THE TREMATODA

To
THE MEMORY
OF
MY FATHER

THE TREMATODA

With Special Reference to British and Other European Forms

BY

BEN DAWES
D.Sc. (London), A.R.C.Sc., D.I.C., F.L.S.
READER IN ZOOLOGY AT KING'S COLLEGE,
UNIVERSITY OF LONDON

CAMBRIDGE
AT THE UNIVERSITY PRESS
1968

PUBLISHED BY
THE SYNDICS OF THE CAMBRIDGE UNIVERSITY PRESS
London Office: Bentley House, 200 Euston Road, London, N.W.1
American Branch: 32 East 57th Street, New York, NY 10022

First Published 1946
Reprinted with Corrections 1956
Reissued 1968

First printed in Great Britain at the University Press, Cambridge
Reprinted in Great Britain by Photolithography by
Unwin Brothers Limited
Woking and London

PREFACE

The growing importance of helminthology during the past two or three decades has been indicated by the publication of several text-books on the subject. Some of these books deal in part with trematodes, but only with a few forms of medical or veterinary interest. For a much longer period of time than that the detailed study of the Trematoda has been sadly neglected in Britain, and zoologists interested in the group have been obliged either to search for information in foreign periodicals, or to take what they could find in a few standard works in German and in the *Oxford Treatise*, the *Cambridge Natural History* and ordinary text-books of Zoology. In writing a book with the hope that it will fill this gap in our scientific literature my aim has been to make available in English, and in a single volume, information that will enable students, teachers and research workers of Zoology in our schools, colleges and universities to identify the trematode parasites of representative animals from the European fauna, and also to provide a broader outline of the structure, modes of life, bionomics and life histories of these animals than can be found in any one book hitherto published in any language. I have not been able to satisfy fully the requirements of specialists, but I have tried as far as possible to indicate where they can find the information they seek.

Parasitism figures prominently in the zoological curriculum at the present time and it is a topic that can be illustrated fully by the study of trematodes, which rarely fails to arouse an enthusiastic interest in students. As a student and university teacher for more than twenty years I have found my wonderment at the phenomenon of parasitism undiminished but, like many other persons, I have been impressed also by the difficulties of naming specimens and finding out what is known about them. The reason for this has been a very extensive literature of original papers written in various languages and dispersed in periodicals so as to be difficult of access. Only about one-half of nearly sixteen hundred papers mentioned in my list of literature are written in English, and most of the many older papers excluded from the list are written in some other language. No doubt many students of Zoology have had their potential interest in the Trematoda stifled at the source by such difficulties as were encountered during the quest for relevant information about the members of this class.

It was perhaps imprudent of me to compile a book such as this during a period of total war, for circumstances during the war years did not favour intensive laboratory work and many journals were especially difficult to consult because put out of reach in places of safety. My task could have been carried out with less trouble and better results in normal times, but it may be taken to represent my unwillingness to relinquish academic research as a supplement to my teaching duties solely on account of enemy activity. Much of the writing was done, many of the figures drawn, during the period when air raids were harassing Bristol and London, and in both these cities. My scientific conscience was consoled by this and other responses of the same kind to enemy action at a time when even teaching work could hardly be guaranteed from one day to the next. Acutely conscious of many defects in my work, I can claim to have striven to attain accuracy in regard to factual matter as far as checking and rechecking can make this so. I have made a special effort also to use scientific names as they ought to be used and to eradicate the confusion that has in some instances resulted from their misuse.

Several friends have given me encouragement and help of some kind or other at all stages of my work, and to all those who have in any way prevented me from relinquishing my set purpose I proffer sincere thanks. Professor D. L. MacKinnon has been patient beyond belief with what must have been my very irritating preoccupation with a special branch of Zoology. Professor C. M. Yonge showed a perennial interest in my work and I was extremely fortunate to have the pleasure of working in his well-regulated Department of Zoology at Bristol. He made me a loan of several specimens and a number of reprints, and he never missed an opportunity to ease my task with appreciative words and actions. In the matter of obtaining periodicals the Librarians of King's College and the University of Bristol gave me unstinted assistance, and both they and their assistants were unbelievably courteous in the face of extortionate requests. My thanks are due also to the Librarian of the Linnaean Society, for both he and his assistant enabled me to consult numerous papers that were unobtainable elsewhere. Dr H. E. Baylis richly deserves my special thanks for the loan of many reprints from his extensive collection and also for much invaluable criticism and advice. He read through the draft of my manuscript at an early stage of its preparation, but in fairness to him and to indicate that he has in no way condoned any errors that may be found in my book I must add that many additions have been made without his cognisance. I am very grateful to him for the time he devoted to my work and for his esteemed counsel. I owe a large debt of gratitude also to my dear friend H. C. Dyer, whose assistance and encouragement were available to me at all times. I am grateful also to the Cambridge University Press for the care they have lavished over the production of my book, and especially for the great skill of their proof-reader, who has saved me from many errors and inconsistencies.

Lastly, I gratefully acknowledge the contributions that have been made to my book by zoölogists too numerous to name here but named in chapter 16, for their researches have provided the fund of knowledge from which I have drawn substantially. I have tried to bear in mind the need to give credit where it is due and if, inadvertently, I have failed to make suitable acknowledgement in any place I can only point to the list of literature given and acclaim my indebtedness to all persons whose names are included therein. Some of my figures are original, but many have been redrawn from those of writers whose names are given, and I am glad to acknowledge here the debt of gratitude I owe to them. My best efforts to do justice to the vast amount of material at my disposal have been inhibited in some instances by the need to keep the size of the book within bounds, but I shall welcome any criticisms, statements and specimens which might improve the book for the working zoologist if and when revision becomes necessary. The need for further research is evident on nearly every page of the book, and it is my earnest hope that British zoologists may be stimulated in one way or another to work towards a better understanding of the interesting animals with which my book deals.

BEN DAWES

DEPARTMENT OF ZOOLOGY
KING'S COLLEGE
LONDON, W.C. 2

26 *April* 1946

CONTENTS

List of Tables in the Text

ANNOTATIONS OF TEXT-FIGURES
ABBREVIATIONS USED

In preparing the text-figures the following conventions have been adopted as consistently as possible in order to facilitate the identification of various organs. *Caeca, excretory canals and vesicle, uterus,* unshaded. *Ovary,* cross-hatching. *Parenchyma,* light stippling. *Receptaculum seminis, seminal vesicle* (distinguished from one another by their positions in the body), irregular wavy-line stippling. *Suckers,* regular line shading. *Testes,* heavier stippling than for the parenchyma. *Vitellaria,* solid black to represent follicles, rarely groups of follicles or the entire compact organs.

A. *aa,* anterior adhesive area. *ab,* 'abdomen' or ecsoma. *ac,* anterior extensions of caeca. *acc,* anterior cephalic cilia. *ag,* apical gland. *ao,* adhesive organ. *ap,* anterior sense papillae. *as,* accessory sac. *as′,* duct of accessory sac. *at,* anterior testis.

B. *b,* brain. *bdy,* body. *be,* bulbus ejaculatorius. *bp,* birth pore. *br,* bristles.

C. *c,* caecum. *c′,* cirrus. c^2, c^3, c^u, secondary, tertiary and ultimate branches of caeca. *ca,* copulatory apparatus. *cc,* cystogenous cells. *cer,* cercaria. *cg,* cephalic glands. *cg′,* caudal glands. *cgp,* common genital pore. *cil,* cilia. *cil′,* especially long cilia. *cx,* caudal cilia. *co,* collar. *cs,* cirrus pouch. *cy,* cyst.

D. *d,* excretory ductule. *dc,* developing cyst. *de,* ejaculatory duct. *dlm,* deep portion of longitudinal muscles. *dm,* diagonal muscles. *dvm,* dorso-ventral muscles. *dp,* dorsal process of opisthaptor.

E. *e,* egg(s). *e′,* embryo. *e″,* eye(s). *eb,* excretory vesicle. *eb″,* lateral excretory canal. *ec,* ectoderm cell of embryo. *ec′,* derivatives of ectoderm cell of embryo. *ed, ej,* ejaculatory duct. *ed′,* excretory duct. *ep,* excretory pore. *eu,* eggs in uterus.

F. *f,* filament(s). *f′,* fin. *fc,* flame cell. *fcs,* false cirrus pouch. *ft,* forked tail.

G. *g,* glands. *g′,* penetration glands. *ga,* genital atrium. *g″,* pore of penetration gland duct. *gb,* germ balls. *gc,* germ cells. *gd,* gland duct. *gic,* genito-intestinal canal. *gp,* genital pore. *gr,* genital rudiment. *gt,* plug of gill tissue. *gs,* genital sucker or gonotyl.

H. *h,* holdfast. *h′, h″, h‴,* first, second and third pair of hooks. *hc,* head collar. *hd,* hermaphrodite duct. *hga,* hooklets in genital atrium. *ho,* head organ(s).

I. *i,* intestine. *i′,* opening of intestine into excretory vesicle. *ic,* inner cyst. *ih,* intestine of host.

L. *l,* lappet(s). *la,* languet. *la′,* larval body. *Lc,* Laurer's canal and/or pore. *Lc′,* blind termination of Laurer's canal. *lh,* larval hooklet(s). *lft,* long forked tail. *lp,* lateral process. *ls,* large spines.

M. *m,* mouth. *m′,* metraterm. *mc,* muscular crest. *mec,* median excretory canal. *mh,* median hook. *mi,* miracidium. *mo,* male organ. *mo′,* marginal organs. *ms,* sheath over mouth. *mu,* muscle in base of haptor.

N. *n,* ventral nerve. *ne,* neck. *nu,* nucleus (nuclei).

O. *o,* ovary. *o′,* ootype. *oc,* outer cyst. *od,* oviduct. *oe,* oesophagus. *op,* operculum. *os,* oral sucker.

P. *p,* pharynx. *p′,* prepharynx. *pa,* papilla(e). *par,* parenchyma. *pb,* posterior region of body. *pb′,* intucked posterior region. *pc,* propagatory cell. *pcc,* post-cephalic cilia. *pd,* posterior disk. *pe,* peduncle. *pg,* prostate gland. *pg′,* pars prostatica. *ph,* pseudohaptor. *pn,* penis. *pn′* and *pn″,* male and female pronuclei. *po,* pores of head glands. *pp,* pharyngeal pouch. *ppa,* posterior sense papilla. *pr,* cup-like posterior region. *ps,* posterior sucker. *ps′,* opening of posterior sucker. *psp,* pigment spot. *pt,* posterior testis.

R. *r*, ciliated ring. *re*, redia. *rh*, rhynchus. *ro*, ring-organ. *rs*, receptaculum seminis. *rsu*. receptaculum seminis uterinum.

S. *s*, sucker(s). *s'*, spine(s). *sbt*, swollen base of tail. *sd*, stalk of opisthaptor. *se*, septum. *sg*, shell glands. *slm*, superficial portion of longitudinal muscle layer. *so*, sense organ(s). *sph*, sphincter. *sps*, skeletal support for posterior sucker. *sq*, squamodisk. *st*, stylet. *st'*, stumpy tail. *sv*, seminal vesicle.

T. *t*, testis(es). *t'*, tail. *tc*, termination of caecum. *th*, tissues of host. *tt'*, trifid tail. *tv*, tail vesicle.

U. *u*, uterus. *u'*, descending limb of uterus. *up*, uterine pore.

V. *v*, vitellaria. *v'*, vestigial vitellarium. *va*, vagina. *vdi*, ventral disk. *vd*, vas deferens. *vd'*, vitelloduct. *vdr*, rudiment of vas deferens. *vp*, vaginal pore. *vp'*, ventral pouch. *vr*, vitelline reservoir. *vm*, vitelline membrane. *vs*, ventral sucker.

W. *wo*, wall of ootype.

Y. *y*, 'yolk' cells. *y'*, fairly mature 'yolk' cells. *y''*, immature 'yolk' cells. *yd*, 'yolk' droplets.

1 and 2 (in Fig. 79 F), first and second polar bodies. ♂ and ♀, male and female pores in Figs. 6 J, 42 C. 60 B and C, but male and female individuals in Figs. 10 I and 59 A.

CHAPTER 1

THE PHYLUM PLATYHELMINTHES

The large and varied assemblage of multicellular animals or Metazoa which constitute the phylum PLATYHELMINTHES is arranged in three classes, Turbellaria or planarians, Trematoda or flukes, and Cestoda or tapeworms. A few forms with indubitable platyhelminth characters do not fit easily into this scheme and are placed in a fourth class called Temnocephalida, which is intermediate between Turbellaria and Trematoda. Some zoologists place the ribbon-worms or Nemertini in the same phylum as the flatworms and, certainly, there is an affinity between the two kinds of worms. But in some respects, notably in their unisexual nature, nemertines are simpler than platyhelminths, and in others, especially the occurrence of a blood-vascular system, they show great advance beyond the typical flatworm organization, so that we are justified in placing them in a different phylum. Roundworms or Nematoda are sometimes grouped with flatworms under the name 'helminths', but they too belong to a phylum with which we are not concerned.

Superficially, the four kinds of flatworms seem to have little if anything in common. They differ so much in appearance, structure and mode of life that it is not easy to formulate the general characters of the phylum as a whole. Sometimes the body is covered with delicate protoplasmic threads or *cilia*, sometimes with a layer of non-living, secreted material or *cuticle*. The digestive system may be elaborate, or not a vestige of it may remain. Male and female reproductive systems may exist in the same individual or, very rarely, in separate individuals. The external surface of the body may be complicated by the existence of various kinds of adhesive organ, or simple as a result of their absence. Differences in the mode of life and life history may suggest not affinity so much as the negation of it.

In spite of these differences, Platyhelminthes have much in common. They are bilaterally symmetrical, i.e. one side of the body resembles the mirror image of the other; the tissues originate from one or another of the three primary germinal layers, ectoderm, endoderm and mesoderm, which develop in the embryos of all animals of higher status than Coelenterata in the animal kingdom; the body lacks a central cavity or *coelom* and a blood-vascular system, the internal organs being enveloped in a spongy packing tissue or *parenchyma* which originates from mesoderm; the principal muscles of the body form a tube several layers thick immediately beneath the integument. Further, the excretory system, by which waste materials are assembled and expelled from the body, comprises in all of them a system of fine canals and finer branches, each of which ultimately terminates in a special kind of cell called a *flame cell*. This has long cilia which project into the lumen of one of the ultimate excretory canaliculi and by their lashings present a flickering appearance, as the name implies. Add to this and the foregoing characters the androgynous nature of most flatworms and the fact that when a digestive system exists it has a mouth, but rarely an anus, and the distinctiveness of the phylum is clear.

(*a*) *Turbellaria.* Planarians are so called because they are flat, unless the name has something to do with Gk. πλανάω, meaning to lead astray or wander. The corresponding scientific name Turbellaria signifies a particular kind of flatworm, the activity of which sets up disturbances or *turbellae* in the watery surroundings. These eddies are created by the lashing motion of innumerable cilia which cover the body, arising from a sheet of cells or *epithelium* at its surface. Other kinds of flatworm lack cilia when adult, but may possess them during the early development. There is no sign of a cuticle in the integument, but the epithelium contains secretory cells which form rod-like bodies called *rhabdites*, and these are extruded and swell in water to form an investing slime which aids ciliary locomotion. Other distinctive characters of the Turbellaria are the position of the mouth behind the brain, the lack of thick-shelled eggs, and the direct nature of the development.

Unlike adult flukes and tapeworms, most planarians live in freedom, many in the sea, others in ponds and streams, and some in damp situations on land. They hunt and devour living prey, the large land planarians thus despatching earthworms, snails and woodlice. Not all planarians preserve such independence, however. Some have come to live in habitual association with molluscs and echinoderms, and a few have acquired the habit of penetrating into and living in the bodies of sea-urchins, holothurians and other invertebrates, thus foreshadowing the parasitic ways of all flukes and tapeworms.

(*b*) *Trematoda.* A trematode or fluke habitually nourishes itself at the expense of another animal, called the *host.* Many trematodes attach themselves to superficial parts of the host, but many others penetrate into the body and settle down in one of the internal organs. The former are called *ectoparasites*, the latter *endoparasites*. The life history provides a more important distinction. Ectoparasitic trematodes develop directly in or on a single type of host and are said to be *monogenetic*; endoparasitic trematodes develop through a sequence of young individuals unlike the parent (*larvae*) with at least one and sometimes more than one change of host, and are said to be *digenetic.* As a rule, digenetic trematodes spend a good part of their larval life in the bodies of molluscs, and some leave the first host and penetrate into another mollusc or a crustacean, or sometimes a fish, before finally settling down in a vertebrate animal to become mature. The vertebrate which harbours the adult fluke is called the *final* host; the invertebrates which succour the larvae and the vertebrates which shelter only juvenile individuals are called *intermediate* hosts, of which a given trematode may thus have one or more than one.

Most trematodes are flattened creatures ranging in size from a minute speck to a length of several inches. They cling to the host by means of special organs of adhesion and anchorage, generally suckers but often hooks, and sometimes both. The cavities of the suckers superficially resemble perforations of the body, a false character to which the name *trematode* refers. Some trematodes browse on mucus extruded from the tissues of the host, others imbibe fluid nutriment by suction from organs like the intestine, and some tackle the tissues themselves, causing their breakdown and resulting in serious injury to the host. Only rarely, however, are the demands of the parasite too heavy to be borne by the host without serious threat to its well-being.

As a rule, trematodes have several other distinctive characters in common. The outermost layer of the integument generally lacks both an epithelium and rhabdite-containing cells, and forms a cuticle. There are no tentacles, though papillae may exist in some regions, notably near the foremost (*anterior*) extremity. The mouth is situated far forward at the tip of the body or near it on the lower (*ventral*) surface. The digestive system is well developed, and generally the intestine is bifurcate. The eggs are encapsulated. The *ensemble* of characters serves to distinguish the trematode from all other kinds of flatworms, including the Temnocephalida.

(*c*) *Cestoda.* Tapeworms and their allies are invariably endoparasitic, and with few exceptions, notably *Archigetes*,* which lives in fresh-water earthworms, attain maturity only in the alimentary canal of a vertebrate animal. When fully grown the cestode may be a mere tenth of an inch long, or, at the opposite extreme, it may attain a length of several yards. This kind of flatworm has undergone extreme modification as a result of long-sustained parasitism in its ancestry, but is simplified in structure rather than degenerate. Only organs which are essential to a comparatively safe and easy existence have been retained. At one end of the worm a *head* or *scolex*, provided with suckers or hooks or both, enables the animal to secure and maintain a hold on the intestine of the host and safeguard itself against a constant threat of expulsion. Behind this a short *neck* of embryonic tissue propagates a long chain or *strobila* of rectangular segments or *proglottides*. Some cestodes are "unsegmented" and without a scolex, and can easily be mistaken for trematodes, but in tapeworms the strobila may consist of more than a thousand proglottides, each of which has at least one set of male and another of female reproductive organs. Apart from this elaboration the structure of the tapeworm is simplified. The body is covered with a cuticle, which generally lacks spines or other appendages. There is no digestive system, all nutriment being imbibed through the integument. Other systems of organs show a low grade of organization, though the nervous and excretory systems are well developed, and the parasite is little more than an efficient machine for producing innumerable eggs. Fecundity is an essential characteristic of Cestoda, as it is of endoparasitic Trematoda, both classes of flatworm being faced with great hazards during the period of larval development, when they have to establish themselves in a sequence of fresh hosts.

(*d*) *Temnocephalida.* Temnocephalids attach themselves to the surface of an invertebrate animal, generally a fresh-water crustacean, but they do not derive nourishment from their 'host'. Instead, they capture and devour insect larvae, rotifers and other small creatures which abound in the surrounding water, and are thus not parasites in the full sense of the term, but merely passengers. In their habits they have more in common with Turbellaria, with which they are linked in some schemes of classification, than with other Platyhelminthes. But they show some structural conformity with Trematoda, having a posterior sucker and a hermaphrodite reproductive system of similar general plan. The gut is simple and saccular, as in some Turbellaria and a few Trematoda, and it opens by a mouth but lacks an anus.

* According to Szidat (1937*b*), *Archigetes* is the neotenic larva of *Biacetabulum*, whose adult lives in fishes.

Despite these resemblances to other classes of flatworms, Temnocephalida are easily distinguished from Trematoda. There is no typical external ciliation in either class, but in the former there is a superficial epithelium containing rhabdite-forming cells and cilia may occur in patches or localized regions in exceptional instances. Eyes occur more frequently than in Trematoda, where they are confined to the larval stages, except in a few monogenetic flukes. The mouth is situated on the ventral surface, more widely removed from the anterior extremity than in trematodes, and the truly distinctive character* is a circlet of at least four and as many as twelve tentacles at the anterior end of the body. Their epithelium contains abundant rhabdite-forming cells. Another distinctive character, though a negative one shared with Turbellaria, is the lack of thick-shelled eggs.

This preamble serves to show that the four classes of the Platyhelminthes can be distinguished by comparatively simple characters of practical importance. The first step in the separation of adult flatworms is to determine the nature of the integument. If it is completely ciliated, there is a clear indication that the worm belongs to the Turbellaria. There is some risk of the loss of the superficial epithelium in a badly preserved turbellarian, but other characters are available and would be utilized in any case to confirm the tentative diagnosis. A flatworm which lacks external ciliation or is ciliated only in patches, but has an epithelial integument, must belong to the Temnocephalida, a diagnosis which is confirmed by the presence of anterior tentacles and a posterior sucker. The existence of suckers on the cuticularized surface of a flatworm with a well-developed digestive system clearly indicates trematode affinities. The lack of a digestive system in an adult flatworm with a cuticle just as clearly signifies cestode affinities. It would be unwise to examine only these elementary characters, but they are obviously the first to be considered.

Mistakes are easily made and the student is forewarned. *Myzostomum* was believed to be a trematode till the great helminthologist Leuckart showed it to be an annelid; *Pentastoma* was included in this class till P. J. van Beneden discovered its embryo and allocated it to the Arthropoda; *Phoenicurus* Rudolphi, 1819† was declared a cestode parasitizing the mollusc *Tethys* till it was proven to be merely an appendage detached from that animal. Perhaps the likeliest source of error parallels that by which *Thysanosoma* was first described as a trematode, though in reality a proglottis detached from the strobila of a cestode. Free cestode proglottides frequently occur in the intestine of fishes, one location in which trematodes are sought. It is good policy to examine the credentials of a suspected parasite whilst it is alive, because the translucency of the living body permits observations to be made of even the finest details of structure (see the Appendix).

* The most notable exception is *Bucephalus polymorphus*, a digenetic trematode belonging to the Bucephalidae, which could never be mistaken for any other kind of flatworm.

† *Phoenicurus* Forster, 1817 is a bird, the redstart.

Monogenea, Aspidogastrea and Digenea, the Orders of the Trematoda

Most trematodes are pale cream in colour, but some acquire a distinctive coloration from the nutriment which they extract from the host. Those which include blood in their diet are tinged red when alive and have brown intestinal contents when preserved. Often the eggs are yellow or golden brown, and when they are numerous the parasite gains further coloration from them. Few trematodes have intrinsic colour, but some larvae possess distinctive pigments, most cercariae of the *Rhodometopa* group being tinged pink.

Size is an important criterion in the classification of Trematoda. Many flukes would be overlooked if we confined our attention to trematodes as large as the well-known species *Fasciola hepatica* or *Polystoma integerrimum*. Most trematodes are less than 30 mm. long, but very large species like *Fasciola gigantica* and *Fasciolopsis buski* may attain a length of 75 mm. and a breadth of 15 mm. At the other extreme, *Heterophyes heterophyes*, the smallest trematode parasite of man, never exceeds 2 mm. in length and 0·7 mm. in breadth. Taking the third dimension into consideration also, evidently some flukes are one thousand times as large as others. The size of a trematode is not known to be correlated with that of the host, but this is not to say that the host is without influence on the growth of its parasites. It seems likely that an internal parasite will be influenced by substances which circulate in the body fluids of its host, but there is a marked hiatus in our knowledge of such matters and little information on the general problem of trematode growth.

Size might be a misleading criterion of distinction were it carelessly applied, because living Trematoda have neither a fixed final size nor a constant shape. The continual play of antagonistic muscles shortens and thickens the body, or alternatively lengthens and attenuates it, considerably modifying its apparent size and shape. Careless fixation may perpetuate one of a number of momentary poses and give a distorted impression of what might be called the mean size and shape. But careful fixation generally provides a specimen of typical form. Difficult as it is to ascertain the sizes of different trematodes, there is no doubt that size is an important character, and in defining it in the following pages I shall use consistently a scheme which was formulated by Lühe, but which has been slightly modified, assuming that the measured parasite was neither unduly extended nor markedly contracted.

Scheme of Size Terminology

Term denoting size	Length of the body (mm.)	Term denoting size	Length of the body (mm.)
Very small	0–1	Fairly large	12–20
Small	1–3	Large	20–35
Fairly small	3–7	Very large	35
Medium	7–12		

Some characters by which the three orders of trematodes are distinguished are so elementary that scrutiny of external structures and apertures on the surface of the body will permit a broad diagnosis to be made. Generally, the mouth is at or near the anterior extremity, though in some trematodes it is near the middle of the ventral surface, as in many Turbellaria. The common genital opening of the reproductive systems is situated in the anterior region of the body, usually in the median plane, but it may occur near the opposite end. A few trematodes have separate openings to the male and female systems and they are generally close together. Some monogenetic trematodes have a ventral opening or a pair of lateral openings into the female system known as vaginal pores, in addition to the genital pore or pores. This is not true for digenetic trematodes, but these have a postero-dorsal opening which is lacking in Monogenea, the pore of Laurer's canal. The excretory system terminates in a single posterior pore in some trematodes, but in a pair of antero-lateral pores in others. Another superficial character which is sometimes useful in diagnosis is the smooth, spiny or scaly nature of the cuticle.

Some of these characters are inconspicuous, but there are more obvious ones. Of greatest importance is the structure of the organs of adhesion, which conform to one of three definite patterns and typify the groups of Trematoda which have been long known to zoologists as the Heterocotylea, Aspidocotylea and Malacocotylea. Some zoologists came to prefer a scheme in which the digenetic trematodes of the two latter groups were linked under the name Digenea and in which the monogenetic Heterocotylea were called Monogenea. More recently, it has been recalled by Faust & Tang (1935) that Aspidocotylea (or Aspidogastrea, a name due to that of the type-genus, *Aspidogaster*) combine the characters of both Monogenea and Digenea but are disqualified from admission to either group because of the equivocal nature of the life history.* For the present it seems preferable to isolate this group and to recognize three orders of Trematoda, namely, the Monogenea, Aspidogastrea and Digenea.

A. *The Order Monogenea*

This group is made up of rather more than a dozen families of Trematoda which infest poikilothermous vertebrates and sometimes, but only rarely, crustaceans, cephalopods and mammals. They are not as well known as digenetic trematodes, because they have only slight economic importance, but they are of at least equal zoological interest and are important in any study of parasites and parasitism. They cling to the skin or gills of fishes, sometimes to the lining of the buccal cavity or pharynx, and they occur in some of these situations in Amphibia and Reptilia. One species at least, *Polystoma integerrimum*, wanders as a larva along the whole extent of the alimentary canal of frogs before settling down in the excretory bladder.

Monogenea are of medium or fairly large size, rarely exceeding 3 cm. in length and often falling short of 1 cm. They are generally flattened dorso-ventrally and disk-, tongue- or leaf-like. The mouth is situated near the anterior tip of the body, where adhesive organs may also be present. Sometimes a pair of small suckers are included in the 'mouth tube' or *prepharynx*; they are called *buccal* suckers.

* E.g. *Lophotaspis vallei*, a parasite of the loggerhead turtle, has a similar larva to typical Monogenea (with anterior and posterior suckers, three patches of cilia and two eye-spots [Manter, 1932]), but this penetrates the conch, *Fasciolaria gigas*, which thus serves as an intermediate host (Wharton, 1939). Other Aspidogastrids, e.g. *Aspidogaster conchicola*, attain maturity in their molluscan hosts.

Similar, but larger, suckers situated at a distance from the mouth near the margins of a somewhat truncated anterior extremity may be called *anterior* suckers. The term *oral* sucker, which is sometimes used indiscriminately, is better reserved for the single sucker which encircles the mouth in a few Monogenea and the majority of Digenea.

The posterior adhesive organs, one or a number of suckers, suckerlets or clamps, may project from the naked ventral surface or may be borne on an outgrowth of it in the form of a disk, which may carry cuticular hooks or hooklets that serve as additional holdfasts. In a few primitive Monogenea the posterior

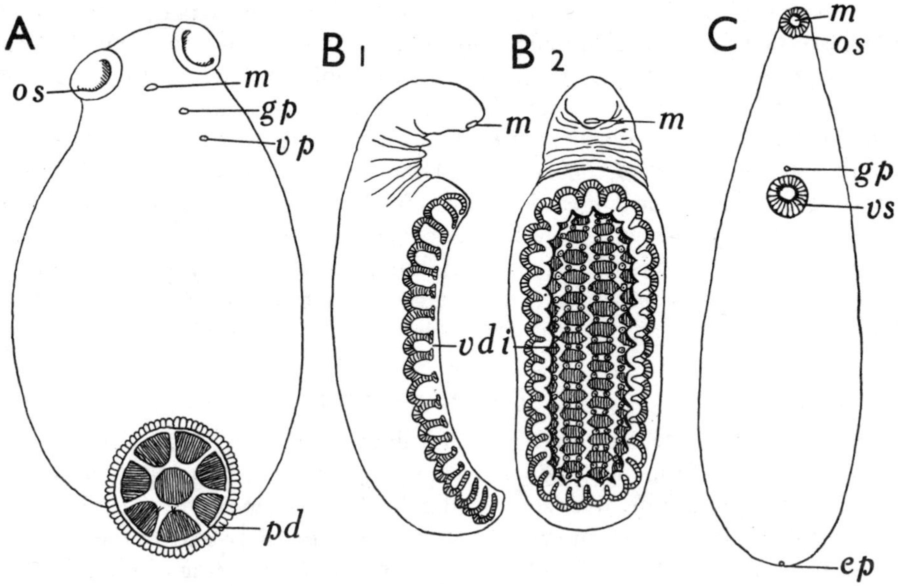

Fig. 1.* The three types of Trematoda. A, Monogenea. B 1, B 2, Aspidogastrea. C, Digenea. (B 1 in side, others in ventral view.)

disk bears only hooks and hooklets, and attachment to the host is less permanent than in other instances, in which the tissues of the host may be deeply scored by the firm, continuous grasp of the suckers.

Other external characters are distinctive when taken together. Eye-spots, when present, are near the anterior end of the body and two or four in number. There is generally a common genital pore, which is close behind the mouth and may be displaced from the median plane. If there is a single vaginal pore, it is close behind the genital pore, but if double pores occur, they are more or less widely separated, sometimes at the lateral margins of the body. The paired excretory pores occupy lateral positions in the anterior region, each communicating by a short canal immediately beneath the integument with a spherical vesicle, which receives the main excretory canal of that side. Thus, the nature of the adhesive apparatus, the positions of various pores on the surface of the body, and a few other and minor characters are sufficient to establish the identity of a monogenetic fluke (Fig. 1 A).

* For a key to the abbreviations used in this and subsequent figures see pp. xv–xvi.

B. *The Order Aspidogastrea*

This order contains the solitary family Aspidogastridae, which has nine genera of distinctive trematodes scattered about the world as parasites of fishes and chelonians, and sometimes of molluscs and crustaceans. All of them penetrate more deeply into the body of the host than do most Monogenea, but the adhesive apparatus is again a character of distinction. It occupies almost the entire ventral surface of the body, consisting of numerous suckerlets or *alveoli* arranged in one row or in three or four rows. In certain elongate Aspidogastrids a single row of alveoli spring from the naked ventral surface, but, as a rule, three or four rows arise from the surface of a large disk which is as conspicuous as the 'foot' of a gastropod mollusc. The maximum number of alveoli which occur in three rows is 32, in four rows 144. With such a formidable adhesive organ accessories are unnecessary, and hooks or hooklets do not occur. The anterior organ of adhesion is feeble or absent in the adult. The mouth is situated at the tip of a very mobile anterior process of conical form. Other external apertures occur in much the same situations as in Digenea. The genital pore is slightly behind the mouth and the excretory pores are posterior. In some Aspidogastrids there is a solitary excretory pore. There is no vaginal pore, but Laurer's canal, which is closed in some species, may open near the posterior extremity, as in Digenea. The Aspidogastrid is thus an easily recognizable trematode (Fig. 1 B1, B2).

C. *The Order Digenea*

Many trematodes which infest man and domesticated animals belong to this group; also many forms which live in vertebrates without injuring them in any way. The dangerous lung, liver and blood flukes are notorious, but the apparently harmless flukes have not commanded such notice. The latter inhabit the stomach or intestine, but a few occur in the coelom, kidneys, ureters and other organs. Every class of the Vertebrata is parasitized by a number of families of Digenea, the class of fishes by families which do not occur elsewhere. The fact that about one-third of the sixty families of Digenea occur only in fishes indicates the value of this class of vertebrate in the study of the Trematoda.

A typical digenetic trematode is easily recognized by the nature of the adhesive apparatus, which is much simpler than in Monogenea or Aspidogastrea, perhaps because of the comparative ease with which the parasite can retain its hold in the sheltered locations of the host's body. The suckers are cup-shaped, muscular organs, generally unaccompanied by hooks or other accessories. One encircles the mouth and is called the *oral* sucker. Another, which may be larger or smaller than the first, occupies a position somewhere on the ventral surface (Fig. 1 C). If far back on the ventral surface, it is called a *posterior* sucker; otherwise, simply the *ventral* sucker. An alternative name for the ventral sucker is *acetabulum*, which is rarely met with outside the American literature. Some Digenea lack the oral, others the ventral sucker, and there are a few forms which are devoid of suckers.

Of other characters which help to characterize the Digenea we may note here the various openings into or out of the body. The genital pore is situated, as a rule, between the suckers on the ventral surface of the body. In rare instances, however, it occurs behind the ventral sucker, sometimes not far from the posterior extremity. Here the excretory pore is almost invariably to be found, the opening of Laurer's canal being a little farther forward on the dorsal surface of the body.

The superficial characters which have been mentioned so far are alone sufficient to enable a broad diagnosis of the three orders of Trematoda to be made. The main characters which serve are:

1. Main adhesive apparatus a postero-ventral disk, or one or a number of suckers set upon a disk or upon the surface of the body, sometimes supplemented by hooks and hooklets. Other adhesive organs, when present, various types of suckers situated near the anterior extremity. Excretory pores paired, situated in the anterior region. Genital pore or pores also anterior. Parasites living on the skin or in other superficial locations of Vertebrata, especially on the gills of fishes. MONOGENEA
2. Main adhesive apparatus a single row of suckerlets or alveoli set upon the ventral surface of the body, or three or four longitudinal rows of alveoli set upon an enormous posterior disk, which lacks hooks or hooklets. Other adhesive organs weakly developed or unrepresented. Excretory and genital pores as in 3. Endoparasites of Vertebrata, especially fishes and chelonians, but also of Mollusca and Crustacea. ASPIDOGASTREA
3. Main adhesive apparatus, when present, a solitary cup-shaped sucker situated somewhere on the ventral surface of the body. Other adhesive organ, when present, an oral sucker encircling the mouth. Excretory pore single and posterior. Genital pore or pores ventral, generally in the anterior region, between the suckers. Vaginal pore or pores absent. Opening of Laurer's canal dorsal, near the posterior extremity. Endoparasites of every class of Vertebrata, and of some Invertebrata. DIGENEA

SOME TREMATODE NAMES AND THEIR DERIVATION

Like other animals, trematodes profit by the law of binomial nomenclature to the extent of receiving two scientific names, one to denote the genus and the other the species to which the animal belongs. If the name was aptly chosen by the person who erected the genus or species, whose privilege it is to devise the name, or if it did not lose its aptness through subsequent discoveries of other animals or more concise knowledge about the same kind of animal, it may provide a distinctive index to the identification of the type. Because very few trematodes have common names, the scientific names are essential to precise delineation. And, like multiple nouns in German, they are apt to be overwhelming unless correctly split and analysed. While it is not absolutely essential to appreciate the derivation of such names, this may aid memory and provide additional interest. A few examples of such derivations might tend to dispel the natural prejudice of persons who are averse to scientific names.

Many scientific names are derived from the Greek, some from Latin. Monogenetic flukes have names many of which end in the suffix *-cotyle* (Gr. cup), and this refers to the character of the main organ of adhesion. Thus arise names like *Monocotyle* (*monos*, alone), *Pseudocotyle* (*pseudes*, false), *Onchocotyle* (*onkos*, barb), *Leptocotyle* (*leptos*, fine), *Merizocotyle* (*merizein*, to divide), *Heterocotyle* (*heteros*, other), *Anthocotyle* (*anthos*, flower), *Acanthocotyle* (*akantha*, thorn), *Megalocotyle* (*megas* (f. *megale*; n. *mega*), great), *Microcotyle* (*mikros*, small), *Lophocotyle* (*lophos*, ridge), *Hexacotyle* (*hex*, six), *Octocotyle* (*okto*, eight), and *Dactylocotyle* (*dactylos*, finger). A less common termination is *-bothrium* (*bothros*, trench), which occurs in names like *Microbothrium*, *Leptobothrium* and *Octobothrium**. Another distinctive name which refers to the adhesive organ is *Trochopus* (*trochos*, wheel). In some names the general appearance of the entire worm is indicated, e.g. in *Entobdella* (*entos*, within; *bdella*, leech), Pteronella (*pteron*, feather or wing) and *Phyllonella** (*phyllon*, leaf).

Some names are inapt because of faulty or insufficient knowledge dating from the time of their inception. Names like *Distoma*, *Tristoma* and *Polystoma* (*dis*, twice; *treis*, *tria*, three; *polys*, many; *stoma*, mouth) refer to imaginary openings into the body which later turned out to be mouths only of the cup-like suckers. But in like instances names may have great significance, e.g. in *Echinostoma* (*echinos*, hedgehog), which has a coronet of stiff spines around the true mouth, and the related *Echinochasmus* (*chasma*, gulf), in which the sequence of spines is broken dorsally.

In some Digenea the name refers to the form of the body or some part of it. Thus, *Schistosoma* (*schistos*, split; *soma*, body), *Gastrodiscus* (*gaster*, belly; *discos*, disk) and *Prosorhynchus* (*proso*, forward; *rhynchos*, snout). Sometimes the name refers to the shape, position or other characters of the gonads, as in *Paragonimus* (*para*, beside; *gonimos*, productive, i.e. gonads), *Metagonimus* (*meta*, with, among, after) and *Opisthogonimus* (*opisthe*, behind). More frequently it refers specifically to the testes, as in *Haplorchis* (*haploos*, single; *orchis*, testis), *Monorchis* (*monos*, alone), *Clonorchis** (*klon*, twig) and *Isoparorchis* (*iso*, equal), as well as *Opisthorchis*, *Schistorchis* and *Metorchis*.

Names are sometimes coined in honour of persons, as in *Lebouria*, *Nitzschia*, *Benedenia*, *Brumptia*, *Bilharziella*. These are generic names, and specific ones more commonly have this origin, as in *Schistosoma mansoni* (after Manson), *Prosthogonimus rudolphii* (after Rudolphi) and *Opisthorchis wardi** (after Ward). Sometimes both generic and specific names serve this purpose, as in *Watsonius watsoni*. Otherwise, the specific name may refer to the host (as in *Alaria canis*), the location of the fluke in the body of the host (as in *Fasciola hepatica* (L. *hepar*, liver) or the geographic locality (as in *Pseudamphistomum danubiense*). Such names tend to lose specificity as the bounds of knowledge widen. More suitable names refer to some special character of the fluke, as in *Dicrocoelium lanceolatum** (L. *lanceola*, a little lance), *Fasciola gigantica*, *Echinochasmus elongatus* and *Haplometra cylindracea**. Even when the specific name is adapted from that of a person, it should be written with a small initial letter, like other specific names.

In classifying animals, categories are established which are known by names such as phylum, class, order, family, genus and species, all but the first of which are components of the category mentioned earlier in the sequence. Thus, a phylum is divided into classes, a genus into species. Intermediate groups are sometimes necessary, in which case the prefix *sub-* is employed to indicate lower status. Thus, a class may be divided into subclasses or an order into suborders. Sometimes a number of similar categories are placed together in a special group, as families are aggregated not into an order, but into a *superfamily*. This particular group is little met with in regard to Trematoda outside the American literature, in which it figures prominently. In what follows we shall refer largely to the particular groups families, subfamilies, genera and species. The first two of these

* Names not in current use.

are named in accordance with their types, particular endings being added for distinction. The family ending is *-idae*, the subfamily *-inae*, and the superfamily, when this appears, has yet another ending, *-oidea*.

A simple example will make clear the practical importance of such distinctive terminology. *Heterophyes* is the name of a genus and happens to be the type of a family, which is thus called the Heterophyidae. This has a number of subfamilies, amongst which are the Metagoniminae, Cryptocotylinae and Apophallinae, respective type genera of which are *Metagonimus*, *Cryptocotyle* and *Apophallus*. There is also a subfamily Heterophyinae, which has as its type the genus which is responsible for the name of the family. This family was included by Price (1940) along with other families in a single superfamily, the type family of which is not the Heterophyidae, but the Opisthorchiidae, which has the genus *Opisthorchis* as its type. This superfamily thus gets the name Opisthorchoidea (which is more correctly written Opisthorchioidea). The value of the endings is obvious; it enables us at a glance to realize with what kind of category or taxonomic unit we are dealing.

The rules which govern nomenclature are deep and wide and need not be gone into here. Suffice it to mention that they have been summarized by Wenyon (1926, 1928), Faust (1930*b*) and Craig & Faust (1940). They have been presented in pamphlet form in another American publication (*Proc. Biol. Soc. Wash.* **39**, 1926, 75–104).

It is necessary, however, to stress the devious ways by which current names sometimes come into being and to emphasize also the need for great care in using names. Students are often puzzled by full names such as *Gyrodactylus elegans* Nordmann, 1832 and *Plagiorchis vespertilionis* (Müller, 1784) Braun, 1900, and it is natural to seek the implication of the brackets and subsidiary designation in one instance and their absence in the other. The names which provide greatest difficulty are those which refer to animals about whose systematic positions there has been most difficulty in the past. A genus or species is named by the person who proposes its erection, defines it and in other ways follows the rules of nomenclature. In the example chosen, the former species (an ectoparasite of fresh-water fishes) was described by Nordmann who, at the same time (1832), named both species and genus, incidentally, in such a way as to make unnecessary any alteration of status or other change. The original name stands to-day as it did more than a century ago. Not so the name of the latter species in the example. Müller in 1784 described this species (an endoparasite of bats) and called it *Fasciola vespertilionis*. But in 1776 Retzius proposed to reserve the generic name *Distoma* for trematodes as distinct from other worms, and Müller's species was referred to this genus by Zeder in 1803. It thus became known as *Distoma vespertilionis* (Müller, 1784) Zeder, 1803, the brackets indicating that the generic title is not that which Müller chose for the species. His original name became a synonym of Zeder's. But in 1899 Lühe proposed the erection of a genus *Plagiorchis*,* and Braun in 1900 referred Müller's species to it. The name of the particular trematode thus became *Plagiorchis vespertilionis* (Müller, 1784)

* In the same year Looss proposed the name *Lepoderma* for this genus, as a result of which some confusion has arisen. According to Hassall (in a letter to Baylis) Lühe's proposal was published a day before Looss's. Consequently, *Plagiorchis* has priority.

Braun, 1900 (Synonyms: *Fasciola vespertilionis* Müller, 1784; *Distoma vespertilionis* (Müller, 1784) Zeder, 1803). Actually, the list of synonyms should include at least one other name, because in 1809 Rudolphi described a species which he called *Distoma lima*, which ultimately proved to be not a distinct species, but merely a form of that we now call *Plagiorchis vespertilionis*. Synonyms may thus represent older names of one and the same species, but also forms which were mistakenly regarded as distinct species but later turned out to belong to the one in question. It will be self-evident that in trematode names which include in parentheses an author's name and a date, and to the right of this a second author's name and a date, there may still be no indication of the authorship or date of the particular genus recognized. Even when the founder of a genus attempts to refer all eligible species to it, which is not invariably the case, he may overlook one or a number, and a second person may be responsible for the inclusion of a particular species within the given genus, hence the necessity for the name and date following the parentheses in the full title. In some instances, and especially in war time when much of the original literature has been placed out of reach, it is difficult to provide all information which is conducive to clear demarcation of a species. But in what follows, the necessary references have been given as consistently as the present, rather difficult, conditions permit.

CHAPTER 2

THE CONTRASTING LIFE HISTORIES OF THE MONOGENEA AND DIGENEA

Having already mentioned the marked contrast between the life histories of the Monogenea and Digenea it remains to amplify what has been stated, and this chapter will be devoted to an outline of the mode of development in one fluke of each kind. A few comments on the eggs of Trematoda must be made before proceeding. Some trematodes are viviparous, and in producing their young alive depart from customary procedure, which is to lay eggs that are enclosed each in a capsule of secretionary material. The manner in which the capsules are moulded from a viscid fluid will be described in a later chapter; suffice it to observe here that they differ in shape and size and that such characters assist broad diagnosis of the two main groups of the Trematoda.

When first formed each capsule as a rule contains an ovum, a number of spermatozoa and a cluster of vitelline cells which provided the materials of which the capsules are formed. Not all the egg capsules (henceforth referred to as 'eggs') of Trematoda are laid in this condition. The ovum may have undergone considerable development by the time the egg is laid, and whole families of Digenea are characterized by eggs which contain at this time a fully developed, ciliated larva. It is important, therefore, to ascertain whether or not eggs still *in utero* show such signs of precocious development, the alternative to which is sometimes a fairly long period of incubation in the external world.

In Digenea the eggs are ovoid and formed of two parts, one of which forms a tiny lid or *operculum* to the other (Fig. 2G, H, I). The emergent larva has the task of pushing this aside at the time of emergence. Some Digenea and practically all Aspidogastrea and Monogenea have non-operculate eggs, and their larvae are liberated by rupture of the capsule. Monogenea frequently have distinctive eggs, one or both poles being drawn out into long filaments, which may be many times as long as the capsule proper and much entangled (Fig. 2E, F). The filaments serve to bring the capsule to an anchorage after it has been laid and to moor it during the period of incubation which follows. This broad distinction between the eggs of Monogenea and Digenea cannot be applied to all Trematoda. *Notocotylus*, a genus of Digenea infecting birds, is characterized by capsules with long, straight filaments (Fig. 2L) which might be mistaken for the eggs of some monogenetic fluke were it not for the unmistakable larva which each contains when laid. Blood flukes belonging to the family Schistosomatidae have non-operculate eggs, and in some species at least each is provided with a short lateral or terminal spine (Fig. 2J, K). With such rare exceptions as these, however, and a few instances of operculate or non-filamentous eggs among the Monogenea (Fig. 2B), the distinction mentioned serves as a means of general diagnosis.

As a rule Monogenea lay fewer and larger eggs than Digenea, but exceptionally large capsules are produced by some of the latter. Thus, the eggs of *Fasciola hepatica* are at least one hundred times as large as those of certain other Digenea.

The most cursory observation will disclose the presence or absence of filaments or an operculum, but important differences of size in the eggs may not become apparent unless careful measurements are made in a manner such as is described in the Appendix of this book.

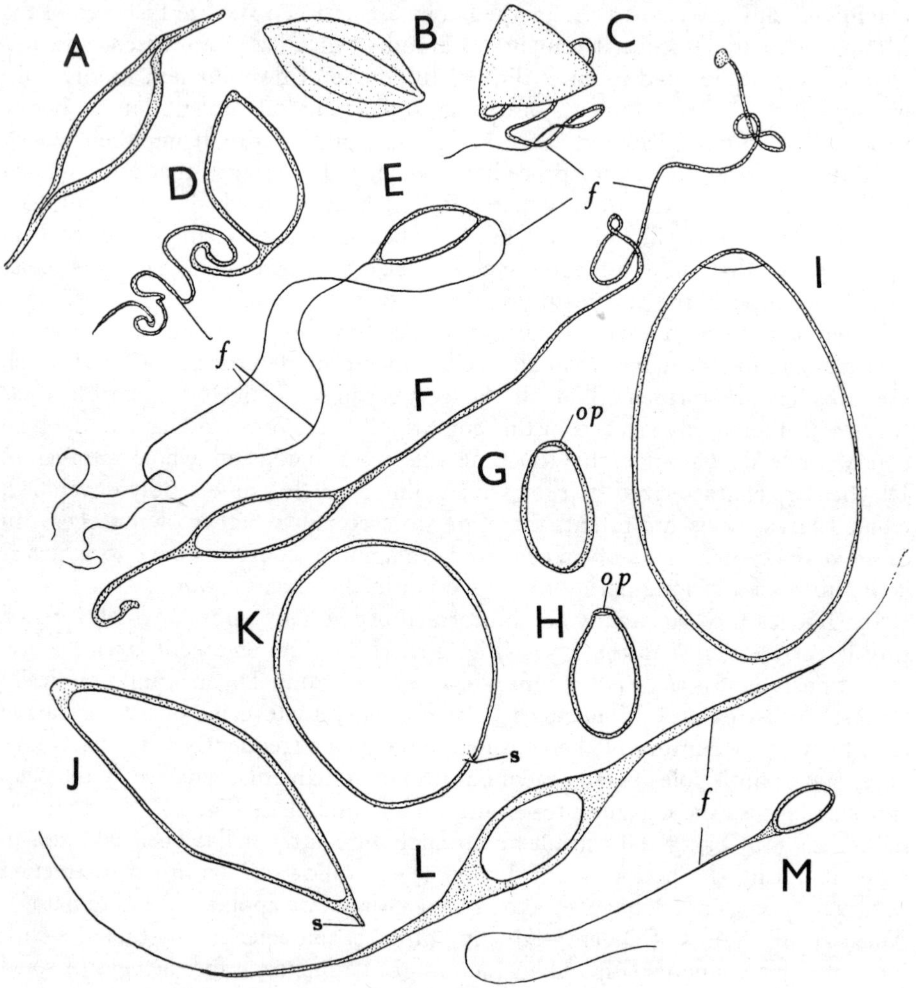

Fig. 2. The eggs of Trematoda. A–F, Monogenea; G–M, Digenea. A, *Hexostoma extensicaudum*. B, *Rajonchocotyle alba*. C, *Nitzschia monticellii*. D, *Acanthocotyle caniculae*. E, *Squalonchocotyle* (=*Erpocotyle*) *canis*. F, *Dactylocotyle carbonarii* (=*Diclidophora denticulata*). G, *Dicrocoelium lanceatum* (=*D. dendriticum*). H, *Opisthorchis felineus*. I, *Fasciola hepatica*. J, *Schistosoma bovis*. K, *Schistosoma japonicum*. L, *Notocotylus attenuatus*. M, *Hemipera sharpei*. (A, original; B–F, after Fuhrmann, 1928; G–L, after Mönnig, 1934; M, after Idris Jones, 1933*b*.)

(*a*) *The life history of a Monogenetic Trematode* (Polystoma integerrimum)

Much of our knowledge on the development of *Polystoma integerrimum* we owe to Zeller's investigation of about 74 years ago, though in recent times Gallien (1932*a*, *b*, 1933, 1935) has reinvestigated the subject with confirmation of

Zeller's results and the acquisition of new knowledge. This is in consequence the best-known monogenetic life history in the literature. For some reason which is obscure, possibly the mistaken idea of simplicity unworthy of serious study, such life histories have been neglected, only about half a dozen having been elucidated. Alvey's researches on the life history of *Sphyranura oligorchis*, which has a much-abbreviated and an embryonic as distinct from a larval development, suggest an interesting field of study which might amply repay willing and able investigators.

During the winter months the reproductive organs of *Polystoma integerrimum* remain dormant, but they reawaken to activity in the spring, when copulation occurs and the flukes again begin to lay eggs. The condition of the reproductive organs is such that when parasitized frogs are brought into the warmth of the laboratory as late in the year as November, the flukes may copulate within a few days and each may produce nearly a gross of eggs a day for more than a week. Temperature also influences the rate of development of the eggs, so that a process which is completed in less than three weeks may take six to twelve if the temperature is lower than 50° F., as is often the case in nature in spring. At the end of this period and when the tadpoles of frogs are so advanced as to have lost their external and acquired internal gills, a larva emerges from the egg of the trematode (Fig. 3A). It is barrel-shaped and has a posterior disk (*pd*) bearing sixteen arrow-like hooklets (*lh*) arranged radially. Five incomplete bands of cilia (*r*) partially encircle the body, three being anterior and ventral, two posterior and dorsal. A tuft of cilia also adorns the anterior extremity. The larva swims about aimlessly in the water in which the eggs were laid, somewhat like a turbellarian, but its behaviour soon becomes remarkable. How it comes to recognize the frog tadpole is uncertain, though it has eye-spots (*e″*) and a brain and seems to have a chemical sense also, but on meeting a tadpole which has reached the internal gill stage of development it moves towards the gill opening, pauses for a moment, and darts into the gill chamber. Once inside, it is attracted to the gill filaments, to which it immediately clings by means of the adhesive apparatus. Feeding upon mucus and detritus with which the filaments are covered, the larva proceeds with its development.

At this stage (Fig. 3A) the mouth (*m*) of the larva is near the anterior end of the body and communicates with a muscular pharynx (*p*), which is capable of suctorial action, and with a relatively simple, unbranched intestine (*i*) which ends blindly in the larval parenchyma. Close behind the intestine are two clusters of unicellular glands (*g*) which elaborate a secretion and pass it into the intestine where, presumably, it serves a digestive purpose. Unwanted waste materials are assembled in a fluid which is collected by flame-cells and swept into one or the other of a pair of longitudinal excretory ducts (*ed*) and out of the body by one of the two lateral excretory pores (*ep*) situated near the middle of the body.

The *Polystoma* larva remains attached to the gills of the tadpole for about eight weeks, till the latter metamorphoses into a tiny frog, when it is faced with a crisis and meets this by itself metamorphosing. As the gills of the tadpole atrophy the young trematode leaves the branchial chamber, moves down the alimentary canal and, when the cloacal bladder develops, enters this chamber and attaches itself to the internal lining. During this migration the cilia of the larva cease to function and atrophy, after which the larva begins to take on some of the characters of the adult. The posterior disk, at first resembling that of primitive Monogenea of the family Gyrodactylidae, begins to develop the suckers, three on either side, and the muscles which bring about their suctorial movements. Some of the larval hooklets are swallowed up by the developing suckers, but others remain on intervening regions of the disk, or all hooklets may be lost, these arrangements varying in different species of the genus. The larval hooklets are too small to be serviceable as grapnels in the adult, and they are superseded by much more formidable hooks which develop in the posterior interval between

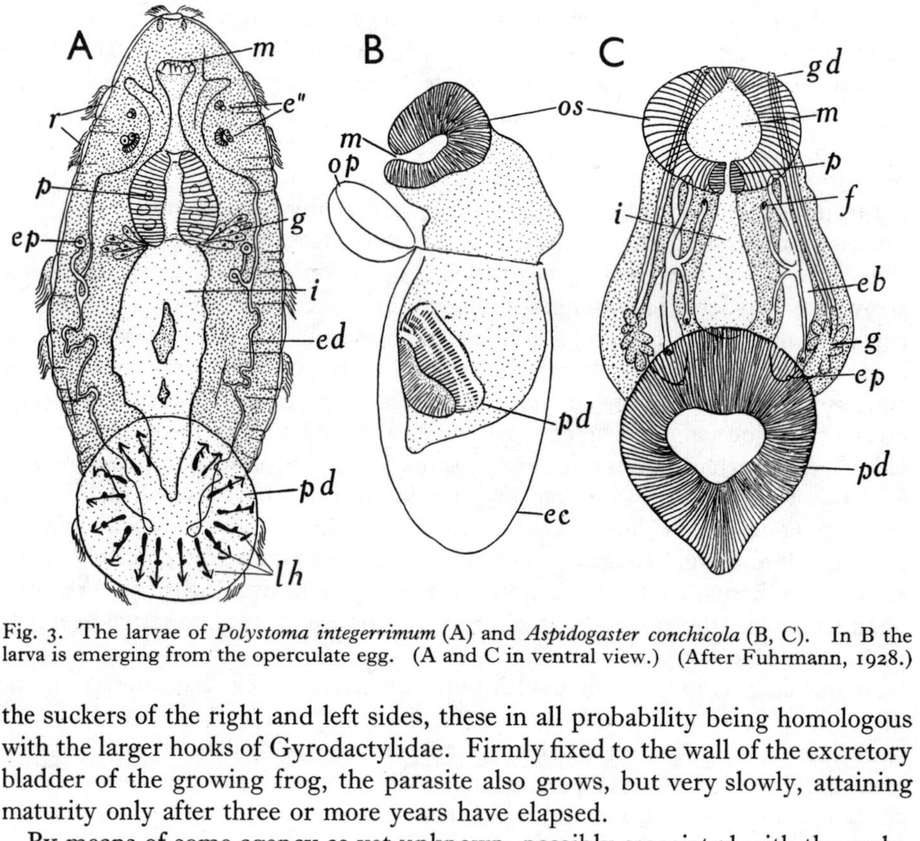

Fig. 3. The larvae of *Polystoma integerrimum* (A) and *Aspidogaster conchicola* (B, C). In B the larva is emerging from the operculate egg. (A and C in ventral view.) (After Fuhrmann, 1928.)

the suckers of the right and left sides, these in all probability being homologous with the larger hooks of Gyrodactylidae. Firmly fixed to the wall of the excretory bladder of the growing frog, the parasite also grows, but very slowly, attaining maturity only after three or more years have elapsed.

By means of some agency as yet unknown, possibly associated with the endocrine activities of the vertebrate host, a *Polystoma* larva which inadvertently attaches itself to the *external* gills of a somewhat younger tadpole than that generally selected, undergoes a greatly accelerated development and, in the short period before the host metamorphoses, is transformed into a miniature fluke which never reaches the bladder of the frog but which produces viable eggs and dies when the tadpole becomes a young frog. It sometimes happens that a larva which becomes attached to the *internal* gills behaves in the same way. The anatomy of the tiny fluke differs greatly from that of the normal one. It has smaller hooks of different shape, a single testis instead of many testes, copulatory organs which are so modified as to be useless, and sperms which lack a head and possess a swollen body. The female organs are even more considerably modified, the vaginae and uterus remaining undeveloped, the vitellaria rudimentary, the ovary small and with far fewer ova, and the ootype opening to the exterior. The modifications render cross-fertilization impossible and the individual fertilizes itself. After about 20 days of development, this modified larva has been transformed into a miniature fluke which is laying eggs that subsequently develop into normal larvae and proceed with their development to the usual adult stage.

Gallien, who confirmed Zeller's discoveries, believes that these facts indicate not merely a chance development along alternative lines according to the degree of development of the host at the time of attachment but a definite alternation of generations. He holds that eggs which are liberated from the parent *Polystoma* in spring yield larvae of the 'external gill' type that come to maturity in the course of a few weeks and produce eggs which in turn provide larvae that cling to the internal gills of older tadpoles and subsequently develop during several years into normal adults. That such an interpretation is possible indicates the need for further research on the problems of development in Monogenea.

(*b*) *The life history of a Digenetic Trematode* (Fasciola hepatica)

(i) *The Miracidium.* The larva which finally pushes aside the operculum of the egg of *Fasciola hepatica* differs greatly from that of *Polystoma* and is known as a *miracidium* (Fig. 4C, D). Neglecting for the moment the mode of its development and the finer details of its anatomy, we can note the general characters of the larva and of its progeny. The pyriform body is completely covered with rather long cilia (*cil*) which facilitate rapid swimming movements in which the blunt end goes foremost. According to Coe (1895) the cilia spring from 21 epithelial cells arranged in five rows: the first row has 6 cells (2 dorsal, 2 ventral and 2 lateral), the second also 6 (3 dorsal, 3 ventral), the third 3 (1 dorsal, 2 ventro-lateral), the fourth 4 (2 right, 2 left) and the fifth 2 (lateral). The cells are connected by non-ciliated matrix forming a layer 1–2 μ thick (Fig. 5). At the anterior end of the body there is a small, conical, somewhat retractile papilla, which later on proves to be an efficient boring organ. No structure even remotely resembling the disk of the *Polystoma* larva exists in the miracidium, whose outline is smooth and streamlined. Internally certain resemblances are seen, however, notably in the eye-spots, the brain and the rudimentary excretory system. But the intestine is rudimentary and there is doubt if it functions as such or may even be represented as a gut at all. A character which is not seen in the larval *Polystoma* is a cluster of large cells which occupies the entire posterior half of the miracidium (Fig. 4C, D, *gc*). This novel character is profoundly important, because the cells are germinal by nature and destined to give rise each to a separate larva of a fresh generation which will continue the life history beyond the miracidial stage.

(ii) *The Sporocyst.* If it is to proceed with its development and continue the life history, the miracidium must enter the body of a suitable species of snail, here in Europe generally *Limnaea truncatula*, but in other continents different species of this genus and other genera. Should it fail to locate and penetrate the snail within 24 hours of the time of its emergence from the egg, the miracidium

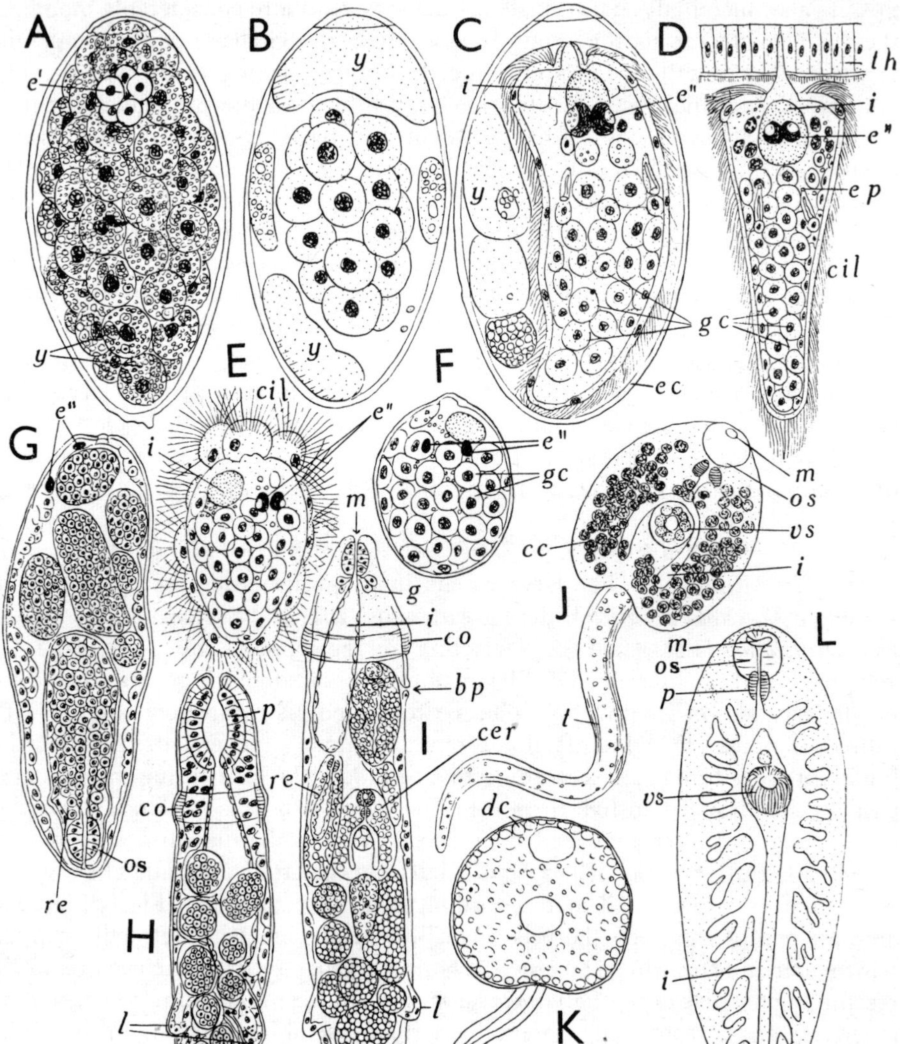

Fig. 4. Stages in the life cycle of *Fasciola hepatica*. A, egg as laid, showing the unsegmented ovum. B, egg at a later stage, during segmentation. C, egg with contained miracidium. D, miracidium represented as penetrating into the tissues of the snail. E, modified miracidium undergoing metamorphosis. F, very young sporocyst. G, mature sporocyst containing germ balls and (at lower end) an almost mature redia. H, young redia (0·5 mm. long) after emergence from sporocyst, containing germ balls in various stages of development. I, mature redia containing one daughter redia and two cercariae, one almost fully developed, as well as germ balls. J, cercaria after emergence from the tissues of the snail. K, cercaria engaged in forming the cyst. L, very young fluke (1·1 mm. long) from the liver of a lamb. (After Thomas, 1883.)

inevitably dies, being unable to survive longer in freedom. Having made an approach to a suitable host, miracidia may be seen to swarm around it and almost immediately to bore into its body, generally choosing a spot near the branchial aperture where the tissues are soft. Once inside the snail these miracidia migrate by way of the blood vessels or the lymph channels to some situation where nutriment is abundant, generally the digestive gland or 'liver' which is situated in the 'hump' beneath the uppermost whorls of the shell.

After penetration into the intermediate host the miracidium loses its ciliated covering, first of all changing to a spherical shape (Fig. 4E, F). This results in the formation of a kind of living cyst, tightly packed with germinal cells but also containing the vestiges of the eyes and 'intestine'. The *sporocyst*, as the young parasite is now called, next increases in size, largely by growth and multiplication of the germinal cells and a stretching of the surface epithelium of the body (Fig. 4G). Each germinal cell has by growth and division given rise to a ball of cells, known as a *germinal ball*, each of which is destined to produce the next larval stage already mentioned. Many such larvae, the *rediae*, distend the sporocyst to the point of rupture, after which they are liberated into the digestive gland of the snail.

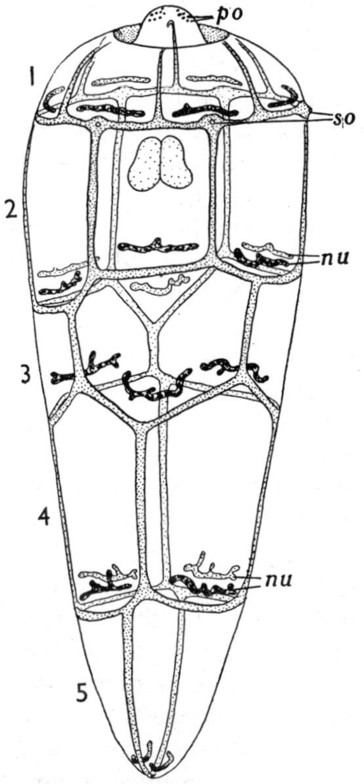

Fig. 5. The miracidium of *Fasciola hepatica*, showing successive tiers of cells antero-posteriorly. Osmium tetroxide, silver nitrate preparation. (After Coe, 1895.)

(iii) *The Redia*. This cylindrical larva (Fig. 4H, I) is readily recognized in light smears of the digestive gland of an infested snail by several characters. It has a pair of marginal processes or lappets (H, *l*), sometimes called *procruscula*, near the posterior and a collar-like structure (H and I, *co*) just behind the anterior extremity of the body. The terminal mouth (I, *m*) communicates with a muscular pharynx (H, *p*) and, beyond this, an unbranched and cylindrical intestine (I, *i*). There is a cluster of unicellular glands at the junction of the two latter organs (I, *g*). Eyes, which persist from the miracidial into the sporocyst stage, are lacking in the redia, but there is an excretory system of delicate canals and flame cells. Antagonistic circular and longitudinal muscles occur in the body wall and their action results in considerable movements, the larva alternating between a short and thick and a very elongate cylindrical form.

The body of the redia is filled with delicate larval parenchyma which is packed with germinal cells resembling those of the miracidium and likewise multiplying to form germinal balls (H, *gb*). These cells are budded off from a central strand of tissue or *rhachis* in the redia rudiment, a mode of origin which is slightly different from that seen in some other digenetic trematodes. There is no doubt that they are segregated early during the formation of the redial body, and it is possible that this happens as soon as the miracidial germinal cell begins to divide, as in other Digenea. The germinal balls to which they give rise are the rudiments of the next larval stage in the life history, the *cercariae*.

The functions of certain organs of the redia were not properly appreciated till recent researches elucidated them. Thomas, who shared with Leuckart the honour of working out the life history of *Fasciola hepatica*, believed the 'collar' to be a muscular ring which serves to maintain the cylindrical form of the anterior end of the redia. Wright (1928) studied rediae both under the microscope and within the tissues of the snail, and from the results of his investigations concluded that the collar is concerned with the difficult problem of locomotion through the tissues of the host. This is effected in the following way.

When the redia moves after a period of quiescence, longitudinal muscles in the collar relax and the collar itself almost disappears. Simultaneously, the redia elongates and pushes its way into the soft tissue of the digestive gland, abrading it and ingesting cellular fragments. While this is happening the posterior region of the redia is becoming swollen and turgid. The procruscula also swell and, pressing against the surrounding tissue, firmly anchor the redia in the burrow it has made in the digestive gland. The turgescence is due to displacement of fluid from the anterior to the posterior region of the body. With the procruscula providing purchase, the anterior region of the larva literally eats its way into the host's tissues.

When the body has become fully extended, the longitudinal muscles of the collar contract, bringing this structure into prominence again and causing it to press against the wall of the 'burrow' as the anterior region shortens and thickens. The larva is thus 'anchored' by its anterior end. The procruscula then lose their turgidity, and as the posterior region of the body shortens it is drawn forward into the 'burrow', to conclude the cycle of events which, oft repeated, produces continual forward progression into the tissues of the host and progressive deepening of the 'burrow'.

The transference of fluid from end to end of the larva during these cyclical changes is probably due to the action of the circular muscles of the body, which effect a kind of peristalsis. The longitudinal muscles also serve by their contraction to draw forward the posterior end of the larva during successive stages of the burrowing process.

From this description it will be evident that the tissues of the host suffer considerable damage. In many infested molluscs the digestive gland, sometimes also the gonad, is almost completely destroyed and replaced by the rapidly growing redial tissue. Sometimes the snails die as a result of their injuries, but not infrequently they show notable powers of resistance to these ravages and survive. Their survival is to some extent due to the fact that the redial stage of development comes to an end with the liberation of the cercariae, but largely to the very great powers of regeneration which the infected and damaged tissues possess.

(iv) *The Cercaria*. As has been stated, this stage develops from a germinal cell inside the redia. When fully developed it escapes by way of the *birth pore* which is situated immediately behind the collar of the redia (Fig. 4I, *bp*). At the time of its emergence it is clearly visible to the naked eye, and microscopic examination shows it to be tadpole-like, with a discoidal body and a long, backwardly projecting tail (Fig. 4J). This resemblance to a tadpole is merely superficial and, as we shall see, the shape may be very different in other Digenea. It already possesses some of the organs of the adult fluke, though in miniature, and we anticipate further description only as far as is necessary to enumerate them. There is an oral sucker (*os*) encircling the mouth (*m*), which pierces the anterior tip of the body, and a ventral sucker which occupies the centre of the body (*vs*). The pharynx, oesophagus and forked intestine are recognizably of trematode pattern though much simpler than in the adult of the species, the last-named organ being devoid of lateral branches. Excretory canals extend along either side of the body and are connected with flame cells by fine branches. The two main canals unite near the posterior end of the body and form a single canal which passes some way along the tail and opens by a median pore on its surface. The gonads and other components of the reproductive systems are not yet developed, though represented by inconspicuous cellular rudiments. At the sides of the pharynx there are clusters of cells (Fig. 4J, *cc*) which contain droplets of fluid. These are destined later to be extruded to form a spherical cyst in which the final period of larval development is spent and for this reason are known as cystogenous cells. We shall find that these are not present in the cercariae of some Digenea.

After 'birth' from the redia along with several of its fellows, the cercaria pushes its way through the tissues of the snail and eventually emerges into the surrounding water. Here in company with others of its kind it enjoys a short period of freedom, but after an hour or two wriggles out of the pool in which the snail lives and clambers a short way up the stem of some plant, usually a blade of grass, which happens to border the pool. Reaching a point just above that at which the herbage will be cropped by some grazing mammal, the cercaria rests. The cystogenous cells then become active and their fluid secretion is passed out on to the surface of the body (Fig. 4K, *dc*). When the cyst begins to harden, the tail is nipped off and the finished cyst contains only the larval body. The cyst soon hardens, in firm contact with the herbage. By the act of casting off the tail the terminal excretory canal is divided and the pore comes to occupy its definitive position at the posterior end of the body. When these changes are complete and the larva, now known as a *metacercaria*, is ensconced in a sheath which has the power of protecting it against the drying agency of wind, all necessary preparations have been made for the infection of the final host, generally a sheep, but sometimes another mammal. All that is needed for the completion of the life history is the ingestion of the cyst and its contained metacercaria.

Before envisaging the establishment of the young parasite in the adult location in the body of the mammal we may recapitulate briefly the main cycle of events which lead up to the entry of the parasite into the final host, adding a few details. The parent fluke living in the bile ducts of the liver of a mammal lays its eggs, which are swept along with the bile into the intestine. Unaffected by the digestive juices of the mammal they pass down the intestine and are cast from the body of the host along with faeces. In damp earth or in shallow pools development proceeds slowly or more quickly in accordance with the dictates of temperature, which may be the principal factor in determining seasonal infestation of snails in spring. At 10° C. development is arrested, but the eggs remain viable for long periods and continue their development when suitable conditions prevail. At 26° C. or slightly lower temperatures a stage of development is soon reached when the miracidium pushes aside the operculum of the egg, emerges, and for a period not longer than about 24 hours swims about freely in its quest for a suitable snail. Unsuccessful miracidia perish, but many are successful in penetrating into snails, lose their ciliated covering and become spherical. Having made their way to the digestive gland of the snail, where nutriment abounds, they grow, while the germinal cells multiply to form germinal balls which enlarge and differentiate into rediae.

After liberation from the enlarged and distorted miracidial body, which is the sporocyst, by rupture of its outer wall, the rediae feed with great destructive effect on the snail's tissues, penetrating ever deeper by the combined action of the collar, the procruscula, the muscles of the body and the feeding organs. Under certain adverse conditions which have not been adequately defined the rediae may produce within themselves a second generation of rediae, but generally they give rise to cercariae which escape by way of the birth pore and leave the snail 4½–7 weeks from the time of its infestation with miracidia. For an hour or two the cercariae swim freely about the pool in which the snail lives, but soon they leave the water and encyst on the stems of grasses and other plants after casting off the tail. At this, the metacercarial, stage they are ready to infect the final host. In laboratory experiments a blade of grass will accommodate up to 1000 cysts (Schumacher, 1939).

The eggs, the larvae which occur in snails and the cercariae can survive the winter and infect the final host during the following spring. But the infestation of snails generally dies down in winter. The encysted metacercariae can remain infective for a few weeks even on dried hay and for many months on moist hay, so that a mammal may be infected with the parasites without going near the pastures if winter feeds of hay are supplied to it.

We may now follow the progress of the parasite from the stage at which it enters the alimentary canal of the mammal as an encysted metacercaria along with herbage. Unaffected by the gastric juice the cysts pass through the stomach into the intestine, where they are digested and the metacercariae freed. These have most of the organs seen in the adult (Fig. 4L), but not the reproductive systems, which develop later from rudiments already laid down. The young fluke has some means of protection against the digestive enzymes of the host, but it does not remain long in the intestine, and soon migrates to the liver. A few young flukes may reach this location by the shortest route, along the main bile duct, but this is not the customary direction in which migration proceeds.

Schumacher (1939) infected guinea-pigs and rabbits in the laboratory with encysted metacercariae of *Fasciola hepatica* and studied the progress of the larvae in their migration to the final location in these hosts. Within half an hour of the cysts reaching the intestine, the metacercariae begin to emerge. Immediately they penetrate the wall of the gut and first appear in the coelom 2 hours after the time of infection. Within 24 hours most of the metacercariae in the coelom assemble and after about 48 hours some of them are boring into the liver capsule, few remaining in the coelom after 4–6 days. For a time the young flukes live in the parenchyma of the liver, eventually seeking out the bile ducts, reaching this situation about 7 weeks after the time of infection, by which time eggs are first appearing in the uterus. Schumacher also fed 2000 cysts to each of five sheep, which were slaughtered respectively 1, 2, 3, 35 and 56 days afterwards. After 24 hours, 125 metacercariae (6 %) were found in the coelom, after 48 hours 186 (9 %), and after 72 hours 116 (6 %). In all, the portal vein and the bile capillaries and ducts were free of parasites. In the coelom of the sheep killed 35 days after infection, about twenty stunted juvenile flukes were found, in the liver about 500 young worms, now 1–4 mm. long. None of them had reached the bile ducts, though eight were found in this situation 56 days after infection, as well as about 400 in the liver tissue. Thus, as might be expected, the young flukes take very much longer to reach their final location in the natural host than in small laboratory animals, though penetration of the gut is equally rapid. The path traversed by the larvae in crossing the coelom remains unknown, but it is certain that they do not use the blood circulatory system, no juvenile flukes having been found in the portal vein.

Details of the life cycle in trematodes other than *Fasciola hepatica* are in some instances very different from those outlined for this fluke, and we shall return to the subject to consider them after having discussed the structure of adult trematodes.

CHAPTER 3

THE MORPHOLOGY OF THE MONOGENEA

Some general characters of the Monogenea have been mentioned already, and in this chapter what has been said will be amplified at least into a detailed impression of the external and internal structure of this order of fluke, so that the 'shorthand' characters of families and genera which are to be given later will be fully intelligible. In dealing with abundant modifications of structure met with, examples of types or families which show them will be given in parentheses so that, in the interests of continuity, the text can be read without reference to the examples but the examples studied if need be.

I. External Structure

Monogenetic trematodes rarely exceed 1 inch in length and often fall short of 1 inch, so that they can be defined as medium to fairly large trematodes. Different forms vary from lancet-shape (Gyrodactylidae) or elliptical (Monocotylidae), through various intermediate shapes of smooth or rugged outline (Diclidophoridae), to a disk-shape (Capsalidae). Sometimes the elongate body is almost cylindrical (Udonellidae), but it may be concave ventrally (Microbothriidae), and when very wide is generally perfectly flat (Capsalidae). The lateral margins may be turned characteristically inwards and ventrally (*Encotyllabe*), but this is not generally the case. Some common shapes are shown in Fig. 6.

The parasite clings to the host while it browses on mucus or abraded tissue or blood, and the anterior end, at or near which the mouth is situated, is closely applied to the tissues of the host, the body sometimes forming a loop encircling organs like the gills (*Hexostoma*). This anterior region is often very mobile, its movements complex. At one time drawn out into a fine, tapering process it may at another be contracted to a degree which buries the mouth deep into the surrounding tissues, lip-like folds rolling into a tube which is formed by inversion of the original process. While extended, the tip of the body may move from side to side and up and down, as if thoroughly exploring the tissues of the host. The whole body in some measure shares these important characters of extensibility and contractility, a point which we are apt to lose sight of if study is confined to preserved flukes.

The cuticle is not generally as spinous as that of some Digenea which live in sheltered nooks and crannies within the host where, no doubt, spines provide a superficial roughness which assists the suckers in locomotion and in maintaining adhesion. Yet, the cuticle is not invariably smooth. It may be raised here and there into papillae and the dorsal and lateral regions of the forepart of the body may have a generous sprinkling of spines (Capsalidae). Distinctive rows of scale-like spines borne on dorsal and ventral cuticular plates or *squamodisks* characterize some genera (*Diplectanum*).

Adhesion is maintained in Monogenea by various modifications of the anterior and posterior regions of the body, especially the latter, where many kinds of holdfasts occur in different species. Sometimes a cup-shaped sucker, the adhesive organ is more often quite different, and Price (1934*a*) proposed the term *haptor* (Gr. *haptein*, to fasten) to indicate an adhesive organ without having

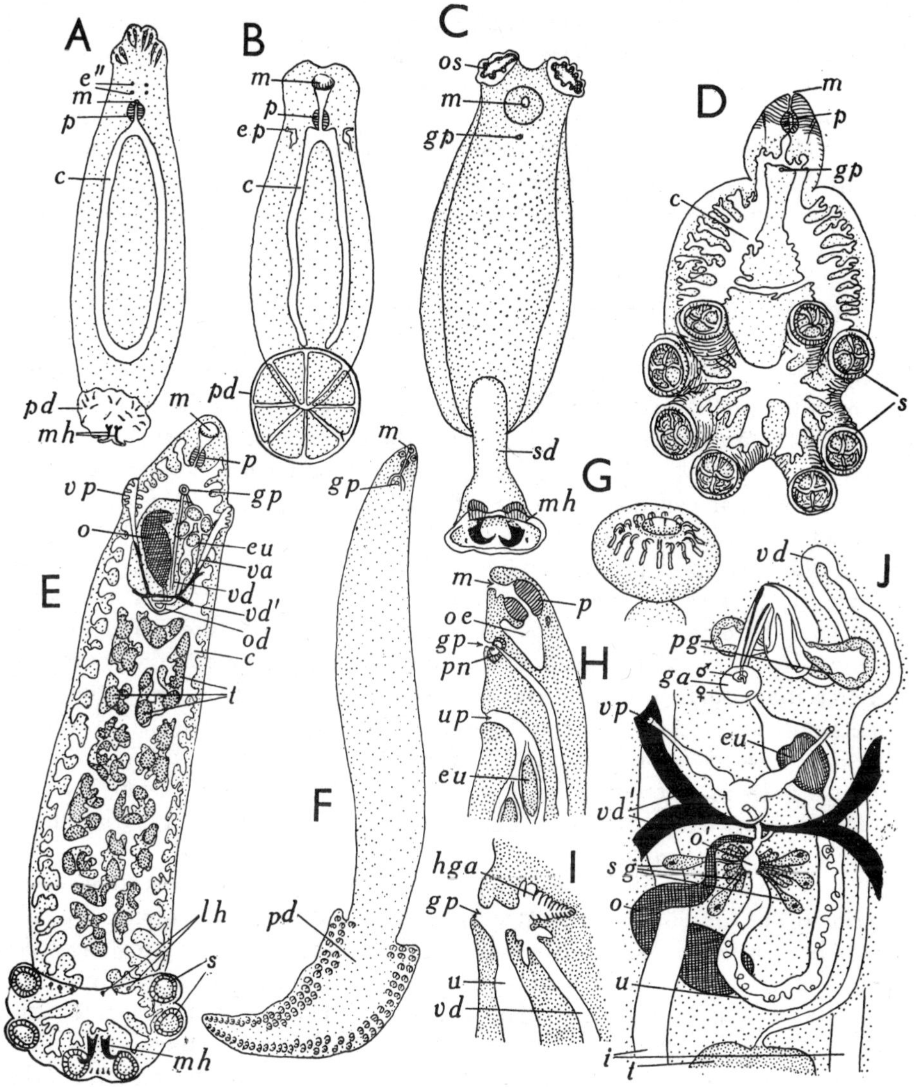

Fig. 6. Various Monogenea (A–F) and diagrams of parts (G–I) or the whole (J) reproductive system. A, *Dactylogyrus amphibothrium*. B, *Monocotyle ijimae* (now *Tritestis ijimae*). C, *Encotyllabe nordmanni*. D, *Cyclocotyla* sp. (probably *C. charcoti* = *C. bellones*). E, *Polystoma integerrimum*. F, *Microcotyle pogoniae*. Note the main types of haptor and (A, B, E) extreme variations in the digestive apparatus. E shows the complete digestive and reproductive systems. G, male organ of *Diclidophora denticulatum*. H, sagittal section of the anterior end of *D. luscae*, showing the separate genital apertures and the armed copulatory organ. I, section similar to H, but of *Microcotyle caudata*, showing common genital pore, unarmed copulatory organ and genital atrium with hooklets. J, general arrangement of the reproductive organs in *Merizocotyle diaphana*. (After Fuhrmann, 1928.)

specific morphological implications. I should like here to modify the nomenclature and propose the name *prohaptor* to denote the anterior adhesive apparatus, and *opisthaptor* for the posterior.

.(i) *The Anterior Adhesive Apparatus or Prohaptor.* In some Monogenea the mouth is encircled by an oral sucker (*Calicotyle*, *Onchocotyle*, *Polystoma*; see Figs. 17B, 18A), i.e. a cup-shaped, muscular organ essentially similar to that of Digenea, but more feebly developed and less sharply set off from the underlying parenchyma. More often there is a pair of suckers, which may be closely associated with the mouth or situated some distance from it. Both types of adhesive organ have been loosely termed oral suckers, but it is preferable to draw a distinction between them. Where the suckers occur in the lumen of the 'mouth tube' or prepharynx, they would be best called *buccal* suckers (*Microcotyle*, Fig. 6F); when situated on the anterior margin of the body at a distance from the mouth, *anterior* suckers (*Acanthocotyle*, *Encotyllabe*, *Tristoma*) (see Figs. 17A, 6C and 1A). The *oral* sucker forms a third type.

Frequently, the prohaptor is ill-defined, not sucker-like, or absent. The latter condition is rare (*Pseudocotyle*) and the former is illustrated by many organs which receive almost as many names. In a few instances there are two grooves which undoubtedly serve a suctorial purpose, one on either side of the anterior extremity (*Nitzschia*); sometimes, instead, there are lateral expansions of this region known as *head lappets* (Calceostomatidae), or papilla-like outgrowths of more concise form called *head organs* with associated multicellular *cephalic glands* and ducts by which a sticky secretion is formed and passed out on to the surface of the prohaptor (Gyrodactylidae, Fig. 14A; Dactylogyridae, Fig. 14B; *Merizocotyle*). Head organs vary in number and are important taxonomic characters; there may be one or two pairs (Gyrodactylidae), three pairs (*Ancyrocephalus*, *Merizocotyle*, *Thaumatocotyle*), several pairs (*Tetraonchus*) or numerous scattered individuals (*Cathariotrema*). In some monogenetic flukes neither head lappets nor head organs occur, but antero-lateral glandular areas alone serve in adhesion (*Entobdella*, Fig. 19A, B; *Macrophyllida*), and in a few instances glandular areas coexist with anterior suckers (*Pseudobenedenia*). In rare instances the mouth is encircled by a somewhat membranous structure which is called a pseudosucker (*Monocotyle*).

Whatever its structure, the primary function of the prohaptor is to apply the anterior tip of the fluke to the substratum during feeding operations. But it is not to be denied a possible locomotor function, because in many instances it is capable of preserving attachment to the host when the worm is seeking a fresh hold with the opisthaptor. The alternate action of the two sets of organs, together with muscular movements of the body, may produce a rudimentary looping movement, though there is little need for the fluke to move far once it has established itself on the host. Most Monogenea are comparatively quiescent, but a few can move very briskly over organs like the gills of fishes (*Microcotyle*).

(ii) *The Posterior Adhesive Apparatus or Opisthaptor.* At its simplest, the opisthaptor is a ventral, disk-like outgrowth of the posterior end of the body such as is seen in some larval Monogenea which later develop more elaborate organs, and a variety of which occurs in some Turbellaria (*Macrostomum*). More complicated types of haptor have probably been derived from this in the phylogeny of Monogenea as they are seen to be in the ontogeny. The disk usually rests on a pad of underlying muscle and is itself muscular. It may be unarmed (Microbothriidae, Fig. 18A, *ps*), but generally it is provided with large and small cuticular hooks and hooklets, as well as accessories by which they are moved after the fashion of anchors or grapnels. The strongly recurved tips can be caused to bite into the tissues of the host and can be abstracted again by the action of muscles which elevate and depress them.

The simpler types of opisthaptor occur in members of the suborder Monopisthocotylea, and the simplest type of all is not sucker-like but just a broad, ventral outgrowth of the body carrying a somewhat complicated armature (Gyrodactylidae, Dactylogyridae, Fig. 14). Other flukes belonging to this group have a simple, undivided, discoidal haptor with a flat or only slightly concave lower surface and a slightly inturned margin, but without hooks or hooklets (Udonellidae, Fig. 15F, *ps*). Many others have an opisthaptor which is little modified from this condition, except in some instances for the presence of hooks and hooklets (*Encotyllabe*; *Entobdella*, Fig. 19A; *Ancyrocotyle*). Some Monopisthocotylea have a discoidal haptor whose ventral surface is divided by slightly raised, muscular, radial septa into a number of sectors, each of which has individual suctorial action and is called a loculus (Figs. 6B, 17B). The number of septa and loculi varies; there may be six or seven (*Megalocotyle*), seven (*Dasybatotrema*), eight (*Monocotyle*, Fig. 6B; *Heterocotyle*, Fig. 16A) or ten (*Trochopus*, Fig. 17F2), and the radial symmetry may be disturbed by the incompleteness of some of the septa and the fusion of others (*Trochopus*) and/or by the bifurcation of the posterior septa (*Capsala*). In more complicated haptors of this type other septa may extend parallel with the margin of the disk. By such modifications there may be, in the simplest cases, one central loculus and a small number of peripheral loculi (*Calicotyle*, Fig. 17B; *Heterocotyle*, Fig. 16A) or, in the most complicated ones, one central loculus, six intermediate and eighteen peripheral loculi (*Merizocotyle*). Radial may give way to bilateral symmetry in such instances, as when the posterior loculus is much larger than the others (*Thaumatocotyle*, Fig. 16B). One other modification of this simpler type of opisthaptor might be mentioned; the tremendous reduction in size of the true disk, which may then occur on the posterior margin of a large secondary disk or *pseudodisk* bearing radial rows of spines (*Acanthocotyle*, Fig. 17A).

A higher grade of organization is seen in the suborder Polyopisthocotylea, the opisthaptor comprising a number of muscular, cup-like suckers or clamps set on a posterior disk or on the naked ventral surface of the body in this region (Fig. 6D, E, F). Generally, the suckers occur in a paired series, but asymmetrical arrangements are met with (4 and 1 in *Grubea*; 4 and 2 in *Hexostoma dissimilis* (Yamaguti, 1937)). The number of suckers or clamps on either side of the body may be one (*Sphyranura*), two (*Platycotyle*), three (*Plectanocotyle*, *Polystoma*) or four (*Diclidophora*, *Mazocraës*), and Poche (1926) used these distinctions for the purpose of separating families, though the validity of such familial characters has been questioned. The most posterior pair of suckers may be of very small size (*Hexostoma*, Fig. 23A), and, besides being of small size, may be borne near the tip of a dorsal outgrowth of the disk which carries the three remaining pairs of large suckers (Hexabothriidae, Fig. 20A).

Hooks and hooklets may occur on all forms of opisthaptor and are generally best developed on the simpler types (Figs. 6A, 14A, B, E*e*). They have been referred to as 'chitinous', but Remley (1942) applied a specific test for chitin with negative results. In some species at any rate, and possibly in most if not all, the hooklets are transferred to the adult from the larval stage, but have been frequently overlooked by investigators. Price (1937*b*) reported their presence in numerous genera of the Capsalidae (*Capsala*, *Tristoma*, *Trochopus*, *Nitzschia*, *Ancyrocotyle*, *Entobdella*, *Benedenia* and *Encotyllabe*) and they occur in other families (Gyrodactylidae and Dactylogyridae). They generally number 14–16, but are fewer in rare instances, and because of their position are known as *marginal* hooklets. The true disk of *Acanthocotyle* bears such larval hooklets, the arrangement of which varies in different species.

In some simple disks which possess marginal hooklets, large hooks are wanting (*Isancistrum*), but generally there is one pair (*Gyrodactylus*, *Dactylogyrus*, *Bothitrema*) or two pairs (*Diplectanum*, *Tetraonchus*). They are sharply pointed and recurved at the tips and are supported by two transverse bars, generally dorsal and ventral in position, sometimes by a single bar only. In forms with a more specialized disk the hooks are different, the supporting bars absent. Elaboration of the disk seems to imply simplification of the grapnels, though perhaps not proportionately. There may be one pair of large hooks (*Polystoma*, *Calicotyle*), (Figs. 6E, 17B 1) or several pairs of much smaller ones (three in *Trochopus*, Fig. 17F 3, 4), and the members of one pair may overlie and thus partially obscure those of another. Mention might also be made of the small stalked appendage bearing hooklets, which occupies the posterior tip of the body in some instances (*Plectanocotyle*, Fig. 17 G 1–3). Also of caudal glands essentially similar to the cephalic glands, but posterior in position which, in some Monogenea, pour their secretion out on the opisthaptor and thus aid adhesion.

Cuticular structures superficially resembling hooks of fantastic shape occur in association with the suckers of the opisthaptor in some Monogenea. They are really deep layers of firm cuticular material providing support for the suckers and attachments for the muscles that bring about their suctorial movements. Before coming to them, apart from thus forewarning students who might be deceived by their appearance in whole mounts, we must consider certain muscular arrangements in the integument and parenchyma. And before thus turning to the internal structures, we might note the more important external apertures of certain organs and make a few remarks on the nature of the cuticle.

(iii) *The External Apertures.* The most obvious opening is, of course, the mouth. It may be prominent or inconspicuous, according to the nature of the prohaptor. Sometimes it is at the anterior tip of the body, or very near it, sometimes much farther back on the ventral surface. It is generally rounded, but may be slit-like (*Microbothrium*).

Other, less prominent openings are the pores of the reproductive and excretory systems. The male and female reproductive systems may open by separate genital pores which may be close together or widely separated, even on different surfaces of the body, or may open together by a common genital pore at the outer end of a genital atrium. Generally, the genital pores occur in the anterior region of the body and their positions are mentioned in diagnoses and descriptions which are reserved for a later chapter. In addition to the genital pores mentioned, others occur in connexion with the female reproductive system, the openings of the vagina or vaginae. These are subject to wide variation and may be single or paired or wanting. When only a single vaginal pore exists it may be ventral and median or slightly, sometimes considerably, lateral. Rarely it may occur on the dorsal surface. Similar variations occur when paired vaginal pores exist; these may be near the median plane or somewhat distant from it. The external apertures of most constant position are the excretory pores, which generally occur near the margins of the body in the anterior region. Other openings relate to glands already mentioned; mostly they are anterior, but may be posterior in position.

(iv) *The Cuticle.* The cuticle is a protective layer of non-living, secreted material, supposedly of a chitinous nature, but not proven so to be, which owes its origin to subcuticular cells that are rarely evident in Monogenea and are believed to have sunk into the parenchyma after producing their secretion. We can discuss the nature of the cuticle of Trematoda in dealing with the Digenea, but might mention here that in some Monogenea it is perforated for the ducts of the cephalic glands and for those of certain other glands, some at least of which produce rhabdites, i.e. rod-like secretionary bodies which are extruded, swell up on coming into contact with water to form an investing slime, and are more characteristic of Turbellaria than of Trematoda.

II. Internal Structure

It is impossible to do justice to the great variety of internal structure seen in Monogenea, and at most only the main anatomical characters can be outlined. The internal organs which show greatest variability are the reproductive systems, especially the copulatory apparatus belonging to the male system. For the most part, the internal anatomy has been insufficiently studied, and the fleshy nature of the body renders study difficult in many instances, but the reproductive organs and alimentary system can generally be made out with comparative ease in whole mounts of the living trematode or slides prepared for permanent use under the microscope.

(i) *The Muscular System.* Immediately beneath the cuticle there are several layers of muscular fibrils which help to form the integument and are responsible for the general movements of the animal. The outermost layer is built up of circularly arranged fibrils, which are wanting in a few Monogenea (*Hexostoma, Hexabothrium*), and is never very thick. Beneath it in turn, or superficially in its absence, are layers of diagonal and longitudinal muscles, the latter generally of considerable thickness. These, the main muscles of the body, function antagonistically. When the circular muscles contract, the longitudinal and the diagonal ones in some measure relax, and *vice versa*. In the first instance the worm becomes slender and attenuated, in the second shorter and thicker. Additional muscular fibrils traverse the parenchyma from the dorsal to the ventral integument, and by their contraction tend to flatten the body.

(ii) *The Parenchyma.* This is a tissue of loosely packed cells and fibrils which fills up all of the available space around the internal organs. In its simplest condition it is a tissue of polygonal cells which link the organs with the parenchyma, but in some instances it is differentiated into ecto- and endoparenchyma, named according to their relative positions, the former consisting of masses of fibrils secreted by cells and the latter of cellular masses. In a few Monogenea the cell boundaries are largely obliterated, so that the parenchyma is syncytial. Fibrillar protoplasm invariably stiffens the ends of the body, the terminal parts of the genital and other ducts, as well as other parts in which rigidity is essential for mechanical reasons.

(iii) *The Mode of Action of the Suckers.* Very little is known about the way in which the suckers produce their adhesive effects. In general, the action resembles that of the leather or rubber sucker with which schoolboys were once glad to play, but doubtless many unknown muscular co-ordinations occur which would well repay close study. In some instances there are two clamp-like jaws capable of gripping gill filaments pincer fashion (*Microcotyle*, etc.). Here we can merely comment briefly on the arrangements which may be seen in *Hexostoma extensicaudum* (Fig. 23 A), a monogenetic fluke with suckers, not clamps, which is found on the gills of the tunny in the North Sea and probably occurs elsewhere in Europe. This trematode has an opisthaptor comprising eight suckers arranged in pairs, of which the most posterior are of very small size. In consequence of this arrangement adhesion may have a different mechanical basis from that seen in other flukes, but its details are instructive. The suckers are composed not of muscular fibrils as in Digenea and some other Monogenea, but of fibrils which at most possess the property of elasticity. They extend perpendicular to the surface of the sucker, their bases firmly inserted in cuticular, plate-like, skeletal pieces which superficially resemble hooks, but are thus completely buried (Fig. 7 C, *ps*, *sps*). Each sucker has three such skeletal supports and is permitted a measure of flexibility by virtue of intervening regions of more pliable tissue.

The large suckers are shallow and saucer-like (Fig. 7 C), but the degree of concavity can be altered by the action of two sets of muscles, both of which arise from the longitudinal muscle layer of the body. One set, originating in the outer part of this layer, is inserted in the rim of the sucker (Fig. 7 C, *slm*), the other, arising in the deep part of this layer, is inserted in the base of the sucker (Fig. 7 C, *dlm*), along with dorso-ventral muscles (*dvm*). All these muscles are firmly attached to the cuticular skeletal structures and their action is readily understood. Contraction of the fibrils inserted in the base of the sucker tends to increase the degree of concavity and produce the necessary vacuum upon which the suctorial effect depends. Contraction of the fibrils inserted in the rim of the sucker tends to raise the rim sufficiently to permit the entry of water and thus dispel the vacuum, thereby serving for release of the sucker. It is especially interesting to find such antagonistic muscles originating from the same muscle layer of the body. More complicated arrangements than these occur in forms like *Diclidophora* which have pedunculate suckers (see Llewellyn, 1941 *c*). Here a powerful ventral muscle is splayed posteriorly into four pairs of muscles which enter the peduncles, one to each, and additional transverse muscles connect the peduncles in pairs. How such muscles work is not known.

The small, posterior suckers of *Hexostoma* belie their appearance, for they are more deeply concave than the large ones and hold the tissues of the host in a stricture-like grip, a plug of the gill tissue being pinched off to a narrow neck (Fig. 7D). They provide a more permanent attachment to the host, as is seen by the enormous extent to which the underlying tissues of the host are scored. The parasite lies in a cavity eroded in the host's tissues and descending almost to the cartilaginous rays of the gill to which it is attached. It seems likely that the small

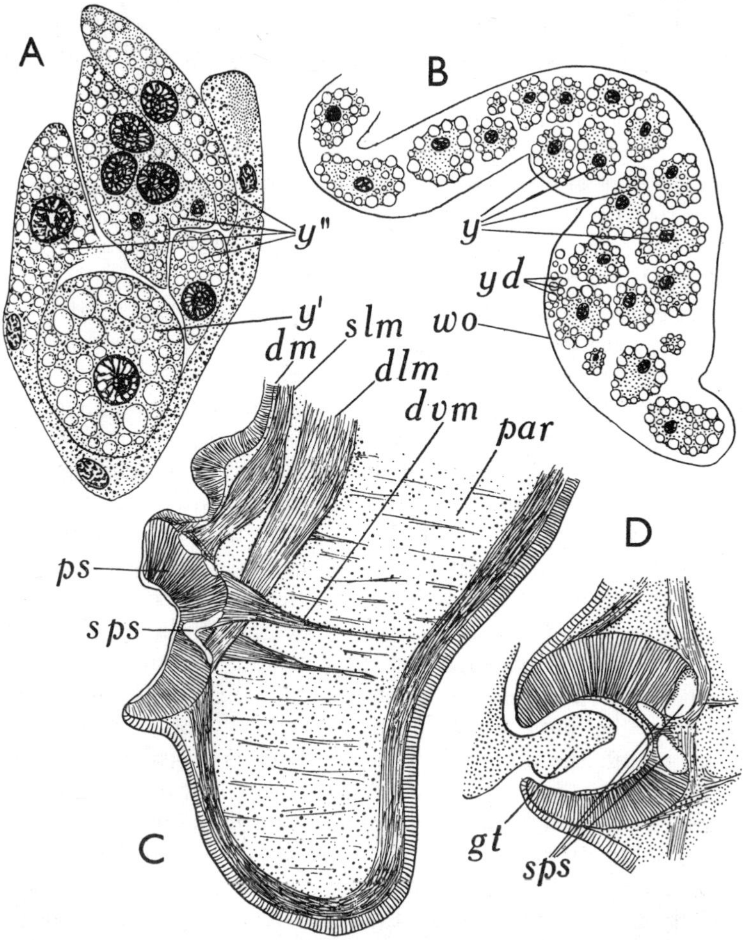

Fig. 7. A, B, the origin of shell-forming material in *Hexostoma extensicaudum*. In A, droplets of fluid are forming in the cells of a nest from the vitellarium (*y″*, early; *y′*, later stages in the process). B showing longitudinal section of the vitelloduct containing cells bulging with droplets of shell substance. C, longitudinal section of posterior region of body showing one large sucker and the muscles that work it. D, longitudinal section of small posterior sucker, showing its firm grasp of the gill tissue. (Original.)

suckers perform the sustained task of adhesion to the host, the large ones perhaps functioning intermittently in emergencies, as when the strong respiratory current in sweeping over the gills threatens the parasite with expulsion from the gill chamber.

(iv) *The Nervous System and Sense Organs.* The nervous system comprises two main clusters of nerve cells or ganglia, linked by a transverse commissure and forming a rudimentary 'brain', and also a number of bundles of fibres or nerves. The brain occupies an anterior position in the body and the nerves extend from it in several directions, anteriorly, laterally and posteriorly for the most part, and are situated in the dorsal, ventro-lateral and ventral regions. In some members of the Polystomatidae, Capsalidae and other families the oesophagus may be enclosed within a nerve ring from which various nerves issue. Variable numbers of nerves emerge from the brain and extend to different parts of the body, but the anterior region is always well supplied, and three pairs of prominent nerves generally proceed posteriorly. Of these, a ventral pair is especially well developed, a ventro-lateral pair delicate and a dorsal pair inconspicuous, sometimes absent, though well formed in *Tristoma* and *Hexabothrium*. The ventral nerves may be connected at intervals by transverse commissures, an arrangement which is reminiscent of the nerve network of some Turbellaria. The various muscles of the opisthaptor are supplied with nervelets from the ventral nerves and the pharynx is another organ with a rich innervation.

Sense organs are more likely to be of use to the monogenetic than to the digenetic fluke, because the former is subjected to the vicissitudes of a more variable environment, and, accordingly, they are better represented, chief amongst them being simple pigmented eye-spots, clusters of pigmented cells and, sometimes, a lens-like structure. Some species of Monogenea have one pair of eye-spots, others two. In *Tristoma* a rudimentary type of taste organ is said to exist, and in most Monogenea the integument is plentifully supplied with fine nerve endings, especially near the tip of the body and in the haptors.

(v) *The Digestive System.* In some Monogenea, as already remarked, there is a mouth tube which may or may not have buccal suckers in its walls. The true mouth is situated at its base. In Digenea the corresponding region is known as a prepharynx, but in Monogenea without a definite oral sucker it is impossible to know whether or not the two regions are homologous. In some instances it seems unlikely, the mouth tube being periodically obliterated by eversion. Beyond this region the alimentary canal proper is divisible into three parts, pharynx, oesophagus and intestine. When buccal suckers do not exist, the pharynx may serve as a sucker. In some Monogenea, this organ also is protrusible (Udonellidae), and it may further resemble the pharynx of Turbellaria by its enclosure within a pharynx sac. The resemblance to rhabdocoele Turbellaria may go further still, as when the intestine is simple and single (*Udonella*, Fig. 15; *Tetraonchus*). This condition may be modified by the development of fenestrations in the region of the gonads.

Both pharynx and oesophagus are variable in size and shape. One characteristic modification of the former is a central constriction (*Capsala*). The intestine is often considerably modified from the simple, sac-like condition. Generally, the root of the intestine bifurcates, forming two tubes which end blindly, the caeca or intestinal crura. These may be simple and unbranched (*Gyrodactylus*, Fig. 14A; *Ancyrocephalus*), and in some instances they merge posteriorly to form a ring-like intestine (*Dactylogyrus*, Fig. 6A). More often, the caeca have branches, both median and lateral in some instances, and the branching may be dendritic (*Entobdella*, Fig. 19A, *c*; *Benedenia*). The finest branches may also coalesce to yield anastomoses (*Polystoma*, Fig. 6E). When the opisthaptor is disk-like, the median posterior anastomosis of the caeca may pass into it to give off branches which ramify in its tissues (*Polystoma*, *Diclidophora*). In some Hexabothriidae branches of the caecal stem may pass into the dorsal process bearing the small suckers (Fig. 20A). Histologically, the lining of the caeca may be a continuous epithelium of cubical or cylindrical cells, or the cells may be discontinuous. Much scope still exists for histological study of this as well as other organ systems in Monogenea.

(vi) *The Excretory System.* We are ill-informed about the details of the excretory system in Monogenea, partly because it is obscured by other organs and difficult to observe, even in living flukes. The ultimate units are flame cells, however, and these will be discussed in connexion with the Digenea. In some Monogenea, fine tubules form networks, but generally, if not invariably, a pair of canals are formed by their union, to course longitudinally through the lateral regions of the body. They begin anteriorly and extend back towards the opisthaptor, where they turn about and, widening considerably, pass forward to the level of the pharynx. Here, each enters a spherical vesicle which communicates with the exterior by a short canal. In distinction to the Digenea, therefore, the excretory pores are paired, lateral and anterior.

(vii) *The Reproductive Systems.* The male components of the hermaphrodite reproductive system are somewhat simpler than the female, though the testes may be modified and the copulatory organ complicated. Many Monogenea possess only a single testis (*Udonella*, *Microbothrium*, *Monocotyle*, *Heterocotyle*, *Leptocotyle*, *Anoplodiscus*), many others have two (*Diplorchis*, *Trochopus*, *Encotyllabe*, *Entobdella*), one genus at least has three (*Tritestis*), and several genera have numerous testes (*Pseudocotyle*, *Tristoma*, *Capsala*). When the testes are numerous they may number less than twenty-seven (*Nitzschia*) or as many as 200 (*Rajonchocotyle*). The primitive number is probably two, and it is doubtful if much larger numbers should be regarded as individual gonads, more probably representing partitioning of the original ones. As a rule, the testes tend to fill up much of the median parenchyma between the caeca, especially when numerous, and they are generally situated behind the ovary, though there are exceptions to this rule (*Diclidophora*, *Tristoma*).

Spermatozoa which develop in the testes pass along fine ductules, the vasa efferentia, which by their union ultimately form a main duct or vas deferens, having or lacking a swollen seminal vesicle, connecting the gonads with the copulatory organ. Often the male organ is comparatively simple, but sometimes it is complicated by accessories whose functions have not been defined adequately. It may be protrusible and eversible (i.e. a *cirrus*) or only protrusible (i.e. a *penis*), so that it is more variable than in Digenea. It projects into an invagination of the ventral integument or genital atrium as a rule, sometimes into a pocket of such a chamber, into which the uterus of the female system may or may not open. The genital pore or pores are more constant in position than in the Digenea, generally in or near the median plane behind the bifurcation of the intestine, but sometimes decidedly lateral (Capsalidae).

The lining of the copulatory organ is frequently a cuticularized ejaculatory duct, and in its more elaborate form the organ may be complicated by a cuticular basal plate and other structures, amongst which are special protractor and retractor muscles (Fig. 18C). The ejaculatory duct sometimes receives the ducts of a pair of large prostate glands (*Merizocotyle*, Fig. 6J, *pg*). At the other extreme, a definite copulatory organ may be wanting, self-fertilization probably being the rule in such instances (*Udonella*). Other monogenetic flukes have a simple organ formed by modification of the terminal part of the vas deferens into a sharply pointed, fibrillar or muscular structure which can be protruded into the genital atrium (*Hexabothrium*, *Microcotyle*). Hooklets may increase the power of the copulatory organ to remain inserted in the genital atrium of another worm during copulation (*Dactylocotyle*, Fig. 6H, *pn*, 6G; *Diclidophora*). In the absence of such an arrangement the same purpose may be served by hooklets in the wall of the genital atrium or a pocket of it (*Microcotyle*, Fig. 6I, *hga*).

Various types of female reproductive organs also occur in Monogenea. The ovary is rarely globular, as it often is in Digenea, but is generally elongate and much folded, sometimes lobed as well. As a rule, it is situated in front of the testes. Ova are gradually developed in all parts of the ovary, but they become mature in the terminal part and pass when ripe into a short tube, the oviduct. In some species (*Hexostoma*) there is a mechanism, comprising a chamber with two sphincters, the ovicapt, for spacing out the ova as they pass into the oviduct and ensuring a steady, continuous flow of them. One other region of the oviduct is specially modified into a chamber in which the ova come into contact with spermatozoa and vitelline cells from the vitellaria and are subsequently encapsulated. This is called the *ootype*, the wall of which is perforated for the ducts of unicellular glands situated outside it and termed *shell glands*. Not infrequently in the literature these two terms are confounded. Several larger ducts open into the ootype or some neighbouring part of the oviduct: (*a*) the *vagina* or *vaginae*, a tube or tubes along which spermatozoa received in copulation from another individual are passed, for storage temporarily in a small chamber, the receptaculum seminis, situated conveniently near the ootype. This part of the female system is sometimes lined with cuticular material and is extremely variable in structure and appearance. The vagina may be absent (*Calceostoma*, *Oculotrema*) and it may be single (*Microbothrium*, *Monocotyle*, *Heterocotyle*) or double (*Leptobothrium*, *Calicotyle*, *Merizocotyle*, *Polystoma*). Such characters may be of great taxonomic importance. The variable positions of the external orifices of the vagina or vaginae have been commented upon already. Other ducts are: (*b*) the *vitelloduct*, a tube formed by the union of lateral ducts along which cells from the vitellaria pass on their way to the ootype; (*c*) the *uterus*, a wider tube which receives the encapsulated ova and conducts them to the genital atrium and so to the exterior. The terminal part of this tube or *metraterm* may be muscular, like the vagina. Finally, (*d*) the *genito-intestinal canal*, a tube which links the oviduct or some part of it with the right caecum, rarely the left, and is probably homologous with Laurer's canal of Digenea. It is absent in all members of the suborder Monopisthocotylea.

Unhappily, the vitellaria were named before their primary function was ascertained. They are paired organs composed of clusters or follicles of cells extending along the body lateral to the caeca, sometimes encroaching on the median parenchyma between the caeca, and of ductules which unite to form ultimately the main vitelloducts. The cells are responsible for the production of fluid secretion out of which the egg capsules are formed and also for its transport to the ootype. In rare instances (some Gyrodactylidae) vitellaria are absent, or may be associated with the ovary to form a germ vitellarium such as characterizes some Turbellaria.

(viii) *The Mode of Formation of the Egg Capsules.* Before proceeding to discuss this subject we must remember that the viviparous habits of some Gyrodactylidae (e.g. *Gyrodactylus elegans*) obviate the need for producing encapsulated eggs. Passing over this interesting point, to return to it later on, we must note also that while some Monogenea produce eggs in considerable numbers, the majority have only one egg or a few eggs in the uterus at any one time. The eggs are relatively large (as much as 0·1–0·3 mm. long), and they may have polar filaments, ten or even fifty times as long as the capsule itself, at one or both ends, though this is not invariably the case. The egg is generally ovoid or spindle-shaped, but we must note that shape varies in different species. A lid or operculum may occupy one pole, but as a rule the eggs are non-operculate. In attempting to outline the mode of formation of the eggs we shall rely on what happens in *Hexostoma extensicaudum*, but the process is probably the same in all trematodes which have non-operculate eggs, and perhaps not very different in those with operculate eggs.

After the trematode has received spermatozoa from another individual in copulation, cells which have undergone a cycle of changes in the vitellaria are split off from the follicles and pass down the vitelloducts towards the ootype. They are crowded with droplets of secretion (Fig. 7A), which eventually cause the cell membrane to bulge and are in process of being extruded when the ootype is reached (Fig. 7B). Meanwhile, ova have come to maturity in the ovary, so that all the preparations for the manufacture of eggs have been made. A precise and beautifully synchronized mechanism now comes into play. An ovum leaves the ovary along with others in a steady stream, impelled in some instances by contraction of the wall of the ovary, and spaced out at the ovicapt adjoining the ootype. As the first ovum enters the ootype, it is followed by a rivulet of spermatozoa from the receptaculum seminis and by a small cluster of vitelline cells laden with secretion. Then, in an instant, a capsule appears, enclosing the three kinds of cells. How this comes about has so far defied the most acute powers of observation, but at any rate we know with certainty that in the process the enclosed vitelline cells lose their droplets of secretion, which must have contributed substantially to the capsule.

Other cells immediately outside the wall of the ootype, the shell glands, certainly secrete a thin fluid into this chamber, but obviously not in sufficient amount to account for all the capsular material laid down. For this fluid several possible functions have been postulated, (*a*) the actual source of shell materials, an idea which is reflected in the name of the glands, (*b*) a medium for the nutrition of the ovum and spermatozoa, (*c*) a lubricant which ensures the smooth passage of the eggs into the uterus, (*d*) an agent which brings about speedy hardening of the capsule. A more recent opinion is that the secretion of the shell glands forms a thin basic layer upon which the material from the vitelline cells is deposited from the inside (Dawes, 1940*b*). If this view is correct, it is difficult to prove its correctness, because the capsule shows no sign of stratification, though when some trematode eggs are treated with NaOH a layer which is recognizable because of external markings is removed. Yet another possibility presents itself. The secretion of the shell glands may, perhaps by virtue of contained electrolytes, induce the vitelline cells to extrude their secretion at a given instant, as they undoubtedly do. Such an effect, which is paralleled in similar physiological circumstances, calls for only a minute amount of secretion, which is all we can expect the insignificant shell glands to produce. Whether or not this view is the correct one, we may be certain that the bulk of the capsule originates in the cells of the vitellaria. It is perhaps significant that in *Sphyranura oligorchis* the ova are encapsulated not in the ootype but in the first part of the uterus (Alvey, 1936), because we may regard this effect as produced by a somewhat delayed action on the part of the shell-gland secretion in evoking liberation of the shell material from the vitelline cells.

No sooner has one egg appeared in the ootype than it is passed into the uterus and another is in process of formation by repetition of the same preliminaries. The shape of the capsule is undoubtedly that of the mould in which it was formed, namely, the ootype. This is generally more elongate than in Digenea, sometimes extending most of the distance across the body (*Hexostoma*). Polar filaments are indubitably formed as threads of viscid secretion like that composing the capsule proper. The clue to their frequency of occurrence in Monogenea and rarity in Digenea is the relationship between the vitellaria and the numbers of eggs produced. Monogenetic trematodes have extensive vitellaria yet produce only few eggs; digenetic trematodes often have puny vitellaria but produce numerous eggs, as a rule. Polar filaments are produced from excess of the viscid secretion of which the capsule is formed, and this must be copious in Monogenea but slight in Digenea. So considerable an excess exists in some Monogenea (e.g. *Erpocotyle eugalei*) that the eggs are joined by continuous filaments, forming a long chain. This implies an abundance of shell material which is not available in Digenea. The rate at which the capsules are formed probably has some bearing on the problem, but it can scarcely be argued that this is greater in Monogenea which produce few eggs than in Digenea which may produce at least a thousand times as many. Scope for further investigation exists, and interesting developments await future work on these problems.

CHAPTER 4

THE STRUCTURE OF *ASPIDOGASTER* AND THE GENERA OF THE ASPIDOGASTRIDAE

The Aspidogastrea contains only a solitary family, the Aspidogastridae, in which there are merely nine genera. These are known, however, as parasites of poikilothermous vertebrates and invertebrates of both salt and fresh waters in five continents. Notably, they occur in gastropod and lamellibranch Mollusca, the larger Crustacea, some fishes and chelonians. They have been known to zoologists since 1827, when Baer discovered *Aspidogaster conchicola* in fresh-water mussels of the genera *Anodonta* and *Unio*. This particular species inhabits various locations in the host and can be seen through the transparent wall of the pericardium. Sometimes, 20–30 individuals are closely packed in the anterior region of the pericardial cavity near the internal opening of the kidney. Juvenile individuals are said to inhabit the intestine and may occasionally be found encysted in the pericardial gland.

The genus *Aspidogaster* is not confined to Mollusca, however, but has been reported from both fresh-water and marine fishes on both sides of the Atlantic Ocean. The fact that the same species may occur in both molluscs and fishes indicates that the former may play the part of both the intermediate and final hosts. Parasitized mussels generally present a starved and shrunken appearance, but whether or not the obvious sickness is attributable to the parasite is not certainly known.

THE STRUCTURE OF *ASPIDOGASTER CONCHICOLA*

(i) *External Characters.* An adult *Aspidogaster conchicola* is generally about 3 mm. long and 1 mm. in greatest breadth, though much smaller individuals than this may reproduce. The body is anvil-shaped, the upper part containing the viscera and the lower forming an enormous adhesive disk. The two parts are separated by a groove which extends along the anterior and lateral regions but is ill-developed in the posterior region. The anterior tip of the body proper is formed of a trumpet-like process which extends well in front of the disk. This part of the animal is very mobile, sometimes extending until it is as long as the disk. The posterior end of the animal is less plastic, but at times it far overreaches the hinder limit of the disk. When the anterior process contracts, lip-like folds of the outer surface roll into a kind of 'mouth tube', until the body is reduced to the length of the disk. The body is covered with a layer of cuticle, which is thinner on the disk than elsewhere and is everywhere spineless.

A transversely oval mouth occupies the tip of the body, which is somewhat curved ventrally (Fig. 1 B1, B2, *m*), and a single median genital pore occurs within the groove separating the anterior process from the disk. Two excretory pores lie side by side where the posterior end of the body joins the disk.

The opisthaptor, perhaps the most distinctive character of this trematode, is a very powerful multiple sucker which can cleave to a sheet of glass so firmly that the animal can be detached only by drastic means such as the scraping action of a scalpel. It is oval in outline, somewhat modified by crenations, and has numerous compartments, the *alveoli*, on its lower surface. These are formed by the intersection of about thirty transverse and three longitudinal ridges. A single alveolus is itself an efficient sucker, and the total adhesive power of the disk may be judged from the fact that it has 118 alveoli (Fig. 1 B 1, B 2, *vdi*). The ridges are formed largely of muscular fibrils perpendicular to the cuticle. Those in the transverse plane lie opposite indentations of the disk laterally, and where they meet the margin there is a series of small retractile papillae, the marginal organs, which are believed but not proven to be sense organs. The ridges mark the limits of the alveoli, each of which is about 0·05 mm. long and 0·15 mm. wide.

(ii) *Internal Characters and Integument.* At the base of the adhesive layer of the disk, which is about 0·05 mm. thick, a membrane separates the disk from the parenchyma. This bears delicate muscle fibrils arranged in two layers, an upper one in which the orientation is transverse and a lower one in which it is horizontal. The muscular arrangements which have been described offer no indication as to how adhesion is achieved and the subject needs further investigation.

The cuticle, which is not more than 0·008 mm. thick, is continued into the various openings which occur on the surface of the body and extends for some distance into the alimentary canal. It is said to be perforated at indeterminate points by the ducts of unicellular glands which secrete on to its surface. At least three kinds of glands have been described, but the differences between them are too vague to warrant a detailed statement. The main type is flask-shaped and granular and occurs in the mouth tube and near the rim of the mouth. Other types exist elsewhere on the body and on the disk.

Beneath the cuticle several strata of muscle fibrils have been identified, a superficial layer of circular, next a layer of *hollow* longitudinal, then two intersecting layers of diagonal and further circular fibrils. The body wall is thus seen to be more complicated than in other Trematoda. The muscles have smooth surfaces, and the ends of individual fibrils are inserted in a deeply staining layer at the base of the cuticle. Myoblasts of two sizes have been identified. Underneath the muscular layers loose parenchyma lends support to the internal organs. In the anterior and posterior regions of the body and laterally this is traversed by vertical (dorso-ventral) muscles.

The parenchyma is also traversed by a thick horizontal partition or *septum* which is not found in other trematodes. This divides the body into an upper and a lower part. It is concave above and merges with the side walls of the body, and incomplete posteriorly, extending back to just beyond the termination of the intestine. The septum is thus a trough-like shelf which extends through the anterior two-thirds of the body. It has an upper layer of transverse and a lower layer of longitudinal muscles, both of which are continued into the general musculature of the body, but its function is not known. It is conceivable that contraction of its muscles decreases the concavity of the septum, relaxation increasing it, and that it is able thus to generate an ebb and flow circulation of fluid in the meshes of the parenchyma.

The septum, as has been said, divides the body anteriorly into an upper and a lower compartment. The upper contains the alimentary canal, the terminal parts of the genital ducts and the vitellaria, which extend along the outer margins of the intestine; the lower contains the ovary, oviduct and ootype and the testis (Fig. 8B, C). The intermediate portions of the genital ducts thus pierce the septum.

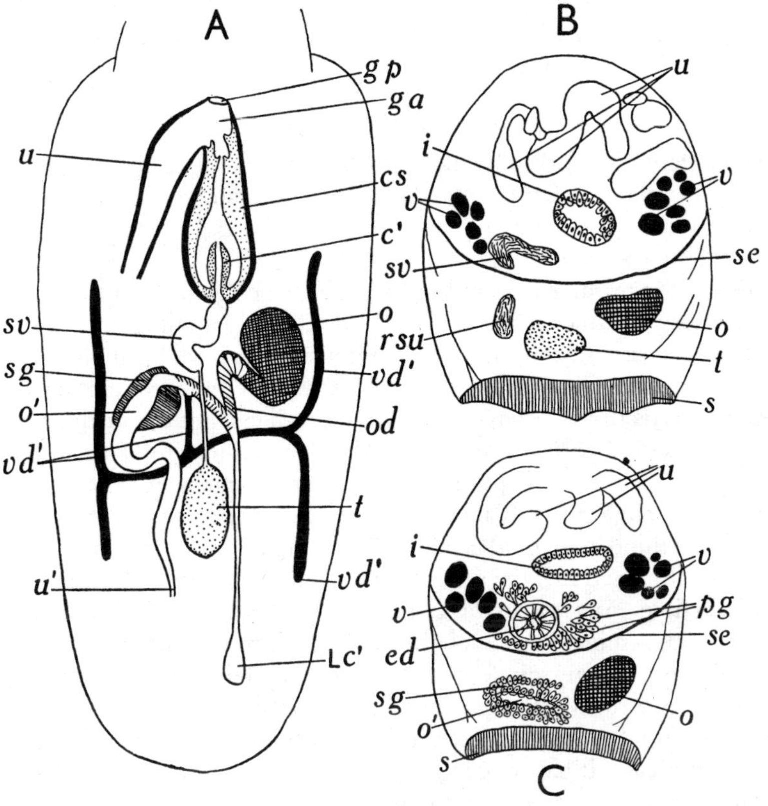

Fig. 8. *Aspidogaster conchicola.* A, schematic diagram of the reproductive system. B, C, transverse sections of the body in the region of the testes (B) and more anteriorly (C). (After Stafford, 1895.)

At the hinder end of the mouth tube or prepharynx, which is muscular but does not possess suckers, there is a strong, muscular pharynx which is longitudinally ovoid. Beyond this there is a simple sac-like intestine. Sections show that the cuticle extends along the alimentary canal to the middle of the pharynx, and this unusual character, which needs reinvestigation, is accompanied by two others. The intestine, which ends blindly in the parenchyma, is lined by an epithelium of tall cells and is provided with a muscular coat comprising outer circular and inner longitudinal muscles. The wall of the pharynx is built up of two thin layers of circular muscles between which there is a thick layer of radial muscles.

Little is known about the nervous system, but the excretory system has been studied in some detail. The two funnel-like posterior pores are said to open into a large, transverse chamber which gives off two wide canals, one on each side of the body. These extend forward in the lower compartment to the level of the anterior border of the disk. Here, the canals narrow and ascend into the upper compartment, proceeding forward to the level of the pharynx, where they turn about and extend back to the posterior end of the body. The recurrent canal on each side of the body gives off three main collecting ducts each of which branches four times, each time into three ductules, finally terminating in flame cells. The number of flame cells on each side of the body is thus $[(3)^4 + (3)^4 + (3)^4] = 243$. The flame-cell pattern, which is repeated on the opposite side of the body, can be represented by the formula

$$2\,[(3)^4 + (3)^4 + (3)^4] = 486.$$

This rather complicated branching is shown diagrammatically in Fig. 12B, in which only 135 flame cells are represented, these being all the flame cells connected with one of the main collecting ducts of the right side of the body (81) and two-thirds of a second (54).

In addition to the flame cells, which occur at the ends of the ultimate branches of the excretory ductules, there are tufts of cilia at various points of the inner lining of the recurrent excretory canal. Similar tufts have been observed in the excretory system of some Monogenea. The flow of fluid containing excretory materials proceeds from the smaller to the larger ducts, but the manner of its propulsion is not known with certainty, the ciliary action not fully accounting for it.

We now turn to the hermaphrodite reproductive organs, which are illustrated diagrammatically in Fig. 8A. Taking the male system first, there is a single testis (*t*), situated just behind the middle of the body in the lower compartment, a vas deferens, which penetrates the septum and then expands into a seminal vesicle (*sv*), an eversible male organ or cirrus (*c'*), and a sac-like sheath, the cirrus pouch (*cs*). The seminal vesicle of the adult parasite may be filled with thread-like spermatozoa. The terminal part of the male system, the copulatory organ or cirrus, can be protruded into the cuticularized genital atrium (*ga*).

The ovary (Fig. 8A, *o*) is situated in front of the testis and slightly to the right of the median plane. It has much in common with the ovary of some Monogenea, being folded several times. Its blind extremity is a large oval mass which gradually tapers to a narrow rhachis. This is folded three or four times behind and to the left of the blind end, so that the entire organ forms a compact mass, though in reality elongate. It communicates with a narrow oviduct (*od*) and this is locally modified to form the ootype (*o'*). Three ducts communicate with the ootype, the vitelloduct (*vd'*), which is formed by the union of paired ducts, the uterus (*u'*) and Laurer's canal (*Lc'*). The first part of the uterus may contain spermatozoa received in copulation from another worm before the eggs were formed, and thus constitutes a receptaculum seminis uterinum. Later on the uterus may be crowded with eggs containing embryos in all stages of development. The uterus shows another unusual character, the presence of outer longitudinal and inner circular muscle fibrils in its walls. The terminal part of this duct (*u*) opens into the genital atrium by a uterine pore.

The vitellaria are formed of clusters of six or more follicles disposed around the main lateral vitelloducts. Cells from them proceed along the ducts to the ootype, conveying their load of shell material to this chamber.

Two characters of a rather puzzling nature remain to be mentioned. They are the ciliated oviduct and blind ending of a posteriorly elongate Laurer's canal. T. H. Huxley discovered the former and no satisfactory explanation has been given of the purpose of either. At one time Laurer's canal was thought to be a yolk reservoir, but the 'yolk cells' are badly named, as I have already remarked. Bearing in mind the true function of these cells, it seems likely that the canal is a depository for unused shell material, which is swept in by ciliary action from the ootype. This arrangement would prevent the delicate female ducts from becoming blocked with hardening fragments.

Self-fertilization has been observed in *A. conchicola* and may be of common occurrence, because the hosts frequently contain a solitary parasite. Cross-fertilization is probably the rule, however. The larvae develop while the eggs are still *in utero* and differ in several interesting particulars from the adult. There is an oral sucker and also a simple posterior sucker (Fig. 3 B, C). The latter enlarges as the young parasite grows, coming to occupy relatively more and more of the ventral surface of the body, and is gradually transformed into the typical disk of the adult by the development of the transverse and longitudinal ridges which mark out the alveoli. Williams (1942) made a special study of the larva of *A. conchicola*, which is gradually transformed into an adult. The eggs measure 0·128–0·134 × 0·048–0·050 mm. and the newly hatched larva is somewhat larger (0·130–0·150 × 0·050–0·055 mm.). It lacks swimming organs and moves very sluggishly. The oral sucker is slightly larger than the ventral (diameters 0·044 and 0·040 mm.), though the latter is deeply sac-like, and maintains its predominance in size until the larva is 0·88–0·96 mm. long and 0·175–0·18 mm. broad, when the ventral sucker becomes slightly the larger (diameters 0·195 and 0·205 mm.). The lateral alveoli appear first, when the larva is 1·2–1·4 mm. long and 0·3 mm. broad and the suckers 0·32 and 0·275 mm. diameter, by which time the excretory system is well developed and the primordia of the reproductive systems have appeared. Williams also noted that when parasitized Unionidae are ingested by cold-blooded vertebrates such as fishes, frogs or turtles, they liberate their parasites in the stomach or intestine of the host, so that a potential second host is annexed to the primitive life cycle. Van Cleave & Williams (1943) discovered that when mature *A. conchicola* are passed by means of a glass tube into the stomach of the turtle, *Pseudemys troosti*, they cling to the wall and will continue to live for periods up to fourteen days.

Species of the Genus *Aspidogaster* Baer, 1827

All the known species of the genus *Aspidogaster* except *A. conchicola* belong to the Old World, two forms from the New World (*A. ringens* Linton, 1907 and *A. kemostoma* MacCallum, 1913) having been transferred by Eckmann (1932*b*) to a new genus, *Lobatostoma*, the chief character of which is the presence of lip-like processes around the mouth. Eckmann recognized three species of the genus and to these she added two new ones, but Bychowsky & Bychowsky (1934), after study of much Aspidogastrid material from the Caspian Sea, reduced the five proposed species to three valid ones. With their proposals, which are indicated below, I am in entire agreement. The principal characters of these three species are as follows:

A. conchicola Baer, 1827. Disk bearing four rows of alveoli of which the two median rows have 14–21 alveoli each and the marginal rows a total of about 32. Laurer's canal ends blindly in a slight dilatation. Parasites of lamellibranch molluscs.

A. limacoides Diesing, 1835 (Syn. *A. donicum* Popoff, 1926). Disk with about 30–34 marginal alveoli, those of the median rows varying between 12 and 18. Laurer's canal opens into the stem of the excretory vesicle. Parasites of fishes, mainly carp, but also many others (see below).

A. decatis Eckmann, 1932 (Syn. *A. enneatis* Eckmann, 1932). Marginal alveoli numbering 20–22. Nature of Laurer's canal undetermined. Parasites of fishes, carp (*Cyprinus carpio*) and barbel (*Barbus* sp.) in Syria and Palestine.

Recent Observations on the Structure of *Aspidogaster limacoides*

Bychowsky & Bychowsky (1934) obtained numerous specimens of *A. limacoides* from the intestine of various fishes inhabiting the Caspian Sea and the Volga Delta, the synonymous form having been found by Popoff in piscine hosts in the Don. Among the hosts are roach, bream, chub, white bream, gudgeon, and a number of non-British fishes. The parasites occurred in the very variable numbers of 1–98 per fish. They were white, tinged with gold or brown according to the degree of maturity attained, and 0·43–2·49 mm. long and 0·2–1·0 mm. broad. Living worms are very contractile and lively. The disk takes up most of the ventral surface of the body and has four rows of alveoli, of which the marginal ones are rounded, the more central or median ones oblong. The shape of the disk varies, being rounded or oval in outline, sometimes heart-shaped, and the dimensions approach that of the body (0·56–2·46 × 0·43–2·0 mm.).

The most interesting and important observations made by Bychowsky & Bychowsky concern the variable structure of the disk and the numbers of alveoli borne thereon. The latter vary between 50 and 74, which are made up of 12–18 in each of the four longitudinal rows and single, median, anterior and posterior alveoli. Further analysis of the observations yielded the following data:

Alveoli of each longitudinal row	12	13	14	15	17	18
Total numbers of alveoli	50	54	58	62	70	74
Numbers of specimens	5	64	105	23	2	1

From statistical considerations it seems likely that fewer than 12 alveoli may constitute a longitudinal row, possibly as few as 10, and Bychowsky & Bychowsky regard *A. decatis* Eckmann as a possible extreme variant of *A. limacoides* Diesing having the minimum number of alveoli.

Although the size of the disk increases in accordance with increase in size of the body, the numbers of alveoli in a longitudinal row seems to be unrelated to the size of the disk. Thus disks measuring 0·80, 0·86, 1·00 and 1·93 mm. in length were found to have 17, 15, 12 and 17 alveoli respectively in each longitudinal row. Such differences can only be due to individual variability.

Internally *A. limacoides* shows certain differences from *A. conchicola* which might be mentioned because they must be taken into consideration in connexion with future re-examination of the latter species. Perhaps the most important difference concerns the horizontal septum which, in *A. limacoides*, begins ventrally underneath the genital pore near the anterior end of the disk and extends to the posterior extremity parallel to the gut, being fused with the margin of the body laterally. According to Voeltzkow (1888) and Stafford (1895) the septum of *A. conchicola* ends posteriorly at the extremity of the gut, which Bychowsky & Bychowsky regard as an error of observation due, perhaps, to displacement of the septum by great development of the uterus. The dorsal compartment of the body in *A. limacoides* contains the digestive system, most of the

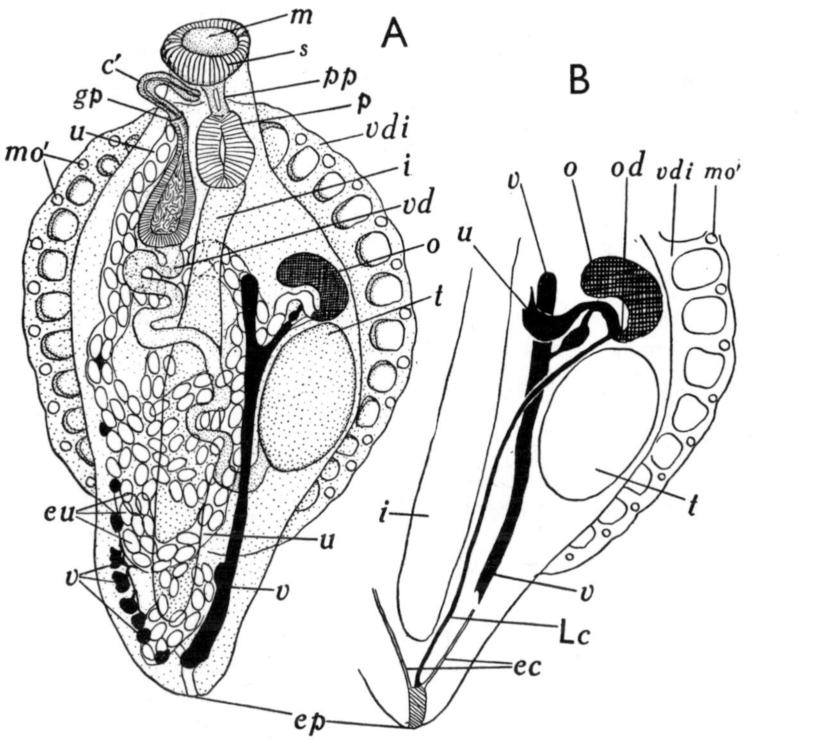

Fig. 9. *Aspidogaster limacoides*. A, entire animal in dorsal view. B, schematic diagram showing connexions between the ducts of the female reproductive system. (After Bychowsky & Bychowsky, 1934.)

uterus, the copulatory organ, the common genital pore and the vitellaria, the ventral compartment the testis, ovary, oviduct, first part of the uterus, vitelline reservoir and a part of Laurer's canal.

Other noteworthy characters of *A. limacoides* are as follows (see Fig. 9). The mouth is encircled by stout circular and longitudinal muscles which form a rudimentary oral sucker (*s*). A short prepharynx (*pp*) and large pharynx (*p*) occupy the slender 'neck' region, and the large, sac-like intestine (*i*) extends almost to the posterior extremity. The pyriform cirrus pouch opens close behind the mouth and contains both a cirrus (*c'*) and a seminal vesicle. The metraterm is muscular and, like the cirrus, contains gland cells. A reniform ovary is situated in front of the ovoid testis near the middle of the body but to one side of the median plane. The vas deferens (*vd*) is long and convoluted and the uterus (*u*) has both descending and ascending limbs. Paired vitellaria (*v*) occur at the sides of the intestine, consisting of elongate rows of follicles and merging posteriorly. Laurer's canal approaches the intestine, runs parallel with it, and opens into the excretory vesicle (see Fig. 9B, *Lc*). The eggs are brown in colour and measure 0·072–0·076 × 0·030–0·042 mm., though extremes of length reach 0·060–0·101 mm.

It is unlikely that other Aspidogastrids will be encountered in this country, but the key here given serves to separate the nine genera of the family Aspidogastridae which are known. Important diagnostic characters are the nature of the adhesive apparatus, the number of testes, the presence or absence of a cirrus pouch and of papillae on the ventral surface of the disk, and the existence of lips bordering the mouth. For further details see Dawes (1941*b*).

KEY TO THE GENERA OF THE FAMILY ASPIDOGASTRIDAE POCHE (after Dawes, 1941*b*)

1. One row of alveoli borne on the disk or on the ventral surface of the body. One or two testes present
   - A. Disk present, suckers confluent; one testis — *Macraspis* Olsson, 1868
   - B. Disk absent, suckers distinct; two testes — *Stichocotyle* Cunningham, 1887
2. Several rows of alveoli borne on the disk. One or two testes present
   - A. Three rows of alveoli present
     - (*a*) One testis
       - (i) Cirrus-pouch present — *Cotylaspis* Leidy, 1857
       - (ii) Cirrus-pouch absent — *Lissemysia* Sinha, 1935
     - (*b*) Two testes — *Cotylogaster* Monticelli, 1892
   - B. Four rows of alveoli present
     - (*a*) One testis
       - (i) Papillae on central part of disk — *Lophotaspis* Looss, 1902
       - (ii) No papillae on central part of disk
         - (*aa*) Mouth with lip-like processes — *Lobatostoma* Eckmann, 1932
         - (*bb*) Mouth without lip-like processes — *Aspidogaster* Baer, 1827
     - (*b*) Two testes. Alveoli numerous (144) — *Multicotyle* Dawes, 1941

CHAPTER 5

THE MORPHOLOGY OF THE DIGENEA

Many digenetic trematodes are flat and leaf-like, but the majority are long and narrow, and circular or oval in cross-section. As a rule, the length of the fluke is greater than its breadth, but the reverse is true of exceptional forms like *Euryhelmis* (Heterophyidae). The body generally tapers more sharply in front than behind but, again, there are many exceptions. So considerable is the variability of structure in Digenea that one finds exceptions to almost any general statement one attempts to make. Some digenetic flukes have powerful oral and ventral suckers, others lack one or the other, even both. In most instances the mouth is near the anterior tip of the body, but in some it is near the middle, in which case great care must be taken to orientate the animal correctly before attempting to examine its structure.

Digenea which have a centrally situated mouth are placed in the suborder GASTEROSTOMATA, all other Digenea in the alternative group PROSOSTOMATA. In the former, the main adhesive organ (if we regard the 'oral sucker' as subsidiary) is situated in front of the mouth; in the latter it occurs behind the mouth. Two simple characters thus suffice in the separation of two very distinctive types of Digenea, though others are needed to clinch the diagnosis.

I. THE SUBORDER GASTEROSTOMATA ODHNER

This group of Digenea contains the solitary family Bucephalidae Lühe, once known as the Gasterostomidae M. Braun, which has adult representatives inhabiting the alimentary canal of fishes and larvae which live in cysts in various parts of the central and peripheral nervous system of fishes, notably in the brain but sometimes in the ear or on the course of the cranial and spinal nerves.

Bucephalopsis gracilescens (Fig. 25A) is a good example of this type of fluke. Hundreds of mature specimens may be found in the pyloric caeca of the angler fish, *Lophius piscatorius*, which infects itself by devouring various gadoid fishes like the cod or whiting which harbour the encysted larvae. It is about 6 mm. long, colourless except for a brown tinge caused by the eggs, and covered with delicate spines (not shown in the diagram). The mouth (*m*) is situated about one-quarter of the distance along the body and is encircled by an oral sucker (*os*) different from the pattern which occurs in other Digenea and probably really the pharynx, which is sunk beneath the integument and about 0·3 mm. in diameter. A slightly larger sucker (0·4 mm. diameter) occupies the anterior tip of the body (*h*), well in front of the oral. In other genera this anterior holdfast may show different characters. It may be a similar organ with retractile tentacles (*Bucephalus*), or a fan-like hood (*Rhipidocotyle*, Fig. 25E), or a somewhat different muscular structure with a beak-like process on its lower margin, variously known as a *rostellum* or *rhynchus* (*Prosorhynchus*, Figs. 25C, D, G, 26A–C). The nature of this anterior holdfast is of generic importance.

Other characters show a fair measure of constancy in all Gasterostomata. True, there are slight differences in the relative positions of the internal organs, but there is no substantial difference in their structural details. The main characters, which stand out boldly from those of Prosostomata, are (Fig. 25 A) a simple sac-like intestine (*i*), a prominent, undivided excretory vesicle (*eb*) with a posterior pore (*ep*), a common genital pore (*gp*) which is also posterior in position, and a characteristic set of hermaphrodite reproductive organs. The male organs include an elongate cirrus (*c'*) lying within its pouch, a short vas deferens, a pair of equally short vasa efferentia, and a pair of rounded testes (*t*). The vas deferens is enlarged at the base of the cirrus to form a seminal vesicle, and this is contained within the cirrus pouch, together with numerous unicellular glands known as prostate glands.

The female organs comprise a rounded ovary (*o*), somewhat smaller than the testes and occupying various positions in relation to them in different species, but in front of them in *Bucephalopsis gracilescens*, a short oviduct, which is slightly enlarged at one point to form the ootype, and several other ducts, (i) the vitelloducts, which pass from the rather coarsely follicular vitellaria (*v*) and converge on the ootype, (ii) the uterus (*u*) which in its simplest form (Fig. 25 B) extends forward to the level of the intestine and then back to the genital pore, but may in the gravid condition possess other, rather indeterminate folds, and (iii) Laurer's canal, a short tube opening dorsally to the exterior. Their ducts stand in the same relation to one another as in Prosostomata and will be discussed in greater detail presently.

The encysted stage of *B. gracilescens* (Fig. 25 F) is perhaps most commonly found in the haddock, the cysts occurring in the spinal canal, or on the nerves, sometimes in the skin. The nerves most frequently bearing them belong to the spinal series, though the auditory nerve is a favourite situation. They are easily recognized by their oval or pear shape and are about 0·6 mm. long. The metacercaria which can be squeezed out of a cyst is colourless and about 2·5 mm. long or smaller. It is readily recognized as a Bucephalid because at this stage it already has most of the organs of the adult (Fig. 25 B).

In reviewing the Digenea of British fishes in a later chapter, more details will be given regarding the commoner species of Bucephalidae, and generic distinctions will be stated in the chapter dealing with the classification of the Digenea. The characters mentioned serve as a ready means of identifying roughly this unique type of trematode.

II. The Suborder Prosostomata Odhner

A. *Some General Types and their Main Characteristics*

The Prosostomata includes several distinctive types of trematodes which are commonly known by such names as Amphistome, Holostome, Monostome, Distome, Echinostome and Schistosome. Not all Prosostomata can be included in such a rough and ready scheme, but most of the important digenetic trematodes which can be found in British Vertebrata are covered by these general terms, for which reason we might examine them more closely. Such types, we must note, are not of equal systematic rank. Amphistomes, Holostomes and Schistosomes represent solitary families (though the tendency in recent times has been to split them), the Paramphistomatidae, Strigeidae (*sensu lato*) and Schistosomatidae respectively, and most Echinostomes belong to the family Echinostomatidae, though other families like Rhopaliadidae and Acanthostomatidae, possibly the Cathaemasiidae, might be included in the general term. Monostomes are more varied. Most of them belong to one or other of two important families, the Cyclocoelidae and Notocotylidae, though other families which might be segregated under this heading, in spite of their different phylogenetic relationships, are the Angiodictyidae*, Mesometridae, Stictodoridae, Eucotylidae, Pronocephalidae, Aporocotylidae, Sanguinicolidae, Spirorchiidae and Heronimidae. Most of the latter are unimportant in that they are unlikely to be met with in British vertebrates, but gain importance in systematic studies on their relationship to other types of Digenea, e.g. Angiodictyids to Amphistomes and Aporocotylids to Schistosomes. Distomes constitute a very formidable group of more than forty families, many of which will be encountered. Some of them have been segregated from others in tentative schemes to be discussed later, but as yet no scheme, simple or complicated, has been formulated for the complete assembly of the families of Prosostomata into higher taxonomic units than families. The great diversity of structure found within the suborder still awaits some unifying influence, which may be said to represent one of the opportunities of future generations of students. Perhaps the biggest stumbling block so far has been the lack of information concerning the life histories, as but few zoologists have seriously concerned themselves with even the adult trematodes of this country.

Some of these general types of Prosostomata can be recognized easily by means of their external characters, as Fig. 10 shows, though internal structures must be taken into account before definite diagnosis can be made. Amphistomes (A and B) have a thick, fleshy body provided with a well-developed posterior sucker. The mouth occurs at the end of a muscular organ which is probably a pharynx, but is commonly spoken of as the oral sucker (*p*). In some Amphistomes it has a pair of diverticula (A, *pp*). The internal organs resemble those of other Digenea, but unusual characters exist in some Amphistomes, notably the large ventral pouch (B, *vp'*) of *Gastrothylax* and related genera. The genital pore is situated in its dorsal wall and it may be regarded as an enormous genital atrium of unknown function. Some Amphistomes parasitize ruminants all over the world, *Paramphistomum* being the genus best known. *Gastrodiscus* (Fig. 59C, D), a parasite of man and other mammals, has a body which is divided into a narrow, cylindrical anterior and a large, expanded posterior region. The latter is covered with regular rows of ventral papillae, and this gives the parasite an *Aspidogaster*-like appearance, but it terminates in a posterior sucker which dispels the illusion. *Diplodiscus subclavatus* (Fig. 10A) occurs in the rectum of frogs and is unusual in having only one testis (*t*). It is of general zoological interest to note that Amphistomes parasitize every class of the Vertebrata.

In Holostomes the body is divided by a constriction into anterior and posterior regions (Fig. 10H). The former is somewhat flattened and carries weak oral and

* The correct name of this family would seem to be Microscaphidiidae Travassos, 1922, because *Angiodictyum* Looss, 1902 is not the type of the subfamily to which it belongs (Microscaphidiinae) and thus cannot be the type of the family (see Stunkard, 1943).

ventral suckers (*os* and *vs*) and, sometimes, also a special adhesive organ provided with glands (*g*). In the figure this is shown embracing one of the villi of the intestine, a characteristic position in the host. The posterior region is generally cylindrical (*pb*), but may be globular or flat, and contains the gonads as well as the main parts of their ducts. The solitary ovary (*o*) is situated in front of the testes (*t*). Holostomes are parasites of the alimentary canal of vertebrates, chiefly birds, though immature individuals commonly occur encysted in the bodies of fishes, being known by such names as *Diplostomum*, *Tylodelphis* and *Tetracotyle* (Fig. 62 K, L).

Monostomes, as the name implies, generally lack one sucker, and this may be the oral or the ventral, but some trematodes, which for convenience are included under this heading, have no suckers. In *Notocotylus*, a well-known genus of Monostomes which parasitizes birds, the ventral sucker is absent. The family to which it belongs, Notocotylidae, is distinctive on account of the three to five rows of glands or clusters of glands on the ventral surface of the body (Fig. 10 F, *g*). Such trematodes are common in the intestine of aquatic birds and some mammals. Lesser known families of Monostomes which lack the ventral sucker are the Mesometridae and Stictodoridae. Of Monostomes which lack the oral sucker perhaps the best known belong to the family Cyclocoelidae, in which the ventral sucker may also be absent (Fig. 10 G). The ring-like intestine is also an important diagnostic character here. These Monostomes are fairly large parasites which inhabit the air sacs, nasal cavities or coelom of aquatic birds. Monostomes which are invariably without suckers belong to several families, one of which is the Microscaphidiidae and occurs in reptiles as well as birds. Others are parasites of the blood-vascular system, the Aporocotylidae and Sanguinicolidae occurring in fishes and the Spirorchiidae in reptiles. Other families of Monostomes are characteristically parasites of sirenians.

Distomes belong to many families in which the mouth is surrounded by an oral sucker, the remaining sucker being situated somewhere on the ventral surface, but not at the posterior end of the body (Fig. 10 C, D). The variable position of the ventral sucker is not due to shifting along the body, as is sometimes supposed, but to modification of the posterior end of the body, which may be regarded as an enormous pouch in forms like *Macrodera longicollis* (Fig. 49 E), and, as such, unrepresented in Amphistomes (Dawes, 1936). Lengthening of the region between the suckers may in some instances be operative also, but modifications which create the illusion of movement of the ventral sucker in different Distomes are mostly such as affect the posterior end of the body. Whole families of Distomes are found only in fishes, and we shall concern ourselves later with the most important of them, namely, the Hemiuridae, Allocreadiidae (members of which occur in bats, e.g. *Crepidostomum*), Acanthocolpidae, Fellodistomatidae, Zoogonidae, Azygiidae and Ptychogonimidae. Others occur only in reptiles or birds or mammals, though many are found in several classes of the Vertebrata. Thus, the Dicrocoeliidae, an important family of liver flukes, occur in all classes except fishes. The importance of Distomes is emphasized in standard works on helminthology, which necessarily ignore many forms of great zoological interest.

In Echinostomes the ventral sucker is close behind the oral and well in the anterior half of the body (Fig. 10 E). The distinctive character of this type of

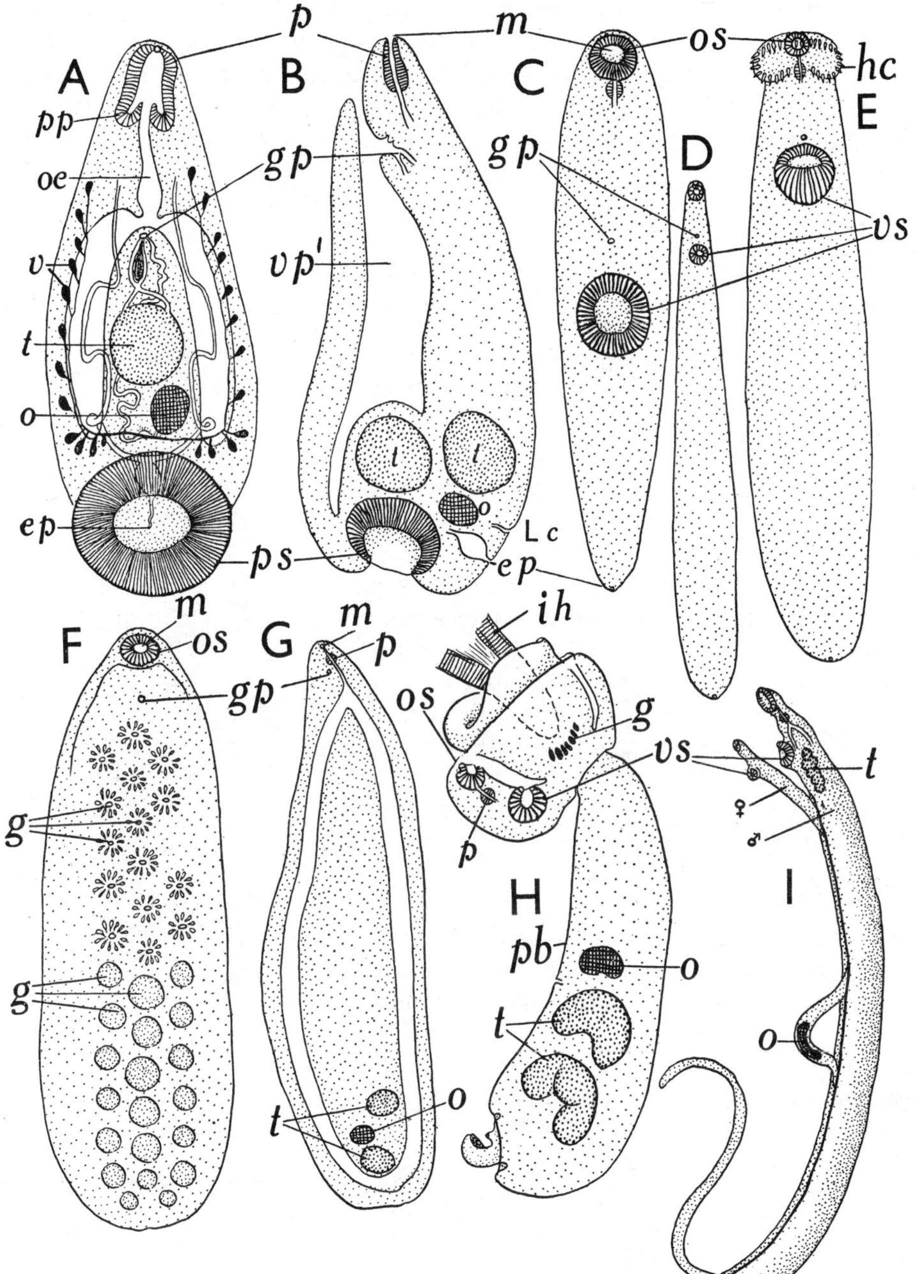

Fig. 10. The main types of Digenea. A, an Amphistome (*Diplodiscus subclavatus*). B, an Amphistome with a ventral pouch (*Gastrothylax*). C, D, Distomes of very different external appearance. E, an Echinostome (*Echinostoma*). F, a Monostome with only an oral sucker (*Notocotylus*). G, a 'Monostome' without suckers (*Cyclocoelum*). H, a Holostome (*Cotylurus*). I, male and female Schistosomes (*Schistosoma*) in association. (A, after Lühe, 1909; F, G, after Fuhrmann, 1928; rest after various writers, altered.)

trematode is a spinous collar surrounding the oral sucker (*hc*). It is called the *head collar*, and the single or double row of stout spines which it bears are known as *head* or *collar spines*. All the well-known Echinostomes belong to the family Echinostomatidae, which are common parasites of birds and mammals. Two families which may be regarded as Monostomes show a remarkable degree of resemblance to Echinostomes in general structure, but lack the head collar. They are the Psilostomatidae and Cathaemasiidae. Both occur in birds. The most unusual Echinostome types, undiscovered in this country, belong to the family Rhopaliadidae, in which the head collar is represented by two anterior, proboscis-like structures armed with hooks.

Schistosomes are baneful, unisexual blood flukes belonging to the family Schistosomatidae, in which the body is long and slender and well suited to life in blood capillaries. They occur in the mesenteries of the intestine and in the liver of birds and mammals. There is sex dimorphism here, the male being larger than the female and habitually carrying her in a ventral groove, or gynaecophoric canal, which is formed by downwardly curved, lateral extensions of his body (Fig. 10 I). Both the oral and the ventral sucker are far forward on the thread-like body, and the gonads (*o* and *t*) occupy very different positions. It might be mentioned that the Schistosomatidae is not the only family of the Prosostomata to show sex dimorphism; another family sharing this unusual character is the Didymozoidae, a family of cyst-dwelling Digenea which infect fishes.

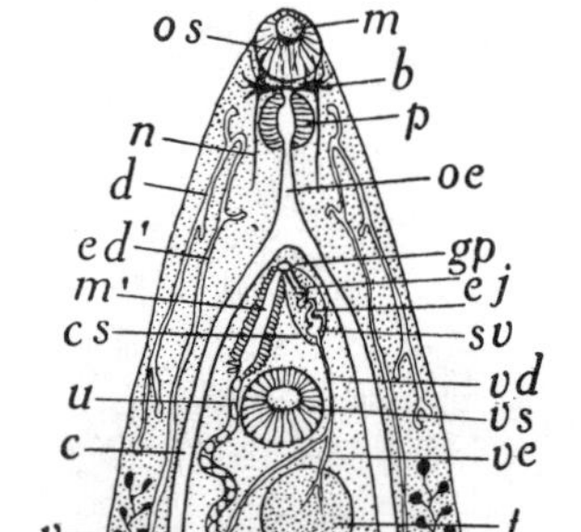

Fig. 11. The chief organs of a 'typical' digenetic trematode; somewhat schematized, but based largely on *Dicrocoelium*.

There is much to be said for the general use of the terms we have mentioned in separating various types of Prosostomata, so long as we regard it merely as a convenience and do not press it beyond useful limits. The types shade off into one another almost imperceptibly, so that real systematic importance is denied them. A knowledge of the location of the trematode in the body of the host is another useful character, whole families of Distomes being characteristically found in the liver, or lungs, or intestine of Vertebrata, but here again we have to avoid over-emphasis.

We can now examine the structure of the prosostomatous Digenea in greater detail, and might note at once that many such trematodes are less specialized and easier to study than *Fasciola hepatica*, the type generally chosen in elementary zoology. Almost every organ of this notorious fluke has suffered excessive branching, whereas the same organs can be studied in a simpler form with more profit in the translucent bodies of many Distomes which the student can easily collect for himself from the fishes, amphibians and other vertebrates which he may be called upon to dissect in the laboratory. This method of teaching has the further advantage of stimulating the student instead of stifling his curiosity by forcing upon his notice an impression of complexity which does not generally exist. The general structure of what might be called a typical digenetic fluke can be summarized by a diagram like Fig. 11. How various Digenea are modified from this somewhat simplified condition will become evident as we proceed.

B. *The Structure of Prosostomata*

I. *External Characters*

Prosostomatous Digenea differ little in size from Monogenea. Most of them are 0·5–10 mm. long, though some attain a length of 20–30 mm. and exceptional thread-like forms 50–60 mm. *Nematobothrium filarina* Beneden, 1858, a member of the Didymozoidae which inhabits the gill chamber of the shadow fish or meagre, is said to attain a length of 1 m.! Most Didymozoids inhabit cysts and some of them neither need nor possess suckers. Some blood flukes have only the rudiments (or vestiges?) of a ventral sucker and rarely show a primary monostomy. Differences between these and other Monostomes show that monostomy is not a criterion of affinity.

Adhesion is generally effected by the ventral sucker, though the oral may also assist and spines may serve the same purpose. Many Digenea have a smooth cuticle, but some intestine and liver flukes have a spiny or a scaly cuticle. The finest cuticular scales occur in some blood flukes. In rare instances the ventral sucker may bear a number of cuticular 'teeth', and spines may occur in other situations, e.g. on the cirrus. The spines of the cuticle generally diminish in size and become sparser posteriorly, indicating a growth gradient of diminishing intensity along the main axis of the body when the spines are being formed, and subsequent lengthening of the body, mainly in the posterior region, which also determines the relative position of the ventral sucker.

The cuticle is strong and elastic, often thicker than in Monogenea, and generally rests on a definite basal membrane. In Monogenea, subcuticular cells rarely exist, but in Digenea they are always sparingly present, if discontinuous. We have no clear conception of the mode of origin of the cuticle, partly because of differences which exist in flukes which have been studied. By some zoologists it is believed to be homologous with the basal membrane of the integument of Turbellaria. This opinion is based on the fact that in some larval Digenea (cercariae) there is at first a cellular epithelium and that this later degenerates. The loss of an embryonic epithelium, however, is no certain criterion of the soundness of this view, because undifferentiated ectoderm cells which underlay the embryonic epithelium may, when the latter disintegrates, give rise to a cuticle by later differentiation.

Monticelli thought that the cuticle was a metamorphosed epithelium which had suffered chemical alteration of its protoplasm and the loss of its nuclei. This opinion is supported by the fact that in some larval and adult Digenea the cuticle contains nuclei, or bodies which resemble them. The cuticle of *Cyclocoelum* has this character. An alternative opinion, which is credible for some Digenea but not others, is that the epithelial cells which formed the cuticle afterwards sank into the parenchyma, retaining only delicate protoplasmic connexions with it. A solution of the problem is difficult to reach, but the Turbellaria throw some light on it. Some of them possess an epithelium and others sunken epithelial cells. Some planarians have a syncytial layer instead of an epithelium, and in a few of them the superficial cytoplasm is cuticularized. This is notably the case in some Turbellaria which have adopted a parasitic mode of life. No doubt the same kind of change has occurred in some Trematoda also, in which case the inclusion of nuclei in the cuticle is intelligible. Monogenea lack sunken epithelial cells, and it may well be that ectoparasitic Trematoda have suffered cuticularization of the entire superficial epithelium or syncytium, as Monticelli believed.

Gland cells occur in the parenchyma of some Prosostomata, mostly in the extremities of the body and particularly in the vicinity of the mouth. They occur in Monorchiidae, Allocreadiidae and some Strigeidae. Special gland cells are developed in relation to the unique adhesive apparatus of the last-named. Such glands are said to be of ectodermal origin, and they open by minute ducts which pierce the cuticle. The ventral glands of the Monostome *Notocotylus* have already been mentioned; they are obvious external characters.

About the suckers we need say little. They are of fairly constant structure, consisting to a large extent of radial fibrils, or bundles of them, which are sharply delimited from the parenchyma. Equatorial and meridional muscle fibrils also exist, especially at the rim of the sucker, where they are thickened and form a sphincter.

II. *The Internal Organs*

(i) *The Muscles and Parenchyma.* In some Prosostomata the body is very muscular, in others its texture is delicate and muscles are feeble. In the former case, the parenchyma may be almost completely filled with longitudinal muscle fibrils. Generally, however, three layers of fibrils are arranged in the form of a tube immediately beneath the cuticle. The outermost layer consists of circularly arranged, the middle of diagonally arranged and the innermost of longitudinal fibrils. The Amphistome *Gastrodiscus* has an additional layer of circular muscles beneath these typical layers. Sometimes, as in *Otodistomum*, longitudinal fibrils occur in the deeper parts of the parenchyma and may be homologous as well as analogous with the deep muscles of Cestoda. Additional vertical muscles traverse the parenchyma in some Digenea, more especially in the lateral regions, where internal organs do not interfere with the arrangement. The anterior tip of the body is usually more muscular than other regions, as can be deduced from the movements of almost any living trematode. Tubular muscles occur in some Echinostomatidae, Paramphistomatidae and Hemiuridae. Muscular structures show various histological characters, but often they are fibrillar. Myoblasts, giant cells with a number of processes each of which is inserted in a muscle fibril or a small group of fibrils, sometimes occur beneath the general musculature, especially in the vicinity of the suckers. Sometimes the myoblasts are grouped in small clusters and may then be syncytial. Some larvae which possess myoblasts show rhythmical pulsatory movements, but these must ultimately depend on special features of the nervous as well as the muscular system. Striated muscle has seldom been identified in Trematoda, but is said to exist in some cercariae.

The parenchyma of Prosostomata is generally open, thereby contrasting with that of Monogenea, and consists of a network of cells and fibrils with irregular cavities in its meshes. Whether or not the spaces represent a schizocoele is uncertain, but the texture of the parenchyma recalls the mesenchyme of embryo Vertebrata. In young trematodes free cells or clusters of them wander through the parenchyma to establish in appropriate situations the rudiments of the gonads and their ducts. Some of the wandering cells resemble lymphocytes and may distribute nutrient substances about the body. Pigment may exist in the parenchyma, indicating that oxidative processes occur there.

(ii) *The Nervous System and Sense Organs.* The nervous system of Prosostomata does not differ significantly from that of Monogenea, consisting of a similar 'brain', with paired ganglia and a wide commissure above the gut, and several nerves. The position of the brain varies slightly, being close to the oral sucker in some forms. Three pairs of nerves generally leave the front of the brain and are distributed to the tip of the body, numerous branches being supplied to the oral sucker and adjacent parts. Three pairs of longitudinal nerves also leave the hindpart of the brain, to supply various regions of the body. These are unequally developed. A ventral pair is conspicuous, and with its commissures forms a network beneath the ventral integument. Commissures also link this with the less conspicuous lateral and dorsal nerves, which are also paired. From both the ring-like commissures, which vary between five and forty in number, and the nerves numerous fibrils arise and branch into abundant endings underneath the muscular layers of the integument. They are mostly motor nerves, as might be expected from the feeble development of sensory structures.

Paired eye-spots exist in the miracidia and sporocysts of some trematodes, and a third eye is present in a few larvae, but such receptors are absent in most adults. Occasionally, scattered granules of pigment occur in the adult, the remnants of the larval eyes. Pear-shaped cells of unknown function and fine hair-like structures occur in the integument in some instances, and possibly serve a sensory purpose. Peculiar 'taste organs' occur on papillae in front of the oral sucker in some trematodes. Obviously, elaborate receptors would scarcely serve a useful purpose, and in fact only simple ones occur.

(iii) *The Digestive System.* Apart from the mouth and the sucker which supports it and assists feeding operations, the digestive system of Prosostomata is closely similar to that of Monogenea. The oral sucker has the same structure as the ventral, consisting of radial and meridional muscles which form a kind of sphincter. In *Bunodera* it has six lobes on the free margin. In Microscaphidiidae, as well as some Paramphistomatidae, it is complicated by the development of lateral pouches in its walls.

The pharynx is a muscular organ situated close behind the mouth. In some Prosostomata a short, non-muscular tube intervenes which, on account of its position, is called the prepharynx. It occurs mainly in elongate forms like the Allocreadiidae and Acanthocolpidae. Generally an almost spherical organ of conspicuous size, the pharynx is rudimentary in some instances and absent in some Gorgoderidae. In Sanguinicolidae both the oral sucker and pharynx are absent, but a thickening of the oesophageal muscles may simulate a pharynx. In most instances a slender tube, the oesophagus, continues the alimentary canal behind the pharynx, but this region is absent in the Ptychogonimidae (parasites of Selachii) and Halipegidae (parasites of frogs), as well as a few more common forms. The oesophagus is usually lined with ectoderm, which is epithelial in some Echinostomes, and sometimes has a cuticular lining. It may also have muscular walls containing transverse and longitudinal fibrils and in some instances receives the ducts of 'salivary glands', which are well developed in parasites of the blood-vascular system.

Beyond the oesophagus there is a pair of more or less elongate blind tubes, the caeca or intestinal crura. Members of the family Syncoeliidae and Cyclocoelidae have a ring-shaped intestine, formed by fusion of the caeca posteriorly. Schistosomes have caeca which fuse farther forward, not far from their anterior ends, the intestine being continued along the body as a single median caecum. Generally, the caeca are simple and unbranched, but where branching occurs the outgrowths may be simple (*Clinostomum*) or complicated by further branching (*Fasciola*, Fig. 55 B, c, c^2, c^3, c^u). In Campulidae (Fig. 56 A–J) and Accacoeliidae (Fig. 36) (parasites of marine mammals and fishes respectively), two forward extensions of the caeca alongside the pharynx impart an H-shape to the intestine. Sanguinicolidae show this character in a modified form, both limbs of the intestine on each side of the body being short, but the forwardly directed ones the larger. Where the caeca are of typical shape they may be short, barely reaching the level of the ventral sucker (*Brachycoelium*, Fig. 49 C), or of medium or great length. Frequently they almost reach the posterior extremity. A most unusual condition is seen in *Haplosplanchnus* (Fig. 35 A), which has a simple, sac-like intestine, and an even more remarkable one in *Haplocladus* (Fig. 45 E), which has a single, asymmetrical caecum. The latter condition is not primary, however, as in Aspidogastridae and Bucephalidae, but secondary. At least one larval trematode is said to lack caeca and may develop into an adult with a similar negative character, though this may merely be a case of delayed development.

The apparently blind caeca of some Echinostomatidae really open into the terminal part of the excretory system, so that the excretory pore also serves as an anus. This also occurs in some cyst-dwelling trematodes like *Balfouria monogama*, where the posterior end of the worm is protruded through a perforation of the host's intestine, so that faecal and excretory materials are cast outside the cyst. This seems to be the characteristic arrangement in Accacoeliidae. In the juvenile fluke a delicate membrane seals the connexion between the excretory and alimentary systems, and is broken through when the adult stage is reached. Some fish trematodes have a true anus. One species of *Acanthostomum* has two anal pores. Double anal tubes and pores occur also in *Schistorchis carneus* and a few other forms. In *Opecoelus* the caeca unite posteriorly, and the ring-like intestine thus formed opens to the exterior by a ventral pore which has no relation to the excretory system. Anal openings are thus commoner than was once supposed, but they provide scanty exceptions to the general rule of a blind intestine, and are not deemed to be of great taxonomic importance.

Circular and longitudinal muscles are included in the wall of the intestine in some Prosostomata, and there is sometimes a simple epithelial lining. The free ends of the epithelial cells may have thread-like processes which project into the lumen, as in Azygiidae, or such free terminations may be brush-like. This type of structure is seen in the intestinal glands of some Hemiuridae.

The food of Digenea is largely digested outside the body of the parasite, though blood cells or serum and epithelial detritus require some internal digestion, which is performed by extracellular means. Blood parasites which lack an alimentary canal or possess only a vestigial one must derive their sustenance from fluid nourishment imbibed through the surface of the body, as occurs in Cestoda.

(iv) *The Excretory System.* Nearly all Prosostomata have a single posterior excretory pore (Fig. 11, *ep*), but *Heronimus* has a dorsal pore in the vicinity of the oesophagus, and *Opisthotrema* shows the primitive arrangement of paired canals and pores. Miracidia, sporocysts and rediae have a pair of pores near the posterior end of the body (Fig. 13 A–C). Cercariae show the same arrangement before the tail develops (D), and when the tail first begins to grow out from the body the two primary pores go with it (E). When the tail process has become well formed, however, the main excretory canals meet and fuse to form a median tube in the hindmost part of the cercarial body (F). And when, finally, the tail is cast off, all signs of the paired origin of the single tube or excretory vesicle are lost (Fig. 13 G). The vesicle thus formed may be small and spherical, as in *Phyllodistomum*, *Asymphylodora* and other forms, or wider than it is long, as in *Paramphistomum*. Fusion of the two primary excretory canals through a great part of the cercarial body otherwise produces a very long excretory vesicle such as is seen in Dicrocoeliidae and Gorgoderidae. If the terminal forks widen, the vesicle becomes V- or Y-shaped, according to whether the stem is short or long. The V-shape is seen in Lecithodendriidae, Ptychogonimidae and other families. The Y-shape with wide forks and stem occurs in Plagiorchiidae, Azygiidae, Hemiuridae, Opisthorchiidae and other families. Further complication takes the form of a branching of the main canals and is seen in some Echinostomes, as well as in *Fasciola*, *Enodiotrema*, *Renifer* and *Styphlodora*.

In most Prosostomata the main excretory canals narrow anteriorly, and the finer canal thus formed turns dorsally near the pharynx and extends back to the posterior end of the body. Numerous fine branches arise from the recurrent limb. The manner of the branching and the ultimate numbers of blind terminations formed are matters of great systematic importance. The system is often rather complicated, but the arrangement can be expressed by means of a formula, the flame-cell formula, which has been applied to some trematodes by Faust and other zoologists. Some characteristic arrangements are shown in Fig. 12, and the method of applying the formula has been outlined already in connexion with the Aspidogastridae. In view of the variety met with in the Prosostomata it is perhaps as well to emphasize that in each formula the number of branches arising from the main canals is indicated by the number of terms in the expression. In each individual term, the bracketed numeral represents the number of branches formed, and the power to which this term is raised represents the number of times branching occurs. Since each of the blind terminal branches ends in a flame cell, the total number of flame cells can be calculated from the formula, which allows for bilateral symmetry. The flame-cell pattern has proved a highly important means of diagnosis, especially for larvae, in which it is simple. The

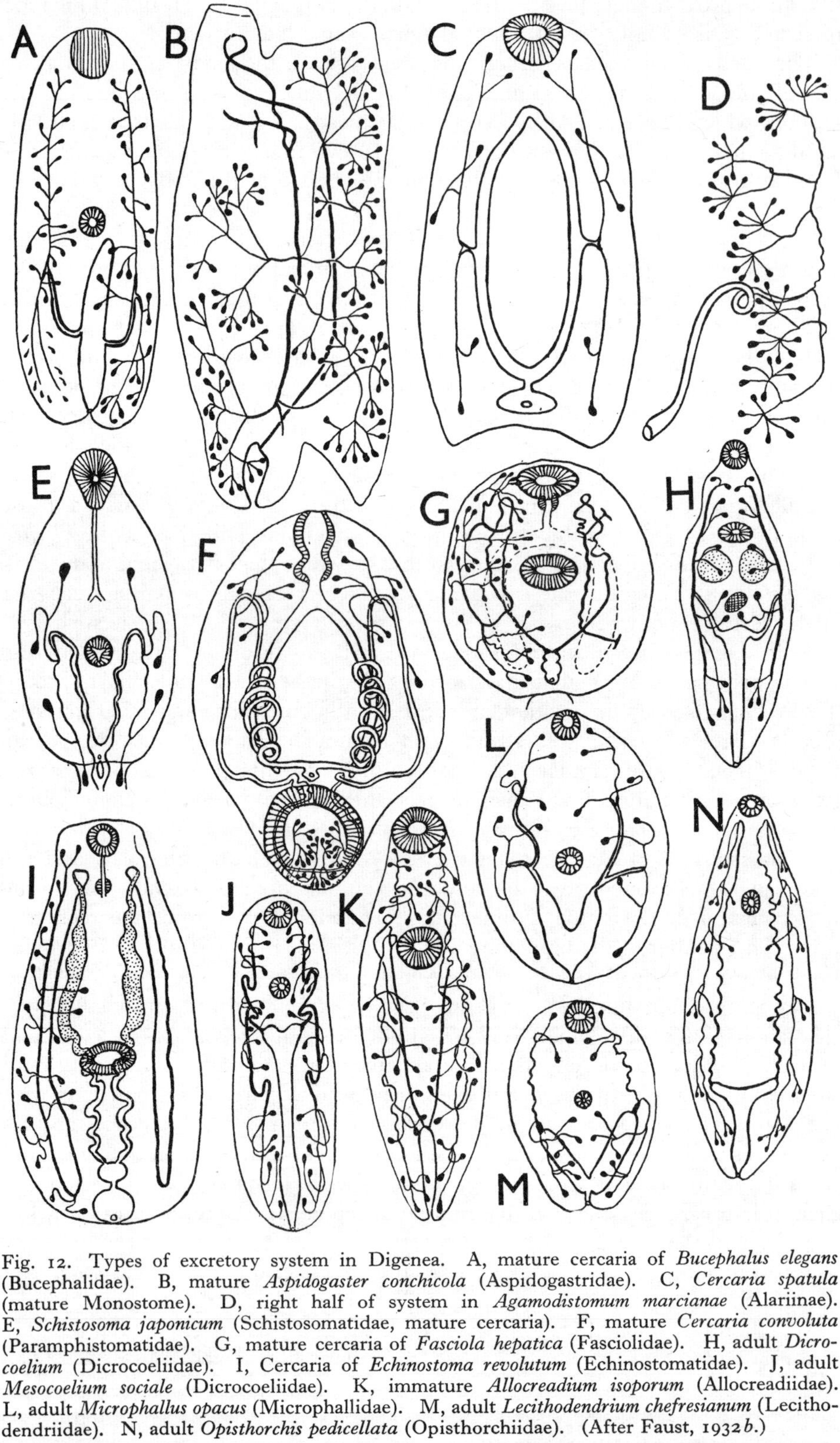

Fig. 12. Types of excretory system in Digenea. A, mature cercaria of *Bucephalus elegans* (Bucephalidae). B, mature *Aspidogaster conchicola* (Aspidogastridae). C, *Cercaria spatula* (mature Monostome). D, right half of system in *Agamodistomum marcianae* (Alariinae). E, *Schistosoma japonicum* (Schistosomatidae, mature cercaria). F, mature *Cercaria convoluta* (Paramphistomatidae). G, mature cercaria of *Fasciola hepatica* (Fasciolidae). H, adult *Dicrocoelium* (Dicrocoeliidae). I, Cercaria of *Echinostoma revolutum* (Echinostomatidae). J, adult *Mesocoelium sociale* (Dicrocoeliidae). K, immature *Allocreadium isoporum* (Allocreadiidae). L, adult *Microphallus opacus* (Microphallidae). M, adult *Lecithodendrium chefresianum* (Lecithodendriidae). N, adult *Opisthorchis pedicellata* (Opisthorchiidae). (After Faust, 1932*b*.)

main drawback in the case of adults is the difficulty of observation in preserved specimens, and even in living trematodes if they happen to be fleshy. Observations should be made whenever possible, because of their great value in systematic work.

The excretory system will be recognized as a protonephridial one, essentially similar to that of Turbellaria. The flame cell (Fig. 13 I) is a cellular unit from which long cilia (*cil'*) project into the blind terminations of the system. The

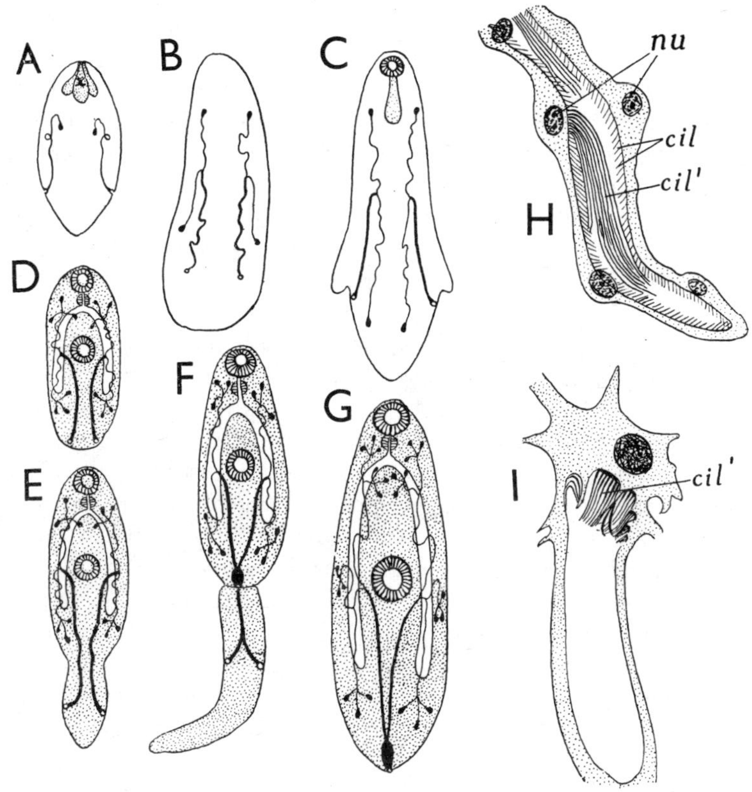

Fig. 13. The excretory system in Digenea. A, miracidium. B, sporocyst. C, redia. D, E, F, stages in development of the cercaria. G, metacercaria. H, tufts of long cilia and large cells forming the ciliated wall of the canal (not seen in the adult). I, young stage flame cell from *Dicrocoelium dendriticum*. (After Fuhrmann, 1928.)

capillaries are cellular or syncytial, and from certain of the cells tufts of cilia may project into the lumen (Fig. 13 H, *cil'*). The way in which flame cells are related to the flow of excreted fluid has never been made clear.

(v) *The Lymphatic System.* A so-called lymphatic system exists in some Prosostomata, being well developed in Paramphistomatidae and Microscaphidiidae. At its simplest, in *Microscaphidium* and *Angiodictyum*, it is made up of three longitudinal canals on either side of the body, respectively dorsal, lateral and ventral. Alternatively, there may be two main canals on either side, as in *Polyangium*, or four on each side, as in *Octangium*. In *Paramphistomum* the system is of cellular origin and in Microscaphidiidae is said to be composed of a single cell. Observation of the living trematode shows these tubes to be contractile, so that a rudimentary circulation of their fluid contents must be brought about, assisted perhaps by movements of the body as a whole. In other Digenea there is no definite lymphatic system, but there may be a circulation of the fluid which fills the spaces in the parenchyma and represents the rudiments of such a system.

(vi) *The Hermaphrodite Reproductive System.* In dealing with the reproductive organs we may leave out of account for the moment unisexual Prosostomata like the Schistosomatidae and Didymozoidae. On the whole, there is a greater multiplicity of sexual structures in Digenea than in Monogenea. The male and female systems generally open close together into a shallow pit or genital atrium, which may be deepened, ultimately giving rise to the unique ventral pouch of forms like *Gastrothylax*. In other trematodes there may be a deep, tubular atrium of complicated structure. The peripheral part may be extended and narrowed to form a hermaphrodite duct, as in some Hemiuridae (Fig. 41 A, *a*, *hd*), and the terminal part may come to resemble the entire atrium of other trematodes. This is so in *Isoparorchis*, *Accacladium* and some Strigeidae. The opening of the genital atrium may be narrow and encircled by feebly or well-developed muscles. The Amphistomes '*Cotylophoron*' and *Cladorchis* and many Heterophyidae have a genital sucker or *gonotyl*, which assists copulation. In a few instances where such a sucker has been lost, the ventral sucker is sunk in the genital atrium and serves in its stead, as in *Metagonimus* and *Apophallus*. Interestingly enough, the converse is sometimes the case, for in some Heterophyidae the ventral sucker has disappeared and the genital sucker takes its place.

The genital pore generally occupies a median position between the oral and ventral suckers, but it may occur on the lateral margin or even the dorsal side of the body. In some Prosostomata it is far forward and at one side of the mouth, in others it is behind the ventral sucker, and in a few it is at or near the posterior extremity. *Prosthogonimus*, *Heterophyes* and *Brachylaemus* respectively illustrate these possibilities. In Strigeidae, Sanguinicolidae and Aporocotylidae the male and female pores are in front of the excretory pore, on the dorsal surface of the body.

The male sexual apparatus consists of a pair of testes (Fig. 11, *t*), vasa efferentia (*ve*), a vas deferens (*vd*), seminal vesicle (*sv*) and a cirrus pouch (*cs*) which encloses the male organ or cirrus, a ductus ejaculatorius and a glandular pars prostatica. As a rule it develops in advance of the female apparatus, protandry being the rule in Digenea and in Cestoda. Many differences from what might be called the typical arrangement are seen in other Prosostomata. There is a solitary testis in the Monorchiidae and Haploporidae, as in some Aspidogastridae, and a variable number of testes in the Schistosomatidae and some Gorgoderidae. Species of *Gorgodera* which have numerous testes develop from cercariae with the rudiments of only two, indicating that two is the primitive number.

The testes are frequently rounded organs of smooth outline, but in some Prosostomata they are lobed, and in *Fasciola* very much branched (Fig. 55 A, *pt*). Shape is of taxonomic importance in some instances. In Distomes the testes are situated behind the ventral sucker, often behind the ovary, seldom in front of it, and very seldom one on either side of it, as in Clinostomatidae and Brachylaemidae. The spermatozoa which develop in the testes are thread-like bodies ranging between 0·02 and 0·2 mm. in length. In species of *Steringophorus*, *Haplocladus* and a few other genera they are enclosed in spermatophores, as they are in the Bucephalid *Prosorhynchus* and in some rhabdocoele Turbellaria. Spermatogenesis will be discussed in a later chapter. Suffice it here to note that the fully formed spermatozoa are swept along the vasa efferentia and vas deferens by ciliary action which is especially evident at the entrance to the seminal vesicle, contraction of the walls of which brings about ejaculation at the proper time.

Several modifications of the copulatory organs occur in Prosostomata, some being of taxonomic importance. The usual copulatory organ is a cirrus, an eversible and protrusible tube contained in a cirrus pouch. Frequently, the cirrus has circular and longitudinal muscles in its walls, and spinelets may project internally into its lumen in the retracted condition, being raised into prominence when the organ is protruded. When included in the cirrus pouch, the seminal vesicle is called a vesicula seminalis *interna*, when outside it the term *externa* is employed. The former type is the more usual, though both occur in a few instances. Sometimes the cirrus pouch contains only the cirrus and ejaculatory duct, the *pars prostatica* terminating in the parenchyma outside. Whole families of Prosostomata lack the cirrus pouch, and the ductus ejaculatorius may open into a genital atrium, as in *Paramphistomum*, or on to the unmodified surface of the body, as in some Holostomes. In Hemiuridae a functional cirrus is lacking and in its place is the hermaphrodite duct, which may be, but is not certainly, formed by fusion of male and female ducts, and may be homologous with the genital atrium of other trematodes (Fig. 41, A*a*). The common duct may be protrusible, like a cirrus, and it is sometimes enclosed in a sac which bears some resemblance to the cirrus pouch, though, obviously, not to be called by this name. Various terms have been used for this structure. Some writers call it a *false cirrus sac* or *pouch*, others a *sinus sac*. The latter is said to be *complete* when the muscles are so considerable as to be continuous, and *incomplete* when the muscles are sparse. This seems to be drawing a rather fine distinction, but it is plain that such a morphologically distinct structure cannot be called simply a cirrus pouch.

The components of the female reproductive system (Fig. 11) are the ovary (*o*), oviduct, ootype (*o′*), receptaculum seminis (*rs*), uterus (*u* and *u′*) and Laurer's canal (*Lc*). As in Monogenea, fertilization and encapsulation of the ova occur in the ootype, which is really a modified section of the oviduct, conveniently situated with respect to the vitelline duct, the opening of the receptaculum seminis and the uterus. In some Prosostomata the uterus proceeds first towards the posterior end of the body by what is here called a descending limb (*u′*), there to turn about and extend towards the genital pore by an ascending limb (*u*). The terminal part or metraterm (*m′*) is muscular to varying degrees in different forms and may serve as a vagina. Frequently it has a thick, cuticular lining, sometimes provided with hooklets, and glands which bear some resemblance to prostate glands sometimes lie outside its walls and open into its lumen. The ovary is often of smooth outline, though lobed in some Heterophyidae, branched in *Fasciola* and a few other genera (Fig. 55 A, *o*) and follicular in the Hemiurid *Otiotrema* and other rare instances. In *Didymozoon* it is extremely long and tubular. It rarely occupies a median position, occurring sometimes on the right and as often on the left.

Ova pass by a short region of unmodified oviduct from the ovary to the ootype, assisted by a ciliary current. Perhaps in more instances than we know at present there is at the junction of ovary and oviduct a muscular chamber or ovicapt, which spaces out the ova as they leave the ovary and passes them in a thin, steady stream along the oviduct during the period of egg production. The manner in which encapsulation occurs is not essentially different from the process already noted for Monogenea, though the formation of an operculate capsule calls for some slight modification of the method we have studied. References to literature bearing on the eggs of Digenea are provided in my paper on those of *Hexostoma* (Dawes, 1940*b*), but special mention may be made here to the papers of Ujiie (1936*a*, *b*), Yumoto (1936), Kouri *et al.* (1936) and Kouri & Nauss (1938). The shell glands contribute little if any material to the egg capsule, but produce a thin secretion, and in some instances this accumulates in the ootype and uterus in copious amount, as in the Aspidogastrid *Stichocotyle*. It may contribute towards the gelatinous capsules of *Pleurogenes* and the similar envelopes which Looss (1902*e*) described for certain Haploporidae, or even the thin sculptured surface layer which can be removed by treatment with NaOH. Only rarely are shell glands absent, however, but this is so in the Sanguinicolidae.

As in Monogenea, shell-producing material develops as droplets of secretion in the vitelline cells, which travel along the vitelloducts to the ootype before discharging their products. The agency by which such movements are carried out has never been disclosed. Only rarely are the vitellaria of Digenea better developed than in Monogenea, so that shell-forming material must be put to more economical use, more so because eggs are produced in far greater numbers. This need for economy explains why polar filaments are of rare occurrence. The vitellaria are generally follicular and occupy the lateral regions of the body. Their extent is fairly constant in a given species, but must not be relied on explicitly as a specific character, being subject to variation as a result of movements of the body as a whole. Compact vitellaria occur in a few Prosostomata such as *Gorgodera*, as well as some Hemiuridae, Zoogonidae and Haploporidae. It is interesting to find that very small, compact vitellaria can produce sufficient material for the manufacture of enormous numbers of eggs, as happens in the first two if not the last two of these examples. Possibly there are qualitative differences in the function of the vitellaria in forms like *Fasciola*, where they are follicular and abundant, and in Hemiuridae, some of which have a single, compact vitellarium and yet produce just as many if not more eggs. This is another matter which has never been investigated. In other instances, reduced vitellaria lead to the production of fewer and thinner, sometimes very delicate eggs, as in Zoogonidae.

Laurer's canal, which is absent in Monogenea but probably represented by the genito-intestinal canal, extends in Digenea from a pore on the dorsal surface of the body to a point near the ootype. In some instances it has a muscular wall, in others it is said to be ciliated. Near, sometimes at, its internal end there is a chamber in which spermatozoa from another individual find temporary accommodation, the receptaculum seminis. This is absent in some Echinostomatidae, Steringophoridae and Azygiidae. Storage of spermatozoa in such circumstances may occur in the proximal part of the uterus, forming what is called a receptaculum seminis *uterinum*. Only rarely is Laurer's canal wanting, as it is in *Derogenes* The purpose of this canal, if we may use the term, is said to be the reception of the cirrus of another individual during copulation and the conveyance of spermatozoa received during this act to the receptaculum seminis. This has been observed in at least two trematodes, *Liolope copulans* and *Collyriclum faba*. From what has been stated already, however, it must be clear that this is not the only means of insemination, and it is unlikely to be the usual means. Self-fertilization by insertion of the cirrus into the metraterm occurs in *Apopharynx bolodes* and has been observed in a number of other trematodes. In many Digenea, cross-fertilization is the rule, and the insertion of the cirrus of one individual into the metraterm of another during copulation has been observed in *Prosotocus confusus*, *Fasciola hepatica* and other forms. When this happens, spermatozoa must travel along the entire length of the uterus to the receptaculum seminis. In forms with a very long uterus this process seems to lack efficacy, but there is no barrier to the transit of spermatozoa in this manner, because copulation occurs before the eggs are formed. The real function of Laurer's canal in many Digenea is thus unconnected with the copulatory act and must be sought elsewhere. In searching for a possible function we must recall the precision with which the eggs are formed in a steady sequence, and realize that any excess of shell-forming material might disrupt the process, were it to accumulate in or near the ootype. So precise is the mechanism, that droplets of secretion left over after the formation of a capsule are few, though irregularities and abnormalities sometimes occur in trematodes which are producing eggs for the first time, e.g. *Fasciola* (see Taylor, 1934). In *Aspidogaster conchicola* and at least one digenetic fluke (*Styphlodora bascaniensis* Goldberger, 1911) Laurer's canal ends blindly and may contain moribund spermatozoa and fragments of shell material. In other Digenea it is distinctly possible, even probable, that this canal serves as a conduit for any excess of shell-forming secretion which may exist as small globules during egg formation, thus serving as a means of maintaining the ootype in a fit condition.

In some instances the folds of the uterus are confined to the region of the body in front of the ovary, a descending limb being absent. In Cyclocoelidae it extends forward from the ovary in wide, transverse folds which may overstep the caeca laterally, a rare condition. When the genital pore occupies a posterior position, as in Brachylaemidae and some Strigeidae, the uterus has a primary ascending limb and a secondary descending one. Another rare condition is that in which only one egg or a few eggs exist in the uterus at any time, the latter being so short that the ootype is virtually in direct communication with the metraterm, as in some Spirorchiidae and Sanguinicolidae. A long uterus is of course correlated with the production of numerous eggs, but is not produced by mere distension, as is perhaps the case in Cestoda. In many juvenile flukes a long and much-convoluted uterus is developed long before the formation of the first egg, e.g. in Bucephalidae (Fig. 25B). On the other hand, we may take it as axiomatic that flukes which produce few eggs possess a short uterus.

Sexual dimorphism has already been referred to, but the previous remarks require amplification. This character, rare in Trematoda, may arise out of hermaphroditism, as seems to be the case in Didymozoidae. *Didymozoon scombri* and *D. faciale*, which live in pairs or small groups encysted in the buccal cavity or beneath the integument of the head region in marine fishes, have male and female reproductive organs of normal, if rather unusual, pattern (Fig. 44). *Wedlia*, species of which also live in cysts, shows a type of dimorphism in which the female organism develops only the rudiments of the male organs, while the male has an ovary, vitellaria, shell gland and receptaculum seminis, though none of these organs is functional. *Nematobothrium filarina* exists in the form of individuals of different thickness. The thicker individual has a body filled with numerous eggs of a golden colour, while the thinner one contains only a few colourless eggs.

In *Schistosoma* sex is determined in the earliest stages of development. The sporocyst of *S. japonicum* can give rise to male or to female cercariae, but not to both, so that the vertebrate host will be infected with cercariae of only one sex if they all came from the same sporocyst. If male cercariae penetrate into the blood-vascular system of the final host, they reach maturity in about 40 days. Female cercariae likewise penetrating into the final host may, in the absence of males, remain immature for periods up to 268 days. When males are present, however, they become mature in about 65 days. An undeveloped female may attain only one-fifth the normal definitive size after a sojourn of months in the blood of the final host. The ovary remains undeveloped and the vitellaria rudimentary. The anomalous females which have remained stunted for 128 days will come to maturity in 19 days, however, if males are introduced into the blood containing them (Severinghaus, 1928). In this instance, it seems possible that the male organism liberates substances, conceivably hormones, which condition the development of the female.

We must now consider the characters of the eggs in Prosostomata. Extending the statement that processes of the egg are rare in Digenea, we must note that they occur in *Notocotylus*, *Hapalotrema*, *Epibathra*, *Ceratotrema*, *Hemipera* and other genera. In Schistosomes the eggs may have each a short lateral or terminal spine, which is really a filament of special kind capable of doing damage to the tissues of the host. For the most part, however, the eggs of Digenea are ovoid and operculate, without spines or filaments. Opercula are lacking in the eggs of Schistosomatidae, Sanguinicolidae and other families, and such capsules must split open before the contained miracidia can emerge. The hatching has been observed closely in operculate capsules, the miracidium obtaining purchase on the rim of the opening while it pushes the lid aside. In non-operculate eggs the details of the hatching process are largely matters for conjecture. We might stress that the form, size and colour of the eggs are often distinctive, and should be noted along with other characters in the identification of Digenea. The largest eggs met with in Prosostomata are those of *Bilharziella polonica*, a Schistosome of birds, being about 0·4 mm. long. Heterophyidae have the smallest eggs, which may be less than 0·025 mm. long. Such considerable differences can only be determined by accurate measurements made under the microscope. The eggs generally pass to the exterior along ducts in the organs of the host, but in some instances (e.g. the eggs of *Schistosoma*) no such easy egress is possible, and they must work their way through the walls of organs like the bladder, and may, while so doing, inflict severe damage on the host.

(vii) *Abnormalities and Artefacts.* In conclusion, we must emphasize the asymmetry which is superimposed on the fundamental bilateral symmetry in Digenea. This chiefly concerns the reproductive systems, but may extend to the alimentary and other systems. At the same time we must note that the asymmetrical organs or parts normally occurring on one side of the body may sometimes be on the other. This condition, known as *situs inversus*, is infrequent and not of any particular importance in taxonomy. More important, because liable to deceive us, are various malformations which may be met with. Thus, Ciurea (1933*b*) observed the atrophy of the left testis and, once, the absence of both the testis and vitellarium of the left side, in *Metagonimus yokogawai*. No doubt such abnormalities are more common than a perusal of the literature would lead us to believe, and they should be recorded. Equally important, for the same reason, are artefacts due to faulty fixation. Trematodes are creatures of variable form, Digenea perhaps more so than Monogenea, and the perpetuation in fixation of abnormal postures may grossly mislead us, not merely as regards what might be considered to be the 'normal' shape but also about the relative positions of the internal organs, i.e. characters upon which taxonomic distinctions are based. Natural variability may thus be greatly extended as far as mere appearances go by mediocre fixation. Many invalid species have been erected for no other reason than that sufficient allowance has not been made for such variability, which should be noted whenever trematodes can be obtained in abundance.

CHAPTER 6

THE TAXONOMY OF THE TREMATODA

I. The Classification of the Monogenea

Numerous diverse trematodes having very distinctive features in common make up this order, the name of which is generally wrongly attributed to Beneden, who, as Price (1937*a*) pointed out, in 1858 used the terms *monogénèses* and *digénèses* in the vernacular sense, the name Monogenea dating back to Carus in 1863. Synonymous names include: Polystomea Leuckart, 1856; Pectobothrii Burmeister, 1856; Cryptocoela Johnston, 1865; Ectoparasitica Lang, 1888; Eterocotylea Monticelli, 1892; Heterocotylea Braun, 1893; Monogenetica Haswell, 1893; and Heterocotylida Lahille, 1918.

Monogenea are ectoparasitic flukes of medium or fairly large size with elongate and cylindrical, or flat and leaf-like bodies. The posterior extremity is modified to form an adhesive organ, the *opisthaptor*, which is generally provided with hooks and hooklets and sometimes suckers or pincer-like clamps as well. Near or at the anterior extremity there may be less conspicuous adhesive organs of a glandular or muscular nature, constituting the *prohaptor*. The mouth is near this end of the body, and the gut comprises a well-developed pharynx and a single sac-like or bifurcate intestine. Eyes may be present or absent. The excretory system is double, opening by a pair of pores symmetrically placed near the level of the genital orifice on the dorsal surface of the body. Hermaphroditism is the rule, and the genital pores usually open into a common atrium. The uterus is short, the eggs few in number and, as a rule, provided with polar filaments. Development is direct, i.e. occurs without an alternation of generations or of hosts, and metamorphosis is incomplete. Such trematodes are mostly parasites of poikilothermous vertebrates, but some live on the crustacean parasites of fishes and one or two on cephalopods and mammals.

Odhner (1912) proposed a division of the order into two suborders, the Monopisthocotylea and Polyopisthocotylea, on the basis of the absence or presence of a genito-intestinal canal, i.e. a tubular connexion between the oviduct and the right (rarely the left) caecum. Most writers have concurred, but Fuhrmann (1928) proposed a tripartite division of the order, two of the groups thus formed representing subdivisions of Odhner's Monopisthocotylea. In this scheme there is little taxonomic advantage, as Price (1937*a*) has pointed out, and Odhner's scheme will be adopted here.

It is hardly practicable to determine the presence or absence of a short and delicate internal canal in an opaque or pigmented trematode, living or preserved, but, fortunately, there are correlated characters, chief amongst which is the structure of the opisthaptor. From a practical point of view, the first thing to do in trying to identify a monogenetic fluke is to scrutinize the opisthaptor. If it is a single disk-like object, whether sucker-like or not, the fluke belongs to the Monopisthocotylea. If, instead, it is made up of two or more suckerlets or clamps set upon a disk, or upon the naked ventral surface of the body, the fluke can be referred forthwith to the Polyopisthocotylea.

Complete classification of the Monogenea has not been satisfactorily achieved, but the sustained efforts of Price (1936*b*, 1937*a–c*, 1938*a, c, d*, 1939*a, b*, 1942*b*, 1943*a, b*) have produced a scheme on which the following arrangements and keys are largely based.* Whether or not the superfamilies have any phylogenetic significance is doubtful, but they provide useful terms for the groups which can be separated. Natural relationships in Monogenea, as in Digenea, cannot be fully elucidated in terms of adult characters only, though at present we have to be satisfied with a scheme which includes some larval characters (e.g. hooklets), but largely excludes larvae from consideration. Studies on the life histories and growth of Monogenea will play their part in the formulation of a final scheme, and will no doubt correct many misconceptions in the groupings which are at present recognized.

KEY TO THE SUBORDERS, SUPERFAMILIES AND FAMILIES OF THE MONOGENEA

I. Opisthaptor a well-developed disk, sometimes sucker-like, with or without radial septa, armed with 1–3 pairs of large hooks and 12–16 marginal hooklets. Genito-intestinal canal absent — MONOPISTHOCOTYLEA

1. Opisthaptor armed, large hooks having supporting bars — *GYRODACTYLOIDEA*
   - A. Viviparous forms — **GYRODACTYLIDAE** (p. 67)
   - B. Oviparous forms
     - (*a*) Anterior extremity not expanded to form head lappets — **DACTYLOGYRIDAE** (p. 67)
     - (*b*) Anterior extremity expanded to form head lappets — **CALCEOSTOMATIDAE** (p. 68)

2. Opisthaptor armed or unarmed; when armed, large hooks lacking supporting bars — *CAPSALOIDEA*
   - A. Prohaptor a pair of suckers or corresponding glandular depressions
     - (*a*) Intestine single — **UDONELLIDAE** (p. 71)
     - (*b*) Intestine double
       - (i) Male and female pores close together — **CAPSALIDAE** (p. 69)
       - (ii) Male and female pores not close together — **ACANTHOCOTYLIDAE** (p. 70)
   - B. Prohaptor not a pair of suckers or glandular depressions
     - (*a*) Opisthaptor provided with hooks — **MONOCOTYLIDAE** (p. 69)
     - (*b*) Opisthaptor not provided with hooks — **MICROBOTHRIIDAE** (p. 70)

II. Opisthaptor a number of suckerlets or clamps borne on a disk-like process or on the ventral surface. Genito-intestinal canal present — POLYOPISTHOCOTYLEA

1. Prohaptor usually an oral sucker, never paired buccal suckers; opisthaptor with three pairs of cup-like suckers (one in *Sphyranura*) — *POLYSTOMATOIDEA*
   - A. Opisthaptor with an appendix-like prolongation — **HEXABOTHRIIDAE** (p. 72)
   - B. Opisthaptor without an appendix-like prolongation — **POLYSTOMATIDAE** (p. 71)

* For comprehensive lists of references the reader is referred to Price's papers.

2. Prohaptor a pair of buccal suckers within the oral cavity; opisthaptor with few or many pairs of suckers or clamps, each organ supported by a number of cuticular pieces or sclerites *DICLIDOPHOROIDEA*
   A. Opisthaptor comprising clamps each with two valves which bite on one another (and function by pinching the gill filament or groups of filaments)
   Sclerites:
      (*a*) Two semicircular marginal pieces (one in each valve) which approach a unilateral basal piece; middle piece, if present, rudimentary (Fig. 22D, *e*) **MAZOCRAËIDAE** (p. 73)
      (*b*) Two pairs of marginal pieces (one pair in each valve making a semicircle) and a well-developed middle piece also of semicircular shape (Fig. 22D, *c*) **MICROCOTYLIDAE** (p. 74)
      (*c*) As in (*b*), but modified; lateral pieces fused, or ventral pieces segmented near the middle, middle piece massive, sometimes having thickenings (Fig. 22D, *b*) **DISCOCOTYLIDAE** (p. 73)
      (*d*) As in (*b*), but modified differently from (*c*); lateral pieces joined (but not fused) and one pair spurred, middle piece meeting one pair of additional pieces (Fig. 22D, *d*) **GASTROCOTYLIDAE** (p. 74)
   B. Opisthaptor sometimes comprising clamps, but generally reinforced suckers which maintain a circular or oval opening (adhesion generally effected by suctorial action)
   Sclerites:
      (*e*) Eight pieces of characteristic complicated shapes (the middle support in two pieces), and sometimes numerous additional pieces of minute size which form a matting of rod-like chains (Fig. 22D, *a*) **DICLIDOPHORIDAE** (p. 74)
      (*f*) Three pieces in each sucker, their shapes indeterminate and variable, but the middle piece the largest and approximately X-shaped (Fig. 22D, *f*) **HEXOSTOMATIDAE** (p. 75)

In reviewing the diagnostic characters of these sixteen families, all of which are represented in British fishes and Amphibia, we shall consider also the distinguishing features of the various subfamilies, noting only the most important genera.

Suborder MONOPISTHOCOTYLEA Odhner, 1912

Syn. Monocotylea Blainville, 1828; Tricotylea and Calicotylea Diesing, 1850; Tristomea Taschenberg, 1879; Oligocotylea Monticelli, 1903; Monopisthodiscinea and Monopisthocotylinea Fuhrmann, 1928.

Superfamily *GYRODACTYLOIDEA* Johnston & Tiegs, 1922

DIAGNOSIS. *Prohaptor:* absent. *Cephalic glands:* generally one group on either side of the pharynx, opening on the surface of one or more than one pair of *head organs*. *Opisthaptor:* disk- or wedge-shaped, bearing one or two pairs of large hooks, supported by one or two, rarely three, cuticular bars. *Gut:* intestine sac-like or having two crura, with or without short diverticula. *Reproductive systems:* genital pore almost or perfectly median. Cirrus simple, cuticular, often with cuticular accessories of complicated form. Vagina present or absent. Genito-intestinal canal almost invariably absent, present only in the Protogyrodactylidae (a family with two genera, *Protogyrodactylus* Johnston & Tiegs, 1922, and *Trivitellina* Johnston & Tiegs, 1922, both Australian).

Family **GYRODACTYLIDAE** Cobbold, 1864 (Fig. 14A)

Syn. Amphibdellidae Carus, 1885, in part.

DIAGNOSIS. *Shape and size:* small, elongate. *Head organs:* one pair. *Opisthaptor:* well developed, generally with one pair of large hooks and 15–16 marginal hooklets. *Gut:* intestine bifurcate, caeca distinct. *Eyes:* absent. *Copulatory organ:* with a row of minute spinelets and, generally, a triangular plate of cuticular material. *Ovary:* V-shaped or lobed, situated behind the testes. *Vitellaria:* absent, or merged with the ovary. *Vagina:* absent. *Habit:* viviparous. Parasites of cephalopods, fishes and amphibians. *Subfamilies:* Gyrodactylinae Monticelli, 1892, emend. Johnston & Tiegs, 1922, and Isancistrinae Fuhrmann, 1928.

KEY TO SUBFAMILIES OF GYRODACTYLIDAE

I. Anterior extremity bilobed, each lobe with a head organ: opisthaptor circular, bearing one pair of hooks and sixteen marginal hooklets. Parasites of fishes and Amphibia
Gyrodactylinae; solitary genus *Gyrodactylus* Nordmann, 1832 (Fig. 14A)

II. Anterior extremity truncated: opisthaptor with fifteen marginal hooklets, but no hooks. Parasites of cephalopods
Isancistrinae; solitary genus *Isancistrum* de Beauchamp, 1912 (only species, *I. loliginis* de Beauchamp, 1912, a parasite of *Loligo media* Linn.)

Family **DACTYLOGYRIDAE** Bychowsky, 1933 (Figs. 6A, 14B, E, F.)

Syn. Gyrodactylidae Cobbold, 1864, in part; Amphibdellidae Carus, 1885, in part.

DIAGNOSIS. *Shape and size:* small, elongate. *Head organs:* two pairs or more, supplied with cephalic glands, which may otherwise be distributed throughout the median region in front of the mouth. *Opisthaptor:* moderately or well developed, with or without accessories in the form of squamodisks, having one pair or two pairs of hooks and, generally, fourteen marginal hooklets. *Vitellaria:* well developed, discrete. *Ovary:* globular, sometimes curved, situated in front of the testes. *Vagina:* present or absent. *Habit:* oviparous. Parasites of fishes. *Subfamilies:* Dactylogyrinae Bychowsky, 1933; Tetraonchinae Monticelli, 1903; Diplectaninae Monticelli, 1903.

KEY TO SUBFAMILIES OF DACTYLOGYRIDAE

I. Opisthaptor with one pair of large hooks — **Dactylogyrinae** (*Dactylogyrus*; *Neodactylogyrus*) (key on p. 109)

II. Opisthaptor with two pairs of large hooks
- A. Opisthaptor with dorsal and ventral squamodisks — **Diplectaninae** (*Diplectanum*)
- B. Opisthaptor without squamodisks — **Tetraonchinae** (*Tetraonchus*, *Haplocleidus*, *Amphibdella*, *Ancyrocephalus*, *Dactylodiscus*, *Linguadactyla*, *Amphibdelloides*) (key on p. 114)

Although the last-mentioned two subfamilies are thus clearly set off from the first, they are, perhaps, better separated by the following characters:

Diplectaninae: Body covered, especially in the posterior half, with forwardly directed spines of scale-like appearance.

Tetraonchinae: Body devoid of scales or spines.

A key to genera of Tetraonchinae is given on p. 114.

Another subfamily is the Bothitrematinae Price, 1936, which was erected for a species of *Bothitrema* Price, 1936 (Syn. *Acanthocotyle* Monticelli, 1888, in part). This is *B. bothi* (MacCallum, 1913) Price, 1936, and need not concern us, except to state that it has one pair of hooks and a circlet of tubular structures on the opisthaptor and is American.

Family **CALCEOSTOMATIDAE** Parona & Perugia, 1890, emend. Price, 1937, and Fischthal & Allison, 1941

DIAGNOSIS. *Anterior extremity:* expanded to form head lappets. *Cephalic glands:* scattered over a considerable area in this region. *Opisthaptor:* sucker-like, not very muscular, with or without hooks and marginal hooklets. *Gut:* intestine having or lacking short diverticula. *Eyes:* present or absent. *Reproductive systems:* testis single. Cirrus simple and cuticularized. Vagina present or absent. *Subfamily:* Calceostomatinae Monticelli, 1892 (as Calceostominae). (*Note:* the Dionchinae Johnston & Tiegs, 1922 was at one time admitted, but is now included in the family Monocotylidae.) The Calceostomatinae consists of five genera, only two of which are represented in Europe, *Calceostoma* Beneden, 1853 and *Anoplodiscus* Sonsino, 1890. The type-species of the former, *Calceostoma calceostoma* (Wagener, 1857), has been found on the gills of *Sciaena* spp. off the Belgian coast, and another species, *Calceostoma inerme* Parona & Perugia, 1889, in Italy and the Mediterranean. Neither of these species has been recorded in this country, but there is every likelihood that one of them may appear. The type-species of the second genus, *Anoplodiscus richiardii* Sonsino, 1890, occurs in Italy; the only other species in Australia.

Superfamily *CAPSALOIDEA* Price, 1936

DIAGNOSIS. *Prohaptor:* present or absent, when present a feeble oral sucker or pseudosucker, or two lateral suckers, or two lateral glandular grooves. *Head organs:* not invariably present. *Opisthaptor:* discoidal, generally large and muscular, with or without hooks, its ventral surface often divided by septa into loculi. Hooks, when present, never supported by bars of cuticular material. *Gut:* intestine single or bifurcate, often having median and lateral diverticula. *Reproductive systems:* genital pore median or lateral. Cirrus sometimes cuticularized, with accessories only in *Anoplodiscus* (Microbothriidae). One testis or a number of testes present. Vagina present or absent. *Habit:* oviparous.

Family **MONOCOTYLIDAE** Taschenberg, 1879 (Figs. 6B, 16A–C, 17B)

DIAGNOSIS. *Shape:* oval or elliptical outline, body flattened. *Prohaptor:* when present, an oral sucker, or several anterior suckers. *Cephalic glands:* present. *Opisthaptor:* discoidal, the ventral surface provided with septa marking out depressed, sucker-like loculi and equipped with one pair of hooks and fourteen marginal hooklets. *Gut:* mouth ventral, near the anterior extremity, pharynx large, oesophagus very short or absent, caeca lacking diverticula, sometimes merging posteriorly. *Reproductive systems:* genital aperture generally median. Cirrus cuticularized. Generally a single testis, rarely three or numerous testes. Ovary curved, sometimes encircling the right caecum. Vagina single or paired, rarely absent. *Subfamilies:* Monocotylinae Gamble, 1896; Calicotylinae Monticelli, 1903; Merizocotylinae Johnston & Tiegs, 1922.

KEY TO SUBFAMILIES OF MONOCOTYLIDAE

I. Prohaptor present as a feeble oral sucker; opisthaptor with seven to eight loculi bounded by radial septa and a central space
  A. Vagina single — **Monocotylinae** (*Monocotyle; Heterocotyle*)
  B. Vaginae paired — **Calicotylinae** (*Calicotyle*)

II. Prohaptor absent, or represented by cephalic glands and head organs; opisthaptor with more than eight loculi bounded by tangential as well as radial septa — **Merizocotylinae** (*Merizocotyle*, *Thaumatocotyle*)

Note. Additional subfamilies are the Loimoinae Price, 1936 and the Dionchiinae Johnston & Tiegs, 1922, with a single genus apiece (U.S.A.).

Keys to genera of Monocotylinae and Merizocotylinae are given on pp. 123 and 128.

Family **CAPSALIDAE** Baird, 1853 (Figs. 1A, 6C, 17F, 19)

Syn. Phyllinidae Johnston, 1846; Tristomidae Cobbold, 1877; Tristomatidae Gamble, 1896; Encotyllabidae Monticelli, 1888.

DIAGNOSIS. *Shape:* oval or elliptical outline, flattened, the anterior region constricted to form a cephalic lobe. *Cuticle:* non-spinous, or bearing papillae, or having spines dorso-laterally. *Prohaptor:* a pair of suckers, or a pair of glandular areas, or suckers and glandular areas, situated at the sides of the cephalic lobe. *Opisthaptor:* disk- or sucker-like, sometimes partitioned by septa to form loculi, equipped with one to three pairs of hooks and fourteen marginal hooklets. *Gut:* mouth ventral, never encircled by an oral sucker, pharynx well developed, caeca with lateral or median diverticula which may be much-branched. *Sense organs:* one pair of sensory papillae on the anterior margin of the cephalic lobe, and two pairs of eyes. *Excretory system:* pore dorso-lateral at or near the level of the pharynx. *Reproductive systems:* genital pores generally lateral, separate or opening into a common genital atrium. Testes numbering two or more. Ovary in front of the testes, median. Vagina present or absent. *Subfamilies:* Capsalinae Johnston, 1929; Trochopodinae Price, 1936 emend.; Benedeniinae Johnston, 1931; Nitzschiinae Johnston, 1931; Encotyllabinae Monticelli, 1892 (all of which have representatives in British Fishes).

KEY TO SUBFAMILIES OF CAPSALIDAE

I. Opisthaptor having a well-developed peduncle arising from the ventral surface of the body posteriorly — **Encotyllabinae** (*Encotyllabe*)

II. Opisthaptor not having a well-developed peduncle
- A. Opisthaptor provided with septa and loculi
  - (*a*) Testes two in number — **Trochopodinae** (*Trochopus, Megalocotyle*) (key on p. 140)
  - (*b*) Testes numerous — **Capsalinae** (*Capsala, Capsaloides, Tristoma*) (key on p. 137)
- B. Opisthaptor not provided with septa and loculi
  - (*a*) Testes two in number — **Benedeniinae** (*Benedenia, Ancyrocotyle, Entobdella*) (key on p. 143)
  - (*b*) Testes numerous — **Nitzschiinae** (*Nitzschia*)

Family ACANTHOCOTYLIDAE Price, 1936 (Fig. 17A)

Syn. Anisocotylidae Tagliani, 1912, in part.

DIAGNOSIS. *Prohaptor:* retractile suckers or feeble suckers encircled by the pores of the cephalic glands. *Opisthaptor:* small, with one pair of centrally situated hooks and fourteen marginal hooklets, sometimes with a large, disk-like pseudohaptor bearing radial rows of spines or radial septa. *Reproductive systems:* genital apertures separate, male pore median or slightly lateral, female pore marginal. Testis single or testes numerous. *Subfamilies:* Acanthocotylinae Monticelli, 1903; Enoplocotylinae Tagliani, 1912.

KEY TO SUBFAMILIES OF ACANTHOCOTYLIDAE

I. Pseudohaptor large, disk-like, with radial rows of spines or with muscular septa; true haptor a very small disk at its margin — **Acanthocotylinae** (*Acanthocotyle, Lophocotyle*) (key on p. 137)

II. Pseudohaptor absent; true haptor large — **Enoplocotylinae** (*Enoplocotyle*)

Family MICROBOTHRIIDAE Price, 1936 (Fig. 18)

Syn. Dermophagidae MacCallum, 1926; Labontidae MacCallum, 1927.

DIAGNOSIS. *Prohaptor:* present or absent, when present in the form of sucker-like structures. *Opisthaptor:* small, without hooks or septa. *Eyes:* rarely present. *Gut:* intestine bifurcate, caeca with or without lateral diverticula. *Reproductive systems:* genital apertures close together, or opening into a common genital atrium. Cirrus cuticularized, or muscular and associated with a cuticular ejaculatory duct. Vagina single, or vaginae paired. *Subfamilies:* Microbothriinae Price, 1938; Pseudocotylinae Monticelli, 1903.

KEY TO SUBFAMILIES OF MICROBOTHRIIDAE

I. One testis, or two testes — **Microbothriinae** (*Microbothrium, Leptobothrium, Leptocotyle*)

II. Numerous testes — **Pseudocotylinae** (*Pseudocotyle*)

A key to genera of Microbothriinae is given on p. 130.

Family **UDONELLIDAE** Taschenberg, 1879 (Fig. 15)

DIAGNOSIS. *Shape:* elongate, cylindrical, or only slightly flattened. *Cuticle:* with ill- or well-marked annuli. *Prohaptor:* present, as two small suckers or sucker-like structures, or absent. Cephalic glands present. *Opisthaptor:* sucker-like, without spines or septa. *Gut:* pharynx well developed, protrusible, intestine sac-like, without diverticula, sometimes with fenestrations in the region of the gonads. *Reproductive systems:* genital aperture median or slightly lateral. Cirrus absent. Testis single. Ovary in front of the testis. Eggs oval or pyriform, each with a filament at one pole. *Type genus: Udonella* Johnston, 1835. (Syn. *Amphibothrium* Frey & Leuckart, 1847; *Nitzschia* Baer, 1826, in part; *Lintonia* Monticelli, 1904; *Calinella* Monticelli, 1910).

Note. The family possibly contains also the *genera inquirenda Echinella* and *Pteronella*, both Beneden & Hesse, 1863.

Suborder POLYOPISTHOCOTYLEA Odhner, 1912

Syn. Polycotyla Blainville, 1828; Octobothrii E. Blanchard, 1847; Eupolycotylea Diesing, 1850; Polycotylea Diesing, 1850, in part.

Superfamily *POLYSTOMATOIDEA* Price, 1936

DIAGNOSIS. *Prohaptor:* an oral sucker situated at or near the anterior extremity. *Opisthaptor:* somewhat disk-like, generally with three pairs of cup-shaped suckers (one pair only in *Sphyranura*), with or without an appendage carrying two small suckers and one to three pairs of small hooks. Large suckers of the cotylophore each with a single hook, which is small in Polystomatidae, but large in Hexabothriidae. *Gut:* prepharynx short, pharynx bulbous, oesophagus short, caeca with or without diverticula or anastomoses. *Eyes:* rarely present. *Reproductive systems:* genital pore common to the male and female systems, ventral. Testis single or testes numerous, situated behind the ovary. Vaginae paired, and, as a rule, with lateral openings. Parasites of fishes, Amphibia and Reptilia, rarely in the eyes of Mammalia.

Family **POLYSTOMATIDAE** Gamble, 1896 (Fig. 6E)

Syn. Polystomidae Carus, 1863; Sphyranuridae Poche, 1926, in part; Dicotylidae Monticelli, 1903.

DIAGNOSIS. *Prohaptor:* an oral sucker. *Opisthaptor:* a disk or cotylophore bearing one pair or three pairs of cup-like suckers, sometimes with a solitary cuticular support or sclerite, and sixteen marginal hooklets; with or without hooks. *Gut:* intestine bifurcate, with or without diverticula, the intestinal crura sometimes merging posteriorly. *Eyes:* as a rule absent in the adults. *Reproductive systems:* genital pore ventral and median. Cirrus generally provided with a coronet of hooklets. Single testis or numerous testes present. Ovary small, on the right or left, in front of the testes. Vaginae present or absent. Parasites of Amphibia and Reptilia occurring in the buccal and nasal cavities, pharynx, oesophagus, urinary bladder and, rarely, in the eyes of aquatic mammals.

Subfamilies: Polystomatinae Gamble, 1896 (Syn. Polystominae Pratt, 1900) (opisthaptor with six suckers) (*Polystoma*) and Sphyranurinae Price, 1939 (opisthaptor with only two suckers). The former has six (non-European) genera besides *Polystoma*, the latter contains the single New World genus *Sphyranura* Wright, 1879, which has not been found in Europe, and which includes three species, two parasitic on the skin of *Necturus*.

Family **HEXABOTHRIIDAE** Price, 1942 (Figs. 17C, 20)

Syn. Onchocotylidae Stiles & Hassall, 1908.

DIAGNOSIS. *Prohaptor:* a more or less well-developed oral sucker or (only in Diclybothriinae) two ventro-lateral bothria. *Opisthaptor:* cotylophore circular or rectangular, bearing six suckers, each with a large crescentic hook, and an appendix provided with a pair of terminal suckers and one to three pairs of hooks. *Gut:* intestine bifurcate, generally with lateral and median diverticula, caeca uniting posteriorly and extending into the opisthaptor. *Eyes:* generally absent, present in Diclybothriinae. *Reproductive systems:* common genital pore ventral and median. Cirrus generally unarmed. Testes numerous. Ovary relatively large, tubular and folded, situated in front of the testes. Vaginae present. Parasites of fishes, notably Selachii. *Subfamilies:* Hexabothriinae Price, 1942; Diclybothriinae Price, 1936; Rajonchocotylinae Price, 1942.

KEY TO SUBFAMILIES OF HEXABOTHRIIDAE

I. Appendix of the opisthaptor having three pairs of hooks
Diclybothriinae (*Diclybothrium*)

II. Appendix of the opisthaptor having one pair of hooks
A. Vaginae uniting to form a single duct which joins the vitelline reservoir
Rajonchocotylinae (*Rajonchocotyle*, *Rajonchocotyloides*)
B. Vaginae opening separately into the transverse vitelloducts
Hexabothriinae (*Hexabothrium*, *Erpocotyle*, *Neoerpocotyle*)

Keys to genera of the last two subfamilies are given on pp. 153 and 149.

Superfamily *DICLIDOPHOROIDEA* Price, 1936

Syn. Dactylocotyloidea Brinkmann, 1942*b*.

DIAGNOSIS. *Prohaptor:* two small buccal suckers within the mouth tube (prepharynx). *Opisthaptor:* generally paired suckers or clamps, pedunculate or sessile, arising from a more or less distinct disk or cotylophore, or from the naked surface of the body. Suckers or clamps typically eight in number, four pairs, but numerous in the Microcotylidae, supported by three or more cuticular rods or sclerites which may terminate in hook-like processes. Hooklets present or absent. *Gut:* prepharynx, bulbous pharynx and short oesophagus present, caeca much-branched, with or without anastomoses, sometimes intestinal crura merge with one another posteriorly. *Reproductive systems:* common genital pore generally present, cirrus sometimes equipped with spines, testes follicular. Ovary elongate and folded, generally in front of the testes, but sometimes with follicles of the testis in front of it. Vaginae present and opening dorsally, or absent. Eggs generally provided with filaments. Parasites on the gills, rarely the skin, of fishes, sometimes on Crustacea which are parasitic on fishes.

Family **MAZOCRAËIDAE** Price, 1936

DIAGNOSIS. *Opisthaptor:* generally four pairs of clamps equally spaced along the lateral regions of an indistinct cotylophore or of the postero-ventral region of the body. Cuticular supports of the clamps as shown in Fig. 22 D, *e*, i.e. relatively simple; in each clamp single dorsal and ventral marginal sclerites with forked tips which approach a straight basal piece, sometimes having a rudimentary middle piece attached to it. Posterior tip of the cotylophore invariably provided with two or three pairs of dissimilar hooklets. *Reproductive systems:* generally numerous testes, but sometimes a single, elongate, lobed testis between the caeca posteriorly. Ovary elongate, folded, near anterior end of testis or testes. Vitellaria follicular, lateral. Vagina, when present, with median dorsal pore. *Genera: Mazocraes* Hermann, 1782; *Ophiocotyle* Beneden & Hesse, 1863; *Kuhnia* Sproston, 1945 (key on p. 156).

Family **DISCOCOTYLIDAE** Price, 1936 (Figs. 17 G, 21)

DIAGNOSIS. *Opisthaptor:* comprising three or four pairs of clamps each having cuticular supports as indicated in Fig. 22 D, *b*, i.e. two pairs of marginal sclerites, an additional pair of basal sclerites, and a middle segment of massive proportions having rows of lacunae extending on either side of a central rib; but variously modified by the fusion of elements and, in the extreme, having two lateral segments and one middle segment, all U-shaped (horseshoe-shaped). *Reproductive systems:* lateral vaginae present or absent. *Subfamilies:* Discocotylinae Price, 1936; Anthocotylinae Price, 1936; Plectanocotylinae Monticelli, 1903; Chimaericolinae n.subfam. (Syn. Chimaericolidae Brinkmann, 1942 *b*).

KEY TO SUBFAMILIES OF DISCOCOTYLIDAE

I. Four pairs of clamps; vaginae generally present
  A. Clamps of the most anterior pair very large, the remaining ones very small
  Anthocotylinae (*Anthocotyle*, *Platycotyle*)
  B. Clamps of the most anterior pair not markedly larger than the others
    (*a*) Posterior region of the body narrow and transversely striated, the body and opisthaptor being connected by a narrow isthmus
    Chimaericolinae (*Chimaericola*)
    (*b*) Posterior region of the body neither narrow nor transversely striated
    Discocotylinae (*Grubea*, *Discocotyle*, *Diplozoon*, *Vallisia*) (key on p. 160)

II. Three pairs of clamps; vaginae absent **Plectanocotylinae** (*Plectanocotyle*)

The sclerites of the clamps differ in the four subfamilies as follows:

1. Discocotylinae: having in each clamp two pairs of lateral sclerites, which may be fused, and a pair of short basal pieces; middle segment of the ventralmost sclerites well developed, sometimes with rows of lacunae extending along either side of a central rib.

2. Anthocotylinae: having in each clamp two pairs of lateral sclerites, the dorsals spurred, and a very well-developed and somewhat U-shaped middle segment having a central rib. (The opisthaptor having a terminal languet bearing two pairs of hooklets.)

3. Plectanocotylinae: having in each clamp two pairs of lateral sclerites, members of the ventralmost pair segmented about their middles, middle segment divided into a curved ventral bar and a perforated dorsal piece (terminal languet present).

4. Chimaericolinae: having in each clamp one pair of lateral sclerites and one median sclerite, all horseshoe-shaped.

Note. Price (1943*b*) excluded the Plectanocotylinae from this family and included a new subfamily, the Vallisinae, made for *Vallisia* Perugia & Parona, 1890.

Family **MICROCOTYLIDAE** Taschenberg, 1879 (Fig. 6F)

DIAGNOSIS. *Opisthaptor:* generally a paired series comprising numerous (in the adult not less than sixteen) clamps, each with cuticular supports as indicated in Fig. 22D, *c*, i.e. two pairs of marginal sclerites, the dorsalmost ones spurred, and a single middle sclerite of U-shape. *Reproductive systems:* single vaginal pore dorsal. Testes numerous. Ovary elongate and folded, situated in front of the testes in the middle third of the body. Eggs equipped with polar filaments. *Subfamily:* Microcotylinae Monticelli, 1892 (key to genera *Microcotyle*, *Axine* and *Pseudaxine* on p. 167).

Family **GASTROCOTYLIDAE** Price, 1943

Syn. Microcotylidae Taschenberg, 1879, in part; Axininae of Nicoll, 1915, in part.

DIAGNOSIS. *Opisthaptor:* generally comprising numerous small clamps arranged in a single row upon a lateral flange-like process. Sclerites somewhat similar to those of Microcotylidae and as indicated in Fig. 22D, *d*, i.e. two pairs of marginal pieces, one pair spurred, and a curved middle piece which meets a pair of submarginal pieces. Cotylophore bearing two to three pairs of dissimilar hooks. *Reproductive systems:* single vaginal pore dorsal. Testes not numerous, posterior. Ovary elongate and folded, situated in front of the testes. Eggs each with two pointed filaments. Type-genus *Gastrocotyle* Beneden & Hesse, 1863.

Family **DICLIDOPHORIDAE** Fuhrmann, 1928 (Figs. 6D, 17D, E, 22A–C)

Syn. Choricotylidae Rees & Llewellyn, 1941; Dactylocotylidae Brinkmann, 1942*a*, in part.

DIAGNOSIS. *Opisthaptor:* comprising four pairs of pedunculate clamp-like suckers having a rather complicated cuticular framework generally consisting of eight sclerites and numerous minute additional pieces which form rod-like chains. Main sclerites arranged as indicated in Fig. 22D, *a*, i.e. four pieces bordering the opening of the sucker ventrally; two others bent at right angles, one part of each bordering the opening dorsally, the other being directed towards its centre; and two others reinforcing the dorsal wall, one T-shaped, the other continuing the stem of the T above its limbs (thus, †). Dorsal wall of the sucker reinforced by the rodlets which are arranged fan-wise in six to seven concentric arcs in which the numbers of pieces vary. *Reproductive systems:* genital hooklets invariably present, each being curved and deeply grooved, but not (as is sometimes stated) necessarily with bifurcate points. Testes numerous. Ovary elongate, folded, generally in front of the testes. Receptaculum seminis generally present. Vaginae generally absent. *Subfamilies:* Diclidophorinae Cerfontaine, 1895 and Cyclocotylinae Price, 1943.

KEY TO SUBFAMILIES OF DICLIDOPHORIDAE

I. Opisthaptor equipped with true cup-like suckers **Cyclocotylinae** (*Cyclocotyla, Diclidophoropsis*)

II. Opisthaptor equipped with clamps (pincer-like organs) **Diclidophorinae** (*Diclidophora, Diclidophoroides, Octodactylus*)

Keys to the genera of Cyclocotylinae and Diclidophorinae are given on pp. 177 and 172.

Family **HEXOSTOMATIDAE** Price, 1936 (Fig. 23)

Syn. Octobothrii Blanchard, 1847, in part; Polystomea Leuckart of Taschenberg, 1879, in part; Octocotylidae Beneden & Hesse, 1863, in part; Octobothriidae Taschenberg, 1879, in part; Plagiopeltinae Monticelli, 1903; *nec* Hexacotylidae Monticelli, 1899.

Diagnosis. *Opisthaptor:* cotylophore bearing four pairs of sessile suckers, those of the morphologically posterior pair being smaller than the remainder, and two pairs of terminal hooks. Sclerites of the larger suckers numbering three (respectively lateral, middle and median topographically) and varying in shape, but the middle piece approximately X-shaped. Sclerites of the small posterior suckers fused, the resultant cuticular structure of indeterminate shape. *Reproductive systems:* vaginal pore dorsal, vagina with a cuticularized lining which is sculptured internally. Testes numerous. Ovary elongate, folded and situated in front of the testes. Eggs each with two short polar filaments. *Type genus: Hexostoma* Rafinesque, 1815, nec *Hexastoma* Rudolphi, 1809, nec *Hexastoma* Kuhn, 1828. (Syn. *Hexacotyle* Blainville, 1828.)

II. The Classification of the Digenea

In the present state of our knowledge, the classification of the Digenea is a difficult undertaking. In addition to more than sixty families, the order includes many genera which have not so far been allocated to families. While the families can be separated from one another without much difficulty, they are not universally agreed upon, and their arrangement in higher taxonomic units cannot be achieved in a comprehensive way. Some writers (Fuhrmann, 1928; Sprehn, 1933) have been content to enumerate families, others have attempted a higher classification into suborders, tribes, subtribes and superfamilies, but many families stand to-day with their nearest relatives undisclosed, and such as have been linked together are not known to be related beyond doubt.

Early schemes of classification of the Digenea were based largely upon the characters of adults, which can be so misleading as to cause the assembly of quite unrelated forms. The difficulty of distinguishing between characters of phylogenetic significance and those of only an adaptive nature seems to be insuperable, and many examples exist of the misinterpretation of convergence in the evolution of the Digenea, e.g. the lack of one or both suckers. We now know that larval characters must be taken into consideration in formulating a natural scheme of classification.

Some characters of both larvae and adults have proved to be of high value. Cort (1917) first drew attention to the importance of the excretory system as a possible basis for classification. Studying several fork-tailed cercariae he saw close similarities in the arrangement of the excretory canals coexisting with great differences in other systems of organs, especially the gut. Faust (1919*a*), Sewell (1922), Szidat (1924), La Rue (1926*a*, *b*, *c*) and others developed his idea, which led to the establishment by Faust of the 'flame-cell formula'.

Making use of the homologies of the excretory systems in cercariae and also of the arrangement of flame cells in miracidia, La Rue first showed the probable relationship between the Strigeidae and the Schistosomatidae, families which could scarcely be linked on the basis of adult characters only. Hunter & Hunter (1934, 1935) further showed that the miracidium of *Clinostomum marginatum* closely resembles the miracidia of Schistosomes and Strigeids in the number of flame cells and in the arrangement of the epidermal cells. Krull (1934) extended such comparative observations to the cercariae. In the modern view, as a result of these and similar studies, the Strigeidae (now split into two families, Strigeidae *sensu stricto* and Diplostomatidae), Schistosomatidae and Clinostomatidae must be grouped together in any scheme of classification which postulates higher groups than the family. To them must be added the Cyathocotylidae, relatives of the Strigeidae, and, according to La Rue (1926*b*), the Bucephalidae will have to be included in the same group. This arrangement seems incredible when adult characters alone are considered, because the four kinds of trematodes seem to have nothing in common.

But the sole use of larval characters and especially single systems of organs can be very misleading, and Stunkard (1929*a*) advocated caution in the use of the excretory system, noting striking differences in it in three genera of trematodes which were by universal agreement placed in the same family, the Heterophyidae. In most trematodes development of the cercaria to the metacercaria and adult is accompanied by increase in the number of excretory ducts and flame cells, and it seems reasonable to suppose that cercariae might be arranged in natural groups on the basis of such changes. Brown (1933) showed, however, that in *Lecithodendrium chilostomum* the same type of excretory system (the $2\,(6 \times 2)$ type) serves all stages of development, and he suggested that an increase in the number of ducts during development might be an expression of the physiological needs of the organism, and similarity in the number of excretory units in a group, or of groups within the system, the result of convergence in evolution and not necessarily an indication of phylogenetic relationship.

Hopkins (1941*b*) has added force to this contention. He noted that in four genera of Monorchiidae which have been studied in sufficient detail the flame-cell formula is 2 [(2 + 2) + (2 + 2)], which may be assumed to be applicable to the entire family, but also characterizes a number of Microphallidae, Heterophyidae and Allocreadiidae, as well as certain Fellodistomatidae and Zoogonidae. These trematodes show such marked variation in adult structure, mode of development and life history that they cannot be very closely related. The conclusion is reached, therefore, that the flame-cell formula does not necessarily denote phylogenetic relationship and that this kind of common structural plan probably represents a primitive arrangement which is superseded in the early development of some cercariae, but persists because of retardation of development in others. It is a plausible theory that this occurs in many otherwise unrelated trematodes, giving a false impression of relationship. On the other hand, there is good reason to doubt the apparent closeness of the relationship between trematodes which show marked differences in the flame-cell formula and other details of the excretory system, especially if the differences are consistently present in larvae and adults.

Price (1940*a*) reviewed the life histories of various members of the Heterophyidae and Opisthorchiidae, in which the cercariae are sufficiently alike to indicate affinity between these families, a conclusion which was shrewdly reached by Witenberg (1929) on the basis of adult characters alone. The cercariae belong to the 'Pleurolophocerca' and 'Parapleurolophocerca' groups of Sewell, have rudimentary ventral suckers and spines around the mouth, and they develop in a redia with short caeca and without either a collar or procruscula. Price also pointed out that some characters of the cercariae are so variable as to be of no more than specific value, even the excretory vesicle varying from a sac-like form to a Y-shape through intermediate shapes. Stressing the opinion that we must depend on the *type* of cercaria, not the characters of any of the larval systems, in assembling families into higher groups, he affirmed that two other families, the Acanthostomatidae and Cryptogonimidae, can be included along with the Heterophyidae and Opisthorchiidae in a single superfamily, the Opisthorchioidea. Adult characters lend support to this judgment, for all members of the four families agree in lacking a cirrus pouch, in possessing a receptaculum seminis, and in the fusion of the terminal parts of the male and female ducts into an hermaphrodite duct. In two of the families there is a genital sucker or *gonotyl*, and some members of the remaining families show a trace of it.

These remarks are intended to make clear the likelihood of taxonomic errors arising out of too implicit a reliance on one system of organs. Common sense suggests, and experience has shown, that in the formulation of a reliable scheme of classification we cannot afford to ignore any characters, and must consider as many characters of both larvae and adults as can be determined. The particular characters or stages which ultimately serve the best purpose will doubtless vary in different families or groups of families, and for this reason and the unsatisfactory taxonomic conditions which prevail comprehensiveness must take precedence over exclusiveness. When all available characters are considered we may come to recognize characters of an adaptive nature in both larvae and adults.

Other groupings of families than those mentioned, the superfamilies Strigeoidea and Opisthorchioidea, have been attempted, one being the Plagiorchioidea, which includes the families Plagiorchiidae, Lecithodendriidae, Dicrocoeliidae and Microphallidae. McMullen (1937*b*) surveyed the literature on larval development in these families, finding further evidence of variability in the excretory systems of cercariae and adults, the vesicle for instance varying from a sac-like to an I-, Y- or V-shape. But all members of these families agree in one important particular; all have Xiphidiocercariae. McMullen admitted three other families, proposed as new, in this superfamily, but it seems very doubtful if they merit such high status.

Other superfamilies, recognized by Poche (1926), include groupings of the Paramphistomatidae, Dissotrematidae and Angiodictyidae*; the Spirorchiidae, Aporocotylidae and Sanguinicolidae; the Pronocephalidae, Notocotylidae, Rhabdiopoeidae and Opisthotrematidae; and the Azygiidae, Hemiuridae, Xenoperidae, Halipegidae and Isoparorchiidae, to which the Ptychogonimidae might be added. One of his superfamilies (named the Fasciolida) was an unwieldy group of thirty-eight families. Faust (1930*b*) divided part of it into superfamilies, of which only the type families and two or three others are mentioned, and Craig & Faust (1940) have put forward the same arrangement. Two of these superfamilies (the Heterophyoidea and Opisthorchioidea) should be merged in view of Price's proposals; one (the Dicrocoelioidea) forms part of the Plagiorchioidea; another corresponds to the 'Hemiurida' of Poche; three others are the Fascioloidea (type family Fasciolidae), Echinostomatoidea (type Echinostomatidae) and Troglotrematoidea (type Troglotrematidae); other families stand aloof, and some of them represent major groups in themselves, e.g. the Didymozoidae and the lesser-known Alcicornidae, Faustulidae and Heronimidae, all of which Poche referred to separate superfamilies. Nicoll (1934) named, but in records only, superfamilies not included in the schemes of Poche, Faust, or Craig & Faust, e.g. the Cyclocoeloidea (family Cyclocoelidae), Allocreadioidea (families Allocreadiidae, Opecoelidae, Coitocaecidae, Sphaerostomatidae, Megaperidae, Lepocreadiidae, Acanthostomatidae and Acanthocolpidae, some of which should scarcely rank as such), the Clinostomatoidea (families Clinostomatidae, Brachylaemidae and Liolopidae) and the Haploporoidea (families Fellodistomatidae, Monorchiidae and Zoogonidae). None of these schemes is satisfactory, and all of them inevitably will be modified in the near future as our knowledge of the life histories of the Digenea grows.

One other consideration must be mentioned as having an important bearing on the working out of a satisfactory taxonomic scheme, the relation between trematode parasites and their hosts. Emphasizing the remarkable degree of host-specificity shown by some Digenea, Szidat (1939*a*) suggested that in each family there are phylogenetic lines, represented by subfamilies or genera, which have followed the evolution of their hosts. In propounding this idea he considered in detail three families, the Paramphistomatidae, Notocotylidae and Schistosomatidae. The first two were regarded as closely related, being derived from the Angiodictyidae*. Of the eleven subfamilies which comprise the Paramphistomatidae in Szidat's scheme, the Dadayatrem(at)inae and Opistholebetinae occur in the gut of fishes, the Diplodiscinae in the rectum of Amphibia, the Schizamphistom(at)inae in the large intestine of Reptilia, the Zygocotylinae in

* [=Microscaphidiidae].

the caeca of birds, and the remainder (Cladorchiinae, Gastrodiscinae, Chiorchiinae, Balanorchiinae, Paramphistom(at)inae and Gastrothylacinae) in various Mammalia. Of the five subfamilies of the Schistosomatidae, the Aporocotylinae* inhabit the heart and truncus arteriosus of fishes, the Spirorchiinae* the heart and arteries of Chelonia, the Bilharziellinae and Schistosomatinae mainly the venous system of birds, and the Eubilharziinae the same system of Mammalia, including man. Interestingly, perhaps significantly, no member of this family, or of the Notocotylidae, is known to occur in Amphibia. Of the eight subfamilies of the Notocotylidae, the Pronocephalinae* occur in aquatic Chelonia (various genera inhabiting different regions of the gut, one genus the bladder), the Notocotylinae in the caeca of birds, the Nudacotylinae in the caecum of rodents, the Hippocrepinae in the large intestine of rodents, the Ogmocotylinae in the small intestine of Cervidae, the Ogmogasterinae in the stomach and lungs of whales, the Rhabdiopoeinae* in the gut and the Opisthotrematinae* in the stomach and lungs of sirenians. The Mesometridae is regarded as a possible subfamily of the Notocotylidae whose members are parasites of teleost fishes.

These opinions are unlikely to meet with the unqualified approval of helminthologists, and host-specificity seems hardly as definite as Szidat maintained. This may be due, as Szidat thought, to substantial errors in the available lists of parasites and their hosts, and to faulty diagnosis. Where formerly we regarded *Notocotylus attenuatus* as a trematode which shows very little host-specificity, he stated, research on the life history and better criteria for diagnosis have revealed whole series of species, each of which shows a decided preference for certain hosts.

That consideration of hosts may lead to modification of our conception of higher taxonomic groups, e.g. families, is shown by Szidat's suggestion that the Fasciolidae, members of which are limited to a relatively small section of the Mammalia, are derived from the much larger family Echinostomatidae through forms like *Cathaemasia*. This genus bears much closer resemblance to Fasciolids than to Echinostomatids when adult, but the cercaria has an indubitable coronet of small head spines, indicating close affinity with the Echinostomatidae.

In concluding this résumé of taxonomic difficulties mention might be made of the inauguration of new possibilities by Wilhelmi (1939, 1940). By the injection of aqueous solutions of lipoid-free antigens he obtained precipitin reactions which indicate closer relationship between Echinostomatidae and Heterophyidae than between either of these families and the Plagiorchiidae or Paramphistomatidae. As far as serological observations go, the Heterophyidae seems to be more closely related to the Plagiorchiidae than to the Paramphistomatidae, while the Echinostomatidae is even less closely related to the latter family. There is the further suggestion (1940) that larval and adult antigens are not serologically distinct in various stages of the life-cycle in some Digenea, so that a possibility exists of the future recognition of larvae corresponding to certain adults by serological methods. It would be premature to predict that such findings will come to serve the ends of taxonomy, but certainly they will provide a means of checking conclusions based on morphological and embryological evidence. Certain cercariae can only be distinguished from one another by their

* These groups are generally given family rank, as in the present book when represented.

behaviour, e.g. the ways in which they swim and rest, though distinct species, and there is some reason for supposing that other physiological and serological characters, if and when they are elucidated, will serve in the definition of higher taxonomic units.

KEY TO FAMILIES OF DIGENEA COMING WITHIN THE SCOPE OF THIS BOOK

Note. Class of vertebrate host referred to as follows: F, fishes; A, Amphibia; R, Reptilia; B, birds; M, mammals. Other references: *b*, bats; *c*, Cetacea; *p*, Pinnipedia. Page numbers refer to Chapter 6, where diagnoses are to be found.

I. Mouth near middle of body (*GASTEROSTOMATA*;
BUCEPHALIDAE (F) (p. 82)

II. Mouth near anterior end of body (*PROSOSTOMATA*)
A. Adults parasitic in blood vascular system
(*a*) Hermaphrodite; parasites of fishes
APOROCOTYLIDAE (F) (p. 93)
(*b*) Unisexual; parasites of birds and mammals
SCHISTOSOMATIDAE (BM) (p. 106)
B. Adults inhabiting cysts or cyst-like cavities in tissues
(*a*) Body thread-like, sometimes expanded posteriorly
DIDYMOZOIDAE (F) (p. 92)
(*b*) Body oval in outline, not thread-like
TROGLOTREMATIDAE (BM) (p. 99)
C. Adults not parasites of blood vascular system, not generally cyst-dwellers
I′ Ventral sucker at posterior end of body
PARAMPHISTOMATIDAE (FARBM) (p. 103)
II′ Ventral sucker, when present, not at posterior end of body
(1) With additional adhesive organ (not a gonotyl) behind ventral sucker
(*a*) Body divided into anterior and posterior regions
(i) Anterior region flat and wide
DIPLOSTOMATIDAE (RBM) (p. 105)
(ii) Anterior region cup- or trumpet-shaped
STRIGEIDAE (RBM) (p. 105)
(*b*) Body not divided into anterior and posterior regions
CYATHOCOTYLIDAE (B) (p. 105)
(2) With a gonotyl, which may be included in genital sinus
HETEROPHYIDAE (BM) (p. 97)
(3) With neither gonotyl nor adhesive organ of Strigeid pattern
[I]″ Ventral sucker absent
(*a*) Testes lateral to caeca
(i) Testes near middle of body
EUCOTYLIDAE (B) (p. 99)
(ii) Testes near posterior end of body
NOTOCOTYLIDAE (BM) (p. 100)
(*b*) Testes median to caeca
(i) Caeca united to form ring-like intestine
CYCLOCOELIDAE (B) (p. 100)
(ii) Caeca terminate blindly
(*a*′) Body elongate, testes one behind the other
MICROSCAPHIDIIDAE (RB) (p. 104)
(*b*′) Body discoidal, testes side by side
MESOMETRIDAE (F) (p. 104)

[II″] Ventral sucker present
(A′) Ovary behind testes, genital pore in front of ventral sucker
(AA) Uterus with ascending limb only
HALIPEGIDAE (FA) (p. 90)
(BB) Uterus with descending and ascending limbs
(*a*) Vitellaria compact, caeca long or short
(i) Vitellaria in a single mass, caeca short
ZOOGONIDAE (F) (p. 88)
(ii) Vitellaria paired, caeca long
HEMIURIDAE (F) (p. 89)
(*b*) Vitellaria follicular, caeca long
(*aa*) Caeca extended anteriorly (intestine H-shaped)
ACCACOELIIDAE (F) (p. 91)
(*bb*) Caeca not extended anteriorly (intestine not H-shaped)
(i) Vitellaria continuous behind ventral sucker
MESOTRETIDAE (M*b*) (p. 102)
(ii) Vitellaria restricted behind ventral sucker
DICROCOELIIDAE (ARBM) (p. 95)
(B′) Ovary between testes, genital pore behind ventral sucker
(*a*) Uterus with ascending limb only
CLINOSTOMATIDAE (RB) (p. 106)
(*b*) Uterus with ascending and descending limbs
BRACHYLAEMIDAE (ARBM) (p. 101)
(C′) Ovary in front of testis or testes, genital pore in front of ventral sucker
(AA) One testis only (two testes in *Monorcheides*)
(*a*) Vitellaria compact or slightly lobed
HAPLOPORIDAE (F) (p. 92)
(*b*) Vitellaria follicular, but not well developed
MONORCHIIDAE (F) (p. 91)
(BB) Numerous testes **ORCHIPEDIDAE** (B) (p. 98)
(CC) Two testes
(A″) Uterus with ascending limb only
(*a*) Prepharynx long, testes one behind the other in posterior region
(i) Cuticle smooth, excretory vesicle funnel- or T-shaped
ALLOCREADIIDAE (F) (p. 86)
(ii) Cuticle spinous, excretory vesicle Y-shaped
(*a′*) Cirrus pouch present, cirrus and metraterm with thorn-like spines
ACANTHOCOLPIDAE (F) (p. 87)
(*b′*) Cirrus pouch absent, cirrus and metraterm without thorn-like spines
ACANTHOSTOMATIDAE (F) (p. 88)
(*b*) Prepharynx short or absent, testes of variable shape and position, gonads in middle or posterior region of body
(i) Head crown and head spines present
ECHINOSTOMATIDAE (BM) (p. 101)
(ii) Head crown and head spines absent
(*a′*) Intestine a single caecum (simple)
HAPLOSPLANCHNIDAE (F) (p. 91)
(*b′*) Intestine H-shaped (anterior extensions)
CAMPULIDAE (M*pc*) (p. 103)
(*c′*) Intestine much-branched **FASCIOLIDAE** (M) (p. 102)
(*d′*) Intestine with two caeca, not H-shaped, unbranched
(i′) Vitellaria tubular (six to seven follicles on either side)
PHILOPHTHALMIDAE (B) (p. 98)
(ii′) Vitellaria follicular, well or fairly well developed
(*a″*) Vitellaria filling lateral regions behind ventral sucker
(i″) Testes rounded **PSILOSTOMATIDAE** (B) (p. 99)
(ii″) Testes branched **CATHAEMASIIDAE** (B) (p. 99)

(*b″*) Vitellaria restricted to a short region behind the ventral sucker
(i″) Body ribbon-like; median excretory canal joining the lateral canals behind the testes **AZYGIIDAE** (F) (p. 89)
(ii″) Body elongate, but not ribbon-like; median excretory canal meeting the lateral canals in front of the testes
OPISTHORCHIIDAE (RBM) (p. 96)
(B″) Uterus with descending and ascending limbs
(*a*) Caeca short, scarcely reaching ventral sucker
MICROPHALLIDAE (B) (p. 96)
(*b*) Caeca long or fairly long, extending far behind ventral sucker
(i) Cirrus pouch absent
(*a′*) Receptaculum seminis absent, pharynx long, oesophagus absent, testes one behind the other, excretory vesicle V-shaped
PTYCHOGONIMIDAE (F) (p. 88)
(*b′*) Receptaculum seminis present, pharynx short or absent, testes obliquely set, excretory vesicle tubular **GORGODERIDAE** (FAR) (p. 93)
(ii) Cirrus pouch present
(*a′*) Oral sucker with six anterior processes
BUNODERIDAE (F) (p. 93)
(*b′*) Oral sucker without anterior processes
(i′) Gonads in front of ventral sucker
STOMYLOTREMATIDAE (B) (p. 98)
(ii′) Gonads behind ventral sucker
(*a″*) Genital pore generally far forward, near pharynx, excretory vesicle Y-shaped
(i″) Oesophagus and pharynx relatively long, prepharynx and receptaculum seminis sometimes absent
PLAGIORCHIIDAE (FARBM) (p. 94)
(ii″) Oesophagus and pharynx relatively short, prepharynx and receptaculum seminis present **CEPHALOGONIMIDAE** (ARB) (p. 95)
(*b″*) Genital pore only slightly in front of ventral sucker, excretory vesicle Y- or V-shaped
(i″) Testes rounded, ovary in front of ventral sucker, receptaculum seminis absent **FELLODISTOMATIDAE** (F) (p. 86)
(ii″) Testes rounded or lobed, ovary sometimes behind testes, receptaculum seminis present
LECITHODENDRIIDAE (ARBM) (p. 94)

The Main Characters of Families of the Digenea

I. *GASTEROSTOMATA*. Diagnosis. Mouth in the middle of the ventral surface. Haptor in front of the mouth, near the anterior extremity. Gut simple, sac-like. Genital pore on the ventral surface of the body, not far from the posterior extremity. Gonads in the posterior, vitellaria in the anterior region. Cercaria furcocercous. Development with exchange of hosts.

Family **BUCEPHALIDAE** Poche, 1907 (Figs. 12A, 24, 25, 26)

Syn. Gasterostomidae Braun, 1893.

Diagnosis. With the characters of the suborder. *Hosts:* marine and freshwater fishes. *Location:* adults in the gut, larvae encysted in the nerves. *Subfamilies* (erected by Nicoll on differences established by Odhner, 1905): Bucephalinae Nicoll, 1914; Prosorhynchinae Nicoll, 1914.

KEY TO SUBFAMILIES OF BUCEPHALIDAE

A. Prae-oral haptor sucker-like — **Bucephalinae** (*Bucephalus*, *Bucephalopsis*, *Rhipidocotyle*) (key on p. 190)

B. Prae-oral haptor a rhynchus — **Prosorhynchinae** (*Prosorhynchus*)

Note. Eckmann (1932*b*) did not recognize these subfamilies, but both Nagaty (1937) and Manter (1940*a*) regarded this as a convenient division of the family.

II. *PROSOSTOMATA.* DIAGNOSIS. Mouth near or at the anterior extremity.

Tabulation of characters (Table 1). In attempting to define the main characters of the families of prosostomatous Digenea, I have worked out a tabular scheme of shorthand characters and a list of explanatory details. Not all characters lend themselves to such abbreviation, but such as do can be read off from one line of the table for a particular family. Where amplification is needed it will be provided in the text which follows. Taking the first line of the table as an example, the following characters can be determined:

EXPLANATION OF ABBREVIATIONS USED IN TABLE 1

HOSTS: F, fishes; A, Amphibia; R, Reptilia; B, birds; M, Mammalia; b, bats; pc, pinnipedes and cetaceans; s, sireniens; ( ), rarely.

SIZE: s, small; vs, very small; fs, fairly small; m, medium; l, large; fl, fairly large; vl, very large.

SHAPE: O, oval; EO, elongate oval; E, elongate; VE, very elongate; sf, slightly flat; f, flat; fa, flat anteriorly; sfa, slightly flat anteriorly; fp, flat posteriorly; c, cylindrical; cp, cylindrical posteriorly; †, shape distinctive.

CUTICLE: s, smooth; w, wrinkled; S, spinous; Sa, spinous anteriorly; S′, scaly; S′a, scaly anteriorly.

PREPHARYNX: −, absent; +, present; ±, present or absent; l, long.

SUCKERS: (1), (2), (3) number present; −o, oral sucker absent; −v, ventral sucker absent; ±, present or absent; p, powerful.

PHARYNX: −, absent; +, present; ±, present or absent; l, large or long.

OESOPHAGUS: −, absent; +, present; ±, present or absent; fl, fairly long; l, long.

CAECA: s, short; m, of medium length; l, long, reaching almost if not quite to the anterior extremity; d, with diverticula, i.e. lobed; b, with branching diverticula; ( ), joined posteriorly; H, H-shaped; o, with anal openings either to exterior or into excretory vesicle; S, gut comprising a single caecum.

GENITAL PORE: A, just in front of ventral sucker; AA, far forward, near mouth; P, behind ventral sucker; PP, far back, near posterior extremity; m, median; l, lateral; d, dorsal; G, with sucker or gonotyl.

CIRRUS POUCH: −, absent; +, present; ±, present or absent; s, short or small; l, large; e, elongate.

TESTES: R, rounded or entire; L, lobed; B, branched; T, one behind the other, i.e. in tandem; O, approximately in same transverse plane, i.e. opposite; D, not in same transverse plane or one behind the other, i.e. diagonal; e, elongate; N, numerous; S, single testis.

OVARY: A, anterior to testes; P, posterior to testes; B, between or at same transverse level as testes; R, rounded or entire; L, lobed; B, branched; m, median; l, lateral; e, elongate.

UTERUS: A, with only ascending loop; DA or AD, with both descending and ascending loops, the order indicating the direction of the first loop; −, absent; ( ), rare.

LAURER'S CANAL: −, absent; +, present.

RECEPTACULUM SEMINIS: −, absent; +, present; s, small; l, large.

VITELLARIA: W, well developed; F, feebly developed; c, compact; s, single.

EXCRETORY VESICLE: F, funnel-like; Y, Y-shaped; V, V-shaped; T, tubular; S, sac-like; MB, much branched; d, with dorsal pore.

EGGS: N, numerous; NN, not numerous; E, each containing an embryo; F, few; s, small; l, large; f1, f2, with one or two terminal filaments; M, each containing a miracidium.

Table 1. *The General Characters of the Main Families of the Digenea*

Family	Hosts	Size	Shape	Cuticle	Prepharynx	Suckers	Pharynx	Oesophagus	Caeca
Fellodistomatidae	F	s-fs	O-EOsf	S	—	2	+	+	m
Allocreadiidae	F	s-m	EO-Esf	s-S	+l	2	+	+	m-l
Acanthocolpidae	F	s-m	E-VEc	S	+l	2	+	+	l
Acanthostomatidae	F	m	EO-Esf	S	+l	2	+	+	l
Zoogonidae	F	s	Osf	S	—	2	±	+l	s
Ptychogonimidae	F	fs-m	O-EOsf	s-w	—	2p	+l	—	lo
Haplosplanchnidae	F	fs	O-EO	.	+	2	+	+	(S)
Azygiidae	F	fs-fl	EO-Esf	s-w	—	2	+l	+	l
Hemiuridae	F(A)	s-m	O-EOc	s-w	—	2	+	+	l
*Syncoeliidae	F	m	EO-Esf	s	—	2	+l	+	(l)
Halipegidae	FA	m-fl	EO-c	s	—	2	+	—	l
Accacoeliidae	F	m-l	EO-Esf	s-w	—	2p	+	+l	lHdo
Didymozoidae	F	l	VE-c	s	—	1-2±v	±	+l	l
Monorchiidae	F	s	O-sf	S	—	2	+	+	l
Haploporidae	F	vs-s	O-EOsf	S-Sa	—	2	+l	+l	s
Bunoderidae	F	s	EO-f†	s	—	2	+	+l	l
Aporocotylidae	F	fs	E	S	—	0	—	+l	m-lH
**Sanguinicolidae	F	vs	Ec	s	—	0	—	+	l
Mesometridae	F	s-fs	O	.	+l	1—v	+	+	m
Gorgoderidae	FAR	s-fs	EO-Ecfp	s	—	2p	±	+fl	l
Plagiorchiidae	FARBM	fs-m	O-EOsf	S-S′	±	2	+	+fl	m-l
Stomylotrematidae	B	fs	O-EO	s	—	2	+l	+	m-l
Dicrocoeliidae	ARBM	s-m	EO-Esf	s-S	—	2	+	+	m-l
Lecithodendriidae	ARBM	s	O-EOsf	s-S	—	2	+	+	m-l
Cephalogonimidae	ARB	s	E-sf	S-Sa	+	2	+	+	s-m
Psilostomatidae	B	s-fs	EO-Ef	s-S	+	2p	+	+	l
Orchipedidae	B	s-m	E	s	—	2	+	±	l
Echinostomatidae	BM	s-vl	EO-Ef	S-Sa	—	2p	+	+	l
Cathaemasiidae	B	fl	EO-E	.	—	2	+	+	l
Clinostomatidae	RB	m-vl	Ef	s	—	2	±	+	ld
Philophthalmidae	B	m-l	EO-Esf	s-S	±	2	+l	+	l
Brachylaemidae	(AR)BM	s-fs	EO-Esf	s-S	+	2	+	+	l
Opisthorchiidae	RBM	s-m	EO-Ef	s-S	—	2	+	+	m-l
Heterophyidae	BM	vs-s	EOsf	S′-S′a	—	1-3	+	+l	l
Microphallidae	B	vs-s	O	S-Sa	+	2	+	+l	s-m-l
Cyclocoelidae	B	fl-vl	EOf	s	—	1-0—0±v	+	+	(ld)
Eucotylidae	B	fs-m	Ef	.	—	1—v	+	±	l
Notocotylidae	BM	s-fs	EOf	Sa	—	1—v	—	+	l(d)
Strigeidae	RBM	s	EOfacp†	s	—	1-2±v	+	+	l
Cyathocotylidae	RB	s	O	s	—	2	+	—	l
Diplostomatidae	BM	s	EOfacp†	s	—	1-2±v	+	+	l
Schistosomatidae	BM	l-vl	VEc†	s-S	—	0-2±0±v	—	+	(l)
Troglotrematidae	BM	m-fl	Osfa	S	--	1-2±v	+	+	m-l
Fasciolidae	M	l-vl	O-EOf†	s-S	--	2	+	+	lb
Campulidae	Mpc	fs-m	Esf	S	+	2	+	+	lHb
Mesotretidae	Mb	m	EO	Sa	—	2	+	+	l
Paramphistomatidae	FARBM	fl-l	EO-Ec	s	—	2p	+	+	m-l

	pore	pouch	Testes	Ovary	Uterus	canal	seminalis	Vitellaria	vesicle	Eggs
Fellodistomatidae	Am-l	+s	RT-O	AR-Ll	DA	+	−	W	YV	NE
Allocreadiidae	Am-l	+l	RT	AR-Ll	A	+	+	W	FT	NN
Acanthocolpidae	Am	+e	RT	ARm	A	.	.	W	Y	NN
Acanthostomatidae	Am	−	RT	ARm	A	.	.	F-W	Y	Ns
Zoogonidae	Al	+e	RO	PRm	DA	+	+	Fcs	S	NNM
Ptychogonimidae	Am	−	R-LT	Am	DA	+	−	W	V	NM
Haplosplanchnidae	AAm	−	SR	AR	A	−	+l	F	Y	M
Azygiidae	Am	+	RT	PRm	A	+	−	W	Y	NNE
Hemiuridae	Am	−	RT-O	PRm	DA	+	+	Fc	Y	Ns
*Syncoeliidae	AAm	−	NRD-O	PL-Bm	DA	−	−	Fc	Y	NN
Halipegidae	AAm	−	RO	PRl	A	+	−	Fc	Y	Nf1
Accacoeliidae	A	−	RT-D	PRm	DA	+	−	W	Y	N
Didymozoidae	AA	−	R-LOe	Bme	A	−	.	Fcs	S	Ns
Monorchiidae	Am-l	+e	SR	ARl	DA	+	±	W	YT	Ns
Haploporidae	Am	+	SR	BRl-m	DA	+	−	W	S	NNf1M
Bunoderidae	Am	+	RT-D	ARl	DA	+	+	W	S	N
Aporocotylidae	PPm-lG	+	N	PR	A	−	−	W	Y	NN
**Sanguinicolidae	PP	−	NRT-O	PLm	--	−	−	W	Y	Fl (one)
Mesometridae	A(A)	−	RO	PL	A	.	.	W	.	NNf1
Gorgoderidae	Am	−	R-LO	ARm-l	DA	+	+	F	Td	Nl
Plagiorchiidae	A(A)m-l	+	R-LT-D-O	AR-Ll	DA	+	±	W	Y	Ns
Stomylotrematidae	AAl	+	RO	ARl	AD	+	+	F	.	Ns
Dicrocoeliidae	Am	+s	RT-O	PRm	DA	+	+s	F-W	TS	N
Lecithodendriidae	Am-l-d	+	R-LO	A-Pl	DA	+	+	F-W	V	N
Cephalogonimidae	AAm-l	+e	RT	ARl	DA	+	+	W	Y	N
Psilostomatidae	Al	+l	RT	ARm	A	+	+s	W	Y	Nl
Orchipedidae	AAm	−	N	AR	A	+	+	W	TY	NNl
Echinostomatidae	Am	+	R-LT	ARl-m	A	+	−	W	Y	NNl
Cathaemasiidae	Am	+	BT	ARm	A	+	−	W	S	NNl
Clinostomatidae	P	+	LT	BRm	A	.	.	W	.	NN
Philophthalmidae	Am	+e	R-LT-D	ARm	A	+	−	F	.	NM
Brachylaemidae	PP	±s	RT	BRm	AD	+	−	W	Y	.
Opisthorchiidae	Am	−	R-BD	ARm	A(D)	+	+	W	Y	Ns
Heterophyidae	Pm-lG	−	R-LT-O	AR-Lm-l	(D)A	+	+	W	Y	NN
Microphallidae	.	±	RO	Am-l	DA	.	±	W	.	N
Cyclocoelidae	AAm	+s	R-LD	A-BRm	A	−	−	F-W	Sd	NM
Eucotylidae	A	−	RO	ARl	AD	.	.	F	.	N
Notocotylidae	AAm	+e	R-LO	BR-Lm	A	+	−	W	Yd	Nf2
Strigeidae	PP	±	RT	ARm	A	+	+	W	MB	Fl
Cyathocotylidae	PP	+l	RD	BRm	D	.	.	W	MB?	Fl
Diplostomatidae	PP	−	RT	ARm	A	.	.	W	MB	Fl
Schistosomatidae	AAm-l	−	NR	-Rm	A	−	−	W	YV	Fl
Troglotrematidae	A-Pm-l	±	L-BOe	ALl	A	+	+	W	TYS	Ns-l
Fasciolidae	Am	+	BT-D	ABl	A	+	+	W	MB	N
Campulidae	Am	+	R-L-BT	AR-Lm-l	A	+	+	W	T	NN
Mesotretidae	Pm	+l	RTe	PRe	AD	.	.	W	.	N
Paramphistomatidae	Am	±	R-LT-D-O	PRm	A	+	.	W	Sd	N or NN
Microscaphidiidae	AA	+	LT	PR	A	+	−	W	.	N

* and ** These groups will be regarded as Subfamilies of the Hemiuridae and Aporocotylidae respectively.

Family **FELLODISTOMATIDAE** Odhner, 1911, emend. Nicoll, 1935 (Figs. 28 A–D, 38 A, B, 39 A–D, 45 D–E)

Syn. Steringophoridae Odhner, 1911.

Hosts: fishes. *Size:* small or fairly small. *Shape:* oval or somewhat elongate, slightly flat. *Cuticle:* spinous. *Suckers:* oral and ventral present. *Gut:* prepharynx absent, pharynx, oesophagus and simple caeca of medium length present. *Genital pore:* in front of the ventral sucker, median or lateral in position. *Cirrus pouch:* present, small. *Testes:* rounded, directly or obliquely one behind the other. *Ovary:* rounded or lobed, in front of the testes, lateral. *Uterus:* with descending and ascending limbs, i.e. extending towards the posterior end of the body before proceeding anteriorly to the genital pore. *Laurer's canal:* present. *Receptaculum seminis:* absent. *Vitellaria:* well developed, follicular. *Excretory vesicle:* Y- or V-shape, i.e. with a long or short median stem. *Eggs:* numerous, each containing an embryo.

These characters can be amplified in the following terms:

Hosts: marine fishes. *Location:* gut, sometimes in gall bladder. *Cirrus pouch:* short and thick, never extending behind the ventral sucker, containing a folded cirrus, a pars prostatica and a bipartite vesicula seminalis. *Uterus:* with folds which tend to fill the posterior region of the body. *Vitellaria:* invariably some distance from the posterior extremity. *Excretory system:* when not Y-shaped, the excretory vesicle tends towards a V-shape by reduction of the median stem. Lateral canals extend forward almost to the anterior extremity. Flame-cell formula undetermined. *Subfamilies:* Fellodistomatinae Odhner, 1911, emend. Nicoll, 1936; Haplocladinae Odhner, 1911.

KEY TO SUBFAMILIES OF FELLODISTOMATIDAE

I. Body oval in outline, testes side by side **Fellodistomatinae** (*Steringotrema, Bacciger, Steringophorus, Fellodistomum, Rhodotrema*) (key on p. 238)

II. Body elongate, testes one behind the other **Haplocladinae** (*Haplocladus, Tergestia, Proctoeces, Ancyclocoelium*) (key on p. 244)

Note. Manter (1940*a*) placed *Tergestia* and *Proctoeces* in the Haplocladinae, but referred this without comment to the family Monorchiidae.

It is assumed that the characters set out in the remaining lines of Table 1 will be taken into consideration along with the following notes. The figures referred to after the headings of the families are those of *this* book.

Family **ALLOCREADIIDAE** Stossich, 1904 (Figs. 12 K, 27 A–D, 28 F, G, I, K, 29 A, B, 35 C)

Diagnosis. *General:* distomes of small or medium size, with well-developed suckers and a mobile anterior extremity. *Hosts:* marine and fresh-water fishes, occasionally mammals. *Location:* gut. *Cuticle:* generally non-spinous, sometimes spinous. *Gut:* prepharynx well marked, caeca long or fairly long, but not quite reaching the posterior extremity. *Cirrus pouch:* large, sac-like, with a stout cirrus, its opening median or lateral in front of the ventral sucker. *Testes:* in the posterior region. *Ovary:* in front of the testes, behind the ventral sucker, generally lateral. *Vitellaria:* follicular, lateral, in the posterior half of the body, generally merging in the median plane posteriorly. *Uterus:* short, with a few loops between the testes and the ventral sucker. *Excretory system:* vesicle funnel-shaped, rarely pyriform, not forked; flame-cell formula $2 [(4 + 4 + 4) + (4 + 4 + 4)]$. *Subfamilies:* Allocreadiinae Looss, 1902; Lepocreadiinae Odhner, 1905; Stephanophialinae Nicoll, 1909; Sphaerostomatinae Poche, 1926.

Note. Stunkard (1932) remarked on the very different types of cercariae met with in this family, which, in his opinion, is heterogeneous and should be partitioned. Cable & Hunninen (1941 *b*) suggested that the Lepocreadiinae and other forms, including *Deropristis* (regarded by some writers as belonging to the Acanthocolpidae), should form the basis of a new family, Lepocreadiidae, restricted to forms having ophthalmoxiphidiocercariae which develop in bivalves. They would include *Lepidauchen*, *Pleorchis* and *Pseudolepidapedon* in the Acanthocolpidae. They also support the proposal of Hopkins (1941 *b*) to restrict a separate family Opecoelidae to include forms having cotylocercous cercariae.

KEY TO SUBFAMILIES OF ALLOCREADIIDAE

I. Cuticle spinous (parasitic in marine fishes) — **Lepocreadiinae** (*Lepidauchen, Opechona, Lepidapedon*) (key on p. 210)

II. Cuticle non-spinous (parasitic in marine or fresh-water fishes)
- A. In marine fishes (excepting *) — **Allocreadiinae** (*Plagioporus, Podocotyle, Helicometra, Peracreadium, Allocreadium*, Cainocreadium*) (key on p. 198)
- B. In fresh-water fishes
  - (*a*) Oral sucker with anterior processes — **Stephanophialinae** (*Crepidostomum*)
  - (*b*) Oral sucker without anterior processes — **Sphaerostomatinae** (*Sphaerostoma*)

Family **ACANTHOCOLPIDAE** Lühe, 1909 (Figs. 27E–H, 28E, 30A–D)

Diagnosis. *General:* elongate, heavily spinous, neither large nor muscular distomes, little flattened and with suckers not far apart. *Cuticle:* bearing groups of stout spines which are sometimes aggregated in definite regions, e.g. round the mouth or 'shoulders'. *Gut:* as in the Allocreadiidae. *Genital ducts:* cirrus pouch elongate; cirrus and metraterm equipped with strong, thorn-like spines; genital atrium tubular. *Gonads, Vitellaria, Uterus:* as in Allocreadiidae. *Excretory system:* vesicle Y-shaped, medium stem short, lateral ducts long. *Eggs:* sculptured, without filaments. *Hosts:* marine fishes, or migratory fishes of fresh water. *Genera: Acanthocolpus* (not in Britain), *Stephanostomum* (Syn. *Stephanochasmus*), *Dihemistephanus*, *Deropristis*, *Neophasis* (Syn. *Acanthopsolus*), *Tormopsolus*.

Note. All these genera have been referred to the subfamily Acanthocolpinae Lühe, 1906, but Ward (1938) erected a new subfamily, Acanthopsolinae (for which the name Neophasinae should be substituted) for *Acanthopsolus* [= *Neophasis*]. In view of the revision of the entire family which is pending we have elected to deal here with the isolated genera (key on p. 214).

Family **ACANTHOSTOMATIDAE** Poche, 1926, emend. Nicoll, 1935 (Fig. 31A–G)

Syn. Acanthochasmidae Nicoll, 1914.

DIAGNOSIS. Somewhat similar to the Acanthocolpidae, but cirrus pouch absent. *Hosts:* marine fishes, notably the wolf-fish. *Location:* intestine. *Subfamily:* Acanthostomatinae, with the genera *Acanthostomum* (Syn. *Acanthochasmus*), *Anisocoelium*, *Anisocladium* (Syn. *Anisogaster*) (key on p. 222).

Family **ZOOGONIDAE** Odhner, 1911 (Figs. 28J, 38C, D, 45F–H)

DIAGNOSIS. *General:* small, slim, spinous, slightly flattened distomes of frail appearance with suckers rarely more than one-third of the body length apart. *Gut:* caeca extending back only to the middle of the body. *Genital ducts:* cirrus pouch elongate, extending towards the median plane from the almost marginal genital pore, pars prostatica well developed. *Gonads:* testes rounded, symmetrical, beside or slightly behind the ventral sucker; ovary between and slightly behind the testes. *Vitellaria:* small, compact, rounded, sometimes a single mass in the posterior region. *Uterus:* tending to fill the posterior region. *Eggs:* fairly large or large (0·035–0·09 mm. long), each containing a miracidium. *Hosts:* marine fishes. *Location:* rectum, rarely the bladder. *Subfamilies:* Zoogoninae, Odhner 1911, Lecithostaphilinae Odhner, 1911.

KEY TO SUBFAMILIES OF ZOOGONIDAE

I. Vitellaria well developed, follicular **Lecithostaphylinae** (*Lepidophyllum*, *Deretrema* (Syn. *Proctophantastes*), *Steganoderma*, *Lecithostaphylus*) (key on p. 250)

II. Vitellaria small and compact **Zoogoninae** (*Diphterostomum*, *Zoogonoides*, *Zoonogenus*, *Zoogonus*) (key on p. 246)

Family **PTYCHOGONIMIDAE** Dollfus, 1936 (Fig. 40A–D)

DIAGNOSIS. *General:* thick, muscular distomes with a wrinkled, non-spinous cuticle and powerful, closely approximated suckers. *Gut:* prepharynx and oesophagus practically absent, caeca simple and long, opening into the median terminal excretory canal posteriorly. *Excretory system:* main canals U-shaped, connected by transverse anastomoses in front of and behind the ventral sucker. *Genital ducts:* genital atrium large and muscular, male and female pores separate on a small, protrusible papilla in its depths. Cirrus pouch absent. Ejaculatory duct short, pars prostatica longer, vesicula seminalis very long and convoluted, extending beyond the posterior border of the ventral sucker. Metraterm extending back only to the middle of this sucker. *Uterus:* with numerous folds passing to left and right of the gonads and extending behind the testes to within a short distance of the posterior extremity. *Laurer's canal:* opening between the levels of the ovary and the gland of Mehlis. *Vitellaria:* follicles numerous, spreading over the excretory vesicle posteriorly, overreaching the ventral sucker anteriorly. *Eggs:* thin, of medium size (0·062–0·067 mm. long). *Miracidium:* carrying rigid 'bristles' anteriorly. *Hosts:* Selachii. *Location:* stomach. *Subfamily:* Ptychogoniminae Dollfus, 1936 with one genus, *Ptychogonimus.*

Family **AZYGIIDAE** Odhner, 1911, emend. Dollfus, 1936 (Figs. 29C, 38E),

DIAGNOSIS. *General:* fairly elongate, little-flattened distomes of variable size with a thick, non-spinous (but wrinkled) cuticle and powerful suckers. *Gut:* oesophagus short, caeca long. *Genital ducts:* genital atrium spacious. Ejaculatory duct, pars prostatica and vesicula seminalis overlie the ventral sucker or are situated in front of it. Uterus with only an ascending limb, extending from the ootype to the genital pore in wide, closely approximated, transverse folds. Metraterm present. *Excretory system:* main canals Y-shaped, lateral canals extending almost to the anterior extremity. *Vitellaria:* lateral, situated in the posterior region, not quite reaching the extremity. *Eggs:* of very variable size (0·045–0·085 mm. long), sometimes sculptured and/or with an albuminous covering. Ripe eggs each containing a non-ciliated embryo. *Hosts:* fresh-water teleosts and Selachii. *Location:* gut. *Genera: Azygia* (in teleosts), *Otodistomum* (in selachians) (key on p. 251).

Note. Ptychogonimus was removed from this family by Dollfus (1936*c*), who retained in the family Azygiidae those genera in which there is a cirrus pouch enclosing the pars prostatica and vesicula seminalis, no uterine folds in front of the ovary, no communication at any time between the caeca and the excretory pore, and no anastomoses connecting the lateral excretory canals *in front of* the oral sucker.

Family **HEMIURIDAE** Lühe, 1901 (Figs. 28L, M, 41A–D, 42A–D, 43A–E)

DIAGNOSIS. *General:* elongate distomes of tapering, cylindrical form in which the cuticle may be ringed, but is not spinous, and the suckers are not widely separated. The posterior part of the body may or may not form an abdomen or *ecsoma*, capable of being retracted into the anterior region or *soma* in a telescoped condition and when fully extended sometimes as long as or longer than the soma. *Genital ducts:* male and female ducts fused into an hermaphrodite duct or *genital sinus*. Cirrus pouch generally absent, but genital sinus sometimes enclosed in a false cirrus pouch or *sinus sac*. This may be well developed (*complete*) or consist of a few muscle fibrils (*incomplete*). Pars prostatica and vesicula seminalis well developed, free in the parenchyma. *Vitellaria:* compact, rounded or lobed, in some instances long and thread-like. *Eggs:* small and thin-shelled. Cercariae cystophorous. *Intermediate hosts:* copepods. *Final hosts:* marine and migratory fishes. *Location:* gut, mainly the stomach, sometimes the gall bladder. *Subfamilies:* Hemiurinae Looss, 1907; Sterrhurinae Looss, 1907; Dinurinae Looss, 1907; Lecithasterinae Odhner, 1905; Derogenetinae Odhner, 1927; Syncoeliinae Looss, 1899.

KEY TO SUBFAMILIES OF HEMIURIDAE

I. Ecsoma present
  A. Mouth overhung by a lip, vitellaria with finger-like or longer lobes
    (*a*) Vitellaria small, with finger-like lobes — **Sterrhurinae** (*Ceratotrema, Synaptobothrium, Lecithochirium, Brachyphallus, Sterrhurus*) (key on p. 261)
    (*b*) Vitellaria with elongate, tubular components — **Dinurinae** (*Dinurus, Ectenurus, Lecithocladium*)
  B. Mouth not overhung by a lip, vitellaria compact or with only very slight lobes — **Hemiurinae** (*Hemiurus, Aphanurus*) (key on p. 257)

II. Ecsoma absent
  A. Vittellaria unpaired, asterisk-like, generally seven-rayed — **Lecithasterinae** (*Lecithaster, Aponurus*) (key on p. 268)
  B. Vitellaria paired
    (*a*) Cirrus pouch absent — **Derogenetinae** (*Derogenoides, Derogenes*) (key on p. 270)
    (*b*) Cirrus pouch present — **Syncoeliinae** (*Hemipera*)

Note. Derogenoides was included by Fuhrman (1928) in the family Syncoeliidae Odhner, which is generally regarded by writers who treat the Hemiuridae in a less restricted sense as a subfamily, the Syncoeliinae. Nicoll (1913*a*), who erected this genus, remarked on the close resemblance between *Derogenoides* and *Derogenes* and, indeed, the characters he gave leads one to believe that the two genera may ultimately prove to be identical. Certainly they cannot be placed in separate subfamilies. Fuhrmann included amongst the characters of the Syncoeliidae a ring-shape intestine formed by fusion of the caeca; this character and others do not fit *Hemipera*, which is debarred from the Derogenetinae by the presence of the cirrus pouch. Other Syncoeliinae have not been found in British waters, so that this is not the place for a revision of the subfamily, which is accepted, however, in a restricted sense.

Family **HALIPEGIDAE** Poche, 1926 (Fig. 47A)

DIAGNOSIS. *General:* elongate, non-spinous, almost cylindrical distomes with widely separated suckers and bearing some resemblance to Hemiuridae which lack an ecsoma. *Gut:* pharynx very small, oesophagus very short, caeca long. *Genital ducts:* genital atrium small. Cirrus small, pars prostatica ill-developed, vesicula seminalis present; cirrus pouch absent. *Excretory system:* main canals Y-shaped, median stem long, lateral branches long and uniting behind and above the oral sucker. *Vitellaria:* ill-developed, each comprising four or five large follicles, situated behind the ovary, one on either side of the median plane. *Uterus:* formed into wide, transverse folds, which may over-step the caeca slightly. *Eggs:* elongate, each having a long filament at the anopercular pole. Cercaria cystocercous, developing in a redia. *Final hosts:* frogs. *Location:* buccal cavity, under the tongue, or in the eustachian recesses. *Genus: Halipegus.*

Family **ACCACOELIIDAE** Dollfus, 1923 (Figs. 36 A–E, 37)

Syn. Accacoeliinae Odhner, 1911; Accacoeliiden Looss, 1912.

DIAGNOSIS. *General:* elongate, more or less cylindrical, muscular distomes of medium or large size having powerful, closely approximated suckers. Ventral sucker invariably larger than the oral, pedunculate, sometimes with a duplicated anterior border or ear-like processes. *Cuticle:* non-spinous, but sometimes bearing papillae anteriorly. *Gut:* prepharynx present, pharynx pyriform or fusiform, oesophagus long and slender, caeca long, with anterior extensions, with or without lateral diverticula, opening into the excretory system posteriorly. *Excretory system:* median stem very short, lateral canals long and united above the pharynx. *Genital ducts:* may open separately, or the terminal parts may fuse to form an hermaphrodite duct. Cirrus pouch absent. Copulatory organ present or absent. *Vitellaria:* very variable, follicles sometimes in small pyriform masses, sometimes branched clusters, sometimes thread-like components. *Eggs:* small, ovoid, 0·026–0·036 mm. long and 0·016–0·022 mm. broad. *Metacercaria:* found in various pelagic invertebrates, principally coelenterates. *Hosts of adults:* pelagic marine fishes. *Location:* gut, sometimes on the gills. *Subfamilies:* Accacoeliinae Odhner, 1911, emend. Dollfus, 1935; Tetrochetinae Dollfus, 1935.

KEY TO SUBFAMILIES OF ACCACOELIIDAE

I. Genital atrium containing a protractile, copulatory organ **Accacoeliinae** (*Accacoelium, Accacladium, Rhyncopharynx, Accacladocoelium*) (key on p. 234)

II. Genital atrium devoid of a copulatory organ **Tetrochetinae** (*Tetrochetus, Orophocotyle, Mneiodhneria*) (key on p. 233)

Family **HAPLOSPLANCHNIDAE** Poche, 1926 (Fig. 35 A)

DIAGNOSIS. *General:* distomes with large, deep, saccular ventral sucker and and intestine formed of a single, simple caecum which is confined to the anterior region of the body. *Reproductive systems:* genital pore slightly behind the oral sucker. Testis single, rounded, not far from the posterior end of the body. Ovary rounded, in front of the testis. Cirrus pouch and cirrus absent. Vas deferens functioning as vesicula seminalis. Vitellaria feebly developed, comprising a few follicles between the gut and the testis. Uterus small. *Eggs:* of medium size, each containing a miracidium. *Hosts:* fishes. *Location:* intestine. *Genus:* *Haplosplanchnus.*

Family **MONORCHIIDAE** Odhner, 1911 (Figs. 35 B, D–G, 45 A, B)

DIAGNOSIS. *General:* small, spinous distomes of variable form, the ventral sucker situated in front of the middle of the body and the anterior region supplied with numerous cutaneous glands. *Gut:* pharynx small, oesophagus short, caeca extending back only to the midbody. *Genital ducts:* genital pore median, in front of ventral sucker or near sucker on left margin of body. Cirrus pouch elongate. Cirrus thick, heavily spinous. Metraterm spinous. *Gonads:* generally only one testis, two testes in *Monorcheides*. Ovary in front of the testis or testes, of variable form, lobed or smooth and elongate. *Vitellaria:* paired, but lightly developed, variable in position. *Eggs:* small. *Hosts:* marine and fresh-water fishes. *Location:* gut. *Subfamilies:* Monorchiinae Odhner, 1911 (flat and oval Monorchiidae with a Y-shaped excretory vesicle) and Proctotrematinae Odhner, 1911 (elongate Monorchiidae with a tubular excretory vesicle), represented in British waters by the genera *Monorchis* and *Monorcheides*, *Asymphylodora* and *Proctotrema* respectively. (Keys on pp. 229 and 230.)

Note. Hopkins (1941*b*) regarded *Proctotrema* Odhner, 1911 as a synonym of *Genolopa* Linton, 1910. If this were proved, the subfamily name would be Genolopinae, but Manter (*Trans. Amer. Micr. Soc.* **61**, 1942) has claimed that both genera are valid on the ground of differences in the spines of the genital atrium, so that the name Proctotrematinae stands.

Family **HAPLOPORIDAE** Nicoll, 1914 (Figs. 32A, B, 33A–E, 34A, B)

DIAGNOSIS. *General:* small or very small, spinous distomes with powerful suckers, the ventral being partly or entirely in the anterior half of the body. *Gut:* prepharynx small, pharynx long, oesophagus fairly long, caeca very short and saccular, generally dorsal and not extending back beyond the ventral sucker. *Excretory system:* vesicle sac-like. *Genital ducts:* uniting in the terminal parts, having a false cirrus pouch or sinus sac, whole sinus long, tubular, non-muscular, but protrusible. Vesiculae seminales *interna* and *externa* present. *Gonads:* testis single, large and rounded, generally on the left side of the body. Ovary small, rounded, median, situated behind the ventral sucker. *Vitellaria:* feebly developed, compact or follicular. *Uterus:* tending to fill the posterior region of the body. *Eggs:* relatively large, thin-shelled, each containing a miracidium with or without eye-spots; sometimes having a mucilaginous envelope. *Hosts:* marine and fresh-water fishes. *Location:* gut. *Genera: Haploporus, Dicrogaster, Saccocoelium, Lecithobotrys.* (Key on p. 224.)

Family **DIDYMOZOIDAE** Poche, 1907 (Fig. 44A, B)

Syn. Didymozoonidae Monticelli, 1888.

DIAGNOSIS. *General:* thread-like Digenea, with or without an expanded posterior region, which live in pairs in cysts or cyst-like cavities within the tissues of the host, generally in the body cavity, buccal cavity or kidney, sometimes in the oesophagus, on the head, in the terminal part of the gut or in the musculature. Some are hermaphrodite, but others show a tendency to unisexuality with rudimentary hermaphroditism in some forms. *Suckers:* oral generally present, ventral absent in some species. *Gut:* pharynx rudimentary or absent, caeca long. *Genital ducts:* male organ rudimentary, genital pore far forward, near the mouth. *Gonads and vitellaria:* generally thread-like and continuous with their ducts. *Eggs:* very small and very numerous. *Classification:* various attempts have been made to divide the family into subfamilies, but incomplete knowledge of forms familiar enough by name make this hazardous. Several European genera are well known, *Nematobothrium*, *Didymozoon*, *Wedlia*, *Köllikeria* and *Didymocystis*. *Hosts:* marine fishes.

Family **APOROCOTYLIDAE** Odhner, 1912 (Fig. 45J)

Syn. Sanguinicolidae Graff of Fuhrmann, 1928, in part.

DIAGNOSIS. *General:* slender, spinous, hermaphrodite Digenea, parasites of the blood-vascular system, lacking suckers. *Gut:* pharynx absent, oesophagus long, caeca with anterior extensions, i.e. intestine H-shaped. *Genital pore:* dorsal, near the posterior extremity, median or lateral. *Gonads:* testes numerous, between the caeca, ovary rounded, behind the testes. *Genital ducts:* cirrus small, uterus short. *Vitellaria:* well developed, with an unpaired vitelloduct. *Eggs:* ovoid or spindle-shaped, not very numerous. *Hosts:* fishes and reptiles. *Subfamilies:* Aporocotylinae (genus *Aporocotyle*) and Sanguinicolinae (genus *Sanguinicola*) (in fishes).

Note. Szidat (1939*a*) proposed to regard this group of parasites of the blood-vascular system as a new subfamily (Aporocotylinae) of the Schistosomatidae.

Family **BUNODERIDAE** Nicoll, 1914 (Fig. 29D)

Syn. Bunoderinae Looss, 1902.

DIAGNOSIS. *General:* small, non-spinous, muscular distomes in which the oral sucker bears a circlet of six short, muscular, anterior processes, sometimes a collar-like expansion of the anterior extremity. *Suckers:* ventral as large as or larger than the oral, slightly in front of the middle of the body. *Cirrus pouch:* membranous. *Vitellaria:* follicular, lateral, extending from the pharynx to the posterior extremity. *Uterus:* with descending and ascending limbs formed into saccular masses in the posterior region, ventral to the testes. *Eggs:* fairly large and numerous. *Hosts:* fresh-water fishes. *Location:* gut. *Genus: Bunodera.*

Family **GORGODERIDAE** Looss, 1901 (Figs. 29E, 47I)

Syn. Gorgoderinae Looss, 1899.

DIAGNOSIS. *General:* small or fairly small, non-spinous distomes having a slender, mobile anterior and a flattened posterior region, also having powerful suckers, the ventral projecting noticeably from the middle of the body. *Cuticle:* sometimes bearing papillae. *Gut:* oesophagus short or long, pharynx present or absent. *Genital ducts:* cirrus pouch and cirrus absent, pars prostatica feebly developed, vesicula seminalis small. *Excretory system:* vesicle simple, tubular, opening on the postero-dorsal surface of the body. *Uterus:* having descending and ascending limbs abundantly folded, the folds tending to fill the posterior region of the body. *Eggs:* fairly large, numerous, increasing in size during progression through the uterus. *Hosts:* fishes, amphibians and reptiles. *Location:* mostly in the ureters and the excretory bladder. *Subfamilies:* Anaporrhutinae Looss, 1901 and Gorgoderinae Looss, 1899, the former with species of *Probilotrema* inhabiting the coelom of Selachii, the latter with genera *Phyllodistomum* in freshwater fishes and *Gorgodera* and *Gorgoderina* in amphibians. *Cercariae:* macrocercous.

Family **PLAGIORCHIIDAE** Lühe, 1901, emend. Ward, 1917
(Figs. 47 B, C, D, F, 48 A–F, 49 A, D, E)

Syn. Lepodermatidae Looss, 1901.

DIAGNOSIS. *General:* slightly elongate, little flattened, spinous or scaly distomes of fairly small or medium size with somewhat extended extremities. *Genital ducts:* cirrus pouch and cirrus well developed as a rule, pars prostatica and seminal vesicle present. Genital pore median or lateral, sometimes far forward and close to the mouth. *Gonads:* shape and position variable, but ovary in front of the testes, generally on the right. *Vitellaria:* well developed, follicular, lateral. *Uterus:* having a few folds between the testes on the descending limb, which may reach almost to posterior extremity, in which case the ascending limb has numerous transverse folds. *Excretory system:* main canals Y-shaped, the unpaired stem much longer than the lateral canals. Flame-cell formula $2\,[(3+3+3)+(3+3+3)]$. *Hosts:* fishes, amphibia, reptiles, birds, sometimes bats. *Location:* gut, sometimes in the buccal cavity of Amphibia and Reptilia. *Cercaria:* belonging to 'Polyadena' group of Xiphidiocercariae. *Genera:* numerous, including *Opisthioglyphe*, *Dolichosaccus* (Plagiorchiinae), *Haematoloechus* and *Haplometra* (Haplometrinae) in Amphibia; *Cercorchis* (Telorchiinae), *Encyclometra* (Encyclometrinae) and *Macrodera* (Haplometrinae) in reptiles, *Prosthogonimus* and *Schistogonimus* (Prosthogoniminae) in birds; *Plagiorchis* (Plagiorchiinae) in all classes of Vertebrata, except fishes.

Note. Many schemes of classification into subfamilies have been put forward, and in the end the family may be split up into a number of families. Stunkard (1924) proposed the erection of one, the Telorchiidae, and McMullen (1937*b*) suggested that three others be formed, Macroderoididae, Reniferidae and Haplometridae, with *Macroderoides* Pearse, 1924, *Renifer* Pratt, 1902, and *Haplometra* Looss, as type genera, his scheme taking into consideration the types of larvae and life histories as well as the excretory system. It is premature at present, however, to attempt a comprehensive analysis of the entire family or group of families.

Family **LECITHODENDRIIDAE** Odhner, 1910
(Figs. 12 M; 47 E, G, H, 51 A, 54 B)

DIAGNOSIS. *General:* small spinous or non-spinous distomes which are neither very compact nor extended, but are of intermediate form. *Suckers:* ventral near the middle of the body. *Gut:* caeca of variable length, in some instances extending almost to the posterior extremity, in others scarcely reaching the middle of the body. *Gonads:* variable in position; testes side by side, sometimes situated in the posterior region, sometimes lateral to the oesophagus; ovary dorsal, anterior or posterior, in front of or behind the testes, generally on the right side. *Genital pore:* anterior, position variable, median, lateral (L.) or dorsal. *Vitellaria:* generally comprising small clusters of follicles laterally, sometimes in the posterior region, or in the anterior. *Uterus:* with descending and ascending limbs much folded in the posterior region. *Eggs:* small (0·017–0·038 mm. long), usually grouped in small masses. *Excretory system:* main canals V-shaped. *Hosts:* Insectivores generally, but well represented in amphibians, reptiles, birds and mammals. *Location:* gut. *Cercariae:* Xiphidiocercariae. *Subfamilies:* Lecithodendriinae Lühe, 1901, emend. Looss, 1902 (Syn. Brachycoeliinae Looss, 1899, in part); Pleurogenetinae Looss, 1899.

KEY TO SUBFAMILIES OF LECITHODENDRIIDAE

I. Vitellaria entirely or mostly in front of the caeca; parasites of Amphibia and bats **Pleurogenetinae** (*Pleurogenes, Prosotocus, Brandesia*)

II. Vitellaria not entirely or mostly in front of the caeca; parasites of birds and bats **Lecithodendriinae** (*Phaneropsolus* and *Eumegacetes* in birds; *Lecithodendrium, Pycnoporus, Parabascus* in bats) (keys on pp. 324 and 383)

Mehra (1935) referred *Phaneropsolus* and *Eumegacetes* to separate new subfamilies, Phaneropsolinae and Eumegacetinae, and proposed other taxonomic amendments to the family.

Family **CEPHALOGONIMIDAE** Looss, 1899

DIAGNOSIS. *General:* small, spinous distomes having a gut which includes a short prepharynx and relatively short caeca, a genital pore which is situated far forward above or on the right of the oral sucker, and a long cirrus pouch which extends back to the ventral sucker. *Gonads:* testes generally in the middle of the body, one behind the other; ovary in front of the testes, near the ventral sucker. *Vitellaria:* well developed, lateral, the follicles mostly distributed in the anterior region. *Uterus:* having descending and ascending limbs which extend almost to the posterior extremity. *Excretory system:* main canals Y-shaped, with or without side branches. *Hosts:* amphibians, reptiles and birds. *Location:* gut. *Genus: Cephalogonimus.*

Family **DICROCOELIIDAE** Odhner, 1910
(Figs. 11, 12H, J, 49B, C, 50B, 57A, 81A–G)

DIAGNOSIS. *General:* elongate, flattened, delicate and translucent distomes of small or medium size with suckers not far apart in the anterior region of the body. *Genital ducts:* cirrus and cirrus pouch small, mainly in front of the ventral sucker, opening in the median plane. *Gonads:* testes close behind the ventral sucker, their relative positions variable, ovary behind the testes. *Vitellaria:* fairly well developed, lateral to the caeca in the middle region of the body. *Uterus:* having descending and ascending limbs abundantly folded, the folds filling most of the body behind the gonads. *Excretory system:* vesicle tubular or sac-like. Flame-cell formula: $2[(2+2+2)+(2+2+2)]$. *Hosts:* amphibians, reptiles, birds and mammals. *Location:* gall bladder and bile ducts, sometimes the pancreatic ducts, rarely in the gut. *Cercariae:* Xiphidiocercariae. *Subfamilies:* Dicrocoeliinae Looss, 1899; Brachycoeliinae Looss, 1899 emend.

KEY TO SUBFAMILIES OF DICROCOELIIDAE

I. Caeca very short, parasites of Amphibia and Reptilia **Brachycoeliinae** (*Brachycoelium*, *Leptophallus*)

II. Caeca long, parasites of birds and mammals* **Dicrocoeliinae** (*Dicrocoelium*, *Eurytrema*, *Oswaldoia*, *Lyperosomum*, *Athesmia* and *Platynosomum*) (keys on pp. 325 and 385)

Family **MICROPHALLIDAE** Viana, 1924
(Figs. 12L, 51C, E, F, G, 52A, B)

DIAGNOSIS. *General:* small or very small distomes, mostly pyriform, having small suckers and a short gut, with caeca scarcely reaching back to the level of the ventral sucker, showing a general resemblance to the Heterophyidae.† *Gonads:* testes at a distance from the posterior extremity. *Genital ducts:* pars prostatica and seminal vesicle situated in front of the ventral sucker. *Uterus:* having descending and ascending limbs, forming folds far behind the region of the testes. *Receptaculum seminis:* absent. *Hosts:* birds. *Location:* gut, sometimes the gall bladder. *Cercariae:* Xiphidiocercariae. *Subfamilies:* Microphallinae Ward, 1901; Maritrematinae Nicoll, 1909; Gymnophallinae Odhner, 1905.

KEY TO SUBFAMILIES OF MICROPHALLIDAE

I. Cirrus pouch present **Maritrematinae** (*Maritrema*)

II. Cirrus pouch absent
- A. Genital atrium in front of the ventral sucker **Gymnophallinae** (*Gymnophallus*)
- B. Genital atrium on left of the ventral sucker **Microphallinae** (*Microphallus*, *Levinseniella*, *Spelophallus*, *Spelotrema*)

Family **OPISTHORCHIIDAE** Lühe, 1901, emend. Braun. 1901
(Figs. 12N, 50C, 54E, 57B)

DIAGNOSIS. *General:* elongate, flattened, lanceolate distomes of delicate appearance, having weakly developed, closely approximated suckers. *Genital ducts:* cirrus small, vesicula seminalis coiled, cirrus pouch absent, genital pore median, slightly in front of the ventral sucker. *Gonads:* testes situated in the posterior part of the body and of variable shape, ovary in front of the testes. *Vitellaria:* follicular, lateral, generally situated in front of the gonads. *Excretory system:* vesicle Y-shaped, with a long stem and short lateral canals which do not extend beyond the level of the ovary anteriorly. Flame-cell formula variable, even in species of the genus *Opisthorchis* (Price, 1940). *Uterus:* having folds mainly confined to the region in front of the ovary. *Eggs:* numerous, small, light brown, each containing a larva. *Hosts:* reptiles, birds and mammals. *Location:* gall bladder and bile ducts. *Cercariae:* Gymnocephalous, of 'Pleurolophocerca' or 'Parapleurolophocerca' type. *Larvae:* encysted in fishes and amphibians. *Subfamilies:* Opisthorchiinae Looss, 1899; Metorchiinae Lühe, 1909.

* But see p. 306 for parasites of reptiles allocated to this group.

† Rankin, jr., J. S. (1940*a*) claimed that the use of Crustacea as second intermediate hosts is a character which helps to distinguish the Microphallidae from the Heterophyidae.

KEY TO SUBFAMILIES OF OPISTHORCHIIDAE

I. Uterus and vitellaria extending anteriorly beyond the level of the ventral sucker **Metorchiinae** (*Metorchis, Holometra, Pachytrema, Pseudamphistomum*) (Keys on pp. 336 and 396)

II. Uterus and vitellaria not extending anteriorly beyond the level of the ventral sucker **Opisthorchiinae** (*Opisthorchis, Amphimerus, Cyclorchis, Cladocystis*) (Key on p. 335)

Family **HETEROPHYIDAE** Odhner, 1914 (Figs. 51 D, 54 F, G)

Syn. Coenogonimidae Nicoll, 1907; Cotylogonimidae Nicoll, 1907; Stictodoridae Poche, 1926.

DIAGNOSIS. *General:* small or very small ovoid or pyriform distomes with feebly developed suckers, the ventral generally enclosed within a genital sinus. *Cuticle:* generally scaly, the scales being more numerous anteriorly. *Genital sinus:* variously modified, containing a cirrus-like body or *gonotyl* (genital sucker), cirrus pouch absent. *Gonads:* one or two testes, situated in the posterior part of body. Ovary in front of the testes or testis. *Uterus:* having transverse folds, mostly between the testes and the genital pore. *Vitellaria:* lateral, follicles arranged on the lateral or the median sides of the caeca, generally in the posterior region. *Excretory system:* vesicle Y-shaped, flame-cell formula variable. *Eggs:* not numerous. *Hosts:* birds and mammals. *Location:* intestine. *Cercariae:* Gymnocephalous, of the 'Pleurolophocerca' or 'Parapleurolophocerca' groups of Sewell. *Metacercariae:* encysted in fishes and amphibians. *Subfamilies:* Heterophyinae Ciurea, 1924; Metagoniminae Ciurea, 1924; Cryptocotylinae Lühe, 1909; Apophallinae Ciurea, 1924; Galactosomatinae Ciurea, 1933; Centrocestinae Looss, 1899; Haplorchiinae Looss, 1899; Adleriellinae Witenberg, 1930; Stellantchasminae Price, 1939, separable and with genera as follows (after Price, 1940*a*):

1. Heterophyinae Ciurea, 1924: ventral sucker not enclosed in genital sinus; gonotyl postero-lateral to the ventral sucker, bearing a row of chitinous rodlets (*Heterophyes*).

2. Metagoniminae Ciurea, 1924: ventral sucker lateral, enclosed in genital sinus; gonotyl inconspicuous, in the form of one or two papilla-like bodies (*Metagonimus:* Syn. *Loossia* Ciurea, 1915; *Yokogawa* Leiper, 1913; *Dexiogonimus* Witenberg, 1929).

3. Cryptocotylinae Lühe, 1909: ventral sucker median, rudimentary, in the anterior wall of a spacious, more or less muscular genital sinus; genital aperture behind the ventral sucker; gonotyl papilla-like, single (*Cryptocotyle:* Syn. *Tocotrema* Looss, 1899; *Dermocystis* Stafford, 1905; *Hallum* Wigdor, 1918; *Ciureana* Skrjabin, 1923).

4. Apophallinae Ciurea, 1924: ventral sucker relatively well developed, enclosed in a small, non-muscular genital sinus; genital pore in front of the ventral sucker; gonotyl papilla-like, double or single (*Apophallus:* Syn. *Rossicotrema* Skrjabin & Lindtrop, 1919; *Cotylophallus* Ransom, 1920).

5. Galactosomatinae Ciurea, 1933: ventral sucker much reduced or absent; gonotyl globular, generally equipped with spines; uterus largely behind the ovary (*Galactosomum:* Syn. *Microlistrum* Braun, 1901; *Cercarioides* Witenberg, 1929; *Tubanguia* Srivastava, 1935).

6. Centrocestinae Looss, 1899: anterior extremity provided with one or two rows of circum-oral spines; ventral sucker relatively well developed, included within a genital sinus; gonotyl, when present, in the form of one or two papilla-like bodies (*Centrocestus:* Syn. *Stamnosoma* Tanabe, 1922; *Stephanopirumus* Onji & Nishio, 1924: *Ascocotyle*: *Phagicola*).

7. Haplorchiinae Looss, 1899: circum-oral spines absent; ventral sucker rudimentary; gonotyle relatively large and equipped with chitinous rodlets or other armature; only a single testis (*Haplorchis:* Syn. *Monorchotrema* Nishigori, 1924; *Kasr* Khalil, 1932).

8. Adleriellinae Witenberg, 1930: ventral sucker absent; gonotyl prominent and with large spines; testes in front of the ovary and vesicula seminalis (*Adleriella:* Syn. *Adleria* Witenberg, 1929, *nec* Rohwer & Fagan, 1917).

9. Stellantchasminae Price, 1939: ventral sucker, when present, small, lateral, enclosed within a genital sinus; gonotyl, when present, relatively large and spinous; one or two testes (*Stellantchasmus* Onji & Nishio, 1916: Syn. *Diorchitrema* Witenberg, 1929) (not represented in this volume).

Family **STOMYLOTREMATIDAE** Poche, 1926 (Fig. 53 D)

Diagnosis. *General:* thick, muscular, non-spinous distomes with widely rounded extremities and very large suckers, the ventral situated behind the middle of the body. *Genital pore:* on the right margin of the body, at the level of the pharynx. *Gonads:* rounded, testes side by side in front of the ventral sucker, ovary lateral, in front of the testes. *Genital ducts:* cirrus and long cirrus pouch present. *Vitellaria:* not well developed, comprising a few large follicles. *Uterus:* having descending and ascending loops which pass on either side of the ventral sucker. *Eggs:* small, thick-shelled, ovoid. *Hosts:* birds. *Location:* cloaca. *Genus: Stomylotrema.*

Family **PHILOPHTHALMIDAE** Looss, 1899 (Fig. 53 F)

Diagnosis. *General:* muscular, non-spinous distomes of medium size having powerful suckers and a very large pharynx. *Genital ducts:* cirrus pouch and metraterm well developed and long, extending well behind the ventral sucker. *Gonads:* one behind the other in the most posterior part of the body, the ovary foremost. *Vitellaria:* not very well developed, tubular, comprising six or seven follicles on either side of the body, lateral or median to the caeca. *Uterus:* folds mostly confined to the region in front of the gonads, sometimes overstepping the caeca laterally. *Eggs:* of medium size, colourless, each containing a miracidium. *Hosts:* birds. *Location:* superficial situations like the orbit and cloaca. *Genera: Philophthalmus* and *Pygorchis.* (Key on p. 346.)

Family **ORCHIPEDIDAE** Skrjabin, 1924 (Fig. 53 B)

Diagnosis. *General:* elongate distomes having well-developed suckers, a conical anterior and a flattened posterior region. *Gut:* prepharynx absent, oesophagus very short, caeca long. *Vitellaria:* well developed, lateral, the follicles distributed along the whole extent of the body behind the ventral sucker. *Genital ducts:* cirrus pouch absent, vesicula seminalis large, long and convoluted, genital pore situated slightly behind the pharynx. *Uterus:* short, containing a few relatively large eggs. *Gonads:* testes numerous and distributed between the caeca and behind the ventral sucker, ovary situated in front of the testes. *Hosts:* birds. *Location:* trachea. *Genus: Orchipedum.*

Family **PSILOSTOMATIDAE** Odhner, 1911, emend. Nicoll, 1935 (Fig. 51 B)

DIAGNOSIS. *General:* elongate, flattened, muscular distomes of small or fairly small size having a general structure closely resembling that of Echinostomatidae, but lacking a crown of head spines. *Gut:* oesophagus epithelial. *Cuticle:* non-spinous or only slightly spinous. *Excretory system:* comprising a subcutaneous network of vessels and two main lateral canals which unite in front of the ventral sucker to form a long median stem. *Eggs:* not very numerous, large, similar to those of echinostomes. *Hosts:* birds. *Location:* gut. *Genera: Psilostomum, Psilochasmus, Sphaeridiotrema, Apopharynx, Psilotrema.* (Key on p. 348.)

Note. All known psilostome life histories resemble one another closely and are essentially echinostome-like, especially in the characters of mother and daughter rediae and miracidia (Beaver, 1939).

Family **EUCOTYLIDAE** Skrjabin, 1924 (Fig. 53 G)

DIAGNOSIS. *General:* monostomes of medium size having a subterminal mouth and an oral sucker. *Genital pore:* situated slightly behind the bifurcation of the intestine. *Cirrus pouch:* absent. *Gonads:* testes elongate-ovoid, near the middle of the body, lateral to the caeca. Ovary situated in front of the testes, sometimes between them. *Vitellaria:* extensive, but comprising only a few coarse follicles on each side of the body. *Uterus:* having descending and ascending loops extending to the posterior extremity. *Eggs:* small, numerous. *Hosts:* birds. *Location:* kidneys and ureters. *Genera: Eucotyle, Tanaisia* and *Tamerlania.* (Key on p. 345.)

Family **CATHAEMASIIDAE** Fuhrmann, 1928, emend. Harwood, 1936 (Fig. 53 E)

DIAGNOSIS. *General:* fairly elongate distomes of oval outline having closely approximated suckers, of which the ventral is the larger, and a cuticle which is spinous ventrally. Internal organs arranged much as in Echinostomatidae, but head crown absent. *Excretory system:* vesicle (in strong contrast with Echinostomatidae and Psilostomidae) having a short, median, terminal canal which is formed on either side into a saccular outgrowth extending anteriorly and laterally. *Gonads:* testes much-branched, one behind the other in the posterior region, ovary slightly in front of the anterior testis. *Vitellaria:* extensive, lateral to the caeca between the ventral sucker and the posterior extremity. *Uterus:* confined to the region in front of the ovary. *Eggs:* fairly numerous, large (0·1 mm. long). *Hosts:* birds. *Location:* gut (oesophagus). *Genus: Cathaemasia.*

Family **TROGLOTREMATIDAE** Odhner, 1914, emend. Braun, 1915 (Figs. 50 G, 54 C, D, 56 K)

Syn. Troglotremidae Odhner, 1914; Collyriclidae Ward, 1917.

DIAGNOSIS. *General:* fleshy, somewhat flattened, spinous distomes of medium or fairly large size having a slightly arched body. Suckers feebly developed, the ventral sometimes absent. Median processes sometimes present at the posterior extremity. *Genital pore:* median or on the left, in front of or behind the ventral sucker when this is present. *Cirrus pouch:* generally absent. *Gonads:* testes elongate, deeply lobed, near the middle of the body or slightly more posterior, ovary in front of the testes and on the right. *Vitellaria:* well developed, the follicles almost filling the dorso-lateral regions. *Uterus:* generally well developed, long and much folded, but sometimes short and of knotted appearance. *Eggs:* small as a rule (0·017–0·025 mm. long), but large in some instances (0·06–0·08 mm. long), size apparently varying inversely as the length of the uterus. *Excretory system:* vesicle Y-shaped or triangular, sometimes tubular. Flame-cell formula: $2([3+(3+3)+(3+3)]+[(3+3)+(3+3)+3])$. *Hosts:* carnivorous birds and mammals. *Location:* cyst-like spaces in various organs such as the frontal sinuses, lungs, kidneys, skin. *Genera: Collyriclum* and *Renicola* in birds; *Paragonimus* and *Troglotrema* in land mammals; *Pholeter* in marine mammals; *Nephrotrema* in water-shrews and moles. (Keys on pp. 361 and 404.)

Family **CYCLOCOELIDAE** Kossack, 1911 (Figs. 10G, 53A)

Syn. Monostomida Kolenati, 1856; Monostomidae Cobbold, 1864; Monostomatidae Gamble, 1896.

DIAGNOSIS. *General:* fairly large or very large monostomes which sometimes lack the ventral as well as the oral sucker, and which have a muscular, slightly flattened body with rounded extremities. *Gut:* intestinal crura merging posteriorly, to form a ring-like intestine, sometimes having short, lateral diverticulae. *Genital pore:* median, far forward in the vicinity of the pharynx. *Genital ducts:* copulatory apparatus and cirrus pouch feebly developed. Vesicula seminalis interna present. *Gonads:* rounded or lobed, grouped together in the posterior region. *Vitellaria:* well developed, the follicles numerous lateral and dorsal to the caeca, sometimes merging posteriorly. *Uterus:* having numerous wide transverse folds which fill up most of the available space within the ring-like intestine, sometimes overstepping it laterally. *Eggs:* thin-shelled, without polar filaments, variable in size, numerous, each containing a miracidium. *Excretory system:* vesicle small, sac-like, situated behind the intestine. *Hosts:* aquatic birds. *Location:* body cavity, nasal cavities, air sacs. *Genera: Cyclocoelum, Hyptiasmus, Typhlocoelum.* (Key on p. 364.)

Family **NOTOCOTYLIDAE** Lühe, 1909 (Figs. 10F, 50H)

DIAGNOSIS. *General:* monostomes of small or fairly small size with an oral sucker, having a superficial resemblance to Cyclocoelidae, but showing marked anatomical differences. *Cuticle:* beset with spinelets, especially on the ventral surface and in the anterior region. *Ventral surface:* bearing three or five rows of groups of unicellular glands. *Gut:* pharynx absent, caeca long and simple, ending blindly near the posterior extremity. *Genital pore:* median, far forward, slightly behind the mouth. *Cirrus pouch:* elongate. *Gonads:* testes slightly lobed, one on either side of the body in the posterior region, lateral to the caeca, ovary between the testes. *Vitellaria:* occupying a short region in front of the gonads laterally. *Uterus:* having a regular series of transverse folds which fill up available space between the caeca from the level of the testes to the posterior end of the cirrus pouch. *Eggs:* each with a long, straight filament at either pole. *Excretory system:* vesicle Y-shaped, the median stem short, the lateral canals long and united above the oral sucker. *Hosts:* birds and mammals. *Location:* the caeca and rectum of birds, especially Anseriformes, the gut of mammals. *Genera:* *Notocotylus*, *Paramonostomum* and *Catatropis*; *Ogmogaster* in marine mammals. (Key on p. 362.) For Szidat's scheme of classification see p. 79.

Family **BRACHYLAEMIDAE** Joyeux & Foley, 1930 (Figs. 50I, 60A–C)

Syn. Harmostomidae Odhner, 1912.

DIAGNOSIS. *General:* somewhat elongate, slightly flattened, smooth or lightly spinous distomes of small or medium size and variable form. *Genital pore:* situated in the posterior region of the body, median, lateral or dorsal, sometimes in front of the gonads, but otherwise terminal. *Genital ducts:* cirrus pouch small and ill-defined, containing only a small cirrus and an ejaculatory duct. Vesicula seminalis *externa* present. *Gonads:* testes one behind the other, posterior, ovary between the testes. *Vitellaria:* follicular, lateral, mostly behind the middle of the body. *Uterus:* having ascending and descending limbs which sometimes reach the level of the oesophagus, but are of variable extent. *Excretory system:* vesicle Y-shaped, comprising a short median stem and long lateral canals extending almost to the anterior extremity. *Eggs:* numerous, small. *Hosts:* birds and mammals, occasionally Amphibia and Reptilia. *Location:* gut; bursa Fabricii of birds. *Genera:* *Brachylaemus* in birds and mammals; *Urotocus* and *Leucochloridium* in birds; *Itygonimus* in moles. (Keys on pp. 366 and 412.)

Note. Dollfus (1934, 1935*a*) retained in this family the genera *Brachylaemus* and *Itygonimus* (subfamily Brachylaeminae [Syn. Heterolopinae Looss; Harmostominae Looss]), but excluded *Leucochloridium* Carus, 1835 (Syn. *Urogonimus* Monticelli, 1888) and *Urotocus* Looss, 1899, both of which were included in a new family, Leucochloridiidae (Syn. Urogoniminae Looss, 1899; Leucochloridiinae Poche, 1907). Other genera which Odhner included in the Harmostomidae (subfamily Liolopinae) are referred to two other families; *Liolope* Cohn, 1902 becomes the type of the Liolopidae Dollfus, 1934; *Hapalotrema* Looss, 1899 is rejected in accordance with the findings of Ward (1921) and Stunkard (*Amer. Mus. Novit.* no. 12, 1921, pp. 1–5), the latter writer having made it the type of a new subfamily, Hapalotreminae, in the Spirorchiidae MacCallum (Syn. Proparorchiidae Ward).

Family **ECHINOSTOMATIDAE** Looss, 1902, emend. Poche, 1926 (Jan.), or Stiles & Hassall, 1926 (Jan.)? (Figs. 10E, 12I, 50D–F, 57D, E)

DIAGNOSIS. *General:* elongate, spinous, muscular distomes of variable size, having closely approximated suckers, of which the ventral may be especially powerful, and a fleshy anterior head collar bearing a single or double coronet of stout 'head' or 'collar' spines. *Cirrus pouch:* generally present, sometimes extending back behind the ventral sucker. *Gonads:* testes ovoid, sometimes slightly lobed, directly or diagonally one behind the other in the posterior region. Ovary generally in front of the testes, median or slightly on the right of the median plane. *Vitellaria:* well developed, follicles mostly in the posterior region, lateral, but sometimes merging in the median plane. *Uterus:* much folded between the ovary and the ventral sucker. *Eggs:* large (0·065–0·12 mm. long) and thin-shelled. *Excretory system:* vesicle Y-shaped, provided with ramifying branches. *Hosts:* birds and mammals. *Location:* intestine, sometimes the bile ducts. *Cercariae:* of echinostome type, entering a second snail or a fish or an amphibian to encyst. *Subfamilies:* Echinostomatinae Looss, 1899; Echinochasminae Odhner, 1910; Himasthlinae Odhner, 1911; and a number of unclassified genera.

KEY TO SUBFAMILIES OF ECHINOSTOMATIDAE

I. Cirrus pouch, if present, extending far behind the ventral sucker; cirrus long and having spines of rose-thorn shape **Himasthlinae** (*Himasthla, Aporchis*)

II. Cirrus pouch and cirrus with other characters
  - A. Head crown with a single or double row of spines which is unbroken dorsally **Echinostomatinae** (*Echinostoma, Echinoparyphium, Euparyphium*) (key on p. 352)
  - B. Head crown with a single row of spines, the sequence being broken dorsally **Echinochasminae** (*Echinochasmus, Heterechinostomum, Stephanoprora*) (key on p. 354)

Isolated genera: Hypoderaeum, Sodalis, Pegosomum, Parorchis, Paryphostomum, Parechinostomum, Petasiger, Chaunocephalus, Echinostephila. (Key to these on p. 356.)

Family **MESOTRETIDAE** Poche, 1926 (Fig. 54A)

DIAGNOSIS. *General:* elongate distomes in which the anterior and posterior regions are sharply marked by a constriction, the anterior region being spinous. *Genital pore:* median, close behind the ventral sucker. *Cirrus pouch:* elongate. *Gonads:* testes large, elongate, diagonally one behind the other in the posterior region, ovary slightly behind the posterior testis. *Vitellaria:* well developed, follicles extending along the entire length of the posterior region. *Uterus:* having descending and ascending limbs very slightly folded. *Eggs:* small, thin-shelled, numerous. *Hosts:* bats. *Location:* gut. *Genus: Mesotretes.*

Family **FASCIOLIDAE** Railliet, 1895 (Figs. 12G, 55A, B)

Syn. Fasciolopsidae Odhner, 1926.

DIAGNOSIS. *General:* large or very large distomes having a broad, flat, muscular, spinous or non-spinous body, and closely approximated suckers. All the important organs are complicated by branching. *Gut:* caeca long and with numerous lateral diverticula with secondary and tertiary branches. *Genital ducts:* cirrus and cirrus pouch well developed. *Gonads:* testes diagonally or directly one behind the other in posterior region, lobed or much-branched, ovary situated in front of the testes, lateral, also much-branched. *Vitellaria:* abundant follicles filling up the parenchyma between the branches of the caeca laterally. *Uterus:* having a few transverse folds in front of the testes, mainly opposite the ovary. *Eggs:* large, thin-shelled, moderate in number. *Excretory system:* main canals much-branched, connected with a subcutaneous network of capillaries. *Hosts:* herbivorous and omnivorous vertebrates, especially Ungulata. *Location:* liver and bile ducts, sometimes the intestine. *Cercariae:* Gymnocephalous. *Genera: Fasciola, Fasciolopsis, Fascioloides.*

Family **CAMPULIDAE** Odhner, 1926 (Fig. 56A–J)

DIAGNOSIS. *General:* elongate, spinous distomes of medium or fairly large size having a body which tapers more gradually posteriorly than anteriorly. *Gut:* caeca generally having anterior extensions, when branched also having median as well as lateral diverticula. *Gonads:* rounded or lobed, one behind another in the posterior half of the body, ovary foremost. *Vitellaria:* follicles abundant in the lateral regions between the bifurcation of the intestine and the posterior extremity, sometimes merging in the median plane behind the gonads. *Uterus:* having a few folds between the ovary and the ventral sucker. *Eggs:* thick-shelled, sculptured, triangular in section. *Hosts:* marine mammals (Cetacea and Pinnipedia). *Location:* liver. *Genera: Lecithodesmus, Zalophotrema, Campula, Orthosplanchnus, Synthesium, Odhneriella.* (Key on p. 388.)

Note. This family is regarded by some writers as a subfamily of the Fasciolidae; distinctive characters such as the less intensive branching of the principal organs, the form of the gut and the distinctiveness of the final hosts would seem to denote worthiness of family rank.

Family **PARAMPHISTOMATIDAE** Fischoeder, 1901, emend. Goto & Matsudaira, 1918 (Figs. 10A, B, 12F, 51I, 59B–D)

DIAGNOSIS. *General:* amphistomatous Digenea having a thick, fleshy, conical body and a powerful posterior sucker. In some forms there is a conspicuous ventral pouch and in others the oral sucker is provided with postero-dorsal out-growths or pockets. *Cuticle:* non-spinous. *Gut:* pharynx absent (sometimes a muscular thickening of the oesophagus occurs), caeca simple and long or only of medium length. *Cirrus pouch:* present or absent. *Gonads:* rounded or slightly lobed, ovary generally behind the testes. *Uterus:* long, but mainly an ascending limb, with folds which occupy the dorso-lateral regions. *Vitellaria:* follicular, lateral, well developed, generally extending along the entire lateral regions. *Eggs:* not very numerous, generally large, sculptured. *Hosts:* all classes of vertebrates. *Location:* gut and bile ducts, sometimes other organs such as the bladder. *Cercariae:* amphistome type. *Subfamilies:** Paramphistomatinae Fischoeder, 1901, emend. Dollfus, 1932; Gastrothylacinae Stiles & Goldberger, 1910; Cladorchiinae Fischoeder, 1901, emend. Hughes, Higginbotham & Clary, 1942; Balanorchiinae Stunkard, 1917; Zygocotylinae Stunkard, 1916; Diplodiscinae Cohn, 1904; Gastrodiscinae Stiles & Goldberger, 1910.

* Szidat (1939*a*) recognized four subfamilies not mentioned here: two occur in fishes, but not in British waters, one in the large intestine of reptiles, and the fourth in the caeca of *Manatus* (his scheme is outlined on p. 78).

KEY TO SUBFAMILIES OF PARAMPHISTOMATIDAE*

1. Body divided into a small, conical, anterior and an expanded, disk-like posterior region **Gastrodiscinae** (*Gastrodiscus*, *Gastrodiscoides*, *Homalogaster*)
2. Body not divided into regions
   - A. Body having a large ventral pouch **Gastrothylacinae** (*Gastrothylax*, *Fischoederius*, *Carmyerius*)
   - B. Body lacking a ventral pouch
     - I. Posterior sucker having an overhanging lip **Zygocotylinae** (*Zygocotyle*)
     - II. Posterior sucker lacking an overhanging lip
       - A′ Cirrus pouch present
         - (*a*) Testes rounded, smooth **Balanorchiinae** (*Balanorchis*)
         - (*b*) Testes lobed or branched **Cladorchiinae** (*Cladorchis*)
       - B′ Cirrus pouch absent
         - (*a*) Vitellaria extensive **Paramphistomatinae** (*Paramphistomum*, *Cotylophoron*, *Pseudodiscus*)
         - (*b*) Vitellaria not extensive, consisting of few large follicles, sometimes a pair of compact organs **Diplodiscinae** (*Diplodiscus*)

Family **MICROSCAPHIDIIDAE** Travassos, 1922

Syn. Angiodictyidae Looss, 1902.

Stunkard (1943) pointed out that as *Angiodictyum* Looss, 1902 is not the type of the subfamily to which it belongs, i.e. the Microscaphidiinae, it cannot be the type of the family, which should be named as shown here.

Diagnosis. *General:* non-spinous monostomes which lack a ventral, but have a small oral sucker, which sometimes has two lateral pouches. *Gut:* pharynx absent, caeca long and simple. *Gonads:* one behind another in the posterior region, the ovary hindmost. *Genital pore:* anterior, not far behind the mouth. *Uterus:* with slight folds between and in front of the gonads. *Vitellaria*: follicles numerous laterally in the posterior region, merging behind the gonads in the median plane. *Excretory system:* pore slightly dorsal, vesicle having eight longitudinal canals provided with connecting vessels. *Lymph system:* comprising a number of longitudinal vessels uniting anteriorly, absent in some instances, e.g. *Dictyangium chelydrae*. *Eggs:* large. *Hosts:* chelonians and birds. *Location:* intestine, sometimes the kidneys. *Genus: Polyangium*, in the kidneys of birds.

Family **MESOMETRIDAE** Poche, 1926 (Fig. 46 A–C)

Diagnosis. *General:* Digenea with an oral, but without a ventral sucker. *Gut:* prepharynx very long, pharynx small, oesophagus very short, caeca wide and long or fairly long. *Gonads:* testes rounded, side by side behind the middle of the body, ovary lobed, median, behind the testes. *Genital pore:* far forward, midway between the bifurcation of the intestine and the mouth. *Vitellaria:* follicles abundant in the dorso-lateral regions behind the pharynx, sometimes merging in the median plane. *Uterus:* having a little folded ascending limb only,

* The subfamily Schizamphistomatinae is excluded, but see p. 313 for notes on the genera in reptiles.

the folds confined within the bounds of the caeca. *Eggs:* large, each generally having a process at one pole. *Excretory system:* vessels forming a much-branched network. *Hosts:* fishes. *Location:* gut. *Genera: Mesometra, Centroderma, Wardula.*

Note. Szidat (1939*a*) referred this group to the suborder Paramphistomata Szidat, 1936, along with the Notocotylidae.

Family **CYATHOCOTYLIDAE** Poche, 1926 (Figs. 53 C, 58 A)

DIAGNOSIS. *General:* distomes of rounded outline with a structure of Strigeid pattern, but in which the body is not divided into anterior and posterior regions. *Genital ducts:* cirrus pouch very large, cirrus large, pars prostatica and seminal vesicle present. *Gonads:* rounded, prominent, ovary between the testes. *Vitellaria:* very well developed, laterally along the whole extent of the body. *Adhesive organ:* very large, occupying the middle of the body behind the ventral sucker. *Glands:* absent. *Excretory system:* flame-cell formula 2 [(2) + (2) + (2) + (2) + (2)]. *Hosts:* reptiles, birds and mammals. *Location:* intestine. *Subfamilies:* Cyathocotylinae, Prohemistomatinae, with genera *Cyathocotyle* and *Mesostephanus* in birds and mammals respectively. Szidat (1936) made *Pharyngostomum* the type genus of a new subfamily, the Pharyngostomatinae, occurring in mammals.

Family **STRIGEIDAE** Railliet, 1919 (Figs. 10 H, 50 J)

Syn. Holostomidae Brandes, 1890.

DIAGNOSIS. *General:* distomes in which the body is divided by a constriction into an anterior, cup- or spoon-shaped and a posterior, ovoid or cylindrical region, the former constituting an adhesive organ, the latter containing most of the reproductive organs. *Suckers:* feebly developed, situated in the anterior region. *Adhesive organ:** posterior to the ventral sucker, when this is present. *Genital pore:* at or near the posterior extremity, sometimes opening into a depression or bursa. *Genital ducts:* cirrus and cirrus pouch absent. *Vitellaria:* well developed, numerous follicles being distributed in both parts of body, or only the posterior. *Uterus:* short, containing a few eggs of large size. *Excretory system:* comprising a subcutaneous network of vessels, pore slightly dorsal, almost terminal. *Hosts:* mainly birds. *Location:* gut. *Miracidium:* having two pairs of flame cells. *Cercariae:* Furcocercous, developing in sporocysts. *Metacercariae:* commonly of the form known as 'Tetracotyle'. *Subfamily:* Strigeinae Railliet, 1919, with genera *Strigea*, *Apharyngostrigea*, *Parastrigea*, *Ophiosoma*, *Cardiocephalus*, *Apatemon*, *Cotylurus*. (Key on p. 370.)

Family **DIPLOSTOMATIDAE** Poirier, 1886, emend. (Fig. 58 B)

DIAGNOSIS. *General:* similar to the Strigeidae, but anterior region of the body more flattened, posterior region cone-like, cylindrical or flat. Ear-like processes sometimes present in the antero-lateral parts of the forebody. *Adhesive organ:* short, circular or elliptical, or elongate, sometimes containing part of the

* Baer, J. G. (*in litt.*) states that this organ contains proteolytic glands which digest the epithelial cells of the host's gut wall; in consequence, he named it the 'tribocytic organ'.

uterus. *Hosts:* mainly birds and mammals. *Location:* gut. *Subfamilies:* Diplostomatinae Monticelli, 1888 (genera, *Diplostomum*, *Neodiplostomum*) (Key on p. 373) and Alariinae Hall & Wigdor, 1918 (genera *Alaria*, *Podospathalium*).

Family **CLINOSTOMATIDAE** Lühe, 1901, emend. Dollfus, 1932

DIAGNOSIS. *General:* flattened distomes of medium to very large size having the suckers close together, the ventral the larger. *Gut:* pharynx absent, caeca long and provided with lateral diverticula. *Genital pore:* situated far behind the ventral sucker and just in front of the anterior testis. *Cirrus pouch:* present. *Gonads:* testes lobed, situated in the posterior region, ovary between the testes. *Uterus:* folded near the ovary and having V-shaped loops towards the ventral sucker. *Vitellaria:* well developed, merging behind the posterior testis. *Hosts:* reptiles and birds. *Location:* buccal cavity, pharynx and oesophagus. *Miracidium:* having two pairs of flame cells and ciliated epidermal plates similar in number and arrangement to those of Schistosomatidae and Strigeidae. *Cercaria:* furcocercous, with a pharynx, penetration glands and an anterior organ. *Genera: Clinostomum*, *Euclinostomum*. (Key on p. 342.)

Family **SCHISTOSOMATIDAE** Looss, 1899, emend. Poche, 1907 (Figs. 10I, 12E, 51H, 59A)

Syn. Schistosomidae Looss, 1899; Bilharziidae Odhner, 1912.

DIAGNOSIS. *General:* greatly elongate, unisexual, dimorphic Digenea parasitic in the blood-vascular system of vertebrates. Female as a rule longer and more slender than the male, sometimes carried in a ventral groove (the gynaecophoric canal) which is formed by ventrally flexed lateral outgrowths of the male body. *Suckers:* poorly developed, sometimes absent in one or both sexes. *Gut:* pharynx absent, oesophagus short, caeca very long and united posteriorly to form a single caecum, sometimes having lateral diverticula. *Genital pore:* situated behind the ventral sucker (when it is present). *Gonads:* male having four or more, sometimes numerous testes situated anteriorly or posteriorly; female having a single, compact ovary, situated in the posterior region, just in front of the union of the caeca. *Vitellaria:* well developed, posterior to the ovary. *Excretory system:* having a short median stem and long lateral canals reaching almost to the anterior extremity. Flame-cell formula: $2\,[(2)^n + (2)^n]$ or $2\,[(3)^n + (3)^n]$. *Eggs:* few, thin-shelled, non-operculate, sometimes having a short lateral or terminal spine. *Hosts:* birds and mammals. *Location:* blood vessels. *Miracidium:* having two pairs of flame cells. *Redia:* absent from the life history. *Cercaria:* furcocercous, entering the final host by way of the skin. No encysted metacercarial stage in development. *Subfamilies:* Bilharziellinae Price, 1929 (in the venous system of birds); Schistosomatinae Stiles & Hassall, 1889 (mainly in the venous system of birds); Eubilharziinae Szidat, 1939 (in the venous system of mammals and man). *Bilharziella*, *Gigantobilharzia*, *Pseudobilharziella* and *Trichobilharzia* belong to the first, *Ornithobilharzia* to the second, and *Schistosoma* to the third of these subfamilies. (Key on p. 374.)

CHAPTER 7

THE MONOGENEA OF BRITISH FISHES AND AMPHIBIA

(*With Notes on their Relatives outside the British Isles*)

Roughly one-third of the known species and one-half of the known genera of Monogenea of the world occur in Europe. About the same fractions of European species and genera have been found in Britain and the seas around. Despite the slight attention given to the subject by British zoologists, nearly seventy species and more than thirty genera are mentioned in our records. Most of the European forms are imperfectly known, existing descriptions being very sketchy in many instances, and there is much scope for restudy of most of them. Increase in our knowledge will add to the number of known species, of course, but we are likely to find that we have been deceived as to the limits of variability in many species. This rather sweeping statement is justified by the existing condition of some genera which contain a multiplicity of species while others have only one. *Microcotyle* has nearly seventy species, *Axine* and *Pseudaxine* together less than a dozen, *Gastrocotyle* only two. This can hardly be due to imperfections in collections of Monogenea, but more probably implies that we have set up highly artificial limits for species in some genera. We cannot hope to devise a satisfactory taxonomic scheme till our knowledge has deepened rather than broadened. We know next to nothing about the variability of cuticular structures and the armature of the opisthaptor, especially about changes due to growth, not to mention elements of the deeper anatomy. Precise study along such lines will remove the vagaries so evident in the diagnoses which follow.

Attempts to formulate a taxonomic scheme have so far resulted in chaos instead of clarity. Price's outstanding contributions to our knowledge have been freely and gratefully drawn upon, but his scheme is not yet complete. Sproston has examined the literature, at my suggestion, and has clarified some issues (*in litt.*), so that to her also my thanks are due. I might explain that Table 2 indicates the Monogenea of fishes which were selected because they provide a representative assemblage of digenetic trematodes (see Chapter VIII), also that page-numbers following the names of families indicate the place in *this* book where diagnoses are given.

Order MONOPISTHOCOTYLEA Odhner, 1912

Syn. Monocotylea Blainville, 1828; Tricotylea and Calicotylea Diesing, 1850; Tristomeae Taschenberg, 1879; Oligocotylea Monticelli, 1903; Monopisthodiscinea Fuhrmann, 1928; Monopisthocotylinea Fuhrmann, 1928.

Superfamily *GYRODACTYLOIDEA* Johnston & Tiegs, 1922, emend. Price, 1937

Family **GYRODACTYLIDAE** Cobbold, 1864 (p. 67)

Syn. Amphibdellidae Carus, 1885 in part and misprints like Gyradactylidae Monticelli, 1888 and Gyrodaktilidae Schneidmühl, 1896

Subfamily **Gyrodactylinae** Monticelli, 1892, emend. Johnston & Tiegs, 1922

With the characters of the family

Genus *Gyrodactylus* Nordmann, 1832 (Fig. 14A)

DIAGNOSIS. *Anterior extremity:* bilobed, each lobe with a head organ. *Opisthaptor:* discoidal, not sucker-like, bearing one pair of hooks and sixteen marginal hooklets. Parasites of fishes and amphibians. There are about fourteen European species and ten non-European, but some are probably invalid. *Type-species:*

Gyrodactylus elegans Nordmann, 1832.

HOSTS. Bream, tench, loach, Crucian carp, pike, three-spined stickleback, ten-spined stickleback, lumpsucker, common goby, minnow, plaice, etc.

LOCATION. Gills.

This species occurs widely on the Continent (Lake Constance, Rhine, Lake of Lucerne) and has appeared in British fresh water.

DIAGNOSIS. *Shape and size:* spindle-like (apart from the opisthaptor), 0·5–0·8 mm. long. *Opisthaptor:* basal part of each of the two hooks with an outer margin curving ventrally to form a space or groove accommodating the two cuticular supports of the hooks. Main part of one support roughly Y-shaped, with plate-like stem arising from a membrane that covers the hooks ventrally and leaves only their points exposed. Limbs of the Y extending forward, dorsally adjacent to basal parts of hooks, each with a short dorsal process. Second supporting piece narrowed at its ends, extending transversely across the space between the limbs of the Y. Hooklets as shown in Fig. 14D, *a*. *Gut:* pharynx with eight conical processes extending into the buccal cavity (prepharynx). *Reproductive systems:* cirrus pouch situated on the left.

Note. A form attributed to this species by Johnstone (1912) was found on the fins of small plaice in the tanks of the aquarium at Piel (Barrow-in-Furness) and can be regarded as an adventitious parasite. Bradley (1861) found specimens on sticklebacks in the Hampstead ponds, and Houghton (1862) recorded the occurrence of the trematode in Shropshire. It also occurs in U.S.A.

OTHER EUROPEAN SPECIES AND THEIR HOSTS (excepting *, unrecorded in Britain):

Gyrodactylus cobitis Bychowsky, 1933. On gills and skin of spined loach (Volga delta, Caspian Sea, Switzerland).

G. gracilis Kathariner, 1894. On gills of common carp, roach, miller's thumb, gudgeon, rudd and other fresh-water fishes.

G. groenlandicus Levinsen, 1881. On gills of short-spined cottus.

†*G. latus* Bychowsky, 1933. On gills and fins of spined loach (Volga delta, Caspian Sea).

**G. medius* Katherine, 1894. On gills of common carp, bream, etc. (minnow near Edinburgh). Also found in Canada.
†*G. parvicopula* Bychowsky, 1933. On gills of bream (Caspian Sea).
†*G. rarus* Wegener, 1910. On gills of three-spined stickleback, ten-spined stickleback and other fishes (Caspian Sea).
†*G. atherinae* Bychowsky, 1933. On gills and fins of an atherine (Caspian Sea).
Also *Gyrodactylus* spp. of Beneden (1870), Lühe (1909) and Wegener (1910).

These species are separable by highly technical differences in the nature and arrangement of parts of the haptorial armature. Bychowsky (1933*b*, p. 24) gave a table of mean measurements for most of them (marked †).

Family **DACTYLOGYRIDAE** Bychowsky, 1933 (p. 67)

Syn. Gyrodactylidae Cobbold, 1864, in part; Amphibdellidae Carus, 1885, in part.

Subfamily **Dactylogyrinae** Bychowsky, 1933

Dactylogyridae with one pair of hooks and fourteen hooklets, a ring-like intestine, eyes, a vagina and rounded gonads.

KEY TO GENERA OF DACTYLOGYRINAE

I.	One cuticular bar supporting hooks of opisthaptor	*Dactylogyrus*
II.	Two similar or dissimilar bars supporting hooks of opisthaptor	*Neodactylogyrus*

Genus *Dactylogyrus* Diesing, 1850, restricted by Price, 1938 (Fig. 14B)

Syn. *Gyrodactylus* Nordmann, 1832, in part.

DIAGNOSIS. Two pairs of head organs. *Opisthaptor:* disk-like, lacking squamodisks, but with one pair of hooks, one clamp-like supporting bar and fourteen marginal hooklets (Fig. 14D, *b*). *Eyes:* present. *Reproductive systems:* ovary in front of the testes. Copulatory apparatus complicated (Fig. 14C, *a*). Vagina sometimes with cuticular supporting structures.

More than twenty European and more than twenty other species are known, but many of them are probably invalid. *Type-species:*

Dactylogyrus auriculatus (Nordmann, 1832) Diesing, 1850

Syn. *Gyrodactylus auriculatus* Nordmann, 1832.

HOSTS. Common carp, bream, goldfish.

LOCATION. Gills.

LOCALITY. Belgian coast, Sweden, and other parts of Europe (not recorded in Britain).

According to Lühe (1909) old descriptions are insufficient for purposes of identification, so that all other species must have been erected without reference to the type. Nybelin (1937) believed that many forms which have been allocated to the type-species are either unintelligible or wrongly placed. He also tried to provide a more complete definition of this species by choosing to regard *Dactylogyrus* sp. Wegener, 1910 and *D. wunderi* Bychowski, 1931 as synonymous forms. Unfortunately, Price (1938*a*) transferred *D. wunderi* to *Neodactylogyrus* out of regard for the two transverse bars which support the hooks of the opisthaptor.

Table 2. *Monogenetic Trematodes of certain Fresh-water and Marine Fishes in Europe*

G, gills; Sk, skin; F, fins; C, cloaca; R, rectum; [Nf] = nasal fossae of sting ray. * Recorded in British waters.

	Roach	Minnow	Common carp	Gudgeon	Perch	Pike	Whiting	Cod	Angler plaice dab†	Various rays	Various dogfish
GYRODACTYLIDAE:											
Gyrodactylus elegans	GSkF	G*	G	—	—	G*	—	—	—	—	—
G. gracilis	—	—	GSk	G	—	—	—	—	—	—	—
G. medius	—	—	G	—	—	—	—	—	—	—	—
DACTYLOGYRIDAE:											
Dactylogyrus anchoratus	—	—	G	—	—	—	—	—	—	—	—
D. auriculatus	—	G	G	—	G	—	—	—	—	—	—
D. cordus	G	—	—	—	—	—	—	—	—	—	—
D. crassus	—	—	G	—	—	—	—	—	—	—	—
D. dujardinianus	G	—	G	—	—	—	—	—	—	—	—
D. elongatus	G	—	—	—	—	—	—	—	—	—	—
D. falcatus	—	—	G	—	—	—	—	—	—	—	—
D. fallax	G	—	—	—	—	—	—	—	—	—	—
D. formosus	—	—	G	—	—	—	—	—	—	—	—
D. major	—	—	—	G	—	—	—	—	—	—	—
D. minutus	—	—	G	—	—	—	—	—	—	—	—
D. similis	G	—	—	—	—	—	—	—	—	—	—
D. sphyrna	G	—	—	—	—	—	—	—	—	—	—
D. trigonostoma	G	—	—	—	—	—	—	—	—	—	—
D. tuba	G	—	—	—	—	—	—	—	—	—	—
D. uncinatus	—	—	—	—	G	—	—	—	—	—	—
D. vastator	—	—	G	—	—	—	—	—	—	—	—
D. wegeneri	—	—	G	—	—	—	—	—	—	—	—
Neodactylogyrus borealis	—	G	—	—	—	—	—	—	—	—	—
N. cryptomeres	—	—	G	G	—	—	—	—	—	—	—
N. crucifer	G	—	—	—	—	—	—	—	—	—	—
N. difformis	—	—	G	—	—	—	—	—	—	—	—
N. gamellus	G	—	—	—	—	—	—	—	—	—	—
N. megastoma	—	—	—	—	—	G	—	—	—	—	—
N. micracanthus	G	—	—	—	—	—	—	—	—	—	—
N. mollis	—	—	G	—	—	—	—	—	—	—	—
N. suecicus	G	—	—	—	—	—	—	—	—	—	—
N. tenuis	—	—	—	—	G	—	—	—	—	—	—
Ancyrocephalus forceps	G	—	—	—	—	—	—	—	—	—	—
A. paradoxus	—	—	—	—	G	—	—	—	—	—	—
Tetraonchus monenteron	—	—	—	—	—	G*	—	—	—	—	—

MONOCOTYLIDAE:											
Calicotyle kroyeri	—	—	—	—	—	—	—	—	—	RCG*	—
Merizocotyle diaphanum	—	—	—	—	—	—	—	—	—	G*	—
'*M. minor*'	—	—	—	—	—	—	—	—	—	G*	—
Thaumatocotyle concinna	—	—	—	—	—	—	—	—	—	[Nf]	—
MICROBOTHRIIDAE:											
Microbothrium apiculatum	—	—	—	—	—	—	—	—	—	—	Sk
M. centrophori	—	—	—	—	—	—	—	—	—	—	F
Leptobothrium pristiuri	—	—	—	—	—	—	—	—	—	—	Sk*
Leptocotyle minor	—	—	—	—	—	—	—	—	—	—	SkG*
Pseudocotyle squatinae	—	—	—	—	—	—	—	—	—	—	Sk*
P. lepidorhini	—	—	—	—	—	—	—	—	—	—	Sk
ACANTHOCOTYLIDAE:											
Acanthocotyle borealis	—	—	—	—	—	—	—	—	—	Sk	—
A. branchialis	—	—	—	—	—	—	—	—	—	G	—
A. elegans, *A. oligoturus* and *Acanthocotyle* sp.	—	—	—	—	—	—	—	—	—	Sk*	—
A. lobianchi	—	—	—	—	—	—	—	—	—	Sk? G	—
A. monticellii	—	—	—	—	—	—	—	—	—	GF*	—
CAPSALIDAE:											
Trochopus brauni	—	—	—	Sk	—	—	—	—	—	—	—
Megalocotyle zschokkei	—	—	—	Sk	—	—	—	—	—	—	—
HEXABOTHRIIDAE:											
'*Hexabothrium appendiculatum*' [=*Erpocotyle* sp.]	—	—	—	—	—	—	—	—	—	—	G*
H. caniculae	—	—	—	—	—	—	—	—	—	—	G*
Erpocotyle abbreviata and *E. canis*	—	—	—	—	—	—	—	—	—	—	G
Neoerpocotyle catenulata	—	—	—	—	—	—	—	—	—	—	G
'*Squalonchocotyle*' *licha* [=*Neoerpocotyle*]	—	—	—	—	—	—	—	—	—	—	G*
N. grisea and '*Squalonchocotyle*' sp.	—	—	—	—	—	—	—	—	—	—	G*
Rajonchocotyle alba, *R. batis*, *R. miraletus* and *R. prenanti*	—	—	—	—	—	—	—	—	—	G*	—
Rajonchocotyloides emarginata	—	—	—	—	—	—	—	—	—	G*	—
DISCOCOTYLIDAE:											
Diplozoon paradoxum	G	G	—	G	—	—	—	—	—	—	—
DICLIDOPHORIDAE:											
Octodactylus morrhua	—	—	—	—	—	—	G*	G*	—	—	—
O. minus	—	—	—	—	—	—	G*	—	—	—	—
Diclidophora merlangi	—	—	—	—	—	—	G*	—	—	—	—
D. palmata	—	—	—	—	—	—	—	G*	—	—	—

† *Gyrodactylus elegans* and *Gyrodactylus* sp. have been recorded on the gills of the plaice and flounder in British waters; *Udonella* sp. on *Caligus* sp. on various gadoid fishes (see pp. 108, 120).

We must conclude that neither this species nor Wegener's unnamed form can be regarded as identical with *Dactylogyrus auriculatus*. At the same time it must be admitted that the distinction between *Dactylogyrus* and *Neodactylogyrus* is unreliable when applied to early and incomplete descriptions. Thus Lühe (1909) credited *Dactylogyrus falcatus* (Wedl) with a solitary transverse haptorial bar, and Price (1938*a*) retained this species in *Dactylogyrus sensu stricto*, although Nybelin (1937, Fig. 10) represented it as having two such bars, which would make it a member of the genus *Neodactylogyrus*.

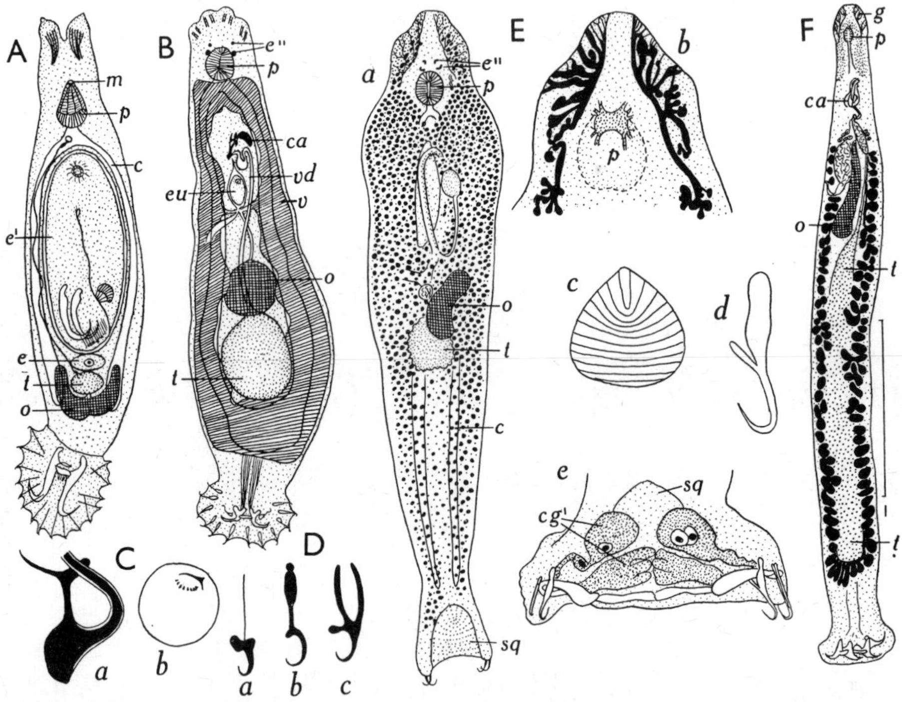

Fig. 14. Some Gyrodactyloidea. A, *Gyrodactylus*. Scheme showing the general structure. B, *Dactylogyrus*. Scheme for comparison with A. C, the copulatory apparatus of *Dactylogyrus* (*a*) and *Gyrodactylus* (*b*). D, the hooklets of *Gyrodactylus* (*a*), *Dactylogyrus* (*b*) and *Tetraonchus* (*c*). E, *Diplectanum aequans*: (*a*) the entire animal; (*b*) the anterior end of the body, showing the distribution of the cephalic glands; (*c*) a squamodisk, showing the lines along which the spines are arranged; (*d*) one of the hooks; (*e*) the posterior end of the body, showing the caudal glands, a squamodisk, the hooks and their supporting bars. F, *Amphibdelloides maccallumi*. (A–D, after Bychowsky, 1933*b*; E, after Maclaren, 1904; F, after Price, 1937*a*.)

OTHER EUROPEAN SPECIES AND THEIR HOSTS (none recorded as yet in Britain)†:

Dactylogyrus amphibothrium Wagener, 1857 (Fig. 6A). On gills of ruffe (Sweden, East Prussia, Russia).

**D. anchoratus* (Dujardin, 1845). On gills of common carp, goldfish (Poland, Austria).

D. chalcalburni Dogiel & Bychowsky, 1934. On *Chalcalburnus chalcoides* (Aral Sea).

D. cordus Nybelin, 1937. On dace and roach (Sweden).

D. crassus Kulwiéc, 1927. On common carp (Poland).

† Excluded from this list are four new species from the River Dnieper recorded by Malevitskaja (1941), namely, *Dactylogyrus bicornis*, *D. chondostomi*, *D. ramulosus* and *D. robustus*.

D. dujardinianus (Diesing, 1850). On common carp, roach, bream, etc. (Belgium, Mediterranean Sea).
D. elongatus Jacob, 1940. On *Leuciscus idus.*
**D. falcatus* (Wedl, 1857). On common carp (East Prussia, Austria, Aral Sea), bream (Sweden).
**D. fallax* Wagener, 1857. On roach, etc. (Switzerland (Lake Constance), Sweden).
D. formosus Kulwiéc, 1927. On carp, Crucian carp (Poland, Sweden).
D. intermedius Wegener, 1910. On Crucian carp (East Prussia, Poland, Russia).
D. minutus Kulwiéc, 1927. On common carp (Poland, Sweden).
**D. major* Wagener, 1857. On gudgeon (locality unspecified).
**D. siluri* Wagener, 1857. On *Siluris glandis* (Switzerland, Middle Europe).
D. similis Wegener, 1910. On roach (Lake Constance, Sweden).
**D. trigonostoma* Wagener, 1857. On roach (locality unspecified).
**D. sphyrna* Linstow, 1878. On roach (Switzerland, Sweden).
**D. tuba* Linstow, 1878. On dace and other fishes (Sweden and elsewhere in Europe).
D. uncinatus Wagener, 1857. On perch (locality unspecified).
D. vastator Nybelin, 1924. On common carp and goldfish (Sweden).
D. wegeneri Kulwiéc, 1927. On carp (Poland, Sweden).

Specific characters include the different shapes of the hooks and their supporting bars, as well as various types of cuticular structures which constitute the copulatory apparatus (see Fig. 14C, *a*). Lühe (1909) gave short diagnoses of the species marked with an asterisk.

Genus *Neodactylogyrus* Price, 1938

Syn. *Dactylogyrus* Nordmann, 1832, in part.

This genus has about forty-five species in Europe and America. Of the following twenty-one European species, of which the first is the type, none has so far been found in British waters, though some may exist there:

**Neodactylogyrus megastoma* (Wagener, 1857) (Syn. *Dactylogyrus megastoma* Wagener, 1857). On white bream and *Rhodeus* (*Cyprinus*) *amarus* (Germany).
**N. alatus* (Linstow, 1878). On white bream and bleak (Germany, Switzerland, Sweden).
N. borealis (Nybelin, 1937). On minnow (Sweden).
**N. cornu* (Linstow, 1878). On *Abramis vimba* (Germany, Middle Europe) and white bream (East Prussia, Sweden).
**N. crucifer* (Wagener, 1857). On rudd, roach, etc. (Aral Sea, Lake Constance, Lake of Lucerne, Sweden).
N. cryptomeres (Bychowsky, 1934). On carp, Crucian carp, gudgeon (Aral Sea).
**N. difformis* (Wagener, 1857). On rudd, chub, carp, etc. (Lake Constance, Sweden).
N. distinguendus (Nybelin, 1937). On white bream (Sweden).
N. fraternus (Wegener, 1910). On bleak (East Prussia, Sweden).
N. gamellus (Nybelin, 1937). On roach (Sweden).
N. gracilis (Wedl, 1861). On *Hydrocyon dentex* (Austria).
N. kulwiéci (Bychowsky, 1931). On barbels (Russia).
N. macracanthus (Wegener, 1910). On tench (East Prussia, Sweden).
**N. malleus* (Linstow, 1877). On barbels (Germany).
N. micracanthus (Nybelin, 1937). On roach (Sweden).
**N. minor* (Wagener, 1857). On bleak, white bream (Germany, Switzerland, Sweden).
**N. mollis* (Wedl, 1857). On carp (Austria).
N. nybelini (Markewitch, 1933). On *Rutilus frisii* (Russia).
N. parvus (Wegener, 1910). On chub, bleak (East Prussia, Lake Constance, Sweden).
N. suecicus (Nybelin, 1937). On roach (Sweden).
**N. tenuis* (Wedl, 1857). On perch (Austria).

Three other species, *propinquus*, *simplicimalleata* and *wunderi* [all (Bychowsky, 1931)], occur in the Aral Sea, five others, *affinis*, *chranilowi*, *frisii*, *haplogonus* and *zandti* [all (Bychowsky, 1933*a*)] in the Caspian Sea (as well as one of the foregoing); the remaining sixteen species occur in U.S.A.

Specific differences parallel those of the genus *Dactylogyrus*. Lühe (1909) wrote notes on species marked with an asterisk.

Subfamily **Tetraonchinae** Monticelli, 1903

Non-spinous Dactylogyridae with two pairs of hooks and sixteen hooklets.

KEY TO THE COMMON GENERA OF TETRAONCHINAE

I. Hooks of the opisthaptor not supported by transverse bars — *Amphibdella*

II. Hooks of the opisthaptor supported by transverse bars or bar
- A. Two transverse bars present
  - (*a*) Hooks of dorsal and ventral pairs about equal in size — *Ancyrocephalus*
  - (*b*) Hooks of the ventral pair much smaller than those of the dorsal — *Haplocleidus*
- B. One transverse bar present
  - (*a*) Dorsal hooks (or one pair) much the larger
    - (i) Opisthaptor lobed and pedunculate — *Dactylodiscus*
    - (ii) Opisthaptor not lobed and pedunculate — *Linguadactyla*
  - (*b*) Dorsal hooks not markedly the larger
    - (i) Intestine comprising a single caecum — *Tetraonchus*
    - (ii) Intestine comprising a pair of caeca — *Amphibdelloides*

Note. About fifteen genera are known outside Europe.* In North America: *Actinocleidus* Mueller, 1937; *Anchoradiscus* Mizelle, 1941; *Cleidodiscus* Mueller, 1934; *Diplectanotrema* Johnston & Tiegs, 1922; *Rhabdosynochus* Mizelle & Blatz, 1941; and *Urocleidus* Mueller, 1934 (which also occurs in Europe). In Japan: *Anchylodiscoides* Yamaguti, 1937; *Ancyrocephaloides* Yamaguti, 1938; *Haliotrema* Johnston & Tiegs, 1922; *Parancyrocephaloides* Yamaguti, 1938; and *Tetrancistrum* Goto & Kikuchi, 1917 (which also occurs in the Philippines and North America). In Australia: *Anchylodiscus* Johnston & Tiegs, 1922; *Daitreosoma* Johnston & Tiegs, 1922; *Empleurosoma* Johnston & Tiegs, 1922; and *Murraytrema* Price, 1937.

Genus *Amphibdella* Chatin, 1874

Syn. *Tetraonchus* of Monticelli, 1903.

DIAGNOSIS. *Shape:* greatly elongate, spindle-like or fusiform. *Head organs:* three pairs. *Opisthaptor:* lobed, distinctly differentiated from the rest of the body, and equipped with two pairs of large hooks of similar appearance and fourteen marginal hooklets. Transverse cuticular bars absent. *Gut:* intestine bifurcate, caeca separate and ending blindly. *Eyes:* absent. *Reproductive systems:* gonads in the anterior region of the body, ovary elongate, curved and not confined between the caeca. Vagina present. Vitellaria confined to the region behind the ootype. Parasites on the gills of the torpedo. The type species, *A. torpedinis*

* Possibly also *Aristocleidus*, *Leptocleidus*, *Onchocleidus* Mueller, 1936 and *Pterocleidus* Mueller, 1937.

Chatin, 1874, has been recorded at Naples and elsewhere in the Mediterranean, *A. flavolineata* MacCullum, 1916, from America and the Irish Sea (Rees & Llewellyn, 1941). The two species differ mainly in the shapes of the hooks which in *A. torpedinis* have a slender blade widening at the roots, and in *A. flavolineata* taper gradually from the point, and in the nature of the copulatory organ.

Genus *Amphibdelloides* Price, 1937

Syn. *Amphibdella* Chatin, 1874, in part.

DIAGNOSIS. Closely similar to *Amphibdella*, but the opisthaptor is not lobed and the hooks are supported by a single cuticular bar. The type and only known species, *Amphibdelloides maccallumi* (Johnston & Tiegs, 1922) (Syn. *Amphibdella torpedinis* Perugia & Parona, 1890, *nec* Chatin; *A. torpedinis* MacCullum, 1916, *nec* Chatin; *A. maccallumi* Johnston & Tiegs, 1922) (Fig. 14F), occurs on the gills of *Torpedo marmorata* in Italy, the torpedo in the Irish Sea (Rees & Llewellyn, 1941) and *Tetranarce occidentalis* in the U.S.A.

Genus *Tetraonchus* Diesing, 1858

Syn. *Gyrodactylus* Wedl, 1857; *Dactylogyrus* of Wagener, 1857, in part; *Monocoelium* Wegener, 1910; *Ancyrocephalus* of Lühe, 1909, in part.

DIAGNOSIS. *Head organs:* two or several pairs, cephalic glands opening on their surfaces. *Opisthaptor:* fairly clearly marked off from the rest of the body, bearing two pairs of hooks supported by a single large cuticular bar, and sixteen marginal hooklets. *Gut:* intestine single and unbranched. *Eyes:* present. *Reproductive systems:* gonads near the middle of the body. Vagina absent. The type species, *Tetraonchus monenteron* (Wagener, 1857), occurs in both Europe and America, the only other species, *T. alaskensis* Price, 1937, in Alaska.

Tetraonchus monenteron (Wagener, 1857) Diesing, 1858

Syn. *Dactylogyrus monenteron* Wagener, 1857; *Gyrodactylus cochlea* Wedl, 1857; *Ancyrocephalus monenteron* (Wagener) of Lühe, 1909.

HOST. Pike.

LOCATION. Gills.

DIAGNOSIS. *Shape and size:* body scarcely tapering posteriorly, 1–2 mm. long. *Opisthaptor:* large, broader than long. Large hooks each with a broad, plate-like base having a stump-like process and a narrow, sickle-like fang. Cuticular supporting bar flattened, dumbbell-shaped. Hooklets as shown in Fig. 14D, *c*. *Gut:* simple, sac-like. *Reproductive systems:* genital hooklets long and slender.

Note. This species has been found on fresh-water fishes in Britain and occurs elsewhere in Europe (Lake Constance, Italy) and in U.S.A. (New York).

Genus *Ancyrocephalus* Creplin, 1839

Syn. *Diplectanum* of authors; *Tetraonchus* Diesing, 1858, in part.

DIAGNOSIS. *Head organs:* generally three pairs, receiving the ducts of the cephalic glands. *Opisthaptor:* fairly distinct from the main part of the body, bearing two pairs of hooks, two cuticular supporting bars and fourteen marginal hooklets. *Eyes:* present. *Gut:* intestine bifurcate, caeca ending blindly. *Reproductive systems:* vagina present. Gonads occupy the middle of the body or a region farther back. Vitellaria generally extending into the posterior third of the body. About six European species are known, including the type, *Ancyrocephalus paradoxus* Creplin, 1839, and about twelve others in America and the Far East.

Ancyrocephalus paradoxus Creplin, 1839

Syn. *Dactylogyrus unguiculatus* Wagener, 1857; *Tetraonchus unguiculatus* (Wagener) Diesing, 1858.

HOST. Perch.

LOCATION. Gills.

DIAGNOSIS. *Shape and size:* elongate, tapering posteriorly, 3–4 mm. long. *Opisthaptor:* small, hooks with curved, sickle-like points and large basal parts, which are flat and plate-like in the larger and more knob-like in the smaller hooks. Each pair of hooks articulating with a beam-like, cuticular connecting bar. *Gut:* intestine bifurcate, caeca terminating blindly. *Reproductive systems:* genital hooklets wide and strong. This species has been found in the Irish Sea, various localities in Europe (Greifswald, Vienna, Gedani) and in Canada.

OTHER EUROPEAN SPECIES AND THEIR HOSTS:

A. bychowskii Markewitch, 1934. On gills of *Hemichromis bimaculatus* (Russia).
**A. cruciatus* (Wedl, 1857). On *Cobitis fossilis* (Austria, also in Canada).
**A. forceps* (Leuckart, 1857). On *Chondrostoma nasus* and roach.
A. vanbenedeni (Parona & Perugia, 1890). On golden mullet (Mediterranean).

Note. Lühe (1909) gave short accounts of the type and species marked with an asterisk.

Genus *Haplocleidus* Mueller, 1937

DIAGNOSIS. *Opisthaptor:* hooks of the ventral pair only about half as large as those of the dorsal. *Reproductive systems:* vagina lateral, on the left. Cirrus long and coiled in a spiral, or having a spiral fin, generally provided with an accessorial piece. *Species:* type, *Haplocleidus dispar* (Mueller, 1936) Mueller, 1937, two others (*H. affinis* Mueller, 1937 and *H. furcatus* Mueller, 1937) in America, and two more in Europe.

Haplocleidus siluri (Zandt, 1924) Price, 1937

Syn. *Ancyrocephalus siluri* Zandt, 1924; *A. vistulensis* Siwak, 1932; *Haplocleidus vistulensis* (Siwak) Price, 1937.

HOST. Catfish, *Siluris glanis* L. (in Lake Constance, Siwak's synonymous form in Poland).

LOCATION. Gills.

According to Zandt, who collected about 200 specimens from a single male fish about 40 cm. long, this trematode is 1·3 mm. long and 0·15–0·175 mm. in greatest breadth, elongate and flattened, with truncated extremities, and has an indistinct cotylophore 0·12 mm. broad. The hooks of the dorsal pair are 0·1 mm. long and have their strongly curved points directed outwards, the rounded root of each bearing a small triangular piece, those of the ventral pair 0·04 mm. long, strongly curved and having two unequal roots. The connecting bar comprises two slender rods 0·025 mm. long set at right angles and forming a V-shaped support. The hooklets are 0·016 mm. long, each having a curved point and a pyriform shaft.

Price (1937*a*) indicated that '*H. vistulensis*' differs in having twelve instead of sixteen hooklets (but suggested that the number should be fourteen in each case) and in the cuticularized nature of the vagina (which Zandt regarded as the oviduct). The suggested synonymy thus seems to be well conceived.

The only other species is *Haplocleidus monticellii* (Cognetti de Martiis, 1925) Price, 1937, which was discovered on the catfish, *Haustor catus*, in Italy.

Genus *Linguadactyla* Brinkmann, 1940

DIAGNOSIS. *Opisthaptor:* indistinctly marked off from the rest of the body, bearing two pairs of hooks, those of one pair much shorter than the other, one cuticular supporting bar and twelve (? fourteen) marginal hooklets. *Eyes:* one pair present in young forms. *Gut:* intestine bifurcate, caeca having branched lateral diverticula. *Reproductive systems:* vagina absent, genital pore median. Gonads near the middle of the body. Vitellaria not extending into the posterior third of the body. Copulatory apparatus comprising two cuticular pieces, one with a sharp point situated in a groove in the other, which is forked. The type-species, *Linguadactyla molvae* Brinkmann, 1940, was discovered on the gills of the blue ling (*Molva byrkelange* Walbaum) in Norway (Bergen?). It is of special interest because young forms only 1·2 mm. long have hooks almost as large (0·12 and 0·06 mm. long) as those of large individuals up to 6 mm. long (hooks 0·143 and 0·07 mm. long), and because it causes a proliferation of the gill epithelium to form an annulus (circular rampart) around the opisthaptor, which seems to be permanently attached to a particular site on the gills.

Genus *Dactylodiscus* Olsson, 1893, *genus inquirendum*, possibly identical with *Ancyrocephalus* (see Price, 1937*a*)

DIAGNOSIS. *Opisthaptor:* lobed, pedunculate, with two pairs of hooks (dorsal pair the larger) and a median cuticular piece of peculiar shape. *Eyes:* present. *Reproductive systems:* gonads rounded, in the middle of the body.

Note. Systematic position doubtful, because information about the cephalic glands, head organs, marginal hooklets and vagina has been excluded from descriptions. The only known species is *Dactylodiscus borealis* Olsson, 1893, which was found on the gills of the grayling and *Coregonus lavaretus* in the lakes of Scandinavia.

Subfamily **Diplectaninae** Monticelli, 1903

Syn. Lepidotreminae Johnston & Tiegs, 1922.

Dactylogyridae with a posterior region covered with scale-like spines and an opisthaptor provided with squamodisks (i.e. disks covered with concentric rows of scale-like spines), two pairs of hooks, basal supporting bars, and fourteen marginal hooklets.

Genus *Diplectanum* Diesing, 1858, emend. Price, 1937

Syn. *Dactylogyrus* of Wagener, 1857, in part; *Acleotrema* Johnston & Tiegs, 1922; *Lepidotes* Johnston & Tiegs, 1922; *Squamodiscus* Yamaguti, 1934.

Some writers identified this genus with *Ancyrocephalus*, but Price (1937*a*) regarded it as valid.

DIAGNOSIS. *Squamodisks:* concentric rows of scale-like spines without groups of accessory spine-like hooks borne upon two plates, dorsal and ventral in position respectively. *Opisthaptor:* equipped with hooks supported by three transverse cuticular bars. *Reproductive systems:* vagina present or absent. Five European species are known, and six others occur in America and the Far East. The type-species is:

Diplectanum aequans (Wagener, 1857) Diesing, 1858 (Fig. 14E, *a–e*)

Syn. *Dactylogyrus aequans* Wagener, 1857.

HOSTS. Bass (also when found at Liverpool and Piel (Barrow-in-Furness)) and *Umbrina cirrosa*.

LOCATION. Gills.

DIAGNOSIS (after Maclaren, 1904). *Size and shape:* 0·5–1·5 mm. long (1–4 mm. according to Stossich), fusiform, tapering posteriorly from the level of the genital pore, slightly flat, greatest breadth about one-quarter of the length. *Cuticle:* with numerous, small papillae. *Head glands:* a large group on either side near the anterior extremity, pores numerous, individual cells pyriform. *Other glands:* two pairs, one behind the other, near the posterior extremity, between the squamodisks. *Opisthaptor:* squamodisks (shield-like plates 0·13 mm. long and 0·15 mm. broad), two in number, on dorsal and ventral sides of posterior region, each with rows of small spines arranged transversely posteriorly, gradually becoming U-shaped rows anteriorly. *Hooks:* two on each side near the posterior extremity, lateral to the squamodisks, 0·06 mm. long (apart from strongly recurved tips), having two unequal roots, connected by a pair of transverse bars which form a joint in the median plane and curve laterally where they meet the roots of the hooks. *Gut:* mouth elongate, subterminal, ventral, with finger-like marginal lobes which are somewhat extensile, pharynx prominent and muscular, oesophagus short, caeca long and unbranched. *Sense organs:* two pairs of eye-spots, slightly in front of the mouth. *Reproductive systems:* genital pore slightly lateral (on the right) some distance behind the mouth, ejaculatory duct long and muscular, ejaculatory bulb bipartite, two prostatic vesicles present (anterior probably the seminal vesicle), penis cuticularized and very long (0·2 mm.). Testis slightly lobed, median, slightly behind the middle of the body, ovary reniform, slightly lateral, immediately in front of the testis on the left. Uterus short and straight. Receptaculum seminis small, a short straight vagina extending from it slightly forward towards the left margin of the body (? opposite side from penis) where the pore is situated. Vitellaria extensive, the follicles scattered laterally along the entire length of the body. *Eggs:* ?.

Note. Stossich found this worm to be common on the bass at Trieste. Maclaren obtained specimens from the same host-species, about one-third of the individuals examined being infested—with two to four worms per fish. A. Scott

(1904, 1906) first recorded this species in Britain, finding it 'sometimes abundant' in the Irish Sea and 'plentiful' at Piel (Barrow-in-Furness). It was mentioned also by T. Scott (1904, 1912). British specimens were relatively large (judging by figures and stated magnifications, size being consistently unspecified), but the descriptions are very brief and incomplete, as well as inaccurate in some particulars (T. Scott confounded the two extremities). The length and breadth seem to have been about 2 and 0·5 mm. respectively and the squamodisks measured about 0·13 × 0·2 mm.

OTHER EUROPEAN SPECIES AND THEIR HOSTS:

Diplectanum aculeatum Parona & Perugia, 1899. On *Corvina nigra* (Italy).
D. echeneis (Wagener, 1857). On gilt-head and Couch's sea bream (Italy).
D. pedatus (Wagener, 1857). On a wrasse (Mediterranean).
D. sciaenae (Beneden & Hesse, 1863). On meagre (English channel).

Note. Maclaren (1904) included in the historical introduction of his paper original diagnoses of the first two and the last of these species.

OTHER GENERA OF THE DIPLECTANINAE:

Lamellodiscus Johnston & Tiegs, 1922. With two species (Australia, Japan).
Lepidotrema Johnston & Tiegs, 1922, emend. Price, 1937. With five species (Australia).
Neodiplectanum Mizelle & Blatz, 1941. With one species (U.S.A.).

Family **CALCEOSTOMATIDAE** Parona & Perugia, 1890, emend. Price, 1937 (p. 68)

Subfamily **Calceostomatinae** Monticelli, 1892

Syn. Anoplodiscinae Tagliani, 1912, in part.

KEY TO THE GENERA OF CALCEOSTOMATINAE

I.	Prohaptor comprising a pair of pseudosuckers	*Anoplodiscus*
II.	Prohaptor not comprising a pair of pseudosuckers	*Calceostoma*

Genus *Anoplodiscus* Sonsino, 1890

Note. According to Price (1938*c*) this genus belongs to the Microbothriidae (subfamily Microbothriinae). Fuhrmann (1928) and Gallien (1937) placed it in the Monocotylidae (subfamily Pseudocotylinae). Johnston & Tiegs (1922) referred it to the family under consideration (subfamily Dionchinae), and Fischthal & Allison (1941), with the concurrence of Price, placed it in the Calceostomatidae emend.

DIAGNOSIS. *Prohaptor:* a pair of pseudosuckers. *Opisthaptor:* sucker-like, devoid of hooks and hooklets. *Eyes:* present. *Gut:* caeca simple, connected by a canal behind the testis. *Reproductive systems:* testis single, in front of the middle of the body. Cirrus cuticularized and with an accessory piece. Ovary in front of the testis. Vagina ventral and on the right. The type-species, *Anoplodiscus richiardii* Sonsino, 1890, occurs on the gills of the sea bream in the Mediterranean, *A. australis* Johnston, 1930 on the fins of *Sparus australis* in Australia.

Genus *Calceostoma* Beneden, 1858

DIAGNOSIS. *Anterior extremity:* expanded to form large, curled, head lappets. *Opisthaptor:* cup-like, armed or unarmed. *Gut:* caeca with numerous short diverticula. *Eyes:* present. *Reproductive systems:* testes elongate, ovary branched, vagina absent. The type-species, *Calceostoma calceostoma* (Wagener, 1857) Johnston & Tiegs, 1922 (Syn. *Dactylogyrus calceostoma* Wagener, 1857; *Calceostoma elegans* Beneden, 1858), is found on the gills of the meagre in northern Europe (and Ostend). The eyes are not mentioned in the original description, and the opisthaptor was supposed to have a single hook, but Johnston & Tiegs (1922) suggested that the second was lost or overlooked, along with cuticular accessories. '*C. elegans*' seems to have been about 10 mm. long, the head lappets taking up one-fifth of this, and the terminal, discoidal episthaptor was about one-third as broad again as the body. Of the three valid species remaining, *C. inerme* Parona & Perugia, 1889, occurs on the gills of *Corvina nigra* in Italy and *Umbrina cirrosa* as well in the Mediterranean, one in Australia and the third in Japan.

OTHER GENERA OF THE CALCEOSTOMATIDAE:

Tricotyle Manter, 1938. With one species (North America).
Acolpenteron Fischthal & Allison, 1940. With two species (North America).
Fridericianella Brandes, 1894. With one species (South America).
Anonchohaptor Mueller, 1938. With one species (North America).

Superfamily *CAPSALOIDEA* Price, 1936

Family **UDONELLIDAE** Taschenberg, 1879 (p. 71)

Genus *Udonella* Johnston, 1835

Syn. *Amphibothrium* Frey & Leuckart, 1847; *Nitzschia* Baer, 1826, in part; *Lintonia* Monticelli, 1904; *Calinella* Monticelli, 1910.

DIAGNOSIS. *Prohaptor:* a pair of small suckers or sucker-like organs. *Opisthaptor:* terminal, sucker-like, without spines, hooks or hooklets. *Gut:* pharynx unarmed, intestine simple and sac-like, but sometimes fenestrated in the region of the gonads. The type-species *Udonella caligorum* is best known, having appeared at Plymouth, in West Scotland and the North Sea and on the other side of both the Channel and the Atlantic (U.S.A. and Canada). Six forms named by Beneden & Hesse (1863) after the hosts of the *Caligus* on which they were found are possibly all synonyms of the type-species, like one unnamed species which Monticelli (1889) obtained from *Caligus* on the flounder. Price (1938*c*) could not find any significant difference between American and European specimens.

Udonella caligorum Johnston, 1835 (Fig. 15)

Syn. *Udonella lupi* Beneden & Hesse, 1863 (from wolf-fish); *U. merlucii* Beneden & Hesse, 1863 (from hake); *U. molvae* (Beneden & Hesse, 1863) (as *Pteronella*) (from ling); *U. pollachii* Beneden & Hesse, 1863 (from pollack); *U. sciaenae* Beneden & Hesse, 1863 (from shadow-fish); *U. triglae* Beneden & Hesse, 1863 (from *Trigla* sp.); *Udonella* sp. Monticelli, 1889 (from flounder); *Nitzschia papillosa* Linton, 1898; *Lintonia papillosa* of Monticelli, 1904; *Udonella socialis* Linton, 1910; *Calinella myliobati* Guberlet, 1936; *Phylline caligi* Kröyer in Beneden, 1858, and *Amphibothrium kröyeri* Frey & Leuckart, 1847.

HOSTS. *Caligus* spp. on halibut, cod, pollack, wolf-fish, common ling, hake, and flounder. Free-swimming *C. rapax* and *C. curtus*. *Anchorella uncinata* on cod. *Alebion carchariae* on *Carcharias milberti*. *Cancerilla vulgaris* on *Amphipholis* sp. *Argulus* sp. on *Neomaenis griseus*. *Alebion carchariae* on *Zygaena malleus*. *Caligus minutus* on wolf-fish. *Anchorella* sp. on shadowfish. *Caligus labracis and C. centrodonti* on ballan wrasse. (Plus other continental hosts.)

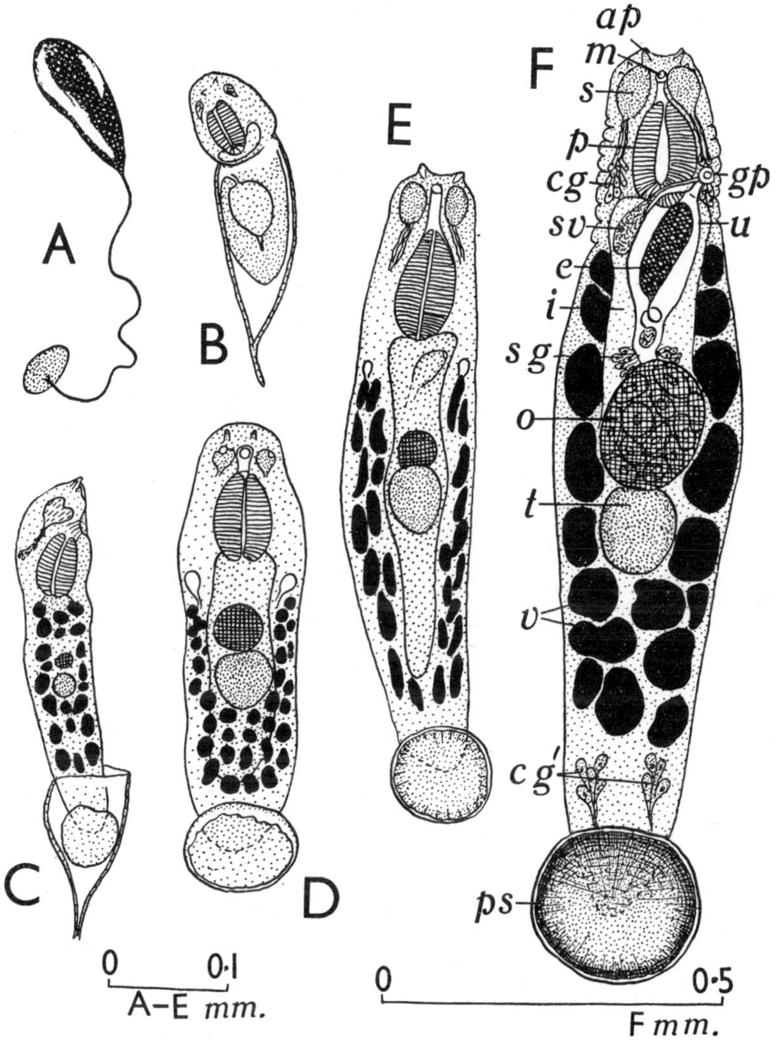

Fig. 15. *Udonella caligorum* (Udonellidae). A, egg. B, C, larva emerging from egg. D, E, young trematodes. F, adult. (After Price, 1938*c*.)

LOCATION. Mainly on the egg sacs of female copepod.

Note. The hatching period is synchronous with that of the copepod, but specimens of various sizes occur on the same egg sac.

DIAGNOSIS. *Shape and size:* elongate, almost cylindrical, 1·1–1·4 mm. long and 0·25 mm. broad. *Cuticle:* annulated anteriorly in the adult. *Prohaptor:* sucker-like, retractile, 0·04–0·06 mm. broad. *Cephalic glands:* present, their ducts leading to the prohaptor. *Opisthaptor:* sucker-like, 0·2 mm. diameter, without hooks or septa. *Gut:* mouth median, subterminal, pharynx ovoid (0·15 × 0·09 mm.), partially protrusible (then showing a papillate surface); intestine simple, sac-like, extending almost to the posterior extremity. *Eyes:* absent. *Sense papillae:* one pair, prominent anteriorly. *Excretory vesicles:* near margin of body at or near base of pharynx. *Reproductive systems:* cirrus absent. Ejaculatory duct slender, connecting with an ovoid vesicula seminalis on the right of the ootype. Testis single, median, in middle of body. Ovary globular, median, in front of testis. Vitellaria formed of few relatively large follicles extending from near the posterior extremity almost to the pharynx. Vagina absent. *Eggs:* elongate, pyriform, 0·133 × 0·042 mm., each with a long, slender filament whose tip is expanded to form an adhesive disk (Fig. 15A).

Note. The worm which escapes from the egg (Fig. 15B, C) is virtually mature, but of small size. The most conspicuous change with growth is reversal of the relative sizes of the testis and ovary; the former is invariably larger than the latter in immature stages, while the reverse is true in maturity (Fig. 15D–F). The smallest individual observed by Price (1938*c*) to escape from the egg was 0·21 mm. long and 0·057 mm. broad.

Other Genera

Genus *Echinella* Beneden & Hesse, 1863, *nec* Swains, 1840—a mollusc (*genus inquirendum* of Price, 1938*c*)

Diagnosis. *Shape:* elongate, cylindrical. *Cuticle:* annulated. *Opisthaptor:* large and sucker-like. *Pharynx:* equipped with two cuticular hooks. *Prohaptor:* absent, represented by two anterior, tentacle-like structures. *Type and only species*:

Echinella hirundinis Beneden & Hesse, 1863

Syn. *Udonella hirundinis* (Beneden & Hesse, 1863), Taschenberg, 1878; *Echinella hirudinis* (Beneden & Hesse). Monticelli, 1888.

Host. *Caligus* sp. on the yellow gurnard (found at Brest).

Diagnosis. *Size and colour:* 2–3 mm. long, pale rose. *Eggs:* of same hue as the body.

Genus *Pteronella* Beneden & Hesse, 1863 (possibly, like *Echinella*, a synonym of *Udonella*)

Diagnosis. *Shape:* elongate. *Cuticle:* annulated in immature individuals. *Anterior extremity:* with cilated ali. *Opisthaptor:* large, sucker-like. *Gut:* pharynx equipped with numerous cuticular stylets. *Type and only species:*

Pteronella molvae Beneden & Hesse, 1863

Syn. *Udonella molvae* (Beneden & Hesse, 1863), Taschenberg, 1878.

Host. *Caligus* sp. (on '*Lota molva*') (found at Brest).

Diagnosis. Of the same size as *Echinella hirundinis*, but white in colour and having dark green eggs, which are laid in clusters with the single filaments grouped together.

Family **MONOCOTYLIDAE** Taschenberg, 1879 (p. 69)

Subfamily **Monocotylinae** Gamble, 1896

Monocotylidae in which the prohaptor is a weak oral sucker, and the opisthaptor has a central loculus bounded by a ridge and seven to eight radial septa marking out peripheral loculi, as well as one pair of hooks and fourteen hooklets.

KEY TO EUROPEAN GENERA OF MONOCOTYLINAE

I.	Mouth encircled by a membranous pseudosucker	*Monocotyle*
II.	Mouth encircled by a weak oral sucker	*Heterocotyle*

Genus *Monocotyle* Taschenberg, 1878

DIAGNOSIS. *Opisthaptor:* ventral surface divided by radial septa into eight equal loculi and equipped with a pair of hooks (possibly hooklets as well). *Prohaptor:* a somewhat membranous pseudosucker, surrounding the mouth. *Eyes:* absent. *Reproductive systems:* testis single. Female genital pore marginal. Vagina and cirrus undescribed. *Type and only species:*

Monocotyle myliobatis Taschenberg, 1878

HOST. Eagle ray (recorded from Italy).

Perugia & Parona (1890) amplified the original description of this species and indicated the marginal position of the uterus which distinguishes the genus from other genera of the same subfamily. This species is 5 mm. long and 2 mm. broad, the prohaptor is 0·5 mm. diameter, the opisthaptor 1·5 mm. Two radial septa of the latter lie in the sagittal plane, the remainder being equally spaced, three on each side. The hooks are 0·46 mm. long. Each has two unequal roots and an outwardly directed point which is bent at right angles to the shaft.

It is as well to mention that some species formerly allocated to this genus were transferred by Price (1938*c*) to the genera *Heterocotyle*, *Tritestis* and *Dasybatotrema*.

Genus *Heterocotyle* T. Scott, 1904

Syn. *Monocotyle* of various writers; *Trionchus* MacCallum, 1916; *Monocotyloides* Johnston, 1934.

DIAGNOSIS. *Prohaptor:* a feeble oral sucker. *Opisthaptor:* ventral surface with a central loculus bounded by a circular septum, giving off eight radial septa to form peripheral loculi. *Eyes:* present or absent. *Reproductive systems:* testis single. Cirrus cuticularized, slender. Vagina present. *Type-species:*

Heterocotyle pastinacae T. Scott, 1904 (Fig. 16A)

Syn. ? *Monocotyle dasybatis minimus* MacCallum, 1916; *M. minima* (MacCallum) Johnston & Tiegs, 1922; *Monocotyloides minima* (MacCallum) Johnston, 1934.

HOST. Sting-ray (at Aberdeen).

LOCATION. Gills.

DIAGNOSIS. *Shape:* slightly elongate, length of body nearly $3\frac{1}{2}$ times the breadth.* *Opisthaptor:* sucker-like, slightly oval in outline, the breadth greater than the length in the ratio 13/11, margin indistinctly crenate, ventral surface

* There is a misprint or an error in Scott's observations on this point.

divided into eight loculi which extend from the periphery to near the middle, where they are interrupted by a small, diamond-shaped space representing the point of attachment of the haptor. The two most posterior loculi are slightly larger than the four anterior, but the two lateral loculi about twice as large as those immediately in front of them. Lateral loculi and the two lower ones between them, each divided by a circular band into two parts (Fig. 16A). At about the middle of the band dividing each lateral loculus from a lower one, there is a short rod which terminates in a strong spanner-shaped hook.

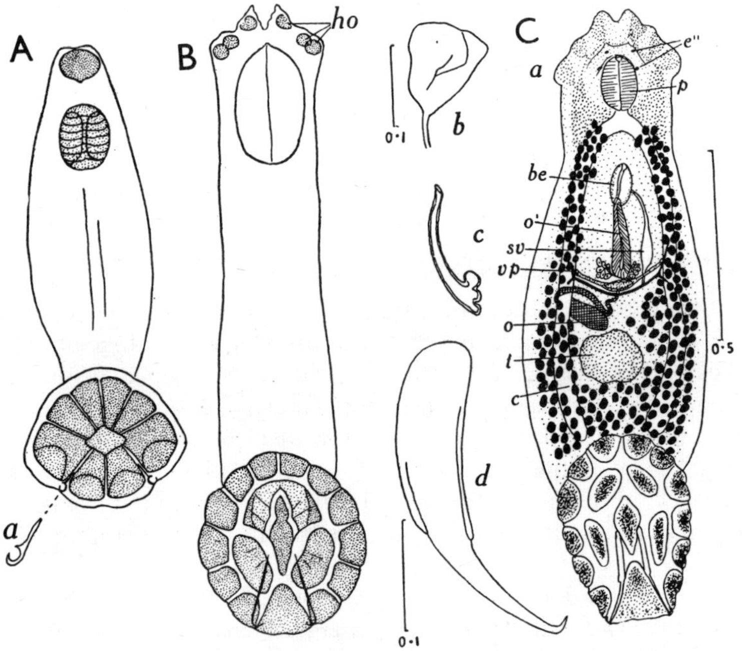

Fig. 16. Some Monocotylidae. A, *Heterocotyle pastinacae*. B, *Thaumatocotyle concinna*. C, *Thaumatocotyle dasybatis*: (*a*) the entire animal; (*b*) the egg; (*c*) the cirrus; (*d*) one of the hooks. (A, B, after T. Scott, 1904; C, after Price, 1938*c*.)

Price (1938*c*) re-examined MacCallum's specimens of *H. minima* and proposed the possible identity of this species with *H. pastinacae*. Brinkmann (1940), unaware of Price's work, reached the same conclusion after scrutiny of the original descriptions, but went further in regarding *Heterocotyle* as a synonym of *Monocotyle*. It is at present preferable to accept Price's opinion that the two genera are distinct, but future study may necessitate such a change.

OTHER SPECIES:

Heterocotyle floridana (Pratt, 1910) Price, 1938 (Gulf of Mexico).
H. minima (MacCallum, 1916) Price, 1938 (U.S.A.).
H. robusta (Johnston & Tiegs, 1922) Price, 1937 (Australia).

OTHER GENERA OF *Monocotylinae*:

Tritestis Price, 1936 (Syn. *Monocotyle* of Goto, 1894) (Fig. 6B) with one species (Japan).
Dasybatotrema Price, 1936 (Syn. *Monocotyle* of MacCallum, 1916) with one species (U.S.A.).

Subfamily **Calicotylinae** Monticelli, 1903

Monocotylidae in which the prohaptor is a feeble oral sucker and the opisthaptor is similar to that of Monocotyle, *likewise having one pair of hooks; cephalic glands opening by a pair of ducts anteriorly. Eyes absent. Testes numerous. Vagina double, as in* Merizocotyle.

Genus *Calicotyle* Diesing, 1850

Syn. *Calycotyle* Diesing, 1850; *Callicotyle* Diesing, 1858; *Callycotyle* Monticelli, 1888; *Callocotyle* of Scott, 1905; *Calycotyle* of St Remy, 1898.

DIAGNOSIS. Having the characters of the subfamily. *Species:* type, *Calicotyle kröyeri* Diesing, 1850 and five others, three European and two of these found in Britain.

Note. At this point a few comments are called for regarding the trematode *Dictyocotyle coeliaca* Nybelin, 1941, one specimen of which was found in the coelom (on the surface of the liver) of *Raia lintea* near Skagen, another in the same location in *R. radiata* at Trondhjem. This trematode bears a remarkable likeness to *Calicotyle*, but differs in the structure of the opisthaptor, the reticular nature of the testes and the branched, follicular blind termination of the ovary. The opisthaptor takes the form of a disk whose surface is divided up into a network of numerous minute loculi, and lacks hooks. The genus *Dictyocotyle* was placed in the Calicotylinae, but the structure of the opisthaptor disqualifies it for admission, unless the characters of the subfamily are drastically altered. Several speculations can be made regarding this trematode, but nothing definite can be done to define its taxonomic status. The location is very unusual and may provide the best solution of the problem which has arisen. In reaching it the trematode may have lost the normal opisthaptor, afterwards making an abortive attempt to regenerate a new one. Or, having reached the location as a larva it may have failed to develop the normal haptor. In any case it seems safe to assume (so wide is the divergence from the structure of any known opisthaptor) that *Dictyocotyle coeliaca* is an abnormal form of a species of *Calicotyle*, probably *C. kröyeri.*

Calicotyle kröyeri Diesing, 1850 (Fig. 17B)

HOSTS. Starry ray, common skate, thornback ray, spotted ray, long-nosed skate, cuckoo ray, painted ray, sandy ray, shagreen ray.

LOCATION. Cloaca (has been found in the rectum; also on the gills; Lebour, 1908*a*).

DIAGNOSIS. *Shape and size:* body oval in outline tapering anteriorly, 4–5 mm. long and almost as broad, much flattened. *Colour:* white or pale yellow. *Prohaptor:* an oral sucker about 0·7 mm. diameter. *Opisthaptor:* sucker-like, about 1·5 mm. diameter in an individual 4·5 mm. long, its ventral margin forming a delicate inturned fringe, the shafts directed medio-anteriorly and the points sharply outwards, not as shown by Lebour (1908*a*, Pl. 5, fig. 1), with seven radial septa (of which one is median and anterior) and the same number of loculi. Hooks curved (Fig. 17B1), attached to the posterior septa. *Gut:* pharynx 0·4 mm. long, oesophagus very short, caeca simple, frequently showing a smaller anterior bow and a larger posterior one, extending to (but not entering) the opisthaptor. *Excretory system:* main canals lateral to the caeca. *Reproductive systems:* common genital pore median, behind the bifurcation of the intestine. Cirrus equipped with a long cuticular piece, not spirally coiled, but bent and 9-shaped. Vaginae paired, opening ventro-laterally slightly behind the level of the genital pore, terminal parts granular. Testes numerous, in the middle of the body and between the caeca posteriorly. Ovary elongate, folded, in front of the testes on the left. Oviduct turning towards the right caecum, recurring and forming a receptaculum seminis, then receiving the vitelloduct, and joining the

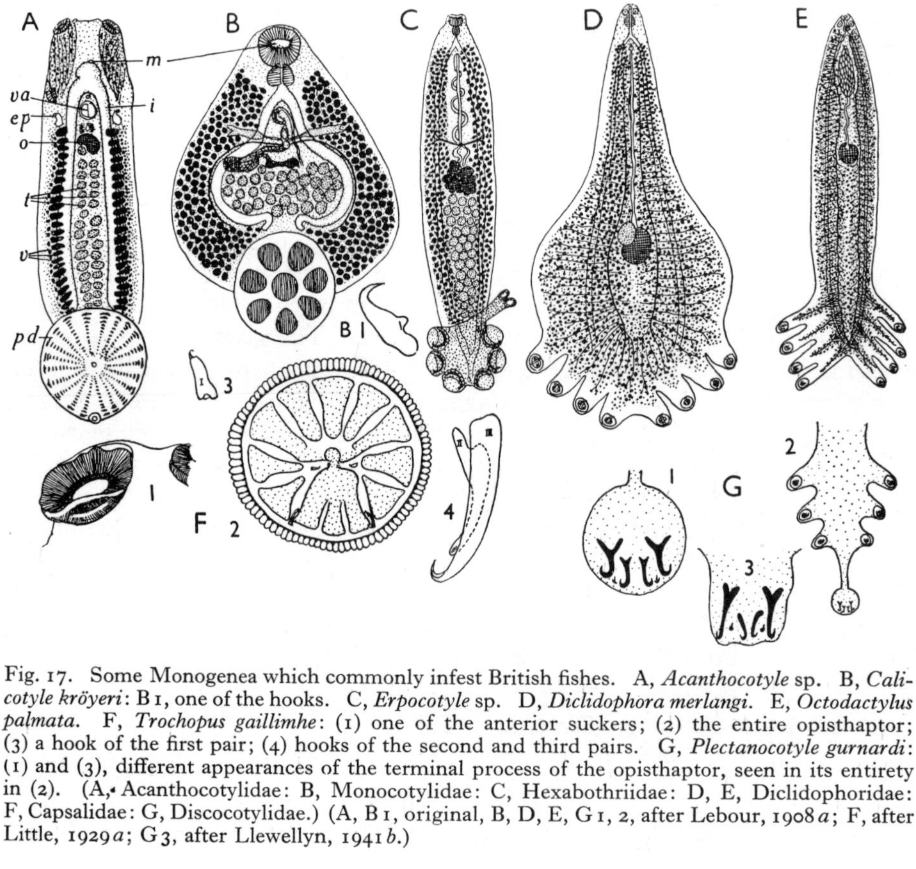

Fig. 17. Some Monogenea which commonly infest British fishes. A, *Acanthocotyle* sp. B, *Calicotyle kröyeri*: B 1, one of the hooks. C, *Erpocotyle* sp. D, *Diclidophora merlangi*. E, *Octodactylus palmata*. F, *Trochopus gaillimhe*: (1) one of the anterior suckers; (2) the entire opisthaptor; (3) a hook of the first pair; (4) hooks of the second and third pairs. G, *Plectanocotyle gurnardi*: (1) and (3), different appearances of the terminal process of the opisthaptor, seen in its entirety in (2). (A, Acanthocotylidae: B, Monocotylidae: C, Hexabothriidae: D, E, Diclidophoridae: F, Capsalidae: G, Discocotylidae.) (A, B 1, original, B, D, E, G 1, 2, after Lebour, 1908*a*; F, after Little, 1929*a*; G 3, after Llewellyn, 1941*b*.)

ootype, beyond which the very short uterus continues to the genital pore. Vitellaria well developed, follicular, lateral between the caeca and the margin of body. Eggs not numerous. (*Note*. Several writers have neglected to mention the characters of the eggs. I have seen only one, *in situ*, in the ootype. It was tetrahedral (triangular in optical section), about 0·100 mm. long, 0·110 mm. broad and had a short filament 0·035 mm. long ending in a small knob at one pole.)

This species has been found at Cullercoats, Liverpool, Galway, Plymouth and in the Irish Sea, as well as in the Kattegat, Mediterranean and Adriatic. Olsson's record of its occurrence on the turbot is probably erroneous (see Nybelin, 1941, footnote, p. 3).

Calicotyle affinis T. Scott, 1912

HOST. Rabbit fish.

LOCATION. Cloaca, even penetrating as deeply as the rectum, rarely on the skin and gills, which are not typical locations.

This species was discovered in the North Sea (Aberdeen). According to Scott, it is larger than *C. kröyeri* (length about 9 mm.) and of a somewhat different, more oval shape. The prohaptor was said to be absent, but was probably overlooked on account of its feeble musculature. Brinkmann (1940) found a number of specimens at Bergen (Norway) and redescribed the species in some detail. It becomes mature when 2·3–2·9 mm. long and attains a length of 7 mm. (to the point of attachment of the opisthaptor). A definite oral sucker is present. The main points of difference, upon which specific identity is claimed, are the oval shape of the body, the form of the opisthaptor, of which the central loculus is transversely elongate, the zoological status of the host, the spherical receptaculum seminis which receives the lateral vaginae separately. Brinkmann, reconstructing *C. kröyeri* from serial sections, found that the receptaculum seminis has two antero-lateral cornua which receive the vaginae. *C. affinis*, if a valid species, which does not seem definitely proven, is more closely related to *C. stossichi* and to the Australian form, *C. inermis* Woolcock, 1936, with which it shares the character of a long, spirally coiled cuticular piece associated with the copulatory apparatus.

OTHER SPECIES

Calicotyle stossichi Braun, 1899

HOST. *Mustelus laevis* (taken from the Mediterranean to the Berlin Aquarium).

In this species the lateral margins of the body are almost parallel, and the central loculus of the opisthaptor transversely elongate.

Two other species, *C. australis* Johnston, 1934 and *C. inermis* Woodcock, 1936, occur in Australia, the sixth and only other species, *C. mitsukurii* Goto, 1894, in Japan.

Nybelin (1941) divided the genus *Calicotyle* into three subgenera, *Calicotyle*, *Calicotyloides* and *Gymnocalicotyle*, on the basis of characters given below, making the specified allocations:

Subgenus *Calicotyle*. *Body:* broadest near the posterior end. *Opisthaptor:* having a rounded central loculus and equipped with two large hooks. *Caeca:* bent strongly inwards near their posterior ends, almost reaching the median plane, then continuing for a short distance posteriorly almost parallel to one another. *Reproductive systems:* cuticularized 'penis tube' short. Vaginae, extending almost directly outwards, opening lateral to the caeca amidst the vitelline follicles. *Species: kröyeri, mitsukurii, australis.*

Subgenus *Calicotyloides*. *Body:* tongue-shaped when extended. *Opisthaptor:* having a transversely oval central loculus and equipped with two large hooks. *Caeca:* extending posteriorly almost parallel to one another. *Reproductive systems:* cuticularized 'penis tube' long. Vaginae long and U-shaped, not overstepping the caeca laterally, opening ventral to them near the pharynx. *Species: affinis, stossichi.*

Subgenus *Gymnocalicotyle*. *Body:* broadest near the posterior end. *Opisthaptor:* having a rounded central loculus; hooks absent. *Caeca:* as in *Calicotylides*. *Reproductive systems:* cuticularized 'penis tube' long. Vaginae long, wide V-shaped, opening lateral to the caeca in front of the foremost vitelline follicles. *Species: inermis.*

Subfamily **Merizocotylinae** Johnston & Tiegs, 1922

Syn. *Anisocotyle* Monticelli, 1903, in part.

Monocotylidae without a prohaptor, but with cephalic glands opening on head organs; opisthaptor with numerous loculi, one pair of hooks and fourteen marginal hooklets. Vagina double.

KEY TO GENERA OF MERIZOCOTYLINAE

I. Opisthaptor with a central loculus, four loculi adjacent to the central and twelve to thirteen marginal loculi *Thaumatocotyle*

II. Opisthaptor with a central loculus, seven loculi adjacent to the central and eighteen marginal loculi *Merizocotyle*

Genus *Thaumatocotyle* T. Scott, 1904

Syn. *Pseudomerizocotyle* Kay, 1942.

Diagnosis. *Head organs:* three pairs, near the anterior extremity. *Opisthaptor:* according to Price (1938*c*), circular in outline, with one central loculus, four loculi adjacent to the central and thirteen marginal loculi, of which the posterior one is largest. (This does not perfectly fit Scott's figure (see below), unless we add that the anterior loculi adjacent to the central are incompletely separated.) *Eyes:* absent. *Testis:* single. *Species:* type, *T. concinna* T. Scott, 1904, and one other, *T. dasybatis* (MacCallum, 1916) (Syn. *Merizocotyle dasybatis* MacCallum, 1916; *Pseudomerizocotyle dasybatis* (MacCallum) of Kay, 1942). (*Note*. Kay (1942), overlooking the paper of Price (1938*c*) and apparently unaware of the existence of Scott's species, transferred the species of MacCallum to a new genus, *Pseudomerizocotyle*, which is unnecessary and accordingly becomes relegated to synonymity with *Thaumacotyle*.)

Thaumatocotyle concinna T. Scott, 1904 (Fig. 16B)

Host. Sting ray.

Location. Nasal fossae.

This species was discovered in the Dornoch Firth, Scotland.

Diagnosis. *Size and shape:* body elongate, about 3 mm. long and approximately one-fifth as broad, anterior extremity terminating abruptly, roughly triangular in outline, with a median indentation. *Head organs:* three pairs, ventral, arranged along the antero-lateral margins of the body. *Opisthaptor:* nearly circular in outline, the ventral surface divided up by septa into loculi in a somewhat complicated way (incompletely described by T. Scott). Thirteen peripheral loculi are bounded by a tangential septum roughly one-fifth of the diameter from the periphery and separated from one another by radial septa, the posterior loculus of the peripheral row being larger than the remainder and confluent with the central space. Central loculus lozenge-shaped, its anterior end approaching closely to but not merging with the tangential septum. Adjacent to the central loculus are three other loculi, of which the anterior is almost divided into two. Hooks moderately slender, arising from the posterior part of the septum bounding the central loculus (Fig. 16B).

Notes. Scott's figures of *Thaumatocotyle* and *Heterocotyle* are wrongly named*; the names should be interchanged. Price (1938*c*) gave the diagnostic characters of *Thaumatocotyle dasybatis* (MacCallum, 1916) (Fig. 16C, *a–d*) and suggested its possible identity with *T. concinna*. Brinkmann independently came to the same conclusion, and proposed to relegate *Thaumatocotyle* Scott to synonymity with *Merizocotyle* Cerfontaine. This proposal cannot be accepted, because Price (*loc. cit.*), whose paper Brinkmann overlooked, gave adequate reasons for recognizing the validity of both genera.

Genus *Merizocotyle* Cerfontaine, 1894

Syn. *Meristocotyle* Rossbach, 1906, a misprint.

DIAGNOSIS. *Head organs:* three pairs, near the anterior extremity. *Opisthaptor:* circular in outline, the septa on its ventral surface forming one central loculus, seven oval loculi adjacent to the central (originally stated as six, but modified by Cerfontaine, erroneously given as five in key to genera by Price, 1938*c*) and eighteen marginal or peripheral loculi. Hooks and hooklets present. *Eyes:* absent. *Testes:* single. *Type-species:*

Merizocotyle diaphana Cerfontaine, 1894, emend.

Syn. *Merizocotyle minor* Cerfontaine, 1898*b*, emend.

HOSTS. Skate, long-nosed skate.

LOCATION. Gills.

DIAGNOSIS (original description). *Size:* 6 mm. long and 1·5 mm. mean breadth. *Shape:* elongate, flattened, broadest near the middle, tapering towards the extremities, concavely truncate in front, broadening in front of the pharynx near the head organs. *Prohaptor:* three pairs of head organs. *Opisthaptor:* large and discoidal, its length about one-third that of the body, its breadth not greatly exceeding that of the body, convex dorsally, slightly concave ventrally, attached to the body by a short peduncle. Loculi as for the genus, but right (or left) anterior intermediate loculus much the smallest and posterior intermediate loculus a profoundly deep invagination almost reaching the dorsal surface of the haptor. Hooks each having a flattened basal plate and a strongly curved point. Hooklets numbering fourteen. *Gut:* mouth not quite terminal, prepharynx small, pharynx fairly large, oesophagus very short, caeca long and simple, not quite reaching the opisthaptor, lined with cubical epithelium. *Reproductive systems:* genital pore median, far behind the bifurcation of the gut, but in front of the uterus and transverse vitelloducts. Testis very large, filling the space between the caeca in the middle third of the body. Seminal vesicle in front of the testis, prostate glands well developed on the right and left of the ejaculatory bulb. Ovary elongate and looped round the right caecum, situated in an antero-

* In the description of the plates.

lateral bay of the testis. Uterus having short descending and ascending limbs. Vitellaria between the pharynx and the ends of the caeca laterally, encroaching on the median region between the pharynx and genital pore and behind the testis. Vaginal pores ventral, vaginae opening separately into a large receptaculum seminis immediately in front of the ootype. Eggs triangular, each having two rounded angles and an acute angle prolonged as a very long filament which, even when coiled, occupies most of the uterus.

Note. '*M. minor*' was found in the latter host and shows only insignificant differences which are explicable in terms of its smaller size and state of contraction, and it is here regarded as a synonym of *M. diaphana*.

Merizocotyle pugetensis Kay, 1942 occurs in America.

OTHER GENERA OF MERIZOCOTYLINAE:

Empruthotrema Johnston & Tiegs, 1922 and *Cathariotrema* Johnston & Tiegs, 1922, neither of which has been found outside U.S.A., and each of which has one species only.

OTHER SUBFAMILIES OF MONOCOTYLIDAE:

Loimoinae Price, 1936 (type and only genus, *Loimos* MacCallum, 1917) and Dionchinae Johnston & Tiegs, 1922 (type and only genus, *Dionchus* Goto, 1899). Both genera occur in America. The former genus has one species, the latter two, of which one, *D. agassizi*, occurs in the West Indies.

Family **MICROBOTHRIIDAE** Price, 1936 (p. 70)

Syn. Dermophagidae MacCallum, 1926; Labontidae MacCallum, 1927.

Subfamily **Microbothriinae** Price, 1938

Syn. *Dermophaginae* MacCallum, 1926; *Labontinae* MacCallum, 1927; *Paracotylinae* Southwell & Kirshner, 1937.

Microbothriidae with an oral sucker, or adoral pseudosuckers, and one or two testes.

KEY TO GENERA OF MICROBOTHRIINAE

I. Caeca with lateral diverticula
  A. Vagina single — *Microbothrium*
  B. Vaginae double — *Leptobothrium*

II. Caeca without lateral diverticula — *Leptocotyle*

Genus *Microbothrium* Olsson, 1868

Syn. *Dermophagus* MacCallum, 1926, *nec* Dejean, 1833; *Philura* MacCallum, 1926; *Pseudocotyle* Beneden & Hesse, 1863 in part; *Labontes* MacCallum, 1927.

DIAGNOSIS. *Prohaptor:* a pair of bothria-like grooves opening into the buccal cavity. *Opisthaptor:* an elliptical groove, markedly cuticularized. *Mouth:* not quite terminal, slit-like. *Gut:* caeca long and equipped with branched lateral diverticula. *Reproductive systems:* genital aperture median. Vagina single, not opening into the genital atrium. Cirrus long and muscular. Ductus ejaculatorius strongly cuticularized. Testis single, near the middle of the body. *Species:* type, *Microbothrium apiculatum* Olsson, 1868, and *M. centrophori* Brinkmann, 1940.

Microbothrium apiculatum Olsson, 1868 (Fig. 18E)

Syn. *Pseudocotyle apiculatum* (Olsson, 1868) Braun, 1890; *Philura ovata* MacCallum, 1926; *Dermophagus squali* MacCallum, 1926.

HOST. Piked dogfish [occurring in the Skagerrack and northern France; also the Orkneys, Canada and U.S.A. (Woods Hole)].

LOCATION. Skin.

DIAGNOSIS (after Price, 1938*c*). *Shape and size:* elliptical outline, convex dorsally, concave or flat ventrally, 1·7–3·2 mm. long and 0·7–1·6 mm. broad.

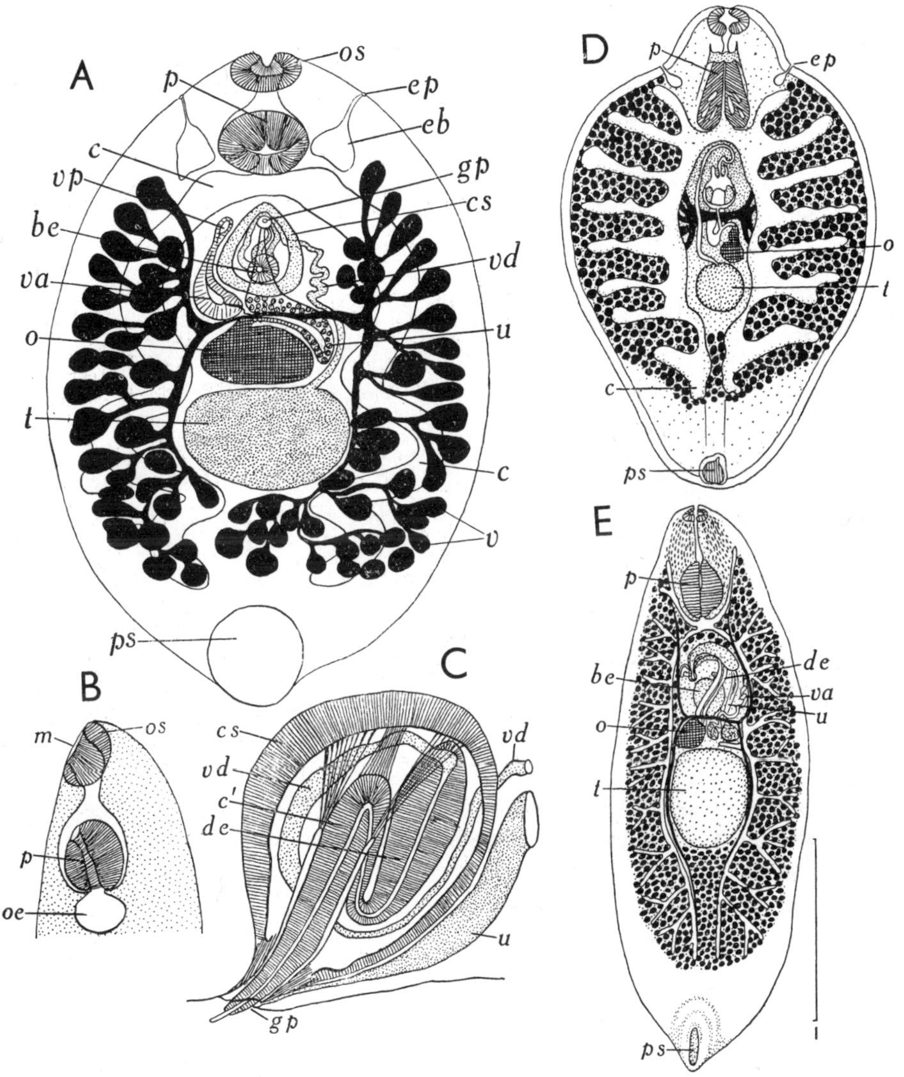

Fig. 18. Some Microbothriidae. A–C, *Leptocotyle minor*: A, scheme of the general structure (dorsal view); B, side view of the anterior end of the body; C, the terminal parts of the genital ducts (side view). D, *Leptobothrium pristiuri*. E, *Microbothrium apiculatum*. (A–C, after Johnstone, 1911; D, after Gallien, 1937; E, after Price, 1938*c*.)

Cephalic glands: present. *Prohaptor:* two sucker-like organs opening into the buccal cavity. *Opisthaptor:* oval, the opening 0·15–0·22 mm. long, terminal, lined with cuticle, unarmed. *Gut:* prepharynx relatively long, pharynx ovoid or pyriform, caeca slender and with somewhat branched lateral diverticula. *Reproductive systems:* genital pore median, about one-third of the distance along the body. Cirrus long, muscular. Genital atrium spacious. Ejaculatory duct cuticularized, sometimes expanded into a bulb. Seminal vesicle globular, at the level of the ovary. Vagina muscular, convoluted, dilated to form a receptaculum seminis between the ovary and the seminal vesicle. Vaginal pore alongside the genital pore, near the left caecum. Testis single, globular, its margins smooth or slightly indented, near the mid-body. Ovary globular, close in front of the testis and on the right. Vitellaria extending from the level of the pharynx to 0·5 mm. from the posterior extremity, merging in the median plane behind the testis. Eggs ovoid, measuring 0·130 × 0·080 mm. each having a short polar process.

Microbothrium centrophori Brinkmann, 1940

Host. *Centrophorus squamosus.*

Location. Caudal fin.

This species was discovered on fishes caught during the *Michael Sars* expedition of 1902 south of the Faröes (lat. 60° 23′ N. long. 8° 55′ W.). According to Brinkmann it is smaller than *M. apiculatum*, being 5·8 mm. long, 3·5 mm. broad and 0·75 mm. thick. It differs also in having a straight vagina, which is situated in and opens on the right side of the body, less compact and more obviously lobulated testes, and a ventrally situated vas deferens, which turns dorsally to join the ejaculatory duct. *Situs inversus* does not necessarily constitute a reliable criterion of specific identity in Trematoda, however, and it is likely that this species will ultimately prove to be synonymous with *M. apiculatum*.

Genus *Leptobothrium* Gallien, 1937

Syn. *Pseudobothrium* Gallien, 1937, *nec* Guiart, 1935.

Diagnosis. *Prohaptor:* a pseudosucker surrounding the mouth. *Opisthaptor:* small, devoid of hooks and hooklets. *Gut:* caeca with unbranched lateral diverticula. *Reproductive systems:* characters largely as in *Microbothrium*, except the vagina which is bifurcate, its branches opening into the genital atrium. *Type and only species:*

Leptobothrium pristiuri (Gallien, 1937) (Fig. 18D)

Syn. *Pseudobothrium pristiuri* Gallien, 1937

Host. Black-mouthed dogfish. (At sea, W. and N. of Ireland, lat. 53° 17′ N., long. 13° 08′ W. and 55° 55′ N., 7° 42′ W.)

Location. Skin.

Diagnosis (after Gallien, 1937). *Shape and size:* broadly lanceolate, tapering somewhat sharply towards the extremities, flattened, up to 1·6 mm. long and 1 mm. broad. *Cuticle:* thick, becoming a rigid membrane near the opisthaptor.

Prohaptor: two circular lips (pseudosuckers). *Opisthaptor:* very small and circular, 0·1 mm. diameter, but very distinct, suckerlike and unarmed. *Gut:* prepharynx short, pharynx large and pyriform, its wall containing lacunae, oesophagus practically absent, caeca diverging round the genital organs, converging without meeting just in front of the opisthaptor, having eight to nine finger-like lateral diverticula which approach the margins of the body. Intestine lined with epithelial cells that become detached and occur free in the lumen. *Reproductive systems:* genital pore close behind the bifurcation of the gut. Genital atrium spacious, receiving a large cirrus anteriorly, the uterus posteriorly and vaginae laterally. Testis large, spherical, tending to fill the space between the caeca at its level, about 0·14 mm. diameter. Vas deferens long, approaching the oesophagus to turn about before meeting the cirrus. Seminal vesicle near the ovary at the proximal end of the vas deferens. Ovary smaller than the testis, slightly in front of it and to one side. Oviduct short to its point of union with the median vitelloduct and vagina. Vitellaria comprising numerous follicles which fill up available space between the diverticula, transverse vitelloducts situated between the genital atrium and the ovary. Lateral vaginae extending from their origin with the median vagina to the genital atrium by short and narrow lateral arcs. Uterus short, straight and wide. Eggs not described.

Genus *Leptocotyle* (Monticelli, 1905) Tagliani, 1912

Syn. *Pseudocotyle* Monticelli, 1888; *Paracotyle* Johnstone, 1911; *Microbothrium* Olsson, 1868 of Baylis & Idris Jones, 1933.

Gallien (1937) and Price (1938*b*) discussed the taxonomy of this genus, correcting previous errors.

Diagnosis. *Prohaptor:* a feeble oral sucker. *Other characters:* similar to those of *Microbothrium*, but caeca without lateral diverticula. *Type and only species:*

Leptocotyle minor (Monticelli, 1888) Gallien, 1937 (Fig. 18A–C)

Syn. *Pseudocotyle minor* Monticelli, 1888; *Paracotyle caniculae* Johnstone, 1911; *Microbothrium caniculae* (Johnstone, 1911) of Baylis & Idris Jones, 1933 and Idris Jones, 1933; *Epibdella* sp. A. Scott, 1906.

Host. Rough hound (Italy as well as at Liverpool and Plymouth in this country and south of the Calf of Man).

Location. Skin, especially dorsally, on the dorsal fins, between the eye and the spiracle, etc.

Diagnosis. *Shape and size:* oval outline, much flattened, slightly pointed anteriorly, about 3 mm. long and half as broad. *Prohaptor:* an oral sucker. *Opisthaptor:* small (0·3 mm. diameter in individual 2 mm. long), circular in outline, sucker-like, devoid of septa and hooks. *Gut:* true mouth at the base of a vestibule or prepharynx, pharynx large (larger than the prohaptor and as large as the opisthaptor), oesophagus very short, caeca simple, extending along a sinuous course to within a short distance of the opisthaptor. *Reproductive systems:* genital pore median, slightly behind the bifurcation of the intestine. (*Note.* Descriptions by Johnstone (1911) and Idris Jones (1933*c*) are at variance regarding the structure of the cirrus pouch and the nature of its contents, that of the former writer seeming to be the more accurate.) Vagina muscular, its pore on the right of the genital pore. Testis single, compact, transversely ovoid, slightly behind the mid-body. Ovary compact, globular, median, slightly in front of the testis. Vitellaria lateral, the follicles spreading over the caeca. Uterus short, only slightly folded, between the ovary and the genital pore. Cirrus pouch muscular, cirrus cuticularized. Eggs few, large, ovoid (0·144 mm. long; Idris Jones (1933*c*) stated 0·0144 mm., obviously a typographical error), pointed and drawn out into a long filament at one pole. In my specimens the capsule measures 0·114–0·145 × 0·042–0·045 mm. and the filament is 0·27–0·32 mm. long.

OTHER MICROBOTHRIINAE:

The possible position of *Anoplodiscus* Sonsino, 1890 in this subfamily has been mentioned. Other genus:

Dermophthirius MacCallum, 1926, with one species, *D. carcharini* MacCallum, 1926 (U.S.A.).

Subfamily **Pseudocotylinae** Monticelli, 1903

Microbothriidae with numerous testes, without a prohaptor, with a small sucker-like opisthaptor, which is devoid of hooks, and with two vaginae which do not open into the genital atrium.

Genus *Pseudocotyle* Beneden & Hesse, 1865

DIAGNOSIS. With the characters of subfamily. *Species:* type, *Pseudocotyle squatinae* Beneden & Hesse, 1865, and *P. lepidorhini* Guiart, 1938, both recorded near our coasts.

Pseudocotyle squatinae Beneden & Hesse, 1865

HOSTS. Monk-fish and '*Squatina angelus*' (at Plymouth, west of Ireland, North Sea, at Roscoff, and at Ostend and Naples).

LOCATION. Skin.

According to the original description this species is 5 mm. long and 3 mm. broad. I have two specimens from the monkfish which have the following characters. *Shape and size:* oval outline, flattened, 4·6 mm. long and 2·6 mm. in greatest breadth (one somewhat narrower than the other, 2·0 mm. broad). *Opisthaptor:* a small sucker-like disk 0·32 mm. diameter situated on a short peduncle at the posterior extremity (or in a slight indentation of the posterior extremity). *Gut:* mouth terminal, pharynx heart-shaped (0·32 × 0·28 mm.) and situated at the base of a capacious prepharynx, oesophagus very short (virtually absent), caeca long and provided with lateral diverticula which branch at most thrice, those of the most anterior pair directed forwards and outwards lateral to the pharynx. *Excretory system:* vesicles large and with dense, parenchymatous walls which are much folded, situated postero-lateral to the pharynx. *Reproductive systems:* genital pore slightly behind the bifurcation of the intestine. Cirrus and ejaculatory duct cuticularized and situated in a parenchymatous pouch. Testes numerous (about 64), occupying an oval zone 1·8 mm. long and 0·95 mm. broad near the middle of the body. Ovary spherical, hollow, immediately in front of the testes and on one side. Receptaculum seminis diagonally behind the ovary and almost contiguous with it. Vitellaria occupying broad lateral zones 0·75 mm. wide which curve round the gonads and their ducts and extend from the pharynx almost to the opisthaptor, leaving triangular anterior and posterior spaces and an oval central space free of follicles, a narrow bridge connecting right and left masses behind the testes. Uterus very short, muscular. Vaginae extending transversely outwards immediately in front of the transverse vitelloducts, opening just in front of the origins of the latter. Egg (only one seen) of irregular ovoid shape, one side more strongly convex than the other, about 0·156 mm. long and 0·08 mm. broad.

Pseudocotyle lepidorhini Guiart, 1938, was discovered on the skin of *Lepidorhinus squamosus* off Finistère. It is said to differ from the type-species in possessing twenty-five testes and in lacking an oesophagus.

Family **ACANTHOCOTYLIDAE** Price, 1936 (p. 70)

Syn. Anisocotylidae Tagliani, 1912, in part.

Subfamily **Acanthocotylinae** Monticelli, 1903

Acanthocotylidae with a large pseudohaptor.

Diagnosis. *Prohaptor:* two retractile suckers, or clusters of openings of the cephalic glands. *Opisthaptor:* true (larval) haptor very small, bearing hooklets and situated at the posterior margin of a large, discoidal pseudohaptor having a ventral armature of spines arranged in radial rows. Testes numerous.

Note. The term '*pseudohaptor*' was devised by Price (1937*a*). That the small disk at its posterior end is really the larval opisthaptor was proved by Bonham & Guberlet (1938) for two American Acanthocotylids, *Acanthocotyle pugetensis* and *A. pacificum*.

Genus *Acanthocotyle* Monticelli, 1888 (Fig. 17A)

Syn. Various misprints: *Achantocotyle* Monticelli, 1888; *Acanthocotile* Parona & Perugia; *Acanthtcotyle* Scott, 1902.

Diagnosis. *Prohaptor:* two retractile sucker-like organs which receive the ducts of the cephalic glands. *Pseudohaptor:* large and discoidal, equipped with curved radial rows of irregular spines. *Gut:* caeca without lateral diverticula. *Species:* type, *Acanthocotyle lobianchi* Monticelli, 1888, and eight others (four European and four American), together with several unnamed forms. The type-species was discovered on the ventral surface of the thornback ray at Naples, and has been twice tentatively recorded in Britain (by T. Scott (1902) in the Clyde; and by Baylis (1939*a*) on the flapper skate in South Devon). Two other species, *A. elegans* Monticelli, 1890, and *A. oligoterus* Monticelli, 1899, seem to have appeared only at Naples, both on the skin of the thornback ray, and on opposite surfaces (dorsal and ventral respectively). The only species known definitely to have appeared in Britain is also a parasite of this host, but occurs on the gills or in the branchial chamber; it is *A. monticellii* T. Scott, 1902, with which *A. branchialis* Willem, 1906 (found on the same host off the coast of Belgium) was shown by Brinkmann (1940) to be identical. The fifth European species is *A. borealis* Brinkmann, 1940, which was discovered on the ventral surface of the starry ray near Bergen (Norway). Early descriptions of some of these species are sketchy in respect of characters which are of use in their separation, and it is possible that some European forms will ultimately be resolved into synonymy with others (e.g. *A. oligoterus* and *A. monticellii* with *A. lobianchi*), but standing for the present they can be separated by certain characters, notably the number of rows of spines on the opisthaptor and the arrangement of spines in the rows, the nature of the cuticle and the number of testes.

KEY TO THE EUROPEAN SPECIES OF *ACANTHOCOTYLE*

A. About twenty rows of spines on the pseudohaptor

I. Margin of the pseudohaptor crenate (spines totalling about 150; cuticle smooth; testes fewer than 30 (about 27)) *A. elegans*

II. Margin of the pseudohaptor not crenate

(*a*) Cuticle smooth (spines totalling more than 200) *A. monticellii*

(*b*) Cuticle bearing small papillae

(i) Spines on the pseudohaptor totalling about 150; testes fewer than 30 (about 22) *A. oligoterus*

(ii) Spines on the pseudohaptor totalling more than 200; testes more than 30 (about 36) *A. lobianchi*

B. More than 30 rows of spines (about 32) on the pseudohaptor; total number of spines more than 250 *A. borealis*

The American species of *Acanthocotyle* show somewhat similar differences. *Acanthocotyle pacifica* Guberlet, 1936 has 40–47 rows of spines on the pseudohaptor; *A. verrilli* Goto, 1899, 30–34 rows; *A. pugetensis* Guberlet, 1936 and *A. williamsi* Price, 1938, 20 rows (rarely 21). The degree of variability within a single species suggests that the number of rows of spines bears some relation to the growth of the pseudohaptor as a whole. This organ is not evident in unhatched larvae, which have a true haptor, and the large size which it attains in adults must be the result of intensive growth that must have a pronounced effect on the development and arrangement of the spines. The testes show correspondingly wide variability in some species, e.g. *A. williamsi*, which has 32–57 (Price, 1938*c*). Such differences within one and the same species far outweigh what have been regarded as specific differences in some European forms. The curious admixture of characters as between European and American species further suggests that forms so far found only on this side of the Atlantic may be identical with others found on the other side. Thus *A. borealis* is so closely similar to *A. verrilli* (as redescribed by Price, 1938*c*) that the two species cannot be separated by the key given by Brinkmann (1940). Possibly, further study will show the two forms to be identical.

Of the five European species, *A. borealis* and *A. monticellii* have a larval opisthaptor equipped with fourteen marginal and two central hooklets. Brinkmann (1940) proved that this holds good for *A. oligoterus*, which was originally credited with only fifteen hooklets. *A. elegans*, also said to have fifteen, very probably has a similar equipment of hooklets. This larval haptor is circular in all species except *A. lobianchi*, in which it is supposed to be tongue-shaped. In certain figures by Monticelli (1899, Pl. I, figs. 5, 6), however, it is shown to be circular, which is probably the characteristic shape.

The only other genus of the subfamily Acanthocotylinae is *Lophocotyle* Braun, 1896, which has one species only, namely, *L. cyclophora* Braun, 1896, discovered at Puerto Toro (South America). The main character of the genus is a large pseudohaptor devoid of radial rows of spines.

Subfamily **Enoplocotylinae** Tagliani, 1912

Acanthocotylidae without a pseudohaptor, but with a large true opisthaptor.

Genus *Enoplocotyle* Tagliani, 1912

DIAGNOSIS. *Prohaptor:* two feeble anterior suckers around which the cephalic glands open. *Opisthaptor:* large, with one pair of central hooks and fourteen marginal hooklets each of which occupies an oval loculus. *Testis:* single, slightly behind the ovary. *Type and only species:*

Enoplocotyle minima Tagliani, 1912

HOST. Murry (from the Mediterranean; not recorded in Britain).
LOCATION. Skin.

Family **CAPSALIDAE** Baird, 1853 (p. 69)

Syn. Phyllinidae Johnston, 1846; Tristomidae Cobbold, 1877; Tristomatidae Gamble, 1896; Encotyllabidae Monticelli, 1888.

Subfamily **Capsalinae** Johnston, 1929

Syn. Tristominae Braun, 1893; Tristomatinae Gamble, 1896.

Capsalidae with a sessile opisthaptor, provided with septa, and with numerous testes.

KEY TO GENERA OF CAPSALINAE

I. Pharynx constricted; testes generally overlapping the caeca laterally *Capsala*

II. Pharynx not constricted; testes confined to the region between the caeca
  A. Posterior rays of the opisthaptor bifid distally, hooks with claw-like tips; dorsal marginal spines crown-shaped, arranged in a single, longitudinal row *Capsaloides*
  B. Posterior rays of the opisthaptor not bifid distally, hooks without claw-like tips; dorsal marginal spines, when present, not crown-like, arranged in numerous short transverse rows *Tristoma*

Genus *Capsala* Bosc, 1811, emend. Price 1939

Syn. *Phylline* Oken, 1815, in part; *Tristoma* Cuvier, 1817, in part; *Tricotyle* Guiart, 1938; *Tristomella* Guiart, 1938; *Capsala* of Guiart, 1938, in part.

DIAGNOSIS. *Opisthaptor:* posterior rays not bifid distally, hooks, when present, simple and without claw-like tips. *Gut:* pharynx constricted. *Reproductive systems:* testes sometimes extending beyond the caeca laterally. *Species:* type, *Capsala martinieri* Bosc, 1811, and about twenty other species, ranging from

U.S.A. to Ceylon, Japan and New Zealand. European species: *Capsala martinieri*; *C. cutanea* (Guiart, 1938); *C. grimaldii* (Guiart, 1938); *C. interrupta* (Monticelli, 1891); *C. onchidiocotyle* (Setti, 1899); *C. pelamydis* (Taschenberg, 1878); and *C. thynni* (Guiart, 1938).

Capsala martinieri Bosc, 1811

Syn. *Phylline diodontis* Oken, 1817; *Tristoma maculatum* Rudolphi, 1819; *T. aculeatum* Couch (quoted by St Remy); *T. coccineum* of Rudolphi, 1819 (Bremser, 1824; Diesing, 1839); *Phylline coccinea* Schweigger, 1820; *Tristoma cephala* Risso, 1826; *Capsala sanguinea* Blainville, 1828; *Tristoma molae* (Blanchard) of Guiart, 1938; *T. coccineum* of Bremser, 1824; *Capsala maculata* (Rudolphi) of Nordmann in Lamarck, 1840; *C. rudolphiana* (Diesing) of Johnston, 1864; *C. cephala* (Risso) of Johnston, 1929; *C. molae* (Blanchard) of Johnston, 1929.

HOST. Sunfish (North Sea, Firth of Forth, West Scotland (Mallaig), Azores, Ireland, Norway, Sicily (Messina) and Italy (Genoa, Naples, Venice), also on east and west coasts of North America).

LOCATION. Skin.

DIAGNOSIS (after Price, 1939*a*). *Shape and size:* outline almost circular, but deeply notched posteriorly, convex dorsally, concave ventrally, 15–21 mm. long and 16–21 mm. broad. *Cuticle:* having numerous minute papillae ventrally, smooth dorsally, except near the margins of the body, where spines (mostly with four cusps) are disposed irregularly throughout a wide zone. *Prohaptor:* a pair of sucker-like organs 1·4–1·8 mm. diameter. *Opisthaptor:* discoidal, 8–10 mm. diameter, surrounded by a pleated marginal membrane 0·5 mm. wide, bearing papillae all over the ventral surface and having an irregular heptagonal central area and seven radiating septa. Hooks absent, but marginal hooklets present. *Gut:* mouth median and close behind the prohaptor, pharynx about 2 mm. long and 1–1·8 mm. broad, having a distinct constriction near the middle, intestine as in other Capsalids. *Reproductive systems:* genital pore slightly on the left of the pharynx. Cirrus pouch club-shaped. Testes very numerous, occupying the central region of the body and extending to within 1·5 mm. of the margins. Ovary lobed, 1·5–2·1 mm. long and 2–3·2 mm. broad, median, in front of the testes and 1·5–2 mm. behind the pharynx. Vitellaria comprising numerous follicles scattered through the body and extending into the cephalic lobe. Vagina slender, opening 0·4–0·45 mm. postero-lateral to the genital pore. Ootype ovoid, immediately behind the cirrus pouch. Eggs not observed.

Specimens of this species which were collected by Prof. C. M. Yonge on the skin of a sunfish at Herdla Fiord, near Bergen (Norway), and handed to me for identification outstrip the size range given above. The largest (preserved) was about 28·5 mm. long and 30 mm. broad. The opisthaptor was nearly 14 mm. diameter, the suckers of the prohaptor about 2 mm. No eggs were found in two specimens which were closely examined.

Genus *Capsaloides* Price, 1936

Syn. *Capsala* Bosc, 1811, in part; *Tristoma* Cuvier, 1817, in part; *Calsaloides* Price, 1936 (typographical error).

DIAGNOSIS. *Opisthaptor:* posterior rays bifid distally, hooks with claw-like tips. *Dorsal marginal spines:* crown-like, arranged in a single longitudinal row.

Gut: pharynx not constricted. *Testes:* confined between the caeca, in a W-shaped mass. *Species:* type, *Capsaloides cornutus* (Verill, 1875); *C. magnaspinosus* Price, 1939; *C. perugiai* (Setti, 1898) (Syn. *Tristoma perugiai* Setti, 1894); *C. sinuatus* (Goto, 1894). The first two named species occur in North America, the last named one in Japan. *C. perugiai* was recorded from the Mediterranean (Spezia) on *Tetrapterus belones*, but has not been found in Britain.

Genus *Tristoma* Cuvier, 1817

Syn. *Capsala* Bosc, 1811, in part.

DIAGNOSIS. *Opisthaptor:* posterior rays not bifid distally and hooks, when present, without claw-like tips. *Dorsal marginal spines:* when present, in numerous, short, transverse rows of similar or dissimilar appearance. *Gut:* pharynx globular or almost so, never constricted. *Testes:* numerous, confined to the region between the caeca. *Species:* type, *Tristoma coccineum* Cuvier, 1817 (Syn. *T. papillosum* Diesing, 1836; *Capsala papillosa* (Diesing) of Nordmann, in Lamarck, 1840) and three other European species only; *T. levinsenii* Monticelli, 1891; *T. integrum* Diesing, 1850 (Syn. *T. coccineum* Cuvier, 1817, in part; *T. coccineum* Cuvier of Taschenberg, 1879; *T. rotundum* Goto, 1894); *T. uncinatum* Monticelli, 1889. None of these species has been recorded in Britain. Price (1939*a*) discussed the taxonomy of the type-species, which occurs on the gills of the sword-fish in Europe and the hammerhead shark in America, and gave a short diagnosis. *Size:* 10–12 mm. long and 7–9·5 mm. broad. *Shape:* bluntly oval, convex dorsally and convex ventrally. *Cuticle:* papillate dorsally, especially behind the ovary, and bearing 43–54 rows of two to four marginal spines, the innermost having one cusp, the second and third with two to seven cusps, and the outermost comb-like. *Prohaptor:* a pair of anterior suckers 1·3–1·7 mm. diameter. *Opisthaptor:* discoidal, 1·8–2·4 mm. diameter, having a marginal membrane 0·17–0·37 mm. wide, a central heptagonal area and seven radiating septa. Hooks straight, 0·13–0·15 mm. long, hooklets 0·015 mm. *Gut:* mouth median, between and behind the anterior suckers, pharynx 1–1·3 mm. long and 1–1·5 mm. broad. *Reproductive systems:* genital pore lateral (behind the left anterior sucker), cirrus pouch club-shaped, its base behind the pharynx on the left. Testes numerous, but confined between the intestinal crura. Ovary lobed, 0·7–0·8 mm. long and 1·1–1·6 mm. broad, median and 0·5–0·7 mm. behind the pharynx. Vitelline follicles scattered mainly lateral to the gut, some along its course. Vagina slender, its base expanded forming a seminal receptacle, its aperture about 1 mm. behind the genital pore. Ootype ovoid and on the left. Metraterm slender. Eggs large (0·114 × 0·095 mm.), more or less triangular, each having four prolongations. For figures see Price (1939*a*).

Tristoma integrum occurs on the sword fish in the Mediterranean (at Genoa, Naples, Messina) and Adriatic (at Venice), also in U.S.A. and Japan. According to Price (1939*a*) it differs from *T. coccineum* in the structure and arrangement of the dorsal marginal spines, which are numerous (more than 300 on each side of the body) and similar in individual rows. See Taschenberg (1879, Pl. 1, figs. 1–2) for two excellent figures, which were merged into one admirable illustration by Braun (1879–93, Pl. 8, fig. 1).

Subfamily **Trochopodinae** Price, 1936

Capsalidae with a sessile, sucker-like opisthaptor having septa, but with only two testes.

KEY TO GENERA OF TROCHOPODINAE
(Forms in which prohaptor is a pair of anterior suckers)

I. Opisthaptor with ten septa *Trochopus*
II. Opisthaptor with six or seven septa *Megalocotyle*

Note. Price (1939*a*) included in this subfamily one other genus, *Macrophyllida* Johnston, 1929, in which the prohaptor takes the form of glandular areas, not suckers. It is represented by one Australian species, *M. antarctica* (Kent Hughes, 1928) Johnston, 1929.

Genus *Trochopus* Diesing, 1850

Syn. *Capsala* of Nordmann; *Tristoma* of Diesing; *Placunella* Beneden & Hesse, 1863, in part.

DIAGNOSIS. *Prohaptor:* a pair of disk-like organs of medium size, varying in form and appearance. *Opisthaptor:* discoidal, scarcely pedunculate, its ventral surface divided into loculi by ten septa and armed with two to three pairs of hooks and fourteen marginal hooklets, the margin having a somewhat plicate membrane. *Gut:* caeca with branched lateral diverticula. *Reproductive systems:* genital pore far forward, at the level of the left anterior sucker. Genital atrium long. Cirrus pouch curved, but not generally crossing the median plane. Testes two in number, ovoid, side by side near the middle of the body. Ovary globular, in front of the testes. Vagina slender, opening immediately behind the common genital pore. *Species:* type, *Trochopus tubiporus* Diesing, 1836 (Syn. *T. longipes* Diesing, 1850) (recorded in Irish waters, the Adriatic (Trieste) and at Brest), and eleven other species, ranging from the British Isles to the Mediterranean, Australia and the Pacific.

Trochopus gaillimhe Little, 1929 (Fig. 17F)

HOST. Yellow gurnard (at Galway).

LOCATION. Gills.

DIAGNOSIS. *Shape, size and colour:* oval outline, flattened, up to 5·85 mm. long and 2·8 mm. broad, milky white, but fairly transparent when extended, and roseate. *Prohaptor:* a pair of deep saucer-like anterior suckers (Fig. 17F 1), each made up of two zones: (*a*) an outer, delicate, semi-transparent, membranous frill, in which numerous dark mucus glands are arranged in radial rows, (*b*) an inner, firmer zone of tissue which is largely muscular, consisting of dorsal and ventral muscles as well as dorso-ventral ones. *Opisthaptor:* discoidal, sucker-like, 1·8 mm. in diameter, having ten radial septa arranged as follows: four anterior ones form one group, three postero-lateral ones form an additional group on each side, the whole forming a horse-shoe figure with the open end posteriorly; in this space are two additional short septa (on right and left respectively) which traverse only half the radius of the haptor (Fig. 17F 2). Peripherally there is a pleated frill, the scallops of which overlap slightly, except posteriorly. Hooks numbering three pairs. First pair 0·1 mm. long, finger-shaped with a small claw, situated at the junction of the three postero-lateral septa on each side, with the points directed towards the centre of the haptor. Second and third pairs much stouter (0·30–0·35 mm. long), close together near the posterior margin of the haptor, on the hindmost septa of the postero-lateral set. Second pair of hooks having four or five rows of serrated ridges distally, third pair of hooks having sharply recurved tips. Adductor and abductor muscles inserted in the main shaft of the second and third pairs of hooks, serve to bring them together and to divaricate them. Third pair of hooks also provided with flexor and extensor muscles.

Trochopus brauni Mola, 1912 (European, but not recorded in Britain)

HOST. Miller's thumb.

LOCATION. Gills.

DIAGNOSIS. *Shape, size and colour:* ellipsoidal, 2·3 mm. long and 0·85 mm. broad, milky white with a slight golden tinge. *Prohaptor:* a pair of anterior suckers 0·3 mm. diameter. *Opisthaptor:* 0·85 mm. diameter, having a marginal membrane and ten muscular septa (arrangement incompletely described) as well as three pairs of hooks. Hooks of the first pair (using Little's terminology, not Mola's) 0·07 mm. long; of the second pair 0·11 mm. long and having recurved tips (like those of first pair); of the third pair 0·17 mm. long, having the tips turned posteriorly. Hooks of the second and third pairs fit together as in *T. gaillimhe.*

Trochopus lineatus T. Scott, 1901

HOST. Streaked gurnard (recorded in the Clyde).

LOCATION. Gill cavities.

This species was said to be one-eighth of an inch long and broad oval in outline. The original description is inadequate, revealing little more than the presence of twelve septa on the opisthaptor, spaced at regular intervals, the two posterior ones with their inner ends free and the remaining ten joined to a central nearly circular ring-like septum. Only one hook on each side is mentioned, situated on the posterior septa. *Eyes:* four, distinct.

Note. It is possible that *T. gaillimhe* is identical with this species, though the formulation of a reliable opinion is impossible without re-examination of Scott's specimens or other similar forms from the streaked gurnard.

OTHER EUROPEAN SPECIES AND THEIR HOSTS:

Trochopus tubiporus (Diesing, 1836). On yellow and red gurnards and black sea bream (Britain, Italy, north-west France.)

T. differens Sonsino, 1891. On black sea bream (Mediterranean).

T. diplacanthus Massa, 1903 (= *Placunella pini* of A. Scott, 1901). On yellow gurnard (Irish Sea and Mediterranean).

T. heteracanthus Massa, 1903. On yellow gurnard (Italy).

T. micracanthus Massa, 1903. On yellow gurnard (Italy).

T. onchacanthus Massa, 1906. Host unknown (Italy).

T. pini (Beneden & Hesse, 1863). On red and yellow gurnards (Italy, north-west France, English Channel).

Genus *Megalocotyle* Folda, 1928

Syn. *Trochopus* of Massa, 1903 and Mola, 1912; in part; *Placunella* Beneden & Hesse, 1863, in part.

DIAGNOSIS. Opisthaptor with six or seven septa; other characters as in *Trochopus*. *Species:* type, *Megalocotyle marginata* Folda, 1928 (in North America) and several others, viz. *M. hexacantha* (Parona & Perugia, 1889) (Syn. *Placunella hexacanthus* Parona & Perugia, 1889; *Trochopus hexacanthus* (Parona & Perugia) Massa, 1906) on gills of *Trigla corax* (Italy); *M. rhombi* (Beneden & Hesse, 1863) (Syn. *Placunella rhombi* Beneden & Hesse, 1863, *Trochopus rhombi* (Beneden & Hesse) Massa, 1903) on turbot (Belgium, France); *M. squatinae* (MacCallum, 1921) on gills of monk-fish (Singapore); *M. zschokkei* (Mola, 1912) on gills of miller's thumb (Italy).

Megalocotyle zschokkei (Mola, 1912) Price, 1939

Syn. *Trochopus zschokkei* Mola, 1912.

DIAGNOSIS. *Shape, size and colour:* ellipsoidal, tapering at the extremities, 2·65 mm. long, pale rose colour. *Prohaptor:* two anterior suckers, cup-like, close together, 0·35 mm. diameter. *Opisthaptor:* discoidal, 0·9 mm. broad, having seven septa which merge. Three pairs of hooks situated on the posterior two septa are close together and similar, except in size; outer hook 0·24 mm. long; middle hook 0·17 mm. long; inner hook 0·06 mm. long.

Note. Mola's figure is small and inadequate, yet the hooks shown bear a close general resemblance to those of *T. gaillimhe*.

OTHER GENUS OF TROCHOPODINAE:

Macrophyllida (Kent Hughes, 1928) Johnston, 1929, with one species in Australia.

Subfamily **Nitzschiinae** Johnston, 1931

Capsalidae with a more or less sessile opisthaptor, without septa or hooks, and with numerous testes.

Genus *Nitzschia* Baer, 1826, *nec* Denny, 1942 (an insect); *nec* Beneden, 1858 (*Entobdella*); *nec* Linton, 1898 (*Udonella*)

Syn. *Hirudo* Abildgaard, 1794, and misprints such as *Nitychia* and *Nityschia* Nordmann, 1833; *Nitschia* Haswell, 1892; *Nitychia* Beneden, 1858.

DIAGNOSIS. *Prohaptor:* two sucker-like grooves, one on either side of the cephalic lobe. *Opisthaptor:* sucker-like, having a well-developed marginal membrane, the ventral surface markedly concave, without septa or papillae, but with three pairs of hooks and fourteen marginal hooklets. *Reproductive systems:* genital aperture median, or on the left, behind the pharynx. Testes numbering up to twenty-seven (or more), confined to the region between the caeca. Ovary rounded. Vagina present. *Species:* type, *Nitzschia sturionis* (Abildgaard, 1794) Kröyer, 1852 (Syn. *N. elegans* Baer, 1826; *N. elongata* of Monticelli, 1909; *Tristoma elongatum* of Nitzsch, 1826; *T. sturionis* (Abildgaard) of Cuvier, and of Blainville; *T. elegans* (Baer) of Beneden & Hesse, 1863), and *Nitzschia superba* MacCallum, 1921. The latter species was found on *Acipenser brevirostrum* of the

Atlantic coast of America; the former occurs on both sides of the Atlantic as a parasite of sturgeons, and has been found in the North Sea, Scotland, the Baltic and Italy. Price (1939*a*) proposed the erection of a third species, *Nitzschia monticellii*, for the 'forma giovani' of '*N. elongata*', which has hooks of unequal length, claiming that two species were represented in Monticelli's specimens.

Subfamily **Benedeniinae** Johnston, 1931

Syn. Capsalidae Baird, 1853, in part; Ancyrocotylinae Monticelli, 1903; Tristominae Braun, 1893 and of Monticelli, 1892, in part.

Capsalidae with a more or less sessile opisthaptor, without septa and hooks, and with only two testes.

KEY TO GENERA OF BENEDENIINAE

I. Prohaptor two sucker-like disks; anterior adhesive organs absent
  A. Caeca with diverticula, much branched — *Benedenia*
  B. Caeca simple, unbranched, not united posteriorly — *Ancyrocotyle*

II. Prohaptor not sucker-like disks; anterior adhesive organs present, opening on a bilobed anterior fold — *Entobdella*

Genus *Benedenia* Diesing, 1858, *nec* Schneider, 1875—a protozoon

Syn. *Epibdella* Beneden, 1856, in part; *Phylline* Oken, 1875, *nec* Abildgaard, 1790, in part; *Tristoma* Cuvier, 1817, in part.

DIAGNOSIS. *Prohaptor:* a pair of sucker-like disks. *Opisthaptor:* sucker-like, lacking septa, but with three pairs of dissimilar hooks and fourteen marginal hooklets. *Reproductive systems:* testes two in number, side by side. Ovary slightly in front of the testes and not separated from them by a belt of vitelline follicles. Vagina present or absent. *Species:* type, *Benedenia sciaenae* (Beneden, 1856) (Syn. *Epibdella sciaenae* Beneden, 1856; *Benedenia elegans* Diesing, 1858; *Phylline sciaenae* (Beneden) Sonsino, 1891; *Tristoma sciaenae* (Beneden), Taschenberg, 1878) and about nineteen others (see Price, 1939*a*), only one of which has been recorded in Europe. This is *Benedenia monticellii* (Parona & Perugia, 1895) (Syn. *Phylline monticellii* Parona & Perugia, 1895; *Epibdella monticellii* of Parona, 1896). The remaining species range from eastern U.S.A. to California, Mexico, Chili and the Galapagos Isles, one occurs in Ceylon and several others in Japan. The type-species has been found on the meagre at Ostend and in Italy, *Benedenia monticellii* on the golden mullet in the Adriatic (Trieste). It is hardly likely that the latter would be encountered in British waters, and the former could be recognized tentatively from the generic diagnosis. According to Beneden (1858) it is 24 mm. long and 12 mm. broad, the opisthaptor 5 mm. diameter.

Genus *Entobdella* Blainville, in Lamarck, 1818 (see Sherborn's *Index Animalium*)

Syn. *Epibdella* Blainville, 1828; *Ertopdella* Rathke, 1843; *Phylline* Oken, 1815, *nec* Abildgaard, 1790; *Phyllonella* Beneden & Hesse, 1863.

DIAGNOSIS. Similar to *Benedenia*, but the prohaptor not comprising anterior suckers or sucker-like disks, instead being represented by a bilobed anterior fold bearing the openings of the cephalic glands. *Species:* type, *Entobdella hippoglossi*

(O. F. Müller, 1776) and five others, two of which occur in America. The other three are *E. diadema* (Monticelli, 1902) (Syn. *Epibdella diadema* Monticelli, 1902; *Phylline diadema* (Monticelli) Linstow, 1903), which occurs on *Trygon violacea* in the Mediterranean, *E. solae* (Beneden & Hesse, 1863), which, like the type-species, occurs in Britain, and *E. steingröveri* (Cohn, 1916), which was discovered on an unidentified fish.

Entobdella hippoglossi (O. F. Müller, 1776) Johnston, 1856

Syn. *Hirudo hippoglossi* O. F. Müller, 1776; *Phylline hippoglossi* (O. F. Müller) Oken, 1815; *Epibdella hippoglossi* (O. F. Müller) Blainville, 1828 and others; *Tristoma hamatum* Rathke, 1843; *Capsala elongata* Baird, 1850, in part; *Nitzschia hippoglossi* (O. F. Müller) Taschenberg, 1878; *Phyllonella hippoglossi* (O. F. Müller) Goto, 1899, Pratt, 1900 and MacCallum, 1927; *Epibdella bumpusii* of Canavan, 1934.

Host. Halibut (T. Scott (1901) found it plentiful on this host at Aberdeen).

Location. Skin, all over the body (Scott).

Price (1939*a*) found that American specimens conform with those described by earlier writers.

Diagnosis. *Shape and size:* oval or elliptical, much flattened, 13–18 mm. long and 3·6–4·8 mm. broad, according to some writers as much as 24 mm. long and nearly half as broad. (I have four specimens which measure about 10–15 × 7–8·5 mm.) Cephalic lobe marked off from the rest of the body by a slight marginal constriction. *Prohaptor:* two elongate, slightly sunken, glandular areas, one on each side of the median plane near the anterior end of the cephalic lobe. *Opisthaptor:* sucker-like, 3·6–4·8 mm. diameter (this probably exaggerates the size, the diameter being much less than the maximum breadth of the body, not equal to it; in my specimens the shape is a broad egg shape, the dimensions about 4·5–7·5 × 4·5–6·5 mm.), surrounded by a membrane rather less than 0·2 mm. wide. Ventral surface concave, its posterior half covered with radiating rows of small tubercles or papillae, and equipped with three pairs of hooks and fourteen marginal hooklets. Hooks of the first pair short, spearhead-like, 0·5–0·65 mm. long; hooks of the second pair slender, 0·8–0·95 mm. long and with recurved tips; hooks of the third pair similar to those of the second but smaller, 0·09–0·12 mm. long. Marginal hooklets about 0·02 mm. long. *Gut:* mouth ventral, between the hinder ends of the cephalic lobe, pharynx large, wider than long, oesophagus very short, caeca with numerous branched diverticula, mainly lateral, united posteriorly(?). *Reproductive systems:* common genital pore far forward, lateral, situated at the posterior end of the left cephalic lobe. Cirrus pouch club-shaped, its base on the right of the median plane, midway between the pharynx and the ovary. Testes globular, side by side in the middle of the body. Ovary transversely ovoid, in front of the testes and separated from them by a broad belt of vitelline follicles. Vitellaria occupying almost the entire body from the mouth to the posterior extremity. Vagina slender, opening behind the genital pore and nearer the median plane. Ootype oval, median, behind the cirrus pouch. Metraterm slender. Eggs tetrahedral, about 0·228 mm. wide, each provided with a long, slender filament.

Note. This species has been found in the North Sea, the Skagerrack, Denmark, the Arctic Ocean and U.S.A., Alaska, Canada and Greenland.

Entobdella soleae (Beneden & Hesse, 1863), Johnston, 1929 (Fig. 19A–C)

Syn. *Phyllonella soleae* Beneden & Hesse, 1863 and of various other authors—Vogt, 1878, 1879; Monticelli, 1888; St Remy, 1892; Perrier, 1897; MacCallum, 1927; Little, 1929; *Epibdella soleae* (Beneden & Hesse, 1863) Monticelli, 1892 and 1902, also of Odhner, 1905 and Nicoll, 1915; *Epibdella producta* Linstow, 1903 (see Odhner, 1906).

HOSTS. Common sole and sand sole (in 60 out of 100 common soles at Galway (Little, 1929)); 57 % common sole in Cardigan Bay; found in the Clyde, off the Lancashire coast, off the Belgian coast and at Brest.

LOCATION. Skin.

DIAGNOSIS. *Shape, colour and size:* oval in outline, flattened, translucent, pale yellow (milky white after death), 2–6 mm. long and 1–3 mm. broad, length

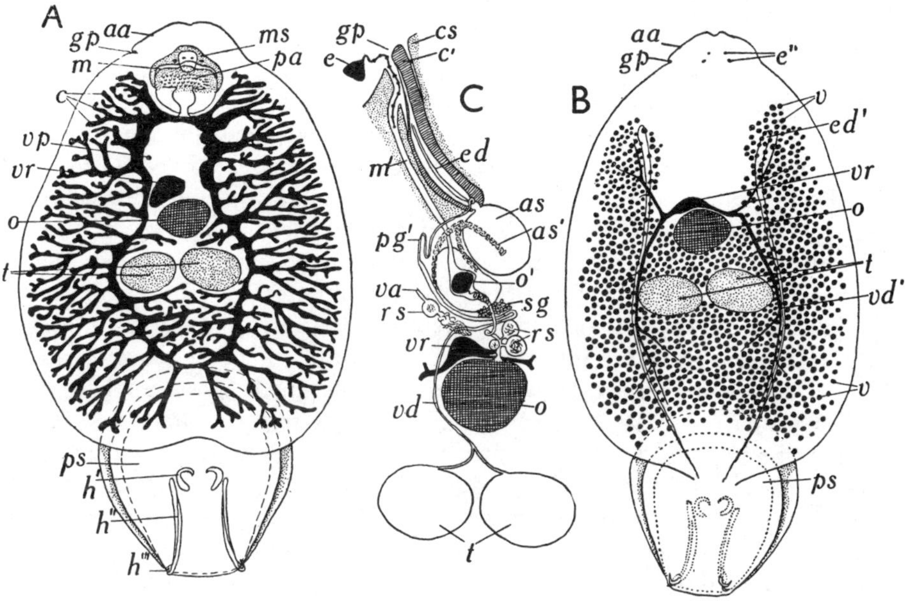

Fig. 19. *Entobdella soleae* (Capsalidae). A, B, entire trematode seen in dorsal view. In A the intestine is represented in solid black. In B the usual convention is followed, vitellaria being shown in black. C, schematic diagram of the reproductive systems. (After Little, 1929c.)

generally slightly greater than twice the maximum breadth (slightly behind the mid-body, opposite the testes). Anterior end having a well-marked cephalic lobe, which is slightly indented in the middle of the anterior border. *Eyes:* two pairs, one behind the other slightly in front of the mouth. Anterior eyes 0·14 mm. apart and 0·37 mm. from the anterior extremity. Posterior eyes 0·26 mm. apart. *Prohaptor:* two lateral adhesive areas along the margins of the cephalic lobe, suckers absent. *Opisthaptor:* sucker-like, the anterior half semicircular in outline, the posterior half slightly extended, then abruptly truncated, 1·5 mm. long and 1·4 mm. broad in an individual 5·9 mm. long. Also having two rays extending along the course of the hooks of the second and third pairs and converging on a point slightly in front of the centre of the sucker, connected by a transverse ray between the hooks of the first pair; also a ridge of inverted horseshoe-shape slightly in front of the others (possibly indicating the position at which the haptor is attached to the body). *Hooks:* three pairs (not two as stated by Beneden & Hesse (1864), T. Scott (1901) & Cunningham (1890)), dissimilar; hooks of the first pair 0·28 mm. long, with a finely bifid base and recurved point, half imbedded in the tissues; hooks of the second pair 0·65 mm. long, only the recurved tip protruding from the tissues, composed of dentine capped with enamel; hooks of the third pair 0·12 mm. long, lateral and close to the recurved tips of the hooks of the second pair, each with a flat base imbedded in the tissues and a minute recurved tip exposed (*not* with two rod-like shafts as in '*Epibdella producta*' Linstow, 1904). *Gut:* mouth ventral, 0·4 mm. from the anterior extremity, a wide, transverse, crescentic slit bordered by anterior and posterior lips and overhung by a dome-like sheath. Pharynx capacious, 0·7 mm. wide, the lining having numerous multicellular papillae with their apices projecting into the lumen and filled with granular secretory materials. Oesophagus absent, or exceedingly short. Two caeca uniting posteriorly slightly in front of the opisthaptor, with branched lateral diverticula in the anterior half of the body, and both median and lateral branched diverticula in the posterior region; the whole intestine lined with large amoeboid cells having prominent pseudopodia. *Excretory system:* two main lateral ducts, lined with cuticle, and dilated (sometimes bifurcate) anteriorly, extending from the anterior end of the vitellaria to the opisthaptor, where they approach the median plane. *Reproductive systems:* common genital pore far forward in the body at the posterior end of the cephalic lobe on the left side. Vaginal pore far behind the genital pore, midway between the ovary and the bifurcation of the intestine, slightly on the left. Testes ovoid, side by side and almost in contact, between the caeca just in the posterior half of the body. Ovary irregularly ovoid, median, slightly in front of the testes. Cirrus pouch present, cirrus long and contractile. Accessory sac (function unknown) communicating with the ejaculatory duct (Fig. 19C, *as*). Vitellaria very well developed, their follicles dispersed dorsal and ventral to the caeca and diverticula, absent only in the median region in front of the ovary. Vitelloducts extending anteriorly and posteriorly, vitelline reservoir prominent, slightly in front of the ovary. Uterus short, metraterm long (1 mm.), and parallel with the cirrus. Vagina narrow, associated with a spherical receptaculum seminis containing sperms. Eggs pyramidal, the apex of each being directed posteriorly and formed into a slender, twisted, nodulated filament. Capsules dark yellow, large, 0·143 × 0·143 mm.; filament 0·44 mm. long. Not more than two eggs have been observed in a single individual, one in the ootype, the other in the metraterm.

Genus *Ancyrocotyle* Parona & Monticelli, 1903

Syn. *Placunella* Beneden & Hesse, 1863, in part.

DIAGNOSIS. *Prohaptor:* two rounded suckers situated on the antero-ventral surface upon large, lateral, almost triangular, fleshy pads. *Opisthaptor:* cup-shaped or discoidal, almost sessile, provided with a marginal membrane and three pairs of unequal hooks as well as fourteen marginal hooklets, third pair of hooks lateral to hooks of second pair. *Eyes:* four in number. *Gut:* pharynx simple, caeca wide and without diverticula. *Reproductive systems:* common genital pore ventral, on the left of the pharynx. Vagina short, opening near the median plane midway between the pharynx and the ovary. Two testes situated side by side in the middle of the body, behind the ovary; or a single testis situated in front of the ovary. *Species:* type, *Ancyrocotyle vallei* (Parona & Perugia, 1895) Parona & Monticelli, 1903 and one other, *A. bartschi* Price, 1934 (from Puerto Rica). *A. vallei* (otherwise known as *Placunella*) has been recorded on the gills of the pilot fish, *Naucrates ductor*, in the Adriatic and Mediterranean, but has not appeared in Britain. *Ancyrocotyle bartschi* occurs on the same host and may be identical with *A. vallei*, but it was based on two immature specimens and shows certain differences.

OTHER GENUS OF BENEDENIINAE:

Pseudobenedenia Johnston, 1931, with one species in New Zealand.

Subfamily **Encotyllabinae** Monticelli, 1892

Syn. Encotyllabidae Monticelli, 1888.

Capsalidae having a pedunculate opisthaptor arising from the postero-ventral surface of the body.

Genus *Encotyllabe* Diesing, 1850 (Fig. 60)

Syn. *Cheloniella* Beneden & Hesse, 1863; *Tristoma* of Taschenberg, 1878 and misprints like *Encotylabe* Gamble, 1896 and *Encotyllahe* Monticelli, 1892.

DIAGNOSIS. *Body:* with thin lateral margins which are turned ventrally. *Prohaptor:* a pair of rounded or elliptical, muscular suckers each having a pleated marginal membrane and projecting somewhat from the antero-lateral margins of the body. *Opisthaptor:* a campanulate disk which is attached to the postero-ventral surface of the body by a long peduncle and is devoid of radial septa, but is equipped with two pairs of hooks (those of one pair very large) and a number of marginal hooklets (possibly fourteen). *Reproductive systems:* genital pore lateral, otherwise similar to *Trochopus* and *Benedenia*. *Species:* type, *Encotyllabe nordmanni* Diesing, 1850 (Syn. *Tristoma excavatum* of Nordmann, quoted '*in litt.*' by Diesing; *T. nordmanni* (Diesing) Taschenberg, 1878; *Encotyllabe normanni* of Braun, 1890) and seven others which range from Britain, Belgium and Italy to Japan, Australia and the Galapagos Isles.

EUROPEAN SPECIES OF *ENCOTYLLABE* AND THEIR HOSTS

Encotyllabe nordmanni (black sea bream) (pharynx) (Fig. 6C).
E. pagelli Beneden & Hesse, 1863 (Syn. *Tristoma pagelli* (Beneden & Hesse) Taschenberg, 1878) (common sea bream) (gills). Found in Ireland and at Brest.
E. paronae Monticelli, 1907 (*Crenilabrus pavo*) (Italy).
E. vallei Monticelli, 1907 (gilt-head) (Italy).

Suborder POLYOPISTHOCOTYLEA Odhner, 1912

Syn. Polycotyla Blainville, 1828; Octobothrii E. Blanchard, 1847; Eupolycotylea Diesing, 1850; Polycotylea Diesing, 1850, in part.

Superfamily *POLYSTOMATOIDEA* Price, 1936

Family **POLYSTOMATIDAE** Gamble, 1896 (p. 71)

Syn. Polystomidae Carus, 1863; Sphyranuridae Poche, 1926; Dicotylidae Monticelli, 1903.

Subfamily **Polystomatinae** Gamble, 1896

Syn. Polystominae Pratt, 1900.

Polystomatidae with an opisthaptor bearing six cup-like suckers.

Genus *Polystoma* Zeder, 1800

Syn. *Linguatula* Frölich, 1798; *Hexathyridium* Treuler, 1793; *Polystomum* Zeder of Diesing, 1850 and others, e.g. *Planaria* Braun; *Fasciola* Gmelin.

DIAGNOSIS. *Opisthaptor:* bearing one pair of hooks. *Eyes:* present or absent. *Reproductive systems:* testes numerous, confined to the region between the caeca. Ovary in front of the testes. Uterus short, in front of the ovary. Vaginae present. Parasites in the urinary bladder of batrachians. *Species:* type, *Polystoma integerrimum* (Frölich, 1791) Rudolphi, 1808 and a number of others.

As restricted by Price, the genus contains six or seven species, including the type. Of these, the type is common in Europe, and one other, *P. gallieni* Price, 1939, was found by Gallien (1938) in the urinary bladder of *Hyla arborea*. The remaining species occur in Africa (*P. africanum* Szidat, 1932), U.S.A. (*P. nearcticum* (Paul, 1935)), and Japan (*P. ozakii* Price, 1939 and *P. rhacophori* Yamaguti, 1936). Other species at one time contained within the genus have been placed in other genera, e.g. *Parapolystoma* Ozaki, 1935; *Polystomoides* Ward, 1917; *Polystomoidella* Price, 1939; and *Neopolystoma* Price, 1939. Of these four, the first genus has a species in Africa (*Parapolystoma alluaudi* (de Beauchamp, 1913)) and another in Australia (*P. bulliense* (Johnston, 1913)); the second has species in North America, India, Japan, Formosa, and Europe (the last, *Polystomoides ocellatum* (Rudolphi, 1819)); the third has species in North America and the fourth species in both North America and Japan. Many such forms have at one time or another been referred to as species of the genus *Polystoma*. Related genera include *Diplorchis* Ozaki, 1931, with species in China, Japan and North America, and *Oculotrema* Stunkard, 1924, which has one species, *O. hippopotami* Stunkard, 1924, in the eye of the hippopotamus (in Egypt).

Polystoma integerrimum (Frölich, 1791) Rudolphi, 1808

Syn. *Linguatula integerrimum* Frölich, 1791; *Planaria uncinulata* Braun, 1790; *Fasciola uncinulata* (Braun, 1790) Gmelin, 1790; *Polystoma ranae* Zeder, 1800; *Hexathyridium integerrimum* (Frölich, 1791) Blainville, 1828; *Hexastoma integerrimum* (Frölich, 1791) Kuhn, 1829; *Polystomum integerrimum* (Frölich, 1791) Diesing, 1850.

HOSTS. Common frog, edible frog, tree frog, common and other toads.

LOCATION. Urinary bladder.

DIAGNOSIS. As for the genus. It is scarcely necessary to outline the characters of this well-known fluke; some are evident in Fig. 6E and others are mentioned in several text-books on zoology.

Polystoma gallieni Price, 1939

Syn. *Polystoma* sp. Gallien, 1938.

HOST. *Hyla arborea* var. *meridionalis* Boettger (in France).
LOCATION. Urinary bladder.
DIAGNOSIS. Differs from *P. integerrimum* in lacking anastomoses of the three branches of the intestine in front of the opisthaptor.

Family **HEXABOTHRIIDAE** Price, 1942 (p. 72)

Syn. Onchocotylidae Stiles & Hassall, 1908.

Subfamily **Hexabothriinae** Price, 1942

Syn. Onchocotylinae Cerfontaine, 1899; Diaphorocotylinae Monticelli, 1903, in part.

Hexabothriidae in which the appendage has one pair of hooks and the vaginae open separately into the transverse vitelloducts.

KEY TO GENERA OF HEXABOTHRIINAE

I. Cirrus armed with spines	*Hexabothrium*
II. Cirrus unarmed	
A. Vitellaria extending into the appendix of the opisthaptor	*Neoerpocotyle*
B. Vitellaria not extending into the appendix of the opisthaptor	*Erpocotyle*

Note. There is a fourth genus, *Heteronchocotyle* Brooks, 1934, characterized by an unarmed cirrus and unequal hooks on the opisthaptor. It has a solitary species *H. hypoprioni* Brooks, 1934, which occurs on the gills of the yellow shark, *Hypoprion brevirostris* at Tortugas, Florida.

Genus *Hexabothrium* Nordmann, 1840

Syn. *Onchocotyle* Diesing, 1850; *Acanthonchocotyle* Cerfontaine, 1899.

DIAGNOSIS. *Opisthaptor* (proper): bearing hooks of approximately equal size. *Cirrus:* armed with spines. *Vitellaria:* not extending into the appendix of opisthaptor. *Eggs:* with one long polar filament. *Species:* type, *Hexabothrium appendiculatum* (Kuhn, 1829) Nordmann, 1840 (Syn. *Polystoma appendiculatum* Kuhn, 1829; *Onchocotyle appendiculatum* (Kuhn) of Diesing, 1850 and various authors; *Acanthonchocotyle appendiculata* (Kuhn) Cerfontaine, 1899), *Hexabothrium canicula* (Cerfontaine, 1899) (Syn. *Onchocotyle appendiculatum* (Kuhn) of Stossich, 1877; *Acanthonchocotyle canicula* Cerfontaine, 1899) and the American form *Hexabothrium musteli* (MacCallum, 1931) Price, 1942*b*.

Price (1942*b*) pointed out that the description of Kuhn, and also a later description by Nordmann (1832) based on parts of Kuhn's material, both fail to supply characters serving to distinguish this type-species, which is established on the characterization provided by Cerfontaine (1899) for specimens taken from the same host and geographical area as Kuhn's. *Onchocotyle appendiculata* (= *Hexabothrium appendiculatum*) already recorded from various hosts is not in some instances the species in question. According to Price, Lebour's (1908*a*) specimens from the piked dogfish should be regarded as *Erpocotyle* sp., those of T. Scott (1901) from the skate and thornback ray probably as *Rajonchocotyle* sp. *Hexabothrium appendiculatum* occurs in the nurse hound, *H. canicula* in the rough hound, these hosts being regarded as types. The differences between these two species are very slight. According to Cerfontaine (1899) *H. appendiculatum* is 6–7 mm. long and has a genital armature of numerous small hooklets which are stronger than those of *H. canicula*, their bases being larger relative to the points. There may be very slight differences in the shape of the haptorial disk, that of the latter species being rounded behind. Guberlet (1933) readily identified *H. canicula* (found on both the rough hound and the nurse hound at Naples) by the characters specified by Cerfontaine, but especially by the size and shape of the genital hooklets and the egg, which has a single polar filament (Fig. 20C).

Genus *Erpocotyle* Beneden & Hesse, 1863 (Fig. 17C)

Syn. *Squalonchocotyle* Cerfontaine, 1899; *Erpetocotyle* Fuhrmann, 1928.

DIAGNOSIS. Similar to *Hexabothrium*, but cirrus unarmed, vitellaria not extending into the appendix of the opisthaptor, and eggs with two polar filaments (Fig. 20D). *Species:* type *Erpocotyle laevis* Beneden & Hesse, 1863 (Syn. *Polystomum appendiculatum* Thaer, 1850, *nec* Kuhn, 1829, in part; *Squalonchocotyle vulgaris* Cerfontaine, 1899), found on *Mustelus laevis* and *M. vulgaris* at Brest, Roscoff, in the North Sea and the Adriatic, and thirteen others, five of them American. The non-American forms (with their hosts) are (new combinations by Price, 1942*b*):

E. abbreviata (Olsson, 1876) (piked dogfish) (Roscoff, Skagerrack).
E. antarctica (Hughes, 1928) (*Mustelus antarctica*).
E. borealis (Beneden, 1853) (Greenland shark, *Scymnus glacialis*) (Greenland, Ostend, Skagerrack).
E. canis (Cerfontaine, 1899) (Syn. *Onchocotyle appendiculata* Beneden, 1858, *nec* Kuhn, 1829 (tope) (Ostend, Roscoff).
E. dollfusi Price, 1942 (Syn. *Squalonchocotyle abbreviata* form D of Dollfus, 1937) (*Echinorhinus brucus*) (Rabat, Morocco).
E. eugalei Price, 1942 (Syn. *Squalonchocotyle abbreviata* form B of Dollfus, 1937) (tope) (Agadir).
E. galeorhini Price, 1942 (Syn. *Squalonchocotyle abbreviata* form A of Dollfus, 1937) (*Galeorhinus mustelus*) (Rio de Oro).
E. torpedinis Price, 1942 (Syn. *Squalonchocotyle abbreviata* form C of Dollfus, 1937) (torpedo) (Morocco).

Price (1942*b*) found that *Erpocotyle laevis* cannot be identified from the original description, but that circumstantial evidence available permits its identity to be established, so that the genus can be revived. On the basis of host specificity Price was satisfied that this species is congeneric with species allocated by Cerfontaine (1899) to *Squalonchocotyle* and, taking two of these (*vulgaris* and *catenulata*) into consideration, decided in favour of the former, making it a synonym of the type-species. Using the synonymous form as described by

Cerfontaine the following diagnosis becomes available. *Size and shape:* about 12 mm. long, but capable of extension to three times this length and then very slender and cylindrical. *Colour:* dark brown. *Cuticle:* covered with small tubercles. *Opisthaptor:* claw of each hook bent in a right angle, its base clearly delimited from the shaft and of smaller girth. Hooklets each having two very unequal roots, the ventral long and curved, the dorsal shorter and narrower. *Gut:* intestinal crura having median and lateral diverticula along their entire length. *Reproductive systems:* vaginal pores slightly behind the level of the genital pore laterally. Cirrus short and almost globular when evaginated. Oviduct (uterus) passing straight from the ootype to the genital atrium. Eggs each having two short and straight polar filaments about as long as the capsule, i.e. 0·200 mm. long (Fig. 20D).

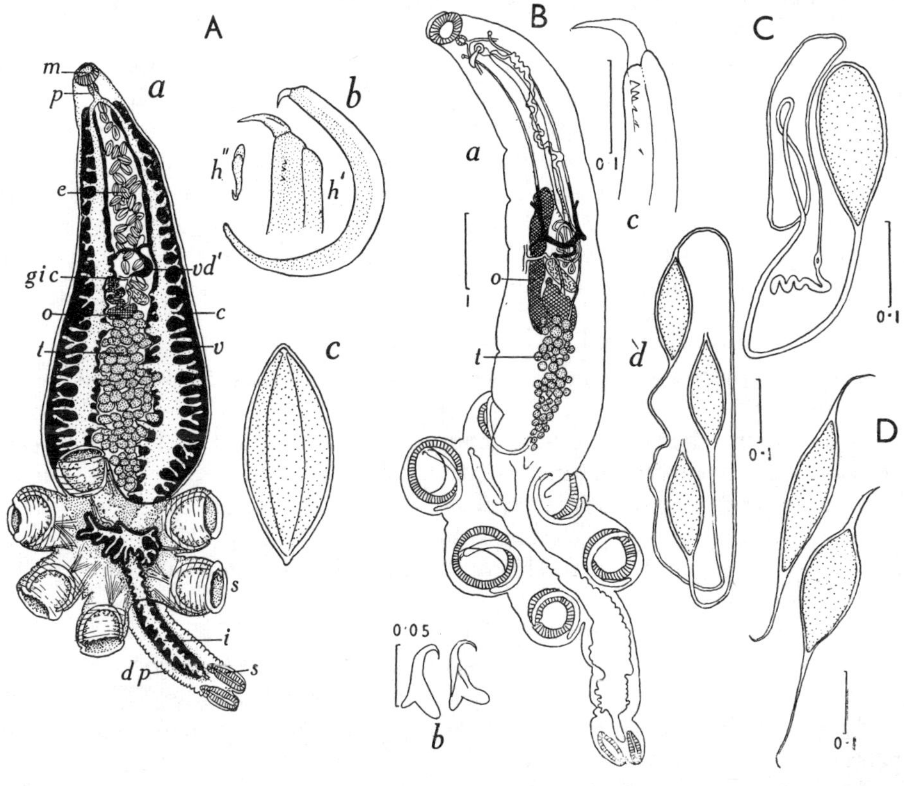

Fig. 20. Some Hexabothriidae and their eggs. A, *Rajonchocotyloides emarginata*: (*a*) the entire animal; (*b*) a hook (*h'*) and its tip enlarged, also a hooklet (*h''*) from the haptorial appendix; (*c*) the egg. B, *Neoerpocotyle catenulata*: (*a*) the entire animal; (*b*) the hooklets of the appendix; (*c*) the tip of a hook; (*d*) part of a chain of eggs. C, the egg of *Hexabothrium canicula*. D, two slightly different eggs of *Erpocotyle vulgaris*. (A, after Price, 1940*b*; B–D, after Guberlet, 1933.)

Note. Guberlet (1933) found several specimens on the gills of *Mustelus laevis* at Naples. They were 10–12 mm. long and 0·8–1·4 mm. broad and had suckers of different sizes, those of the third pair smaller than the others and having much smaller hooks. (This, in my experience, is not a constant character.) The eggs measured about 0·22 × 0·06 mm. and the filaments were 0·13 mm. long.

Erpocotyle abbreviata (Olsson, 1876) Price, 1942

Syn. *Onchocotyle abbreviata* Olsson, 1876; *Squalonchocotyle abbreviata* (Olsson) Cerfontaine, 1899.

According to Cerfontaine (1899) this species has some characters in common with *E. canis*, but note the hooks and hooklets. *Size:* 7–8 mm. long. *Opisthaptor:* claw of each hook distinct from the shaft and relatively long, bent in a right angle, but having a straight point. Hooklets having straight roots. *Gut:* median diverticula of the intestinal crura ill-developed. *Reproductive systems:* cirrus long and cylindrical, uterus somewhat convoluted. Eggs equipped with two long polar filaments and about 0·100 mm. long.

Erpocotyle canis (Cerfontaine, 1899) Price, 1942

Syn. *Onchocotyle appendiculata* of Beneden, 1858, *nec* Kuhn; *Squalonchocotyle canis* Cerfontaine, 1899.

According to Cerfontaine: *Size:* 7–8 mm. long. *Cuticle:* tuberculate, the tubercles occurring in the cavities of the oral sucker and the suckers of the opisthaptor. *Opisthaptor:* claw of each hook curved in a right angle, its base almost of the same girth as the shaft. *Suckers:* having cushion-like rims. *Hooklets:* with somewhat triangular roots, the point of each displaying two sharp bends. *Gut:* as in *E. laevis*. *Reproductive systems:* uterus and cirrus as in *E. abbreviata*. Eggs about 0·100 mm. long, each having two very long filaments (rarely, only one filament).

Erpocotyle borealis (Beneden, 1853) Price, 1942

Syn. *Onchocotyle borealis* Beneden, 1853; *Squalonchocotyle borealis* (Beneden) Cerfontaine, 1899.

This is a large form 20–30 mm. long and 3–4 mm. broad. The claw of each hook is regularly curved and its base passes almost imperceptibly into the shaft. The dorsal root of each hooklet (opposite the point) is longer and more delicate than the ventral. The vas deferens is long and convoluted. The eggs are very large (0·250 mm. long), but the filaments relatively short and sometimes folded down on the capsule.

Note. It seems rather unlikely that the various forms of *Squalonchocotyle abbreviata* recognized by Dollfus are distinct species. Price (1942*b*) bases specific identity on the following characters, apart from host affiliations.

Form A: differences in the size of the body and the large hooks of the opisthaptor.

Form B: differences in the hooks, the position of the vaginal pores lateral to the caeca and posterior to the genital pore, eggs arranged in chains.

Form C: differences in the hooks, vaginae short, their pores posterior to the genital pore.

Form D: marked differences in the size of body, terminations of vaginae club-shaped.

Genus *Neoerpocotyle* Price, 1942

Syn. *Squalonchocotyle* Cerfontaine, 1899, in part.

Diagnosis. Similar to *Erpocotyle*, but vitellaria extending into the appendix of the opisthaptor. *Species:* type, *Neoerpocotyle maccallumi* Price, 1942 and seven others, only three of which occur outside America. One of these, *N. licha* (Rees & Llewellyn, 1941) is known by name only, the other two occur in Italy, one of them west of Ireland.

Neoerpocotyle grisea (Cerfontaine, 1899) Price, 1942

Syn. *Onchocotyle appendiculata* Taschenberg, 1879, *nec* Kuhn, 1829; *Squalonchocotyle grisea* Cerfontaine, 1899.

HOST. Six-gilled shark (at Naples and Trieste and off west Ireland).

LOCATION. Gills.

DIAGNOSIS. According to Cerfontaine (1899): *Size:* about 15 mm. long. *Opisthaptor:* claw of each hook not clearly differentiated from the shaft and curved at right angles to it. Hooklets Y-shaped, the roots and point being of equal size, the ventral root bent towards its own side and terminating in a small knob. *Gut:* median and lateral diverticula present. *Reproductive systems:* vaginal pores near the genital pore and the level of the excretory pores. Eggs about 0·175 mm. long, each having two short filaments about half as long as the spindle-shaped capsule.

Neoerpocotyle catenulata (Guberlet, 1933) Price, 1942 (Fig. 20B, *a–d*)

Syn. *Polystomum appendiculatum* Thaer, 1850, *nec* Kuhn, 1829, in part; *Squalonchocotyle catenulata* Guberlet, 1933.

HOSTS. Smooth hound, *Mustelus laevis* (at Naples).

LOCATION. Gills.

DIAGNOSIS. *Size:* 7–11 mm. long and about 1·5 and 1 mm. in greatest breadth and thickness. *Shape:* fusiform, tapering towards the extremities. *Prohaptor:* an oral sucker 0·5 mm. diameter having a margin free of tubercles. *Opisthaptor:* rectangular, the suckers being arranged in two parallel rows of three. Hooks having a shaft bearing a slight lateral ridge and six or more spinelets on either side and a long, sharply bent point. Hooklets curved, 0·07–0·08 mm. long. Appendage almost cylindrical, about 2 mm. long and 0·75 mm. thick. *Other characters:* as in the figures, eggs measuring 0·127–0·141 × 0·052–0·066 mm., each having a filament 0·3–0·7 mm. long at either pole, these being continuous with the filaments of adjacent eggs.

Note. Price (1942*b*) found that the vitelline follicles extend into the appendage of the opisthaptor, contrary to Guberlet's observation. He further credited Guberlet with an otherwise complete and accurate description, which only goes to show that the main character on which the genus *Neoerpocotyle* was founded is ill-conceived and that the genus may have to be relegated to synonymity with *Erpocotyle*.

Neoerpocotyle licha (Rees & Llewellyn, 1941) n. comb. was found on the gills of *Scymnorhinus licha* on Porcupine Bank (Atlantic seaboard).

Subfamily **Rajonchocotylinae** Price, 1942

Hexabothriidae in which the appendix of the opisthaptor has one pair of hooks and the vaginae unite to form a single duct opening into the vitelline reservoir.

KEY TO GENERA OF RAJONCHOCOTYLINAE

I.	Vitellaria extending into the appendix of the opisthaptor	*Rajonchocotyloides*
II.	Vitellaria not extending into the appendix of the opisthaptor	*Rajonchocotyle*

Genus *Rajonchocotyle* Cerfontaine, 1899

DIAGNOSIS. With the characters of the subfamily, but vitellaria as in the key above. *Species:* type, *Rajonchocotyle batis* Cerfontaine, 1899, and seven others, three of which occur in North America and one in Japan.

Rajonchocotyle batis Cerfontaine, 1899

Syn. *Onchocotyle appendiculata* Olsson, 1867, *nec* Kuhn, 1829.

HOST. Skate (at Plymouth, Ostend, off west Ireland and in the Skagerrack).

LOCATION. Gills.

DIAGNOSIS. *Size:* 12–15 mm. long. *Opisthaptor:* circular, bearing very large suckers. Shaft of each hook chamfered near the base of the point, the surfaces of contact being of equal girth. Hooklets having three delicate axes, two forming roots (dorsal much the longer), the other bearing the point being very short and humped on the dorsal (convex) side. *Reproductive systems:* vaginal pores far behind the genital pore near the median plane, median vaginal canal long. Eggs lacking polar filaments, but generally bearing a small tubercle at one pole, and about 0·175 mm. long.

Rajonchocotyle alba Cerfontaine, 1899

HOST. Burton skate (at Roscoff).

LOCATION. Gills.

DIAGNOSIS. *Size:* 8–9 mm. long. *Opisthaptor:* hooks long and of uniform girth, tapering near the base of the claw, but not in a straight chamfer. Dorsal root of each hooklet more massive than the ventral. *Reproductive systems:* median vaginal canal short. Eggs as in *R. batis* and only slightly larger, 0·180 mm. long.

Rajonchocotyle prenanti (St Remy, 1890) Cerfontaine, 1899

Syn. *Onchocotyle prenanti* St Remy, 1890; *O. borealis* Stossich, 1885, *nec* Beneden, 1853.

HOST. Long-nosed skate (at Roscoff and Trieste).

LOCATION. Gills.

DIAGNOSIS. Similar to *R. alba*, but hooks shorter and of more uniform girth, tapering markedly to the base of the point, and hooklets showing a deep space (not a shallow one) between the roots. *Eggs:* as in other species, but very large (about 0·220 mm. long).

Rajonchocotyle miraletus Rees & Llewellyn, 1941

HOST. Cuckoo ray (west of Ireland).

LOCATION. Gills.

This species has not been described.

Genus *Rajonchocotyloides* Price, 1940

DIAGNOSIS. With the characters of the subfamily, but vitellaria extending into the appendix of the opisthaptor. (*Note.* Perhaps in this case, as in that of *Neoerpocotyle*, Price has chosen an inconstant character. In many hexabothriids the vitelline follicles enter or fail to enter the appendage in different individuals undoubtedly belonging to the same species.) *Type and only species:*

Rajonchocotyloides emarginata (Olsson, 1876) Price, 1940 (Fig. 20*a–c*)

Syn. *Onchocotyle emarginata* Olsson, 1876; *O. appendiculata* Sonsino, 1891, *nec* Kuhn, 1829.

HOST. Thornback ray (at Plymouth and Roscoff).

LOCATION. Gills.

DIAGNOSIS. *Size:* body proper 5 mm. long and 1·8 mm. broad in the posterior region. *Prohaptor:* an oral sucker 0·28 mm. diameter. *Opisthaptor:* 2·5 mm. diameter, having six suckers arranged more or less in a circle, and an appendix 1·4 mm. long and 0·5 mm. broad which terminates in a pair of muscular suckers. Suckers of the haptor proper 0·6 mm. diameter, each containing a large hook about 1·4 mm. long. Suckers of the appendix constricted at their bases, 0·32 mm. long and 0·15 mm. broad, the region between them occupied by a pair of hooklets 0·04 mm. long. *Gut:* pharynx of medium size (0·15 mm. long and 0·115 mm. broad), oesophagus very short, caeca uniting posteriorly, with lateral and median diverticula and with several short diverticula extending into the haptor. Single caecum extending into the appendix almost to the bases of the small suckers. *Reproductive systems:* genital pore near the bifurcation of the intestine, 0·47 mm. from the anterior extremity. Testes numerous, confined between the caeca and between the ovary and the posterior end of the body proper. Ovary with several loops, on the right of the median plane, near the mid-body. Receptaculum seminis large, on the left near the ovary. Vitellaria extending from the bifurcation of the intestine to the tip of the caecum in the appendix, follicles distributed along the caeca. Uterus long, slightly folded. Eggs very large, lemon-shaped (Rugby-football-shaped), measuring 0·170–0·197 × 0·080–0·097 mm., provided with meridional ridges.

Subfamily **Diclybothriinae** Price, 1936

Hexabothriidae in which the appendage has three pairs of hooks.

Genus *Diclybothrium* F. S. Leuckart, 1835

Syn. *Diclibothrium* Leuckart, 1836; *Diplobothrium* Leuckart, 1842 (=*Diclybothrium* renamed); *Diklibothrium* Leuckart, in Kollar, 1836.

DIAGNOSIS. With the characters of the subfamily. *Species:* type, *Diclybothrium armatum* F. S. Leuckart, 1835 (on the gills of various sturgeons in Russia, U.S.A. and Canada) and one other, *D. hamulatum* (Simer, 1929), an American parasite on the gills of the paddle fish, *Polyodon spathula*.

Diclybothrium armatum F. S. Leuckart, 1835

Syn. *Diclibothrium armatum* Leuckart, 1836; *Diklibothrium crassicaudatum* Leuckart, in Kollar, 1836; *Diplobothrium armatum* (Leuckart, 1835) Leuckart, 1842; *Hexacotyle elegans* Nordmann, 1840; *Polystoma* (*Hexacotyle*) *armatum* (Leuckart, 1835) Dujardin, 1845; ? *Erpocotyle circularis* Linstow, 1904; ? *Diclibothrium circularis* (Linstow, 1904) Skwortzoff, 1928.

DIAGNOSIS (after Price, 1942*b*). *Shape and size:* slender, elliptical in cross-section, 2·5–13 mm. long and 0·22–1·1 mm. broad. *Prohaptor:* two ventro-lateral bothria of oval shape, 0·08–0·15 mm. broad. *Opisthaptor:* circular or rectangular, 0·32–1·0 mm. long and 0·32–0·80 mm. broad, with six sessile suckers and a somewhat triangular appendix 0·22–0·32 mm. long and 0·12–0·24 mm. wide. Suckers 0·20–0·37 mm. diameter, each with a large hook 0·40–0·47 mm. long. Appendix bearing one pair of relatively small suckers (0·06 mm. long) and three pairs of large hooks (those of the *outer* pair 0·44–0·54 mm. long, directed posteriorly and ventrally: those of the second pair 0·34–0·35 mm. long, directed posteriorly and dorsally: those of the third pair not curved like the others, but straight, 0·10–0·12 mm. long, directed ventrally or anteriorly). *Gut:* mouth ventral, 0·14–0·40 mm. from the anterior extremity, pharynx oval. *Reproductive systems:* genital pore median, 0·24–0·68 mm. from the anterior extremity. Cirrus muscular, its tip armed with short, thickly set spines. Seminal vesicle pyriform. Testes very numerous (300 or more). Ovary tubular, much convoluted, median, about one-quarter of the distance along the body, in front of the testes. Vitellaria lateral, extending from the level of the vaginal pores to near the posterior extremity. Ootype conspicuous, in front of the ovary. Uterus long, slender, sinuous, median. Eggs without polar filaments, measuring 0·208–0·224 × 0·088–0·140 mm.

Note. Skwortzoff's specimens were small (3·25–5·88 × 0·41–0·85 mm.), but had a large opisthaptor (measuring 0·55×0·57 mm.) and showed other characters of apparent distinction, including relatively small eggs 0·11–0·14 mm. long. Further studies are likely to show, however, that these differences merely indicate wide limits of variability in a single, highly host-specific trematode species.

Superfamily *DICLIDOPHOROIDEA* Price, 1936

Syn. Dactylocotyloidea Brinkmann, 1942*a*.

Family **MAZOCRAËIDAE** Price, 1936 (p. 73)

Syn. Octobothriidae Taschenberg, 1879; Octobothridae of Monticelli, 1888; Octocotylidae Beneden & Hesse, 1863; Mazocriidae Southwell & Kirshner, 1937: all *partly* identical with Mazocraëidae.

KEY TO GENERA OF MAZOCRAËIDAE HAVING A COTYLOPHORE

(*Note.* This qualification excludes the genera *Mazocraëoides* Price, 1936, which has one species (*M. georgei* Price, 1936) in U.S.A. and *Neomazocraës* Price, 1943, emend., which has one species (*N. dorosomatis* (Yamaguti, 1938)) in Japan.)

I. Cotylophore of the opisthaptor with two pairs of small, accessory suckers *Ophiocotyle*

II. Cotylophore of the opisthaptor without small, accessory suckers
   A. Genital hooklets in two transverse rows; dorsal vagina present *Mazocraës*
   B. Genital hooklets in two longitudinal rows; vagina absent *Kuhnia*

Genus *Mazocraës* Hermann, 1782

Syn. *Octobothrium* Leuckart, 1827; *Octostoma* Kuhn, 1829; *Octocotyle* Diesing, 1850; *Octoplectanum* Diesing, 1858; *Octobothrium* (*Octocotyle*) of St Remy, 1891; *Glossocotyle* Beneden & Hesse, 1863.

Diagnosis. *Shape:* body narrow, flattened, tapering gradually to a bluntly rounded anterior extremity. *Opisthaptor:* cotylophore almost triangular, having four pairs of clamp-like suckers (Fig. 22D, *e*) along the lateral margins, terminating in a short, truncated edge which bears two pairs of dissimilar hooklets, those of the outer pair the larger. *Reproductive systems:* testes numerous, filling the region between the caeca in the posterior half of the body. Ovary elongate, in front of the testes and on the left. Vaginal pore in the mid-dorsal line. Genital pore transversely elongate, provided with twelve bipartite hooklets; one in each lateral angle and a horizontal row of five along both the anterior and the posterior lip. Eggs generally with two polar filaments. Parasites on the gills of Clupeoid fishes. *Species:* type, *Mazocraës alosae* (Beneden & Hesse, 1863), three others in Europe (one a *species inquirendum*) and an American form (*M. cepedianum* Kimpel, 1938) from fresh water, so far known only by name.

Mazocraës alosae Hermann, 1782

Syn. *Octobothrium lanceolatum* Leuckart, 1827, of various writers; *Octostoma alosae* (Hermann, 1782) Kuhn, 1829; *Octocotyle lanceolatum* (Leuckart, 1827) Diesing, 1850; *Glossocotyle alosae* Beneden & Hesse, 1863; *Octoplectanum lanceolatum* (Leuckart) Diesing, 1858; *Octobothrium alosae* (Hermann) Cerfontaine, 1896.

Note. Confusion is manifest in the literature, where as many as three of the above names have been used in a single paper to denote one and the same species.

HOSTS. Allis shad, twaite shad (the former both in fresh water and the sea).

LOCATION. Gills.

DIAGNOSIS. *Shape:* tapering acutely from a little behind the middle of the body towards the anterior extremity, gradually narrowing a little towards the opisthaptor, middle region of the body flattened. *Size:* 10–12 mm. long, but possibly larger. *Colour:* greyish yellow modified by a dark or greenish gut. *Reproductive systems:* genital hooklets as for the genus.

Note. In this and the following species of *Mazocraës* host-specificity has been depended on implicitly, but this is incomplete and the species require redefinition. It is possible that not all are valid. This species was discovered in north-west France, but has been recorded in Germany (Rhineland) and Britain (at Aberdeen, in the Irish Sea and at Plymouth).

Mazocraës harengi (Beneden & Hesse, 1863)

Syn. *Octocotyle harengi* Beneden & Hesse, 1863; *Octobothrium harengi* (Beneden & Hesse) Taschenberg, 1879; *Octoplectanum harengi* (Beneden & Hesse) of Nicoll, 1915 and Linstow, 1889.

HOSTS. Herring, allis shad.

LOCATION. Gills.

DIAGNOSIS. *Opisthaptor:* hooks, two pairs, borne on an ovoid posterior extension of the cotylophore, those of the lateral pair larger and having broad, tapering bases and very slender points. *Gut:* intestine extending back as far as the posterior pair of clamps, which project slightly from the margin of the cotylophore. *Eggs:* relatively narrow, fusiform and equipped with two delicate polar filaments only slightly longer than the capsule (Beneden & Hesse). This species was discovered in the same locality as the preceding, but has since appeared at Plymouth and in the Solway Firth.

Mazocraës pilchardi (Beneden & Hesse, 1863)

Syn. *Octocotyle pilchardi* Beneden & Hesse, 1863; *Octobothrium pilchardi* (Beneden & Hesse) Taschenberg, 1879; *Octoplectanum pilchardi* (Beneden & Hesse) of Linstow, 1889 and Nicoll, 1915.

Host. Pilchard.

Location. Gills.

Diagnosis. Similar to *M. harengi*, but smaller and having progressively smaller clamps situated on short peduncles, and rust-coloured eggs each having a long filament with a terminal plate at one pole and a stumpy process at the other. Found at Brest.

Mazocraës heterocotyle (Beneden, 1870) *sp. inq.*

Syn. *Octostoma heterocotyle* Beneden, 1870; *Octobothrium heterocotyle* (Beneden) Taschenberg, 1879; *Octoplectanum heterocotyle* (Beneden) of Linstow, 1885 and of Nicoll, 1915.

Host. Sprat.

Location. Gills.

This species was reported on the coast of Belgium, but has not appeared in Britain.

Genus *Ophiocotyle* Beneden & Hesse, 1863 *genus inq.*

Syn. *Octobothrium* of Taschenberg, 1879 and of St Remy, 1891; *Ophycotyle* Monticelli, 1888.

Diagnosis. *Shape:* body tapering gradually towards the anterior extremity, having a narrow neck-like anterior region. *Opisthaptor:* comprising a cotylophore bearing four pairs of clamps and terminating in a process bearing two pairs of minute accessory suckers and equipped with two pairs of hooks, those of the external pair the larger. *Reproductive systems:* genital hooklets as in *Mazocraës*. *Type and only species:*

Ophiocotyle fintae Beneden & Hesse, 1863

Syn. *Octobothrium fintae* (Beneden & Hesse) St Remy, 1891.

Host. Twaite shad.

Location. Gills.

This species has not appeared in Britain, though discovered across the English Channel at Brest. It might be mentioned that in the work by Beneden & Hesse (1864) the specific diagnosis of this trematode seems to have been confounded with that of *Glossocotyle* (now *Mazocraës*) *alosae* and interchange of the diagnoses does not dispose of all the difficulties, because the differences in size do not then fit the absolute sizes given in Pl. IX, figs. 11 and 19. (The names of the two species are confounded (by reference to the headings in the text) in the captions for Pl. IX on p. 134.)

Genus *Kuhnia* Sproston, 1945

Syn. *Octostoma* Kuhn, 1829, *nec* Otto, 1823; *Octobothrium* Leuckart, 1827 of Leuckart, 1842, in part; *Octocotyle* Diesing, 1850, in part, *nec* Goto, 1894; *Octoplectanum* Diesing, 1858, in part.

DIAGNOSIS. *Shape:* lanceolate, flattened, posterior half of the body with lateral margins tending to curve ventrally. *Opisthaptor:* cotylophore with four pairs of almost sessile clamps, arranged in two lateral rows, and at least two pairs of hooks of twisted, sigmoid shape, the outer with spurs near the middle and a ridged shaft, the inner small. Small larval hooklets may be present also. *Gut:* caeca not extending into the cotylophore. *Reproductive systems:* cirrus with a longitudinal, slit-like aperture, guarded by a row of inwardly curved hooks on each side, and with a pair of antero-lateral masses of attached tissue, each bearing a hook with a ventrally and outwardly curved blade and two wing-like basal processes resembling broad spines; vagina absent. *Other characters:* as in *Mazocraës*. Parasites on the gills of Scombridae. In addition to the type-species, *Kuhnia scombri*, there are possibly two or three others, e.g. *K. minor* (Goto, 1894) and *K. macracantha* (Meserve, 1938) in the Galapagos Isles.

Kuhnia scombri (Kuhn, 1829)

Syn. *Octostoma scombri* Kuhn, 1829; *Octobothrium scombri* (Kuhn) of Nordmann, 1832 and other writers; *Octocotyle scombri* of Dujardin, 1845; *O. truncata* Diesing, 1850; *Octoplectanum truncatum* of Diesing, 1858; *Pleurocotyle scombri* of Taschenberg, 1878; *Octocotyle major* Goto, 1894; *Octocotyle scombri* of Nicoll, 1915.

HOSTS. Mackerel, Spanish mackerel (and in houttyn at Vladivostok and '*Scomber japonicus*' in Japan).

LOCATION. Gills.

LOCALITY. North Sea, English Channel, eastern Atlantic, Mediterranean, Russia, Japan.

DIAGNOSIS. *Shape and size:* lanceolate, broadest near the mid-body, tapering towards the narrow anterior quarter of the body and towards the opisthaptor, 1·3–6·5 mm. long, the breadth increasing with increase in size, 10–17 % of body length. *Colour:* white, but the gut showing bluish grey or greenish brown, punctuated with dark spots. *Prohaptor:* elongate oval buccal suckers. *Opisthaptor:* heart-shaped, distinct from the body, 0·23–0·62 mm. long, the relative size diminishing during growth from 17 to 9 % of body length. Clamps in two diverging rows. Large hooks each with a stout, hollow blade and a ridged shaft bearing a spur near its middle; small hooks sigmoid, each with a twisted shaft and a sickle-like point. *Gut:* caeca unbranched, extending into the opisthaptor to a region between the third and fourth clamps, generally unequal in length. *Reproductive systems:* five genital hooks in each median row about 0·017 mm. long, each having a semicircular blade which is sharply bent on the shaft. Hooks of the lateral pair each having a broad blade and a wide base, only slightly larger than the more median hooks. Eggs (laid singly) fusiform, operculate, with a filament at each pole, very variable in size and shape, measuring 0·194–0·312 × 0·049–0·126 mm.; the anterior filament about as long as the egg, the posterior shorter.

Family **DISCOCOTYLIDAE** Price, 1936 (p. 73)

Subfamily **Discocotylinae** Price, 1936

Discocotylidae with a posterior region not narrow or striated, four pairs of clamps, those of the anterior pair not markedly larger than the others, and (generally) vaginae.

KEY TO EUROPEAN GENERA OF DISCOCOTYLINAE

I. Body asymmetrical, flexed laterally and with an eminence on the convex, sometimes on the concave side *Vallisia*

II. Body not asymmetrical (excluding the opisthaptor from consideration)
- A. Adults united in permanent copula, each pair X-shaped *Diplozoon*
- B. Adults not united in permanent copula
  - (*a*) Cotylophore of the opisthaptor having four pairs of clamps *Discocotyle*
  - (*b*) Cotylophore of the opisthaptor having four similar clamps on the right side and a solitary small sucker on the left *Grubea*

Genus *Vallisia* Perugia & Parona, 1890

DIAGNOSIS. *Shape:* as given in the key above. *Opisthaptor:* cotylophore slightly delimited from the rest of the body by a slight constriction, bearing four pairs of sessile clamps near its margins. Sclerites of the suckers being of typical Discocotylid type. Terminal part of the cotylophore bearing one pair of hooks. *Gut:* caeca united in front of the cotylophore, a median canal extending through the latter and having (?) an anus. *Reproductive systems:* gonads in the posterior quarter of the body, the ovary behind the testes. Eggs very large, each equipped with two polar filaments. *Species:* type only, *Vallisia striata* Parona & Perugia, 1890 (Syn. *Octocotyle arcuata* Sonsino, 1890), discovered on the gills of the horse mackerel in the Adriatic Sea, but unrecorded elsewhere.

Note. The original figure suggests to me a damaged worm with cytolysing protoplasm still confined within a bulging and newly regenerated surface membrane.

Genus *Grubea* Diesing, 1858 (erected for '*Octobothrium scombri* Nordmann' of Grube, 1855)

DIAGNOSIS. *Shape:* body symmetrical, but equipped with an asymmetrical cotylophore having an ear-like lobe and bearing four similar clamps on the right side and one small clamp on the left, as well as two pairs of hooks in a terminal position. *Reproductive systems:* lips of the genital pore provided with two rows of eight hooks (one row on the anterior, the other on the posterior lip) and two larger, bipartite hooklets, one in each lateral angle. *Species:* type only, *Grubea cochlear* Diesing, 1858 (Syn. *Pleurocotyle scombri* of Pratt, 1900), discovered on the gills of the mackerel and horse mackerel in Italy (Naples) and unrecorded in Britain.

Genus *Diplozoon* Nordmann, 1832

Syn. *Diporpa* Dujardin, 1845; *Diplozoum* (Nordmann, 1832), of Burmeister, 1835.

DIAGNOSIS. Adults united in pairs in permanent copula. *Opisthaptor:* four pairs of clamps (Fig. 21 *b*) set close together along the postero-lateral margins of a rectangular cotylophore (larvae with one, two, three and four pairs at successive stages of development). Terminal part of cotylophore with two pairs of hooks, one of which is recurved, the other needle-like. *Gut:* intestine not bifurcate, but long and having numerous branched diverticula. *Reproductive systems:* genital pore situated in the posterior region of each member of the pair of individuals, surrounded by a gonotyl or genital sucker which fuses at maturity with a dorsal papilla bearing the vaginal pore of the other individual (corresponding change occurring in the inverse sense, so that male duct of each individual is contiguous with the female duct of the other). Testis single, compact, slightly in front of the opisthaptor. Ovary formed into a loop a little in front of the testis. Vitellaria comprising numerous follicles which fill up the anterior region in front of the point of union of the individuals. Uterus little folded, its pore near the point of union of the vagina of one individual with the male pore of the other. Eggs ovoid, each having a single, long, coiled filament at the anopercular pore. Parasites on the gills of fresh-water fishes. *Species:* type, *Diplozoon paradoxum* Nordmann, 1832 and one other, *D. nipponicum* Goto, 1891, which was discovered in Japan.

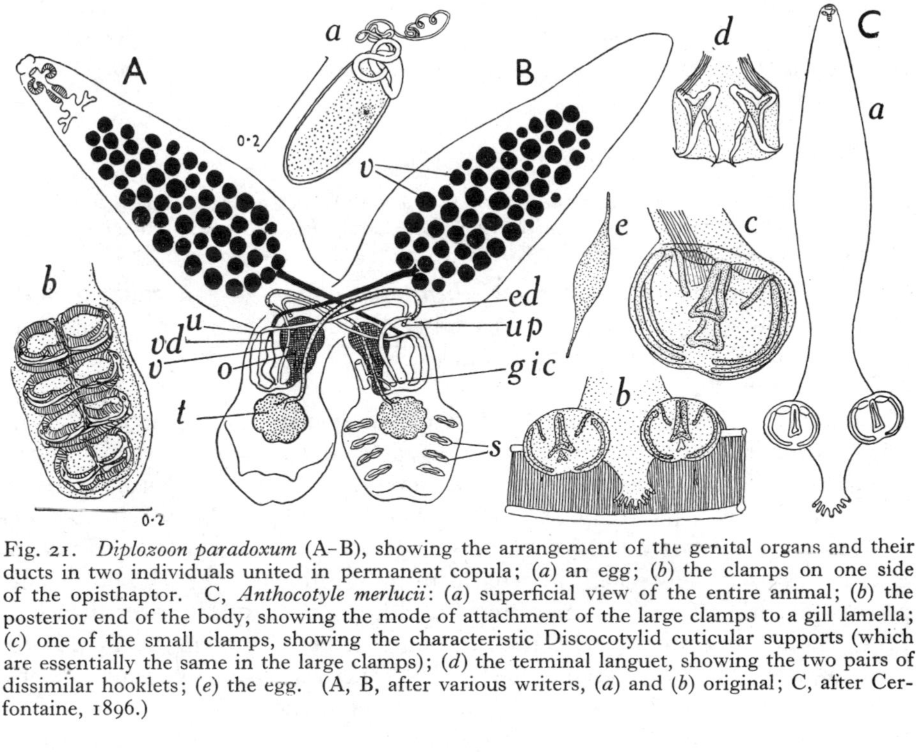

Fig. 21. *Diplozoon paradoxum* (A–B), showing the arrangement of the genital organs and their ducts in two individuals united in permanent copula; (*a*) an egg; (*b*) the clamps on one side of the opisthaptor. C, *Anthocotyle merlucii*: (*a*) superficial view of the entire animal; (*b*) the posterior end of the body, showing the mode of attachment of the large clamps to a gill lamella; (*c*) one of the small clamps, showing the characteristic Discocotylid cuticular supports (which are essentially the same in the large clamps); (*d*) the terminal languet, showing the two pairs of dissimilar hooklets; (*e*) the egg. (A, B, after various writers, (*a*) and (*b*) original; C, after Cerfontaine, 1896.)

Diplozoon paradoxum Nordmann, 1832 (Fig. 21 A–B and *a*, *b*)

Syn. *Diporpa dujardinii* Diesing, 1850.

HOSTS. Miller's thumb, burbot, crucian carp, gudgeon, roach, rudd, silver (white) bream, bream, bleak, minnow, three-spined stickleback, etc. (apparently unrecorded in Britain, but undoubtedly occurring here, also in Finland, Russia, Germany, France, Switzerland, Italy).

LOCATION. Gills.

DIAGNOSIS. With the characters of the genus. (*Note.* This trematode is described and figured in several text-books on Zoology (see Gamble, 1896; Benham, 1901) and Braun gave good figures (Pl. 13, figs. 1–5).)

Genus *Discocotyle* Diesing, 1850

Syn. *Placoplectanum* Diesing, 1858; *Discotyle* Braun, 1890.

DIAGNOSIS. *Shape:* body elongate, truncated posteriorly. *Prohaptor:* buccal suckers well developed. *Opisthaptor:* cotylophore somewhat rectangular, with four pairs of clamps, generally sessile, arranged along the lateral margins. Cuticular bars of the clamps of uniform type, all elements directed inwards towards the centre; two pairs of lateral pieces articulating with a well-developed middle segment, dorsal laterals spurred. *Reproductive systems:* lateral vaginae in the anterior third of the body, uniting to form a long, median duct. Genital pore equipped with spines having each a single point. Parasites of marine and fresh-water fishes. *Species:* type, *Discocotyle sagittata* (Leuckart, 1842) and three others, *D. sybellae* (T. Scott, 1909) (discovered on the gills of the trout at Aberdeen), *D. dorosomatis* Yamaguti, 1938 (in Japan) and *D. thyrites* (Hughes, 1928) (in Australia).

Discocotyle sagittata (Leuckart, 1842) Diesing, 1850

Syn. *Cyclocotyla lanceolata* Zäringer, 1829 (name only); *Octobothrium sagittatum* Leuckart, 1842; *Placoplectanum sagittatum* (Leuckart) Diesing, 1858; *Mazocraës sagittatum* (Leuckart) Southwell & Kirshner, 1937; *Discocotyle salmonis* Shaffer, 1916.

HOSTS. Trout, salmon (as *Mazocraës sagittatum* on the 'brown trout' in North Wales (Southwell & Kirshner, 1937*c*) and in Scotland (Friend, 1939), Cambridgeshire, the Black Forest, Switzerland, and U.S.A.).

LOCATION. Gills.

DIAGNOSIS. *Shape and size:* elongate or lanceolate, tapering anteriorly to a blunt extremity, slightly constricted anteriorly at the level of the vaginal pores, about 6–9 mm. long and one-quarter or one-third of this in greatest breadth posteriorly. (*Note.* In attenuated specimens of my own the body is barely one-sixth as broad as long.) *Colour:* off-white. *Prohaptor:* a pair of buccal suckers. *Opisthaptor:* relatively small (its length in my specimens 0·1–0·15 body length and its breadth 0·4–0·6 the maximum breadth of the body), broader than long (in the ratio of 1·3–1·6 : 1), sometimes having a posterior margin indented in the median plane, narrowing slightly posteriorly, otherwise as for the genus. *Reproductive systems:* genital pore median, at the level of the vaginal pores and close behind the bifurcation of the gut. Testes numerous (about forty). Ovary elongate, curved, slightly in front of the testes. Vitellaria extending along most of the lateral regions between the lateral vaginae and the opisthaptor. Uterus little folded. Vagina bifurcating close behind the genital pore, lateral vaginae opening to the exterior at the anterior constriction of the body. Eggs few (as few as four *in utero* together, but such as have imperfect shells difficult to observe sometimes), yellow and very large. Southwell & Kirshner gave the dimensions of the eggs as 0·25 × 0·13 mm., which seems an excessive size, but in my own specimens these figures were surpassed, the largest normal egg measuring about 0·30 × 0·14 mm., though one abnormal egg was less than half this size. *Notes*. Southwell & Kirshner (1937*c*) found more than one hundred specimens on the gills of a host, observing that these were pallid and covered with mucus, death of the host being caused, presumably, by the parasites. These writers aimed very wide of the mark in affirming that the three species erected by MacCallum (1917), namely, *Diclidophora merlangi*, *D. prionoti* and *D. cynoscioni*, are all synonyms of *Discocotyle sagittata*. All three belong to a different family, the Diclidophoridae; the first named belongs to the Diclidophorinae and is now known as *Diclidophoroides maccallumi*, the second and last named to different genera of the Cyclocotylinae, being known now as *Cyclocotyla prionoti* and *Neoheterobothrium cynoscioni* (see Price, 1943*b*, pp. 37, 49 and 51). Price was somewhat doubtful about the validity of '*Discocotyle salmonis*' Shaffer, 1916, but allowed the species to stand in the absence of recent descriptions of *D. sagittata*. In my opinion it is clearly identical with the type-species and must be regarded as a synonym of *D. sagittata* (Leuckart, 1842), coming well within its range of variability in Europe.

Subfamily **Chimaericolinae** n. subfamily

Syn. Chimaericolidae Brinkmann, 1942.

Discocotylidae with a narrow striated posterior region, four pairs of clamps, those of the anterior pair not markedly larger than the others, and (generally) vaginae.

Genus *Chimaericola* Brinkmann, 1942

Syn. *Discocotyle* Diesing, 1850, in part; *Placoplectanum* Diesing, 1858, in part.

Diagnosis. *Shape:* body elongate, elliptical anteriorly, attenuated posteriorly to the fan-like opisthaptor and here transversely striated. *Prohaptor:* absent. *Opisthaptor:* cotylophore distinctly set off from the rest of the body, bearing four pairs of bowl-like muscular clamps set on short peduncles. Sclerites as for the subfamily. *Gut:* intestine forming a large median anterior sac and long, simple caeca extending into the cotylophore and terminating each in an irregular pouch. *Reproductive systems:* genital pore median. Cirrus equipped with numerous hooklets. Testes numerous. Ovary lobed, situated in front of the testes in the middle third of the body proper. Oviduct slightly convoluted, uterus very long and forming several ascending and descending limbs between the ootype and the genital pore. Vaginal pores situated in the anterior third of the body, midway between the median plane and the lateral margins, the ends of the vaginae having much-folded walls forming a glandular complex including unicellular glands. Eggs large, lemon-shaped, operculate, each containing a segmenting ovum. *Type and only species:*

Chimaericola leptogaster (Leuckart, 1830)

Syn. *Octobothrium leptogaster* Leuckart, 1830; *Discocotyle leptogaster* (Leuckart) Diesing, 1850; *Placoplectanum leptogaster* (Leuckart) Diesing, 1858; *Octocotyle* (*Octobothrium*) *leptogaster* (Leuckart) Parona & Perugia, 1892; *Neoheterobothrium leptogaster* (Leuckart) Price, 1943.

Host. Rabbit fish.
Location. Gills.

This species has been recorded in the North Sea, at Oslo and Trondhjem (Norway), Kristineberg (Sweden) and in the Skagerrack. T. Scott (1911) provided the only British record, and his figures indicate a trematode with an extended, transversely striated posterior region, terminating in a fan-like opisthaptor comprising four pairs of clamp-like suckers set on short peduncles. Brinkmann (1942*b*) has redescribed the trematode in great detail.

DIAGNOSIS. As for the genus. *Additional characters: Size:* up to 50 mm. long, about 10–12·5 mm. in maximum breadth and 4–5 mm. in maximum thickness (Brinkmann gave separately the measurements of several writers). *Reproductive systems:* hooklets on the cirrus about 0·018 mm. long, comprising a slender shaft 0·0035 mm. broad and a stouter base 0·07 mm. broad. Uterus forming seven limbs (counting ascending and descending ones) between the ootype and the genital pore. Vitellaria extending throughout most of the intestinal region as far back as the attenuated posterior region, connected above the anterior half of the uterine region by a bridge of follicles and having medio-lateral extensions underlying the testes. Vitelloducts connected by a transverse canal slightly behind the vaginal pores, this having lateral openings into the vaginae. Eggs lemon-shaped, measuring 0·185 × 0·103 mm. (0·201 × 0·091 mm. according to Parona & Perugia, 1892), operculate at the blunter end.

Subfamily **Anthocotylinae** Price, 1936

Discocotylidae having four pairs of clamps, those of the most anterior pair very large, the remaining ones very small, and (generally) lateral vaginae.

Genus *Anthocotyle* Beneden & Hesse, 1863

DIAGNOSIS. With the characters of the subfamily. *Shape:* elongate, the anterior region broad and bluntly rounded in front, but with a narrow, pointed termination. *Type and only species:*

Anthocotyle merlucii Beneden & Hesse, 1863 (Fig. 21C, *a–e*)

Syn. *Anthocotyle merlucii americanus* MacCallum, 1916; *A. americanus* (MacCallum) Price, 1943*b*.

HOSTS. Hake, coal-fish (at Plymouth, Aberdeen, and in the North Sea, off west Ireland, at Brest and Genoa, and in Canada and U.S.A.).

LOCATION. Gills.

DIAGNOSIS. The original description was framed in very general terms, but according to Cerfontaine (1895–6): *Size:* 10–15 mm. long. *Shape:* tapering to an obtuse point anteriorly, narrowing behind the mid-body, but attaining maximum breadth just in front of the opisthaptor. *Opisthaptor:* large clamps each having a dorsal and a ventral valve, hinged anteriorly, and a wide, slit-like opening bordered by a pleated membrane. Sclerites of discocotylid type augmented by rod-like pieces which line the cavity of each clamp. Small clamps each borne on a short peduncle and not significantly different from the others as regards sclerites. Two pairs of hooks, situated on a terminal lanquet about as large as a small clamp, dissimilar and unequal in size, those of the outer pair the larger and more curved and equipped with median and lateral roots, of which the latter is the smaller. (*Note*. Price (1943*b*) thought that a third pair of hooks of minute size was overlooked by Cerfontaine.) *Gut:* mouth transversely oval, terminal, pharynx small, oesophagus short, intestine bifurcate, the crura having a few short median and more as well as larger lateral diverticula, but sending a single canal to the base of the opisthaptor, into which several simple diverticula extend. *Reproductive systems:* genital pore median and just behind the bifurcation of the gut. Testes numerous, but confined between the intestinal crura in the posterior half of the body. Cirrus having forty hooklets with simple points, vas deferens median and folded. Ovary ∩-shaped with folded limbs, situated in front of the testes in the second quarter of the body. Uterus comparatively straight. Vaginal pores ventro-lateral, vaginae continuous with the main vitelloducts and connected immediately in front of the ovary by two anastomoses. Vitellaria extending laterally along the intestine between the vaginal pores and the posterior limit of the testes. Eggs clear yellow, each having two polar filaments about as long as the capsule.

Note. In claiming validity for *A. americanus* Price (*loc. cit.*) depended upon inequality of size between the two large clamps, but this is seen to be an unreliable criterion when account is also taken of specimens found by Linton (1940) at Woods Hole, the asymmetry being the opposite of that noted by Price. It is more than probable that American and European forms are identical, therefore, and that the synonymy given above is justified.

The only other genus of the Anthocotylinae is *Winkenthughesia* Price, 1943, in which the clamps of the opisthaptor are of about equal size. The type and only species, *W. thyrites* (Hughes, 1928) Price, 1943 occurs in Australia.

Subfamily **Plectanocotylinae** Monticelli, 1903

Discocotylidae with only three pairs of clamps and without vaginae.

Genus *Plectanocotyle* Diesing, 1850

Syn. *Plectanophorus* Diesing, 1858; *Phyllocotyle* Beneden & Hesse, 1863; *Plectanocotyle* Monticelli, 1888 (in the family Octocotylidae).

DIAGNOSIS. *Shape:* elongate, anterior end of the body tapering, very extensile. *Opisthaptor:* cotylophore rather narrow, bearing three pairs of clamps and an extensile and contractile terminal process which terminates in a disk bearing three pairs of dissimilar hooklets. *Reproductive systems:* copulatory organ with a number of delicate cuticular rods (of which there may be five on each side). Testes numerous, in the posterior half of the body. Ovary elongate, curved, in front of the testes. Vaginae absent. *Species:* type, *Plectanocotyle elliptica* Diesing, 1850 and one other.

Plectanocotyle elliptica Diesing, 1850

Syn. *Plectanophorus ellipticus* Diesing, 1858.

HOST. '*Labrax mucronatus*' (recorded in Europe, but found by Kollar in America; not recorded in Britain).

LOCATION. Gills.

DIAGNOSIS. *Shape:* outline oval or broadly elliptical. *Ophisthaptor:* suckers in line along the margin of a rounded posterior extremity.

Plectanocotyle gurnardi (Beneden & Hesse, 1863) Llewellyn, 1941 (Fig. 17G)

Syn. *Phyllocotyle gurnardi* Beneden & Hesse, 1863; *Plectanocotyle lorenzii* Monticelli, 1899; *Plectanocotyle caudata* Lebour, 1908.

HOSTS. Yellow, red, streaked and grey gurnards (at Plymouth, Liverpool, Galway, Aberdeen, Cullercoats, also Brest and in Italy).

LOCATION. Gills.

DIAGNOSIS. *Shape:* elongate, tapering at both extremities. *Size:* 4–7 mm. long and about 0·8 mm. broad at the level of the ovary. *Prohaptor:* a pair of buccal suckers. *Opisthaptor:* somewhat shield-shaped cotylophore terminating in a narrow process or a disk bearing hooklets. (*Note.* Sometimes merely a small languet: see Fig. 17G, 3.) Clamps, three pairs, each borne on a short peduncle, having characteristic Discocotylid sclerites. Hooklets of the outer pair the largest, Y-shape, the stem having a curved tip; those of the inner pair similar to the outer, but smaller; those of the innermost pair very small, anchor-like. *Gut:* intestine bifurcate, caeca obscured by the vitellaria. *Reproductive systems:* genital pores side by side slightly behind the bifurcation of the gut. Copulatory organ equipped with five pairs of slender cuticular rods, those of the central pair being smaller and having knob-like terminations. Uterus little folded, opening on the left of the male pore. Testes numerous, situated in the posterior half of the body. Ovary elongate and curved, slightly in front of the testes. Vitellaria well developed, the follicles occupying the lateral regions along almost the entire length of the body and extending into the cotylophore. Eggs ovoid, pale yellow, 0·065 mm. long, each having a single polar filament about as long as the capsule.

OTHER GENUS:

Octoplectanocotyle Yamaguti, 1937, with the species *O. trichuri* Yamaguti, 1937 in Japan.

Family **GASTROCOTYLIDAE** Price, 1943 (p. 74)

Syn. Microcotylidae Taschenberg, 1879, in part; Axininae of Nicoll, 1915, in part.

Genus *Gastrocotyle* Beneden & Hesse, 1863

DIAGNOSIS. *Opisthaptor:* cotylophore a simple lateral flange bearing a single row of small clamps. Cuticular elements (sclerites) arranged as in Fig. 22D, *d*, comprising two pairs of lateral pieces (joined, but not fused distally, those of the more dorsal pair with curved, inwardly directed spurs), one pair of basal pieces (which meet the laterals), a single, curved middle piece which is bifurcate ventrally, and additional pieces which form a pair and extend from the ventral lateral towards the middle sclerite. Cotylophore also equipped with two to three pairs of dissimilar terminal hooks. *Species:* type, *Gastrocotyle trachuri* Beneden & Hesse, 1863 and one other species, *G. japonica* Ishii & Sawada, 1938, in Japan.

Gastrocotyle trachuri Beneden & Hesse, 1863

HOST. Horse mackerel.

LOCATION. Gills.

According to Beneden & Hesse (1864) the cotylophore of this species bears thirty-one to thirty-eight suckers. Parona & Perugia (1890*a*) gave the number as about thirty-five and Yamaguti (1938), who found a specimen in the same host species in Japan, twenty-seven. Idris Jones (1933 *d*) redescribed the trematode in the following (modified) terms: *Size:* 4·7 mm. long and 1·2 mm. in greatest breadth posteriorly. *Shape:* elongate, hatchet-shaped (the narrow anterior third corresponding to the 'handle', the posterior two-thirds, together with the cotylophore to the 'blade'). *Prohaptor:* a pair of oval buccal suckers measuring 0·023 × 0·015 mm. *Opisthaptor:* a cotylophore bearing a single series of thirty-two to forty clamps along one side, each about 0·08 mm. diameter, and three pairs of hooks, those of the most lateral pair the largest. *Gut:* pharynx ovoid, measuring 0·046 × 0·030 mm., oesophagus 0·2 mm. long, its walls pouched, intestine bifurcate and provided with numerous lateral diverticula, the crura uniting posteriorly. *Reproductive systems:* genital pore median, a short distance from the anterior extremity (ventral to the posterior end of the oesophagus), genital atrium muscular, spherical, 0·023 mm. diameter. Genital hooklets numbering twelve, arranged in a circlet, each having a bifurcate root and a small curved point. Testes not very numerous and situated near the posterior extremity. Ovary elongate, U-shaped, in front of the testes. Vaginal pore median, dorsal, a short distance behind the level of the genital pore. Vitellaria comprising numerous follicles distributed along the courses of the intestinal crura and their diverticula. Uterus not folded. Eggs not observed.

Note. Idris Jones did not describe the sclerites of the clamps, but gave figures (1933 *d*, Figs. 1–4) in which the shapes and arrangement of the pieces were misinterpreted. According to Beneden & Hesse (1864) the egg has two pointed filaments, each (as their figure shows) about as long as the fusiform capsule. Yamaguti observed a solitary egg in the Japanese specimen, and this was 0·25 mm. long, 0·1 mm. broad and its filaments were 0·315 mm. long. This species has been found at Plymouth, Brest, in the eastern Atlantic and in the Mediterranean (Genoa).

Family **MICROCOTYLIDAE** Taschenberg, 1879 (p. 74)

Syn. Axininae of Nicoll, 1915, in part.

Subfamily **Microcotylinae** Monticelli, 1892

With the characters of the family.

KEY TO GENERA OF MICROCOTYLINAE

I. Clamps distributed along both margins of the cotylophore, which generally lacks hooks posteriorly
  A. Clamps approximately equal in number on the two sides — *Microcotyle*
  B. Clamps decidedly unequal in number on the two sides — *Axine*

II. Clamps distributed along one margin of the cotylophore, which has hooks posteriorly — *Pseudaxine*

Note. A number of other genera are known in North America, Japan, China and the Galapagos Isles, including *Bicotylophora* Price, 1936 (with one species in U.S.A.), *Protomicrocotyle* Johnston & Tiegs, 1922 (with one species in U.S.A.

and another in Costa Rica), *Cestrocolpa* Meserve, 1938 (with one species in the Galapagos Isles, another off the coast of Mexico) and *Axinoides* Yamaguti, 1938 (with two species in Japan and two in North America).

Genus *Axine* Abildgaard, 1794, *nec* Oken, 1835, a crustacean

Syn. *Heteracanthus* Diesing, 1836; *Axime* Moulinie, 1856.

DIAGNOSIS. *Body:* asymmetrical in regard to the opisthaptor. *Prohaptor:* a pair of buccal suckers. *Opisthaptor:* cotylophore extending along one postero-lateral margin of the body, bearing numerous small clamps arranged obliquely in relation to the longitudinal axis. *Gut:* intestine bifurcate, caeca long and provided with median and lateral diverticula. *Reproductive systems:* genital pore median, near the bifurcation of the intestine. Testes numerous, in the posterior half of the body. Ovary elongate, curved, in front of the testes. Vagina, generally with a lateral pore on the left, and without hooklets. Eggs with two polar filaments. *Species:* type, *Axine belones* Abildgaard, 1794, three species in Japan, one in America and three in the Galapagos Isles.

Axine belones Abildgaard, 1794

Syn. *Heteracanthus pedatus* Diesing, 1836; *Axine triglae* Beneden & Hesse, 1863; *Axine orphii* Benèden & Hesse, 1863.

HOST. Gar-fish (at Plymouth, in the North Sea and Baltic, at Brest and Trieste).

LOCATION. Gills.

DIAGNOSIS. A good description of this species was given by Lorenz (1878). *Size:* 4–8 mm. long. *Shape:* very elongate, tapering gradually from the opisthaptor to the anterior extremity. *Colour:* translucent, milky white, clear grey laterally. *Prohaptor:* a pair of buccal suckers. *Opisthaptor:* terminal, but obliquely set, the left margin the more anterior, clamps (Lorenz pointed out the mistake made by Beneden & Hesse in deeming them suckers) numbering fifty to seventy, arranged in a single row. Median sclerite of each clamp having unequal limbs, the shorter on the left, marginal sclerites numbering four on each side, supplemented by a small 'Spange' (? a process of one of the lateral marginals) on each side. *Gut:* pharynx ovoid, smaller than the buccal suckers and situated behind and between them, oesophagus long, intestine having numerous short median and lateral diverticula along its entire length. *Reproductive systems:* genital pore median, slightly in front of the bifurcation of the intestine, genital atrium muscular, its lining bearing two lateral groups of twelve to twenty hooklets arranged in a double row, and a ventral hemispherical elevation provided with a small circlet of eight to twelve hooklets. Cirrus fusiform and muscular, with a thick basal annulus with sixteen to twenty-four hooklets in its ventro-lateral part. Vas deferens long and convoluted, seminal vesicle not far from the base of the cirrus. Testes (represented as a single testis) numerous, filling the space between the caeca behind the ovary, which is J-shaped and situated on the left, the long limb continuous with the oviduct. Uterus thick-walled, almost median, entering the genital atrium dorsal to the cirrus. Vagina with three regions, a thick and muscular basal, a capacious middle, and a trumpet-shaped terminal part, vaginal pore lateral or slightly dorsal. Vitellaria lateral between the middle segment of the vagina and the opisthaptor. Eggs apparently laid singly, each having two polar processes, the shorter sometimes represented by a knob-like structure.

Note. '*Axine triglae*', which was found on the yellow gurnard, resembles *A. belones*, but showed *situs inversus* of the opisthaptor, which is directed towards the left. This is not a reliable character of specific distinction. Lorenz had no misgivings about the identity of '*Axine orphii*' with *A. belones*, and this too is added to the synonymy.

Genus *Pseudaxine* Parona & Perugia, 1890

DIAGNOSIS. *Opisthaptor:* cotylophore unilateral, with a single row of clamps along its oblique lower edge, terminating in a short posterior appendage bearing two or more pairs of hooks. *Species:* type, *Pseudaxine trachuri* Parona & Perugia, 1890 and three others, two in Japan and one in Mexico.

Pseudaxine trachuri Parona & Perugia, 1890

HOST. Horse mackerel (Italy, Japan; in Britain at Plymouth).

LOCATION. Gills.

Note. Specimens of this species from Japan and measuring 2·7–3·1 mm. long have an opisthaptor bearing twenty to twenty-two clamps, but still fewer than in European specimens, and two pairs of hooklets with bifurcate bases on the terminal process. Members of the pair at the base of the appendage are twice as large as and more strongly curved than those of the pair near the tip; length about 0·024 mm.

Genus *Microcotyle* Beneden & Hesse, 1863 (Fig. 6F)

DIAGNOSIS. *Opisthaptor:* cotylophore more or less distinct from the body proper, or forming a frill around its posterior half, generally symmetrical, but sometimes having more clamps on one side than on the other, frequently with its anterior part projecting from and parallel with the ventral surface of the body, occasionally set at an angle to the main axis of the body. Clamps very numerous. Hooklets absent. *Reproductive systems:* genital atrium sometimes very complex, having an anterior, or several lateral, or paired posterior outgrowths, sometimes unarmed, sometimes with one type or several types of hooks or spines. Vaginal pore with or without spines, frequently associated with paired dorsal suckers. Ovary elongate, curved and with a concavity facing posteriorly, in front of the testes. Eggs ovoid and with two polar filaments. *Species:* type, *Microcotyle donovani* Beneden & Hesse, 1863 and about sixty-seven others. Meserve (1938) compiled a list of fifty-eight species (two of them also included in *Gotocotyle* Ishii, 1936) and added a new one. At least twenty-two species occur in Japan, about twenty in North America, about seventeen in Europe, six in Australia, and different single species in Egypt, Java, the Galapagos Isles and the Philippines. It is very improbable that all these species are valid, and about many of them our knowledge is very scanty.

EUROPEAN SPECIES AND THEIR HOSTS (all are parasitic on the gills):

Microcotyle donavini Beneden & Hesse, 1863; comber wrasse (France); ballan wrasse (Scotland; T. Scott, 1905) (Plymouth Aquarium, 1940; Sproston, *in litt.*).
M acanthurum Parona & Perugia, 1890; *Brama rayi* (Italy).
**M. alcedinis* Parona & Perugia, 1890; '*Smaris alcedo*' and '*Moena vulgaris*' (Italy).
M. canthari Beneden & Hesse, 1863; black sea bream (France, Italy).
M. centrodonti Brown, 1929; common sea bream (London, Zool. Soc. Aquarium).
M. chrysophryi Beneden & Hesse, 1863; gilt-head (Italy).
M. draconis Briot, 1904; greater weever (English Channel and North Sea).
M. erythrini Beneden & Hesse, 1863; pandora (France) and bogue and axillary bream (Italy).
**M. fusiformis* Goto, 1894; *Centronotus rubulosus* (Japan), butter-fish (Britain; Cullercoats (Crofton, 1940) and Plymouth).
M. labracis Beneden & Hesse, 1863; bass (France); also Roscoff (Vogt, 1878), Scotland (T. Scott, 1905) and Liverpool (A. Scott (1904), also Johnstone).
**M. lichiae* Ariola, 1899; horse mackerel (Italy).
**M. mormyri* Lorenz, 1878; *Pagellus mormyrus* (Italy).
**M. mugilis* Vogt, 1878; grey mullet (Roscoff (Vogt, 1878) and Japan (Yamaguti, 1938)).
M. pancerii Sonsino, 1891; *Umbrina cirrhosa* (Italy).
**M. salpae* Parona & Perugia, 1890; *Box salpa* (Italy).
**M. sargi* Parona & Perugia, 1889, 1890; *Sargus rondeletii*, *S. vulgaris*, *S. annularis* and *S. salviani* (Italy).
**M. trachini* Parona & Perugia, 1889; *Trachinus radiatus* (Italy).

MacCallum, G. A. and MacCallum, W. G. (1913) compiled a table showing the characters of thirty species of *Microcotyle*, including those marked with an asterisk in the list given above. They differ in size and shape, length and shape of the cotylophore and number of clamps borne thereon,* absence or presence and nature of the genital hooklets, and the shapes and sizes of the eggs. According to Beneden & Hesse (1864), *M. donavini* is characterized by a relatively large mouth surrounded by a fleshy lip having a median incision, numerous genital hooklets which are flat, triangular and sharp, an oval cotylophore (number of clamps not specified, but twenty-four pairs figured) and reddish eggs, each of which has a delicate pointed filament about as long as the capsule at one pole and a slightly longer and stouter process having the shape of a crozier at the other. Of the other species described by Beneden & Hesse, *M. canthari* is very similar to the type-species (and may prove to be identical with it), but has thirty to forty genital hooklets having curved points. The same type of genital hooklet was described for *M. chrysophryi*, which was admitted to be similar to *M. canthari*, but said to differ in the character of the mouth and the buccal suckers. It is certainly identical with *M. canthari* and, if this species falls, with the type-species also. One indubitable synonym of *M. donavini* is the form *M. erythrini*, which has similar genital hooklets and eggs. *M. labracis* seems to be a valid species, and is said by the authors to have genital hooklets with triple points of unequal sizes and reddish eggs each of which has a simple filament at one pole and a peculiar process terminating in an anchor-like structure with a central perforation at the other.

* The number of clamps cannot be accepted as a character of specific distinction, Remley (1942) having shown that it increases during development in *Microcotyle spinicirrus* by more or less continual lengthening of the cotylophore and the formation of new clamps posteriorly.

Species of this genus most likely to be met with in Britain are *M. donavini*, *M. centrodonti*, *M. fusiformis*, *M. draconis* and *M. labracis*. The second of these species has been described in considerable detail (Brown, 1929) and can be used for representing the genus here.

Microcotyle centrodonti Brown, 1929

HOST. Common sea bream.

LOCATION. Gills.

DIAGNOSIS. *Shape, colour and size:* elongate, tapering towards the extremities, more rounded anteriorly, transparent, or whitish, 2·5–4·5 mm. long. *Movements:* a looping movement effected by the use of the suckers carries the fluke swiftly across the gills. *Cuticle:* thin, with a thicker underlying subcuticle having a slightly striated appearance. *Prohaptor:* a pair of transversely ovoid buccal suckers, 0·06 mm. along the major axis. *Head (sticky) glands:* two lateral and one median and dorsal, situated at the anterior extremity. *Opisthaptor:* numerous clamps arranged in a paired series on a prominent cotylophore, which is invariably longer than the body (it may be twice as long) and is continued forward some distance in front of its origin on the ventral surface of the body. *Clamps:* sixty to eighty pairs, seventy-three pairs common in larger specimens, each clamp transversely oval or rectangular in outline, largest (0·08 mm. broad) where the cotylophore joins the body, smallest (0·03 mm. broad) at posterior end of cotylophore. *Sclerites:* five pieces; a hollow (?), unpaired middle piece of irregular U-shape (occupying the middle of the clamp) and two pairs of lateral pieces. Median sclerite having unequal anterior and posterior parts, the former bifurcate at its tip, the latter wedge-shaped. Anterior lateral sclerites extending from the central sclerite along the antero-ventral surface of the clamp, then curving dorsally to terminate on its posterior surface. Posterior lateral sclerites situated in the ventral part of the posterior wall, curving dorsally to terminate on the dorsal wall. *Gut:* funnel-like mouth tube (prepharynx) not quite terminal, pharynx small and spherical, oesophagus of medium length, caeca long, terminating near cotylophore, with numerous long lateral and shorter median diverticula. Lining of the intestine not a continuous epithelium, but instead a membrane containing scattered cells. *Reproductive systems:* genital atrium large (0·05 × 0·07 mm.), common genital pore situated at its anterior end, slightly behind the bifurcation of the gut. Vas deferens opening on a conical papilla having rod-like bodies in its walls. Testes numerous (14–23) situated in the posterior quarter of the body proper. Ovary elongate, much convoluted, a little in front of the testes. Vitellaria well developed, the follicles distributed laterally along the courses of the intestine and the diverticula. Uterus straight and wide, narrowing as it approaches the genital atrium, having a non-ciliated lining. Vagina narrow, opening in a shallow pit in the median dorsal line slightly behind the genital atrium, devoid of spines. Eggs ovoid, measuring 0·050 × 0·015 mm., each with two polar filaments (that at the anterior pole very long and convoluted, terminating in a point, that at the posterior pole much shorter, but longer than egg proper and terminating in a knob-like button with eight points along its edge). *Dorsal suckers:* one pair of openings of the sucker-like pits, having walls containing rod-like bodies and a tongue-like retractile and protractile process bearing twelve simple spines, immediately in front of vaginal pore (probably an aid to adhesion during copulation; present only in this species and in *M. alcedinis* and *M. canthari*).

Notes. This species was discovered on hosts in captivity in the aquarium of the Zoological Society, London. It is presumably a mucus feeder, no blood being found in the gut. Infected fishes seem to have been suffocated by excessive secretion of mucus, which choked up the gills. Considerable host-specificity was shown; the gills of Couch's sea bream in the same tank with heavily infected common sea bream were free of this species, though one specimen bore two individuals of another species.

Other species likely to be found in Britain can be identified tentatively by the number of clamps on the opisthaptor. It must be made clear, however, that much individual variation occurs in the numbers found in members of one and the same species; the following numbers are significantly different from those given above, however, except in two instances. These two species can be separated from *M. centrodonti* by their lack of the dorsal suckers: *M. draconis* (11 pairs of clamps); *M. fusiformis* (30 pairs); *M. labracis* (63 pairs); *M. chrysophryii* (78 pairs).

Family **DICLIDOPHORIDAE** Fuhrmann, 1928 (p. 74)

Syn. Choricotylidae Rees & Llewellyn, 1941; Dactylocotylidae Brinkmann, 1942*a*, in part.

Subfamily **Diclidophorinae** Cerfontaine, 1895, *sensu* Price, 1943

Syn. Dactylocotylinae Brinkmann, 1942*a*, in part.

Diclidophoridae in which the suckers are closed and clamp-like.

KEY TO GENERA OF DICLIDOPHORINAE

I. Opisthaptor comprising a cotylophore which is sharply marked off from the rest of the body — *Diclidophoroides*

II. Opisthaptor not comprising such a cotylophore as the above
   A. Testes situated exclusively behind the ovary — *Octodactylus*
   B. Testes situated in front of and behind the ovary — *Diclidophora*

Genus *Diclidophora* Diesing, 1850

Syn. *Dactycotyle* Beneden & Hesse, 1863; *Dactylocotyle* Marschall, 1873; *Dactylocotyle* 'Parona & Perugia, 1889' of Brinkmann, 1942*a*, in part.

DIAGNOSIS. *Opisthaptor:* comprising an indistinct cotylophore equipped with four pairs of pedunculate, clamp-like suckers (reinforced suckers). *Reproductive systems:* testes numerous and distributed mainly behind, but also in front of the ovary. Eggs with polar filaments. *Species:* type, *Diclidophora merlangi* Kuhn, in Nordmann, 1832 (Syn. *Diclidophora longicollis* Diesing, 1850) and several others, *D. denticulata* (Olsson, 1876), *D. luscae* (Beneden & Hesse, 1863) and *D. pollachii* (Beneden & Hesse, 1863) all occurring in Britain.

Diclidophora merlangi (Kuhn, in Nordmann, 1832) Kroyer, 1834–40 (Fig. 17 D)

Syn. *Octostoma merlangi* Kuhn, in Nordmann, 1832 (manuscript name); *Octobothrium platygaster* Leuckart, 1842; *Diclidophora longicollis* Diesing, 1850; *Octoplectanum longicolle* (Diesing) Diesing, 1858; *Octobothrium merlangi* (Kuhn) of Beneden, 1856, 1858, Monticelli, 1888, Scott, 1900 and Lebour, 1908; *Dactylocotyle merlangi* (Kuhn) of Cerfontaine, 1895, and possibly others, including *Diclidophora luscae* (Beneden & Hesse, 1863) and *D. polachii* (Beneden & Hesse, 1863), which are supposed to parasitize the bib and poor cod and the pollack respectively.

HOST. Whiting.

LOCATION. Gills.

The validity of the specific name *merlangi* has been put beyond dispute by Price (1943 *a*, footnote 5, p. 45). The species is common in this country and has been found at Plymouth, Galway, in the Irish Sea and the Firth of Forth, at Aberdeen, Liverpool and Cullercoats. It occurs also in the Mediterranean (at Naples). I have specimens which answer to the following diagnosis. *Size:* up to 9 mm. long and 2·6 mm. in greatest breadth posteriorly (Lebour's specimens were up to 11 × 4 mm.). *Shape:* anterior third of the body narrow (0·8 mm. broad behind, tapering to a point in front), posterior region broadening sharply towards the end of the second third, then tapering gradually towards the opisthaptor, about 2 mm. broad at the anterior end of the latter. *Prohaptor:* a pair of buccal suckers about 0·012 mm. diameter. *Opisthaptor:* not clearly marked off from the rest of the body, occupying the posterior five-eighteenths, tapering gradually towards the origins of the most posterior suckers (1·5 mm. behind those of the most anterior) and here 0·9 mm. broad as against 2 mm. anteriorly. The four pairs of suckers borne on peduncles of about equal length (0·8–0·9 mm. long) which are directed obliquely outwards and backwards. Sclerites resembling those of *D. denticulata*, but spines of the anterior valve absent and median sclerite having a small number of tubercles arranged in about three longitudinal rows, sometimes on keel-like ridges. *Gut:* mouth not quite terminal, pharynx much larger than the buccal suckers (0·23 mm. diameter), caeca with branched lateral diverticula, some of which pass into the peduncles of the suckers. *Reproductive system:* genital pore median, 0·6 mm. from the anterior extremity and 0·22 mm. behind the posterior end of the pharynx. Genital hooklets numbering sixteen, with ridged points (appearing like a fork with two closely situated prongs, but strictly unforked). Gonads as for the genus. Vitellaria very well developed, practically filling the body with their follicles from a level 0·2 mm. behind the genital pore to the posterior end of the opisthaptor, some follicles extending into the proximal halves of the peduncles. Eggs not observed, but probably having two filaments, the longer terminating in a button-like swelling.

Diclidophora denticulata (Olsson, 1876) Price, 1943

Syn. *Octobothrium denticulatum* Olsson, 1876; *Dactylocotyle denticulatum* (Olsson) of Cerfontaine, 1895; *D. carbonarii* Cerfontaine, 1895 (in relation to figures).

HOSTS (in Britain). Coal-fish, poor cod, hake.

LOCATION. Gills.

This species was discovered on the coal-fish in the Skagerrack, but occurs outside Europe (at Woods Hole, U.S.A. and in Nova Scotia, Canada). In Britain it has been found at Plymouth, Aberdeen and off western Ireland. Cerfontaine (1895) found that the trematodes attach themselves to the second or third gills, the opisthaptor directed towards the branchial arch and the clamps grasping groups of filaments. When detached from the host the parasite alters its shape, the clamps curving on their peduncles towards the ventral surface of the body. Cerfontaine also described the structure of the clamps in considerable detail. Each clamp has a dorsal and a ventral valve, but during attachment the whole clamp is rotated through about 90° so that the ventral valve is anterior in topographical position. The rather complicated structure need not be described: suffice it here to state that the species can be recognized by the fact that centrally the external half of the anterior (ventral) valve bears about thirty (possibly as many as forty) small spines which occupy a superficial position and are directed diagonally forward and inwards. The trematode is rather small (about 7 mm. long).

Diclidophora luscae (Beneden & Hesse, 1863) Price, 1943

Syn. *Dactycotyle luscae* Beneden & Hesse, 1863; *Dactylocotyle luscae* (Beneden & Hesse) Cerfontaine, 1895; *Octobothrium luscae* (Beneden & Hesse) Taschenberg, 1879.

HOSTS. Bib, poor cod (in Britain and at Brest and Roscoff).

LOCATION. Gills.

DIAGNOSIS. The original description was sketchy and without a figure, and it gave the impression that this species could easily be confounded with *D. pollachii*. Cerfontaine at first (1895) failed to find it, but later (1898*a*) identified it twice at Roscoff and gave its characters as: *Size:* 4.7 mm. long. *Shape:* resembling *D. pollachii*, but more squat, the attenuated anterior end shorter, enlargement to maximum breadth occurring sharply near the opisthaptor. *Opisthaptor:* trapezoidal and broadest anteriorly, the peduncles robust, each bearing a characteristic clamp having irregular nodules on the ventral valve. *Gut:* pharynx spherical, oesophagus fairly long, bifurcation of the intestine slightly behind the male pore, but far in front of the female pore, median and lateral diverticula not numerous, anastomoses very evident, ramifications extending into the peduncles of the opisthaptor. *Reproductive systems:* male pore decidedly in front of the female pore. Testes numerous, situated between the components of the intestine. Genital hooklets generally numbering ten, each with a short, straight base and a curved blade. Ovary folded into an N-shape, situated between the peduncles of the first pair. Oviduct and uterus straight and median, ootype club-shaped and situated behind the ovary. Receptaculum seminis anterior to the ovary. Vitellaria as in other species. Vagina short and opening ventrally. Eggs having each a relatively short, crozier-like filament at one pole and a long filament terminating in a funnel-shaped structure having a milled margin at the other. *Note.* The short, bent filaments project anteriorly *in utero* and the long filaments become entangled.

Diclidophora pollachii (Beneden & Hesse, 1863) Price, 1943

Syn. *Dactycotyle pollachii* Beneden & Hesse, 1863; *Dactylocotyle pollachii* (Beneden & Hesse) Cerfontaine, 1895; *Octobothrium pollachii* (Beneden & Hesse) Taschenberg, 1879.

HOST. Pollack (at Liverpool, Aberdeen, in Irish waters, at Brest and Roscoff).

LOCATION. Gills.

DIAGNOSIS. According to Cerfontaine (1895): *Size:* 8–13 mm. long. *Shape:* pointed anteriorly, but broadening abruptly in the anterior half of the body and uniformly broad in the posterior half. *Opisthaptor:* clamps and their peduncles not as robust as in *D. denticulata*, but larger than in *D. pollachii*. *Gut:* intestinal crura having fewer median and lateral diverticula than in either of the above species especially in the anterior region, and a few anastomoses near the opisthaptor. *Reproductive systems:* genital hooklets generally numbering fourteen (rarely eleven, ten or, even, nine). Testes numerous, many of them lateral to the intestine. Ovary only slightly in front of the opisthaptor. Eggs each having two polar filaments, the shorter crozier-like, but more irregular than in *D. denticulata*.

Genus *Octodactylus* Dalyell, 1853

Syn. *Pterocotyle* Beneden & Hesse, 1863.

DIAGNOSIS. *Opisthaptor:* as in *Diclidophora*. *Reproductive systems:* testes numerous and situated entirely behind the ovary. Eggs generally without polar filaments. *Species:* type, *Octodactylus palmata* (F. S. Leuckart, 1830) and two others, *O. minor* (Olsson, 1876) and *O. morrhuae* (Beneden & Hesse, 1863), all three of which occur in Britain.

Octodactylus palmata (F. S. Leuckart, 1830) Price, 1943 (Fig. 17E)

Syn. *Octodactylus inhaerens* Dalyell, 1853; *Octobothrium digitatum* Rathke, 1843; *Pterocotyle palmata* Beneden & Hesse, 1863, and of Lebour, 1908; *Dactylocotyle molvae* Cerfontaine, 1895; *D. palmatum* (F. S. Leuckart) Cerfontaine, 1895.

HOST. Ling (Dalyell stated that twenty-nine worms were removed from part of the gills of one host).

LOCATION. Gills.

This species has appeared in Irish waters, the Moray Firth, the Clyde and at Aberdeen and Cullercoats. Cerfontaine (1895) gave the length as 10–20 mm., up to 30 mm. after death, the breadth as fairly uniform throughout the length of the body, the shape of the anterior extremity as very obtuse. The clamps and peduncles are said to be sturdy, the cotylophore rounded behind so that the posterior clamps are close together and the entire haptor has a palmate appearance. There are sixteen genital hooklets and the eggs are large and devoid of filaments. I have a specimen (obtained from a ling at Plymouth) which has the following characters. *Size:* 22·5 mm. long, 4·2 mm. in greatest breadth. *Shape:* tapering from the mid-body towards both extremities, but of uniform girth (2·3 mm. broad) in a region 0·2–0·45 mm. from the anterior extremity. *Prohaptor:* a pair of buccal suckers 0·18 mm. long and 0·14 mm. broad. *Opisthaptor:* cotylophore indistinct, 2 mm. long, tapering and 0·54 mm. broad posteriorly. Clamps uniform and of similar sizes (0·6 × 0·4 mm.), the peduncles 0·6–0·8 mm. long and 0·5 mm. broad, those of the posterior pair slightly the largest. Sclerites in general as in *Diclidophora*. *Gut:* mouth 0·18 mm. wide and 0·1 mm. from the anterior end of the body, oesophagus very short, intestinal crura with lateral diverticula and median anastomosing branches posteriorly, one main branch serving each peduncle. *Reproductive systems:* genital pore median, about 1 mm. from the anterior extremity. Genital hooklets numbering eighteen, each with a plate-like root and a curved and deeply grooved (but not bifurcate) point. Uterus with many folds in the narrow anterior region. Eggs very numerous (more than 200 *in utero*), devoid of filaments, 0·22 × 0·09 mm., yellow, having an inner lining with thread-like processes, an embryo with eye-spots developing before it is laid.

Octodactylus minor (Olsson, 1876) Price, 1943

Syn. *Octobothrium palmatum* forma *minor* Olsson, 1868; *O. minor* Olsson, 1876; *Dactylocotyle minus* (Olsson) of Gallien, 1937.

HOSTS (in British waters). Poutassou, whiting.

This species occurs in Scandinavian waters (Bergen), but was found off the south-west of Ireland by Gallien (1937), who described it in the following terms. *Size:* up to 5·2 mm. long and 1·1 mm. broad at the middle of the body. *Colour:* brown. *Shape:* elongate, very flat, tapering towards the extremities. *Opisthaptor:* cotylophore one-tenth as long as the body and indistinct, bearing four pairs of pedunculate clamps and a terminal languet. *Reproductive systems:* genital pore just behind the pharynx, cirrus equipped with twelve sickle-shaped hooklets. Testes not very numerous, situated posteriorly (not anteriorly as Olsson stated). Vagina present. Uterus containing a single fusiform egg having two short polar prolongations.

Octodactylus morrhuae (Beneden & Hesse, 1863) Price, 1943

Syn. *Pterocotyle morrhuae* Beneden & Hesse, 1863; *Dactylocotyle morrhuae* (Beneden & Hesse) St Remy, 1898.

HOSTS. Cod, whiting.

LOCATION. Gills.

This species has been found in Irish waters and at Aberdeen as well as elsewhere in Europe (Brest). According to Beneden & Hesse (1863) it is 14–15 mm. long, the most anterior peduncles are shortest, the most posterior the longest, the cotylophore is deeply cleft behind and the clamps of the two sides diverge, and the eggs are large, ovoid and devoid of filaments.

Octodactylus macruri (Brinkmann, 1942) n. comb.

Syn. *Dactylocotyle macruri* Brinkmann, 1942.

HOST. *Macrurus rupestris* Gunnerus (caught in the Skagerrack).

LOCATION. Gills.

DIAGNOSIS (after Brinkmann, 1942*a*). *Size:* 1·25–3·5 mm. long (mean 2·25 mm.) and 0·5–0·75 mm. in greatest breadth posteriorly. *Shape:* tongue-shaped or lanceolate, tapering anteriorly. *Opisthaptor:* cotylophore broad, merging imperceptibly with the rest of the body and equipped with four pairs of uniform clamps set on short peduncles laterally. Clamps each comprising an anterior (ventral) and a posterior (dorsal) valve, the former with three sclerites, the latter with six, together with a number of delicate rod-like bodies. *Reproductive systems:* genital pore median, close behind the pharynx. Testes numerous (twenty) and posterior. Cirrus bulbous, about 0·06 mm. diameter and equipped with a variable number (ten to fourteen) hooklets. Eggs not observed.

Genus *Diclidophoroides* Price, 1943

DIAGNOSIS. *Opisthaptor:* clearly set off from the rest of the body. *Reproductive systems:* testes situated behind the ovary. *Other characters:* similar to *Octodactylus*. *Species:* type, *Diclidophoroides maccallumi* Price, 1943, and possibly one other, *D. phycidis* (Parona & Perugia, 1889) (Syn. *Dactylocotyle phycidis* Parona & Perugia, 1889; *Dactycotyle phycidis* (Parona & Perugia) of Rees & Llewellyn, 1941).

Diclidophoroides maccallumi Price, 1943

Syn. *Diclidophora merlangi* MacCallum, 1917; *Dactylocotyle minor* (Olsson, 1876) of Manter, 1926; *D. phycidis* Parona & Perugia, 1889 of Stafford, 1904; *Choricotyle merlangi* (MacCallum, 1917) of Llewellyn, 1941.

This species is confined to the North American continent and has been redescribed by Price (1943 *a*), who gave adequate reasons for regarding the original name of the species as a homonyn and accordingly provided a new name.

The species *Diclidophoroides phycidis* (Parona & Perugia, 1889), which was tentatively assigned to this genus by Price (1943 *a*), was discovered on the greater fork-beard in Italy. According to Price the original description is not adequate for making a definite generic assignment.

Subfamily **Cyclocotylinae** Price, 1943

Syn. Diclidophorinae Cerfontaine, 1895, in part; Dactylocotylinae Brinkmann, 1942*a*, in part.

Diclidophoridae equipped with true cup-like suckers.

KEY TO EUROPEAN GENERA OF CYCLOCOTYLINAE

I. Vaginae absent	*Cyclocotyla*
II. Vaginae present	*Diclidophoropsis*

Note. Price (1943*a*) included five other genera in the subfamily: *Cyclobothrium* Cerfontaine, 1895 (with three species in Japan); *Heterobothrium* Cerfontaine, 1895 (which is monotypic and also confined to Japan); *Neoheterobothrium* Price, 1943 (with two species, both American); *Cyclocotyloides* Price, 1943 (monotypic) and *Pedocotyle* MacCallum, 1913 (monotypic; American). Price suggested that *Octobothrium leptogaster* Leuckart, 1830 possibly belongs to this genus, but we have seen that it forms the type of a subfamily of Discocotylidae.

Genus *Cyclocotyla* Otto, 1823

Syn. *Octostoma* Otto, 1823, *nec* Kuhn, 1829; *Cyclostoma* Otto, 1823, *nec* Lamarck, 1799; *Cyclobothrium* Cerfontaine, 1895, in part; *Choricotyle* Beneden & Hesse, 1863; *Diclidophora* Diesing, of Goto, 1894, in part; *Mesocotyle* Parona & Perugia, 1889.

DIAGNOSIS. *Opisthaptor:* clearly marked off from the rest of the body, the suckers pedunculate or almost sessile and fairly equally spaced out. *Reproductive systems:* genital atrium not muscular, cirrus equipped with hooklets, testes situated behind the ovary, vitellaria extending into the opisthaptor, vaginae absent. *Species:* type, *Cyclocotyla bellones* Otto, 1823 and about eleven others, several of which occur in Europe, two in America, two in Japan and one in the Galapagos Isles.

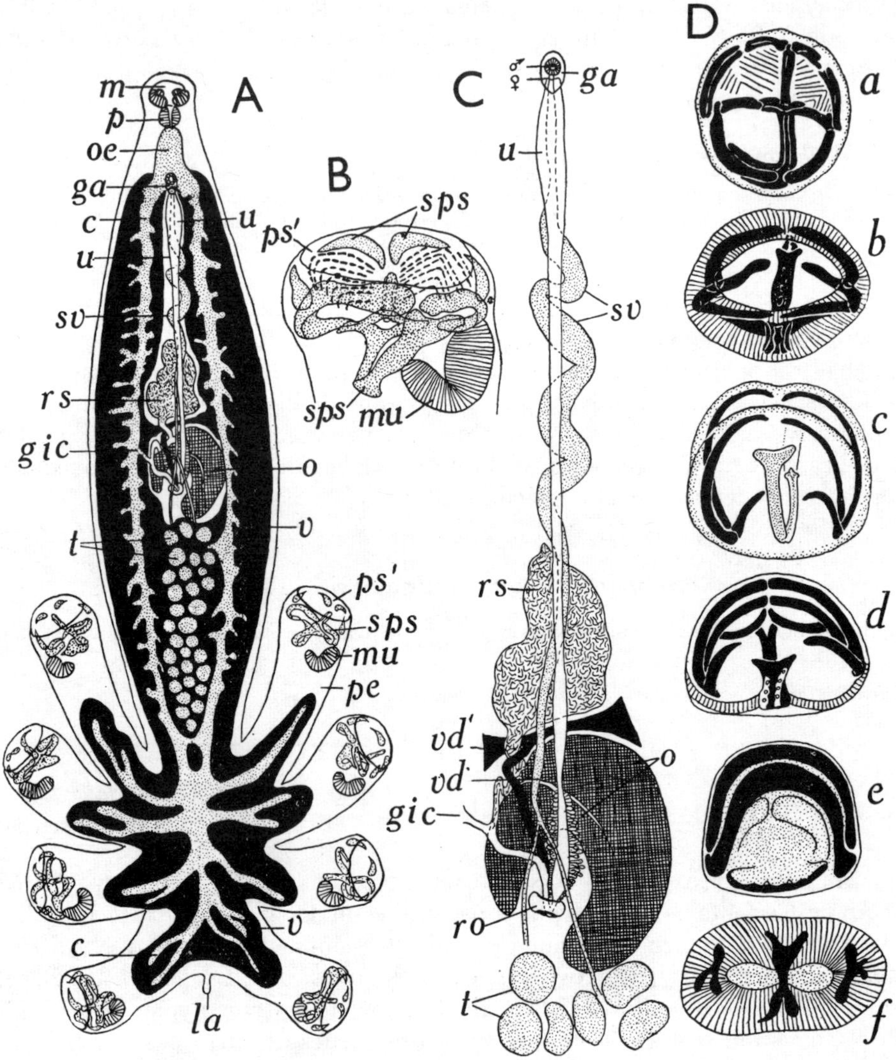

Fig. 22. A–C, *Cyclocotyla chrysophryi* (Diclidophoridae): A, the entire animal (ventral view); B, one of the suckers of the left side (ventral view); C, diagram of the reproductive systems. D, types of suckers or clamps occurring in the Diclidophoroidea: (*a*) Diclidophoridae; (*b*) Discocotylidae; (*c*) Microcotylidae; (*d*) Gastrocotylidae; (*e*) Mazocraëidae; (*f*) Hexostomatidae. (A–C, after Llewellyn, 1941*c*; D, after Price, 1943*a*.)

Cyclocotyla bellones Otto, 1823

As Price pointed out (1943*a*), this species, discovered on the dorsal skin of a gar-fish at Naples, was described only superficially and is similar to *Cyclocotyla charcoti* (Dollfus, 1922) (probably the species represented in my Fig. 6D). The latter species (Syn. *Cyclobothrium charcoti* Dollfus, 1922; *Choricotyle charcoti* (Dollfus) of Llewellyn, 1941*a*; *Diclidophora* sp. Fuhrmann, 1928), which Price regards as possibly identical with the type, is (according to Dollfus) 3–4·7 mm. long and almost as broad when contracted. Anteriorly a constriction marks off a cephalic lobe, which is capable of great extension, however, so that the body becomes flask-shaped, the narrow anterior region occupying two-fifths of the total length. The cotylophore is almost circular, the suckers uniform and borne on short peduncles. Each sucker has eight sclerites arranged as in a typical diclidophorid, and numerous rod-like bodies such as occur in species of *Diclidophora*. The internal organs are incompletely known, but the genital hooklets number six and the eggs measure 0·354–0·375 × 0·062–0·076 mm. (i.e. are exceptionally large) and have a short filament at either pole. *Cyclocotyla charcoti* was found on the body of a crustacean (*Cymothoa* (*Meinertia*) *oestroides*), itself parasitic in the buccal cavity of the horse mackerel and the bogue (off Gijon, northern Spain).

Cyclocotyla chrysophryi (Beneden & Hesse, 1863) Price, 1943 (Fig. 22A–C)

Syn. *Choricotyle chrysophryi* Beneden & Hesse, 1863; *C. chrysophris* (Beneden & Hesse) Monticelli, 1888; *C. chrysophrii* (Beneden & Hesse) Cerfontaine, 1898.

HOSTS. Gilt-head, common sea bream.

LOCATION. Gills.

This species was described in a very general manner from specimens about 6 mm. long by Beneden & Hesse, but has been redescribed in considerable detail by Llewellyn (1941*c*) in the following terms. *Size:* about 5 mm. long and 1 mm. in greatest breadth. *Shape:* as in Fig. 22A. *Colour:* brown. *Prohaptor:* a pair of buccal suckers 0·10 mm. diameter. *Opisthaptor:* four pairs of pedunculate suckers (cup-like and with circular or oval apertures) and a terminal languette 0·11 mm. long and 0·03 mm. broad. Peduncles diminishing in length in a posterior direction (1·0, 0·75, 0·50 and 0·46 mm. long respectively), the most anterior directed forwards, the most posterior backwards and intervening ones in general conformity. Sclerites eight in number (arranged as in Fig. 22B), excluding numerous rod-like bodies. The base of each sucker containing a muscular cup. *Gut:* intestine bifurcate, the crura uniting near the opisthaptor (into which diverticula penetrate, one main branch entering each peduncle) and bearing median and lateral diverticula. *Reproductive systems:* genital pore 0·37 mm. behind the almost terminal mouth. Cirrus spherical, 0·03 mm. diameter, equipped with eight bifid (?) hooklets, rarely nine. Testes numerous (about thirty), ovary folded and situated just in front of the testes. Eggs not observed. Found at Brest and on the Irish Atlantic Slope (lat. 53° 30′–54° N., long. 11° 40′ W.) at an average depth of 138 fathoms.

Cyclocotyla pagelli (Gallien, 1937) Price, 1943

Syn. *Diclidophora pagelli* Gallien, 1937.

HOST. Common sea bream.

LOCATION. Gills.

This species was discovered off south-west Ireland (lat. 53° 17′ N., long. 13° 08′ W.) and described from a single specimen, which has the following characters. *Size:* 3·4 mm. long and 1 mm. broad. *Shape:* broadly rounded anteriorly, gradually narrowing posteriorly, thick and fleshy (but not very muscular). *Opisthaptor:* cotylophore elongate (0·9 mm. long) tapering posteriorly, bearing four pairs of suckers on thick lateral peduncles which diminish in length posteriorly. Sclerites numbering eight, relatively thick, the median pieces arranged in an arc which gives the entire organ great depth. Base of each sucker comprising a thick, muscular cup. *Reproductive systems:* genital pore near the pharynx. Cirrus ill-developed, but larger than in *C. chrysophryi* (0·06 mm. diameter), hooklets present, but their number doubtful (more than 8?). Eggs not observed.

Note. The most noteworthy difference between this and the preceding species is the clearly delimited cotylophore and almost contiguous origins of the peduncles of the most anterior suckers.

Other Species of Cyclocotyla

Cyclocotyla labracis (Cerfontaine, 1895) Price, 1943

Syn. *Diclidophora labracis* Cerfontaine, 1895; *Choricotyle labracis* (Cerfontaine) Llewellyn, 1941.

This species was discovered on the gills of the bass in the North Sea. It is oval, broadly rounded anteriorly and has an indistinct cotylophore with laterally arranged peduncles of equal size (short) and a terminal languette. The cirrus is equipped with eight hooklets.

Cyclocotyla squillarum (Parona & Perugia, 1889) Price, 1943

Syn. *Mesocotyle squillarum* Parona & Perugia, 1889.

This species was discovered on *Bopyrus squillarum* at Trieste.

Cyclocotyla taschenbergii (Parona & Perugia, 1889) Price, 1943.

Syn. *Diclidophora taschenbergi* Parona & Perugia, 1889; *Dactylocotyle taschenbergi* (Parona & Perugia) Stossich, 1898.

A parasite of *Sargus rondeletii* at Genoa. This species may have to be transferred to the genus *Diclidophoropsis*.

Genus *Diclidophoropsis* Gallien, 1937

Diagnosis. *Opisthaptor:* comprising pedunculate suckers. *Reproductive systems:* genital atrium not muscular, cirrus equipped with hooklets, testes situated behind the ovary, vitellaria extending into the opisthaptor, vaginae present. *Type and only species:*

Diclidophoropsis tissieri Gallien, 1937

Host. *Macrurus laevis* (caught off south-west Ireland at lat. 53° 17′ N., long. 13° 08′ W.).

Location. Skin near (but not on) the gills.

Gallien (1937) found most of thirteen hosts parasitized with five to six individuals each. The characters according to his description are: *Size:* 6–7 mm. long when fairly extended and 1 mm. broad (immature individuals 3·5 × 0·7 mm.). *Colour:* white with yellow margins, clear centrally. *Shape:* flat and elongate, narrowing anteriorly. *Prohaptor:* buccal suckers measuring 0·25 × 0·175 mm. *Opisthaptor:* cotylophore small, bearing four pairs of suckers radially arranged, their peduncles long and narrow, extensile and variable. Sclerites slender, eight in number, arranged in the form of a cross with a circle around it and maintaining the sucker in an open state. *Reproductive systems:* cirrus muscular, fleshy and equipped with 128 (± 5) simple hooklets 0·025 mm. long and flexed at an obtuse angle. Eggs not observed.

Diclidophorid of Uncertain Position

Genus *Platycotyle* Beneden & Hesse, 1863

Diagnosis. *Opisthaptor:* comprising two pairs of pedunculate suckers borne on a transversely rectangular cotylophore and arranged like a diagonal cross. *Type and only species:*

Platycotyle gurnardi Beneden & Hesse, 1863

Host. Grey gurnard.
Location. Gills.

This species was found once only (at Brest) and is very imperfectly known. According to the authors it is 5 mm. long, flat, elongate, attenuated at the extremities. The four '*crochets*' (*sclerites*?) with which each of the suckers are said to be equipped probably represent an incompletely described Diclidophorid skeletal arrangement.

Family **HEXOSTOMATIDAE** Price, 1936 (p. 75)

Syn. Octobothrii Blanchard, 1847, in part; Polystomeae Leuckart of Taschenberg, 1879, in part; Octocotylidae Beneden & Hesse, 1863, in part; Octobothriidae Taschenberg, 1879, in part; Plagiopeltinae Monticelli, 1903, *nec* Hexacotylidae Monticelli, 1899.

Genus *Hexostoma* Rafinesque, 1815 *nec Hexastoma* Rudolphi, 1809 *nec Hexastoma* Kuhn, 1828

Syn. *Hexacotyle* Blainville, 1828.

Diagnosis. With the characters of the family. *Species:* type, *Hexostoma thynni* (Delaroche, 1811) Rafinesque, 1815 and six others; two European (one in British waters), three in Japan and one in the Galapagos Isles.

Hexostoma extensicaudum (Dawes, 1940) n. comb. (Fig. 23A, B).

Syn. *Hexacotyle extensicauda* Dawes, 1940.

Host. Tunny (North Sea: lat. 54° 17′ N., long. 1° 52′ E.).
Location. Gills.

Diagnosis. *Shape and size:* elongate, 11 mm. long, body divided into anterior, middle and posterior regions whose respective dimensions are 1·5 × 1·3,

5·5 × 3·3 and 4·0 × 2·6 mm.; anterior region passing abruptly into the middle region, which is separated from the posterior region by a narrow waist 1·7 mm. broad; anterior extremity with a median papilla 0·25 mm. long and 0·3 mm. broad. *Prohaptor:* a pair of minute buccal suckers 0·1 mm. diameter. *Opisthaptor:* an ill-defined posterior cotylophore with four pairs of suckers, those of the most posterior pair being very small, and two pairs of hooklets. Suckers of

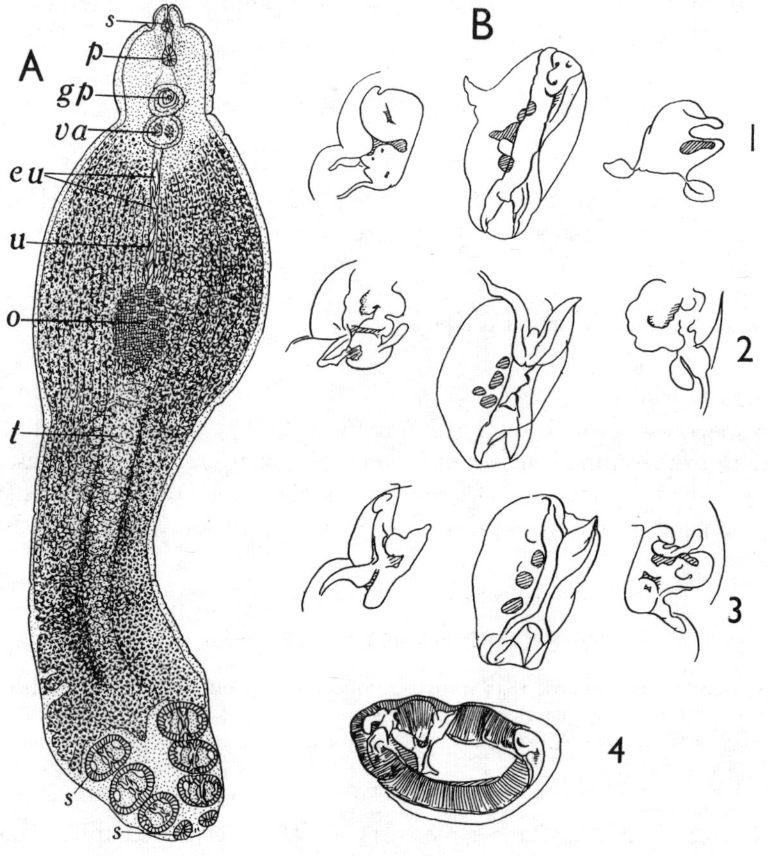

Fig. 23. *Hexostoma extensicaudum* (Hexostomatidae). A, entire trematode in ventral view. B 1–4, skeletal supports of the 1st–4th suckers, the last *in situ*. Median, middle and lateral pieces are shown from left to right. (After Dawes, 1940*a*.)

first three pairs oval in outline, cup-like, having their long axes in the transverse plane and about 0·067 of the body length. Suckers of the posterior pair more strongly concave than preceding ones, 0·03 of the body length in breadth. Ratio of the breadths of small and large suckers 0·45. Hooklets with strongly recurved points and with a 'guard' which approaches the tip in pincer fashion. Hooklets of the lateral pair 0·075 mm. long, those of the median pair 0·015 mm. long. Cuticular supports of the large suckers comprising three dissimilar pieces, lateral, middle and median respectively, the middle piece approximately X-shaped, differing in shape in different suckers, sunk deeply in musculature (Fig. 23 B, 1–4).

Supports for the small suckers rudimentary (vestigial?), showing some signs of a tripartite nature. *Musculature:* outer circular muscles of the body absent or vestigial, oblique and longitudinal muscles present, the latter thicker dorsally than ventrally and posteriorly than anteriorly, divided ventrally into superficial and deep parts which respectively assist in the releasing and adhering action of the suckers, being set in the rim and the centre of the suckers respectively. *Gut:* pharynx small and non-muscular, oesophagus short and narrow, caeca long and with profusely branched median and lateral diverticula, extending into the cotylophore. *Reproductive systems:* common genital pore near the bifurcation of the intestine at the junction of the anterior and middle regions of the body. Genital atrium capacious, 0·6 mm. long, 0·3 mm. broad and 0·4 mm. deep. Vaginal pore median and dorsal, close behind the level of the genital pore. Distal part of the vagina bilobed, capacious, as large as the genital atrium, lined with cuticle having numerous tubercles. Copulatory organ conical, fibrillar, 0·35 mm. long and 0·2 mm. broad. Vas deferens long and convoluted. Prostate glands well developed. Testes numerous, confined to the region between the caeca in the waist connecting middle and posterior divisions of the body. Ovary U-shaped with limbs folded in the transverse plane about a dozen times, situated in front of the testes. Vitellaria well developed, extending laterally throughout the middle division of the body, consisting of 'nests' of vitelline cells in various stages of secretory activity, one cell of each nest at least being large and distended with droplets of shell material. Uterus little folded, its terminal part (metraterm) straight. Ootype elongate in the transverse plane, situated between the testes and the ovary. Oviduct with an ovicapt having two sphincters. Eggs not numerous, ovoid, very large, 0·25 × 0·15 mm., each with two polar filaments about 0·25 mm. long and hollow in their basal parts.

Note. This trematode causes considerable erosion of the host's tissue, especially in the upper parts of the gills, but also at the point of attachment near the free ends of the gills.

Other Species

Hexostoma thynni (Delaroche, 1811)

Syn. *Polystoma thynni* Delaroche, 1811; *P. duplicatum* Rudolphi, 1819; *Hexacotyle thynni* (Delaroche) Blainville, 1828; *Plagiopeltis duplicata* (Rudolphi) Diesing, 1850.

Host. Tunny (Mediterranean (Naples and the Balearic Isles) and Atlantic coast of North America; not recorded in British waters).

Location. Gills.

Diagnosis. *Shape:* body narrow and pointed anteriorly, broad and truncated posteriorly. *Opisthaptor:* three pairs of larger and one (posterior) pair of small suckers, all set on the transversely truncated posterior extremity. Large hooklets with recurved tips and a 'guard', smaller hooklets straight and situated anterior to the larger. *Reproductive systems:* genital atrium with two lateral cuticular pieces. Vagina equipped with hooklets.

Hexostoma thunninae (Parona & Perugia, 1889)

Syn. *Octocotyle thunninae* Parona & Perugia, 1889; *Octobothrium thunninae* (Parona & Perugia) St Remy, 1891; *Hexacotyle thunninae* (Parona & Perugia) Goto, 1899.

HOST. Tunny (Mediterranean (Genoa); not recorded in British waters).

LOCATION. Gills.

DIAGNOSIS. *Opisthaptor:* suckers set on the rounded posterior extremity much as in *H. extensicaudum*, their long axes in the transverse plane, but relatively of much smaller size (0·031 and 0·019 of the body length respectively in the large and small suckers). Ratio of sizes in small and large suckers 0·61. Hooklets much larger than in this species, 0·128 and 0·048 mm. long respectively in the two pairs.

In three species discovered in Japan (one of which, *H. acuta* (Goto, 1894) has been recorded on the tunny in the North Sea (Baylis, 1939*a*), but the British form is possibly *H. extensicaudum*) the large suckers and the hooklets are intermediate in size between those of *H. thunninae* and *H. extensicaudum*. Measurements show that relatively large suckers coexist with relatively small hooklets and vice versa (Dawes, 1940*a*).

CHAPTER 8

SOME DIGENETIC TREMATODES OF BRITISH AND SOME OTHER FISHES

The trematode parasites of fishes have been studied in some detail in the British Isles and nearby waters, and this chapter contains brief descriptions of more than one hundred species of Digenea belonging to seventeen families, as well as information regarding the hosts, locations and localities in which the parasites are found. Our knowledge rests fundamentally on the pioneer work of continental zoologists, but valuable contributions to it have been made by Lebour (1905–1935), Nicoll (1906, 1907*a*, *b*, 1910*b*, 1912*b*, 1913*a*, *b*, 1914*a*, 1915*a*, 1924*a*) and a few other British zoologists, whose work is duly and gratefully acknowledged by meagre references in the text.

Trematodes differ from one another substantially in their relations to the host. Some Digenea of fishes are cosmopolitan and occur in a very wide range of fishes that are not very closely related; others show a remarkable degree of host-specificity. In the case of some widespread parasites, infection may be accidental. Thus, a voracious feeder like the angler acquires the parasites of its victims and, at the same time, provides suitable conditions for maintaining the infection. Delicate feeders like the red-band fish do not run the same risk of infection and, more often than not, are devoid of trematodes. We shall return to this question in a later chapter. Meanwhile, we can survey the taxonomic characters of the commoner and some of the less common Digenea which occur in British fishes. Many of the species mentioned have been recorded in British waters, and some of the others may reasonably be expected to occur therein.

The whiting is the most heavily and diversely infected fish in British seas. The cod and the pollack harbour fewer trematodes than the whiting, but have parasites which are somewhat characteristic of all gadoid and a few other fishes. It is noteworthy that, as far as records go, the cod, bib, haddock, poor cod, hake and pollack together have acquired only two common species of Digenea which are not found in the whiting; these are *Lepidapedon rachion* and *L. elongatum*.

Next to the whiting, perhaps the grey gurnard has the greatest variety of digenetic trematodes, some species of which have not been found in gadoids, though the genera may be represented, e.g. *Stephanostomum triglae* and *Prosorhynchus triglae*. The genus *Rhipidocotyle* is represented only in three gurnards (grey, red and yellow) and in the greater and lesser weevers, but Lebour found it represented in practically every fish she examined, sometimes in fair abundance. It is thus one of the best examples of host-specificity in the British trematode fauna.

The dab is another widely distributed fish having a varied array of digenetic flukes, and a good second to it among flatfishes is the plaice. Both of them succour members of the Fellodistomidae and Zoogonidae, as well as the much commoner Allocreadiidae and Hemiuridae. The ballan wrasse and the angler also are notable hosts of Digenea; the latter harbours no less than six hemiurids and the former both Hemiurids and Allocreadiids. The horse mackerel and the conger are other common fishes with a varied assemblage of trematodes. To student and teacher alike, such fishes as these provide trematodes which are far superior to the notorious *Fasciola hepatica* as zoological types.

Seven families are commonly represented in British fishes; they are the Hemiuridae, Allocreadiidae, Acanthocolpidae, Fellodistomatidae, Azygiidae, Zoogonidae and Bucephalidae. Several others are less commonly represented, e.g. the Ptychogonimidae, Monorchiidae, Acanthostomatidae and Didymozoidae, and larval forms belonging to other families occur in an encysted condition. From this preamble it is evident that fishes like the cod, whiting, plaice, dab and angler are useful sources of trematode specimens, for which reason their trematode parasites are shown in Table 3, together with those of skates, rays and dogfishes, which also have a distinctive fauna of digenetic flukes.

The commonest Digenea of our marine fishes belong to the Hemiuridae. *Derogenes varicus* occurs in more than forty species of fishes in British waters, *Hemiurus communis* in more than thirty, *Lecithaster gibbosus* and *Podocotyle atomon* in more than twenty, and *Zoogonoides vivipara* in more than a dozen, of which about eight are flatfishes. With such possibilities as exist for obtaining these parasites, there is surely no excuse for their exclusion from teaching collections of schools and colleges. *Derogenes varicus* is of such comparatively simple structure as to be worthy of first place, and has the additional interest of being cosmopolitan. It infests altogether more than fifty species of fishes whose distribution ranges from our waters into the Russian Arctic, to Puget Sound and the Gulf of Maine. It even occurs in the Gulf of Mexico, though only in deep water, and is found in the Far East.

The trematode parasites of fresh-water fishes have not been systematically studied by British zoologists, and most of the records mentioned in this chapter refer to such as inhabit continental waters. Here, then, is a whole field of research scarcely touched upon. Assemblages of trematodes which can be found in fresh-water fishes differ from those of marine fishes. The most common and characteristic of them are members of the families Bucephalidae, Allocreadiidae, Azygiidae and Gorgoderidae, namely, *Bucephalus polymorphus*; *Allocreadium isoporum* and *Sphaerostoma bramae*; *Azygia lucii*; *Phyllodistomum folium*. Digenea of less frequent occurrence belong to the genera (and families) *Crepidostomum* (Allocreadiidae), *Haploporus*, *Saccocoelium* and *Lecithobotrys* (Haploporidae), *Hemiurus*, *Derogenes*, *Lecithochirium* and *Lecithaster* (Hemiuridae), and *Bunodera* (Bunoderidae). Familial characters would go a long way towards establishing the identity of such species as may be expected to occur in fishes inhabiting our rivers and lakes.

In addition, larval trematodes are of common occurrence in fresh-water fishes, encysted beneath the integument or within the orbit, and have been wont to go by such names as '*Diplostomum*', '*Tetracotyle*' and '*Tylodelphis*' (all closely related larval Strigeidae or Diplostomatidae). Less common larval forms belong to the families Opisthorchiidae, Heterophyidae and Clinostomatidae. A whole chapter would be needed to deal with these and other larval forms. Suffice it here to say that Nicoll (1924*a*) included forms known as '*Diplostomum cuticola*' and '*D. volvens*' amongst the characteristic trematodes of fresh-water fishes.

Migratory fishes like the salmon and fishes which, like the three-spined stickleback, live in brackish water may harbour mixed assemblages of trematodes which are characteristically found in marine and fresh-water fishes respectively. It is self-evident that a migrant fish which happens to be infested with any of the characteristic trematodes of fresh-water fishes probably acquired its parasites whilst living in fresh water. Some fishes which, like the flounder, shad, sturgeon and bass, habitually live in both fresh and salt water may harbour trematodes which are typically marine. The salmon shows this feature also, but, perhaps accidentally, includes amongst its parasites *Distoma miescheri* Zschokke, 1890, a form which has not been found in any truly marine fish. The common eel is more tolerant of fresh-water trematodes, being parasitized commonly by *Bucephalus polymorphus*, *Sphaerostoma bramae* and *Azygia lucii* in continental waters. Yet it may include amongst its parasites three of the best-known trematodes of marine fishes, *Derogenes varicus*, *Hemiurus communis* and *Lecithaster gibbosus*, indicating that on entering fresh water it does not rid itself of trematode parasites, but merely substitutes one aggregate for another.

Were we to search the records for the half-dozen fresh-water fishes likely to provide the most varied and abundant collection of trematodes, we should be bound to select the roach, minnow, common carp, perch, gudgeon and pike. Unfortunately, we know very little about their trematode fauna in British waters, but on the Continent they harbour all the characteristic forms and several others as well. In fact, out of ninety-eight species of trematodes listed by Nicoll (1924*a*), no fewer than forty-one have been found in these six fishes. These represent twenty-two genera out of a total of forty-one (thirty-four only if we exclude such as occur also in marine fishes). All the families to which European fresh-water trematodes belong are represented. Accordingly, the trematode parasites of the six fishes are included along with those of certain marine fishes in Table 3.

Of the common trematodes of fresh-water fishes, *Sphaerostoma bramae* is essentially a parasite of cyprinids, though it occurs also in the eel, pike and perch. The same is true of *Allocreadium isoporum*, *Phyllodistomum macrocotyle*, the larval form '*Diplostomum cuticola*' and *Bucephalus polymorphus*, though other fishes are parasitized as well. On the other hand, *Azygia lucii* has not been found in any cyprinid. On the Continent at least, the roach seems to be more heavily infested than any other fresh-water fish; it harbours five of the seven characteristic forms (including '*D. cuticola*' and '*D. volvens*'), lacking only *Azygia lucii*, and *Bucephalus polymorphus*. At the other extreme, many fresh-water fishes, especially the chars, are not known to be parasitized by trematodes. Of the trematodes themselves, *Sphaerostoma bramae* is the most widespread form in fresh water (as *Derogenes varicus* is in the sea), occurring in at least fourteen species of fishes.

In a more or less comprehensive account of our trematode fauna it is impossible to do full justice to the Digenea of fishes in particular. Many forms which deserve more are treated cursorily in this chapter and are not included in the figures. It might be mentioned that this defect will be remedied in the near future by the publication of a Ray Society Monograph in which I shall give a fuller account of the trematode parasites of British fishes.

Table 3. *Digenetic Trematodes of certain Fresh-water and Marine Fishes in Europe*

Few of the fresh-water forms have been found in British waters, but most of the marine forms are known in the seas round our coasts.
B, blood vessels; b, bladder; C, pyloric caeca; Co, coelom; D, duodenum; E, ear; G, gills; I, intestine; O, oesophagus; R, rectum; S, stomach; s, subcutaneous; U, ureter. ( ), encysted; † juvenile.

	Roach	Minnow	Common Carp	Gudgeon	Perch	Pike	Cod	Whiting	Angler	Plaice	Dab	Various rays	Various dogfish
BUCEPHALIDAE:													
Bucephalus polymorphus	—	—	—	I	I	I	—	—	—	—	—	—	—
Bucephalopsis gracilescens	—	—	—	—	—	—	(E)	(E)	SIC ( )	—	—	—	—
Prosorhynchus crucibulum	—	—	—	—	—	—	(G)	—	† I	—	—	—	—
'*P. grandis*' (=*P. squamatus*)	—	—	—	—	—	—	C	CI	—	—	—	—	—
ALLOCREADIIDAE:													
Allocreadium isoporum	I	I	I	—	—	I	—	—	—	—	—	—	—
Sphaerostoma bramae	I	I	I	—	I	I	—	—	—	—	—	—	—
Plagioporus varia	—	—	—	—	—	—	—	—	—	I	—	—	—
Lepidapedon rachion	—	—	—	—	—	—	I	—	—	—	—	—	—
L. elongatum	—	—	—	—	—	—	D	—	—	—	—	—	—
Opechona retractilis	—	—	—	—	—	—	—	CI	—	—	—	—	—
Podocotyle atomon	—	—	—	—	—	—	—	C	—	CI	I	—	—
ACANTHOCOLPIDAE:													
Stephanostomum baccatum	—	—	—	—	—	—	—	—	† ( )	—	(s)	—	—
S. caducum	—	—	—	—	—	—	—	D	—	—	—	—	—
S. pristis	—	—	—	—	—	—	CI	CI	—	—	—	—	—
S. rhombispinosum	—	—	—	—	—	—	—	CI	—	—	—	—	—
FELLODISTOMATIDAE:													
Steringophorus furciger	—	—	—	—	—	—	—	—	D	D	DI	—	—
Steringotrema cluthense	—	—	—	—	—	—	—	—	—	—	† (I)	—	—
Haplocladus minor	—	—	—	—	—	—	—	—	—	—	I	—	—

ZOOGONIDAE:													
Diphterostomum betencourti	—	—	—	—	—	—	—	—	—	—	—	—	I
Zoogonoides viviparus	—	—	—	—	—	—	—	—	—	R	R	—	—
Zoogonus rubellus	—	—	—	—	—	—	—	—	—	R†	R†	—	—
AZYGIIDAE:													
Azygia lucii	—	—	—	—	I	S	—	—	—	—	—	—	—
Otodistomum veliporum	—	—	—	—	—	—	—	—	—	—	—	OS	—
PTYCHOGONIMIDAE:													
Ptychogonimus megastoma	—	—	—	—	—	—	—	—	—	—	—	—	S
HEMIURIDAE:													
Hemiurus appendiculatus	—	—	—	—	SIC	—	—	—	—	—	—	—	—
H. communis	—	—	—	—	—	—	S	S	S	S	S	—	—
H. levinsini	—	—	—	—	—	—	S	—	—	—	—	—	—
H. ocreatus	—	—	—	—	—	—	—	S	—	—	—	—	—
Derogenes varicus	—	—	—	—	—	—	S	S	S	S	SI	—	—
Lecithaster gibbosus	—	—	—	—	—	—	—	I	—	I	I	—	—
Lecithochirium rufoviride	—	—	—	—	—	—	—	—	S	—	—	—	—
Sterrhurus fusiformis	—	—	—	—	—	—	—	—	S	—	—	—	—
Synaptobothrium caudiporum	—	—	—	—	—	—	—	S	S	S	—	—	—
BUNODERIDAE:													
Bunodera luciopercae	—	—	—	—	I	I	—	—	—	—	—	—	—
GORGODERIDAE:													
Phyllodistomum macrocotyle	U	—	—	U	—	b	—	—	—	—	—	—	—
Probilotrema richiardii	—	—	—	—	—	—	—	—	—	—	—	Co	Co
APOROCOTYLIDAE:													
Sanguinicola inermis	—	—	B	—	—	—	—	—	—	—	—	—	—
Aporocotyle simplex	—	—	—	—	—	—	—	—	—	—	BG	—	—
UNCLASSIFIED:													
Centrovarium lobotes	—	—	—	—	—	S	—	—	—	—	—	—	—
Heterostoma luteum	—	—	—	—	—	I	—	—	—	—	—	—	—

Suborder GASTEROSTOMATA Odhner, 1905

Family **BUCEPHALIDAE** Poche, 1907 (p. 82)

Syn. Gasterostomidae Braun, 1893.

Subfamily **Bucephalinae** Nicoll, 1914

Pre-oral haptor sucker-like.

KEY TO GENERA OF BUCEPHALINAE

1. Anterior extremity provided with a muscular sucker having a circlet of six or seven muscular, retractile tentacles or fimbriae *Bucephalus*
2. Anterior extremity provided with a simple, globular, muscular sucker only *Bucephalopsis*
3. Anterior extremity provided with a feeble, shallow sucker surmounted by a fan-shaped hood *Rhipidocotyle*

Another genus, *Dolichoenterum* Ozaki, 1924, differs from the three preceding genera in the position of the ovary, which is situated between the testes, not in front of them. Its type species, *D. longissimum* Ozaki, 1924, occurs in the intestine of *Leptocephalus myriaster* in Japan.

Genus *Bucephalus* Baer, 1827

Syn. *Gasterostomum* Siebold, 1848; *Eubucephalus* Diesing, 1855.

The characteristic tentacles are conspicuous when extended, but may seem to be very inconspicuous papillae when retracted; they vary in shape and number in different species. *Type species:*

Bucephalus polymorphus Baer, 1827 (Fig. 24A–D)

Syn. *Distoma campanula* Dujardin, 1845; *Gasterostomum fimbriatum* Siebold, 1848; *G. laciniatum* Molin, 1859; ? *Bucephalus elegans* Woodhead, 1930; *B. varicus* Manter, 1940.

Hosts. Adults in pike, perch, burbot, gudgeon, etc.; cercariae in fresh-water mussels; metacercariae in houting, dace, rudd, bleak, silver bream.

Location. Adults in the intestine. Metacercariae encysted beneath the skin.

Diagnosis. *Size:* about 1 mm. long and 0·2 mm. broad (0·62–2·29 × 0·11–0·35 mm. according to Nagaty, 1937). *Cuticle:* spinous, spines minute, most numerous anteriorly. *Haptor:* longitudinally oval, much larger than the pharynx (0·08–0·19 × 0·07–0·17 mm.) and equipped with six tentacles (seven according to Nagaty (1937), three dorsal, two lateral and two ventro-lateral) with broad bases and tapering extremities. Basal part of each tentacle having two ventral processes, sometimes only one process. (*Note.* The completely retracted tentacles may be so insignificant as to be overlooked.) *Gut:* mouth slightly behind the mid-body, pharynx (in the literature often referred to as an oral sucker) almost spherical (0·03–0·08 × 0·03–0·10 mm.), intestine simple and sac-like, directed anteriorly or posteriorly. *Excretory system:* vesicle simple and saccular, extending forward almost to the haptor, the pore being situated at the posterior extremity. *Reproductive systems:* genital pore ventral and near the excretory pore, genital atrium present. Cirrus pouch large (0·17–0·57 mm. long) and elongate, directed towards the left side, containing a pars prostatica, an oval seminal vesicle and a stout cirrus. Testes globular, one behind the other on the right posteriorly adjacent to the cirrus pouch, diameters 0·06–0·19 mm. Ovary spherical, 0·05–0·13 mm. diameter, situated in front of the testes near the mid-body, also on the right. Vitellaria comprising nine to twenty-four coarse follicles on either side of the median plane in front of the gonads. Uterus rarely extending in front of the vitellaria, much folded. Eggs very numerous, golden yellow, ovoid and operculate, measuring 0·021–0·027 × 0·013–0·023 mm.

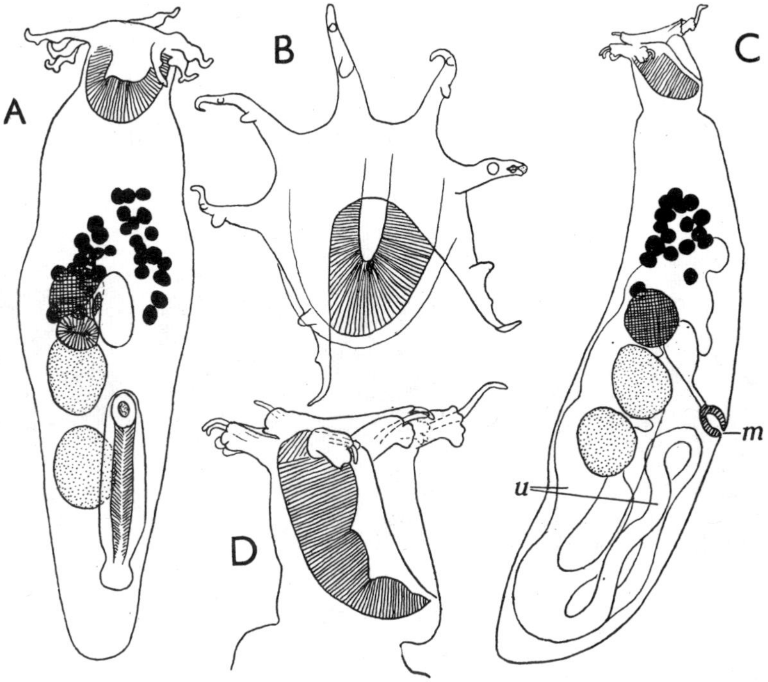

Fig. 24. *Bucephalus polymorphus* (Bucephalidae). A, typical specimen in ventral view. B, anterior end of A showing the extended tentacles. C, mature specimen in side view (from right). D, anterior end of C showing the tentacles partially retracted. (After Nagaty, 1937.)

Notes. In early descriptions tentacles are not mentioned, but Wagener (1858)* described them as having each a single ventral process, and other writers have concurred, '*Bucephalus elegans*' having this character. Nagaty (*loc. cit.*) showed that some specimens have one ventral process, but that two are typically present. Manter (1940*a*) referred Nagaty's specimens to a new species characterized by relatively large eggs, but, as the size of the eggs shows extreme variability in European and American specimens, such a change is unwarranted, and *B. varicus* Manter, 1940 is relegated to synonymity with *B. polymorphus*, which has occurred in East Prussia, France, Italy, the Red Sea and America, and (in the pike) in Wiltshire and Berkshire.

* (*Arch. Naturg.* 24. Jahrg., II, 250–6.)

OTHER SPECIES:

Bucephalus gorgon (Linton, 1905) (U.S.A.).
B. marinus Vlasenko, 1931 (Black Sea).
B. introversus Manter, 1940 (America: Mexico and Colombia).
B. tridenticularia Verma, 1936; *B. aoria* Verma, 1936 (India).
B. uranoscopi Yamaguti, 1934 (Japan).

Genus *Bucephalopsis* (Diesing, 1855) Nicoll, 1914

Syn. *Prosorhynchoides* Dollfus, 1929.

Bucephalopsis gracilescens (Rudolphi, 1819) *nec* Tennent, 1906 (Fig. 25 A, B)

Syn. *Distoma gracilescens* Rudolphi, 1819; *Gasterostomum gracilescens* (Rudolphi, 1819).

HOSTS. Adults in the angler and the conger. Metacercariae in the haddock, cod, whiting, pollack, common ling, greater forkbeard. Cercariae in the cockle.

LOCATION. Adults in pyloric caeca, stomach and intestine. Larvae encysted in nerves, especially the spinal nerves near the tail and the auditory nerves.

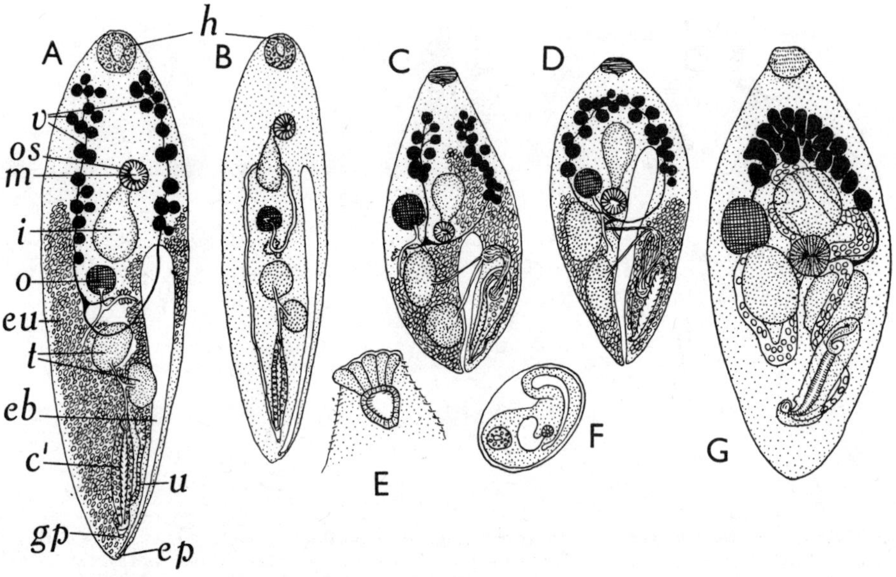

Fig. 25. Some Gasterostomata (Bucephalidae) from British marine fishes. A, *Bucephalopsis gracilescens*; mature. B, same species; immature, extracted from a cyst in which it is shown in F. C, *Prosorhynchus squamatus*. D, *P. grandis* (=*P. squamatus*). E, anterior end of *Rhipidocotyle viperae* (=*R. galeata*) showing the anterior holdfast. G, *P. squamatus*. (E, after Nicoll, 1914*a*; G, after Markowski, 1933*a*; rest after Lebour, 1908*a*.)

DIAGNOSIS. *Size:* about 6 mm. long. *Shape and colour:* elongate-oval outline, broader anteriorly, but variable in life, colourless except for the eggs. *Cuticle:* spinous. *Haptor:* larger than the pharynx, diameter 0·4 mm. *Gut:* pharynx 0·3 mm. diameter, intestine simple and sac-like, mouth situated about one-quarter of the distance along the body. *Excretory system:* vesicle long and narrow, extending almost to the mid-body, its contents granular and highly refractive. *Reproductive systems:* genital pore near the posterior extremity. Cirrus pouch long and narrow, containing a narrow pars prostatica, a pyramidal vesicula seminalis and a long, spinous cirrus. Testes ovoid, diagonally one behind the other alongside the excretory vesicle. Ovary spherical, in front of the testes, but behind the gut and slightly on the right. Vitellaria lateral, in the anterior half of the body, comprising fifteen to eighteen coarse follicles on each side. Uterus with ascending and descending limbs, much convoluted at full maturity, filling the posterior two-thirds of the body. Eggs golden brown, 0·026 mm. long (according to Mathias (1934) 0·018–0·024 × 0·01–0·015 mm.).

Note. This species has been found at Plymouth, Liverpool, Aberdeen and Cullercoats, also Trieste, Banyuls and U.S.A. (Woods Hole).

OTHER SPECIES:

Nagaty (1937) listed fourteen other known species of *Bucephalopsis* and added four new ones. The full list, which includes some species which will probably fall when the genus is reviewed in the light of modern knowledge, is:

Bucephalopsis haimeanus (Lacaze-Duthiers, 1854) (Syn. *Bucephalus haimeanus* Lacaze-Duthiers, 1854; *Bucephalus cuculus* McCrady, 1874; *Gasterostomum* sp. Linton, 1900 and 1901; *Gasterostomum gracilescens* Linton, 1906 *nec* Rudolphi, 1919) (Balearic Isles; U.S.A.).
B. triglae (Beneden, 1870) *nec* Nicoll, 1909 (Syn. *Gasterostomum triglaè* Beneden, 1870). (Belgium).
B. tergestinum (Stossich, 1883) (Syn. *Gasterostomum tergestinum* Stossich, 1883) (Italy).
B. arcuatus (Linton, 1900) (Syn. *Gasterostomum arcuatum* Linton, 1900; *G. pusillum* (Stafford, 1904); *Bucephalopsis pusilla* (Stafford, 1904)) (U.S.A.).
B. ovatus (Linton, 1900) (Syn. *Gasterostomum ovatum* Linton, 1900; *Prosorhynchoides ovatus* (Linton, 1900) Dollfus, 1929) (U.S.A.).
B. exilis Nicoll, 1915 (North Queensland, Australia).
B. latus Ozaki, 1928 (Japan).
B. elongatus Ozaki, 1928 (Japan).
B. ozakii Nagaty, 1937 (Syn. *B. ovatus* Ozaki, 1928 *nec* Linton, 1900) (Korea).
B. basaringi Layman, 1930 (Russia).
B. pleuronectis Layman, 1930 (Russia).
B. fusiformis Verma, 1936 (India).
B. garuai Verma, 1936 (India).
B. magnum Verma, 1936 (Syn. *B. confusus* Verma, 1936; *B. minimus* Verma, 1936) (India).
B. southwelli Nagaty, 1937 (Red Sea).
B. lenti Nagaty, 1937 (Red Sea).
B. longicirrus Nagaty, 1937 (Red Sea).
B. megacetabulus Nagaty, 1937 (Red Sea).

B. triglae is known only by a figure. *B. tergestinum* was described by Stossich from materials obtained in Trieste and was regarded by Eckmann (1932*b*) as a synonym of *B. gracilescens*, but it is a much smaller form (about 1 mm. long) which Nagaty chose to regard as a valid species, though possibly identical with *B. haimeanus*.

Genus *Rhipidocotylė* Diesing, 1858

Syn. *Nannoenterum* Ozaki, 1924.

Rhipidocotyle galeata (Rudolphi, 1819) Eckmann, 1932 (Fig. 25 E)

Syn. *Monostomum galeatum* Rudolphi, 1819; *Gasterostomum galeatum* (Rudolphi) of Stossich, 1898; *G. minimum* Wagener, 1852; *G. triglae* (Beneden) of Nicoll, 1909; *Rhipidocotyle minimum* (Wagener) of Diesing, 1858; *Rhipidocotyle viperae* Nicoll, 1914 *nec* Beneden, 1870.

Hosts (British waters; Plymouth, St Andrews, Galway). Greater weever, lesser weever, red, yellow and grey gurnards.

Location. Intestine.

Diagnosis (based on Nicoll's form *viperae* which, according to both Eckmann (1932*b*) and Nagaty (1937), is identical with *R. galeata*). *Shape and size:* oval outline, anterior extremity square-cut, posterior extremity pointed, 0·7–1·2 mm. long and about 0·4 mm. broad. *Cuticle:* spinous. *Haptor:* terminal, 0·13 mm. diameter in largest specimens, musculature feeble, with five-rayed, fan-like hood surmounting it. Oral sucker (pharynx) near the mid-body. *Gut:* simple, sac-like. *Reproductive systems:* genital pore posterior. Gonads close together in front of the mouth (on the right), ovary 0·25 mm. from the anterior extremity. Vitellaria along the lateral margins of the second quarter of the body. Uterus much folded, filling up the space between the vitellaria, extending back between the testes and the gut. Cirrus pouch long, slender, extending from the posterior extremity to a short distance in front of the mouth. Pars prostatica long, cirrus short. Eggs measuring 0·036–0·037 × 0·018–0·021 mm.

Note. '*R. minima*' was recorded from the last four of the hosts mentioned above and, according to Nicoll (1914*a*), differs from '*R. viperae*' in the topographical arrangement of the gonads and in the shorter cirrus pouch. Nagaty is probably right in regarding it as a synonym of *R. galeata*.

Other Species (none of these has been found in Europe):

Rhipidocotyle baculum (Linton, 1905) Eckmann, 1932 (Syn. *Gasterostomum baculum* Linton, 1905; *Nannoenterum baculum* (Linton, 1905) Manter, 1931) (U.S.A.).

R. papillosum (Woodhead, 1929) Eckmann, 1932 (Syn. *Bucephalus papillosum* Woodhead, 1929; *B. pusillus* Cooper, 1915 *nec* Stafford, 1904 (U.S.A.).

R. pentagonum (Ozaki, 1924) Eckmann, 1932 (Syn. *Nannoenterum pentagonum* Ozaki, 1924) (Japan).

R. septapapillata Krull, 1934 (U.S.A.).

R. transversale Chandler, 1935 (U.S.A.).

R. elongatum McFarlane, 1936 (Canada).

R. khalili Nagaty, 1937 (Red Sea).

R. eckmanni Nagaty, 1937 (Red Sea).

Subfamily **Prosorhynchinae** Nicoll, 1914

Pre-oral haptor a rhynchus.

In addition to the well-known genus *Prosorhynchus* this subfamily may contain two valid genera, *Neidhartia* Nagaty, 1937 and *Alcicornis* MacCallum, 1917, as well as several somewhat doubtful ones, namely, *Gotonius* Ozaki, 1924; *Skrjabiniella* Issaitschikow, 1928; *Mordvilkovia* Pigulewsky, 1931 and *Dollfusina* Eckmann, 1932 (renamed *Dollfustrema* Eckmann, 1934), all of which are regarded by Nagaty (1937) as synonyms of *Prosorhynchus*. In *Neidhartia* the ovary is on the left side of the body opposite the testes, in *Alcicornis* and *Prosorhynchus* on the right side in front of the anterior testis. *Alcicornis* differs from *Prosorhynchus* chiefly in having tentacles associated with the rhynchus. Species of *Neidhartia*, namely, *N. neidharti* and *N. ghardagae*, both Nagaty, 1937, occur in *Serranus* sp. in the Red Sea. *Alcicornis carangis* MacCallum, 1917 was found in the intestine of *Caranx ruber* in the U.S.A. and *Alcicornis baylisi* Nagaty, 1937 in the bayad, *Caranx* sp., in the Red Sea. Manter (1940*c*) believed that *Dollfustrema* and *Mordvilkovia* should be excluded from the synonymy proposed by Nagaty, holding that the cuticular folds on the rhynchus in the latter are suggestive of the spines in the former. Accordingly, he considered *Mordvilkovia* a valid genus with *Dollfusina* and *Dollfustrema* as synonyms. He further regards *Pseudoprosorhynchus* Yamaguti, 1938 as a synonym of *Neidhartia.* Jones (1943), who studied only *Prosorhynchus aculeatus* and overlooked the papers of Nagaty (1937) and Manter (1940*c*), considered *Skrjabiniella* to be a valid genus, containing those species of *Prosorhynchus* in which the body is elliptical, not elongate; the rhynchus oval, not conical; the mouth situated behind the anterior testis, not in front of it; the testes symmetrical, not 'in tandem'; and the vitelline follicles arranged in the form of an arc, not two lateral groups. *P. crucibulum* and *P. ozaki* will not fit into either scheme based unilaterally on these alternatives, and differ from one another in four out of five of them. This, and certain discrepancies (e.g. the 'symmetrical' disposition of the testes, negatived by Nicoll's observations (1910*b*, p. 351)), suggest that the combination of the five characters cannot be taken as an index of generic status, and that separately they represent no more than specific differences. Accordingly, it seems preferable at present to retain *Skrjabiniella* as a synonym of *Prosorhynchus.*

Genus *Prosorhynchus* Odhner, 1905

Syn. *Gotonius* Ozaki, 1924; *Skrjabiniella* Issaitschikow, 1928; *Mordvilkovia* Pigulewsky, 1931; *Dollfusina* Eckmann, 1932 (preoccupied); *Dollfustrema* Eckmann, 1934; *Neidhartia* Nagaty, 1937; *Pseudoprosorhynchus* Yamaguti, 1938.

The most familiar species of this genus are *P. crucibulum* (Rudolphi, 1819), *P. aculeatus* Odhner, 1905 and *P. squamatus* Odhner, 1905. Nagaty regarded the first two as valid and the last as a synonym of the penultimate species. This opinion, due primarily to Eckmann (1932*b*), was not shared by Manter (1940*c*), who was satisfied with the distinctions disclosed by Odhner and referred Eckmann's specimens of *P. crucibulum* to a new species, *P. caudovatus*, on account of the distinctive polar processes of the eggs. Other forms which have been found in this country are *P. grandis* Lebour, 1908 and *P. triglae* sp. inq. Nicoll, 1914. The former is closely similar to *P. squamatus* (see Fig. 25C, D) (differing mainly as a result of general contraction), and the latter is a juvenile form (Fig. 26C) which, in the disposition of the gonads, shows a similar possible relationship.

Prosorhynchus aculeatus Odhner, 1905 (Fig. 26B)

Syn. *Gasterostomum crucibulum* Beneden, 1870; *G. armatum* Olsson, 1876; *Skrjabiniella aculeatus* (Odhner, 1905) Issaitschikow, 1928.

Host. Conger.

Location. Intestine.

Diagnosis. *Shape, colour and size:* body oval in outline, pointed at the extremities, yellowish brown, 1–2·5 mm. long and about half as broad. *Cuticle:*

spinous, spines scale-like, arranged in regular diagonal rows. *Haptor:* a rhynchus, varying in appearance according to state of retraction or protraction from a saucer-like depression to a small button, about 0·27 mm. broad and 0·15 mm. deep. *Gut:* oesophagus short, surrounded by unicellular gland cells, intestine sac-like, extending forward not quite to the mid-body, pharynx 0·14 mm. diameter, situated behind the mid-body. *Excretory system:* pore terminal, vesicle extending slightly in front of the mouth. *Reproductive systems:* genital pore almost terminal. Cirrus pouch thick and muscular, occupying the posterior third of the body, directed towards the left side. Large vesicula seminalis *interna* present, but no vesicula seminalis *externa* (Jones, 1943), pars

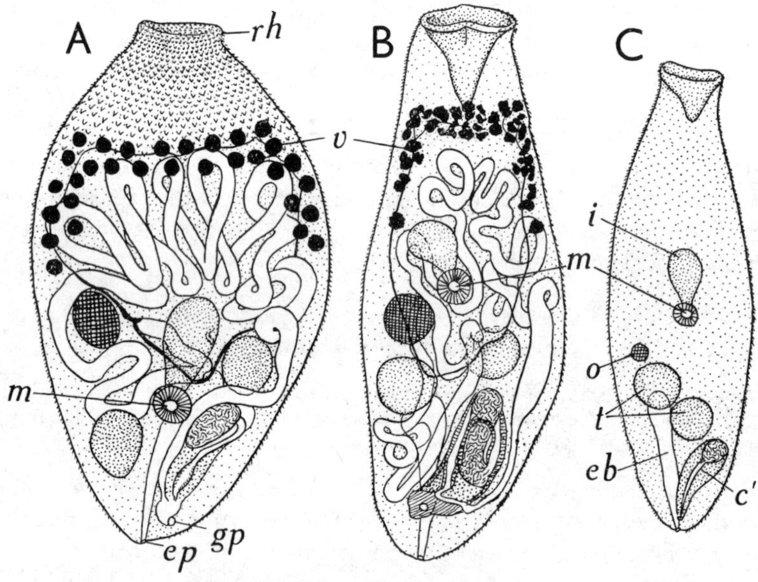

Fig. 26. Other Bucephalidae from British marine fishes. A, *Prosorhynchus crucibulum*. B, *P. aculeatus*. C, *P. triglae* (=*P. squamatus*?). (A, B, after Nicoll, 1910*b*; C, after Nicoll, 1914*a*.)

prostatica forming a loop. Testes ovoid, one on either side of the mouth, but variable in position. Ovary ovoid, near the mid-body on the left. Vitellaria in the form of an arc of about twelve to fifteen large follicles on each side in the anterior region. Uterus with ascending and descending limbs, formed into numerous folds, tending to fill up all space at the sides and in front of the gonads. Eggs small, numerous, measuring 0·026–0·031 × 0·016–0·020 mm. (mean, 0·085 × 0·0185 mm.), containing segmenting ova.

Note. This species occurs in the Far East and in arctic waters.

Prosorhynchus crucibulum (Rudolphi, 1819) (Fig. 26A)

Syn. *Monostomum crucibulum* Rudolphi, 1819; *Gasterostomum crucibulum* (Rudolphi, 1819) *nec G. crucibulum* Beneden, 1870; *G. armatum* Molin, 1859; *Prosorhynchus costai* Travassos, Artigas & Pereira, 1928; *Mordvilkovia elongata* Pigulewsky, 1931; *Prosorhynchus scalpellus* McFarlane, 1936.

HOST (British). Adults in conger. Larvae in angler and cod.

LOCATION. Adults in the intestine.

DIAGNOSIS. *Size:* larger than *P. aculeatus*, 2–6 mm. long. *Shape:* elongate, flat, truncated anteriorly, rounded posteriorly. *Colour:* adults yellowish brown, coloured by the eggs. *Cuticle:* spinous, spines more abundant anteriorly. *Haptor:* a very large rhynchus measuring 0·62 × 0·57 mm. in an individual 3·15 mm. long. *Gut:* pharynx near the mid-body. Other characters as shown in Fig. 26A.

Prosorhynchus squamatus Odhner, 1905 (Figs. 25C, D, G, 26C)

Syn. *Bucephalus crux* Levinsen, 1881; *Prosorhynchus grandis* Lebour, 1908; ? *Prosorhynchus triglae* sp. inq. Nicoll, 1914.

HOSTS. Short-spined cottus, long-spined cottus, Montagu's sucker ('*P. grandis*' in cod and whiting; '*P. triglae*' in grey gurnard).

LOCATION. Pyloric caeca ('*P. triglae*' in stomach and intestine).

DIAGNOSIS. *Size and shape:* somewhat smaller than *P. aculeatus*, 1–1·5 mm. long (but '*P. grandis*' 2–2·5 × 0·7 mm.), pyriform, pointed anteriorly. *Cuticle:* spinous. *Haptor:* a rhynchus 0·09 mm. long and 0·12 mm. broad. *Gut:* pharynx centrally placed and about 0·10 mm. diameter. Eggs numerous and small, 0·026–0·033 mm. long (0·024–0·038 × 0·022–0·027 mm. in specimens 1–1·5 mm. long according to Markowski, 1933*a*). Other characters as shown in the figures. Found at Plymouth, Millport, St Andrews, Cullercoats, Galway, in Italy, the Polish Baltic, the Red Sea and U.S.A.

The encysted form of *P. squamatus* was found outside the pyloric caeca and in the skin and muscles of the bullhead by Levinsen (1881), who discovered the curious larva, *Bucephalus crux*, which Odhner regarded as the larval form of this species and which is therefore included in the synonymy.

Note. There are reasons for supposing that both *Prosorhynchus aculeatus* and *P. squamatus* are identical with *P. crucibulum*, the type-species.

OTHER SPECIES:

Prosorhynchus facilis (Ozaki, 1924) (Syn. *Gotonius facilis* Ozaki, 1924; *G. platycephali* Yamaguti, 1934; *Prosorhynchus apertus* McFarlane, 1936) (Japan, Canada, Red Sea).
P. uniporus Ozaki, 1924 (Syn. *Skrjabiniella uniporus* (Ozaki, 1924) Issaitschikow, 1928) (Japan).
P. vaneyi Shen, 1930 (Syn. *Dollfusina vaneyi* (Shen) Eckmann, 1932; *Dollfustrema vaneyi* (Shen) Eckmann, 1934) (China).
P. freitasi Nagaty, 1937 (Red Sea).
P. ozakii Manter, 1934 (Tortugas, Mexico, Galapagos Isles).
P. rotundus Manter, 1940 (Galapagos Isles).
P. gonoderus Manter, 1940 (Galapagos Isles).
P. pacificus Manter, 1940 (Galapagos Isles).
P. platycephali (Yamaguti, 1934) (as *Gotonius*) (Japan).
P. manteri Srivastava, 1937 (India).
P. arabiana Srivastava, 1937 (India).

Suborder PROSOSTOMATA Odhner, 1905

Family **ALLOCREADIIDAE** Stossich, 1904 (p. 86)

Sub-family **Allocreadiinae** Looss, 1902

Cuticle non-spinous; parasites of marine fishes, excepting *.

KEY TO GENERA OF ALLOCREADIINAE

1. Genital pore slightly lateral, on the left; oesophagus short; testes rounded
   A. Ovary rounded; vitellaria extending in front of the ventral sucker; cirrus pouch not extending back behind the middle of the ventral sucker *Plagioporus*
   B. Ovary three-lobed; vitellaria extending forward only to the ventral sucker; cirrus pouch extending back behind the ventral sucker *Podocotyle*
2. Genital pore median; oesophagus short or long; testes rounded or lobed
   A. Testes irregularly lobed; eggs each having a filament at the anopercular pole *Helicometra*
   B. Testes not lobed; eggs without filaments
      (*a*) Cirrus pouch projecting behind the ventral sucker *Peracreadium*
      (*b*) Cirrus pouch not projecting behind the ventral sucker
         (i) Oesophagus long; pars prostatica present; vitellaria absent from the narrow anterior region (parasitic in fresh-water fishes) *Allocreadium**
         (ii) Oesophagus short; pars prostatica absent; vitellaria present in anterior region (parasitic in marine fishes) *Cainocreadium*

Genus *Plagioporus* Stafford, 1904

Syn. *Lebouria* Nicoll, 1909; *Caudotestis* Yamaguti, 1934.

Miller (1940, see 1941*a*) indicated the validity of the genus *Plagioporus* and discussed briefly the synonyms stated above. Several species of the genus, *P. alacer*, *P. varius*, *P. idonea* and *P. tumidulus*, have been recorded in British waters, but their validity is not beyond question and only one, at most two, species are clearly defined. Issaitschikow (1928) proposed the erection of the subgenera *Caudotestis*, *Mediantestis* and *Lebouria*, with the types *nicolli* n.sp., *tumidulus* and *idoneus* respectively, an apparently unnecessary procedure as regards British species.

Plagioporus varius (Nicoll, 1910) (Fig. 81)

Syn. (*Lebouria*) *alacris* (Looss, 1901) Nicoll, 1909; *Lebouria varia* of Nicoll, 1910; *L. idonea* Nicoll of Johnstone, 1910.

Hosts. Dragonet, ballan wrasse, cuckoo wrasse, plaice, flounder.

Location. Intestine.

Diagnosis. *Shape and size:* oval, flat, 1·25–1·75 mm. long and about one-third as broad. *Suckers:* transversely oval, ventral larger than the oral and slightly in front of the mid-body, dimensions 0·35 × 0·29 mm. and 0·18 mm. diameter in an individual 1·5 mm. long. *Gut:* prepharynx and oesophagus short, pharynx large (half as broad as the oral sucker), caeca long, not quite reaching the posterior extremity. *Reproductive systems:* genital pore slightly in front and to the left of the bifurcation of the intestine. Cirrus pouch club-shaped, ejaculatory duct short, seminal vesicle convoluted. Pars prostatica not distinct. Testes diagonal and contiguous slightly behind the mid-body. Ovary rounded, in front of the testes, slightly lateral. Receptaculum seminis pyriform, between the ovary and the testes. Vitellaria extending from the level of the oesophagus to the posterior extremity, overlapping the caeca behind the ventral sucker. Uterus restricted to the region between the gonads and the ventral sucker. Eggs few (not more than thirty at any time), measuring 0·085–0·093 × 0·038–0·051 mm. (mean, 0·088 × 0·045 mm.), each having a knob-like process at the anopercular pole. This species has appeared in various parts of Britain (Plymouth, Liverpool, Millport, St Andrews) and at Galway.

Plagioporus alacer (Looss, 1901)

Syn. *Distomum alacre* Looss, 1901; *Lebouria alacris* (Looss) of Nicoll, 1910, 1914, 1915; *Distomum* sp. Johnstone, 1907.

HOSTS. Ballan wrasse, cuckoo wrasse, gilthead, gold-sinny, rock cook.

LOCATION. Intestine.

DIAGNOSIS. Said to be distinguishable from *P. varius* by the paler colour, more delicate texture, smaller size (0·8–1·45 × 0·46 mm.), longer oesophagus, smaller testes and anteriorly more extensive vitellaria, which merge in the median plane near the bifurcation of the intestine. So difficult is the separation of this species (reported at Millport, Morecambe Bay, Aberystwyth and Plymouth) and *P. varia* that the latter is almost certainly synonymous with it.

Plagioporus idonea (Nicoll, 1909)

Syn. *Lebouria idonea* Nicoll, 1909; *L.* (*Lebouria*) *idonea* of Issaitschikow, 1928.

This species, found in the intestine of the wolf-fish, is said to be distinguishable from *P. varius* by its larger size (1·5–2·5 × 0·7–1·0 mm.), stockier build, and distinct yellow colour. The knob-like processes of the eggs are more pronounced than in *P. varius*. This species may also prove to be merely a synonym of *P. alacer*. Another species, *P. tumidulus* (Rudolphi, 1819) Syn. *Distoma tumidula* Rudolphi, 1819; *Lebouria tumidula* (Rud.) of Nicoll, 1909; *L.* (*Mediantestis*) *tumidula* of Issaitschikow, 1928), is said to occur in the intestine of the ocean pipe-fish in Europe. Pigulewski (1931) described (as *Lebouria*) a species, *acerinae*, from the ruffe in the Ukraine.

Mention can be made here of the genus *Coitocaecum* Nicoll, 1915, which Poche (1926) made the type of a subfamily, Coitocaecinae. This genus bears a close resemblance to *Plagioporus* in the general anatomy and the relationship between the suckers, but differs notably in the structure of the intestine, the intestinal crura uniting posteriorly to form a ring-shaped intestine. The best-known species of the genus is *Coitocaecum anaspidis* Hickmann, 1934, which exhibits progenesis in the Tasmanian shrimp, *Anaspides tasmaniae*, and about half a dozen others occur in Japan, one (see Manter, 1940*a*) in the Galapagos Isles, Mexico and other localities. Pigulewski (1931) described two species, *Coitocaecum macrostomum* and *C. ovatum*, from the gut of *Siluris glanis* and that of the pike and ruffe in the Sodsch-Dnieper region.

Genus *Podocotyle* (Dujardin, 1845) Odhner, 1905

Syn. *Sinistroporus* Stafford, 1904; *Podocotyloides* Yamaguti, 1934.

The history of this genus has been discussed fully by Odhner (1905) and by Manter (1926).

Podocotyle atomon (Rudolphi, 1802) Odhner, 1905 (Fig. 27D)

Syn. *Fasciola atomon* Rudolphi, 1802; *Distoma atomon* Rudolphi, 1809; *D. simplex* Rudolphi, 1809 of Olsson, 1868; *D. angulatum* Dujardin, 1845; *Allocreadium atomon* (Rudolphi) of Odhner, 1901; *Sinistroporus simplex* Stafford, 1904, in part; *Psilostomum redactum* Nicoll, 1906; *Distomum vitellosum* Linton of Johnston, 1907; ? *Fasciola aeglefini* Müller, 1877, in part; *Podocotyle atomon* var. *dispar* Nicoll, 1909.

Hosts. Short-spined cottus, long-spined cottus, Montagu's sea-snail, two-spotted goby, butter-fish, viviparous blenny, three-spined stickleback, fifteen-spined stickleback, bib, whiting, coal-fish, pollack, five-bearded rockling, three-bearded rockling, Norwegian topknot, Dover sole, plaice, flounder, dab, ocean pipe-fish, common eel, gilt-head.

Location. Intestine and pyloric caeca.

This is a common parasite of the fishes of rock pools in many British localities (Millport, St Andrews, Firth of Clyde, Cullercoats, Irish Sea, North Wales, Galway, Plymouth), and is widely distributed elsewhere (Sweden, Canada, Greenland, U.S.A.).

Diagnosis. *Shape, colour and size:* elongate-oval outline, translucent and colourless, 1·9–3·5 mm. long. *Suckers:* ventral much larger than the oral, diameters 0·30 and 0·16 mm. in an individual 2·5 mm. long. *Gut:* prepharynx short, oesophagus twice as long as the pharynx, caeca long. *Reproductive systems:* genital pore on the left of the oesophagus. Cirrus pouch club-shaped, extending behind the ventral sucker. True pars prostatica absent. Testes one behind the other in the middle of the posterior half of the body. Ovary three-lobed, in front of the testes. Vitellaria of variable extent, generally from the ventral sucker to the posterior extremity, follicles small and compact. Uterus folded between the ovary and the ventral sucker. Eggs few (not more than sixty at any time), brownish yellow, variable in size, 0·06–0·08 mm. long.

Note. Hunninen & Cable (1941*a*) claim that this species has a cotylomicrocercous and not a trichocercous cercaria as was declared by Palombi (1938*b*). It develops in *Littorina rudis* and has a double-pointed stylet, three pairs of penetration glands and a flame-cell formula $2\,[(2 \times 2) \times (2 \times 2)]$. *Gammarus* spp. serve as intermediate hosts.

Other European Species (mainly British) and their Hosts (all are parasitic in the intestine):

Podocotyle furcata (Bremser, in Rudolphi, 1819). Red mullet (Italy), Dover sole.

P. reflexa (Creplin, 1825). Three-bearded rockling, fifteen-spined stickleback, lump-sucker, grey gurnard (Sweden and at Plymouth) (possibly identical with *P. atomon*).*

P. olssoni Odhner, 1905 (Syn. *Distoma simplex* of Olsson, 1868). *Gadus melanostomus*; '*Lumprenus*' *maculatus* (Sweden) (possibly identical with *Podocotyle atomon*).

P. syngnathi Nicoll, 1913. Greater pipe-fish, ocean pipe-fish (Plymouth).

* Rees (1945) found *P. reflexa* in the five-bearded rockling at Aberystwyth.

P. atherinae Nicoll, 1913–15 sp. inq. Sand smelt (probably only an abnormal specimen of *P. atomon*) (Plymouth).
P. odhneri Issaitschikow, 1928. *Gymnacanthus tricuspis* (Russian Arctic).
P. levinseni Issaitschikow, 1928. Sea snail, lumpsucker (Russian Arctic).

Park (1937*b*), who described eight new American species of the genus, disagreed with Issaitschikow's use (1928) of the distribution of the vitellaria and the direction of the lobes of the ovary in the separation of species, finding that the

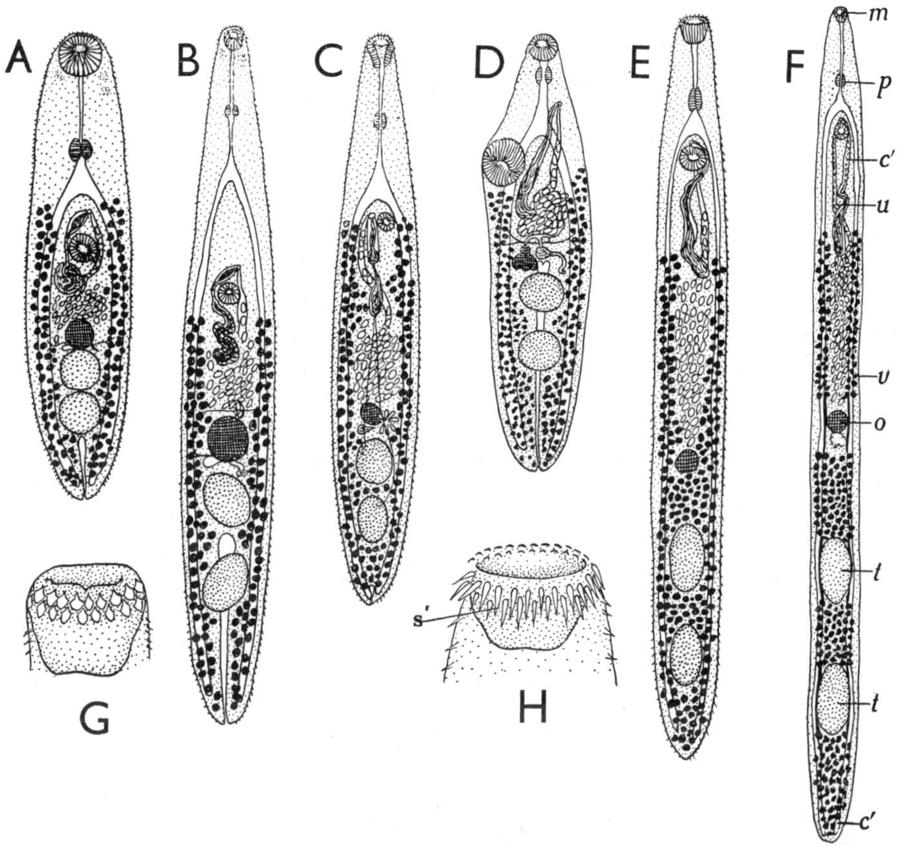

Fig. 27. Allocreadiidae (A–D) and Acanthocolpidae (E–H) from British marine fishes. A, *Lepidapedon rachion*. B, *L. elongatum*. C, *Opechona bacillaris*. D, *Podocotyle atomon*. E, *Stephanostomum caducum*. F, *S. pristis*. G, anterior end of *S. rhombispinosum*. H, anterior end of *S. caducum*. (After Lebour, 1908*a*.)

vitellaria vary widely in this respect within a given species. Certain other characters are said to be less variable, notably the flat or cylindrical form of the body, the straight, sinuous or coiled condition of the seminal vesicle, the extent of the cirrus pouch in front of or behind the ventral sucker, the morphological characters of the genital organs, and the relative positions of the testes. In his key to nineteen species (which excluded *P. furcata* and *P. atherinae*, but included *P. atomon* var. *dispar* Nicoll, 1909) it is notable that all the European forms (and several others) have a coiled seminal vesicle. It is claimed that the European species can be separated by characters modified to form the key:

I. Testes very close together
  A. Cirrus pouch extending back to the mid-point of the ventral sucker *P. levinseni*
  B. Cirrus pouch extending back to the posterior end of the ventral sucker *P. odhneri*

II. Testes considerably apart, or separated by vitelline follicles
  A. Cirrus pouch not extending behind the ventral sucker *P. syngnathi*
  B. Cirrus pouch extending behind the ventral sucker
    (*a*) Oesophagus twice as long as the pharynx
      (i) Vitelline follicles not reaching the ventral sucker *P. atomon*
      (ii) A few vitelline follicles extending in front of the ventral sucker *P. atomon* var. *dispar*
    (*b*) Oesophagus about as long as the pharynx
      (i) Body flattened *P. olssoni*
      (ii) Body cylindrical *P. reflexa*

It must be emphasized that *P. atomon* is a very variable species. As long ago as 1909 Nicoll found that variability affects certain characters upon which Odhner depended for the determination of *P. reflexa*, e.g. shape, flattening, relative positions of the suckers. More importantly, he showed that artefacts produced by inappropriate methods of fixation modify specimens so that they answer to the descriptions of other species. Several species which occur outside Europe will certainly prove to be invalid, and the same might be said of some European forms.

Genus *Helicometra* Odhner, 1902, *nec* Travassos, 1928 (a misprint for *Helicotrema*; Brachylaemidae)

Syn. *Loborchis* Lühe, in Stossich, 1902; *Halicometra* Pratt, 1902.

Palombi (1929*a*) reviewed this genus and recognized three species, claiming that they can be separated by the following characters:

I. Testes lobed *H. fasciata*

II. Testes globular, not lobed
  A. Testes diagonally one behind the other; vitellaria mainly confined to the lateral margins of the body *H. pulchella*
  B. Testes directly one behind the other; vitellaria lateral, but tending to fill the posterior region behind the testes *H. sinuata*

According to this writer the main characters of the three species (with identical forms) are as indicated in the following diagnoses:

Helicometra pulchella (Rudolphi, 1819) Odhner, 1902

Syn. *Distoma pulchellum* Rudolphi, 1819; *Distomum labri* Stossich, 1886, *nec* Beneden, 1870; *Allocreadium labri* (Stossich) of Odhner, 1901; *Distoma* (*Dicrocoelium*) *pulchellum* of Barbagallo & Drago, 1903; *Allocreadium alacre* Stossich of Palombi, 1929.

Hosts. Cuckoo wrasse, greater weever, butterfly blenny and *Gobius jozo* (in Italy and Sicily).

Location. Intestine.

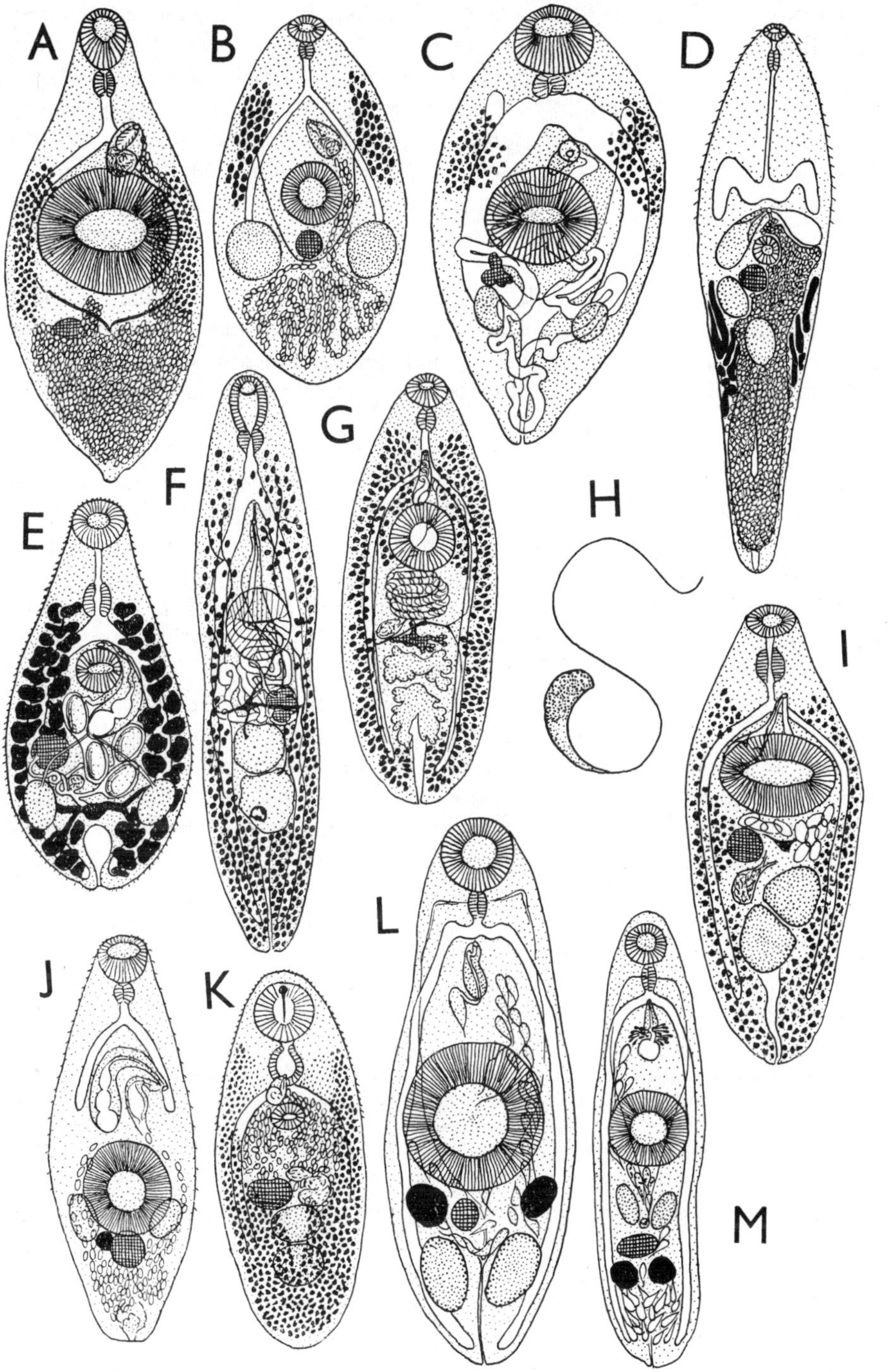

Fig. 28. Fellodistomatidae (A–D), Acanthocolpidae (E), Allocreadiidae (F, G, I, K), Zoogonidae (J) and Hemiuridae (L, M) of British marine fishes. A, *Steringotrema pagelli*. B, *Bacciger bacciger*. C, *Fellodistomum agnotum* (=*F. fellis*). D, *Ancylocoelium typicum*. E, *Neophasis lageniformis*. F, *Cainocreadium labracis*. G, *Helicometra fasciata* (recorded by Nicoll as *H. pulchella*) (H, its egg). I, *Plagioporus varius*. J, *Zoonogenus vividus*. K, *Lepidauchen stenostoma*. L, *Hemipera ovocaudata*. M, *Derogenoides ovacutus*. (A, B, after Nicoll, 1914*a*; C, after Nicoll, 1909; D, J, after Nicoll, 1913*c*; E, after Lebour, 1910; F, after Johnstone, 1908; G, H, I, after Nicoll, 1910*b*; K, L, M, after Nicoll, 1913*a*.)

DIAGNOSIS. *Shape and size:* oval outline, tapering anteriorly, 1·35–2·0 mm. long and 0·65–0·75 mm. in greatest breadth. *Cuticle:* smooth. *Suckers:* ventral larger than the oral, diameters 0·22–0·25 mm. (mean 0·24 mm.) and 0·15–0·20 mm. (mean 0·18 mm.). *Gut:* pharynx 0·066–0·086 mm. long and 0·075–0·13 mm. broad. *Reproductive systems:* genital pore median, near the middle of the oesophagus. Cirrus pouch extending back to a point near the middle of the ventral sucker. Testes globular, entire, diagonally one behind the other posteriorly. Ovary slightly lobed (having three lobes in Mediterranean forms, many in northern forms). Receptaculum seminis globular. Vitellaria extending laterally from the pharynx to the posterior ends of the caeca, merging in the middle line posteriorly. Uterus coiled into a spiral form. Eggs variable in number and size, 0·050–0·058 mm. long, each having a single polar filament about 0·20 mm. long.

Note. Found at Plymouth and in Irish waters, but records doubtful.*

Helicometra sinuata (Rudolphi, 1819)

Syn. *Distoma sinuatum* Rudolphi, 1819; *Allocreadium sinuatum* (Rudolphi) of Odhner, 1901; *Helicometra fasciata* Stossich of Palombi, 1929, *nec* Rudolphi.

HOSTS. Greater weever, *Ophidion barbatum*, *Fierasfer imberbis* (in Italy).

LOCATION. Intestine.

DIAGNOSIS. *Shape and size:* more elongate than *H. pulchella*, 1·75–2·38 mm. long (mean 2·06 mm.; according to Odhner, 2·3 mm.) and 0·54–0·6 mm. in greatest breadth (mean 0·56 mm.). *Suckers:* ventral the larger, diameters 0·29–0·34 mm. (mean 0·31 mm.) and 0·19–0·23 mm. (mean 0·21 mm.). *Gut:* pharynx 0·087 mm. diameter, oesophagus very short, but discrete. *Reproductive systems:* testes globular, entire, directly one behind the other in the posterior region. Ovary multilobate. Vitellaria enormously developed, follicles completely filling the region behind the testes and extending laterally to the oesophagus, but interrupted slightly near the posterior part of the ventral sucker. Eggs 0·042–0·050 mm. long (mean 0·047 mm.), each equipped with a filament 0·2 mm. long. (*Note.* The sizes of the eggs are misprinted in one place in Palombi's paper (1929*a*).)

Helicometra fasciata (Rudolphi, 1819) Odhner, 1902 (Fig. 28G, H)

Syn. *Distoma fasciatum* Rudolphi, 1819, *nec* Stossich, 1885, 1886, 1892, 1898; *Allocreadium fasciatum* (Rudolphi) of Odhner, 1901; *Distoma* (*Dicrocoelium*) *fasciatum* of Barbagallo & Drago, 1903; *Distomum gobii* Stossich, 1883; *Helicometra gobii* (Stossich) of Wallin, 1910; *Loborchis mutabilis* Stossich, 1902; *Helicometra mutabilis* of Stossich, 1904; *H. flava* of Stossich, 1903 and Wallin, 1910; *H. pulchella* (Rudolphi) of Nicoll, 1910; *H. epinepheli* Yamaguti, 1934; *H. hypodytis* Yamaguti, 1934.

HOSTS. Cuckoo wrasse, gold-sinny, comber, eel, conger, rock goby, shanny, gattorugine, Cornish sucker, common topknot (in Britain, according to Palombi; at Plymouth and Millport); cuckoo wrasse, eel (at Bergen); *Labrus turdus*, *L. merula*, *Crenilabrus coeruleus*, *C. pavo*, *Serranus scriba*, *S. hepatus*, *Gobius jozo*, and the eel (in Italy).

LOCATION. Intestine.

This species is said to occur in more than twenty species of fishes in Europe, as well as others in various parts of America (Florida, Mexico). It also occurs in

* Excepting Rees (1945) who found *H. pulchella* in the shanny and *H. fasciata* in the long-spined cottus at Aberystwyth.

the Far East. According to Palombi it is the species recorded in Britain by Nicoll. Baylis & Idris Jones (1933) recorded the species in the gilt-head at Plymouth, which seems to be the only record in Britain. Palombi gave the characters as follows. *Shape and size:* elongate-oval outline, flattened, 1·5–3 mm. long, 0·4–0·75 mm. in greatest breadth and 0·33 mm. thick. *Suckers:* ventral 0·22–0·30 mm. diameter, oval 0·15–0·20 mm. *Gut:* prepharynx flask-shaped, but variable, pharynx 0·09 mm. diameter, oesophagus, 0·15–0·3 mm. long and S-shaped in living specimens. *Reproductive systems:* cirrus pouch pyriform, cirrus 0·15 mm. long. Testes lobed, diagonally one behind the other, shape variable. Ovary variably lobed, situated immediately in front of the testes. Receptaculum seminis pyriform, 0·11–0·20 mm. long and 0·07 mm. broad. Vitellaria mainly lateral, extending to the posterior extremity, merging behind the testes. Uterus spirally coiled, as in other species. Eggs 0·060–0·075 mm. long and 0·025–0·029 mm. broad, each having a filament 0·20 mm. long.

Nicoll's specimens (recorded as parasites of the comber, yellow gurnard, red gurnard, streaked gurnard, rock goby, gattorugine, shanny, ballan wrasse, cuckoo wrasse, gold-sinny, common topknot, common eel and conger) were somewhat larger (1·3–4·3 mm. long and about one-third as broad) and had suckers of relatively slightly different sizes (0·35 and 0·23 mm. diameter). Other characters conform to those described by Palombi, and the eggs measured 0·063–0·084 × 0·032–0·037 mm., were dark brown, convex on one side and concave on the other, and equipped each with a filament six to eight times as long as the capsule.

According to Palombi (1929*a*), who also studied the life history of this species, encysted metacercariae occur in shrimps of the species *Leander serratus* (Penn), *L. squilla* L. and *L. xiphias* (Risso).

Two other species have been recorded in Europe: *Helicometra gurnardus* Thapar & Dayal, 1934, was found in a grey gurnard at the Aquarium of the Zoological Society, London; *H. plovmornini* Issaitschikow, 1928 in several (three) fishes in the Russian Arctic. In the latter the vitellaria do not extend anteriorly as far as the ventral sucker.

American species include *H. torta* and *H. execta*, both Linton, 1910. A related form having numerous (nine) testes arranged in two longitudinal rows in the posterior region was referred by Linton (1910) to a new species and genus, *Helicometrina nimia*. The species *Helicometra azumae* Layman, 1930 has ten testes and was found in *Azuma emmnion* in Peter the Great Bay.

Genus *Peracreadium* Nicoll, 1909

Peracreadium genu (Rudolphi, 1819) Nicoll, 1909

Syn. *Distoma genu* Rudolphi, 1819; *Distomum fasciatum* Stossich, 1892; *Allocreadium genu* (Rudolphi) Odhner, 1901.

Hosts. Shanny, ballan wrasse (at Millport and Plymouth).*

Location. Posterior part of the intestine.

Diagnosis. *Shape, colour and size:* elongate, somewhat flat, neutral grey, 1·5–2·4 mm. long and 0·6 mm. broad at the level of the ventral sucker in an individual 2 mm. long. *Suckers:* ventral larger than the oral and slightly in front

* Rees (1945) found *P. genu* in the five-bearded rockling at Aberystwyth.

of the mid-body, dimensions 0·32×0·35 mm. and 0·2 mm. diameter. *Gut:* prepharynx short, pharynx large and rounded, oesophagus short, caeca long. *Reproductive systems:* genital pore median, near the bifurcation of the intestine. Cirrus pouch elongate (not club-shaped as in most Allocreadiidae), extending far behind the ventral sucker to the level of the ovary. Ejaculatory duct and seminal vesicle convoluted. Pars prostatica distinct. Cirrus frequently protruding and observed inserted into the female pore. Testes one close behind the other in the final third of the body, outline oval but may be irregular. Ovary rounded, in front of the testes, on the right. Vitellaria extending from the posterior extremity to the ventral sucker, where the sequence of follicles is broken, plus a wedge-shaped mass in front of the ventral sucker on each side of the body. Uterus short. Metraterm well differentiated. Eggs few (never more than thirty at one time), pale yellow, measuring 0·080–0·088 × 0·044–0·056 mm. (mean 0·0845 × 0·051 mm.).

Peracreadium commune (Olsson, 1867) Nicoll, 1909

Syn. *Distomum commune* Olsson, 1867; *Allocreadium commune* (Olsson) of Odhner, 1901; *Distomum labri* Beneden, 1870, *nec* Stossich, 1886.

HOSTS. Ballan wrasse, corkwing.

LOCATION. Posterior part of the intestine.

DIAGNOSIS. Differs from *P. genu* in the following characters: broader and flatter, brown in colour, having a more elongate (almost fusiform) pharynx, smaller testes, and vitellaria which are not interrupted at the level of the ventral sucker. The eggs are of variable size, measuring 0·076–0·098 × 0·026–0·048 mm. This species also occurs at Plymouth and Millport and is probably a synonym of *P. genu*. Another species, *P. perezi* Mathias, 1926, occurs in the ballan wrasse and gilt-head, but has not appeared in Britain.

Genus *Cainocreadium* Nicoll, 1909

Cainocreadium labracis (Dujardin, 1845) Nicoll, 1909 (Fig. 28F)

Syn. *Distomum (Dicrocoelium) labracis* Dujardin, 1845; *Echinostoma labracis* (Dujardin) Beneden, 1870; *Allocreadium labracis* (Dujardin) of Johnstone, 1908.

HOST. Bass (in the Irish Sea, Cardigan Bay, Morecambe Bay).

LOCATION. Intestine.

DIAGNOSIS. *Shape and size:* elongate, almost cylindrical, unusually large (for an allocreadiid), 10 mm. long and 2 mm. broad. *Suckers:* ventral sucker near the mid-body. *Gut:* prepharynx absent or very short, oesophagus short, caeca long, bifurcation of the intestine far in front of the ventral sucker. *Reproductive systems:* genital pore close behind the oesophagus. Cirrus pouch elongate, but not extending behind the ventral sucker. Testes one close behind the other in the posterior region. Ovary three-lobed, slightly in front of the testes, on the right. Ejaculatory duct long, vesicula seminalis convoluted. Vitellaria extending from the posterior extremity to the level of the pharynx. Uterus mostly confined between the gonads and the ventral sucker. Eggs measuring 0·07–0·10 × 0·04–0·06 mm.

Genus *Allocreadium* Looss, 1900

Syn. *Creadium* Looss, 1899, *nec* Vieill., 1816, a bird.

Allocreadium isoporum (Looss, 1894) Looss, 1900 (Figs. 12 K, 29 A)

Syn. *Distomum isoporum* Looss, 1894; *Creadium isoporum* of Looss, 1899.

HOSTS. Pike, common carp, barbel, tench, minnow, chub, roach, bream.

LOCATION. Intestine.

DIAGNOSIS. *Shape, colour and size:* elongate, white, yellowish or reddish, 3–5 mm. long and less than 0·75 mm. broad. *Pigment:* aggregates of brown

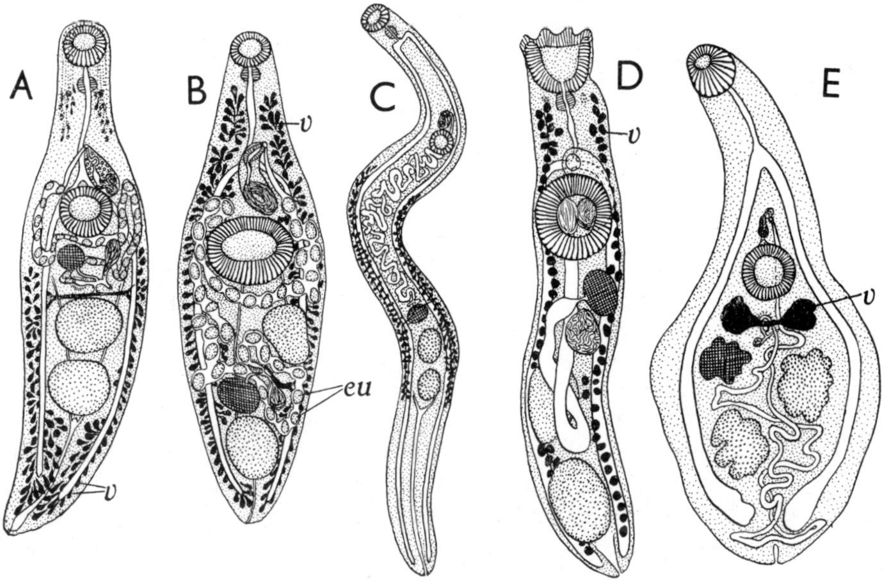

Fig. 29. Some Digenea of British fresh-water fishes. Allocreadiidae (A, B), Azygiidae (C), Bunoderidae (D) and Gorgoderidae (E). A, *Allocreadium isoporum*. B, *Sphaerostoma bramae*. C, *Azygia lucii*. D, *Bunodera luciopercae*. E, *Phyllodistomum folium*. (B, C, after Fuhrmann, 1928; rest after Lühe, 1909; all drawn with reference to the originals by Looss, 1894*a*.)

granules, vestiges of larval eye-spots, occur at the sides of the pharynx. *Suckers:* of about equal size, the ventral very slightly the larger (in the ratio of 14/13) and slightly in front of the mid-body. *Gut:* prepharynx short, pharynx present, oesophagus and caeca long. Bifurcation of the intestine near the ventral sucker. *Reproductive systems:* genital pore in front of the bifurcation of the intestine. Cirrus pouch large, pars prostatica present, seminal vesicle and ejaculatory duct folded. Testes one behind the other in the posterior half of the body. Ovary in front of the testes. Vitellaria overlying the caeca and overstepping their inner margins posteriorly, extending from the ventral sucker to the posterior extremity, their follicles large. Uterus with few folds confined to the region in front of the testes. Eggs large, 0·09 × 0·06 mm., and few, thin and brownish yellow, each containing an ovum and ten to fifteen vitelline cells. According to Looss (1894) the trematode is generally found free in the lumen of the intestine, not attached to its wall. It has been found in Cambridgeshire and Berkshire in the roach and chub respectively.

Allocreadium angusticolle (Hausmann, 1896) Odhner, 1901

Syn. *Distomum angusticolle* Hausmann, 1896; *Creadium angusticolle* of Looss, 1899; *Pera creadium angusticolle* of Nicoll, 1909.

HOST. Miller's thumb.

LOCATION. Intestine.

DIAGNOSIS. *Size:* 1·5 mm. long and 0·3 mm. in greatest breadth. *Suckers:* ventral much larger than the oral, diameters 0·22 and 0·09 mm. *Vitellaria:* extending forward from the posterior extremity to the level of the pharynx. *Genital pore:* near the bifurcation of the intestine, slightly lateral. Eggs 0·055–0·060 mm. long.

Note. This species occurs at Basle and in Lake Constance, but has not appeared in Britain.

Allocreadium transversale (Rudolphi, 1802) Odhner, 1901

Syn. *Fasciola transversalis* Rudolphi, 1802; *Distoma transversale* of Rudolphi, 1819.

HOST. Spined loach (in East Prussia).

LOCATION. Intestine.

DIAGNOSIS. *Shape, colour and size:* elongate, slightly flat, rounded posteriorly, narrowing anteriorly, white, reddish posteriorly, about 2·3 mm. long and 0·75 mm. broad. *Suckers:* oral smaller than the ventral (in the ratio 2/3), dimensions 0·25 mm. diameter and 0·43 × 0·32 mm. *Vitellaria:* of the same extent as in *A. isoporum*. Eggs horn-yellow, 0·115 × 0·083 mm. (Odhner).

Note. Szidat (1939*b*) redescribed this species from a single specimen found in East Prussia, the only one collected for more than 130 years. It was 2 × 0·8 mm., the oral and ventral suckers measured 0·25 mm. diameter and 0·34 × 0·40 mm. respectively, the vitellaria extended forward to the bifurcation of the intestine, and the eggs measured 0·115 × 0·065 mm. This species has not been found in Britain.

Allocreadium polymorphum Layman, 1933

This species was found in the intestine of *Cottus kneri* (in Russia).

Subfamily **Sphaerostomatinae** Poche, 1926

Thapar & Dayal (1934) proposed the elevation of this group to family rank, an apparently unwarranted alteration.

Genus *Sphaerostoma* Rudolphi, 1809

Syn. *Spaerostoma* Pratt, 1902.

Sphaerostoma bramae (O. F. Müller, 1776) Lühe, 1909 (FIG. 29B)

Syn. *Fasciola bramae* Müller, 1776; *Distoma globiporum* Rudolphi, 1802; *Sphaerostomum globiporum* (Rudolphi) Looss, 1899.

HOSTS. Perch, ruffe (pope), pike, common carp, barbel, tench, minnow, chub, dace, roach, rudd, silver bream, bleak, loach (in Germany, Switzerland, Poland, Norway, Denmark). (Found in the roach in Cambridgeshire.)

LOCATION. Intestine.

DIAGNOSIS. *Shape and size:* oval outline, attenuated at the extremities, especially the anterior, 1–4·2 mm. long and about three-eighths as broad. *Cuticle:* non-spinous, but having regular transverse rows of delicate papillae, perforated for the numerous openings of the integumentary glands. *Suckers:* ventral situated near the mid-body and almost twice as large as the oral, 0·3–0·45 mm. diameter. *Gut:* pharynx small, oesophagus long and slender, caeca long. Bifurcation of the intestine well in front of the ventral sucker. *Reproductive systems:* genital pore near the bifurcation of the intestine, sometimes in front of it. Cirrus pouch large, but confined to the region in front of the ventral sucker. Cirrus small and thick, seminal vesicle large and coiled. Testes sometimes rounded, but more often lobed, diagonally one behind the other in the posterior half of the body. Ovary situated in a rather wide space between the testes, lateral. Vitellaria extending from the pharynx to the posterior extremity, overlying the caeca and tending to overstep their inner margins. Uterus with short descending and long ascending limbs, formed into folds between the testes and at the sides of the ventral sucker. Eggs large, pale golden yellow, dimensions about 0·076 × 0·060 mm.

Note. Markowski (1933*a*) found this species in the common eel in the Polish Baltic. In his smaller specimens (1–1·5 mm. long) the vitellaria were ill developed, barely reaching the ventral sucker anteriorly, and the eggs were much smaller (0·057 × 0·049 mm.).

Subfamily **Stephanophialinae** Nicoll, 1909

Cuticle smooth, oral sucker with anterior processes; parasitic in freshwater fishes.

Genus *Crepidostomum* Braun, 1900

Syn. *Acrodactyla* Stafford, 1904; *Stephanophiala* Nicoll, 1909; *Acrolichanus* Ward, 1917; *Crepidastomum* Pratt, 1902.

Crepidostomum farionis (Müller, 1784) Lühe, 1909 (Figs. 35 C, 76)

Syn. *Fasciola farionis* Müller, 1784; *Distoma farionis* of Blanchard, 1891; *D. laureatum* Zeder, 1800 of various writers; *Crepidostomum laureatum* (Zeder) of Braun, 1900, Stafford, 1904, Cooper, 1916; *Stephanophiala laureata* (Zeder) of Nicoll, 1909; *S. transmarina* Nicoll, 1909; *S. vitelloba* Faust, 1918; *Crepidostomum ussuriense* Layman, 1930, emend.; *C. vitellobum* (Faust) Hopkins, 1931; *C. faronis* Linton, 1940.

HOSTS. Sea trout, char, houting, grayling (in Britain).

LOCATION. Intestine.

DIAGNOSIS. *Shape and size:* oval outline, 2–6 mm. long and up to 1·5 mm. broad. *Pigment:* vestiges of the larval eye-spots occurring at the sides of the oesophagus. *Suckers:* oral 0·4–0·5 mm. diameter, having six separate muscular processes, ventral about half as large again (0·6–0·75 mm. diameter) and situated in front of the mid-body. *Gut:* prepharynx short, pharynx small, oesophagus short, caeca long, bifurcation of the intestine slightly in front of the ventral sucker. *Excretory system:* pore slightly dorsal, median canal extending to the level of the anterior testis, lateral canals to the oral sucker. *Reproductive systems:* genital atrium absent, or very shallow, cirrus pouch club-shaped, confined to the region in front of the ventral sucker, seminal vesicle elongate oval, pars prostatica globose, cirrus sometimes formed into a single loop, but variable. Testes globular, one behind the other in the middle of the hindbody. Ovary almost globular, close behind the ventral sucker and far in front of the testes, sometimes slightly lateral. Vitellaria extending laterally from the oesophagus to the posterior extremity, merging in the median plane behind the testes. Uterus much folded between the testes and the ventral sucker, metraterm straight, as long as the cirrus pouch, muscular, but lined with cuticle. Eggs comparatively few (less than one hundred), bright yellow or slightly darker, containing unsegmented ova when laid, measuring 0·065–0·085 × 0·040–0·044 mm. (see Hopkins, 1934).

Notes. This species occurs in France, Germany, Austria, Sweden, Denmark, Finland, Russia, U.S.A., Canada, Alaska, has appeared in Hampshire and is common in some Scottish and Yorkshire rivers. It was studied in detail under the name *Stephanophiala laureata* by Nicoll (1909). Brown (1927), who made a study of the life history, found all trout fished from the River Warfe infected, likewise 84 % of grayling. Hopkins (1934) examined many specimens from U.S.A., Alaska and Britain, also studying available descriptions of specimens from other localities, concluding that minor differences noted come well within the range of individual variability. The synonymy given above was largely determined by him.

OTHER EUROPEAN SPECIES:

C. auriculatum (Wedl, 1857) Lühe, 1909 (Syn. *Distoma auriculatum* Wedl, 1857; *Acrodactyla auriculata* (Wedl) Odhner, 1910; *Acrolichanus auriculatus* (Wedl) Skwortzoff, 1927; *Acrolichanus similis* Wisniewski, 1933) (in *Acipenser ruthenus* L. in the Danube, Volga and Oka).

C. latum (Pigulewski, 1931) (Syn. *Stephanophiala lata* Pigulewski, 1931) (in the rudd in the Ukraine).

C. baicalensis Layman, 1933 (in *Thymallus arcticus* and *Cottus kneri* in Lake Bajkal).

C. suecicum Nybelin, 1932 (in grayling, sea trout, burbot, miller's thumb and char in Sweden).

Note. Nybelin thought this form identical with *C. metoecus* (Braun, 1900), a parasite of bats; Hopkins (1934) regarded fishes as the natural hosts, bats as 'accidental' hosts.

Subfamily **Lepocreadiinae** Odhner, 1905

Cuticle spinous; parasites of marine fishes.

KEY TO GENERA OF LEPOCREADIINAE

1. Pharynx relatively very large (10 % body length), prepharynx short, oesophagus absent; oral sucker larger than the ventral — *Lepidauchen*

2. Pharynx not relatively large, prepharynx and oesophagus present, lengths variable; oral sucker not significantly larger than the ventral
   A. Prepharynx long, oesophagus short — *Lepidapedon*
   B. Prepharynx short, oesophagus (or pseudo oesophagus) long — *Opechona*

Genus *Lepidauchen* Nicoll, 1913

Lepidauchen stenostoma Nicoll, 1913 (Fig. 28K)

HOST. Ballan wrasse (at Plymouth).

LOCATION. Intestine.

DIAGNOSIS. *Shape and size:* oval outline, moderately flat, 2·9–3·25 mm. long and 1·3 mm. in greatest breadth. *Cuticle:* spinous anteriorly. *Suckers:* oral having a longitudinal, slit-like aperture, much larger than the ventral, diameters 0·55 and 0·27 mm. *Gut:* prepharynx very short, pharynx very large (0·31 × 0·34 mm.), oesophagus absent, caeca long, not quite reaching the posterior end of the body. Bifurcation of the intestine slightly in front of the ventral sucker. *Reproductive systems:* genital pore median, slightly in front of the bifurcation of the gut. Cirrus pouch small but stout, reaching back only to the ventral sucker. Seminal vesicle large and globular. Testes rounded, one behind the other, contiguous, situated in the posterior region. Ovary transversely ovoid, in front of the testes on the right. Vitellaria comprising abundant follicles, extending from the level of the pharynx to the posterior extremity, encroaching on the median plane behind the gonads, approaching the median plane antero-dorsally. *Receptaculum seminis uterinum* takes the place of the more usual structure. Uterus not very long, its folds occupying available space between the gonads and the ventral sucker. Eggs not numerous (fewer than 100 at any time), brownish yellow, measuring 0·078–0·084 × 0·046–0·050 mm.

Genus *Lepidapedon* Stafford, 1904

Syn. *Lepodora* Odhner, 1905.

Stafford erected this genus for *Distomum rachion* and the name *Lepidapedon* has the right of priority over *Lepodora*.

Lepidapedon rachion (Cobbold, 1858) Stafford, 1904 (Fig. 27A)

Syn. *Distomum rachion* Cobbold, 1858; *Lepodora rachiaea* Odhner, 1905 and of Lebour, 1908; *Distoma increscens* Olsson, 1868, in part.

HOSTS. Haddock, cod, coal-fish, pollack.

LOCATION. Intestine.

This species has been found at St Andrews, Cullercoats, Millport and Plymouth and occurs in U.S.A. and Canada.

DIAGNOSIS. Generally occurring in groups of two to six individuals. *Shape, colour and size:* elongate, opaque, pale yellow, 2–4 mm. long and about one-quarter or one-fifth as broad. *Suckers:* oral larger than the ventral, which is situated slightly in front of the mid-body, diameters 0·36 and 0·20 mm. in individuals 3·5 mm. long. *Gut:* prepharynx fairly or very long (sometimes several times as long as the pharynx), pharynx conspicuous (0·2 mm. diameter), oesophagus shorter than the pharynx, caeca wide and long. Bifurcation of the intestine behind the mid-point between the suckers. *Reproductive systems:* genital pore near the bifurcation of the intestine, displaced slightly to the left. Cirrus pouch elongate, enclosing a folded seminal vesicle, a pars prostatica and a short cirrus.

Vesicula seminalis *externa* present. Testes rounded, one behind the other, contiguous, in the posterior region. Ovary rounded, in front of the testes. Vitellaria extending from the posterior extremity to a point midway between the bifurcation of the intestine and the ventral sucker, merging in the median plane behind the testes. Uterus with few folds, confined between the ovary and the ventral sucker. Eggs few, yellowish brown, about 0·06–0·07 mm. long (Manter (1934) gave the range of length as 0·059–0·065 mm., Linton (1940) as 0·06–0·07 mm., for American forms). Issaitschikow (1928) described (as *Lepodora*) a subspecies, *gymnacanthi*, from the Arctic.

Lepidapedon elongatum (Lebour, 1908) Nicoll, 1915 (Fig. 27B)

Syn. *Lepodora elongata* Lebour, 1908.

HOST. Cod (at Cullercoats).

LOCATION. Duodenum and intestine.

DIAGNOSIS. *Shape, colour and size:* more elongate than *L. rachion*, colourless, fairly translucent, 3·5 mm. long and 0·54 mm. broad. *Suckers:* small, ventral larger than the oral, diameters 0·12 and 0·10 mm.; ventral well in front of the mid-body. *Gut:* prepharynx very long, pharynx small, oesophagus as long as the prepharynx, caeca long and wide. Bifurcation of the intestine far in front of the ventral sucker, midway between it and the oral sucker. *Reproductive systems:* genital pore slightly in front of the ventral sucker and far behind the bifurcation of the intestine, slightly on the left. Gonads as in *L. rachion*, but more distant from one another. Vesicula seminalis *externa* much longer than in *L. rachion*, vitellaria more restricted anteriorly, not extending in front of the ventral sucker. Uterus with few folds, mostly between the ovary and the vesicula seminalis externa. Eggs not numerous, but generally more than 100 *in utero* at any time, more pointed at one pole than at the other, measuring 0·066 × 0·036 mm.

Note. Manter (1934) found this fluke in several fishes which were trawled at fair depths (140–367 fathoms) near Tortugas. Linton (1940) found that the eggs of specimens from Woods Hole measured 0·054–0·07 × 0·03–0·04 mm.

Genus *Opechona* Looss, 1907

Syn. *Pharyngora* Lebour, 1908.

Ward & Fillingham (1934) pointed out that Looss used the name *Opechona* for a supposititious genus with *Distomum bacillare* (later *Pharyngora bacillaris*) as its type about one year before Lebour erected the genus *Pharyngora* for the species *P. retractilis*. Accordingly, it must stand as the valid name.

Opechona retractilis (Lebour, 1908)

Syn. *Pharyngora retractilis* Lebour, 1908.

HOST. Whiting (at Cullercoats) (rare).

LOCATION. Intestine.

DIAGNOSIS. *Shape, colour and size:* elongate, transparent and colourless, very contractile, 4–6 mm. when at rest, but extensile beyond these limits, about 0·5 mm. broad. Girth uniform, but the anterior end tapering, the posterior rounded. *Suckers:* oral larger than the ventral, which is situated not far in front of the mid-body, diameters 0·2 and 0·12 mm. *Gut:* prepharynx very long, as much as three times the length of the pharynx, but very contractile, pharynx small, oesophagus very long (0·4–0·8 mm.), caeca both wide and long. *Reproductive systems:* genital pore slightly in front or to one side of the ventral sucker. Cirrus pouch cylindrical. Vesicula seminalis *interna* spherical. Vesicula seminalis *externa*, long and pyriform. Testes one behind the other, in the posterior part of the body. Ovary heart-shaped, in front of the testes. Vitellaria extending from near the posterior extremity to the level of the ventral sucker. Uterus with few loops between the ovary and the vesicula seminalis *externa*. Metraterm somewhat inflated. Eggs few (20–40 at any time) and large (about 0·08–0·10 mm. long).

Opechona bacillaris (Molin, 1859) Looss, 1907 (Fig. 27C)

Syn. *Distoma bacillare* Molin, 1859; *D.* (*Dicrocoelium*) *bacillare* (Molin) Stossich, 1886; *Pharyngora bacillaris* of Nicoll, 1914 and other writers.

HOSTS. Mackerel, boar-fish, lumpsucker.

LOCATION. Intestine and pyloric caeca.

DIAGNOSIS. Differs from *O. retractilis* in the following particulars: body much less elongate, ventral sucker more anteriorly situated, prepharynx and oesophagus much shorter, vitellaria less extensive anteriorly (not reaching the ventral sucker), vesicula seminalis *externa* absent.

Notes. According to Nicoll (1910*b*, but not 1914), *O. retractilis* is identical with *O. bacillaris* and *Distoma increscens* Olsson, 1868. Ward & Fillingham (1934) preferred to regard it as a valid species, pointing out that if the structure of this form is allowed as coming within the range of variability of *Opechona bacillaris* the validity of several other species of the genus will not have been well established. Study of specimens from the whiting might impart some finality to one or the other of these opinions. Unfortunately they are rare, having been found in Britain only once, during December. The trematode occurs in Italy and Sicily, and Markowski (1933*a*) found numerous specimens in the mackerel in the Polish Baltic and figured a well-developed vesicula seminalis *externa*. Perhaps it was for this reason that he included *O. retractilis* in the list of synonyms of this worm. A brief diagnosis based on his findings includes the following characters. *Size:* 2–4·5 mm. long. *Colour:* old individuals brownish, young ones colourless. *Shape:* elongate. *Cuticle:* spinous, spines small. *Zones:* three recognized: (*a*) oral sucker to ventral sucker, 1·045 mm. long; (*b*) ventral sucker to ovary, 0·365 mm. long; (*c*) ovary to posterior extremity, 1·56 mm. long. *Suckers:* oral 0·27–0·285 mm. long and 0·195–0·21 mm. broad; ventral 0·12–0·195 mm. diameter. *Gut:* prepharynx very short, pseudo-oesophagus long (0·3–0·6 mm.), oesophagus short. *Reproductive systems:* genital pore near the ventral sucker, cirrus pouch elongate, vesicula seminalis *externa* and *interna* and pars prostatica present. Testes one behind the other in the posterior third of the body, ovary three-lobed, situated slightly in front of the anterior testis. Receptaculum seminis ovoid. Vitellaria extending from the level of the vesicula seminalis *externa* to the ends of the caeca (near the posterior extremity). Uterus confined near the median plane between the ventral sucker and the ovary. Eggs measuring about 0·082 × 0·035–0·041 mm.

Note. The species *Opechona orientalis* (Layman, 1930), discovered in *Scomber japonicus* in Peter the Great Bay, also has a vesicula seminalis *externa*, but relatively small eggs, measuring 0·054–0·062 × 0·032–0·035 mm. (breadth misprinted in the German translation) in specimens 2·1–3·3 mm. long.

Family **ACANTHOCOLPIDAE** Lühe, 1909 (p. 87)

Cable & Hunninen (1941 *b*), have concluded, on the basis of a study of the life history of *Deropristis inflata* (Molin, 1859), that the aggregate of genera at present comprising this family is not a natural group of closely related trematodes. They claim that a redefinition of the family in the light of recent work would result in the exclusion of *Deropristis* and *Dihemistephanus*, whilst other genera (with little interest for us) would be included for the first time. Postulating that the closest relatives of *Deropristis* are members of the subfamilies Anallocreadiinae and Lepocreadiinae (family Allocreadiidae), they suggest that this genus, together with these subfamilies, should form the basis of a new family, Lepocreadiidae, distinct from Allocreadiidae. It would be premature at this stage, however, to follow these suggestions, which depend entirely upon the cercarial types. Accordingly, *Deropristis* and *Dihemistephanus* are included, as hitherto, in the family under consideration. It might be as well to mention that M. J. Miller (1941 *a*) removed *Neophasis* (*Acanthopsolus*) from the Acanthocolpidae (as seems desirable) and referred it to the subfamily Lepocreadiinae of the Allocreadiidae, thus further indicating the need for a revision of both these families. In view of pending changes we must be satisfied with an array of genera.

KEY TO GENERA OF ACANTHOCOLPIDAE

1. Oral spines arranged in a double row around the mouth
   - A. Sequence of oral spines unbroken ventrally — *Stephanostomum* (*Stephanochasmus*)
   - B. Sequence of oral spines broken ventrally — *Dihemistephanus*
2. Oral spines not differentiated
   - A. Testes far back in the posterior region of the body
     - (*a*) Anterior region of body expanded to form shoulder-like processes, and here beset with especially large spines — *Deropristis*
     - (*b*) Anterior region not expanded to form shoulder-like processes, and not beset with large spines — *Neophasis* (*Acanthopsolus*)
   - B. Testes not far back in the posterior region of the body — *Tormopsolus*

Genus *Stephanostomum* Looss, 1899, *nec Stephanostoma* Danielsen, 1880

Syn. *Stephanochasmus* Looss, 1900; *Lechradena* Linton, 1901; *Echinostephanus* Yamaguti, 1934.

Manter (1934) has pointed out that the generic name *Stephanostoma* Danielsen, 1880 (an insect) does not invalidate Looss's first choice of a name for this genus, so that *Stephanostomum* must stand. Incidentally, it seems to be the more appropriate name.

In this genus Looss (1901 *b*) arranged five species showing close general resemblance in spite of variation in size, but differing in the numbers, sizes and arrangement of the oral spines, the relative sizes of the suckers and the absolute sizes of the eggs. They are *Stephanostomum cesticillum* (Molin, 1858), *S. bicoronatum* (Stossich, 1883), *S. pristis* (Deslongchamps, in Lamouroux, 1824), *S. caducum* (Looss, 1901) and *S. minutum* (Looss, 1901). The first two differ from the others in showing a ventral hiatus in the rows of oral spines. Of the five species, two have not appeared in Britain (*S. bicoronatum* and *S. minutum*), but three other species have been discovered in this country, namely, *S. baccatum* (Nicoll, 1907), *S. rhombispinosum* (Lebour, 1908) and *S. triglae* (Lebour, 1908). Another European species is *S. sobrinum* (Levinsen, 1881), also unrecorded in Britain.

Stephanostomum pristis (Deslongchamps, in Lamouroux, 1824) Looss, 1899 (Fig. 27F)

Syn. *Distoma pristis* Deslongchamps, 1824; *Echinostoma pristis* (Deslongchamps) Dujardin, 1845 and of Cobbold, 1860; *Anoiktostoma pristis* (Deslongchamps) of Stossich, 1899; *Stephanochasmus pristis* of Looss, 1901.

HOSTS. Cod, whiting.

LOCATION. Intestine and pyloric caeca.

DIAGNOSIS. *Shape and size:* ribbon-like, about 7 mm. long (4·5 mm. according to Looss) and 0·35 mm. broad. *Cuticle:* spinous. *Oral spines:* a double, unbroken row of thirty-six strong, thick spines. *Suckers:* of about equal size, or oral smaller than the ventral, diameters 0·13 and 0·19 mm., the ventral relatively far forward on the body (about one-sixth of the distance along it). *Gut:* prepharynx long, pharynx small (length 0·1 mm.), oesophagus short, caeca long. Bifurcation of the intestine very slightly in front of the ventral sucker. *Reproductive systems:* genital pore near the anterior margin of the ventral sucker. Cirrus pouch very elongate, curving round the ventral sucker and extending back to a level midway between this sucker and the ovary. Vesicula seminalis club-shaped, cirrus long and spinous, the spines irregularly arranged. Testes ovoid, one behind the other, 0·5 mm. apart, the posterior testis about 1 mm. from the posterior extremity. Ovary spherical, about 1 mm. in front of the anterior testis. Vitellaria extending from the posterior end of the cirrus pouch to the posterior extremity, tending to fill up available space behind, between and in front of the gonads, but interrupted opposite them. Uterus with only a few folds between the ovary and the ventral sucker. Metraterm long and lined with spinelets similar to those of the cirrus. Eggs not numerous, about 0·06 mm. long.

Note. This species was discovered in Normandy and has been found at Plymouth, Galway, Millport and Cullercoats. Lebour (1908*a*) did not see the oral spines, but Nicoll (1910*b*) observed them in immature specimens from the cod.

Stephanostomum rhombispinosum (Lebour, 1908) (Fig. 27G)

Syn. *Stephanochasmus rhombispinosus* Lebour, 1908; *S. pristis* (Deslongchamps) Stossich, 1886.

HOST. Whiting (at Cullercoats), five-bearded rockling (at Aberystwyth).

LOCATION. Pyloric caeca.

DIAGNOSIS. *Shape and size:* ribbon-like, 5–10 mm. long and about 0·3 mm. broad. *Cuticle:* spinous to the posterior extremity. *Oral spines:* 36–38 very flat, broad, rhombic spines arranged in a double row. *Suckers:* of equal size, diameters about 0·16 mm. *Gut:* as in *S. pristis*, but oesophagus broad. *Reproductive systems:* as in *S. pristis*, but vitellaria slightly more extensive anteriorly, extending along the sides of the seminal vesicle. Eggs about 0·08 mm. long.

Stephanostomum triglae (Lebour, 1908)

Syn. *Stephanochasmus triglae* Lebour, 1908.

HOST. Grey gurnard (at Cullercoats, St Andrews and Galway).

LOCATION. Pyloric caeca and intestine.

DIAGNOSIS. *Shape and size:* less ribbon-like than *S. pristis* and *S. rhombispinosus*, 3·2 mm. long and 0·74 mm. broad when contracted, 4·8 mm. long and 0·50 mm. broad when extended. *Cuticle:* spinous, spines thickly clustered anteriorly. *Oral spines:* number uncertain, but more than forty-two and far fewer than fifty-six (the number in *S. baccatus* (Nicoll, 1907)), spines of the anterior (oral) row slightly larger than those of the posterior (aboral), lengths 0·04 and 0·036 mm. respectively. *Suckers:* oral smaller than the ventral, dimensions 0·26 × 0·14 mm. and 0·28 mm. diameter. *Gut:* prepharynx long, but short in contracted individuals, pharynx twice as large as in the foregoing species (0·2 mm. long), oesophagus very short, caeca long. *Reproductive systems:* similar to those of foregoing species, but vitellaria continuous from the ventral sucker to the posterior extremity. Eggs measuring 0·099 × 0·056 mm.

Note. Nicoll (1906) observed fifty spines in a single immature specimen, 2·9 mm. long; those in the anterior row were 0·03 mm. long, those in the posterior row 0·035 mm., thus reversing the order of sizes noted by Lebour.

Stephanostomum caducum (Looss, 1901) (Fig. 27E, H)

Syn. *Stephanochasmus caducus* Looss, 1901; *S. caducus* var. *lusci* Nicoll, 1914.

HOSTS. Bib, poor cod, whiting (at Plymouth, Galway and Cullercoats).

LOCATION. Intestine and pyloric caeca.

DIAGNOSIS. *Shape and size:* less ribbon-like than *S. pristis* and *S. rhombispinosum*, about 4·4 mm. long and 0·34 mm. broad. *Cuticle*: spinous. *Oral spines:* forty-eight, in two unbroken rows of twenty-four each, spines of the anterior row larger than those of the posterior, lengths 0·039 and 0·033 mm. respectively. *Suckers:* equal in size, about 0·3 mm. diameter (0·17 mm. diameter in specimens 4 mm. long, according to Looss). *Gut:* prepharynx about 0·34 mm. long. *Reproductive systems:* similar to those of *S. triglae*, but vitellaria extending from the posterior extremity to the hinder part of the vesicula seminalis. Eggs measuring 0·066 × 0·036 mm. Nicoll (1914*a*) recorded the variety *lusci* from the first two of the specified hosts. In his specimens the 'head-spines' were arranged in two rows of twenty-five, those of the posterior row being slightly the larger.

Stephanostomum baccatum (Nicoll, 1907)

Syn. *Stephanochasmus baccatus* Nicoll, 1907; *Distomum valdeinflatum* Stossich of Johnstone, 1905.

HOSTS. Adults in short-spined cottus, halibut, and witch; larvae in angler, long rough dab, dab, and lemon sole.

LOCATION. Adults in the intestine and rectum; larvae encysted under the skin and in the muscles.

DIAGNOSIS. *Size:* 3·34 mm. long and 0·75 mm. in greatest breadth. *Oral spines:* fifty-six in two alternating rows of twenty-eight each, spines of the anterior row shorter than those of posterior, lengths 0·031 and 0·037 mm. respectively. *Suckers:* ventral much larger than the oral and 0·87 mm. from the anterior extremity, diameters 0·33 and 0·23 mm. *Pigment:* traces of the vestigial eye-spots of the larva present.

Note. Adults have been found at Aberdeen, Millport and St Andrews, larvae in the Irish Sea and at Cullercoats.

OTHER EUROPEAN SPECIES:

Stephanostomum cesticillum (Molin, 1858) Looss, 1899 (Syn. *Distoma cesticillum* Molin, 1858; *Stephanochasmus cesticillus* (Molin, 1858) Looss, 1901; *Echinostoma cesticillus* (Molin) Monticelli, 1893; *Anoiktostoma cesticillus* (Molin) Stossich, 1899) (in the stomach and intestine of the John Dory (rare)).

Stephanostomum sobrinum (Levinsen, 1881) Looss, 1899 (Syn. *Distoma sobrinum* Levinsen, 1881; *Stephanochasmus sobrinum* (Levinson, 1881) Odhner, 1905) (in the intestine of the short-spined sea scorpion).

Linton (1940) gave accounts of several American species, viz. *S. dentatum* (Linton, 1900), *S. tenue* (Linton, 1898) and *S. filiforme* Linton, 1940, and two unnamed forms.

Note. Immature specimens of species of *Stephanostomum* have been found in Britain by Johnstone (1905), Lebour (1908*a*), Nicoll & Small (1909) and Nicoll (1910*b*), in America by Linton (1898) and Stafford (1904), in India by Lühe (1906) and in Japan by Yamaguti (1934*a*, 1937). In these instances encysted metacercariae were found in the tissues of fishes, suggesting that the latter serve as the second intermediate host. In Britain, flatfishes seem to serve in this capacity to a marked extent, and voracious feeders like the angler are probably facultative hosts. Martin (1938*b*, 1939*b*) extended our knowledge by completing the life history of *Stephanostomum tenue* (Linton) experimentally. Ophthalmoxiphidiocercariae develop in rediae in the digestive gland of *Nassa obsoleta*. Taken into the gut of the second intermediate host (small fishes), the cercariae migrate through the wall of the intestine, metacercariae encysting in the liver or mesenteries, to await ingestion of this by the definitive host. Martin showed that in this species the flame-cell formula is 2 (3 + 3 + 3 + 3 + 3 + 3 + 3).

Genus *Dihemistephanus* Looss, 1901

Dihemistephanus sturionis Little, 1930 (Fig. 30 A–C)

HOST. Sturgeon.

This species was discovered in the Bristol Channel, west of Lundy Island.

LOCATION. Intestine (?).

DIAGNOSIS. *Shape and size:* very elongate, oval in section, frail and extensile, 3·5–6·2 mm. long and 0·49–0·69 mm. broad. *Cuticle:* spinous, except posteriorly. *Oral spines:* forty-four flat, almond-shaped spines in two rows, twenty-one to twenty three in the anterior and twenty one in the posterior row, those in the anterior row only half the size of the others. *Suckers:* either of equal size or ventral slightly smaller than the oral, ventral situated in the first fifth or sixth of the body. *Gut:* prepharynx short, pharynx large, oesophagus short, caeca long. *Reproductive systems:* genital pore near the anterior margin of the ventral sucker. Genital atrium extending back to the posterior margin of this sucker. Cirrus

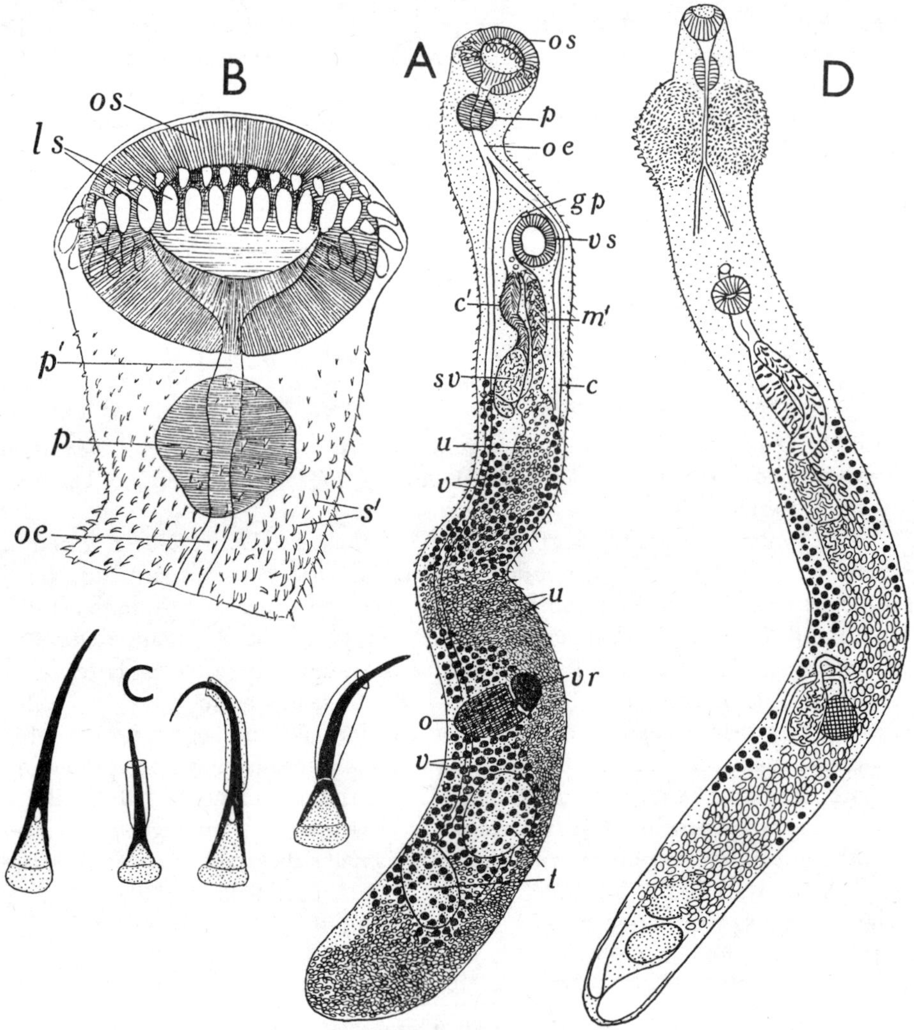

Fig. 30. Acanthocolpidae. A–C, *Dihemistephanus sturionis* a parasite of the sturgeon. A, entire trematode in ventral view. B, anterior end in dorsal view. C, 'hooklets' from the metraterm. D, *Deropristis inflata*. (A–C, after Little, 1930; D, after Markowski, 1933*a*.)

and metraterm equipped with spines about 0·05 mm. long, those of metraterm interspersed with cilia. Testes ovoid, one behind the other in the most posterior part of the body. Ovary in front of the testes, spherical. Vitellaria extending from the base of the cirrus pouch to the posterior extremity, mainly lateral and dorsal. Uterus with descending and ascending limbs, abundantly folded, occupying much of the available space in the posterior half of the body. Eggs numerous, thick-shelled, yellow, measuring 0·043–0·049 × 0·034–0·043 mm.

Dihemistephanus lydiae (Stossich, 1896) Looss, 1901

Syn. *Echinostoma lydiae* Stossich, 1896; *Stephanostomum lydiae* (Stossich) of Looss, 1899.

HOST. Sunfish (at Trieste; also recorded in the Firth of Forth by Nicoll, 1909).

LOCATION. Intestine.

DIAGNOSIS. *Size:* about 4 mm. long and 0·8 mm. broad. *Oral spines:* thirty-nine long, cylindrical, spike-like spines of approximately equal size arranged in an unbroken row of twenty and a posterior row of nineteen. *Suckers:* of about equal size, the ventral situated near the mid-body.

Note. Nicoll's specimens were 1·8–2·4 mm. long and had oral spines numbering eighteen in the anterior row and seventeen in the posterior. The eggs were much larger than in *D. sturionis*, but very variable in size, measuring 0·060–0·072 × 0·032–0·040 mm.

Genus *Deropristis* Odhner, 1902

Ward (1938) reviewed the literature bearing on this genus in some detail.

Deropristis hispida (Abildgaard, in Rudolphi, 1819)

Syn. *Distoma hispidum* Abildgaard, in Rudolphi, 1819; *Echinostoma hispidum* of Beneden, 1870, *nec* Cobbold, 1858.

Rudolphi first described this species, but cited Abildgaard as the authority for the name. Odhner (1902) claimed Rudolphi as the author and some writers have followed him in this. Ward (1938) pointed out that as the first description bore the alternative name, no violence is done to the original publication in which the species was founded by attributing the species to Abildgaard.

HOSTS. Sturgeon, smelt.

LOCATION. Intestine.

This species has not been recorded in Britain, but was found by Beneden (1870) off the Belgian coast, and also occurs in the Mediterranean and in North America.

DIAGNOSIS. *Size:* up to 12 mm. long and 0·5–0·65 mm. broad, American forms less than half as long (Ward). 'Shoulders' about 0·7 mm. broad, bearing ten to twelve thick spines arranged in a single row, the anterior spines the smallest. Above the pharynx there is a cluster of six to seven similar spines, irregularly arranged. *Note.* Discrepancies occur in descriptions of the numbers and arrangement of the spines. Ward's observations may be summarized as follows: entire body covered by closely set rows of sharply pointed spines, largest anteriorly, becoming smaller towards both extremities, absent behind the testes. Margins of the 'shoulders' beset with rows of flat, ovate spines, the humped dorsal region bearing still heavier spines. Exact number and arrangement undetermined. *Suckers:* small, ventral slightly the larger, diameters 0·19 and 0·17 mm. (in American forms 0·15 and 0·12 mm.). Eggs 0·038–0·043 mm. long (in American specimens 0·036–0·045 × 0·016–0·025 mm.).

Deropristis inflata (Molin, 1858) (Fig. 30 D)

Syn. *Distomum inflatum* Molin, 1858; *Echinostomum hispida* of Beneden, 1870.

This species, unrecorded in Britain, was discovered in the eel at Padua, and has been found in the same host in the Polish Baltic (Markowski, 1933*a*). It also occurs in Canada and U.S.A. Looss (1902*a*, *b*) gave full descriptions and figures. Specific differences from the previous species include the dimensions of various organs and the eggs, which are larger, measuring 0·043–0·049 × 0·023–0·027 mm. (mean 0·045 × 0·025 mm.) (0·046–0·052 × 0·019–0·024 mm. according to Markowski). Specimens from Poland have the following characters. *Size:* 3·6 mm. long. *Cuticle:* spinous back to the middle of the body. '*Shoulders*': about 0·375–0·39 mm. broad, bearing on each lateral margin about twenty-eight spines (measuring 0·024–0·027 × 0·011–0·013 mm.) arranged in two longitudinal rows. *Suckers:* of almost equal sizes, the oral measuring about 0·10–0·11 × 0·12 mm., the ventral about 0·10–0·11 × 0·13–0·14 mm. *Other characters* as in Fig. 30 D.

Note. Cable & Hunninen (1942*a*) showed that the cercaria is of a modified trichocercous type, the tail having a ventral fin fold and six pairs of ventro-lateral tubercles, each carrying a short bristle. In America the cercaria develops in a redia in the branchial region and digestive gland of *Bittium alternatum* and later penetrates the polychaete *Nereis virens*, encysting in the axial region or in the parapodia. The eel becomes infected after devouring annelids containing metacercariae. These writers also gave good figures of the adult which show the peculiar spination admirably. The large spines develop in the metacercaria and may become worn, giving the dorsal hump a ragged appearance, or lost in the adult.

Genus *Tormopsolus* Poche, 1926

Tormopsolus osculatus (Looss, 1901) Poche, 1926

Syn. *Distomum osculatum* Looss, 1901.

HOST. Five-bearded rockling.

LOCATION. Intestine.

DIAGNOSIS. According to Looss (1901*b*): *Size:* about 4 mm. long and 0·4 mm. broad. *Shape:* ribbon-like, rounded posteriorly, tapering anteriorly from the ventral sucker. *Cuticle:* spinous, spines small and fine around the mouth, becoming larger laterally towards the cirrus pouch and ventrally towards the pharynx, then smaller again farther back. *Eyes:* remains of the cercarial eyes present as clusters of dark pigment at the sides of the pharynx. *Suckers:* very close together, their centres 0·6 mm. apart, the oral very small (0·09 mm. diameter), the ventral twice as large (0·02 mm. (*sic*, misprint for 0·2 mm.)). *Gut:* prepharynx narrow, pharynx large and urn-shaped, 0·13 mm. broad (i.e. larger than the oral sucker), caeca long. *Reproductive systems:* genital organs topographically arranged much as in *Stephanostomum*. Genital pore close in front of the ventral sucker, cirrus pouch club-shaped, overreaching the ventral sucker by about twice its breadth, seminal vesicle sac-like, occupying one-third of the cirrus pouch, cirrus about the same length and lined with spinelets. Testes large and ovoid, ovary rounded, situated near the mid-body. Receptaculum seminis absent. Vitellaria extending anteriorly to the end of the cirrus pouch, follicles filling the body behind the gonads. Uterus mainly between the ovary and the cirrus pouch, metraterm about as long as the cirrus and similarly lined with spinelets. Eggs large and few, thin-shelled, yellow, with a flattened opercular pole, measuring about 0·07 × 0·046 mm.

Notes. This species has not appeared in Britain. Manter (1934) pointed out the very close resemblance which it shows to species of *Stephanostomum*, especially *S. caducum*, and the rarity of its occurrence (only two specimens are known to exist) also suggests that it may be a trematode belonging to the genus, but one which had lost the oral spines.

Genus *Neophasis* Stafford, 1904

Syn. *Acanthopsolus* Odhner, 1905.

Neophasis lageniformis (Lebour, 1910) M. J. Miller, 1941 (Fig. 28E)

Syn. *Distomum* sp. Lebour, 1908; *Acanthopsolus lageniformis* Lebour, 1910.

HOST. Wolf-fish (at Cullercoats).

LOCATION. Upper intestine.

DIAGNOSIS. *Shape and size:* very small, flask-shaped, slightly flat (especially in front of the ventral sucker), colourless except for the eggs, 0·5–1·3 mm. long. *Cuticle:* spinous. *Suckers:* both small, ventral smaller than the oral and just in front of the mid-body. *Pigment:* one eye-spot situated on each side of pharynx. *Gut:* prepharynx long, but highly contractile, pharynx relatively large, oesophagus very short and wide, caeca wide and long. Bifurcation of the intestine close in front of the ventral sucker. *Reproductive systems:* genital pore a median, transverse slit near the anterior margin of the ventral sucker (may be seen to open and shut frequently in the living worm). Cirrus pouch elongate, curving round the ventral sucker. Cirrus spinous, vesicula seminalis hour-glass-shaped, pars prostatica feebly developed. Testes side by side, widely separated, in the posterior region. Ovary in front of the right testis. All gonads rounded. Vitellaria consisting of large follicles in the lateral regions, extending from the pharynx to the posterior extremity. Uterus with only two or three folds, containing as a rule four eggs (never more than eight). Eggs very large (as large as the testes), 0·08–0·10 mm. long, occurring in individuals little more than 0·5 mm. in total length.

Infected wolf-fish may contain hundreds of individuals of this species. Small specimens occur in the stomach, even as far forward in the alimentary canal as the pharynx and buccal cavity. The cercariae occur in rediae in the digestive gland of *Buccinum undatum*.

Notes. The type-species *Neophasis pusilla* Stafford, 1904 is a parasite in the urinary bladder of the wolf-fish in Canada.

Another species, *N. oculatus* (Levinsen, 1881), was discovered off the east coast of Greenland in the intestine and pyloric caeca of the herring and the short-spined sea scorpion.

Family **ACANTHOSTOMATIDAE** Poche, 1926, emend. Nicoll, 1935 (p. 88)

Syn. Acanthochasmidae Nicoll, 1914.

Subfamily **Acanthostomatinae** Nicoll, 1914, emend. Hughes, Higginbotham & Clary, 1942

Syn. Acanthochasminae Nicoll, 1915; Anoiktostominae Nicoll, 1915.

With the characters of the family.

KEY TO GENERA OF ACANTHOSTOMATINAE

I. Vitellaria restricted to a short region near the ventral sucker, scarcely extending behind it and mainly in front of it — *Anisocoelium*

II. Vitellaria situated in the mid-body or hind-body, their anterior terminations far behind the ventral sucker

- A. Body ribbon-like, suckers very close together and of small and uniform sizes, vitellaria situated near the mid-body — *Anisocladium*
- B. Body elongate, but not ribbon-like, suckers not very close together and of dissimilar sizes (oral larger than ventral), vitellaria mainly situated in the hind-body — *Acanthostomum*

Genus *Acanthostomum* Looss, 1899

Syn. *Acanthochasmus* Looss, 1900; *Caimanicola* Freitas & Lent, 1938.

Acanthostomum imbutiforme (Molin, 1859) (Fig. 31 B, C)

Syn. *Distoma imbutiforme* Molin, 1859; *Echinostoma(um) imbutiforme* (Molin) of Stossich, 1898 and Johnstone, 1906; *Anoiktostoma imbutiforme* (Molin) of Stossich, 1899; *Acanthochasmus imbutiformis* (Molin) of Nicoll, 1915.

HOST. Bass, wolf-fish.

LOCATION. Intestine.

Johnstone (1906) briefly described specimens found in the bass off the Lancashire coast.

DIAGNOSIS. *Shape and size:* elongate, tapering from the ventral sucker towards both extremities, but truncate anteriorly, about 6·5 mm. long (7·5 mm. according to Looss, 1901 *b*). *Cuticle:* spinous, spines small and straight. *Suckers:* oral larger than the ventral, about one-sixth of the body-length apart (1·1 mm.), diameters 0·39 and 0·33 mm. (0·27 and 0·2 mm., Looss). *Oral spines:* mouth encircled by eighteen or nineteen, rarely seventeen (even sixteen according to Johnstone), stout, straight spines 0·09 mm. long, arranged in a single circlet. *Gut:* prepharynx short, pharynx small, oesophagus short, caeca long. *Reproductive systems:* genital pore a transverse slit-like aperture on the tip of a papilla midway between the bifurcation of the intestine and the ventral sucker. Testes ovoid, one behind the other in the extreme posterior region of the body. Ovary large and rounded, some distance in front of the testes and about one-third of the distance along the body. Vitellaria extending as two lateral bands of follicles from a point midway between the ventral sucker and the ovary to the level of the testes. Uterus with descending and ascending limbs, slightly folded. Eggs measuring 0·023–0·024 × 0·012 mm. (Looss, 1901 *b*)

OTHER SPECIES:

Acanthostomum praeteritum (Looss, 1901) has been reported in the intestine of the wolf-fish, but outside British waters. Fig. 31 A shows *A. spiniceps* Looss, 1890, which was found in the gut of '*Bagrus bayad*' in Cairo. A fourth species is *A. absconditum*. The four species differ in the sizes of the suckers and numbers of oral spines. Looss (1901 *b*) counted the latter in more than 230 individuals of the four species. They mainly number nineteen in *absconditum*, eighteen in *imbutiforme*, twenty-one to twenty-two in *praeteritum* and twenty-eight to twenty-nine in *spiniceps*.

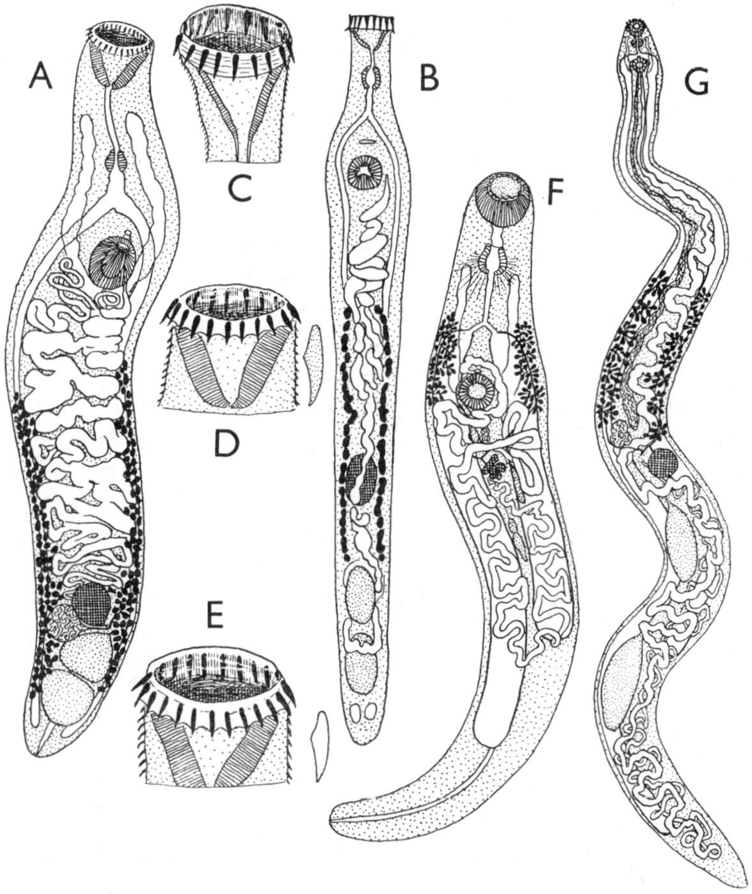

Fig. 31. Acanthostomatidae. A, *Acanthostomum spiniceps*. B, *A. imbutiforme*. C, D, anterior end of B. E, anterior end of *A. praeteritum*. In D and E an isolated spine is shown. F, *Anisocoelium capitellatum*. G, *Anisocladium fallax*. (B, C after Johnstone, 1906; rest after Looss, 1901 *b*.)

Genus *Anisocoelium* Lühe, 1900

Anisocoelium capitellatum (Rudolphi, 1819) (Fig. 31 F)

Syn. *Distoma capitellatum* Rudolphi, 1819.

HOST. *Uranoscopus scaber*, *Sparus salpa* (Italy; found by Mathias (1934) in former host and the same location at Banyuls).

LOCATION. Gall bladder.

DIAGNOSIS. *Size:* 3–5 mm. long (according to Looss, up to 8 mm.). *Cuticle:* spinous, spines in the anterior region short and wide, having eight to nine sharp points at the free margin, in the posterior region longer and narrower and with fewer points, true spines only in the most posterior region. *Oral spines:* a circlet of twenty-four, rarely twenty-five to twenty-six. *Other characters:* as in the figure.

Note. Mathias (1934) determined that the oesophagus is never longer than the pharynx to the extent indicated by Looss (1901 *b*), the caeca never reach the posterior extremity as claimed by Stossich (1896), the testes are not oval and disposed as stated by Monticelli (1893), and the uterus is not folded posteriorly as was maintained by Lühe (1900). The eggs are numerous and very small (0·012–0·015 × 0·007–0·010 mm.).

Genus *Anisocladium* Looss, 1902

Syn. *Anisogaster* Looss, 1901.

Anisocladium fallax (Rudolphi, 1819) (Fig. 31 G)

Syn. *Distoma fallax* Rudolphi, 1819; *Anisogaster fallax* of Looss, 1901; *Anisogaster gracilis* Looss, 1901.

HOST. *Uranoscopus scaber* (at Naples).

LOCATION. Gall bladder.

DIAGNOSIS. *Shape and size:* very elongate, 8–10 mm. long when contracted, up to 16 mm. when extended, broader in the posterior than in the anterior region. *Cuticle:* thickly spinous, spines small, becoming sparse at the level of the ovary. *Oral spines:* twenty-five, rarely twenty-four, in a single row. *Other characters:* as in the figure.

Looss (1901 *a*) described as *A. gracilis* smaller and more contracted specimens (found in the same host at Trieste) which were said to differ from *A. fallax* in having unequal suckers, the oral being much more than twice as large as the ventral, slightly larger eggs and fewer vitelline follicles. This form must be considered identical with *A. fallax*, which Vlasenko (1931) also found (as *gracile*) in the Black Sea.

Family **HAPLOPORIDAE** Nicoll, 1915 (p. 92)

KEY TO GENERA OF HAPLOPORIDAE

1. Vitellaria so close together as to seem single, their borders curved, shaped like a clover leaflet — *Dicrogaster*
2. Vitellaria obviously paired
   - A. Vitellaria close together between the caeca and spherical — *Haploporus*
   - B. Vitellaria widely separated and lateral to the caeca
     - (*a*) Vitellaria compact, irregularly three-cornered — *Saccocoelium*
     - (*b*) Vitellaria formed of a small number (seven) of separate follicles — *Lecithobotrys*

Genus *Haploporus* Looss, 1902

Haploporus benedenii (Stossich, 1887) Looss, 1902 (Fig. 32 A, B)

Syn. *Distoma benedenii* Stossich, 1887.

HOST. Thick-lipped grey mullet.

LOCATION. Intestine.

DIAGNOSIS. *Shape and size:* oval outline, more pointed anteriorly, 1·25–2·5 mm. long and 0·5–0·6 mm. broad, contracted specimens sometimes more thick set (Stossich). *Cuticle:* spinous, spines arranged in regular rows, becoming sparse posteriorly. *Suckers:* oral larger than the ventral, diameters 0·25 and 0·19 mm. Ventral sucker barely in the middle third of the body. *Gut:* prepharynx short (Looss does not mention it, but it is evident in his excellent figures), pharynx spherical, oesophagus long, caeca very short and wide, extending only to the level of the testis, i.e. slightly beyond the ventral sucker. Bifurcation of the intestine dorsal to the ventral sucker. *Reproductive systems:* genital pore in front of the bifurcation of the intestine. False cirrus pouch very large, but not

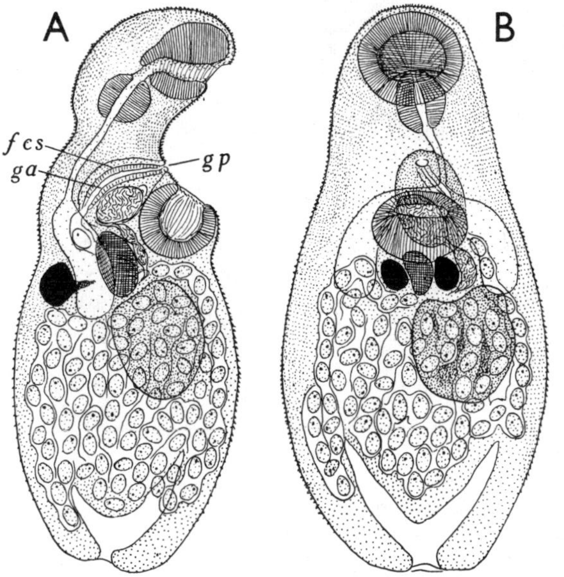

Fig. 32. *Haploporus benedenii* (Haploporidae). A, side view. B, ventral view. (After Looss, 1902*e*.)

extending back as far as the posterior margin of the ventral sucker. Single testis rounded, close behind the ventral sucker, decidedly lateral, on the right. Ovary in front of the testes. Vitellaria almost spherical, between the caeca and the ovary, but dorsal to both. Uterus with descending and ascending limbs, formed into a number of folds in the posterior region. Eggs fairly numerous, measuring 0·045–0·053 × 0·030–0·034 mm., apparently without gelatinous envelopes, each containing a miracidium with a relatively small, X-shaped eye-spot.

Note. Nicoll (1914*a*) obtained a few specimens of this species at Plymouth from the host mentioned above.

Haploporus lateralis Looss, 1902 (Fig. 33A, B)

HOSTS. Golden grey mullet, thick-lipped grey mullet.

LOCATION. Intestine.

DIAGNOSIS. *Shape and size:* oval outline, pointed at the extremities, but truncated posteriorly, very small, 0·8–0·95 mm. long, 0·38 mm. in greatest breadth and

0·27 mm. in greatest thickness. *Cuticle:* spinous. *Suckers:* oral slightly smaller than the ventral, diameters 0·11 and 0·12 mm. *Gut:* caeca extending back to the level of the middle of the testis. *Reproductive systems:* genital pore in front of the bifurcation of the intestine. Testis decidedly lateral and at the same level as the

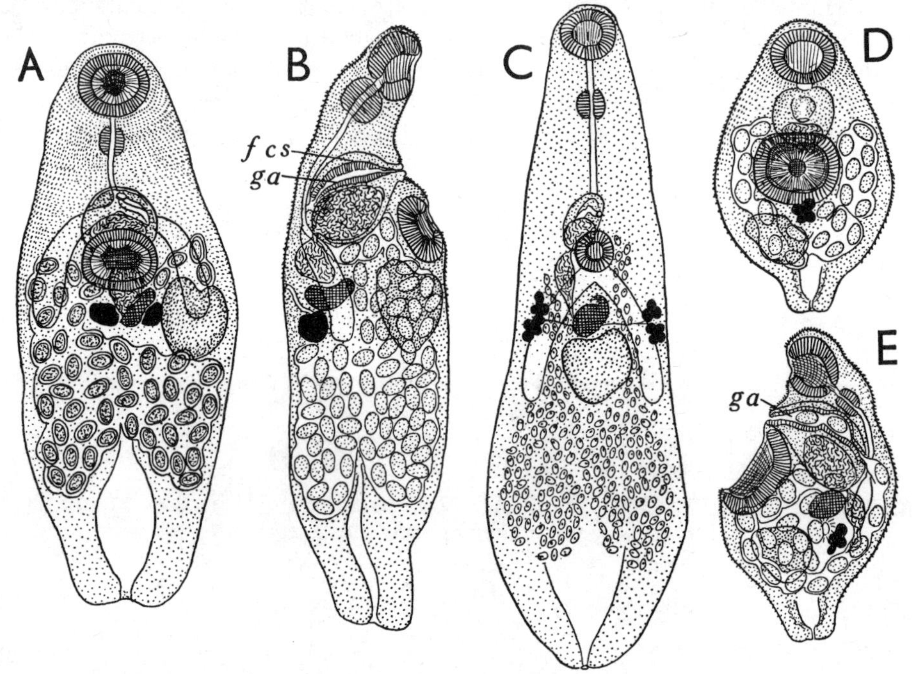

Fig. 33. Other Haploporidae from British fishes. A, B, *Haploporus lateralis* in ventral and side views. C, *Lecithobotrys putrescens* in ventral view. D, E, *Dicrogaster contractus* in ventral (A) and lateral (B) views. (After Looss, 1902*e*.)

ovary. Eggs smaller than in *H. benedinii* (0·042–0·045 × 0·023–0·026 mm.), having gelatinous envelopes 0·008 mm. thick, each containing a miracidium without eye-spots.

Note. This species will probably prove to be identical with *H. benedenii.*

Haploporus longicollum Vlasenko, 1931 was found in the intestine of *Mugil cephalus* in the Black Sea.

Genus *Dicrogaster* Looss, 1902

This genus does not seem to have been recorded for fishes inhabiting British waters, but there is every likelihood of its being found in this country. Looss (1902*e*) described the two species for which brief diagnoses are given below. Both members of the genus are small haploporids with short, tubular caeca which lie immediately beneath the *dorsal* surface of the body and do not reach back beyond the ventral sucker. The vitellaria are so close together near the median plane as to appear like a single mass of three or four follicles, and they are equally near the tips of the caeca and the dorsal surface of the body.

Dicrogaster perpusilla Looss, 1902, emend.

HOST. Thick-lipped grey mullet.

LOCATION. Rectum.

DIAGNOSIS. *Shape and size:* very small, oval outline, tapering anteriorly from behind the mid-body, rounded posteriorly, but with a papilla-like extremity, 0·3 mm. long, 0·18 mm. in greatest breadth, and 0·15 mm. thick. *Suckers:* oral smaller than the ventral, diameters 0·07 and 0·1 mm. Ventral sucker in the mid-body. *Gut:* pharynx relatively small (0·018 mm. long). *Reproductive systems:* testis only slightly lateral. Eggs few (twelve to twenty at any time), measuring 0·053 × 0·025 mm., having gelatinous coverings, each containing a miracidium with eye-spots.

Dicrogaster contracta Looss, 1902, emend. (Fig. 33D, E)

HOST. Thick-lipped grey mullet.

LOCATION. Intestine.

DIAGNOSIS. *Size:* 0·45 mm. long and 0·24 mm. in greatest breadth. *Shape:* as in *D. perpusilla*. *Suckers:* relatively large, oral smaller than the ventral, diameters 0·1 and 0·125 mm. *Gut:* pharynx relatively large, cylindrical, 0·047 mm. long. *Reproductive systems:* testis on the right side of the body. Eggs relatively small (0·035–0·040 mm. long), apparently without gelatinous envelopes, each containing a miracidium with a relatively small, X-shaped eye-spot.

Note. This species was based on a single specimen and is almost certainly identical with *D. perpusilla*.

Genus *Saccocoelium* Looss, 1902

Saccocoelium obesum Looss, 1902 (Fig. 34A, B)

HOSTS. Golden grey mullet, thick-lipped grey mullet, thin-lipped grey mullet.

LOCATION. Intestine.

DIAGNOSIS. *Shape and size:* pyriform, very small, 0·6–0·8 mm. long, 0·33–0·37 mm. broad and almost as thick. *Cuticle:* spinous. *Suckers:* oral slightly smaller than the ventral, 0·1–0·12 mm. diameter. Ventral sucker near the mid-body. *Gut:* prepharynx of medium length, pharynx large and elongate (0·11 × 0·08 mm.), oesophagus long, caeca very short, egg-shaped and sharply set off from the oesophagus. Entire gut situated close beneath the dorsal surface of the body (see Fig. 34B). *Reproductive systems:* genital pore close to the anterior margin of the ventral sucker. False cirrus pouch not appreciably larger than the ventral sucker, extending back slightly beyond it. Testis single, lateral. Ovary in front of the testis, slightly behind the ventral sucker. Vitellaria compact, postero-lateral to the tips of the caeca. Uterus with descending and ascending limbs formed into a few folds in the posterior region. Eggs measuring 0·045 × 0·026 mm., having gelatinous envelopes, each containing a miracidium with large eye-spots.

Note. Nicoll (1914*a*) found a few specimens of this species in the intestine of the thick-lipped grey mullet at Plymouth.

Saccocoelium tensum Looss, 1902

HOST. Thick-lipped grey mullet.

LOCATION. Intestine.

DIAGNOSIS. *Size:* 0·6 mm. long and 0·2 mm. broad and thick. *Suckers:* oral decidedly smaller than the ventral, diameters 0·075 and 0·095 mm. *Gut:* pharynx shaped as in *S. obesum*, but relatively smaller (0·077 × 0·051 mm.). *Reproductive systems:* similar to *S. obesum*, but false cirrus pouch relatively larger, testis only slightly lateral, eggs measuring 0·045 × 0·028 mm., and having gelatinous envelopes, each containing a miracidium with eye-spots.

It is very probable that further study will prove *S. tensum* to be identical with *S. obesum*. The differences specified by Looss were insufficient for the erection of two species.

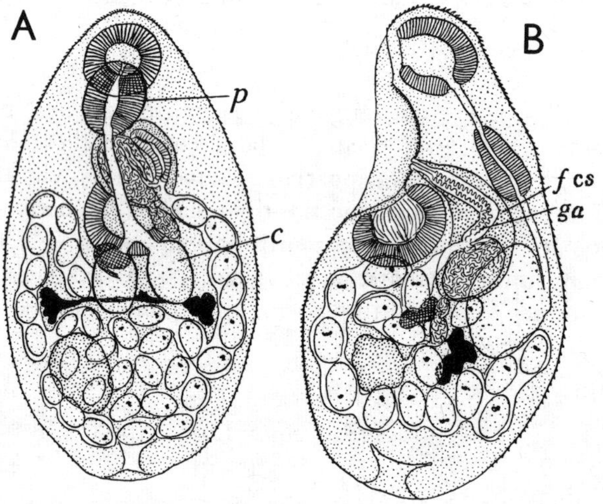

Fig. 34. *Saccocoelium obesum* (Haploporidae) in ventral (A) and lateral (B) views. (After Looss, 1902*e*.)

Genus *Lecithobotrys* Looss, 1902

Lecithobotrys putrescens Looss, 1902 (Fig. 33C)

HOST. Golden grey mullet.

LOCATION. Rectum.

DIAGNOSIS. *Shape and size:* fusiform, tapering anteriorly from the more rounded posterior extremity, 2·3 mm. long and 0·75 mm. in greatest breadth. *Suckers:* oral larger than the ventral, the latter nearly half-way along the body, diameters 0·21 and 0·15 mm. *Gut:* prepharynx fairly long, pharynx pyriform, 0·1 mm. diameter, oesophagus long, caeca short and wide, but longer than in other haploporids, extending back far beyond the ventral sucker to the posterior border of the testis. Bifurcation of the intestine dorsal to the ventral sucker. *Reproductive systems:* genital pore in front of the bifurcation of the intestine. Testis median, not far behind the ventral sucker, between the caeca. Ovary median, in front of the testis. Vitellaria formed of seven spherical follicles on each side, lateral to the middle of the caeca. Uterus with descending and ascending limbs, much folded in the posterior region behind the caeca and the gonads. Eggs very numerous, measuring 0·044–0·047 × 0·026–0·028 mm., each containing a miracidium with eye-spots which develop relatively slowly.

Family **MONORCHIIDAE** Odhner, 1911 (p. 91)

Hopkins (1941*b*) has provided a key to thirteen genera belonging to this family. Only five genera are reported from Europe, however, namely: *Monorchis* Looss, 1902; *Monorcheides* Odhner, 1905; *Physochoerus* Poche, 1926; *Genolopa* Linton, 1910 (Syn. *Proctotrema* Odhner, 1911); and *Asymphylodora* Looss, 1899. Only the first and last of these have been reported from fishes in British waters. Two other genera, additional to the thirteen already mentioned, *Lasiotocus* Looss and *Pristisomum* Looss (in Odhner, 1911), have not been adequately described. The genera with which we are more concerned can be separated by the following keys:

Subfamily **Monorchiinae** Odhner, 1911

Flattened forms of circular or short oval outline having a Y-*shaped excretory vesicle, generally with vitellaria confined to the region in front of the ventral sucker.*

KEY TO GENERA OF MONORCHIINAE

I.	Two testes present	*Monorcheides*
II.	One testis present	*Monorchis*

Genus *Monorchis* Looss, 1902, *nec* Clerc, 1902—a cestode

Monorchis monorchis (Stossich, 1890) Looss, 1902 (Fig. 45 A)

Syn. *Distomum monorchis* Stossich, 1890; *D. tartinii* Stossich, 1899; *Monorchis parvus* Looss, 1902.

HOST. Gattorugine (in Britain; Plymouth).

LOCATION. Stomach.

DIAGNOSIS. *Size:* small, 1–1·2 × 0·8 mm. (Looss, 1902). *Cuticle:* spinous. *Suckers:* oral larger than the ventral (diameters 0·23 and 0·13 mm.), the latter near the mid-body. *Gut:* prepharynx short, pharynx small (0·08 mm. diameter), oesophagus short, caeca long. *Reproductive systems:* genital pore median, in front of the ventral sucker, near the bifurcation of the intestine. Cirrus pouch elongate, extending back well behind the ventral sucker. Cirrus spinous. Testis in the posterior region. Ovary in front of the testis, lateral. Vitellaria lateral, in front of the ventral sucker, lateral to the anterior regions of the gut. Uterus with descending and ascending limbs, much folded, mainly in the middle of the body, lateral to the other organs. Eggs yellowish brown, very small (0·21–0·023 × 0·012–0·013 mm.).

Looss described small specimens from *Sargus annularis* and *S. rondeletii* at Trieste as '*Monorchis parvus*', which is evidently synonymous with *M. monorchis*.

Genus *Monorcheides* Odhner, 1905

Monorcheides diplorchis Odhner, 1905 (Fig. 45B)

HOST. *Lumprenus* (*sic*) *medius* (on west side of King's Bay, Spitzbergen).

LOCATION. Intestine.

This species has not been recorded in British waters, but would be easily recognizable on account of having two testes. According to Odhner it is 0·45–0·8 mm. long, 0·3–0·5 mm. broad, and the eggs (which contain embryos) measure 0·026–0·028 × 0·014–0·015 mm.

Another species is *Monorcheides soldatovi* Issaitschikow, 1928, which occurs in *Aspidophoroides olriki* in Russia. Yet another, *Monorcheides cumingiae* (Martin, 1938) was reared experimentally in flounders and eels and small puffers fed on clams containing encysted metacercariae in U.S.A. Martin (1939*a*, 1940) described the cercariae, which develop in sporocysts in the clam (*Cumingia tellinoides*) and leave the body of the mollusc, later re-entering it to encyst.

Subfamily **Proctotrematinae** Odhner, 1911, emend. Yamaguti, 1934

Body somewhat elongate, little flattened, excretory vesicle tubular, vitellaria generally confined to the region behind the ventral sucker.

KEY TO GENERA OF PROCTOTREMATINAE

I. Oral sucker larger than the ventral; oesophagus very short; genital pore median *Proctotrema*

II. Oral sucker smaller than the ventral; oesophagus fairly long; genital pore lateral *Asymphylodora*

Genus *Asymphylodora* Looss, 1899

Asymphylodora tincae (Modeer, 1790) Lühe, 1909 (Fig. 35B)

Syn. *Fasciola tincae* Modeer, 1790; *Asymphylodora ferruginosa* (Linstow, 1877); *A. ex-spinosa* (Hausmann, 1896); *A. immitans* (Mühling, 1898); *A. tincae* var. *kubanicum* Issaitschikow, 1923; *A. macrostoma* Ozaki, 1925; *Distoma punctatum* (Zeder, 1800) Rudolphi, 1809 (see Witenberg & Eckmann, 1934); *Asymphylodora tincae* var. *donicum* Issaitschikow, 1923.

HOSTS (continental). Barbel, common bream, tench, carp.

LOCATION. Intestine.

DIAGNOSIS. *Shape, colour and size:* oval outline or spindle-shaped, extremities tapering, colourless or reddish, up to 1·3 mm. in length. *Cuticle:* scaly, scales with their free ends produced into three-cornered processes. *Suckers:* oral smaller than the ventral, which is one-third of the distance along the body, diameters 0·15–0·26 and 0·26–0·34 mm. *Gut:* prepharynx short, pharynx globular, oesophagus long and folded, caeca fairly long and extending to the last quarter of the body. *Reproductive systems:* genital pore at the level of the ventral sucker, near the lateral margin of the body. Genital atrium well developed. Vesicula seminalis large and folded, pars prostatica short, cirrus and metraterm spinous. Vitellaria lateral, not very well developed, occupying a short region of the body near its middle. Uterus with descending and ascending limbs, formed into double U-shaped folds, extending to the posterior region of the body. Eggs small, brownish red, with a small nodule or short filament asymmetrically placed near the anopercular pole, measuring 0·023–0·027 × 0·012–0·015 mm. (according to Witenberg and Eckmann (1934), 0·021–0·026 × 0·011–0·013 mm.). The cercariaeum develops in a redia.

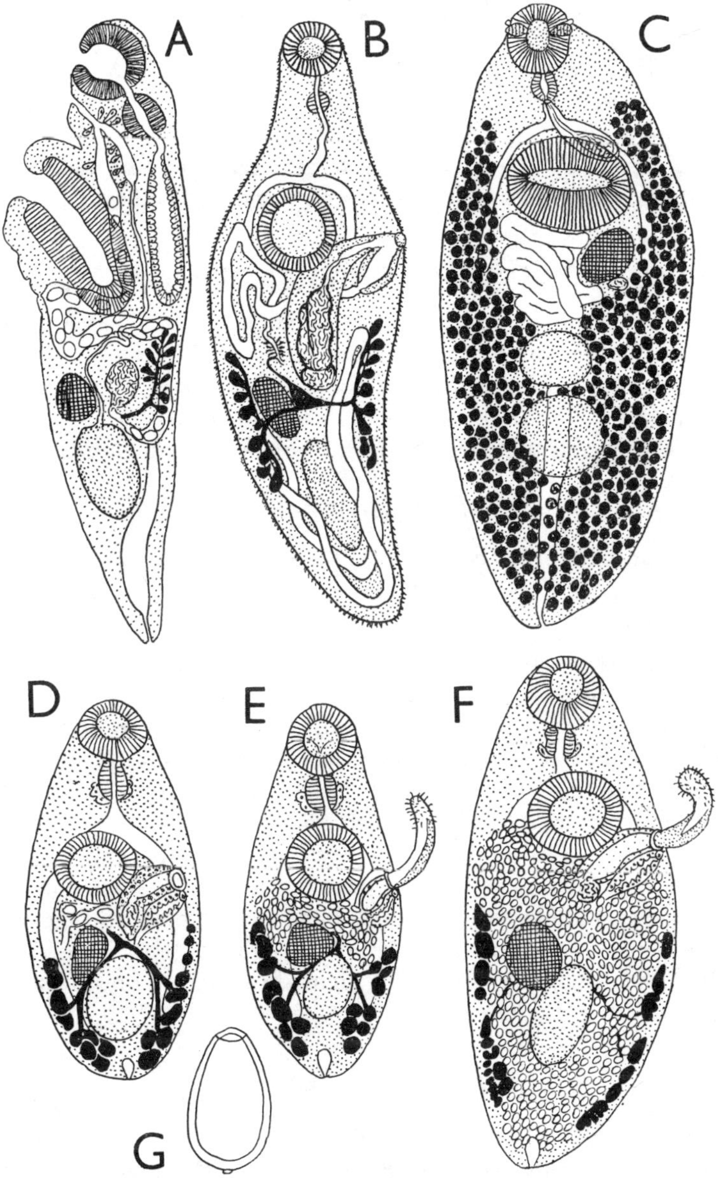

Fig. 35. Haplosplanchnidae (A), Monorchiidae (B) and Allocreadiidae (C) from British fresh-water fishes. A, *Haplosplanchnus pachysoma*. B, *Asymphylodora tincae*. C, *Crepidostomum farionis*. D–G, a Monorchiid, *Asymphylodora demoli* (= *A. tincae*?), from the common goby in the Polish Baltic. D–F, different stages in sexual maturity; G, the egg. (A, after Fuhrmann, 1928; B after Lühe, 1909; C, after Nicoll, 1909; D–G, after Markowski, 1935.)

Note. Since Witenberg & Eckmann affirmed that this is the only species of the genus, Markowski (1935) has described *A. demeli* (Fig. 35 D–F), a parasite of the common goby in the Polish Baltic. It is said to differ from *A. tincae* in the position of the vitellaria, the small size of the cirrus pouch, the shortness of the oesophagus and the symmetrical form of the pyriform eggs, but is probably identical with this species.

Genus *Proctotrema* Odhner, 1911

Proctotrema bacilliovatum Odhner, 1911

HOST. Red mullet.

LOCATION. Rectum.

This species was discovered at Trieste and can easily be identified by the bacilli-like eggs, which measure 0·031–0·033 × 0·008–0·009 mm.

Family **HAPLOSPLANCHNIDAE** Poche, 1926 (p. 91)

This family contained till recently only a single species of the type and only genus *Haplosplanchnus*, but Srivastava (1939*a*) described *H. purii* and a new genus and species, *Laruea caudatum*, both of which are parasites of a food-fish, *Mugil waigiensis*, in India. *Laruea* is said to be distinguished by the shape of the body and ventral sucker and the positions of the gonads and vitellaria.

Genus *Haplosplanchnus* Looss, 1902

Haplosplanchnus pachysoma (Eysenhardt, 1829) Looss, 1902 (Fig. 35 A)

Syn. *Distoma pachysoma* Eysenhardt, 1829; *Podocotyle pachysomum* (Eysenhardt) of Stossich, 1898.

HOSTS. Thin-lipped grey mullet, thick-lipped grey mullet.

LOCATION. Intestine.

This trematode occurs in Europe (Italy, Sicily) and Japan, though unrecorded in Britain. It can be recognized by the familial characters, especially the shape, sac-like gut and single testis. Looss (1902*b*) gave a detailed account of the anatomy. It is brownish yellow and up to 3·3 mm. long, but mature at 1·2 mm. According to Yamaguti (1934*a*) the eggs may hatch *in utero* and the miracidia have uniform ciliation, a large X-shaped eye-spot with a lens-like body on either side, and are slightly larger than the egg which contained them, measuring 0·06–0·07 × 0·027–0·03 mm. The vacated egg shells were about 0·051 mm. long and 0·027 mm. broad, but Looss specified eggs measuring 0·040–0·055 × 0·026–0·031 mm.

Family **ACCACOELIIDAE** Dollfus, 1923 (p. 91)

Syn. *Accacoeliinae* Odhner, 1911; Accacoeliiden Looss, 1912.

Members of this family occur for the most part in the intestine of the sun-fish (Molidae) and are uncommon in British waters. One species has been found at Salcombe, Devon, and others occur in Europe as close to our shores as the Atlantic coast of France. In briefly reviewing the family I have freely interpreted the work of Dollfus (1935*c*).

Subfamily **Tetrochetinae** Dollfus, 1935

Syn. Tetrochetinen Looss, 1912.

Genital atrium without a papilla, i.e. copulatory organ.

KEY TO GENERA OF TETROCHETINAE

1. Anterior wall of the ventral sucker double; vitellaria composed of small pyriform masses of follicles — *Orophocotyle*
2. Anterior wall of the ventral sucker not double; vitellaria composed of long tubules with ramifying branches — *Tetrochetus*
3. Ventral sucker very prominent, with four ear-like outgrowths; vitellaria composed of slender cords between the pharynx and the ovary — *Mneiodhneria*

Genus *Orophocotyle* Looss, 1902

Two species of this genus have been found in the oblong sun-fish (*Ranzania truncata*) at Trieste: *O. planci* (Stossich, 1899) Looss, 1902 (Syn. *Podocotyle planci* Stossich, 1899) and *O. divergens* Looss, 1902 (Fig. 45 C), which is probably a synonym of the former. Looss (1912) claimed that members of this genus differ from those of the genus *Tetrochetus* in lacking diverticula near the bifurcation of the intestine. Odhner (1927) showed that such diverticula occur. Adults are 3–3·4 mm. long and produce eggs measuring 0·028–0·030 × 0·016–0·020 mm.

Genus *Tetrochetus* Looss, 1912

Tetrochetus raynerianus (Nardo, 1827) Looss, 1912

Syn. *Distoma raynerianum* Nardo, 1827.

Host. *Luvarus imperialis* (essentially a Mediterranean fish which sometimes visits British waters).

Location. Intestine.

Locality. Venice, Trieste, Isle of Elba.

Another species, *T. proctocolus* Manter, 1940, occurs in the Galapagos Isles.

Genus *Mneiodhneria* Dollfus, 1935

Syn. *Odhnerium* Yamaguti, 1934.

Mneiodhneria foliata (Linton, 1898) Dollfus, 1935

Syn. *Orophocotyle foliata* (Linton) of Looss, 1902; *Distoma foliatum* Linton, 1898.

Host. Sun-fish.

Location. Intestine.

Locality. U.S.A. (Massachusetts, California) and Canada, also in Europe (see Nicoll, 1915).

Linton (1940) gave a table of separate measurements of five specimens of this species. There is great variability in the shapes of the suckers, but the ventral is invariably much the larger, has accessory lobes and is pedunculate. The vitellaria extend from the oral sucker to the testes and the eggs measure 0·030–0·033 × 0·018–0·021 mm.

Mneiodhneria calyptrocotyle (Monticelli, 1891) Dollfus, 1935 (Fig. 36D, E)

Syn. *Distomum calyptrocotyle* Monticelli, 1891; *Accacoelium calyptrocotyle* of Monticelli, 1893; *Orophocotyle calyptrocotyle* (Monticelli) of Looss, 1902.

HOST. Sun-fish.

LOCATION. Intestine.

LOCALITY. Trieste, Roscoff (Finistère), Rabat (Morocco) (see Dollfus), also in Sweden (host *Mola nasus*).

This species can be recognized by the generic characters established by Dollfus. American forms were mentioned by Sumner, Osburn & Cole (*Bull. U.S. Bureau of Fisheries*, 1911, pp. 31, 583) and described by Linton (1940).

DIAGNOSIS. *Shape:* very elongate, cylindrical, attenuated posteriorly, often formed into an **S** with the anterior region convex dorsally. On the antero-dorsal surface a series of six to eight transverse, muscular crests occur. *Cuticle:* with or without small papillae in front of the ventral sucker. *Suckers:* ventral larger than the oral and prominent, with or without a peduncle, with four large ear-like outgrowths. *Gut:* prepharynx very short, pharynx elongate, caeca long and with anterior prolongations, neither the caeca nor their extensions having diverticula. *Reproductive systems:* genital pore slightly behind the ventral sucker, genital atrium elongate, but not deep. Testes ellipsoidal, diagonally one behind the other in the middle of the region behind the ventral sucker. Ovary ellipsoidal, slightly behind the posterior testis. Vitellaria tubular, with a number of folds, extending between the ovary and the oral sucker. Uterus with descending and ascending limbs, extending to within a short distance of the posterior extremity, formed into folds which occur as far forward as the level of the pharynx. Eggs extremely numerous, 0·030–0·040 mm. long.

Subfamily **Accacoeliinae** Odhner, 1911, *sensu* Dollfus, 1935

Genital atrium with a papilla, i.e. copulatory organ.

KEY TO GENERA OF ACCACOELIINAE

1. Ventral sucker without muscular expansions
   - A. Caeca and their anterior extensions without diverticula; vitellaria composed of small dendritic groups of follicles, extending from near the testes to a short distance in front of the posterior extremity *Accacoelium*
   - B. Caeca with diverticula, their anterior extensions without diverticula; vitellaria elongate, with small ramified lateral branches, extending between the ovary and the ventral sucker *Accacladium*
   - C. Caeca and their anterior extensions with diverticula (the latter with only two); vitellaria in the form of tubes and short, dendritic branches extending between the middle of the ovary and that of the ventral sucker *Rhyncopharynx*
2. Ventral sucker with muscular expansions: caeca and their anterior extensions with diverticula, the latter with several (six); vitellaria in cords or in an arborescent mass anteriorly *Accacladocoelium*

Genus *Accacoelium* Monticelli, 1893 (*sensu* Odhner, 1928)

Accacoelium contortum (Rudolphi, 1819) Monticelli, 1893 (Fig. 36A)

Syn. *Distoma contortum* Rudolphi, 1819; *Podocotyle contortum* (Rudolphi) Stossich, 1896.

HOST. Sun-fish.

LOCATION. Gills.

LOCALITY. Mediterranean and Atlantic, also in Sweden (host *Mola nasus*).

DIAGNOSIS. The characters already given will suffice. This is the only known species. See Odhner (1928) and Dollfus (1935*c*). American specimens were described by Linton (1898, 1940) and by Sumner, Osburn & Cole (*loc. cit.*).

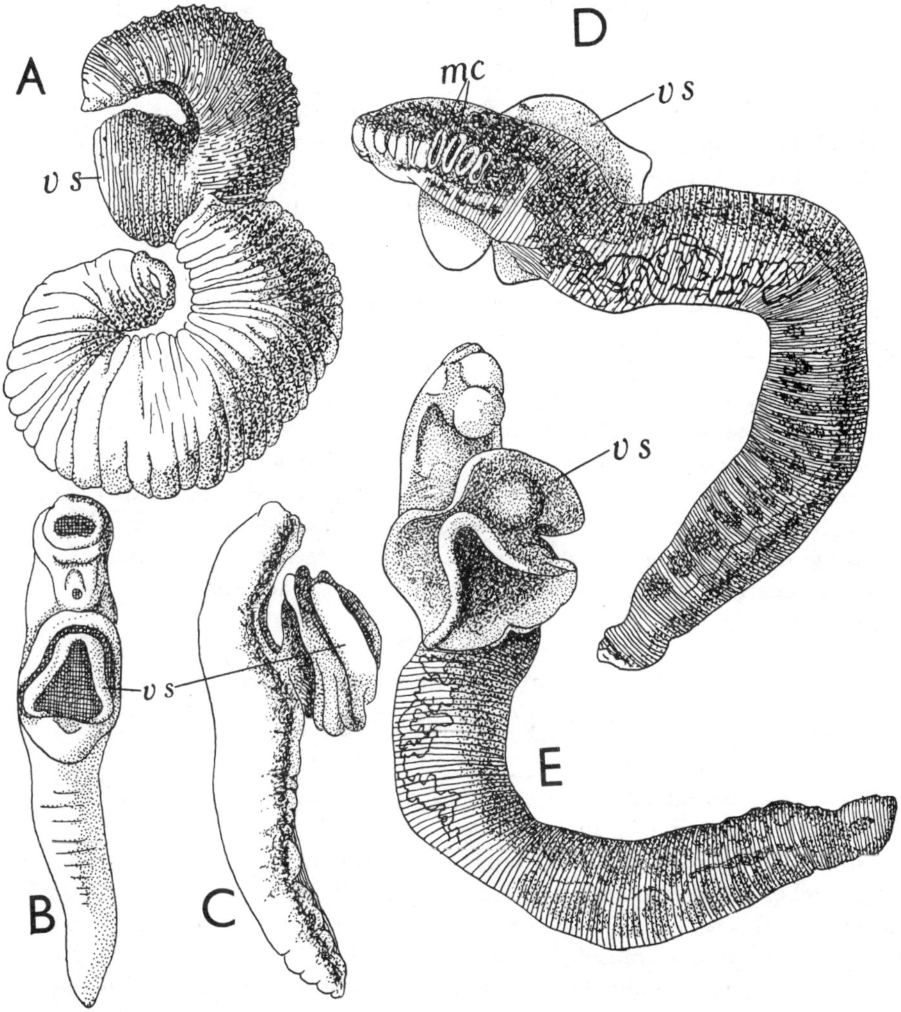

Fig. 36. Accacoeliidae from the sunfish. A, *Accacoelium contortum*. B, C, *Accacladocoelium petasiporum*. D, E, *Mneiodhneria calyptrocotyle*. A and C in side, B and E in ventral and D in dorsal view. (After Dollfus, 1935*c*.)

Genus *Accacladium* Odhner, 1928

Accacladium serpentulus Odhner, 1928

HOST. Sun-fish.

LOCATION. Intestine.

LOCALITY. Naples, Bergen, U.S.A. and Japan.

DIAGNOSIS. The characters already given will suffice. For descriptions of American specimens see Linton (1898, 1940) and Sumner *et al.* (*loc. cit.*).

Another species, *A. nematulum* Noble, 1937, was discovered in the gut of a sun-fish at Monterey Bay (California).

Genus *Rhyncopharynx* Odhner, 1928

Rhyncopharynx paradoxa Odhner, 1928

HOST. Sun-fish.

LOCATION. Intestine.

LOCALITY. Naples, Trieste, also the Pacific coast of Japan.

DIAGNOSIS. *Shape and size:* elongate, 20–31·5 mm. long, 1·5–2 mm. broad in the posterior region, forebody short and slender, 2–2·75 mm. long and 1–1·2 mm. thick at the base. *Suckers:* oral invariably elongate-oval (0·48–0·57 × 0·34–0·45 mm.), ventral large (1·25–1·4 mm. diameter) and prominent, stalked or not according to the state of contraction or extension of the process on which it is situated. *Gut:* anterior region much modified. *Pharynx:* comprising a main (posterior) part 0·5–0·65 mm. diameter, immediately in front of which a second vesicle 0·33–0·48 mm. diameter and thin-walled (the 'Rüsselblase') is connected on its ventral side with a bent and tubular structure (the 'Rüsselgang') terminating just behind the oral sucker in a narrower pointed tube (the 'Rüssel') which is enclosed within a sheath (the 'Rüsselscheide') whose lumen is continuous with that of the oral sucker. Odhner presumed that this rather complex arrangement (resembling rubber-bellows of the 'double' type) serves as a kind of aspirator. *Oesophagus:* also modified, its anterior part forming a bulb. *Intestine:* characteristically H-shaped, each of the two anterior extensions having a bifid termination.

Genus *Accacladocoelium* Odhner, 1928

Accacladocoelium alveolatum Robinson, 1934 (Fig. 37)

HOST. Sun-fish.

LOCATION. Intestine.

LOCALITY. Britain (Salcombe, Devon).

DIAGNOSIS. *Shape and size:* elongate-oval outline, 6·5 mm. long and 1·5 mm. broad. *Cuticle:* much wrinkled, especially at the base of the ventral sucker. *Suckers:* oral smaller than the ventral, about 1 mm. apart, diameters 0·5 and 1·0 mm. *Integument:* muscles thick and well developed. *Posterior extremity:* may be invaginated, giving a false impression of a posterior sucker. '*Honeycomb*': in posterior region, a zone of closely set papillae and cuticle-lined spaces which stain intensely, presenting the appearance of a 'honeycomb' about 0·25 mm. broad with cells about 0·08 mm. deep. The free edges of the partitions are here and there swollen into flanges in which the cuticle is vesicular. Nature of 'honeycomb' undetermined; no observed connexion between the papillae and subcuticular structures; gland cells absent from the adjacent tissues. *Gut:* pharynx small, oesophagus somewhat S-shaped (extending posteriorly, then dorsally to the origin of caeca), caeca complex. Main caeca long, extending back to the posterior extremity, having numerous lateral diverticula. Forward extensions of the caeca at the sides of the pharynx, having six short, lateral diverticula. The entire intestine is thus H-shaped, both forward and hinder crura having diverticula. *Excretory system:* main canals Y-shaped. *Reproductive systems:* genital pore median, far forward, just behind the mouth. Genital atrium narrow, a papilla in its cavity bearing a single pore. Testes transversely ovoid, slightly behind the ventral sucker, contiguous. Ovary spherical, a little behind the posterior testis. Vitellaria highly characteristic; one vitellarium is very small (vestigial), close behind the ootype (which is lateral to the ovary); the remaining vitellarium is very well developed, follicular or dendritic, confined to the region in front of the ventral sucker, completely occupying all available space above and between the suckers. Uterus with descending and ascending limbs, formed into numerous transverse folds which tend to fill up the posterior region of the body. Eggs not described.

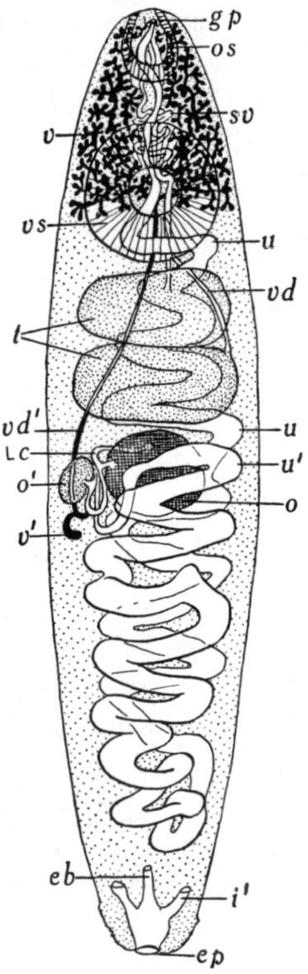

Fig. 37. Another Accacoeliid from the sunfish, *Accacladocoelium alveolatum*. (After Robinson, 1934.)

Other Species

Accacladocoelium nigroflavum (Rudolphi, 1819) Odhner, 1928

Syn. *Distoma nigroflavum* Rudolphi, 1819; *D. megninii* Poirier, 1885.

Host. Sun-fish.

Location. Intestine.

Locality. Atlantic coast of Europe, Naples, Isle of Elba and Sweden (host *Mola nasus*), also in Canada.

Diagnosis. *Vitellaria*: in the form of very long, sinuous threads which anastomose, extending from the pharynx to the posterior border of the ventral sucker (which is devoid of muscular expansions).

Accacladocoelium macrocotyle (Diesing, 1858) *sensu* Monticelli, 1893

Syn. *Distoma macrocotyle* Diesing, 1858; *Podocotyle macrocotyle* (Diesing) of Stossich, 1896

Host. Sun-fish.

Location. Intestine.

Locality. Naples, Sweden (host *Mola nasus*), also U.S.A. and Canada.

Diagnosis. *Vitellaria:* in the form of fine, delicate, very sinuous threads, extending from the vicinity of the pharynx to slightly behind the ovary. *Note.* The original specimens were found in Irish waters and the original description was based on the findings of Bellingham (1844). For descriptions of American specimens see Linton (1898, 1940) and Sumner *et al.* (*loc. cit.*).

Accacladocoelium petasiporum Odhner, 1928 (Fig. 36B, C)

Host. Sun-fish.

Location. Intestine.

Locality. Italy (Naples and Trieste).

DIAGNOSIS. *Ventral sucker:* unlike that of other species, having muscular expansions similar to those of *Mneiodhneria*. *Copulatory organ:* present. *Gut:* caeca and their forward extensions provided with diverticula. (In the last two characters, *A. petasiporum* differs from *Mneiodhneria* spp.). *Vitellaria:* in the form of an assemblage of delicate, sinuous threads situated between the peduncle of the ventral sucker and the genital pore.

Note. The genus *Hirudinella* Garsin, 1730 was referred by Nicoll (1915 *a*) to the Accacoeliinae, which was placed in the Hemiuridae. Manter (1926) maintained this allocation, but Dollfus (1935 *c*) affirmed that it was erroneous for the reason that Accacoeliinae are characterized by an H-shaped intestine and lack a special chamber (the gland-stomach) between the oesophagus and the origin of each crus of the intestine, while *Hirudinella* lacks the former, but possesses the latter character. Previously (1932 *b*) Dollfus gave several reasons for excluding *Hirudinella* from the Hemiuridae (in which family Poche (1926) also placed it) and erected for it a new family Hirudinellidae (superfamily Hemiuroidea).

Family **FELLODISTOMATIDAE** Odhner, 1911, emend. Nicoll, 1935 (p. 86)

Syn. *Steringophoridae* Odhner, 1911.

The opinion of Stiles, quoted by Stunkard & Nigrelli (1930), justifies the use of Nicoll's name for this family, though the synonymous form is frequently used.

Subfamily **Fellodistomatinae** Nicoll, 1909 emend.

Body oval, testes side by side.

KEY TO GENERA OF FELLODISTOMATINAE

1. Genital pore well in front of the ventral sucker, near the bifurcation of the intestine
   - A. Cirrus pouch globular or ovoid
     - (*a*) Ventral sucker behind the middle of the body, genital pore lateral, anterior region of the body more pointed than the posterior — *Steringotrema*
     - (*b*) Ventral sucker not behind the middle of the body, genital pore almost exactly median, anterior region not more pointed than the posterior
       - (i) Vitellaria almost entirely confined to the region in front of the ventral sucker, oesophagus much longer than the pharynx — *Bacciger*
       - (ii) Vitellaria almost entirely confined to the region behind the ventral sucker, oesophagus not longer than the pharynx — *Steringophorus*
   - B. Cirrus pouch elongate — *Rhodotrema*
2. Genital pore close to the anterior margin of the ventral sucker — *Fellodistomum*

Genus *Steringophorus* Odhner, 1905

Syn. *Leioderma* Stafford, 1904

Steringophorus furciger (Olsson, 1868) Odhner, 1905 (Fig. 38A)

Syn. *Distoma furcigerum* Olsson, 1868; *Distomum furcigerum* of Levinsen, 1881; *Leioderma furcigerum* (Olssen) of Stafford, 1904.

HOSTS. Angler, long rough dab, plaice, lemon sole, dab, witch.

LOCATION. Stomach and intestine.

Diagnosis. *Shape, size and colour:* spindle-shaped, pyriform when moving (blunt end foremost), 1·5–3·5 mm. long and almost half as broad, bright red and further tinged by eggs. *Cuticle:* smooth. *Suckers:* ventral larger than the oral (in the ratio 5/3), slightly in front of the mid-body, diameters 0·35 and 0·2 mm. in an individual 3 mm. long. *Gut:* prepharynx short, pharynx present, oesophagus short, caeca fairly long, extending back beyond the testes, but not reaching the posterior extremity. Bifurcation of the intestine far in front of the ventral sucker.

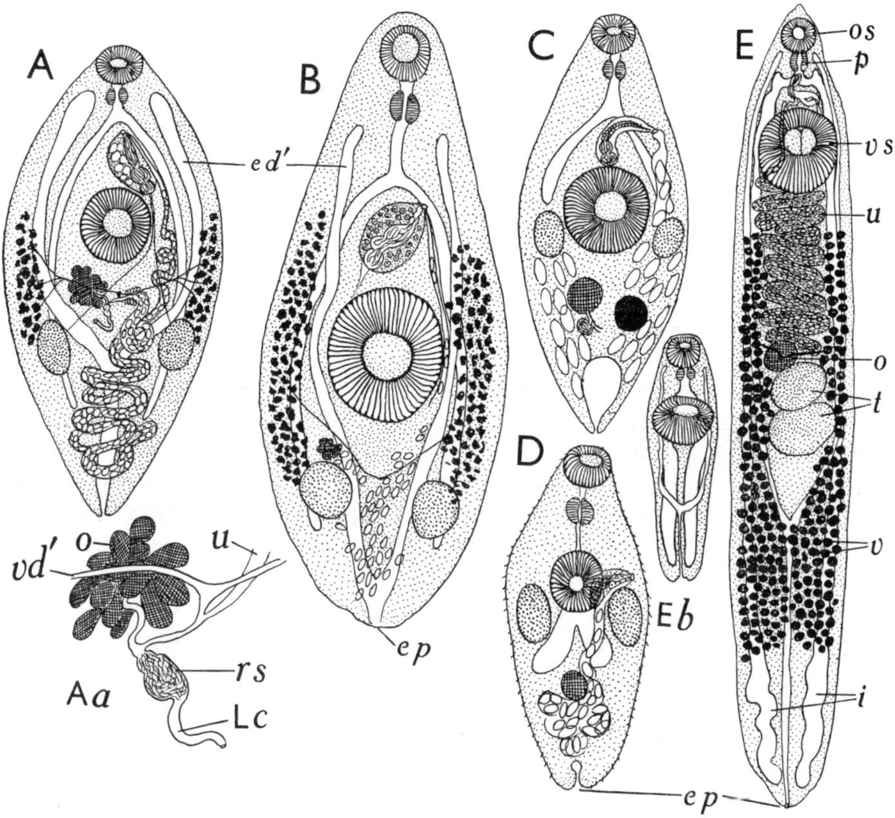

Fig. 38. Common Fellodistomatidae (A, B), Zoogonidae (C, D) and Azygiidae (E) of British marine fishes. A, *Steringophorus furciger*. B, *Steringotrema cluthense*. C, *Zoogonoides viviparus*. D, *Zoogonus rubellus*. E, *Otodistomum cestoides* (= *O. veliporum*). A*a*, ovary and main female ducts of A. E*b* juvenile specimen of species E. (B, D E*b*, after Nicoll, 1909, 1909 and 1913*c* respectively, the rest after Lebour, 1908*a*.)

Excretory system: main canals Y-shaped, bifurcation between the testes, lateral branches reaching the level of the pharynx. (The canals contain highly refractive materials and appear conspicuously dark when viewed by transmitted light.) *Reproductive systems:* genital pore median or slightly on the left, close behind the bifurcation of the intestine. Cirrus pouch large, ovoid, containing a very small cirrus, a pars prostatica and a vesicula seminalis of hour-glass shape. Testes side by side, widely separated, ventral to the caeca in the posterior region. Ovary lobed, in front of the testes, slightly on the right. Receptaculum seminis absent

(Lebour (1908*a*) thought it present). Uterus with descending and ascending limbs, forming a number of transverse folds between and behind the testes. Vitellaria lateral, in the middle third of the body. Eggs numerous, colour golden brown, measuring 0·046 × 0·019 mm. (according to Miller (1941*a*) 0·055 × 0·029 mm.).

Notes. Only immature individuals have been found in the stomach; mature flukes occur throughout the intestinal region, are most usually found in the pyloric caeca, but may occur in association with *Zoogonoides viviparus* in the rectum. Little (1929*a*) observed the formation of spermatophores in this species, which occurs in the Far East, Greenland, U.S.A., Canada and Sweden and has appeared at Cullercoats, St Andrews, Galway and Plymouth. It is said to be equally common in winter and summer.

Genus *Steringotrema* Odhner, 1911

Syn. *Steringophorus* Odhner, 1905, in part

Steringotrema cluthense (Nicoll, 1909) Odhner, 1911 (Fig. 38B)

Syn. *Steringophorus cluthensis* Nicoll, 1909; *Leioderma cluthense* of Nicoll, 1910.

HOSTS. Dab, lemon sole (Firth of Forth, Galway, Plymouth).

LOCATION. Duodenum and pyloric caeca.

DIAGNOSIS. *Shape and size:* oval outline, more pointed anteriorly, flattened, 1·5–2 mm. long and 0·6–0·8 mm. broad. (Living worm capable of great extension.) *Cuticle:* smooth. *Suckers:* oral slightly behind the anterior tip of the body, 0·22 mm. long and 0·15 mm. broad. Ventral almost twice as large as the oral, just in the posterior half of the body, 0·44 mm. diameter in an individual 2 mm. long. *Gut:* prepharynx absent, pharynx contiguous with the oral sucker (Nicoll's statements and diagrams disagree on these points), oesophagus twice as long as the pharynx (about 0·17 mm. long), caeca long, extending slightly behind the testes, but not to the posterior extremity. *Excretory system:* main canals Y-shaped, but with a very short main stem; lateral branches extending forward to the level of the pharynx. *Reproductive systems:* genital pore near the bifurcation of the intestine on the left. Cirrus pouch almost globular, entirely in front of the ventral sucker. Testes side by side, somewhat separated, midway between the ventral sucker and the posterior extremity. Ovary multilobate, small, immediately in front of the right testis. Vitellaria lateral, extending from the level of the testes to that of the cirrus pouch, i.e. well in front of the ventral sucker anteriorly. Uterus as in *Steringophorus furciger*, but with fewer folds. Eggs relatively thick shelled, measuring 0·044–0·056 × 0·028–0·032 mm.

Steringotrema pagelli (Beneden, 1870) Odhner, 1911 (Fig. 28A)

Syn. *Distoma pagelli* Beneden, 1870; *Distomum tergestinum* Stossich, 1889; *D. actaeonis* Pagenstecher, 1863.

HOST. Common sea bream (Galway and Billingsgate [? North Sea or English Channel]).

LOCATION. Intestine.

DIAGNOSIS. Similar to *S. cluthense*, but larger (3·6–4·1 mm. long) and with a relatively much larger ventral sucker (three times as large as the oral), more restricted vitellaria (extending only slightly in front of the ventral sucker), a rounded ovary, a genital pore which is situated in front of the bifurcation of the intestine and a more abundantly folded uterus. Eggs very numerous, measuring 0·048–0·051 × 0·028 mm. according to Odhner (1911-13), but much larger (0·057–0·063 × 0·033–0·037 mm.) according to Nicoll (1914*a*).

Steringotrema divergens (Rudolphi, 1809) Odhner, 1911 (Fig. 39D)

Syn. *Distoma divergens* Rudolphi, 1809; *Fasciola blennii* Müller, 1776.

HOSTS. Butterfly blenny (at Plymouth), other blennies on the Continent.

LOCATION. Stomach and intestine.

DIAGNOSIS. *Shape and size:* more elongate than *S. cluthense* and *S. pagelli*, little flattened, 0·8–1·3 mm. long and 0·35–0·5 mm. in greatest breadth. *Suckers:* ventral near (slightly in front of) the mid-body, transversely elongate, larger than the oral, dimensions 0·27–0·38 × 0·21–0·27 mm. and 0·15–0·16 mm. diameter. *Gut:* pharynx 0·09–0·12 mm. diameter, oesophagus longer than in other species, caeca extending back to the posterior limits of the testes. *Reproductive systems:* genital pore close to the bifurcation of the intestine, cirrus pouch almost globular. Gonads rounded, the ovary larger than in other species and situated in front of the right testis, testes not widely separated. Vitellaria comprising an array of follicles between the pharynx and the ends of the caeca on each side, the sequence being broken near the ventral sucker. Uterus much folded, filling the hind-body. Eggs of golden colour, measuring 0·045–0·048 × 0·031 mm.

Notes. Rudolphi's specimen was obtained from the gattorugine, but the trematode has been found at Naples, Kiel and elsewhere in the butterfly blenny. Mathias (1934) described specimens 0·97–2 mm. long and 0·5–1 mm. broad which were obtained near Banyuls. The dimensions of the suckers were 0·35–0·48 × 0·28–0·40 and 0·15–0·20 mm. diameter, of the eggs 0·040–0·055 × 0·022–0·032 mm. The arrangement of vitelline follicles which Odhner described (vide diagnosis) provided an unequivocal character of distinction.

Genus *Bacciger* Nicoll, 1914

Bacciger bacciger (Rudolphi, 1819) Nicoll, 1914 (Fig. 28B)

Syn. *Distoma baccigerum* Rudolphi, 1819.

HOST. Sand smelt (at Plymouth).

LOCATION. Stomach.

DIAGNOSIS. *Shape and size:* oval outline, flattened, about 0·95 mm. long and 0·52 mm. broad. *Cuticle:* smooth. *Suckers:* ventral near the mid-body, about half as large again as the oral, diameters 0·15 and 0·105 mm. *Gut:* prepharynx absent, pharynx small, oesophagus two or three times as long as the pharynx, caeca long and narrow, their terminations obscured by the uterus. Bifurcation of the intestine far in front of the ventral sucker. *Reproductive systems:* genital pore median, slightly behind the bifurcation of the intestine and far in front of the ventral sucker. Cirrus pouch small, confined to the region in front of the ventral sucker. Pars prostatica inflated, ejaculatory duct folded, cirrus small. Testes rounded, side by side, widely separated, close behind the ventral sucker. Ovary rounded, between the testes and directly behind the ventral sucker.

Vitellaria lateral, extending from the level of the oesophagus to that of the ventral sucker. Uterus with descending and ascending limbs formed into a number of folds, mostly longitudinal, in the region behind the gonads. Eggs very small and numerous, measuring 0·020–0·024 × 0·014–0·017 mm.

Palombi (1932*a*, 1933, 1934*a*, *b*) identified *Cercaria pectinata* Huet, 1891 with *B. bacciger* and elucidated the life history of this species, which occurs also at Naples.

Genus *Rhodotrema* Odhner, 1911

Rhodotrema ovacutum (Lebour, 1908) Odhner, 1911 (Fig. 39A, B)

Syn. *Steringophorus ovacutus* Lebour, 1908.

HOST. Long rough dab (on west coast of Sweden, also at Aberdeen and Cullercoats).

LOCATION. Intestine.

DIAGNOSIS. *Shape:* oval outline, sometimes pyriform and with an extended anterior extremity. *Colour:* bright red, rapidly fading at death, when the colour is yellow with a greenish brown patch due to the eggs. *Cuticle:* smooth. *Size:* 1·8–2·3 mm. long, extensible beyond the upper limit, 1·2–1·3 mm. in greatest breadth (at the level of the ventral sucker). *Suckers:* ventral about twice as large as the oral and situated at the mid-body, diameters 0·6–0·8 and 0·32–0·4 mm. *Gut:* prepharynx very short, pharynx well formed, oesophagus short and narrow, caeca fairly long, extending slightly behind the testes. *Reproductive systems:* genital pore far forward, close to the pharynx, on the left of the median plane. Testes slightly postero-dorsal to the ventral sucker, ventral to the caeca, one on either side of the body. Cirrus pouch well developed (0·3 mm. long), obliquely oriented, containing a large pars prostatica, the cells of which almost fill it. Uterus with descending and ascending limbs, its fold tending to fill the posterior region. Vitellaria in front of the mid-body, comprising small groups of large follicles laterally, between the oesophagus and the anterior part of the ventral sucker, with three main ducts on either side. Eggs bright brown with a green tinge, pointed at one end, rounded at the other (hence the specific name), measuring 0·046 × 0·026 mm.

Rhodotrema skrjabini and *R. problematicum*, both Issaitschikow, 1928 occur in the Russian Arctic, *R. quinquelobata* Layman, 1930 in Peter the Great Bay.

Genus *Fellodistomum* Stafford, 1904

Fellodistomum fellis (Olsson, 1868) Nicoll, 1909

Syn. *Distoma fellis* Olsson, 1868; *Fellodistomum incisum* (Rudolphi, 1809) of Stafford, 1904; *F. agnotum* Nicoll, 1909.

HOST. Wolf-fish.

LOCATION. Gall bladder.

DIAGNOSIS. *Colour, shape and size:* when fresh dull green with a pinkish sucker, oval outline, thick and fleshy, 2·5–3·3 mm. long and 1·1–1·5 mm. broad. *Cuticle:* smooth, or wrinkled anteriorly, equipped with rod-like bodies which stand out vertically from the surface. *Suckers:* globular, large and prominent, ventral more than twice as large as the oral and situated slightly behind the mid-body. *Gut:* prepharynx short, pharynx fairly large, oesophagus very short (practically absent), caeca fairly long and very wide, extending to the level of the testes. Bifurcation of the intestine well in front of the ventral sucker. *Excretory system:* main canals Y-shaped, but with a short median stem. Lateral branches extending to the level of the pharynx. *Reproductive systems:* genital pore on a

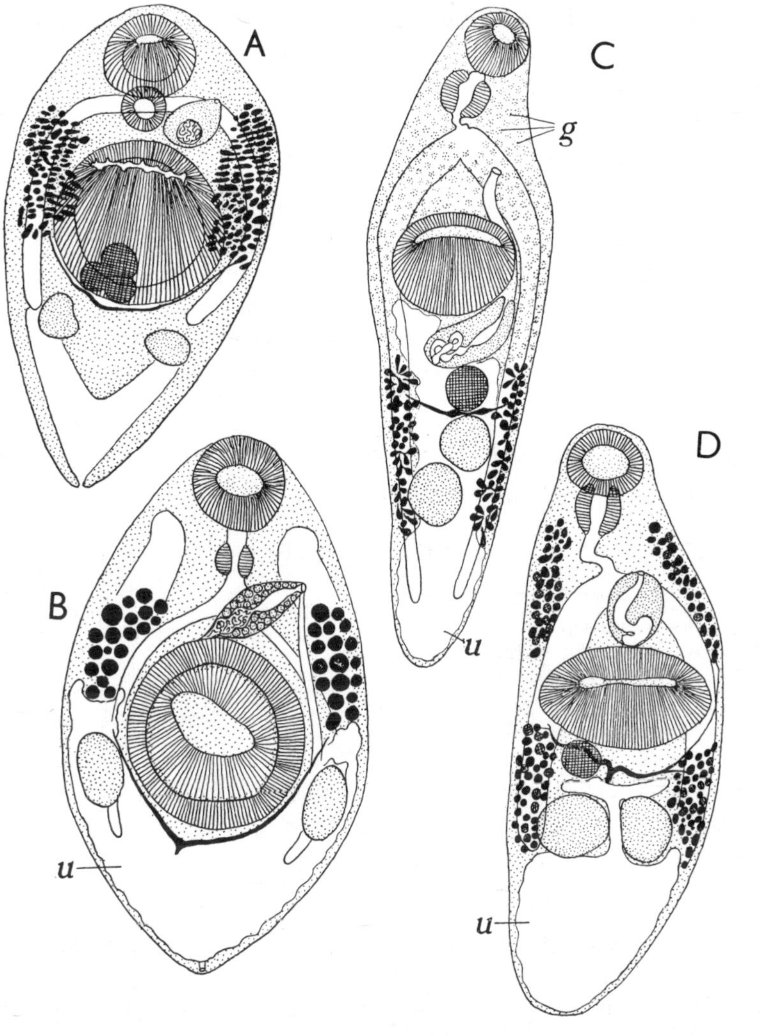

Fig. 39. Other Fellodistomatidae. A, B, *Rhodotrema ovacutum*. C, *Proctoeces maculatus*. D, *Steringotrema divergens*. (A, D, after Odhner, 1911–13; B, after Lebour, 1908*b*; C, after Looss, 1901*a*.)

large papilla close to the anterior margin of the ventral sucker, well to the left of the median plane. Cirrus pouch containing a short ejaculatory duct, a small bipartite vesicula seminalis and a well-developed pars prostatica. Testes ovoid, side by side, somewhat separated, midway between the ventral sucker and the posterior extremity. Ovary lobed, in front of the right testes, close to the ventral sucker. Receptaculum seminis absent (?). Vitellaria of very limited extent, follicles few, confined to short lateral regions in front of the ventral sucker. Part of uterus proximal to the ootype forming a receptaculum seminis uterinum. Uterus with descending and ascending limbs, formed into a number of rather small folds. Eggs measuring about 0·042 × 0·023 mm.

Notes. The form '*Fellodistomum agnotum*' (Fig. 28C) comprised small specimens found in the gall bladder and duodenum of the wolf-fish along with *F. fellis* and in the proportion of 1 to 30. They were mature when little more than 1 mm. long, whereas the length of *F. fellis* at maturity was said to be 2·5 mm. It is difficult to believe that such specificity of host and location could be shown by two distinct species of trematode occurring in various parts of Europe (Trieste, west Sweden, Cullercoats) and in Canada.

Pigulewsky (1938) erected a new genus, *Ovotrema*, for a Fellodistomatid (*O. pontica*) discovered in the gut of the anchovy in the Black Sea. The original description is very brief, but the genus would seem to necessitate an inclusion 1 A (*c*) in the key given above, this to read: 'Ventral sucker not behind the middle of the body, genital pore decidedly lateral (on the left).' According to the figure given, *O. pontica* lacks both a prepharynx and an oesophagus.

Subfamily **Haplocladinae** Odhner, 1911

Body elongate, testes one behind the other.

KEY TO GENERA OF HAPLOCLADINAE

I. Only a single caecum developed	*Haplocladus*
II. Two caeca present	
A. Caeca very short, ending in front of the ventral sucker	*Ancylocoelium*
B. Caeca not very short	
(*a*) Oral sucker having thirteen lancet-like processes	*Tergestia*
(*b*) Oral sucker devoid of lancet-like processes	*Proctoeces*

Note. Manter (1940*a*) referred *Tergestia* and *Proctoeces* to the Haplocladinae, but placed this subfamily in the Monorchiidae.

Genus *Haplocladus* Odhner, 1911

Odhner recognized three species of this genus, two of them new. None has appeared in British waters. They are:

H. typicus Odhner, 1911 (Fig. 45E). This species was said to be common in the gut of the horse mackerel at Palermo and Trieste.

H. filiformis (Rudolphi, 1819). Found in the gut of the red-band fish at Trieste and said to differ from *H. typicus* by the more anterior position of the testes and the smaller eggs (0·034–0·037 mm. long as against 0·04 mm.).

H. minor Odhner, 1911. Found (rarely) in the gut of the dab in Sweden and said to have a furcocercaria which was discovered in an aquarium containing *Nucula nucleus* and *Syndosmya alba*.

It is distinctly possible that further study would resolve these three forms into one and the same species, the specific characters cited by Odhner being insufficient to warrant the erection of separate species.

Genus *Tergestia* Stossich, 1899

Syn. *Theledera* Linton, 1910.

Tergestia laticollis (Rudolphi, 1819) Stossich, 1899 (Fig. 45 D)

Syn. *Distoma laticolle* Rudolphi, 1819; *D. polonii* Molin of Olsson, 1869.

HOST. Horse mackerel (in Britain).

LOCATION. Intestine.

This is a widely distributed trematode (occurring in Italy, the Black Sea, Tortugas, Costa Rica and Japan) which has appeared at Aberdeen and Plymouth.

DIAGNOSIS. *Shape and size:* elongate, up to 4 mm. long and 0·2–0·3 mm. broad, mature when 1·5 mm. long. *Suckers:* ventral larger than the oral and near the end of the first third of the body. *Gut:* pharynx very elongate (0·17–0·25 × 0·08–0·10 mm.). *Reproductive systems:* vitellaria ending close behind the ventral sucker. Eggs brownish yellow, measuring 0·021–0·023 × 0·015 mm.

Note. According to Odhner (*loc. cit.*) the larva of this species is *Cercaria dichotoma* Müller (figured by La Val., 1855), but not the form dealt with by Pelseneer (1906) and Lebour (1908*b*), which is *C. fissicauda* Villot, 1879, *nec* La Val., 1855, an entirely different furcocercaria.

Nicoll (1913*c*) briefly described a single adult specimen of this species found at Aberdeen and (1914*a*) reported its frequent occurrence at Plymouth. His specimens showed certain slight differences from those of Odhner, attaining a length of 4·6 mm., at which the suckers are relatively smaller (diameter of the ventral $= \frac{1}{13}$th body length).

Odhner (1911–13) briefly described another species, *Tergestia acanthocephala* (Stossich, 1887), which was said to have a shorter pharynx, more extensive vitellaria and more dissimilar suckers, and seems to have been confined largely to Italy and Sicily, but has been recorded in the mackerel in Devon. Other species occur in America and Japan.

Genus *Proctoeces* Odhner, 1911

Proctoeces maculatus (Looss, 1901) Odhner, 1911 (Fig. 39 C)

Syn. *Distomum maculatum* Looss, 1901; *Proctoeces erythraeus* Odhner, 1911.

HOST. Butterfly blenny ('*P. erythraeus*' in *Chrysophrys bifasciata* and *Iulis lunaris*).

LOCATION. Intestine.

This species has not been found in Britain.

DIAGNOSIS. *Size:* up to 3 mm. long, but mature at 1 mm. *Suckers:* ventral much larger than the oral and about one-third of the distance along the body. *Eggs:* measuring 0·072–0·079 × 0·027 mm. '*P. erythraeus*' was said to differ in the smaller size of the ventral sucker, less extensive vitellaria and smaller eggs (0·045 mm. long). Only the last character is significant, but this is marred by the fact that only a solitary mature specimen was found.

Note. Looss remarked on the gland cells which are scattered throughout the anterior region of this trematode. Odhner substantiated the statement, but indicated that similar cells occur in *Steringophorus furciger*.

Manter (1940*a*) described another species, *P. magnorus*, from Cerros Island, Mexico.

Genus *Ancylocoelium* Nicoll, 1912 (provisionally in Haplocladinae)

Ancylocoelium typicum Nicoll, 1912 (Fig. 28D)

HOST. Horse mackerel (at Aberdeen).

LOCATION. Intestine.

DIAGNOSIS. *Shape and size:* spatulate, widest in the anterior region, flat, 2·25 mm. long and 0·48 mm. broad. *Colour:* when alive, coloured a rich yellow by the eggs. *Cuticle:* spinous, spines scale-like, sparse posteriorly. *Suckers:* not conspicuous, diameters: oral 0·1 mm., ventral 0·13 mm. Ventral sucker slightly in front of the mid-body and 0·92 mm. from the anterior extremity. *Gut:* pharynx small and elongate (0·08 × 0·04 mm.), oesophagus long (0·5 mm.) and narrow, caeca very short and wide, turning sharply anteriorly, then posteriorly, terminating only a short distance behind the bifurcation of the intestine and well in front of the ventral sucker. *Reproductive systems:* genital pore median, close behind the bifurcation of the intestine. Cirrus pouch club-shaped, confined to the region in front of the ventral sucker. Vesicula seminalis large, ovoid, pars prostatica and cirrus short. Testes diagonally one behind the other. Ovary in front of the testes. All gonads in the middle of the body, close together, slightly behind the ventral sucker. Vitellaria lateral, slightly asymmetrical, the left a little in advance of the right, extending from the anterior border of the foremost testis to within 0·7 mm. of the posterior extremity. Follicles of the vitellaria arranged in the form of one continuous tubule thrown into a rosette-shaped knot. Uterus with descending and ascending limbs, voluminous and little folded, filling all available space between the genital pore and the posterior extremity. Eggs very numerous, small, bright yellow, measuring 0·024 × 0·013 mm.

Note. Nicoll mentioned an 'accessory genital sac', arising from the genital atrium; it is a thin-walled sac of undetermined function and measures about 0·24 × 0·11 mm.

Family **ZOOGONIDAE** Odhner, 1911 (p. 88)

Subfamily **Zoogoninae** Odhner, 1911

Vitellaria small and compact.

KEY TO GENERA OF ZOOGONINAE

1. Vitellaria paired	*Diphterostomum*
2. Vitellaria unpaired	
A. Bifurcation of the intestine in front of the ventral sucker	
(*a*) Cirrus with needle-like spinelets	*Zoogonoides*
(*b*) Cirrus without needle-like spinelets	*Zoonogenus*
B. Bifurcation of the intestine behind the ventral sucker	*Zoogonus*

Genus *Diphterostomum* Stossich, 1904

Syn. *Diphtherostomum* of Stafford, 1905.

Diphterostomum brusinae (Stossich, 1889) Stafford, 1905 (Fig. 45F)

Syn. *Distomum brusinae* Stossich, 1889, and of Looss, 1901; *Pleurogenes brusinae* of Stossich, 1899; *Brachycaecum brusinae* (Stossich) Barbagallo & Drago, 1903.

HOSTS. Various Labridae (notably the gilt-head), and butterfly blenny.

LOCATION. Rectum.

This trematode has not appeared in Britain, but Odhner (1911–13) gave a clear diagnosis of specimens discovered in Trieste. *Shape and size:* elongate, posterior end rounded, anterior end tapering and more flattened, 0·4–0·85 mm. long and 0·16–0·21 mm. broad. *Colour:* yellowish, but with reddish suckers. *Texture:* frail. *Suckers:* ventral larger than the oral and transversely ovoid, dimensions 0·15–0·17 × 0·08–0·10 mm. and 0·075–0·095 mm. diameter. (*Note.* The lips of the ventral sucker described by Looss (1901*a*), but not by Stossich (1899), were regarded by Odhner as an artefact.) *Gut:* pharynx very small (0·03 mm. diameter), oesophagus very long. *Reproductive systems:* genital pore slightly dorsal (ventral according to Looss). Cirrus fairly large, pars prostatica spherical. Testes longitudinally ovoid, situated close behind the ventral sucker. Metraterm two-thirds as long as the cirrus pouch. Vitellaria very small. Eggs increasing in size *in utero* from 0·043–0·046 × 0·028 to 0·057 × 0·023 mm., each containing a lively miracidium.

Palombi (1929*b*, 1930*a*, *b*) has studied the life cycle of this species.

Diphterostomum betencourti (Monticelli, 1893) Odhner, 1911

Syn. *Distomum betencourti* Monticelli, 1893; *D. luteum* Beneden, 1870, *nec* Baer, 1827; *Pleurogenes betencourti* (Monticelli) Stossich, 1899.

This species has appeared in the nurse hound at Wimereux and in the rough hound off the coast of Belgium. Stossich (1899) found it in both hosts at Boulogne. According to Odhner (*loc. cit.*) it is larger and more robust than the previous species (2–2·5 × 1·1–1·3 mm.), but has relatively smaller suckers (ventral 0·6–0·65 × 0·45 mm. diameter, oral 0·3 mm.), as well as a long cirrus pouch, but relatively short cirrus, a very elongate pars prostatica and thick-shelled eggs 0·034–0·036 mm. long.

A third species, *D. sargus annularis* Vlasenko, 1931, occurs in the host after which it was named in the Black Sea.

Genus *Zoogonoides* Odhner, 1902

Zoogonoides viviparus (Olsson, 1868) Odhner, 1902 (Fig. 38C)

Syn. *Distoma viviparum* Olsson, 1868; *Zoogonus viviparus* (Olsson) of Looss, 1901; *Zoogonoides subaequiporus* Odhner, 1911.

HOSTS (in Britain). John Dory, dragonet, wolf-fish, gattorugine, butterfly blenny, long rough dab, turbot, Dover sole, thickback, plaice, dab, lemon sole, witch.

LOCATION. Intestine, especially rectum.

This species has been found at Plymouth, Liverpool, St Andrews, Millport, Aberdeen, Cullercoats, Galway and in Scandinavia.

Diagnosis. *Shape, colour and size:* spindle-shaped, very small, bright red, 0·8–1·6 mm. long and 0·31–0·42 mm. broad. *Cuticle:* spinous, spines minute, arranged in alternating rows, most abundant anteriorly. *Suckers:* oral 0·12–0·16 mm. diameter, ventral almost twice as large and situated in the middle of the body. *Gut:* prepharynx short, oesophagus twice as long as the pharynx, caeca short, extending to the ventral sucker, but never beyond its posterior end. Bifurcation of the intestine approximately midway between the suckers. *Excretory system:* vesicle small, oval. *Reproductive systems:* genital pore well in front of the ventral sucker, on the left of the bifurcation of the intestine. Cirrus pouch sickle-shaped. Vesicula seminalis hour-glass-shaped. Cirrus equipped with needle-like spinelets. Testes ovoid, one on either side of the ventral sucker or slightly further back. Ovary almost spherical, close behind the ventral sucker, slightly on the left. Receptaculum seminis immediately behind the ovary. Vitellarium compact, on the left of the ovary. Uterus with descending and ascending limbs, slightly folded in the posterior part of the body. Eggs few, thin-shelled, 0·07–0·08 mm. long (according to Nicoll (1907*a*), measuring 0·086–0·094 × 0·042–0·044 mm.), each containing a miracidium.

Note. 100 or more specimens may occur in an infected fish, often associated (in the dab) with *Steringophorus furciger*. *Zoogonoides subaequiporus* Odhner, 1911, occurs in the rectum of the wolf-fish in European waters, but has not been found in Britain. The size relations of the suckers did not justify the erection of this species (as Odhner believed), and it must be relegated to synonymity with *Z. viviparus*. Its fate may be shared ultimately by the American species *Z. laevis* Linton, 1940, the life history of which was described recently (1943*b*) by Stunkard.

Genus *Zoonogenus* Nicoll, 1912

Zoonogenus vividus Nicoll, 1912 (Fig. 28J)

Host. Common sea bream (at Aberdeen and Plymouth).

Location. Rectum.

Diagnosis. *Shape, colour and size:* oval outline, broadest in the posterior region, blotchy blood red (redder than the gut contents of the Crustacea-eating host), about 1·4 mm. long and 0·46 mm. broad. *Cuticle:* spinous in the anterior region. *Suckers:* ventral much larger than the oral and situated behind the mid-body (0·94 mm. from the anterior extremity), diameters 0·34 and 0·16 mm. Both suckers are globular, and in some instances the ventral is nearly three times as large as the oral. *Gut:* prepharynx apparently absent, oesophagus somewhat shorter than the pharynx, caeca narrow and short, their terminations in front of the ventral sucker. *Excretory system:* vesicle short and bulbous, containing yellowish green or brown concretions. *Reproductive systems:* genital pore lateral, underlying the left caecum. Cirrus pouch curved, situated in front of the ventral sucker. Vesicula seminalis hour-glass-shaped. Ejaculatory duct long, its walls folded. Cirrus non-spinous. Testes ovoid, one situated over each postero-lateral quadrant of the ventral sucker. Ovary spherical, between the testes, close to the posterior margin of the ventral sucker. Vitellarium small, compact, to the right of the ovary or behind it. Uterus with descending and ascending limbs, folded in the posterior region of the body. Metraterm muscular. Eggs relatively small, thin-shelled, increasing in size during passage through uterus from 0·016 × 0·012 mm. to 0·036 × 0·018 mm. (i.e. very much smaller than those of *Zoogonoides viviparus*).

Note. This is a delicate trematode, which dies and becomes macerated under conditions which do not affect *Derogenes varicus* and *Hemiurus communis*.

Genus *Zoogonus* Looss, 1901

Zoogonus rubellus (Olsson, 1868) Odhner, 1902 (Fig. 38D)

Syn. *Distoma rubellum* Olsson, 1868.

HOST. Wolf-fish (as a juvenile in the dab and cuckoo wrasse).

LOCATION. Intestine and rectum (juveniles in the intestine).

This species was discovered by Olsson in the ballan wrasse in west Sweden, but it is rare in British waters. According to Odhner (1911–13) it is 0·9–1·2 mm. long and 0·25 mm. broad, has suckers of unequal sizes (oral 0·10–0·12 mm. diameter, ventral 0·13–0·14 mm.), relatively long caeca, a cirrus pouch which does not cross the median plane, and eggs containing miracidia 0·1 mm. long.

Zoogonus mirus Looss, 1901

HOSTS. Various Labridae (at Trieste and Rovigno).

LOCATION. Rectum.

This species occurs in the Mediterranean and has not appeared in Britain. Looss (1901*a*) seems to have dealt with rather unusual specimens, Goldschmidt (1902) with more normal ones. According to Odhner (1911–13) this species never exceeds 0·6 mm. in length and 0·2 mm. in breadth, has suckers of nearly equal sizes (oral 0·08–0·09 mm. diameter, ventral 0·09–0·11 mm.), a cirrus pouch which extends diagonally across the median plane, and a miracidium which by comparison with that of *Z. rubellus* more nearly fills the egg.

Notes. Odhner remarked on the close resemblance between the northern and the Mediterranean forms, but affirmed that they are clearly distinct species when seen side by side. The cercaria which was described under the name of *Distomum lasium* Leidy, 1891, and which develops in *Nassa obsoleta* on the east coast of North America, was found by Stunkard (1938*b*) to encyst in Polychaeta, especially in *Nereis virens*, and to develop to maturity in eels and toadfish, belongs to a species of *Zoogonus* which Stunkard believed to be identical with both *Z. rubellus* and *Z. mirus*. Later studies (1940*b*, 1941) of larvae obtained at Wimereux produced such discordant results, however, that he reconsidered the question of specific identities and came to regard American specimens as forming a distinct species, *Z. lasius* (Leidy, 1891). At the same time he indicated that the morphology of adults is so variable that it is impossible to decide whether the European forms represent two distinct species, or should be regarded as identical.

Metacercariae of a species of *Zoogonus* occur plentifully near Wimereux, encysted in the muscles and connective tissue of Aristotle's lantern of the echinoid *Psammechinus miliaris*, and Stunkard (1941) examined numerous molluscs suspected of serving as the first intermediate host, without success. The metacercariae did not differ significantly from those discovered by Timon-David

(1933*a*, 1934, 1936) in sea urchins near Marseilles, probably representing the same species (*Zoogonus mirus*). They could not be found by Stunkard in a number of polychaetes (*Eunereis longissima* and *Nereis errorata*) examined, just as the American species was undiscovered in available sea urchins (*Arbacia punctatula* and *Strongylocentrotus drobachiensis*) at Woods Hole. It is odd that two species of the same genus of Trematoda should utilise on the east and west shores of the Atlantic second intermediate hosts belonging to distinct phyla of the Invertebrata.

Subfamily **Lecithostaphylinae** Odhner, 1911

Vitellaria well developed, follicular.

Members of this subfamily have not appeared in Britain, but most of them occur in Europe.

KEY TO GENERA OF LECITHOSTAPHILINAE

I. Suckers small and feeble; caeca extending back only to the middle of the body. (Parasites in the urinary bladder of fishes) *Lepidophyllum*

II. Suckers large or fairly large; caeca extending back to the posterior third of the body. (Parasites in the intestine of fishes)
  A. Suckers especially large, the ventral situated in the posterior half of the body; caeca not extending behind the testes *Deretrema*
  B. Suckers not especially large, the ventral situated in front of the mid-body; caeca extending behind the testes
    (*a*) Pharynx very small *Steganoderma*
    (*b*) Pharynx not very small *Lecithostaphylus*

Note. Yamaguti (1934*a*) regarded *Lecithostaphylus* as a synonym of *Steganoderma* (but gave no reasons) and adopted the name Steganoderminae for the subfamily in accordance with priority requirements.

Genus *Lepidophyllum* Odhner, 1902

Lepidophyllum steenstrupi Odhner, 1902 (Fig. 45 G)

Hosts. Wolf-fish (in Norway), *Anarrhichas pantherinus* (in Iceland), *Zoarces anguillaris* (in Canada).

Diagnosis. *Size:* up to 5 mm. long and two-thirds as broad. *Cuticle:* scaly. *Suckers:* oral larger than the ventral in the ratio 4/3, diameters 0·27–0·28 and 0·20–0·22 mm. *Gut:* oesophagus much longer than the pharynx, caeca ending near the mid-body. *Eggs:* 0·04–0·043 × 0·021–0·023 mm.

Genus *Deretrema* Linton, 1910

Syn. *Proctophantastes* Odhner, 1911.

Price (1934*d*) affirmed that as *Deretrema fusillus* Linton, 1910 is congeneric with *Proctophantastes abyssorum* Odhner, 1911, Odhner's genus must fall.

Deretrema abyssorum (Odhner, 1911) Price, 1934 (Fig. 45 I)

Syn. *Proctophantastes abyssorum* Odhner, 1911.

Hosts. Haddock and *Coryphaenoides rupestris* (at Trondhjem).

Location. Rectum.

DIAGNOSIS. *Size and shape:* 0·8–1·1 mm. long and 0·35–0·4 mm. in greatest breadth, anterior region narrowed and pointed when extended, posterior widely rounded. *Suckers:* oral spherical, 0·25 mm. diameter, ventral of the same length, but 0·35–0·4 mm. broad. *Gut:* oesophagus about twice as long as the pharynx, which is 0·06 mm. diameter. *Eggs:* 0·034–0·037 × 0·018 mm.

Note. Unrecorded in Britain.

Genus *Lecithostaphylus* Odhner, 1911

Lecithostaphylus retroflexus (Molin, 1859) (Fig. 45H)

Syn. *Distoma retroflexum* Molin, 1859; *Podocotyle retroflexum* (Molin) Stossich, 1898.

HOST. Greater pipe-fish (at Trieste and Palermo).

LOCATION. Intestine.

DIAGNOSIS. *Size:* 1·5–2·5 mm. long and 0·4–0·55 mm. in greatest breadth. *Suckers:* ventral slightly the larger, diameters 0·18–0·23 and 0·15–0·2 mm. *Reproductive systems:* vitellaria lateral, between the ventral sucker and the testes. Eggs 0·038–0·041 × 0·02 mm.

This species was said by Odhner to be common at Trieste and Palermo. The larva, *Cercaria thaumanthiatis*, is a bristle-tail cercaria, first mentioned in Graeffe (1860) and discovered on the disk of a Hydromedusa of the genus *Eucope*.

Genus *Steganoderma* Stafford, 1904

Steganoderma formosum Stafford, 1904

This species occurs in the intestine and caeca of the halibut, but has not appeared in Europe.

Family **AZYGIIDAE** Odhner, 1911 (p. 89)

Fairly large trematodes, up to and even exceeding 30 *mm. in length.*

KEY TO GENERA OF AZYGIIDAE

1. Suckers of approximately equal size, oral slightly larger than ventral; parasites of fresh-water fishes *Azygia*
2. Ventral sucker approximately twice as large as oral; parasites of selachians *Otodistomum*

Genus *Azygia* Looss, 1899

Syn. *Mimodistoma* Stafford, 1904; *Megadistoma* Stafford, 1904; *Hassalius* Goldberger, 1911.

Azygia lucii (Müller, 1776) Lühe, 1909 (Fig. 29C)

Syn. *Fasciola lucii* Müller, 1776; *Distoma tereticolle* (Rudolphi) of Looss, 1894; *Azygia loossii* Marshall & Gilbert, 1905; *Ptychogonimus volgensis* Linstow, 1907; *Distomum volgense* (Linstow) of Lühe, 1909; *Azygia tereticollis* (Rudolphi) of Looss, 1899 and Odhner, 1911; *A. volgensis* (Linstow) of Odhner, 1911; *A. robusta* Odhner, 1911; and, possibly, forms discovered outside Europe.

HOSTS. Perch, burbot, trout, grayling, chub, pike (various Salmonidae in parts of Europe).

LOCATION. Stomach and intestine, sometimes oesophagus (Looss).

Lühe (1909) considered this species to be very common in the pike on the Continent, and it has appeared in this host in Britain (Norfolk). It also occurs in North America. Manter (1926) has pointed out that contraction of the body greatly alters the topography of the internal organs, as well as the shape, and that the dimensions of the pharynx are subject to considerable variation. This explains the extensive synonymy.

DIAGNOSIS. *Shape, colour and size:* very elongate, almost cylindrical, flesh or rose colour (fading rapidly to pure white after death), 10–30 mm. long (according to Looss (1894*a*) up to 54 mm. in length) and about 1·5 mm. broad. *Cuticle:* relatively thick (0·02 mm.), ringed when the body is contracted, non-spinous. *Suckers:* oral slightly larger than the ventral, which is situated well within the anterior third of the body. *Gut:* pharynx cylindrical, oesophagus very short, caeca long and, in the living worm, showing lively peristalsis. Contents of the gut never coloured, according to Looss, but including fat-droplets. *Reproductive systems:* genital pore slightly in front of the ventral sucker. Cirrus pouch almost spherical, cirrus short, vesicula seminalis long and coiled. Testes ovoid, rather small, one behind the other well in the posterior half of the body, slightly in front of the origin of the lateral excretory canals. Ovary rounded, slightly in front of the anterior testis. Vitellaria lateral, extending from the level of the posterior testis to a point slightly behind the ventral sucker. Uterus with an ascending limb only, formed into slight transverse folds which fill the available space between the caeca from the ovary to the ventral sucker. Eggs clear yellowish brown, measuring 0·045 × 0·023 mm., those in the terminal part of the uterus each containing a fully developed non-ciliate miracidium, bearing bristles which arise from basal plates ('Borstenplatten').

For descriptions of the American species, *A. longa* (Leidy) see Manter (1926) and Linton (1940).

Genus *Otodistomum* Stafford, 1904

This genus was erected for trematodes (found in the barndoor skate in Canadian Atlantic waters) which Stafford identified as *Distoma veliporum* Creplin, 1837. Lebour (1908*a*) found a similar fluke in the stomach of the starry ray and assigned it to this species. Odhner (1911–13) examined specimens found in Europe and recognized two species of *Otodistomum*, *O. veliporum* (Creplin, 1837) and *O. cestoides* (Beneden, 1870), also expressing the opinion that Stafford's specimens belonged to the latter. Although the two species were difficult to separate, Odhner listed certain characters by which they could perhaps be identified. Other helminthologists came to agree that the separation of the two species is practically impossible on purely morphological grounds, but some faith persisted in differences in the sizes of the eggs and the thickness of their shells, which Odhner gave as 0·086 × 0·06–0·063 and 0·006 mm. for *veliporum* and 0·065–0·072 × 0·043 and 0·003 mm. for *cestoides*. Manter (1926) determined a clear-cut difference between the sizes of the eggs in specimens collected in Alaska and Maine respectively, giving the mean measurements of length as 0·0855 and 0·0694 mm. Accordingly, he regarded the Pacific forms as belonging to the species *veliporum*, the Atlantic forms to *cestoides*. Dollfus (1937) made a very detailed study of the genus, however, and decided that all North American forms belong to the species *cestoides*. He defined the geographical limits and hosts of the two species precisely, viz.:

O. cestoides: in the North Atlantic, north of 50° N. for Europe and 40° N. for America and in the North Pacific of America; parasites of *Chlamydoselachus* and *Raja*.

O. veliporum: in the Atlantic of south Europe, north Africa (50–20° N.) and in the Mediterranean; parasites in various Squaliformes and Torpediniformes.

Combining published descriptions and his own observations Dollfus (*loc. cit.*) concluded that the genus can be divided up into six forms according to the sizes of the eggs, the thickness of their shells, and host affiliations, viz.:

1. Eggs with shells not more than 0·004–0·005 mm. thick (American forms)
   - A. Adults in *Raja* and *Chlamydoselachus* of North Atlantic and North Pacific (California). Eggs measuring 0·065–0·072 × 0·043–0·046 mm.; shells 0·0025–0·0045 mm. thick — *O. cestoides cestoides*
   - B. Adults in *Raja* of North Pacific. Eggs measuring 0·085 × 0·0575 mm.; shells 0·004 mm. thick — *O. cestoides pacificum*
2. Eggs with shells 0·006–0·010 mm. thick (European and Australian forms)
   - C. Adults in *Pristiophorus* in Australia — *O. pristiophori*
   - D. Adults in Squaliformes and Torpediniformes of Mediterranean and Atlantic, South Europe to North Africa. Eggs measuring 0·085–0·097 × 0·062–0·070 mm.; shell 0·006–0·007 mm. thick — *O. veliporum leptotheca*
   - E. Adults in Hexanchidae of Europe. Eggs intermediate in dimensions and thickness between (D) and (F) — *O. veliporum veliporum*
   - F. Adults in Torpediniformes of Atlantic, South Europe to North Africa — *O. veliporum pachytheca*

While bestowing deserved credit on a thoroughgoing attempt to settle an extremely difficult problem at no matter what cost in labour, it must be admitted that these findings will be both difficult and hazardous to put into practice. Abundant evidence has come to light regarding the variability of size in the eggs of Trematoda. Manter (1926) concluded that there is no overlap between the sizes of the eggs in Atlantic and Pacific forms of *Otodistomum*, but this has not been borne out by the recent work of Van Cleave & Vaughn (1941), who found instead complete overlap. Consequently, these writers concluded that the criterion of size of the eggs no longer holds good as a basis for the separation of the subspecies *O. cestoides cestoides* and *O. cestoides pacificum* as proposed by Dollfus and that, accordingly, all American forms should be regarded as belonging to a single highly variable species bearing the name *Otodistomum cestoides*. Miller (1941 *a*) independently reached the conclusion that *O. veliporum* and *Xenodistomum melanocystis* of Stafford, 1904 really belong to this species. If eggs show this extent of variability in two of the six subspecies proposed by Dollfus, then the remaining subspecies are invalidated. Pending further inquiries, which will not provide a very attractive problem, it seems logical to conclude that a single species of the genus *Otodistomum* infects cartilaginous fishes all over the world, and that it should be known not as *O. cestoides* but as *O. veliporum*.

Otodistomum veliporum (Creplin, 1837) Stafford, 1904 (Fig. 38E, E*b*)

Syn. *Distoma veliporum* Creplin, 1837; *D. insigne* Diesing, 1850; *Fasciola squali grisei* Risso of Diesing, 1850; *Distoma microcephalum* Baird, 1853; *D. cestoides* Beneden, 1870; *D. nigrescens* Olsson, 1876; *Agamodistomum chimaerae* Ariola, 1899; *Xenodistomum melanocystis* Stafford, 1904; *Otodistomum cestoides* (Beneden) of Odhner, 1911 and other authors; *Cercaria cestoides* Nicoll, 1913; *Otodistomum cestoides cestoides* of Dollfus, 1937; *O. c. pacificum* of Dollfus, 1937; *O. veliporum leptotheca* Dollfus, 1937; *O. v. veliporum* of Dollfus, 1937; *O. v. pachytheca* of Dollfus, 1937; and, probably, *O. pristiophori* (Johnston, 1902) (Syn. *Distoma pristiophori* Johnston, 1902).

Hosts. Various Selachii (in Britain, flapper skate, starry ray).

Location. Stomach, rarely oesophagus, branchial cavity, intestine (? accidental). In Britain, this species has been found in Northumberland, Aberdeen and Galway. It occurs also in Belgium, France, Italy, Sweden, Norway, Russia, several regions of the American continent (Maine, Woods Hole, Washington, California, Canada, Alaska, Chili) and in New Zealand.

Diagnosis. *Shape and size:* very elongate (breadth about one-twelfth of the length), slightly flattened, up to 80 mm. long (Stafford, 1904). In view of the extreme variability resulting from differences in size and degree of contraction, we might consider the characters of the specimens described by Lebour (1908*a*), which were rather small, about 32 mm. long and about 2·6 mm. broad. *Colour:* opaque white, coloured yellow, brown or deep purple by the eggs (according to age) and yellowish brown by the gut contents. *Cuticle:* non-spinous, sometimes slightly wrinkled. *Suckers:* well developed, ventral twice as large as the oral and close behind it, diameters 2·4 and 1·2 mm. *Gut:* prepharynx absent, pharynx conspicuous (0·4 mm. long and 0·35 mm. broad), caeca long and slightly sinuous, extending almost to the posterior extremity. Bifurcation about midway between the suckers. *Excretory system:* main canals Y-shaped, with a fairly long median stem and lateral canals extending to the pharynx. *Reproductive systems:* genital pore near the pharynx, slightly on the left. Cirrus pouch rounded, narrowing towards the spacious genital atrium, vesicula seminalis small and club-shaped. Testes ovoid, one behind the other and contiguous, near the mid-body. Ovary in front of and contiguous with the anterior testis, on the right. Receptaculum seminis absent. Vitellaria well developed, extending from close behind the ventral sucker to the final fifth of the body, merging in the median plane behind the posterior testis. Uterus with an ascending limb only, formed into transverse folds between the ovary and the ventral sucker. Eggs thin-shelled, broad oval, about 0·06 mm. long.

Notes. Exclusively a parasite of Selachii, but metacercariae are sometimes found in large, globular, orange-yellow cysts in the gastric mucosa of the witch. Adults are usually found attached to the wall of the stomach, not free in its lumen. According to Manter (1926) the trematode becomes mature when about 11 mm. long. The ratio of the sizes of the oral and ventral suckers does not change with general growth, but remains at about 3/4 or 3/5 (the ratio 1/2 determined by Lebour being unusual). The terminal parts of the genital ducts (unobserved by Lebour) penetrate a conical muscular papilla which projects into the genital atrium and is covered with cuticle. The massing of vitelline follicles across the median plane posteriorly which was observed by Lebour also seems to be unusual; generally, the lateral rows remain distinct, though a strict sequence is not invariably maintained. Manter (*loc. cit.*) also made a serious attempt to determine the growth characteristics of more than 200 specimens ranging in length between 2·3 and 65 mm. While the relative sizes of the suckers to one another remain constant, their size relation to the body as a whole changes considerably, and the position of the ventral sucker becomes progressively more anterior, by increase in the length of the region behind it. The miracidium is non-ciliate and its cuticle bears numerous fine bristles anteriorly. 'Borstenplatten' such as occur in the miracidium of *Azygia lucii* are absent, but the bristles are carried by five areas radiating from the anterior end of the body. These, like the bristles, are shed eventually, and during the transition period the larva presents a curious appearance, the strips of tissue resembling appendages.

Family **PTYCHOGONIMIDAE** Dollfus, 1936 (p. 88)

Subfamily **Ptychogoniminae** Dollfus, 1937

With the characters of the family.

Genus *Ptychogonimus* Lühe, 1900

Ptychogonimus megastoma (Rudolphi, 1819) Lühe, 1909 emend. (Fig. 40)

Syn. *Distoma megastomum* Rudolphi, 1819; *D.* (*Brachylaimus*) *megastomum* Rudolphi of Parona, 1896; *D. soccus* Molin, 1858 of Stossich, 1883; *D. lymphaticum* Linstow, 1903.

HOST. Smooth hound (in Britain); many other Selachii (tope, nurse hound, blue shark) on the Continent. Unencysted metacercariae in Brachyura.

LOCATION. Adults in the stomach; larvae in the tubules of the testes, the vas deferens and the coelom (near the ovary) of the crustacean.

Dollfus (1937) gave a detailed account of the history of this species, which occurs in Italy, France, Belgium and North Africa and has appeared at Plymouth. It is a very distinctive trematode, which is recognizable from the familial characters, but a brief diagnosis can be given.

DIAGNOSIS. *Shape:* slightly elongate, the breadth about one-third of the length, thick and flattened dorso-ventrally. *Cuticle:* non-spinous, but having large, irregular, transverse folds, 0·009–0·014 mm. thick, its surface finely granular and having groups of small vacuoles. *Suckers and musculature:* well-developed suckers, powerful, fairly close together, oral slightly larger than the ventral. *Gut:* prepharynx very short, practically absent, pharynx large, oesophagus absent, caeca long. Bifurcation of the intestine dorsal to the ventral sucker, proximal parts of caeca extending anteriorly, then turning and continuing posteriorly to open into the excretory vesicle (Fig. 40D). *Excretory system:* main canals U-shaped plus a short, straight, muscular canal opening to the exterior. *Reproductive systems:* genital pore between the suckers (slightly nearer the ventral). Genital atrium large and muscular, at its base a small protrusible papilla on which are the male and female pores. Cirrus pouch absent. Ejaculatory duct short, pars prostatica slightly longer, vesicula seminalis very long and convoluted. Testes smooth or slightly lobed, one behind the other in the posterior region. Ovary in front of the testes. Receptaculum seminis absent. Vitellaria well developed, extending from slightly in front of the ventral sucker to the posterior extremity, spreading over the excretory vesicle ventrally; follicles discrete. Uterus with descending and ascending limbs, numerous (ten to fourteen) folds passing to right and left of the gonads and filling up much of the space behind them. Metraterm short, not extending back beyond the level of the centre of the ventral sucker. Eggs numerous, thin-shelled, measuring about 0·062–0·067 × 0·036–0·038 mm. (according to Jacoby (1899) 0·057 × 0·024 mm.; according to Linstow (1903) 0·073 × 0·044 mm.).

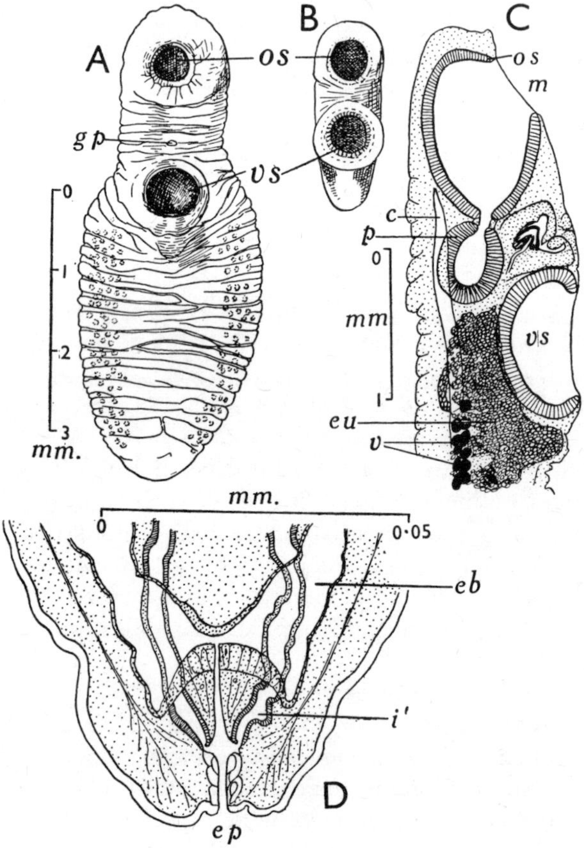

Fig. 40. *Ptychogonimus megastoma* (Ptychogonimidae). A, adult (6 × 2·5 mm.). B, juvenile (2 × 0·6 mm.). C, anterior end of adult. D, posterior end of adult, showing connexions between caeca and excretory vesicle. A, B, D in ventral, C, in side view. (After Dollfus, 1936*c*.)

Notes. Each egg contains a non-ciliate miracidium equipped with rigid bristles anteriorly. According to Dollfus, metacercariae in *Carcinus* attain a length of 2·8–3 mm. and the smallest specimens found in the smooth hound were 1·6 mm. long. The juvenile fluke differs considerably from the adult as regards the proportions of the body (see Fig. 40B). Monticelli (1890) found an adult specimen in the body cavity of *Maia* sp. (near the ovary), but this and other instances of progenesis require confirmation.*

* Palombi (*Nota prev. Riv. Parasit. Roma*, **5**, 1941, 127–8) named *Cercaria dentalii* Pelseneer as a larva of this species.

Family **HEMIURIDAE** Lühe, 1901 (p. 89)

Subfamily **Hemiurinae** Looss, 1899

Ecsoma present; mouth not overhung by a lip; vitellaria compact, or only slightly lobed.

KEY TO GENERA OF HEMIURINAE

I. Vitellaria paired *Hemiurus*
II. Vitellaria unpaired *Aphanurus*

Genus *Hemiurus* Rudolphi, 1809

Syn. *Parahemiurus* Vaz & Pereira, 1930.

The greatest difference which exists between *Hemiurus* and '*Parahemiurus*' is the bipartite nature of the seminal vesicle in the former and its undivided state in the latter. Manter (1934) was doubtful about the validity of the new genus

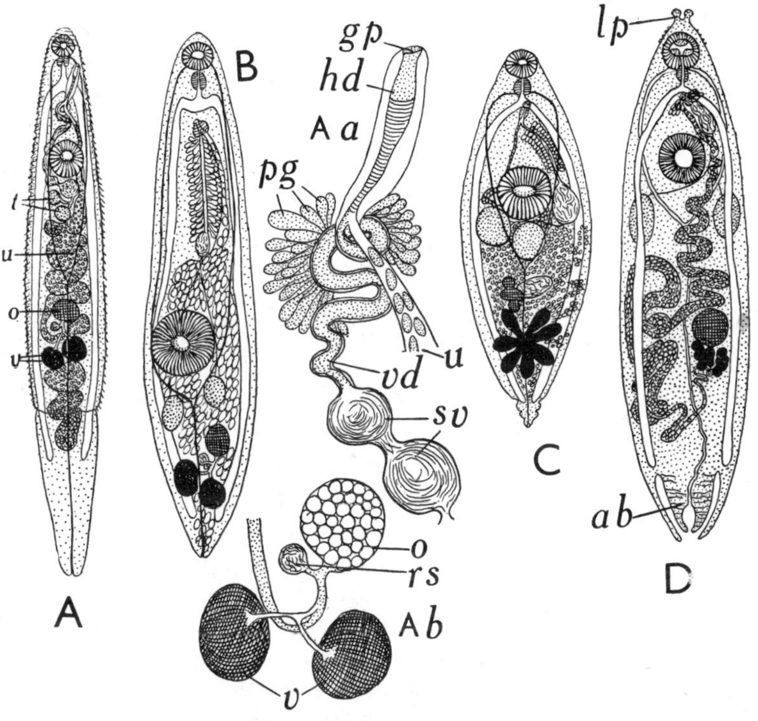

Fig. 41. Hemiuridae which (excepting D) are among the commonest Digenea of British marine fishes. A, *Hemiurus communis*. B, *Derogenes varicus*. C, *Lecithaster gibbosus*. D, *Ceratotrema furcolabiatum*. A*a*, terminal parts of genital ducts of A. A*b*, female organs of A, showing relationship between ovary, vitellaria, receptaculum seminis and various ducts. (D, after Idris Jones, 1933*b*; rest after Lebour, 1908*a*.)

(erected for *P. parahemiurus* Vaz & Pereira, 1930, which is identical with a species described by Linton (1910) as *Hemiurus merus* and previously (1898) as *Distomum appendiculatum*), but Woolcock (1935), Lloyd (1938) and others accepted it, as did Manter himself ultimately (1940*a*), then stressing the wide distribution of *Parahemiurus merus* (Linton) in the Gulf of Mexico, south American Atlantic, American Pacific and, probably, Japanese seas, and regarding various more recent species of the genus (*P. parahemiurus* Vaz & Pereira, 1930; *P. platichthyi* Lloyd, 1938; *P. atherinae* Yamaguti, 1938; and *P. harengulae* Yamaguti, 1938) as identical with it. At the same time he accepted the validity of four other species of the genus, but regarded two of them (*P. sardinae* and *P. seriolae*) as possible synonyms of *P. merus*. Thus, he pruned the new 'genus' of more than half its growth of species, but referred to the 'rapid growth of the genus' as dispelling doubts regarding its validity. This setting up of a parallel series to species of *Hemiurus* has been due entirely to comparison of newly discovered forms with '*Parahemiurus*', but not *Hemiurus*, as a result of easy acceptance of this trifling character of distinction. All species of Vaz & Pereira's 'genus' are under suspicion of being species of *Hemiurus*, and many of them will prove to be known species insufficiently described. In *Hemiurus levinseni* the seminal vesicle has a non-muscular wall, and no doubt under certain conditions of extension or contraction of the body it may assume a variety of shapes, including that attributed to '*Parahemiurus*'. The extent to which muscles are developed in the wall of the vesicle seems to vary in other species, and it is likely that in some the entire vesicle is muscular.

Hemiurus appendiculatus (Rudolphi, 1802) Looss, 1899 (Fig. 43 D, E)

Syn. *Fasciola appendiculata* Rudolphi, 1802; *Distoma appendiculatum* Rudolphi, 1809 and Rudolphi, 1819, in part; *Distomum ventricosum* Wagener, 1860, in part, *nec* Rudolphi, 1819; *Apoblema appendiculatum* of Juël, 1899 and Monticelli, 1891, in part; *A. appendiculatum* of Looss, 1896, *nec* Mühling, 1898.

HOSTS. Perch, burbot, river trout, houting, twaite shad, allis shad, river lamprey.

LOCATION. Stomach, sometimes intestine and pyloric caeca.

DIAGNOSIS. *Size:* soma up to 3·5 mm. long and 0·65 mm. broad and thick, ecsoma when fully extended about three-quarters of the length of the soma. *Suckers:* ventral about twice as large as the oral and close behind it, diameters 0·33–0·5 and 0·17–0·23 mm. *Cuticle:* striated dorsally only to a point slightly behind the pharynx. *Reproductive systems:* genital atrium short, not longer than the breadth of the sinus sac. Anterior part of the seminal vesicle having a thick, muscular wall. Vitellaria irregularly rounded. Limbs of the uterus extending into the ecsoma and almost coextensive with the caeca. Eggs measuring 0·020–0·023 × 0·010–0·012 mm. (0·026–0·027 × 0·010–0·012 mm. according to Patzelt (1930)).

Notes. According to Lühe this species is common in the Rhine, and Looss found it equally common in the Nile, specimens from this locality in no way differing from others found at Trieste. The trematode also occurs in U.S.A. and Canada. Patzelt (1930) made a detailed study of specimens found in *Sardinella aurita* at Palma (Majorca). Linton (1940) gave diagnoses and measurements of specimens which have been found in nineteen species of fishes in the Woods Hole district.

Hemiurus communis Odhner, 1905 (Fig. 41 A, *Aa*, *Ab*)

Syn. *Distoma appendiculatum* of Olsson, 1868 and Juël, 1889, in part, *nec* Rudolphi, 1802; *D. appendiculatum* of Johnstone, 1907.

HOSTS (in Britain). Common sea bream, short-spined cottus, long-spined cottus, yellow gurnard, red gurnard, grey gurnard, streaked gurnard, angler, greater weever, horse mackerel, boar-fish, rock goby, butter-fish, cod, bib, poor cod, whiting, coal-fish, pollack, common ling, torsk, greater sand eel, lesser sand eel, halibut, long rough dab, common topknot, plaice, flounder, dab, ocean pipe-fish, conger, common eel.

Location. Stomach.

Diagnosis. According to Looss: *Shape and size:* plumper than *H. appendiculatus*, 1·3 mm. long when the ecsoma is retracted, 2 mm. long when it is extended, 0·4–0·45 mm. broad and thick. *Cuticle:* annulated ventrally almost

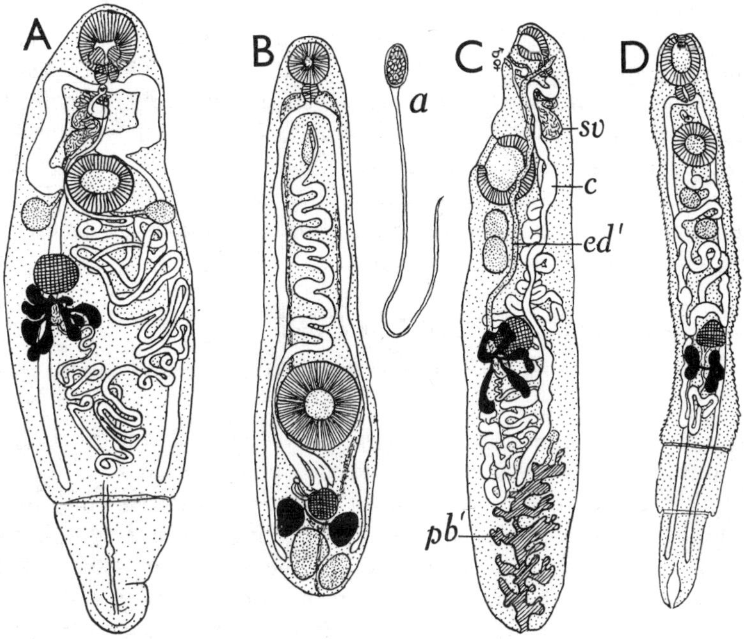

Fig. 42. Less common Hemiuridae. A, *Lecithochirium rufoviride*. B, *Hemipera sharpei*, (*a*), its egg. C, *Sterrhurus fusiformis*. D, *Brachyphallus crenatus*. (A, after Johnstone, 1907; B, (*a*) after Idris Jones, 1933*b*; C, D, after Fuhrmann, 1928.)

to the end of the soma, dorsally hardly at all, striae faint on the ecsoma. *Suckers:* smaller than in the previous species, ventral almost twice as large as the oral, diameters about 0·3 and 0·14 mm. *Other characters:* hermaphrodite duct fairly thick, about as long as the ventral sucker is broad. Seminal vesicle bipartite, the anterior part having a thin wall. Vitellaria irregularly rounded. Eggs measuring 0·019–0·021 × 0·010–0·012 mm. Uterus with ascending and descending limbs extending into the ecsoma, abundantly folded between the ecsoma and the testes.

Notes. Other descriptions of this trematode occurring in the literature differ from the above diagnosis in details. Specimens found by Lebour (1908*a*) at Cullercoats were about 2·8 mm. long and 0·5 mm. broad (the ecsoma taking up two-thirds of total length) and had suckers 0·28 and 0·14 mm. diameter and eggs measuring 0·026 × 0·01 mm. The trematode is widely distributed and shows a decided preference for gadoid fishes. As a rule, several specimens occur in one host, but the fluke is sometimes solitary, or associated with other hemiurids. In Britain it has been found at Plymouth, New Brighton, Piel (Barrow-in-Furness), Galway, St Andrews, Millport, Aberdeen and Cullercoats. It also occurs in Scandinavia.

Hemiurus levinseni Odhner, 1905

Syn. *Distomum appendiculatum* of Olsson & Juël, in part.

HOSTS. Cod, short-spined cottus, greater fork-beard.

LOCATION. Stomach.

This trematode has been found in Greenland, Denmark, Canada and U.S.A. According to Odhner (1905) and Looss (1908) it is 1–1·6 mm. long and 0·3–0·5 mm. broad and thick, the ecsoma, which is invariably retracted (?), 0·1–0·13 mm. long. The oral sucker is somewhat larger than the ventral, diameters 0·17 and 0·14 mm. The sinus sac is about as long as the oral sucker is broad, the pars prostatica is little folded, the vitellaria are almost spherical, and the eggs measure 0·026–0·028 × 0·012–0·013 mm. For descriptions of American specimens see Manter (1926) and Linton (1940).

Hemiurus ocreatus (Rudolphi, 1802) Looss, 1899

Syn. *Distoma ocreatum* Molin, of Olsson, 1868; *Hemiurus lühei* Odhner, 1905 and of Looss, 1907.

HOSTS. Herring, sprat, pilchard (west Sweden), salmon (at Leipzig) and (?) grey gurnard, mackerel, boar-fish, whiting, pollack, hake.

This species seems to take the place of *H. communis* in clupeoid fishes and is frequently the only trematode parasite of the herring. According to Looss (1908, as *H. lühei*) it is 1·5–1·7 mm. long when the ecsoma is retracted, 2·5–2·8 mm. when it is extended, the breadth under these circumstances being 0·3 and 0·23–0·27 mm. *Cuticle:* with annuli extending on the dorsal surface to a point midway between the ventral sucker and the testes. *Suckers:* ventral nearly twice as large as the oral, diameters 0·17–0·21 and 0·1–0·12 mm. Eggs measuring 0·02–0·022 × 0·011–0·012 mm.

Note. Markowski (1933*a*) described specimens from the herring in the Polish Baltic. They were reddish in colour, 2–3·5 mm. long when partially contracted posteriorly, annulated to the level of the ovary, had suckers 0·18–0·24 and 0·165–0·24 mm. diameter respectively and eggs measuring 0·025–0·027 × 0·011–0·014 mm.

Hemiurus rugosus Looss, 1907

Syn. *Hemiurus stossichi* of Lühe, 1901, *nec Apoblema stossichii* Monticelli, 1891.

HOSTS. Pilchard, sardine (at Trieste), turbot.

LOCATION. Stomach.

This species was discovered at Trieste. According to Looss (1908) it is not larger than the previous species, about 3 mm. long and 0·4–0·6 mm. broad and thick when contracted, 4 mm. long when extended. The annuli extend back on the dorsal surface to a point above the testes, generally to near the ovary. The ventral sucker is half as large again as the oral, diameters 0·24–0·28 and 0·17–0·19 mm. The sinus sac is notably more slender than in *H. ocreatus*, the vitellaria have a variable number of deep clefts and the eggs measure 0·019–0·021 × 0·011–0·012 mm. Host affiliations would suggest that this species is closely related to *H. ocreatus*, perhaps identical with it.

Genus *Aphanurus* Looss, 1907

Aphanurus stossichii (Monticelli, 1891)

Syn. *Distomum ocreatus* Monticelli, 1887, Stossich, 1888, 1898 *nec* Rudolphi, 1819, *nec* Olsson, 1867; *Apoblema stossichii* Monticelli, 1891; *Hemiurus stossichi* (Monticelli) of Lühe, 1901; *Aphanurus virgula* Looss, 1907.

HOSTS. Pilchard, horse mackerel, bogue, etc. (at Naples and Trieste) (anchovy and red-band fish as *A. virgula*), *Sardinella aurita* at Palma (Majorca) (see Patzelt, 1930).

LOCATION. Stomach and oesophagus.

DIAGNOSIS. *Size:* up to 0·9 mm. long and 0·3 mm. broad. *Suckers:* ventral at least twice as large as the oral, diameters 0·13–0·16 and 0·06–0·07 mm. *Reproductive systems:* sinus sac never longer than the breadth of the ventral sucker and ending far in front of it when the body is extended. Vitellaria immediately behind the ovary, transversely elongate or reniform, generally irregular in shape. Eggs measuring 0·023–0·025 × 0·011–0·012 (confirmed by Patzelt).

Notes. Looss remarked on discrepancies in the original description of this trematode, e.g. the position of the genital pore and the occurrence of a small ecsoma. Looss came to regard the latter as an artefact, perhaps a local constriction of the integument posteriorly. The form '*Aphanurus virgula*' was based on insignificant differences from *A. stossichi* and was found in the anchovy and red-band fish only in the absence of this species.

Subfamily **Sterrhurinae** Looss, 1907

Ecsoma present; mouth overhung by a lip; vitellaria small and having finger-like lobes.

KEY TO GENERA OF STERRHURINAE

I. Anterior extremity produced into a pair of processes — *Ceratotrema*

II. Anterior extremity not produced into a pair of processes
- A. Presomatic pit (an invagination in front of the ventral sucker) present
  - (*a*) Presomatic pit a small, circular depression
    - (i) Oral sucker having two ventro-lateral thickenings and a well-defined dorsal lip — *Lecithochirium*
    - (ii) Oral sucker devoid of ventro-lateral thickenings and having an ill-defined dorsal lip — *Synaptobothrium*
  - (*b*) Presomatic pit a transverse, slit-like depression — *Brachyphallus*
- B. Presomatic pit absent — *Sterrhurus*

Note. Excluded from the key is the genus *Plerurus* Looss, 1907, which would separate along with *Sterrhurus*, but is said to differ by having a concavity (not to be confounded with the presomatic pit, which is absent) between the suckers.

The type-species, *Plerurus digitatus* (Looss, 1899) (Syn. *Hemiurus digitatum* Looss, 1899; *Lecithochirium digitatum* (Looss) of Lühe, 1901), was discovered in the stomach of *Sphyraena vulgaris*, seems to be very rare, and has not appeared in northern Europe.

Genus *Sterrhurus* Looss, 1907

Syn. *Lecithochirium* Lühe, 1901, in part.

Several species of this genus occur in Europe. The type-species, *S. musculus* Looss, 1907, is parasitic mainly in the eel and the dentex, but also in the sturgeon, turbot, angler and many other fishes, isolated specimens having been found in the greater weever, pandora, bass and comber. This species was discovered at Trieste and has not appeared in Britain. According to Looss, it is 1–1·5 mm. long, has ventral and oral suckers respectively 0·2 and 0·1 mm. diameter, vitellaria with finger-like lobes and eggs which measure 0·019–0·021 × 0·011–0·013 mm. *S. grandiporus* (Rudolphi, 1819) (Syn. *Distoma grandiporum* Rudolphi, 1819; *Lecithochirium grandiporum* (Rudolphi) of Lühe, 1901), a parasite of the murry at Naples, is said to be slightly larger, 1·6–1·7 mm. long, but to have much larger suckers (diameters 0·4 and 0·2 mm.), vitellaria divided almost to the bases of the processes, and almost spherical eggs measuring 0·02 × 0·015–0·017 mm. *Sterrhurus imocavus* Looss, 1907 is a parasite of the tunny in Egypt and is larger than the previous two species (up to more than 3 mm. long), has more dissimilar suckers (diameters 0·35–0·45 and 0·12–0·17 mm.), and comparatively small eggs (dimensions 0·016 × 0·01 mm.). A fourth species, *S. fusiformis* (Lühe, 1901) (Syn. *Lecithochirium fusiformis* Lühe, 1901) (Fig. 42C), and the only one to appear in Britain, has the following characters (based on Looss (1907), but with findings due to Jones (1943) in parentheses). *Size:* 3·65 mm. long and 1·12 mm. broad when the body is contracted (5 × 1·7 mm., the ecsoma being as long as the soma in life, but contracting more strongly in fixation). *Suckers:* ventral about twice as large as the oral, diameters 0·5–0·6 and 0·25–0·3 mm. (0·60 and 0·28 mm.). *Vitellaria:* lobes long, thin, tapering, sometimes forked at the tips. *Eggs:* fairly thick-shelled, dimensions 0·02–0·023 × 0·017–0·019 mm. (0·024 × 0·018 mm.). The hosts of this species in Britain are the angler and conger. For descriptions of the American species, *S. monticellii* (Linton, 1898), see Linton (1898, 1910, 1940). Other species in America include *S. laevis* (Linton, 1898); *S. floridensis* Manter, 1934; *S. praeclarus* Manter, 1934; *S. robustus* Manter, 1934; *S. profundus* Manter, 1934; *S. branchialis* Stunkard & Nigrelli, 1934; and *S. magnatestis* Park, 1936. It is unlikely that all are valid species.

Genus *Ceratotrema* Idris Jones, 1933

Ceratotrema furcolabiatum Idris Jones, 1933 emend. (Fig. 41D)

HOST. Five-bearded rockling (at Wembury, near Plymouth).

LOCATION. Coelom, attached to the parietal peritoneum, liver and intestine, also along the course of the portal vein.

DIAGNOSIS. *Shape, colour and size:* elongate, reddish brown mottled with green, 7·7 mm. long and 1·7 mm. in greatest breadth. *Ecsoma:* short, invariably invaginated. *Suckers:* ventral slightly larger than the oral and close behind it;

oral with three cushion-like projections into its lumen, the postero-lateral ones round and prominent, the anterior one small. *Lip:* a process of the dorsal surface near the anterior extremity, having a broad base and a pair of outgrowths which are capable of great extension and contraction, presumably the homologue of lip of other Sterrhurinae, but more muscular, better developed and not overhanging the mouth even when contracted. *Gut:* pharynx present, oesophagus short, caeca long, but not extending into the ecsoma. *Excretory system:* main canals Y-shaped, with a long, median canal having a dilatation near the pore. Point of union of the three main canals close behind the ventral sucker, at the level of the testes. *Reproductive systems:* genital pore well in front of the ventral sucker, near the bifurcation of the intestine. Testes elongate oval in outline, side by side and widely separated, overlying the caeca, close behind the ventral sucker. Ovary rounded, some distance behind the testes, in the posterior region of the soma, slightly lateral. Vitellaria slightly lobed, close behind the ovary. Uterus with descending and ascending limbs confined to the soma, not reaching far behind the ootype, the ascending limb wide and formed into a few, wide folds which sometimes overstep the caeca laterally. Eggs greenish, numerous, measuring 0·023 × 0·011 mm.

Genus *Lecithochirium* Lühe, 1901

Looss (1908) gave as a diagnostic character of this genus the presence of two lateral elevations extending into the lumen of the oral sucker. Lloyd (1938), as well as Manter (1931, 1934), showed that they are not invariably present, and regarded the presence of the presomatic pit as the best diagnostic character. Incidentally, Lloyd thought this pit might serve as a chemoreceptor playing some part in controlling movements of the ecsoma.

Lecithochirium rufoviride (Rudolphi, 1819) Lühe, 1901 (Fig. 42A)

Syn. *Distoma rufoviride* Rudolphi, 1819; *Hemiurus rufoviride* (Rudolphi) of Looss, 1899; *Distomum ocreatum* Molin of Johnstone, 1907.

Hosts. Angler, conger, common eel (at St Andrews, Millport, Aberdeen, Irish Sea, Galway, Plymouth and Naples).

Location. Stomach.

Diagnosis. *Shape and size:* elongate-oval outline, cylindrical, 2·5–8 mm. long and about one-third as broad, having a well-developed ecsoma. *Colour:* brownish orange, sometimes blood red when alive. *Cuticle:* smooth, thickest in the antero-ventral region. *Suckers:* of nearly equal size and about 0·9 mm. diameter in specimens 8 mm. long (oral 0·7 mm., ventral 0·79 × 0·83 mm. in specimens 4–6 mm. long according to Jones, 1943). (*Note.* The size relations (diameters) of the suckers in small, medium and large specimens was given by Looss as 0·3–0·4 and 0·4–0·5 mm., 0·6–0·7 and 0·7–0·8 mm., and 0·8 and 1·0 mm. The oral sucker is invariably slightly the smaller, but I have specimens about 5 mm. long in which equality is more nearly attained.) *Reproductive systems:* vitellaria having three to five lobes which are short, irregular and as broad as long. Eggs measuring 0·022 × 0·013 mm. (0·018 × 0·009 mm. according to Jones).

Notes. An individual host may harbour more than 100 flukes of this species (see Johnstone, 1907). Larvae also occur in brown cysts attached to the viscera (generally the intestine) of the shanny. The cysts may be loosely attached to the intestine, or deeply imbedded in its tissue. Encysted individuals may have well-developed gonads, even eggs *in utero*, though they do not exceed about 2 mm. in length. Jones (*loc. cit.*) gave a detailed redescription of specimens from the Conger.

Lecithochirium gravidum Looss, 1907 (Fig. 43A)

Syn. *Apoblema rufoviride* of Juël, 1889.

HOSTS. Conger, common eel, bass, etc. (at Plymouth, Catania, Trieste, west Sweden).

LOCATION. Stomach.

According to Looss (1908) this species can be distinguished from *L. rufoviride* by its smaller size and the different size relations of the suckers. It is said to be 2·5–2·7 mm. long (2·9 mm. when the ecsoma is extended) and 0·5–0·8 mm. broad. The oral sucker is smaller than the ventral, dimensions in small individuals 0·17–0·25 and 0·3–0·43 mm., in larger individuals 0·25–0·33 and 0·45–0·53 mm., and in the largest 0·45 and 0·75 mm. *Other characters:* similar to *L. rufoviride.*

Notes. Looss affirmed the validity of this species, but admitted that it is very similar to *L. rufoviride.* Neither of the criteria on which specific identity was claimed, are reliable, however, and further study is likely to prove the two forms identical. The trematode has been found encysted and exhibiting progenesis in various Labridae and other fishes.

Two other species of *Lecithochirium*, *L. conviva* Lühe, 1901 and *L. physcon* Lühe, 1901, occur respectively in the conger (Berlin Coll.) and the angler at Trieste, but have not appeared in Britain. *L. copulans* (Linstow, 1904) (Syn. *Synaptobothrium copulans* Linstow, 1904) was discovered in *Arnoglossus laterna* at Louvain.

Genus *Synaptobothrium* Linstow, 1904

By some writers (e.g. Sprehn, 1933) this genus is regarded as identical with *Lecithochirium.* According to Looss it differs in the structure of the terminal parts of the genital ducts, which are cylindrical instead of pyriform, and include a long metraterm and a long, tubular pars prostatica which is not clearly marked off from the hermaphrodite duct. The eggs are supposed to be reniform in profile.

Synaptobothrium caudiporum (Rudolphi, 1819) Looss, 1907 (Fig. 43B)

Syn. *Distoma caudiporum* Rudolphi, 1819; *Lecithochirium caudiporum* (Rudolphi) of Lühe, 1901.

HOSTS. Adults in the John Dory, plaice, angler, yellow gurnard, cod, bib, brill, conger (at Plymouth and Galway); larvae in the shanny, ballan wrasse, corkwing.

LOCATION. Adults in the stomach; larvae encysted in the viscera.

DIAGNOSIS (after Looss, 1908). *Size:* up to 2·7 mm. long when the ecsoma is half retracted and 0·7 mm. in greatest breadth. *Suckers:* oral about half as large as the ventral, diameters 0·13–0·2 and 0·25–0·4 mm. *Reproductive systems:* metraterm long, extending back to a level above the middle of the ventral sucker. Vitellaria with short and thick, finger-like lobes which invariably diverge. Eggs thin-shelled, reniform in profile, measuring 0·032 × 0·013 mm., each containing when laid a distinctive miracidium with a small anterior space (?) with a bright central dot.

Note. My somewhat larger specimens (total length 4·4 mm., ecsoma 1·2 mm. long) are clearly recognizable by the tapering anterior extremity, the relative sizes of the transversely oval suckers (dimensions 0·12 × 0·24 and 0·44 × 0·50 mm.), which have their centres about 0·8 mm. apart. The eggs are smaller than Looss specified, measuring 0·025 × 0·015 mm.

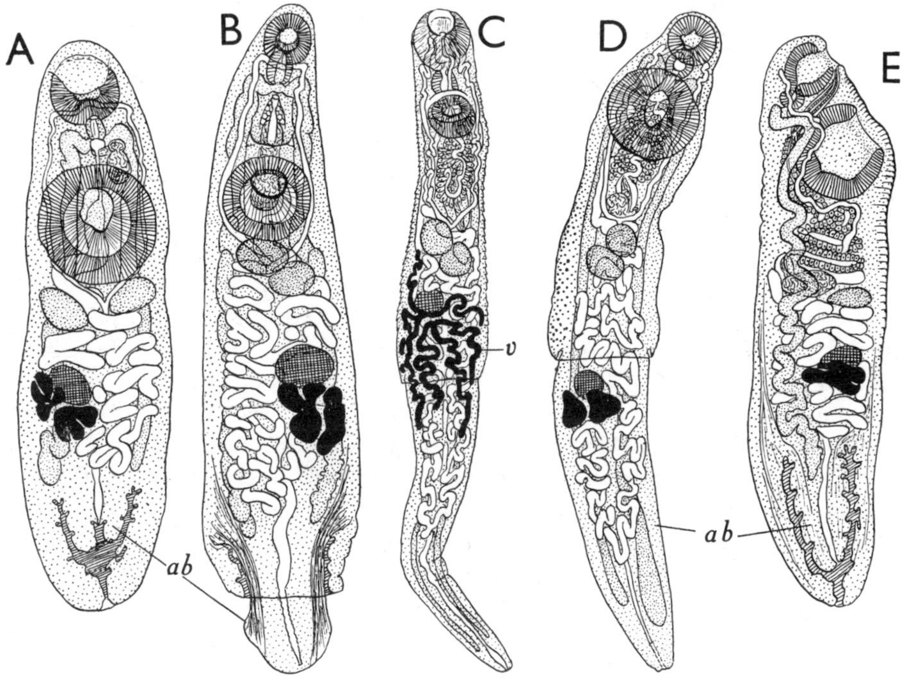

Fig. 43. Other Hemiuridae. A, *Lecithochirium gravidum*. B, *Synaptobothrium caudiporum*, C, *Lecithocladium excisum*. D, E, *Hemiurus appendiculatus* with the ecsoma respectively extended and retracted. (After Looss, 1907*a*.)

Genus *Brachyphallus* Odhner, 1905

This genus, as Lloyd (1938) pointed out, combines the characters of Hemiurinae and Sterrhurinae, having the ringed cuticle of the former, but the presomatic pit and vesicular parenchyma of the latter. Odhner (1905) and Fuhrmann (1928) include the genus in the Hemiurinae. Looss (1908) placed it in the Sterrhurinae because of the lack of a complete sinus sac, which occurs, however, in *Brachyphallus crenatus* (see Lander, 1904; Odhner, 1905; Lloyd, 1938), so that the terminal parts of the genital ducts have the characters of the Hemiurinae, although the deeper parts (ejaculatory duct and short pars prostatica) have those of Sterrhurinae.

Brachyphallus crenatus (Rudolphi, 1802) Odhner, 1905 (Fig. 42 D)

Syn. *Fasciola crenata* Rudolphi, 1802; *Distoma crenatum* Rudolphi, 1809, *nec* Rudolphi, 1810, *nec* Molin, 1859; *D. appendiculatum* Rudolphi, 1819, in part; *Distomum ventricosum* Wagener, 1860, in part; *D. ocreatum* Olsson, 1867, *nec* Rudolphi, 1819, *nec* Molin, 1861; *Apoblema ocreatum* Juël, 1899; *A. appendiculatum* Monticelli, 1892, in part, and Mühling, 1898, *nec* Rudolphi, 1802; ? *Fasciola serratulata* Müller, 1780; ? *Distomum ocreatum* of Linton, 1900; ? *Fasciola salmonis* Müller, 1780; *Hemiurus ocreatus* (Rudolphi) of Lühe, 1901.

HOSTS. Lesser sand eel (in Britain); father lasher, dab, three-spined stickleback, lesser sand eel, salmon, sea trout, smelt (elsewhere in Europe (North Sea and Baltic)); greater sand eel and salmon in the Polish Baltic (Markowski, 1933 *a*).

LOCATION. Stomach.

DIAGNOSIS. *Shape and size:* elongate, cylindrical, 3·2–3·6 mm. long, 0·5–0·65 mm. broad near the ventral sucker, 0·7–0·75 mm. broad near the end of the soma (Nicoll). Ecsoma about 1·1–1·2 mm. long (one-third of total length). Mature, according to Lühe, when 1·25–1·35 mm. long, the ecsoma then three-quarters the length of the soma. *Cuticle:* striated, striae present from the mouth to the end of the soma (Looss); according to Nicoll (1907) ecsoma markedly striated. *Suckers:* of almost equal size, ventral generally larger than the oral in the ratio of 7/6, diameters 0·28–0·3 and 0·24–0·26 mm. in individuals more than 3 mm. in total length. *Gut:* as in *Hemiurus*, caeca extending almost to the extremity of the ecsoma. *Reproductive systems:* genital pore midway between the suckers. Hermaphrodite duct very short. Pars prostatica in the same straight line as the seminal vesicle, which is opposite the middle of the ventral sucker. Testes diagonally one behind the other, close behind the ventral sucker. Ovary rounded, far behind the testes. Vitellaria compact, rounded or very slightly lobed, close behind the ovary. Uterus confined to the soma. Eggs of variable size; measuring 0·021–0·029 × 0·011–0·016 mm., according to Markowski (1933) 0·016–0·025 × 0·011–0·014 mm. For an account of American specimens see Linton (1940).

Another species of this genus, *B. anuriensis* Babaskin, 1928 occurs in the intestine of *Onchorhynchus keta* in Russia.

Subfamily **Dinurinae** Looss, 1907

Ecsoma present; mouth overhung by a lip; vitellaria having elongate, tubular outgrowths.

Looss (1908) included three genera in this subfamily, two of them new. Neither of them has appeared in Britain, for which reason short diagnoses will suffice here.

Genus *Dinurus* Looss, 1907

DIAGNOSIS. Ecsoma very large, pars prostatica and vitelline tubules long and convoluted, ejaculatory duct devoid of prostate gland cells for only a short region near the seminal vesicle.

Dinurus tornatus (Rudolphi, 1819) Looss, 1907

Syn. *Distoma tornatum* Rudolphi, 1819; *Dinurus barbatus* (Cohn, 1903) Looss, 1907; *D. breviductus* Looss, 1907; *D. longisinus* Looss, 1907.

HOSTS. Pelamid, *Coryphaena equisetis*, *C. hippuris*.

LOCATION. Stomach.

This trematode occurs in the Red Sea, and has been found in the Atlantic (10° S.) and U.S.A. The specific characters of *D. tornatum* Rudolphi are, according to Looss: *Size:* up to 3·7 mm. long and 1·2 mm. broad when the ecsoma is retracted, 2–4·5 × 0·8–1·3 mm. when it is extended. *Cuticle:* annulated dorsally behind or near the ventral sucker, finely striated in small specimens. *Suckers:* of almost equal sizes, oral 0·5–0·7 mm. diameter, ventral 0·6–0·8 mm. *Reproductive systems:* sinus sac and hermaphrodite duct reaching the anterior and posterior borders of the ventral sucker respectively. False cirrus provided with hooklets. Eggs measuring 0·018–0·020 × 0·012–0·015 mm.

Notes. The differences on which the last three species listed in the synonymy above were erected are insufficient to establish their specific identity. Host affiliations also suggest that they are synonyms of *D. tornatus.* Dubois (1933) found this species in the dolphin. Linton (1940) gave an account of this and described a new species, *D. pinguis*, at Woods Hole.

Genus *Ectenurus* Looss, 1907

Diagnosis. Distinguished from *Dinurus* mainly by the limitation of prostate gland cells to a short region of the ejaculatory duct, of which the main part is free of them, and by the short, only slightly convoluted vitelline tubules.

Ectenurus lepidus Looss, 1907

Hosts. Horse mackerel, Spanish mackerel, angler, etc.

Location. Stomach.

This species was discovered at Trieste and has appeared in Britain (at Aberdeen) in the horse mackerel (see Nicoll, 1915*a*). According to Looss it is 2 mm. long and 0·25–0·3 mm. broad and thick, having a pair of dorsolateral outgrowths of the anterior extremity and markedly dissimilar suckers (diameters 0·08–0·12 and 0·2–0·3 mm.). The almost colourless eggs measure about 0·02 × 0·01 mm. *E. virgula* Linton (1910) occurs in America.

Genus *Lecithocladium* Lühe, 1901

Lecithocladium excisum (Rudolphi, 1819) Lühe, 1901 (Fig. 43 C)

Syn. *Distoma excisum* Rudolphi, 1819; *Lecithocladium excisiforme* Cohn, 1903; *Distoma cristatum* Rudolphi, 1819; *Lecithocladium cristatum* (Rudolphi) of Looss, 1907; ? *Distomum crenatum* Molin, 1859; *D. gulosum* Linton, 1901 of Johnstone, 1906; ? *Lecithocladium crenatum* (Molin) of Looss, 1907; ? *L. gulosum* (Linton, 1901) of Looss, 1907, and Linton, 1940.

Hosts. Mackerel, Spanish mackerel, etc. Immature flukes in horse mackerel, bogue, angler, red-band fish, etc.

Location. Stomach.

This trematode was discovered at Trieste, and occurs at Millport, Aberdeen, Plymouth, Walney Island (Barrow-in-Furness), also at Naples, in America and Japan. Markowski (1933*a*) found a specimen 6 mm. long in the mackerel in the Baltic Sea. According to Looss (1908) the form '*L. excisum*' has the following characters: *Size:* 3–4 mm. long when the ecsoma is retracted, 6–8 mm. when it is extended, 0·5–1·0 mm. broad and thick. *Cuticle:* annuli characteristically curled or crinkled above the pharynx (but not in specimens less than 1–1·5 mm.

long). *Neck-hump* ('Nackenbuckel'): feebly developed. *Suckers:* oral, funnel- or flask-shaped and much larger than the ventral, dimensions 0·65 × 0·5 and 0·35 × 0·45 mm. diameter. *Gut:* pharynx 0·5 mm. long and 0·25 mm. broad. *Reproductive systems:* genital pore near the mouth. Pars prostatica long and convoluted, seminal vesicle spindle-shaped; both situated between the ventral sucker and the testes. Vitellaria with long, thread-like processes, three on one side, four on the other. Uterus having long limbs which are much folded and extend into the ecsoma. Eggs measuring 0·02–0·022 × 0·010–0·012 mm. (0·019–0·027 × 0·011 mm. according to Markowski (*loc. cit.*)).

Notes. Looss himself believed the form '*L. excisiforme*' to be a contracted form of '*L. excisum*' and stated that '*L. cristatum*' differs from the latter in the more equal sizes of the suckers (of which the oral is cup-shaped), slightly smaller pharynx, constant 'neck-hump' not having crinkled annuli, and slightly smaller eggs, all of which characters suggest (when examined in detail) that the two forms are synonymous. Looss also regarded both *L. crenatum* and *L. gulosum* as *species inquirenda* because incompletely described. The latter is an American form which Linton (1940) referred to the genus *Ectenurus*. Johnstone's specimens from the mackerel were very large (10 mm. long and 1·1 mm. in greatest breadth), but had relatively small suckers (diameters 0·6 and 0·5 mm.) with their centres 1·8 mm. apart. The cuticle was described merely as 'unarmed' and the 'neck-hump' was not mentioned, but annuli were indicated in a figure both dorsally and ventrally in the anterior region and there is no doubt about the identity of the specimens with this species.

Subfamily **Lecithasterinae** Odhner, 1905

Small forms without a typical ecsoma and with a 7-lobed, star-shaped vitellarium

KEY TO GENERA OF LECITHASTERINAE

I. Body spindle-shaped; folds of the uterus mostly close behind the ventral sucker *Lecithaster*

II. Body broad and cylindrical behind the ventral sucker; folds of the uterus mainly in the posterior region of the body *Aponurus*

Genus *Lecithaster* Lühe, 1901

Syn. *Leptosoma* Stafford, 1904; *Mordvilkoviaster* Pigulewsky, 1938.

Lecithaster confusus Odhner, 1905

Syn. *Distomum mollissimum* of Stossich, 1899, *nec* Levinsen, 1881; *Apoblema mollissimum* of Looss, 1896, in part; *Hemiurus bothryophorus* of Looss, 1899, *nec Distoma bothryophoron* Olsson, 1869; *Derogenes cacozelus* Nicoll, 1907.

Hosts. Herring, twaite shad.

Location. Rectum.

This species occurs at Trieste, in Egypt and northern European waters, and on the Atlantic coast of north America, but has not appeared in Britain. According to Looss (1908) it is 1·2 mm. long and 0·4 mm. in greatest breadth and thickness, has a distinct, but non-muscular lip, ventral and oral suckers 0·25–0·27 and 0·14–0·15 mm. diameter, a long pars prostatica having few spaced gland cells, a four-lobed ovary, vitellaria with relatively short and slightly lobed processes and eggs measuring 0·015–0·017 × 0·009 mm.

Notes. Linton (1940) listed the hosts of this trematode near Woods Hole and gave details regarding the structure of American specimens. Hunninen & Cable (1943*b*) showed that it has small cystophorous cercariae which develop in long, constricted rediae in the digestive gland of *Odostomia trifida* in U.S.A., also that copepods of the genus *Acartia* ingest the cercariae, which develop into metacercariae in the haemocoele. Experimental infection of fishes which serve as the final hosts was also achieved. See this paper also for illustrations (Figs. 15–20) of the general structure and the details of the excretory system in the adult.

Lecithaster gibbosus (Rudolphi, 1802) Lühe, 1901 (Fig. 41 C)

Syn. *Fasciola gibbosa* Rudolphi, 1802; *Distoma gibbosum* of Rudolphi, 1809; *D. bergense* Olsson, 1868; *D. mollissimum* of Levinsen, 1881, *nec* Stossich, 1899; *D. botryophoron* Olsson, 1868; *Apoblema mollissimum* (Levinsen) of Looss, 1896; *Lecithaster bothryophorus* (Olsson) of Stafford, 1904.

Hosts. According to Looss, whiting, mackerel, herring, gar-fish. Comber, red gurnard, grey gurnard, lesser weever, horse mackerel, John Dory, dragonet, two-spotted goby, red-band fish, ballan wrasse, lesser sand eel, halibut, long rough dab, Norwegian topknot, common sole, flounder, dab, plaice, trout, smelt, gar-fish, sprat (in Britain).

Location. Intestine.

This species has been found at Plymouth, Foulney Island (Barrow-in-Furness), St Andrews, Millport, Aberdeen, Cullercoats and Galway, also at Egedesminde, in Greenland and Canada, and on the Atlantic coast of U.S.A. According to Looss it differs from the preceding in being of larger size, and having smaller suckers, a shorter pars prostatica and seminal vesicle, longer vitelline lobes and larger eggs. The specific characters are: *Size:* 1·75 mm. long and 0·5 mm. in greatest breadth and thickness. *Suckers:* 0·18–0·25 and 0·1–0·14 mm. diameter. *Reproductive systems:* pars prostatica so short that the seminal vesicle lies above and slightly behind the ventral sucker. Ovary four-lobed, the lobes hardly longer than their breadth. Vitellaria having processes which are clearly longer than wide. Eggs (according to Odhner) measuring 0·025–0·027 × 0·013 mm. Lebour (1908*a*) described specimens from the whiting and grey gurnard having the following characters: *Size and shape:* about 1·2 mm. long and 0·4 mm. in greatest breadth, elongate or spindle-shaped, having a constriction posteriorly which simulates a short ecsoma. *Suckers:* ventral twice as large as the oral and situated in front of the mid-body, diameters 0·20 and 0·10 mm. *Gut:* pharynx 0·06 mm. long, oesophagus very short, caeca long and narrow. *Excretory system:* median canal meeting the lateral canals near the mid-body, lateral canals uniting behind the pharynx. *Reproductive systems:* genital pore median, near the oesophagus. Testes globular, above and slightly behind the ventral sucker (figured slightly differently), vasa deferentia short and joining the seminal vesicle separately, pars prostatica well developed, the gland cells large, seminal vesicle spherical and situated on the left of the ventral sucker. Ovary four-lobed, receptaculum seminis large and ovoid, vitellaria forming a seven-lobed mass behind the ovary, eggs pale greenish brown, of undetermined size.

Note. For descriptions of American specimens see Manter (1926) and Linton (1940).

Another species, *L. stellatus* Looss, 1907, was found in the dentex and *Maena vulgaris* at Trieste, and occurs in Japan. Yet another species, *Lecithaster galeatus* Looss, 1907, discovered in the golden grey mullet off the coast of Egypt, was referred by Pigulewsky (1938) to the new genus *Mordvilkoviaster*, one distinctive character of which was said to be the non-lobed, oval or spherical vitellaria. The erection of this genus was unwarranted and *Mordvilkoviaster* is relegated to synonymity with *Lecithaster*. Pigulewsky (1938) also described a new species of *Lecithaster*, *L. tauricus* (discovered in the anchovy in the Black Sea), and other species are known in America and Japan.

Genus *Aponurus* Looss, 1907

Aponurus laguncululus Looss, 1907

HOSTS. Gar-fish and other fishes (anchovy, rarely greater weever, red mullet).

LOCATION. Stomach and oesophagus.

This species was discovered at Trieste and is said to be common on the Continent, but is unknown in Britain. According to Looss it is about 1 mm. long and 0·25 mm. in greatest breadth. The oral sucker is only half as large as the ventral (diameters 0·1 and 0·2 mm.), the vitellaria comprise large rounded follicles and the eggs measure about 0·027 × 0·016 mm.

Notes. Another species, *A. tschugunovi* Vlasenko, 1931, occurs in the red mullet in the Black Sea, *A. trachinoti* Manter, 1940 occurs in Mexico and *Aponurus* sp. in the Woods Hole district of U.S.A.

Two other genera of Lecithasterinae might be mentioned briefly: *Lecithophyllum* Odhner, 1905 (type *L. botryophoron* (Olsson, 1868) Odhner, 1905 (Syn. *Distoma botryophoron* Olsson, 1868; *D.* (*Brachylaimus*) *botryophoron* (Olsson) Stossich, 1886)) which Looss distinguished from *Lecithaster* mainly by the elongate genital sinus, Odhner by the size of the egg; and *Lecithurus* Pigulewsky, 1938 (type *L. lindbergi* (Layman, 1930), discovered in Peter the Great Bay). The latter genus is said to be characterized by a ventral sucker which is half as large again as the oral, caeca extending into the posterior region and vitellaria with long, finger-like lobes. Both can only be regarded as *genera inquirendae*.

Subfamily **Derogenetinae** Odhner, 1927

Ecsoma absent, vitellaria compact, but paired, cirrus-pouch absent.

KEY TO GENERA OF DEROGENETINAE

I. Ventral sucker situated in the posterior part of the body; eggs ovoid — *Derogenes*

II. Ventral sucker situated slightly in front of the mid-body; eggs each drawn out into a sharp point at one pole — *Derogenoides*

Genus *Derogenes* Lühe, 1900

The type species, *D. ruber* Lühe, 1900, was discovered in the gall bladder of the streaked gurnard.

Derogenes varicus (Müller, 1784) Looss, 1901 (Fig. 41 B)

Syn. *Fasciola varica* Müller, 1784 and of Rudolphi, 1802; *Distoma varicum* Zeder of Rudolphi, 1809; *Distomum dimidiatum* Creplin, 1829, in part; *Derogenes varicum* (Müller) of Olsson, 1868 and Levinsen, 1881, *nec* Monticelli, 1890; *D. minor* Looss, 1901; *D. plenus* Stafford, 1904; *D. fuhrmanni* Mola, 1912; *D. crassus* Manter, 1934.

HOSTS (in Britain). Common sea bream, short-spined cottus, long-spined cottus, yellow gurnard, red gurnard, grey gurnard, angler, greater weever, lesser weever, horse mackerel, boar fish, John Dory, dragonet, lumpsucker, butterfly blenny, cod, bib, poor cod, whiting, coalfish, pollack, common ling, hake, three-bearded rockling, torsk, hallibut, long rough dab, turbot, brill, Norwegian topknot, common sole, flounder, dab, lemon sole, plaice, salmon, trout, houting, grayling, sprat, conger, sturgeon.

LOCATION. Stomach, sometimes intestine; may be regurgitated into the buccal cavity.

DIAGNOSIS. *Shape, size and colour:* spindle-shaped, broadest in the posterior region, 1·5–7 mm. long (usually 1·5–3 mm.), colourless except for the eggs. *Cuticle:* non-spinous, but generally wrinkled anteriorly, and showing transverse striae. *Suckers:* ventral about twice as large as the oral and situated far back behind the mid-body, diameters 0·33–0·4 and 0·19–0·24 mm. in individuals 1·5–2 mm. long (about 0·6 and 0·3 mm. in individuals 4·5 mm. long). *Excretory system:* median and lateral canals meeting at a point close behind the ventral sucker, lateral ones extending to the level of the oesophagus, above which they unite. *Gut:* prepharynx absent, oesophagus shorter than the pharynx, caeca long, extending almost to the posterior extremity. *Reproductive systems:* genital pore far in front of the ventral sucker, close behind the bifurcation of the intestine. Hermaphrodite duct short, its wall containing a few isolated muscles (i.e. forming a rudimentary incomplete sinus sac of Lloyd, 1938). Pars prostatica long (extending to the midpoint between the genital pore and the ventral sucker), straight, with very large and conspicuous cells, free in the parenchyma. Vesicula seminalis dilated. Testes ovoid, diagonally side by side close behind the ventral sucker. Ovary spherical, between and behind the testes, lateral. Vitellaria spherical or ovoid, diagonally placed in the extreme posterior region. Uterus with descending and ascending limbs, abundantly folded, filling up the posterior region. Eggs golden yellow, 0·050–0·062 × 0·025–0·034 mm. (possibly larger, up to 0·06 × 0·038 mm.), always largest in that part of uterus in front of the ventral sucker.

Notes. This is the most widely distributed digenetic trematode of fishes, having been found at St Andrews, Millport, Aberdeen, in the Tweed (in the salmon; see Tosh, 1905), off the coast of Cumberland, at Galway, Plymouth, Trieste, near Banyuls, as well as in the Russian Arctic, Canada, U.S.A. and the Galapagos Isles. Forms not mentioned above which have been regarded as species of *Derogenes* include *D. affinis* (Rudolphi, 1819) Lühe, 1901 (in *Scorpaena cirrosa* at Arimini) and *D. urocotyle* (Parona, 1899) Odhner, 1905 (in *Scorpaena scrofa* at Portoferrajo).

Mention can be made at this point also of the genus *Genarches* Looss, 1902, especially of *Genarches mülleri* (Levinsen, 1881) Looss, 1902 (Syn. *Distoma mülleri* Levinsen, 1881; *Progonus mülleri* (Levinsen) Looss, 1899), which was discovered in *Cottus scorpius* and *Gadus ovak* at Egedesminde. This species bears a remarkable resemblance to *Derogenes varicus*, differing chiefly in the structure of the terminal parts of the genital ducts. It occurs also in America, where there is at least one other species, *Genarches infirmus* Linton, 1940.

Genus *Derogenoides* Nicoll, 1913

Derogenoides ovacutus Nicoll, 1913 (Fig. 28M)

HOST. Greater weever (at Plymouth).

LOCATION. Stomach.

DIAGNOSIS. Closely similar to *Derogenes varicus*, but showing several differences which Nicoll deemed of generic value, viz. rarely exceeds 0·6 mm. long when mature, extremities rounded, suckers of the same proportions (0·123 and 0·066 mm. diameter), but the oral with a small projecting anterior lobe, pars prostatica smaller, prostate cells fewer, genital sinus cylindrical instead of globular, seminal vesicle smaller, ventral sucker relatively farther forward, eggs much smaller (0·033–0·042 × 0·015–0·019 mm.; mean 0·038 × 0·018 mm.) and with a pointed anopercular pole. *Derogenoides skrjabini* Vlasenko, 1931 occurs in the intestine of the three-bearded rockling in the Black Sea.

Subfamily **Syncoeliinae** Looss, 1899

Syn. Syncoliinae Fuhrmann, 1904.

Similar to Derogenetinae, but cirrus pouch present.

Genus *Hemipera* Nicoll, 1913

Hemipera ovocaudata Nicoll, 1913 (Fig. 28L)

HOST. Cornish sucker, three-bearded and five-bearded rocklings (at Plymouth).

LOCATION. Stomach.

DIAGNOSIS. *Shape and size:* elongate-oval outline, almost cylindrical, 1·5 mm. long and 0·56 mm. broad. *Cuticle:* non-spinous. *Suckers:* ventral much larger than the oral and distinctly behind the mid-body, 0·9 mm. from the anterior extremity, diameters 0·4 and 0·22 mm. *Gut:* prepharynx absent, pharynx small (0·066 mm. diameter), caeca long. *Reproductive systems:* genital pore far in front of the ventral sucker, close to the bifurcation of the intestine. Cirrus pouch enclosing only the ejaculatory duct and pars prostatica. Vesicula seminalis *externa* present. Testes ovoid, side by side near the posterior extremity. Ovary in front of the testes, on the right. Vitellaria compact, rounded, one on either side of the ovary. Uterus little folded. Eggs large, not numerous (largest specimens with less than thirty in uterus at any time) slightly curved, measuring about 0·100 × 0·027 mm. and having a filament about 0·2 mm. long at the anopercular pole.

Hemipera sharpei Idris Jones, 1933 (Fig. 42B)

HOST. Red-band fish (at Plymouth).

LOCATION. A solitary specimen was found under the gill cover.

DIAGNOSIS. *Shape and size:* elongate, more bluntly rounded posteriorly than anteriorly, 4·8 mm. long and 0·85 mm. broad. *Suckers:* ventral twice as large as the oral and situated far back in the body (in the posterior third), diameters 0·74 and 0·37 mm. *Gut:* as in *H. ovocaudata*. *Reproductive systems:* gonads and vitellaria more compacted and nearer the posterior extremity than in *H. ovocaudata*, testes slightly diagonal and contiguous; ovary in front of the more anterior testis and median; vitellaria between the testes and the ovary, lateral and compact; uterus formed into short transverse folds in front of the ventral sucker. Eggs not curved, measuring 0·100 × 0·038 mm., each having a filament about twenty times as long as the capsule proper at the anopercular pole. Other differences from *H. ovocaudata*: presence of Laurer's canal and prominent receptaculum seminis.

Note. In the genus *Syncoelium* Looss, 1899 the crura of the gut unite posteriorly to form a ring-like intestine (see Fuhrmann, 1928, Fig. 55).

Family **BUNODERIDAE** Nicoll, 1914

Syn. Bunoderinae Looss, 1902.

Genus *Bunodera* Railliet, 1896

Bunodera luciopercae (Müller, 1776) Lühe, 1909 (Fig. 29D)

Syn. *Fasciola luciopercae* Müller, 1776; *F. percae cernuae* Müller, 1776; *Planaria lagena* Braun, 1788; *Fasciola percae* Gmelin, 1790; *F. percina* Schrank, 1790; *F. nodulosa* Frölich, 1791; *Distoma nodulosum* Zeder, 1800, and of Looss, 1894; *Crossodera nodulosa* of Cobbold, 1860; *Bunodera nodulosa* of Looss, 1899.

HOSTS. Perch, pope, barbel, pike, sea trout (and others on the Continent).

LOCATION. Intestine.

This species has been recorded in the perch in Hertfordshire and also occurs in Scotland, Denmark, Sweden, France, Russia, Germany, as well as in Canada and U.S.A.

DIAGNOSIS. *Shape and size:* lancet-shape, with a tapering anterior region (more attenuated than in the figure), 1–4·5 mm. long. *Suckers:* oral with six anterior processes (two dorsal, two lateral and two ventral); ventral just in front of the mid-body. *Gut:* pharynx small, oesophagus long, caeca long. Bifurcation of the intestine well in front of the ventral sucker in extended, but slightly behind it in contracted specimens. *Reproductive systems:* genital pore slightly in front of ventral sucker. Cirrus pouch membranous, not muscular, containing a small pars prostatica and a cirrus which is little differentiated from the ejaculatory duct. Testes ovoid, one behind the other in the extreme posterior region, lateral. Ovary a short distance in front of the testes. Vitellaria extending from the pharynx to near the posterior extremity, follicles rather coarse. Uterus with short descending and long ascending limbs, underlying the gonads, slightly folded, but occupying available space between the caeca posteriorly. Eggs brown, measuring 0·100 × 0·05 mm., each containing (in larger specimens, but not smaller ones, according to Looss) a miracidium.

Notes. Remains of the larval eye-spots persist for a time as two very small masses of pigment, slightly behind the oral sucker; more conspicuous in young than in older flukes.*

Hopkins (1934), after examining both European and American specimens of this species, accepted the description of Looss (1894*a*) with three slight reservations: (i) the excretory vesicle is tubular, not forked, and extends forward to the level of the receptaculum seminis; (ii) the vitellaria are situated lateral to the uterus and both ventral and lateral to the caeca, generally extending to the posterior extremity; and (iii) the eggs are of smaller size, measuring 0·063–0·082 × 0·038–0·051 mm. This writer gave a diagnosis based on specimens from both continents, but regarded this species as one of the most variable he had studied, mainly because of changes in form consequent upon enlargement of the uterus, emphasizing the fact that were it not for the existence of forms indicating intermediate conditions extreme forms could well be regarded as distinct species. The size of the eggs seems to vary even more than Hopkins realized, because Linton (1940) gave the range as 0·045–0·069 × 0·030–0·045 mm. in American specimens (named '*Bunodera nodulosa* (Froelich)') 1·12 mm. long, and specified the size 0·077 × 0·042 mm. for slightly larger specimens 2·25 mm. long.

One other species of the genus is *Bunodera sacculata* Van Cleave & Mueller, 1932, a parasite of *Perca flavescens* in U.S.A.

Family **GORGODERIDAE** Looss, 1901 (p. 93)

Syn. *Bunoderinae* Looss, 1902.

Subfamily **Gorgoderinae** Looss, 1901

Genus *Phyllodistomum* Braun, 1899

Syn. *Spathidium* Looss, 1899; *Catoptroides* Odhner, in Looss, 1902; *Microlecithus* Ozaki, 1926.

Looss (1902*c*, p. 862) credited Odhner with the genus '*Catoptroides*' and gave as its diagnostic characters the sharp demarcation between the fore-body and the spatulate hind-body and the almost symmetrical arrangement of the testes. Odhner (1910*a*) used a different combination of characters, the grouping of the gonads slightly behind the ventral sucker, the symmetrical arrangement of the testes and the position of the vitellaria beside and a little behind the ovary, not in front of it, attaching little importance to the general shape. Soon afterwards (1911) he transferred '*Catoptroides macrocotyle* Lühe, 1909' (Syn. *Phyllodistomum folium* of Sinitsîn, 1905) to *Phyllodistomum* and expressed the opinion that *P. angulatum* Linstow, 1907 (Syn. *Catoptroides angulatus* (Linstow) Lühe, 1909) is an immature form of the same species. Species remaining in *Catoptroides* were *C. spatula* (Odhner, 1902) and *C. spatulaeforme* (Odhner, 1902), which were described as species of *Phyllodistomum* discovered in the urinary bladder of catfish in the Sudan, and which are probably identical. Neither they nor other forms since allocated to *Catoptroides* can be distinguished from species of

* For accounts of the life cycle, see Layman (*Bull. Soc. Nat. Moscow* (Biol.) **49,** 1940, 173–80) and Komarova (*C.R. Acad. Sci. Moscow*, **31**, 1941, 184–5).

Phyllodistomum by any generic characters put forward, despite attempts like that of Loewen (1929) to revive the genus *Catoptroides*, which, once and for all, must fall as a synonym of *Phyllodistomum*.

Phyllodistomum folium (v. Olfers, 1817) Braun, 1899 (Fig. 29E)

Syn. *Distoma folium* v. Olfers, 1817, and of Rudolphi, 1819 and Looss, 1894; *Spathidium folium* (Olfers) Looss, 1899; *Phyllodistomum pseudofolium* Nybelin, 1926.

HOSTS. Three-spined stickleback (Cambridgeshire), ruffe (pope), miller's thumb, grayling, pike (in central Europe).

LOCATION. Urinary bladder and ureters.

DIAGNOSIS. *Shape and size:* spatula-like, broad posteriorly, tapering gradually towards the anterior end, up to 2 mm. long. (According to Lühe the forebody is 0·25–0·33 mm. broad, the hindbody three times this, but Zandt (1924) found the differences less marked and gave the breadths 0·3 and 0·43 mm.) *Suckers:* ventral only slightly larger than the oral and situated slightly in front of the mid-body, diameters about 0·16 mm. (0·112–0·247 and 0·185–0·25 mm. according to Zandt). *Reproductive systems:* testes rounded, only slightly indented, obliquely set and fairly close together, about twice as large as the ovary, which is of similar shape and is situated on the right. Vitellaria compact and of similar shapes, one on either side close behind the ventral sucker. Uterus profusely coiled in the posterior region, mainly between the caeca. Eggs numerous, measuring 0·035 × 0·018 mm. when newly formed, but much larger (0·053 × 0·031 mm.) in some parts of the uterus (according to Zandt, 0·058–0·06 × 0·031–0·035 mm.).

Notes. This species has also been found in Sweden and Canada. The specimens described by Nybelin as *P. pseudofolium* come within the range of variability of the species, so that this name must be regarded as a synonym of *P. folium*. According to Lühe (1909) the cercaria is of rhopalocercous type and is *Cercaria duplicata* (Baer), six to eight cercariae developing at a time in egg-like sporocysts 0·66–1·0 mm. long in fresh-water mussels, each of which may harbour up to five thousand larvae.

Species not recorded in Britain, but occurring in Europe:

Phyllodistomum conostomum (Olsson, 1876)

Syn. *Distoma conostomum* Olsson, 1876.

HOST. Houting (in Sweden).

LOCATION. Oesophagus and on the gills.

DIAGNOSIS. *Shape and size:* elongate, broad posteriorly and tapering towards the anterior end, 5 mm. long and 1·5 mm. broad. *Suckers:* ventral slightly the larger and situated in the anterior third of the body, diameters 0·44 and 0·38 mm. *Reproductive systems:* testes lobed, two or three times as large as the ovary, well separated in the middle of the hind-body. Ovary trilobate, slightly larger than the vitellaria and well behind them, almost opposite the anterior testis. Vitellaria and uterus as in *P. folium*, eggs measuring about 0·041 × 0·027 mm.

Note. Probably identical with *P. folium*.

Phyllodistomum acceptum Looss, 1901

HOSTS. *Crenilabrus pavo*, *C. griseus* (at Trieste).

LOCATION. Bladder.

DIAGNOSIS. *Shape and size:* fore-body and hind-body sharply delimited in life, otherwise broad posteriorly and tapering anteriorly, up to 7·2 mm. long and 3·4 mm. in greatest breadth. *Colour:* golden red. *Cuticle:* beset with irregular nodules ('Knötchen'), especially in the fore-body. *Suckers:* of about equal size (0·6 mm. diameter), the ventral slightly in front of the mid-body. *Reproductive systems:* testes small, only slightly lobed, diagonally set and widely separated, about twice as large as the globular or slightly indented ovary. Vitellaria larger than the ovary, deeply lobed and separated from the ventral sucker by folds of the uterus, which overreach the caeca laterally. Eggs measuring 0·038 × 0·024 mm. (misprinted by Looss as 0·38 × 0·24 mm.).

Note. Probably identical with *P. folium*.

Phyllodistomum macrocotyle (Lühe, 1909) Odhner, 1911

Syn. *Phyllodistomum folium* of Sinitsîn, 1905, *nec* Olfers, 1817; *Catoptroides macrocotyle* Lühe, 1909; ? *Phyllodistomum angulatum* Linstow, 1907; ? *Catoptroides angulatus* (Linstow) Lühe, 1909.

HOSTS. Crucian carp, barbel, gudgeon, rudd, roach, chub, bream, pike, white bream (near Warsaw).

LOCATION. Ureters, rarely the bladder.

DIAGNOSIS. *Shape and size:* hind-body four to six times as broad as the fore-body in juveniles, three and a half times as broad (hardly one-third of the body length) in adults, pointed at the posterior extremity, 2–4 mm. long, rarely 5 mm. *Suckers:* ventral at or a little in front of the mid-body and about twice as large as the oral, but the ratio of the diameters very variable (1 : 1 in individuals less than 1 mm. long; 3 : 2 in individuals 1 mm. long, thereafter increasing to 11 : 6 and even 5 : 2). *Reproductive systems:* ovary having three to six lobes, but generally four, situated slightly behind one of the vitellaria. Testes deeply lobed, symmetrically or diagonally set. Eggs increasing in size as they pass through the uterus, from 0·032 × 0·018 to 0·054 × 0·036 mm.

Notes. Five or six specimens were taken from a single host as a rule, but in some instances ten to fifteen and, exceptionally, as many as forty. The cercariae are of the microcercous type (the tail hardly one-tenth as long as the body) and develop twelve to fourteen at a time in sporocysts, in which they later encyst. The intermediate host is *Dreissena polymorpha*. Sporocysts containing encysted metacercariae leave the mussel by way of the gills and are eaten by Cyprinid fishes serving as the definitive hosts.

Phyllodistomum simile Nybelin, 1926

Syn. *Phyllodistomum megalorchis* Nybelin, 1926.

HOSTS. Miller's thumb, grayling, burbot, sea trout (in Sweden).

LOCATION. Urinary bladder.

DIAGNOSIS (data for *P. megalorchis* in parentheses). *Shape:* fore-body narrow, sharply separated from the broadly oval hind-body (the same). *Size:* 0·82–

2·2 mm. long (3·5–5·4 mm.) and 0·54–1·2 mm. broad (1·6–2·3 mm.). *Suckers:* ventral slightly larger than the oral, diameters 0·21–0·34 and 0·16–0·25 mm. (0·33–0·44 and 0·26–0·35 mm.), ratio of the diameters 1 : 1·3–1 : 1·36 (1 : 1·25–1 : 1·26). *Reproductive systems:* testes slightly lobed (more deeply lobed), up to twice as large as the ovary (more than twice as large), diagonally one behind the other (closer together). Ovary irregularly oval (slightly lobed). Vitellaria oval, less than half as large as the ovary (approximately half the size) and near the ventral sucker. Uterus much folded, the folds overstepping the caeca laterally (the same). Eggs measuring 0·033–0·037 × 0·022–0·026 mm. (0·059–0·063 × 0·040–0·048 mm.).

Note. The slightly different characters of *P. megalorchis* are correlated with differences in size, and the two forms should be regarded as identical. In the ultimate analysis *P. simile* is likely to prove identical with *P. folium.*

Phyllodistomum elongatum Nybelin, 1926 *sp. inq.*

Hosts. Tench, bream (in Sweden).

Location. Ureters.

Diagnosis. Differing from *P. simile* mainly in shape, the elongate hind-body being little broader than the fore-body and separated from it by a constriction, and the small size of the gonads, particularly the testes. *Size:* 1·65–3·5 mm. long and 0·38–0·84 mm. broad. *Suckers:* of almost equal size, the oral a trifle the smaller, diameters 0·165–0·23 and 0·165–0·255 mm. *Eggs:* 0·055 × 0·033 mm.

Note. This form is almost certainly identical with *P. simile* and will no doubt ultimately prove to be nothing but an extended form of *P. folium.*

Other Species of *Phyllodistomum*:

About three dozen species unmentioned so far have been referred to this genus, more than half of them American, others scattered throughout the world in Japan, China, India, Russia and Africa. Two or three of them occur in amphibians, others in fresh-water shrimps, but the great majority occur in fishes. Nybelin (1926), Holl (1929), Lewis (1935), Wu (1938), Fischthal (1942*b*, 1943) and Meserve (1943) have made attempts to review the genus, Meserve confining his attention to the parasites of Amphibia, Lewis compiling a table of measurements and drawing or redrawing figures of those known to his time (though, unfortunately, about nineteen new species have been described since then). Some of these 'species' have fallen as synonyms of *P. folium* or other species, but a long list of vaguely defined forms still remains. Many of the species still standing will undoubtedly share a similar fate when someone reviews the genus as Rankin (1938) reviewed species of *Brachycoelium.* It is safe to infer from the available descriptions that *Phyllodistomum folium* will closely rival trematodes like *Otodistomum veliporum* in the wideness of its distribution and *Bunodera luciopercae* in the extent of its variability. For what it is worth, the following alphabetical list of species not included in the diagnoses above is given, together with localities, where these are not U.S.A. or elsewhere in north America. Species marked * occur in amphibians, that marked † in shrimps:

**P. almorii* Pande, 1937 (India); **P. americanum* Osborne, 1903; *P. brevicaecum* Steen, 1938; *P. catostomi* Wu, 1938; *P. caudatum* Steelman, 1938; **P. coatneyi* Meserve, 1943; *P. cotti* Wu, 1938; *P. etheostomae* Fischthal, 1942; *P. fausti* Pearse, 1924; *P. kajika*

(Ozaki, 1926) (Japan); *P. lacustri* (Loewen, 1929); †*P. lesteri* Wu, 1938 (China); *P. linguale* Odhner, 1902; *P. lohrenzi* (Loewen, 1935); *P. lysteri* Miller, 1940; *P. macrobranchicola* Yamaguti, 1934 (Japan); *P. marinum* Layman, 1930 (Peter the Great Bay); *P. mogurndae* Yamaguti, 1934 (Japan); *P. nocomis* Fischthal, 1942; *P. parasiluri* Yamaguti, 1934 (Japan); *P. patellare* (Sturges, 1897) Braun, 1899 (Syn. *P. enterocolpum* Holl, 1930); *P. pearsei* Holl, 1929; *P. sinensis* Wu, 1937 (China) ? *nomen nudum*; *P. singulare* Lynch, 1936; *P. semotili* Fischthal, 1942; *P. shandrai* Bhalerao, 1937 (India); *P. solidum* Rankin, 1937; *P. spatula* Odhner, 1902 (Sudan); *P. spatulaeforme* Odhner, 1902 (Sudan); *P. staffordi* Pearse, 1924 (Syn. *P. carolini* Holl, 1929; *P. hunteri* (Arnold, 1934); *Catoptroides hunteri* Arnold, 1934); *P. superbum* Stafford, 1904; *P. umbrae* Wu, 1938; *P. undularis* Steen, 1938; *P. unicum* Odhner, 1902 (Red Sea).

Subfamily **Anaporrhutinae** Looss, 1901

Genus *Probilotrema* Looss, 1902

According to Dollfus (1935*c*) this genus contains two species, *P. richiardii* (Lopez, 1888) and *P. capense* Looss, 1902, which are readily distinguished by the size relationships of the suckers; in the latter species the suckers are of approximately equal sizes, in the former the ventral is much the larger. *P. richiardii* is known only in the Mediterranean, interestingly, as a parasite in the coelom of various Selachii (piked dogfish, smooth hound, eagle ray). *P. capense* seems to have been found once only, in the coelom of a dogfish near the Cape of Good Hope. Dollfus (*loc. cit.* and 1937) gave the characters of a specimen of *P. richiardii* which was found in the coelom of the piked dogfish at Algiers. Other species of the genus exist elsewhere; Woolcock (1935) described two new species, *P. philippi* and *P. antarcticus*, discovered in sharks in Port Phillip Bay (Australia). The genus is clearly distinguishable by the broad oval shape, numerous testes (twenty-six to thirty on each side of the body) lateral to the caeca, and small vitellaria (two lateral groups of follicles immediately in front of the testes).

Family **DIDYMOZOIDAE** Poche, 1907 (p. 92)

Syn. Didymozoonidae Monticelli, 1888.

Usually cyst-dwellers with a tendency towards unisexuality.

Notable attempts have been made to classify the unusual trematodes which belong to this family (Braun, 1879–93; MacCallum, 1917; Dollfus, 1926, 1935*c*; Ishii, 1935*b*), but there is little to be gained by attempting to define species or even genera at the present time. As Baylis has pointed out (1938*a*), early studies were incomplete and our knowledge is still limited, our ignorance profound, on the structure of forms well known by name. Thus, in describing *Nematobothrium filarina* Beneden (1858) did not see either gut or oral sucker, and the species still awaits redescription. Of the well-known genera in Europe, three are true hermaphrodites; they are *Nematobothrium* Beneden, 1858, *Didymozoon* Taschenberg, 1878 and *Didymocystis* Ariola, 1902, and the last is distinguishable from the former two by the wide posterior region of the body. Other genera which are well known have individuals which are male or female. Amongst them are *Wedlia* Cobbold, 1860, emend. Odhner, 1907 and *Köllikeria* Cobbold, 1860, of which only the latter is said to have a ventral sucker. Both have a wide, ovoid, globular or reniform posterior region.

Genus *Didymozoon* Taschenberg, 1879, *sensu* Odhner, 1907

Didymozoon scombri Taschenberg, 1879 (Fig. 44A)

Syn. *Nematobothrium* (*Benedenozoum*) *scombri* of Ishii, 1935.

HOST. Mackerel (in Britain), (Spanish mackerel also in Italy).

LOCATION. In cysts on the roof of the buccal cavity (between the pharyngobranchial elements), also on the external surfaces of the basi-branchials and other parts of the gill-bar, as well as on the internal surface of the operculum (Johnston, 1914).

This species was found in Italy, and occurs in west Sweden, the Irish Sea, at Plymouth and in U.S.A.

DIAGNOSIS. *Cysts:* generally isolated, rarely in groups, pale yellow (due to eggs showing through the body of the fluke and the wall of the cyst), about 4–5 × 2–3 mm., sometimes slightly larger. Some cysts contain one pair of individuals, some as many as sixteen worms, which lie with their thread-like bodies intertwined. *Worm extracted from cyst: Shape and size:* 15–20 mm. long and 0·5–1 mm. broad and thick (according to Baylis, up to more than 55 mm. long), anterior region with a narrow neck, posterior region thicker than the neck, but not expanded. *Suckers:* ventral absent; oral very small, 0·3–0·4 mm. diameter in individuals 25–35 mm. long (Odhner) or about 0·5 mm. diameter in specimens 15–20 mm. long (Baylis, based on Johnstone's figure). *Gut:* pharynx minute, oesophagus very slender, caeca probably long. *Reproductive systems:* genital pores separate, on a papilla close to the mouth. Cirrus pouch and pars prostatica absent, vasa deferentia and vas deferens short, testes short, tubular, anterior, continuous with their ducts. Ovary very elongate, tubular, overlapping the first and second thirds of the body, continuous with the oviduct. Vitellaria elongate, extending through the posterior region. Uterus with an initial ascending limb extending to the level of the testes, a descending limb extending to the posterior extremity, and a second ascending loop proceeding to the metraterm, which is short, narrow and thick-walled. Eggs small, rounded, thick-shelled, measuring 0·15 × 0·01 mm. according to Odhner, 0·016–0·0175 × 0·010–0·0105 mm. according to Baylis (1938*a*). Embryo as in *D. faciale*. Linton (1940) gave an account of American specimens of this species and of *D. sardae* (MacCallum & MacCallum, 1916).

Didymozoon faciale Baylis, 1938 (Fig. 44B)

HOST. Mackerel (in the English Channel).

LOCATION. In cysts beneath the outer skin of the head, behind the eye.

DIAGNOSIS. *Cysts:* yellowish, 3–7 mm. long and 1–1·5 mm. wide, each containing one to four worms. Cyst wall tough, worms very delicate and inseparable in preserved material. *Worms extracted from the cysts: Shape:* anterior region lancet-like, slightly flat, joined by 'narrow neck' with wider, cylindrical, posterior region. *Size:* 0·71–16·3 mm. long and 0·08–1·4 mm. broad (under coverslip pressure). Specimens less than 3 mm. long are generally without eggs, most of them having undifferentiated female organs. Testes well developed and functional even in specimens only 1·1 mm. long, but specimens 0·7 mm. long having undifferentiated testes. Proportionate sizes of fore- and hind-body changing with age, the latter undergoing much more rapid growth. *Suckers:* oral small, ovoid or pyriform, 0·004–0·008 mm. long and 0·003–0·005 mm. broad, ventral absent. *Gut:* pharynx small, oesophagus narrow, caeca probably long, but obscured by the uterus in the posterior region. *Reproductive systems:* genital pore situated on a rounded, muscular papilla ventral to the oral sucker. Metraterm median, equipped with circular muscles. Vas deferens dorsal to the metraterm and parallel

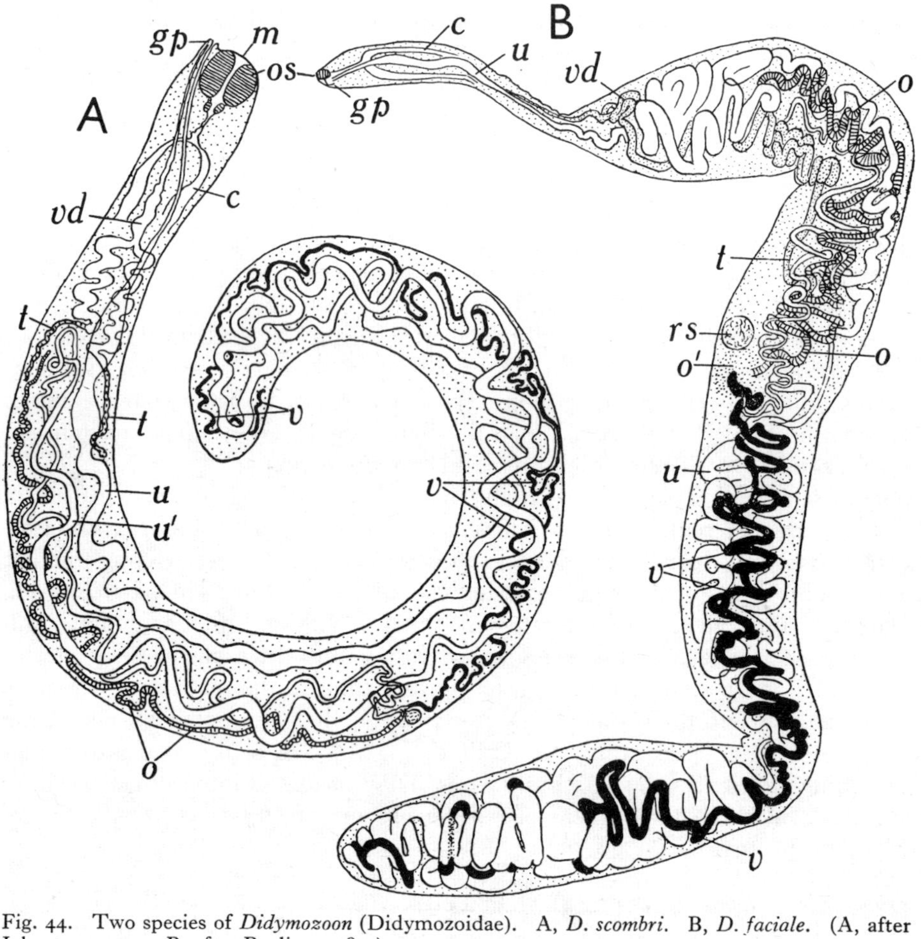

Fig. 44. Two species of *Didymozoon* (Didymozoidae). A, *D. scombri*. B, *D. faciale*. (A, after Johnstone, 1914; B, after Baylis, 1938*a*.)

with it, but narrower and more convoluted. Paired vasa deferentia situated slightly behind the neck region, continuous with the testes, which are long, tubular and convoluted, occupying most of the body in young individuals, but extending only about one-third of the distance along it in full maturity. Ovary very long and thread-like, extending from near the posterior ends of the testes to just in front of the mid-body. Receptaculum seminis globular, oviduct and ootype also situated just in front of the mid-body. Uterus extremely long and very much convoluted, completely filling the body, comprising an ascending limb extending from the ootype to just behind the neck region, a descending limb passing back to the posterior extremity, and a second ascending limb passing forward to the metraterm. Eggs extremely numerous, very small and thick-shelled, operculate, yellow, measuring 0·018–0·020 × 0·011 mm. Embryos hatched *in utero* having fine spinelets (long and short alternating) radially arranged in an anterior coronet.

OTHER EUROPEAN SPECIES OF *DIDYMOZOON*:

Didymozoon pelamydis Taschenberg, 1879

Syn. *Nematobothrium* (*Benedenozoum*) *pelamydis* (Taschenberg) of Ishii, 1935.

HOST. Pelamid (in Italy and near Algiers).

LOCATION. Between the gill lamellae.

DIAGNOSIS. *Size:* 30 mm. long. According to Dollfus (1935*c*) differs from *D. scombri* mainly in having a very long oesophagus, part of which is surrounded by distinct cells, the merging of the intestinal crura posteriorly, the position of the genital pore behind the oral sucker, and the existence of a solitary testis, which may fail to develop or disappear in one member of the encysted pair.

Didymozoon sphyraenae Taschenberg, 1879

Syn. *Didymozoum sphyraenae* Taschenberg of Ishii, 1935.

HOST. *Sphyraena vulgaris* (in Italy and Sicily).

LOCATION. Underneath the buccal mucous membrane.

DIAGNOSIS. *Size:* 15 mm. long and 1·5 mm. broad. *Other characters:* body short and compact, rounded posteriorly, having a well-marked neck region and an only slightly broadened anterior end. Pharynx and intestine present.

Didymozoon auxis Taschenberg, 1879

Syn. *Didymozoum auxis* Taschenberg of Ishii, 1935.

HOST. Plain bonito (in Italy).

LOCATION. On the outer side of the gill lamellae.

DIAGNOSIS. *Size:* 12 mm. long. Body long and narrow, bent in a right angle, sometimes having curved lateral margins, the neck region pen-shaped and very sharply marked off. Pharynx present, intestine absent.

Didymozoon taenioides Monticelli, 1888

Syn. *Monostoma molae* Rudolphi, 1819, in part; *Didymozoum taenioides* Monticelli of Ishii, 1935; *Nematobothrium* (*Didymozoon*) *taenioides* Monticelli of Maclaren, 1904.

HOST. Sunfish (in Italy and Belgium).

LOCATION. Muscles.

Didymozoon tenuicolle (Rudolphi, 1819)

Syn. *Monostoma tenuicolle* Rudolphi, 1819; *Didymozoon lampridis* Lönnberg, 1891; *Didymozoum tenuicolle* (Rudolphi) of Ishii, 1935.

HOST. Opah (in the North Sea and Sweden).

LOCATION. Gills and muscles.

Didymozoon pretiosus Ariola, 1902

Host. Tunny (in Italy).

Location. Gills.

This species is said to have a characteristic swelling at the junction of the anterior and posterior regions of the body, but, like the previous four species, is not adequately defined.

Genus *Nematobothrium* Beneden, 1858

Several species of this genus occur in Europe, but none in Britain.

Nematobothrium molae Maclaren, 1904

Syn. *Nematobothrium* (*Maclarenozoum*) *molae* Maclaren of Ishii, 1935.

Host. Sunfish (at Naples, Roscoff and near Algiers).

Location. Gills.

Diagnosis. *Cysts:* largest 4·5 mm. long, containing two inextricably intertwined worms forming a U-shaped mass 7 cm. long, one end containing the anterior ends of the worms being thicker than the other. *Trematode: Shape and size:* anterior end narrow and pen-like, 0·0675 mm. diameter at the pharynx, total length estimated at 1–1·5 m. *Suckers:* oral absent, ventral very small (0·01 mm. diameter) and situated about 1 mm. behind the mouth. *Gut:* pharynx pyriform, 0·055 mm. long, oesophagus short, caeca commencing just in front of the ventral sucker and ending just behind the origin of the vas deferens (i.e. relatively short). *Reproductive systems:* genital pores separate and near the mouth, the female pore behind the male. Gonads tubular, continuous with their ducts, extending throughout the body. Vitellaria very thread-like. Uterus capacious, its folds filling the hind-body. Eggs measuring about 0·02 × 0·015 mm.

Note. Dollfus found four cysts on the specified host near Algiers and gave the dimensions as 20–25 × 3–5 mm. He was unable to extricate the worms entire, but showed that the sizes of the eggs vary both in these specimens (0·0175–0·018 × 0·012–0·0125 mm.) and in others at Roscoff (0·0155–0·0195 × 0·011 mm.).

Other Species:

Nematobothrium filarina Beneden, 1858

Syn. *Nematobothrium* (*Benedenozoum*) *filarina* Beneden of Ishii, 1935.

Host. Shadow fish (in France and Belgium).

Location. Gills.

Nematobothrium benedeni (Monticelli, 1893) Maclaren, 1904

Syn. *Didymozoon benedeni* Monticelli, 1893.

Host. Sunfish (in Italy).

Location. Gills.

This species is similar to *N. molae*, indeed Maclaren showed his specimens to Monticelli believing them to belong to the present species, but Monticelli expressed the opinion that they represent a distinct species.

Nematobothrium guernei Moniez, 1891

Syn. *Nematobothrium (Benedenozoum) guernei* Moniez of Ishii, 1935.

HOST. Albacore (in Gascony).

LOCATION. Gills, muscles, intestine.

Genus *Didymocystis* Ariola, 1902

Syn. *Didymocistis* Ariola, 1902; *Didymostoma* Ariola, 1902.

Didymocystis wedli Ariola, 1902

Syn. *Monostoma bipartitum*, third form of Wagener, 1858; *Didymozoon thynni* Taschenberg of Braun, 1879–93; *Didymocystis kobayashii* Dollfus, 1926; *Wedlia katsuwonicola* Okada, 1926.

HOST. Tunny (in France, Italy, near Algiers and in Japan).

LOCATION. Gills and gill lamellae.

According to Yamaguti (1934*a*) and Dollfus (1935*c*) the shape of this trematode is distinctive, the slender anterior region projecting from one end of the swollen hind-body, which is comma-like, relatively slender, and of somewhat beaded appearance. Dollfus showed that the eggs measure about 0·015–0·0185 × 0·007–0·015 mm.

Didymocystis thynni Taschenberg, 1879

Syn. *Didymocystis reniformis* Ariola, 1902; *Monostoma bipartitum*, second form of Wagener, 1858.

HOST. Tunny (in France, Italy and at Syracuse).

LOCATION. Gills or operculum.

In this species the slender anterior region projects from one end of a wider and reniform hind-body.

Genus *Köllikeria* Cobbold, 1860

Köllikeria filicollis (Rudolphi, 1819) Cobbold, 1860

Syn. *Monostoma filicolle* Rudolphi, 1819; *Distoma okenii* Kölliker, 1846; *D. filicolle* (Rudolphi) Beneden, 1858; *Köllikeria okenii* (Kölliker) Ariola, 1906 and of Parona, 1912; *K. (Köllikerizoum) filicollis* of Ishii, 1935.

HOST. Ray's bream (in the North Sea, Irish Sea, France, Italy and Finistère).

LOCATION. Gill arches, gill rakers, buccal cavity, etc.

According to Johnstone (1911) the cysts are 12–25 mm. diameter, the largest containing more than one pair of worms. His Fig. 5 represents one of the females removed from a cyst, and clearly indicates the filiform anterior and reniform posterior regions. A male from the same cyst was 5–15 mm. long and thread-like.

Genus *Wedlia* Cobbold, 1860

Wedlia bipartita (Wedl, 1855) Cobbold, 1860

Syn. *Monostomum bipartitum* Wedl, 1855; *Didymostoma bipartitum* (Wedl) Ariola, 1902; *Didymozoon micropterygis* Richiardi, 1902; *Didymostoma micropterygis* (Richiardi) Ariola, 1902; *Köllikeria (Wedlia) bipartita* (Wedl) of Ishii, 1935.

HOSTS. Tunny, *Seriola dumerili* (in Italy and France), *Thunnus secundodorsalis* in U.S.A.

LOCATION. Gills, gill arches, skin of head.

This, the only European species of the genus, is said to be distinguished by the presence of a pharynx and by the light attachment between individuals of the two sexes. For figures see Fuhrmann (1928, Figs. 85, 97). Linton (1940) gave details about American specimens of this species and of a species, *Wedlia xiphiados* (MacCallum & MacCallum, 1916) from the swordfish, *Xiphias gladius*.

Family **APOROCOTYLIDAE** Odhner, 1912 (p. 93)

Syn. Sanguinicolidae Graff, of Fuhrmann, 1928, in part.

Subfamily **Aporocotylinae**

Genus *Aporocotyle* Odhner, 1900

Aporocotyle simplex Odhner, 1900 (Fig. 45J)

HOSTS. Flounder, dab.

LOCATION. Gills.

Odhner discovered this parasite of the blood-vascular system in Sweden, but it has not appeared in Britain. Generally, it is solitary, but two to seven individuals may occur together in the same host.

DIAGNOSIS. *Shape and size:* elongate, flat, 3·5–5 mm. long, 0·45–0·75 mm. broad, and 0·10–0·15 mm. thick. *Cuticle:* beset with fine spines arranged in small groups, abundant in the antero-ventral and lateral regions. *Suckers:* absent. *Gut:* mouth small, slightly ventral, pharynx absent, oesophagus very long (1 mm.), intestine H-shaped, comprising simple caeca extending to the posterior extremity and anterior extensions extending alongside the posterior two-thirds of the oesophagus. (*Note.* Caeca generally filled with blood.) *Excretory system:* pore slightly dorsal, median canal very short, lateral canals extending forward beneath the caeca. *Reproductive systems:* common genital pore dorsal, near the posterior extremity. Cirrus pouch pyriform, but curved. Testes numerous (about 130), situated between the anterior two-thirds of the caeca. Ovary globular or ovoid. Uterus formed into a compact mass of small folds between the ovary, cirrus pouch and testes. Receptaculum seminis *uterinum* present. Metraterm extending transversely across the body immediately behind the testes. Vitellaria lateral to the caeca and oesophagus, especially abundant follicles anteriorly on median and lateral sides of the anterior diverticula. Egg (found once only by Odhner) spindle-shaped, very large (measuring about 0·125 × 0·033 mm.), containing an ovum and vitelline cells.

Subfamily **Sanguinicolinae**

Genus *Sanguinicola* Plehn, 1905

Syn. *Janickia* Rasin, 1929.

Sanguinicola inermis Plehn, 1905

HOSTS. Cyprinids.

LOCATION. Blood-vascular system.

This species was first described as a turbellarian, then (by Plehn, 1908) as a cestode. Odhner (1911 *b*) first classified the trematode and showed its relationship to *Aporocotyle*, also correcting certain observations made by Plehn. Scheuring (1923) discussed the historical associations and redescribed both the adult and its larvae. He credited the adult with the following characters: *Size:* up to 1 mm. long and 0·3 mm. broad, but of variable proportions, though generally of lanceolate shape. *Cuticle:* beset with diagonal rows of spinelets. *Gut:* mouth terminal, pharynx absent, oesophagus long, intestine represented by four to five

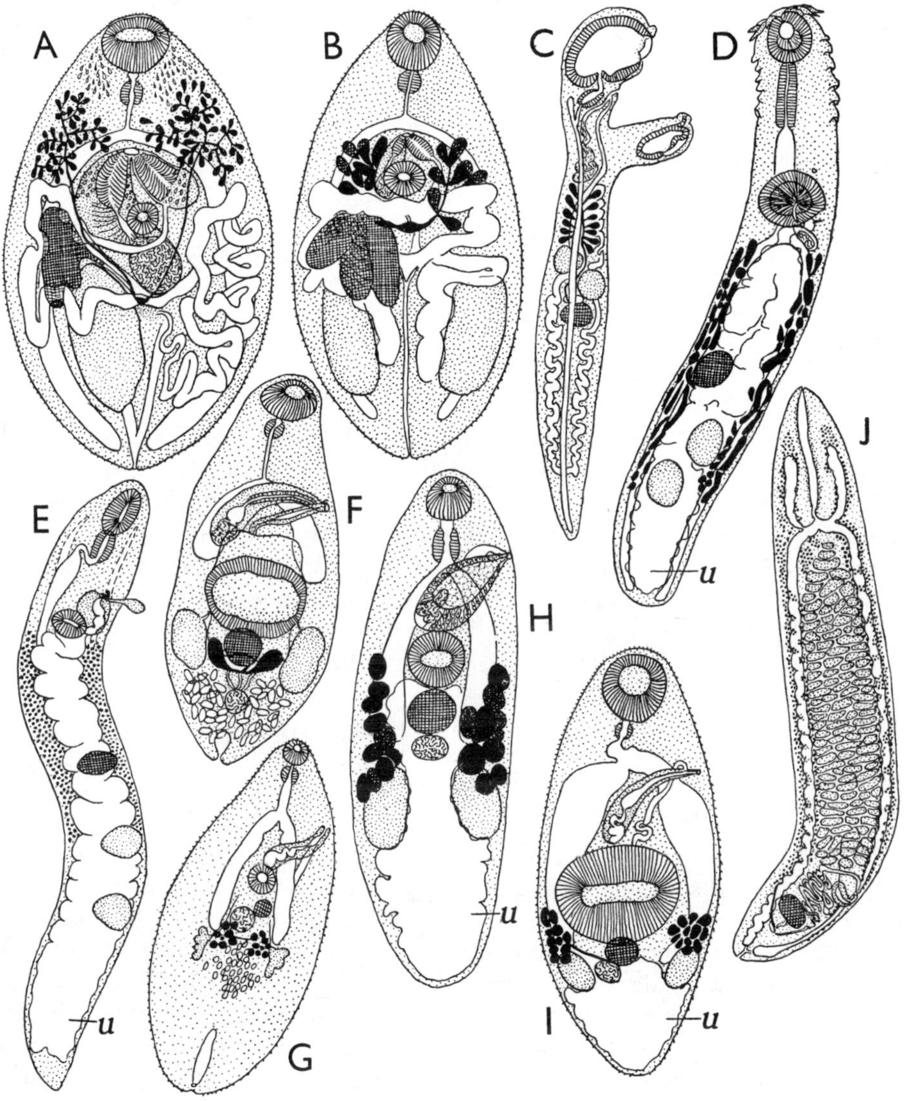

Fig. 45. Some rare Digenea of marine fishes, most of them so far unrecorded in British waters. A, *Monorchis monorchis*. B, *Monorcheides diplorchis* (Monorchiidae). C, *Orophocotyle divergens* (Accacoeliidae). D, *Tergestia laticollis*; E, *Haplocladus typicus* (Fellodistomatidae). F, *Diphterostomum brusinae*; G, *Lepidophyllum steenstrupi*; H, *Lecithostaphylus retroflexus*; I, *Deretrema abyssorum* (Zoogonidae). J, *Aporocotyle simplex* (Aporocotylidae). (A, C, after Looss, 1902*b*, *a*; F, after Looss, 1901*a*; G, after Odhner, 1902; B, after Odhner, 1905; D, E, H, I, J, after Odhner, 1911–1913.)

short, lobed sacs (sometimes six). *Reproductive systems:* genital atrium absent, male and female pores separate, slightly lateral (the former the more lateral) at the posterior fifth of the body. Testes numerous, arranged in a paired series on either side of the median plane in the middle third of the body, commencing immediately behind the gut. Cirrus pouch feeble, but muscular, cirrus represented by an acorn-like retractile papilla, seminal vesicle absent. Ovary with two large lateral lobes (butterfly-shaped), situated immediately behind the testes. Vitellaria almost filling the anterior region, but mainly lateral to the gut and testes, extending as far back as the ovary. Vitelloduct single from its origin anteriorly, situated on the left side. Receptaculum seminis absent. Uterus virtually absent, represented by a short metraterm containing a solitary egg. Eggs characteristic in shape (elliptical in transverse, triangular in longitudinal section) measuring 0·030–0·069 mm. in length (according to Odhner about 0·040 mm. long in the metraterm of the fluke, but up to 0·070 mm. long in the blood of the host). Miracidium having two anterior processes, a 'stylet', eye-spots and a gut. Sporocysts and cercariae of very variable size. The cercaria is the well-known form *C. cristata* La Valette, 1852, or a closely related Lophocercaria.

Family **MESOMETRIDAE** Poche, 1926 (p. 104)

Monostomes are of rare occurrence in fishes. Rudolphi twice found specimens in the gut of *Box salpa* at Naples, calling them *Monostomum orbiculare*. Stossich (1883) gave a fresh description of similar specimens found in the same host at Trieste, and a year later Parona described the net-like lymph system in the same trematode. Brandes (1892*b*) also worked on it, and Stossich (1898) again dealt with it. Lühe (1901*b*) examined Rudolphi's specimens, as well as specimens collected by Wagener and those studied by Parona, founding the genus *Mesometra* for forms which he believed to belong to two species.

Genus *Mesometra* Lühe, 1901

Mesometra orbicularis (Rudolphi, 1819) (Fig. 46A, B)

Syn. *Monostomum orbiculare* Rudolphi, 1819.

Diagnosis. *Shape and size:* flat, disk-like, oval or circular outline with wavy margins, smallest specimens a little more than 1 mm. long, largest 3 mm. long and 2·7 mm. broad, most oval forms measuring 2·25 × 1·65 mm. *Cuticle:* non-spinous. *Suckers:* ventral absent, oral not quite terminal, not rounded, but with a small lug on each side, 0·145–0·25 mm. long and 0·17–0·30 mm. broad. *Gut:* oesophagus feebly muscular, thickened in front of the bifurcation of the intestine to form a spherical 'pharynx bulb' 0·05–0·07 mm. diameter, caeca large, horseshoe-shaped, enclosing the genital organs. *Reproductive organs:* genital pore median, near the middle of the oesophagus. Testes side by side, separated by a space in which the uterus lies, ovary lobed, median, behind the testes. Receptaculum seminis absent, Laurer's canal present. Vitellaria comprising numerous follicles in the dorsal part of the body, mainly above the caeca, but spreading; the anterior third of the body and a narrow marginal zone devoid of follicles. *Excretory system:* pore on the dorsal surface of the body, near the posterior extremity. Eggs without filaments, extremes of size 0·062–0·083 × 0·025–0·042 mm., but mainly 0·078–0·083 × 0·035–0·040 mm. Palombi (1937*a*) studied the development of the cercaria to the metacercaria stage.

Mesometra brachycoelia Lühe, 1901 (Fig. 46 C)

DIAGNOSIS. *Shape:* as in *M. orbicularis*. *Size:* 1·35–1·7 mm. long and 1·1–1·2 mm. broad. *Cuticle:* non-spinous. *Suckers:* ventral absent, oral rounded. *Gut:* caeca only half as long as in *M. orbicularis*, ending in the region of the testes. *Reproductive systems:* similar to *M. orbicularis*, but vitellaria mainly dorsal, region above caeca, testes, uterus and ovary relatively free of follicles, also more numerous follicles in front of the bifurcation of the intestine and at the sides of the oesophagus. *Excretory system:* excretory pore as in *M. orbicularis*. Eggs 0·078–0·080 × 0·035–0·040 mm. In all probability this species is identical with *M. orbicularis*.

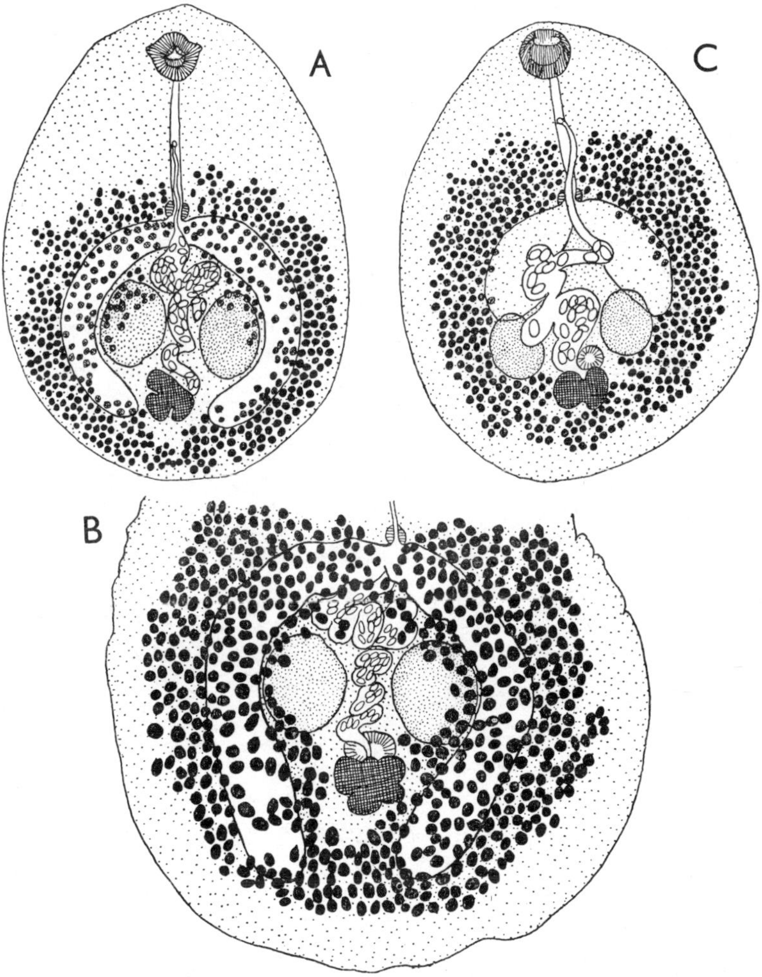

Fig. 46. Mesometridae from *Box salpa* in Italy. A, *Mesometra orbicularis*. B, posterior half of A. C, *M. brachycoelia*. A, C in ventral, B in dorsal view. (After Lühe, 1901*b*.)

Family **PARAMPHISTOMATIDAE** Fischoeder, 1901, emend. Goto & Matsudaira, 1918 (p. 103)

This family is well represented in fishes in various parts of the world, e.g. India, Japan, Australia, though not in Europe. In this country no amphistome parasite of fishes has ever been found, but it would be wrong to allow this negative to pass without comment. In many parts of the world fishes are regularly parasitized by Amphistomes. In his scheme of classification based on the zoological status of the hosts, as well as on the structural characters of the trematodes, Szidat (1939*a*) recognized two subfamilies of amphistomes which live exclusively in the gut of fishes. The Dadayatrem(at)inae Fukui, 1929 contains seven genera, *Dadayatrema* Travassos, 1932; *Dadayus* Fukui, 1929; *Microrchis* Daday, 1907; *Pseudodiplodiscus* Szidat, 1939; *Pseudocladorchis* Daday, 1907; *Travassosinia* Vaz, 1932; and *Parabaris* Travassos, 1922 (for the first, second, fourth and seventh of which synonyms are given). The second subfamily, Opistholebetinae Fukui, 1929, contains three genera:* *Opistholebes* Nicoll, 1925; *Gyliauchen* Nicoll, 1915; and *Paragyliauchen* Yamaguti, 1934. In addition there are eight unclassified genera; *Kalitrema* Travassos, 1933; *Flagellotrema* Ozaki, 1937; *Helostomatis* Bhalerao, 1937; *Neocladorchis* Bhalerao, 1937; *Heterolebis* Ozaki, 1937; *Petalocotyle* Ozaki, 1937; *Nicollodiscus* Srivastava, 1937; and *Orientodiscus* Srivastava, 1937. This implies a diversity of amphistome structure second only to that found in mammals; and greater than that seen in amphibians, reptiles and birds together.

* Szidat was misled by Southwell & Kirshner (1937*a*) into including in this subfamily the genus *Paracotyle* Johnstone, 1911, which belongs to the Monogenea (and is more correctly called *Leptocotyle*).

CHAPTER 9

SOME COMMON DIGENEA OF AMPHIBIA AND REPTILIA

In this chapter are assembled the diagnostic characters of some of the commoner digenetic trematodes of frogs, newts, grass-snakes and lizards. Such hosts yield a valuable (if small) collection of flukes, any one of which far surpasses *Fasciola hepatica* as a subject for study. The ease with which trematodes like *Dolichosaccus* (*Opisthioglyphe*) can be obtained, hundreds of individuals from the intestine of a single frog, and the clarity with which the structure of a typical trematode can be demonstrated, make such flukes ideal types for laboratory work, far superior to most trematodes of veterinary interest. When the locations of such parasites are taken into account, their value is greatly enhanced, for they are found in the intestine, rectum, urinary bladder, lungs and buccal cavity, as well as encysted in the skin and the subcutaneous tissues, the muscles, the brain and the nerves. Representatives of at least half a dozen families of flukes can be collected and studied with a minimum of effort and, for the most part, they are complementary to the families which are to be found in fishes. Again, some of these trematodes are amongst the oldest known and, therefore, of some historical interest.

Reptiles are insignificant in the British fauna, and their trematode parasites have been little studied. The grass snake may not be a consistently reliable source of material, but it is possible sometimes to collect dozens of *Macrodera longicollis* when this vertebrate is being dissected. Few as are the trematodes dealt with in this chapter (some common trematodes of amphibians and reptiles in Europe are shown in Table 4), they create more enthusiasm in students than the best preparations of *Fasciola*, *Paragonimus*, *Schistosoma* and other notorious parasites. One great advantage is that they can be studied alive and not regarded as mere microscopic objects.

Family **PLAGIORCHIIDAE** Lühe, 1901, emend. Ward, 1917 (p. 94)

Syn. Lepodermatidae Looss, 1901.

Subfamily **Plagiorchiinae** Lühe, 1901, emend. Pratt, 1902

Genus *Plagiorchis* Lühe, 1899

Syn. *Lepoderma* Looss, 1899.

Plagiorchis mentulatus (Rudolphi, 1819) Stossich, 1904 (Fig. 47F)

Syn. *Distoma mentulatum* Rudolphi, 1819; *D. lacertae* Rudolphi, 1819; *Lepoderma mentulatum* (Rudolphi) of Looss, 1899.

HOSTS. Common frog, grass snake, common lizard, sand lizard.

LOCATION. Intestine.

LOCALITY. Europe.

DIAGNOSIS. *Shape and size:* outline oval, slightly elongate, both extremities tapering, 1·7–2 mm. long. *Suckers:* ventral much smaller than the oral and

Table 4. *Some Common Digenetic Trematodes of Amphibia and Reptilia*

B, buccal cavity; b, bladder; C, cloaca; GB, gall bladder; I, intestine; L, lungs; m, muscles; O, oesophagus; p, peritoneum; R, rectum; s, subcutaneous; Sf, spinal fluid; Sc, spinal cord; ( ), encysted.

	Edible frog	Common frog	Common toad	Natter-jack	Crested newt	Common newt	Grass snake	Viper	Slow worm	Common lizard	Sand lizard
PARAMPHISTOMATIDAE:											
Diplodiscus subclavatus	R	R	R	—	—	R	—	—	—	—	—
Opisthodiscus diplodiscoides	R	—	—	—	—	—	—	—	—	—	—
PLAGIORCHIIDAE:											
Opisthioglyphe ranae	I	I	I	I	I	—	—	—	—	—	—
Dolichosaccus rastellus	I	I	—	—	—	—	—	—	—	—	—
Haematoloechus variegatus	L	L	L	—	—	—	—	—	—	—	—
H. similis	L	L	—	—	—	—	—	—	—	—	—
H. asper	L	—	—	—	—	—	—	—	—	—	—
Haplometra cylindracea	L	L	L	—	—	—	—	—	—	—	—
Plagiorchis mentulatus	—	I	—	—	—	—	I	—	—	I	I
Macrodera longicollis	—	—	—	—	—	—	L	—	—	—	—
Cercorchis nematoides	—	—	—	—	—	—	I	—	—	—	—
‘*C. aculeatus*’ [=*linstowi*]	—	—	—	—	—	—	I?	—	—	—	—
C. ercolanii	—	—	—	—	—	—	I	I	—	—	—
Encyclometra caudata	—	—	—	—	—	—	I	—	—	—	—
LECITHODENDRIIDAE:											
Brandesia turgida	p	—	—	—	—	—	—	—	—	—	—
Pleurogenes claviger	I	I	I	I	—	I	—	—	—	—	—
P. loossi	I	—	—	—	—	—	—	—	—	—	—
Pleurogenoides medians	I	I	I	I	—	—	—	—	—	—	—
Prosotocus confusus	I	I	I	I	—	—	—	—	—	—	—

DICROCOELIIDAE:											
Brachycoelium salamandrae	—	I	I	—	I	R	—	—	I	—	—
Leptophallus nigrovenosus	—	—	—	—	—	—	OI	—	—	—	—
Paradistomum mutabile	—	—	—	—	—	—	—	—	—	—	GB
CEPHALOGONIMIDAE:											
Cephalogonimus europaeus	I	—	—	—	—	—	—	—	—	—	—
C. retusus	I	I	—	—	—	—	—	—	—	—	—
Cephalogonimus sp.	—	—	—	—	—	—	I	—	—	—	—
HALIPEGIDAE:											
Halipegus ovocaudatus	B	B	—	—	—	—	—	—	—	—	—
GORGODERIDAE:											
Gorgodera cygnoides	b	b	—	—	—	—	—	—	—	—	—
G. pagenstecheri	b	b	—	—	—	—	—	—	—	—	—
G. varsoviensis	b	b	—	—	—	—	—	—	—	—	—
Gorgoderina vitelliloba	b	b	b	—	—	—	—	—	—	—	—
OPISTHORCHIIDAE:											
Ratzia parva (larva)	(m)	—	—	—	—	—	—	—	—	—	—
HETEROPHYIDAE:											
Euryhelmis squamula (larva)	—	(s)	—	—	—	—	—	—	—	—	—
'STRIGEID' larvae:											
'*Tetracotyle colubri*' [=*Strigea strigis*]	—	—	—	—	—	—	(s)	—	(p)	—	—
'*T. crystallina*'	(p)	(p)	—	—	—	—	—	—	—	—	—
'*Tylodelphis rachiaea*'	(Sf)	(Sc)	—	—	—	—	—	—	—	—	—
'*Codonocephalus urniger*'	(p)	—	—	—	—	—	—	—	—	—	—

about one-third of the distance along the body, diameters 0·12 and 0·2 mm. *Gut:* prepharynx and oesophagus very short or absent, pharynx large, but smaller than the ventral sucker (0·1 mm. diameter), caeca long. Bifurcation of the intestine far in front of the ventral sucker. *Reproductive systems:* genital pore slightly in front of the ventral sucker, cirrus pouch long and curving round the sucker. Testes approximately spherical, diagonally situated in the third quarter of the body. Ovary slightly behind the cirrus pouch, on the right. Vitellaria extending throughout the lateral regions of the body from the level of the pharynx to the posterior extremity. Uterus with long descending and ascending limbs, only slightly folded. Eggs yellowish, measuring 0·032–0·036 × 0·020 mm.

Note. This species would be readily identified by means of the generic characters and the hosts. Other species occur in birds and mammals, but one, *P. ramlianus* (Looss, 1896) Stossich, 1904, was found in the intestine of the frog and chamaeleons in Alexandria (Egypt). *Distoma arrectum* Molin, 1859, a parasite of the common lizard in Europe, has been named *Plagiorchis molini* by Lent & Freitas (1940*b*).

Genus *Dolichosaccus* Johnston, 1912

Syn. *Opisthioglyphe* Looss, 1899, of authors, in part; *Lecithopyge* Perkins, 1928.

Perkins (1928) showed that the trematode known as *Opisthioglyphe rastellus* does not belong to the genus *Opisthioglyphe* Looss, 1899, erecting for it the new genus *Lecithopyge*. At the same time he commented on several genera, including *Dolichosaccus* Johnston, 1912, which was erected for certain trematodes of Australian frogs which closely resemble '*Opisthioglyphe rastellum*'. From his diagnoses it is clear that the two genera are very closely similar, and Travassos (1930*b*) seems to have been justified in regarding *Lecithopyge* as a synonym of *Dolichosaccus*.

Dolichosaccus rastellus (Olsson, 1876) Travassos, 1930 (Fig. 48C–E)

Syn. *Distomum rastellus* Olsson, 1876; *D. endolobum* Linstow, 1888 *nec* Dujardin, 1845; *Opisthioglyphe rastellus* Looss, 1907; *O. histrix* of Nicoll, 1926; *Lecithopyge rastellus rastellus* Perkins, 1928; *L. rastellus subulatum* Perkins, 1928; *L. rastellus cylindriforme* Perkins, 1928.

Hosts. Common frog, edible frog.

Location. Intestine.

Diagnosis (based on Lühe, with ranges of measurements given by Travassos shown in parentheses). *Shape and size:* outline oval, wide and rounded anteriorly tapering posteriorly, 3·5–4·0 mm. long (2–4·1 mm.) and 0·8–1·0 mm. broad (0·8–1·2 mm.). *Cuticle:* spinous. *Suckers:* oral larger than the ventral, dimensions 0·33–0·35 mm. (0·24 × 0·21–0·40 × 0·30 mm.) and 0·24–0·27 mm. (0·20–0·24 mm.), ventral about one-quarter of the distance along the body. *Gut:* oesophagus short, scarcely longer than the stout pharynx, which is 0·2 mm. broad (0·14 × 0·21–0·40 × 0·35 mm.), bifurcation of the intestine slightly in front of the ventral sucker, caeca long. *Reproductive systems:* genital pore situated between the ventral sucker and the bifurcation of the intestine, median. Cirrus pouch long and sinuous, extending behind the ventral sucker (for about

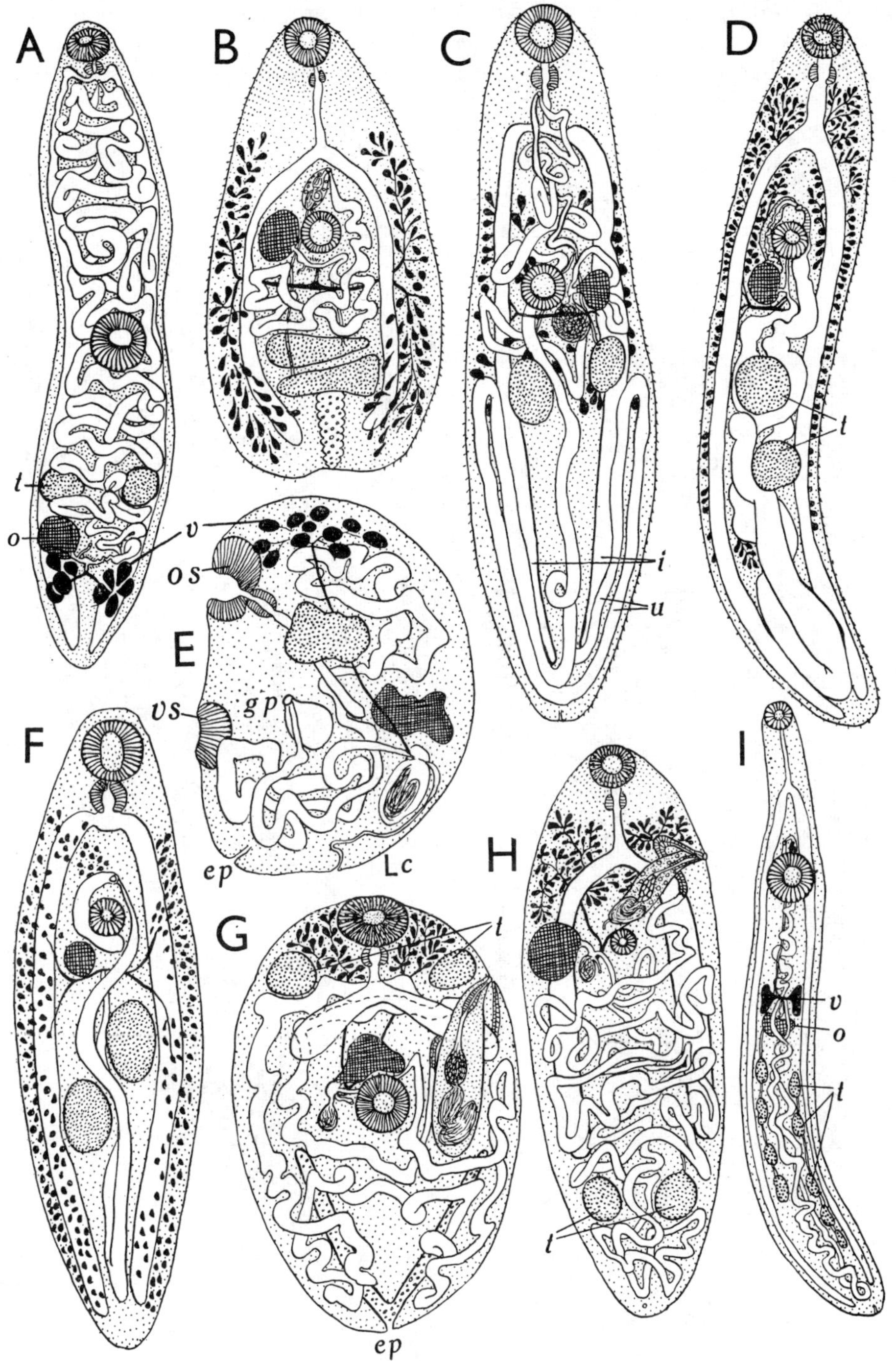

Fig. 47. Halipegidae (A), Plagiorchiidae (B, C, D, F), Lecithodendriidae (E, G, H) and Gorgoderidae (I) from British Amphibia. A, *Halipegus ovocaudatus*. B, *Opisthioglyphe ranae*. C, *Haematoloechus variegatus*. D, *Haplometra cylindracea*. E, *Brandesia turgida*. F, *Plagiorchis mentulatus*. G, *Prosotocus confusus*. H, *Pleurogenes claviger*. I, *Gorgodera cygnoides*. (After Lühe, 1909, but with due regard for the originals of Looss, 1894*a*, in all except E and F.)

its own width) containing the cirrus, the pars prostatica and a voluminous seminal vesicle. Testes somewhat distant from the posterior extremity, rounded, but with slightly indented borders, orientated diagonally one behind the other.

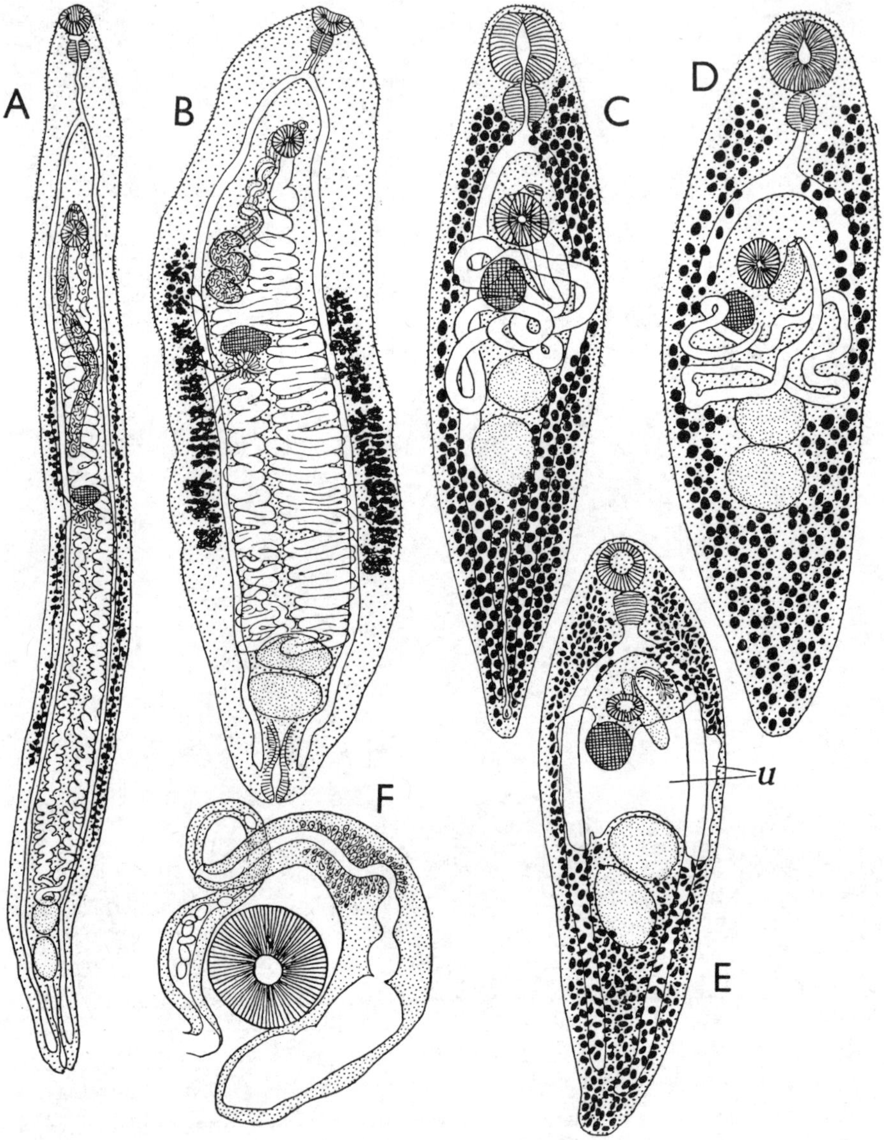

Fig. 48. Plagiorchiidae from the grass snake (A), mud-puppy (B) and frogs (C–F). A, *Cercorchis ercolanii*. B, *C. necturi*. C–E, *Dolichosaccus rastellus*: C, '*rastellus subulatum*'; D, '*rastellus rastellum*' (Molin's specimen); E, specimen from Hamburg. F, metraterm, ventral sucker and cirrus pouch of E. (A–D, after Perkins, 1928; E, F, after Travassos, 1930*b*.)

Ovary situated behind the ventral sucker on the right, near the posterior end of the cirrus pouch. Vitellaria comprising numerous follicles above and lateral to the caeca, distributed throughout the body between the oral sucker and the posterior extremity. Uterus with a descending limb reaching to the anterior testis and an ascending limb, when gravid formed into transverse folds which may slightly overstep the caeca laterally. Eggs measuring 0·042 × 0·025 mm. (0·045–0·051 × 0·021–0·024 mm.). The cercaria is *C. limnaeae ovatae*, which develops in sporocysts in *Limnaea ovata* and later encysts in tadpoles (see p. 489), the thin cysts being about 0·18–0·3 mm. in diameter.

Perkins's observations clearly indicate that *Dolichosaccus rastellus* is an exceedingly variable species. Apart from the shape, differences in which are hardly significant, greatest variation occurs in the dimensions of the eggs and the length of the cirrus pouch. In *D. rastellus rastellus* (represented by Olsson's specimens from the frog and the toad in Sweden) the cirrus pouch is short (not much longer than the ventral sucker) and the eggs small (measuring 0·036 × 0·02 mm.). In *D. rastellus subulatum* the eggs measure 0·040 × 0·023 mm. (frog) or 0·043 × 0·025 mm. (toad), in *D. rastellus cylindriforme* 0·055 × 0·033 mm., and in both of these forms the cirrus pouch is two or three times as long as the ventral sucker. Considerable as is the variation, it is consistent with findings in regard to other trematodes, and it is unlikely to serve the purpose of defining geographical races, as Perkins thought.

OTHER SPECIES:

The only other species of *Dolichosaccus* known at present are parasites of Anura in Australia, namely, *D. trypherus*, *D. ischyrus* and *D. diamesus*, all Johnston, 1912. They do not differ among themselves any more than do the various forms of *D. rastellus* (Travassos, *loc. cit.*, has given comparative data and figures) and probably represent at most a single species. Indeed, they may ultimately prove to be identical with *D. rastellus*.

Genus *Opisthioglyphe* Looss, 1899

Syn. *Brachysaccus* Johnston, 1912; *Dolichosaccus* Johnston, 1912, in part.

The genus *Brachysaccus* was erected for two species of trematodes found in Australian frogs, *anartius* and *symmetrus*, both Johnston, 1912. Perkins (1928) accepted the genus, noting that it differs from *Dolichosaccus* in the larger uterus and more completely diagonal arrangement of the testes, but Travassos (1930*b*) is more probably correct in regarding it as a synonym of *Opisthioglyphe*, which Perkins incorrectly excluded from consideration on the grounds that species other than *rastellus* belong to the Psilostominae rather than the Telorchiinae. Travassos also transferred two species of *Dolichosaccus*, *D. juvenilis* Nicoll, 1918 (which Perkins regarded as *Brachysaccus*) and *D. amplicava* Travassos, 1924, to *Opisthioglyphe*, other species of which are the type *O. ranae* (Froelich, 1791), *O. siredonis* (Poirier, 1886), *O. locellus* Kossack, 1910, a parasite of water-shrews, *O. adulescens* Nicoll, 1914, from *Vipera aspis* in the London Zoo, and *O. magnus* Szidat, 1932, a parasite of reptiles in West Africa.

Opisthioglyphe ranae (Frölich, 1791) Looss, 1907 (Fig. 47B)

Syn. *Fasciola ranae* Frölich, 1791; *Distoma endolobum* Dujardin, 1845; *D. retusum* Beneden, 1861; *Monostomum hystrix* Molin, 1861; *Opisthioglyphe endoloba* (Dujardin) of Looss, 1899; *O. hystrix* of Kossack, 1910.

HOSTS. Common and edible frogs, common toad, natterjack toad, crested newt.

LOCATION. Intestine.

DIAGNOSIS (based on Lühe, with ranges of measurements given by Travassos (1930*b*) shown in parentheses). *Shape, colour and size:* outline oviform, tapering anteriorly and broad, but notched posteriorly, yellowish brown, 1·6–2·5 mm. long (1·6–4·5 mm.) and about half as broad behind the middle (0·9–2 mm.). *Cuticle:* spinous. *Suckers:* oral somewhat larger than the ventral, dimensions 0·15 mm. (0·14–0·20 mm.) and 0·13 mm. (0·09–0·16 mm.), the ventral situated near (slightly in front of) the mid-body. *Gut:* prepharynx very short, pharynx small (measuring 0·06–0·09 × 0·11 mm.), oesophagus of medium length, caeca long, extending almost to the posterior extremity. Bifurcation of the intestine approximately midway between the suckers. *Reproductive systems:* genital pore slightly behind the bifurcation of the intestine. Cirrus pouch pyriform, situated entirely in front of the ventral sucker, containing the cirrus, the pars prostatica and a coiled seminal vesicle. Testes transversely elongate, of irregular shapes, situated one behind the other in the posterior region. Ovary far in front of the testes, near the ventral sucker, on the right. Vitellaria follicular, lateral to the caeca, extending from the bifurcation of the intestine to the posterior extremity and slightly beyond the ends of the caeca. Uterus having the descending and ascending limbs formed into folds, mainly between the anterior testis and the ventral sucker. Eggs clear yellowish brown, measuring 0·049 × 0·038 mm. (0·045 × 0·027–0·051 × 0·032 mm.). *Excretory system:* main canals Y-shaped, the median stem extending to a point above the anterior testis, the lateral canals only slightly the shorter. *Cercariae*: Xiphidiocercariae developing in rediae, which occur in *Limnaea stagnalis* and *L. palustris*.

One other genus of Plagiorchiinae might be mentioned, *Astiotrema* Looss, 1900, members of which are typically parasites of chelonians. About eleven species are known, four in India, two in Korea, two others in Africa and one in Japan. Remaining species are European, *A. emydis* Ejsmont, 1930, occurring in *Emys orbicularis* in Poland, *Astiotrema monticellii* Stossich, 1904 in '*Natrix viperina*' in Italy.

Another Plagiorchiid which has been found in this country, but (presumably) does not belong to the fauna, is *Pneumotrema travassosi* Bhalerao, 1937*b*, which was discovered in the lung of a Brazilian snake, *Amphisbaena alba*, which died in the London Zoological Gardens.

Subfamily **Haplometrinae** Pratt, 1902

Genus *Haplometra* Looss, 1899

Haplometra cylindracea (Zeder, 1800) Looss, 1899 (Fig. 47D)

Syn. *Distoma cylindraceum* Zeder, 1800; *D.* (*Dicrocoelium*) *cylindraceum* of Dujardin, 1845.

HOSTS. Common frog, edible frog, common toad.

LOCATION. Lungs.

This was probably the first trematode ever to be observed in frogs, almost certainly being the form seen by Swammerdam and recorded in 1737. After this, and before Zeder named it, at least two zoologists (Pallas, 1760; Goeze, 1787) saw it without recognizing it as a distinct species.

DIAGNOSIS. *Shape and size:* elongate, cylindrical, sometimes 20 mm. long but generally less than half this length, attaining maturity when 3–4 mm. long. *Colour:* opaque, milky white when seen by reflected light, reddish in transmitted light. *Cuticle:* scaly, more especially anteriorly. *Suckers:* ventral smaller than the oral in the ratio of 3/4 and situated a little behind it and about one-third of the distance along the body. *Gut:* pharynx small, oesophagus of medium length, caeca long. Bifurcation of the intestine far in front of the ventral sucker. *Reproductive systems:* genital pore median, much nearer the ventral sucker than the bifurcation of the intestine. Cirrus pouch long and narrow, containing a stout cirrus, a well-differentiated pars prostatica and a much convoluted seminal vesicle. Testes spherical, situated one behind the other in the third quarter of the body. Ovary somewhat anterior to the testes, slightly behind or beside the ventral sucker. Vitellaria well developed between the pharynx and the posterior extremity, conspicuous near the oesophagus, overlying the caeca and extending towards the median plane. Uterus with wide descending and ascending limbs extending back between the testes almost to the posterior extremity and only slightly folded. Eggs very numerous, dark brown, small, measuring 0·04 × 0·022 mm.; 0·040–0·044 × 0·022–0·026 mm. according to Travassos (1930*a*). *Cercaria:* a well-known 'Cystocercous' form developing in sporocysts in the 'liver' of *Limnaea ovata*. A larval or imaginal insect serves as the second intermediate host, metacercariae encysting in the haemocoele. The host may harbour a dozen mature flukes in each lung, but as a rule the parasites are fewer.

Genus *Haematoloechus* Looss, 1899

Syn. *Pneumonoeces* Looss, 1902; *Ostiolum* Pratt, 1902.

Harwood and Ingles have pointed out, separately (1932), that the name first chosen by Looss for this genus was not invalidated by the hemipteran genus *Haematoloecha* Stäl, 1874, and should stand in accordance with the International Code of Zoological Nomenclature. Reviewing the genus, Travassos & Darriba (1930) listed eighteen species (together with three of *Ostiolum*) which occur mainly in North and South America, but also in Europe, India and Australia.

EUROPEAN SPECIES:

Haematoloechus variegatus (Rudolphi, 1819) Looss, 1899 (Fig. 47C)

Syn. *Distoma variegatus* Rudolphi, 1819; *Pneumonoeces variegatus* (Rudolphi) of Looss, 1902, *nec D. variegatum* Looss, 1894.

HOSTS. Common frog, common toad, edible frog.

LOCATION. Lungs.

DIAGNOSIS. *Shape and size:* elongate, almost cylindrical, more pointed and broader posteriorly than anteriorly, 7–18 mm. long and almost one-third as broad. *Cuticle:* non-spinous. *Suckers:* ventral smaller than the oral and situated slightly in front of the mid-body. *Gut:* prepharynx very short, pharynx not large, oesophagus of medium length, caeca long and relatively wide, almost reaching the posterior extremity. Bifurcation of the intestine far in front of the ventral sucker. *Reproductive systems:* genital pore far forward, lateral to the pharynx. Cirrus pouch elongate, tubular, but confined to the region in front of the ventral sucker. Testes ovoid, sometimes having indented margins, situated side by side (or slightly diagonally) just in the third quarter of the body. Ovary longitudinally ovoid, sometimes of irregular shape, situated in front of the testes and to one side of the ventral sucker. Vitellaria extending as ten to twelve rosette-like groups of six to seven follicles on each side from the level of the oesophagus almost to the posterior extremity, somewhat hidden posteriorly by the folds of the uterus. Uterus with descending and ascending limbs, somewhat irregularly folded in front of the gonads and arranged in double lateral folds on each side of the body between the testes and the posterior extremity. Eggs dark brown, measuring 0·025–0·032 × 0·013–0·019 mm., mostly 0·029 × 0·016 mm. (according to Travassos & Darriba (1930) 0·024–0·032 × 0·016–0·020 mm.).

The subspecies *abbreviatus* Bychowsky, 1932 occurs in the lungs of the fire-bellied toad in the Ukraine.

Haematoloechus similis (Looss, 1899)

Syn. *Distoma simile* Looss, 1899, *nec* Sonsino, 1890; *Haematoloechus similigenus* of Stiles & Hassall, 1902; *Distoma variegatum* of Looss, 1894, in part; *Pneumonoeces similis* of Klein, 1905; Lühe, 1909; and other writers; *P. similigenus* of Cort, 1915; Nicoll, 1926.

HOSTS. Edible frog, common frog.

LOCATION. Lungs.

DIAGNOSIS. *Size:* 7–10 mm. long and up to 2 mm. broad. *Cuticle:* spinous. *Other characters:* said to differ from *H. variegatus* further by the regular rounded or ovoid shape of the ovary and the larger eggs, which measure 0·034–0·042 × 0·017–0·021 mm. (according to Travassos & Darriba (1930), 0·032–0·040 × 0·016–0·026 mm.). The vitellaria are relatively short, extending only half-way between the suckers in an anterior direction and to the hinder testis posteriorly.

Haematoloechus asper Looss, 1899

Syn. *Distoma variegatum* of Looss, 1894, in part; *Pneumonoeces asper* of Klein, 1905; Lühe, 1909; and other writers.

HOST. Edible frog.

LOCATION. Lungs.

DIAGNOSIS. *Size:* comparatively small, about 7 mm. long. *Cuticle:* spinous. *Vitellaria:* extensive, not formed into rosette-like folds. *Eggs:* relatively large (0·055 × 0·029 mm., but according to Travassos & Darriba (1930), 0·054–0·064 × 0·024–0·032 mm.).

Haematoloechus schulzei (Wundsch, 1911) n. comb.

Syn. *Pneumonoeces schulzei* Wundsch, 1911.

HOSTS. Frogs.

LOCATION. Lungs.

DIAGNOSIS. *Size:* large, up to 18 mm. long. *Other characters:* similar to *H. asper* in the general anatomical features, but distinguished from it by the rosette-like clusters of vitelline follicles and the small size of the eggs (0·025–0·027 × 0·019 mm. according to Wundsch, 0·032–0·035 × 0·016–0·018 mm. according to Travassos & Darriba).

Haematoloechus sibericus (Issaitschikow, 1927) n. comb.

Syn. *Pneumonoeces sibericus* Issaitschikow, 1927.

Occurs in the lungs of *Rana* spp. in Russia.

Genus *Macrodera* Looss, 1899

Syn. *Saphedera* Looss, 1902.

This genus seems to have been renamed because of *Macroderes* Westwood, 1843 (Coleoptera), a change which was unnecessary under Art. 35 of the Rules of Nomenclature, so that *Macrodera* must stand.

Macrodera longicollis (Abildgaard, 1788) Lühe, 1909 (Fig. 49E)

Syn. *Fasciola longicollis* of Frölich, 1792; *Distomum attenuatum* Rudolphi, 1814; *D. naja* Rudolphi, 1819; *D. longicolle* of Cobbold, 1860; *Macrodera naja* of Looss, 1899; *Saphedera naja* of Looss, 1902; *Macrodera longicollis* (Abildgaard) of Lühe, 1909; *Saphedera longicollis* of authors.

Hosts. Grass snake and *Natrix chrysarga*.

Location. Lung-sacs.

Diagnosis. *Shape and size:* very elongate, little flattened, having a short and rounded anterior region bearing the suckers, a very narrow 'neck' and a long, gradually widening, sac-like posterior region, about 12 mm. long and 1 mm. broad. *Cuticle:* spinous. *Suckers:* very close together, the ventral larger than the oral, diameters 0·67–0·73 and 0·45 mm. *Gut:* pharynx and oesophagus of medium length, caeca wide and long, extending almost to the posterior extremity. *Reproductive systems:* genital pore immediately behind the pharynx, but still close to the ventral sucker. Cirrus pouch very elongate, cylindrical, thickened in its terminal part, because of a well-developed pars prostatica, containing also a seminal vesicle. Testes spherical, one behind the other and somewhat separated in the posterior region. Ovary spherical, smaller than and situated some distance in front of the testes. Vitellaria fairly well developed, lateral and distributed mainly in front of the testes, tending to enter the 'neck', more extensive on one side posteriorly than on the other, the follicles arranged in about six clusters on either side of the body. Uterus very long and well developed, both the descending and the ascending limb richly folded in the posterior region between and behind the gonads, the folds reaching back to the extremity. Eggs very numerous, brown, measuring 0·036 × 0·021 mm.

There seems to be only one other species of this genus, *Macrodera cantonensis* (Wallace, 1936), and it is also parasitic in reptiles, occurring in *Natrix piscator* in China.

Subfamily **Encyclometrinae** Nicoll, 1932

Syn. Encyclometriinae Mehra, 1931.

Genus *Encyclometra* Baylis & Cannon, 1924

Syn. *Odhneria* Baer, 1924 *nec* Travassos, 1921.

The type-species of this genus, *E. caudata*, occurs in various snakes in both Europe and Asia. Two other species occur in Japan and a third in China.

Encyclometra caudata (Polonio, 1859) Joyeux & Houdemer, 1928 (Fig. 49A)

Syn. *Distomum caudatum* Polonio, 1859; *D. subflavum* Sonsino, 1892; *Distomum* sp. no. 1 Timotheev, 1900; *Paraplagiorchis timotheevi* Dollfus, 1924; *Odhneria bolognensis* Baer, 1924; *Encyclometra bolognensis* (Baer) of Baylis & Cannon, 1924; *E. natricis* Baylis & Cannon, 1924; *Orthorchis natricis* Mödlinger, 1924; *Encyclometra subflava* (Sonsino) of Dollfus, 1924.

Hosts. Grass snake and other snakes (*Natrix chrysarga*, *N. piscator*, etc.).

Location. Intestine.

According to Joyeux & Houdemer (1928), who quote Dollfus (*in litt.*), the correct name of this species, for some time in dispute, is as shown. *Distomum colubri murorum* (spelt by them *colubrimurorum*) Rudolphi, 1819, Dollfus thought, is probably the same species, but its type is lost. Nevertheless, Hughes, Higginbotham & Clary (1942) have boldly used the name *Encyclometra colubri-murorum* (*sic*), with *caudatum* as a synonym.

Diagnosis. Based on '*Odhneria bolognensis*' of Baer: data of Baylis & Cannon (1924*a*) in parentheses. *Shape and size:* slightly elongate, about 6 mm. long (3·2–4·5 mm.) and 1·2 mm. broad (1·1–1·75 mm.) at the level of the ventral sucker. *Cuticle:* thick and non-spinous. *Suckers:* ventral larger than the oral in the ratio of 1·4/1* and situated slightly in front of the middle of the body (diameters 0·49–0·78 and 0·46–0·66 mm.). *Gut:* prepharynx distinct (short), pharynx well developed (pyriform, 0·26–0·4 × 0·22–0·3 mm.), oesophagus short (absent), caeca long (extending to final eighth of body). Numerous glands occur at the base of the pharynx and open into the oesophagus. Caeca in living worms of variable length, sometimes asymmetrical. *Excretory system:* vesicle Y-shaped, having a long median stem and short lateral canals. *Reproductive systems:* genital pore situated between the anterior margin of the ventral sucker and the anterior regions of the caeca on the left side. Cirrus pouch elongate, curving towards the median plane and thus both in front of and lateral to the ventral sucker, containing a short cirrus, a long pars prostatica and a bulbous seminal vesicle (constricted in the middle). Testes ovoid (figured as slightly lobed) situated one behind the other in the posterior region. Ovary spherical, situated in front of the testes, behind and above the ventral sucker and on the right. Vitellaria composed of isolated follicles distributed laterally between the ventral sucker and the posterior extremity (not discontinuous, but variable). Uterus with long descending and ascending limbs, only slightly folded (with many convolutions). Eggs large, thin-shelled, yellow, measuring 0·095 × 0·057 mm., each containing a miracidium showing some activity.

* Baylis & Cannon (1924*b*) mentioned that Baer re-examined his specimens and confirmed their measurements of the suckers.

Subfamily **Telorchiinae** Looss, 1899

Genus *Cercorchis* Lühe, 1900

Perkins (1928) outlined the confusion and difficulties which resulted from the simultaneous publication (on 28 December 1899) of revisions of the Trematoda by Looss and Lühe, both of whom erected genera under the name *Telorchis*, but arranged in them assemblages of species which prevented their acceptance as synonymous genera. In 1900 Lühe showed that the species *clava* Diesing, 1855, which he designated as the type of *Telorchis*, did not conform to the genus as set up by Looss. Accordingly he made *clava* the type of his genus *Telorchis* and founded a new genus, *Cercorchis*, for the other species, separation being based on the characters:

1. Oesophagus absent. Uterus not confined between the caeca *Telorchis* Lühe, 1900
2. Oesophagus present. Uterus confined between the caeca *Cercorchis* Lühe, 1900

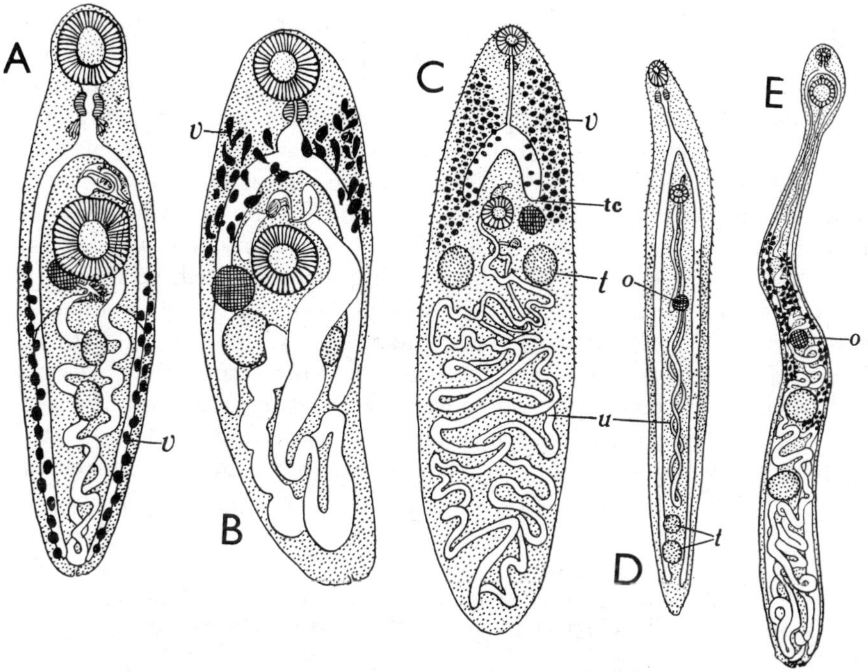

Fig. 49. Plagiorchiidae (A, D, E) and Dicrocoeliidae (B, C) from British Reptilia. A, *Encyclometra caudata*. B, *Leptophallus nigrovenosus*. C, *Brachycoelium salamandrae*. D, *Cercorchis nematoides*. E, *Macrodera longicollis*. C occurs also in Amphibia. (A, after Baer, 1924; E, after Fuhrmann, 1928; rest after Lühe, 1909.)

Stunkard (1915) refused to accept this scheme, and for some time the genera were regarded as synonymous under the name *Telorchis*. Perkins (1928) agreed that neither of the characters used in the scheme is reliable, but insisted that two genera are needed, and modified the definitions not only of these and other genera which later arose, but also that of the Telorchiinae to which they belong. Bennett & Tobie (1936) accepted *Cercorchis* as redefined by Perkins, but Wharton (1940) rejected it. Only two species mentioned by Perkins belong to the genus *Telorchis*, namely, *clava* and *anacondae* MacCallum, 1918, both of which are parasites of snakes in America. Species of the large genus *Cercorchis*, on the other hand, occur mostly in American water tortoises, but two species live in snakes and two in Amphibia.

The characters which, according to Perkins, apply to this genus are as follows: Genital aperture situated near and in front of the ventral sucker. Cirrus pouch very long and slightly coiled, the metraterm being about as long and parallel with it. Testes one behind the other in the posterior region. Vitellaria confined to the region between the ventral sucker and the testes and distributed lateral to the caeca, which bound the folds of the well-developed uterus.

Cercorchis nematoides (Mühling, 1898) (Fig. 49D)

HOST. Grass snake.

LOCATION. Intestine.

LOCALITY. Europe.

DIAGNOSIS. *Shape and size:* very elongate, 3–4 mm. long and about one-tenth as broad (probably variable, because Perkins gives the dimensions 3·5 × 0·85 mm.). *Cuticle:* spinous back to the level of the ovary, spines fine and numerous. *Suckers:* ventral smaller than the oral and situated about one-quarter of the distance along the body, diameters 0·125 and 0·135 mm. *Gut:* prepharynx short, pharynx spherical (0·068 mm. diameter), oesophagus about 0·2 mm. long, caeca extending almost to the posterior extremity. Bifurcation of the intestine some distance in front of the ventral sucker. *Reproductive systems:* genital pore situated slightly in front of the ventral sucker. Cirrus pouch very long (more than 0·7 mm.), extending posteriorly behind the ovary. Testes rounded, one behind the other near the ends of the caeca. Ovary far in front of the testes, slightly in front of the middle of the body, near the base of the cirrus pouch. Vitellaria lateral, extending from a point midway between the ventral sucker and the ovary to within a short distance of the anterior testis. Uterus with long descending and ascending limbs (the folds of which mingle) confined to the region in front of the testes. Eggs dark brown, measuring 0·033 × 0·019 mm.

OTHER EUROPEAN SPECIES:

Cercorchis linstowi (Stossich, 1890) (=*aculeatus* of Braun, 1891) in *Testudo graeca*.
C. ercolanii (Monticelli, 1893) (=*D. signatus* Ercolani *nec* Diesing) in the grass snake and viper.
C. poirieri (Stossich, 1895) (=*gelatinosum* Poirier *nec* Rudolphi) in *Emys orbicularis*.
C. parvus (Braun, 1901) in *Emys orbicularis*.
C. solivagus (Odhner, 1902) in *Chlemmys caspica*.
C. stossichi Goldberger, 1911 (=*poirieri* Stossich, 1904 *nec* 1895) in *Emys orbicularis*.
C. shelkownikowi Skrjabin & Popoff, 1924 in *Emys orbicularis*.

Cercorchis ercolanii (Monticelli, 1893) (Fig. 48A)

Perkins remarked on the doubt which has been felt regarding the validity of this species which, occurring in the intestine of the grass snake, might otherwise be a synonym of *C. nematoides*. He found it to be very common in the more humid districts of Cambridgeshire, and regarded it as distinct from *C. nematoides* before becoming aware of the existence of *C. ercolanii* as a described species. It bears a close resemblance to *C. medius* (Stunkard, 1916), which is found in American tortoises. It differs from *C. nematoides* in being broader, in having a smaller oral sucker and a larger pharynx, as well as a shorter cirrus pouch, which is separated from the ovary by folds of uterus, and in having vitelline follicles arranged in groups. Perkins has supplied the following ranges of measurements (and, in parentheses, means). *Size:* 4·3–6·5 mm. × 0·4–0·7 (4·96 × 0·54 mm.). *Distance between suckers:* 0·6–1·25 mm. (0·926 mm.). *Suckers:* oral, 0·09–0·16 × 0·09–0·14 mm. (0·122 × 0·199 mm.); ventral, 0·11–0·15 × 0·10–0·17 mm. (0·122 × 0·126 mm.). *Pharynx:* 0·06–0·105 mm. (0·087 mm.). *Eggs:* small (0·027 × 0·016 mm.), but variable (0·022–0·036 × 0·014–0·018 mm.).

There seems to be no reasonable doubt regarding the validity of the species.

Cercorchis necturi Perkins, 1928 (Fig. 48B)

This species was discovered in the intestine of *Necturus maculatus* which were sent to this country from the Biological Laboratory at Woods Hole. A brief description is desirable here, because such material sometimes reaches this country for purposes of dissection. The only other species of *Cercorchis* which occurs in an amphibian is *C. stunkardi* Chandler, 1923, which parasitizes *Amphiuma* and is closely similar to *Cercorchis necturi* in that the ventral sucker is larger than the oral, a character which does not hold good for any other species.

Perkins gave the following ranges of measurements (means in parentheses). *Size:* 3·5–5·0 × 0·9–1·15 mm. (4·25 × 1·01 mm.). *Distance between suckers:* 0·43–0·62 mm. (0·56 mm.). *Suckers:* oral, 0·15–0·19 × 0·11–0·13 mm. (0·172–0·115 mm.); ventral, 0·17–0·20 × 0·17–0·20 mm. (0·19 × 0·185 mm.). *Pharynx:* 0·09–0·12 × 0·08–0·11 mm. (0·1 × 0·099 mm.). *Eggs:* 0·020–0·034 × 0·016–0·026 mm. (0·0278 × 0·0185 mm.). He also defined the species in terms which indicate that it contains broad, spinous forms, with suckers in the size ratio of about 3/4, without a prepharynx but having an oesophagus, with a genital aperture situated slightly on the left of the median plane, with a well-developed and slightly coiled cirrus pouch and metraterm, and somewhat asymmetrical vitellaria, that on the right side being less developed than that on the left and placed farther forward. The ovary is separated from the base of the cirrus pouch by one or two folds of the uterus, which is confined between the caeca, and the testes are transversely compressed, the anterior being slightly the smaller.

Genus *Cercolecithos* Perkins, 1928

Perkins erected this genus to include *Cercorchis arrectus* (Molin, 1859) (Syn. *Plagiorchis molini* Lent & Freitas, 1940), a form found in *Lacerta viridis* and *Podarcis muralis* in Europe, but not rediscovered until 1940. It agrees with *Cercorchis* in all particulars except the character of the vitellaria, which extend behind and above the testes, a condition which does not arise in any species of *Cercorchis* proper. This solitary species from European lizards is about 3·5 mm. long and 1·0 mm. broad, spinous, without a prepharynx but with an oesophagus. The genital pore is median, the cirrus pouch extends behind the ovary, the vitellaria are joined in a tract of folicles above the testes, and the descending and ascending limbs of the uterus are distinct.

OTHER PLAGIORCHIIDAE:

More than twenty genera and 100 species of reptilian trematodes which bear an obvious relationship to plagiorchiids but are referred by some writers to a separate family, Reniferidae Baer, 1924, occur in various parts of the world, especially America, chiefly in snakes. The following genera and species have been recorded in or near Europe:

Genus *Enodistrema* Looss, 1900

Syn. *Enodia* Looss, 1899, *nec* Huebn. 1816.

Enodiotrema acariaeum Looss, 1902; *E. instar* Looss, 1901; *E. megachondrum* (Looss, 1899) Looss, 1901; *E. reductum* Looss, 1901. All in the loggerhead turtle in Egypt.

Genus *Pachypsolus* Looss, 1901

Pachypsolus irroratus (Rudolphi, 1819) Looss, 1901. In the loggerhead turtle and *Chelonia mydas* near Europe and Africa; also in Australia.

Genus *Styphlodora* Looss, 1899

Styphlodora serrata Looss, 1899. In *Varanus niloticus* (Egypt).

Genus *Styphlotrema*

Styphlotrema solitaria (Looss, 1899) Odhner, 1910. In the loggerhead turtle near Egypt; also in Florida.

Genus *Rhytidodes* Looss, 1901

Rhytidodes gelatinosus (Rudolphi, 1819) Looss, 1901. In the loggerhead turtle and *Chelonia mydas* near Egypt.

Genus *Lechriorchis* Stafford, 1905

Lechriorchis inermis Lebour, 1913

Family **CEPHALOGONIMIDAE** Nicoll, 1915 (p. 95)

Subfamily **Cephalogoniminae** Looss, 1899

Genus *Cephalogonimus* Poirier, 1886

Cephalogonimus retusus (Dujardin, 1845)

Syn. *Distoma retusum* Dujardin, 1845.

HOSTS. Edible frog, common frog.

LOCATION. Intestine.

DIAGNOSIS. *Size:* small. *Cuticle:* spinous. *Suckers:* ventral only about half as large as the oral and about one-third of the distance along the body. *Gut:* prepharynx short, pharynx small, oesophagus and caeca short and extending back only slightly behind the ventral sucker. Bifurcation of the intestine slightly in front of the ventral sucker. *Excretory system:* median vesicle long and wide, extending forward to a point near the middle of the body, lateral canals short, failing to reach the level of the ventral sucker. *Reproductive systems:* genital pore very far forward, at the level of the mouth, on the *dorsal* surface of the body. Cirrus pouch relatively long, extending back to, sometimes behind, the bifurcation of the intestine. Testes slightly irregular in shape, one behind the other in the body slightly behind the ventral sucker. Ovary in front of the testes, dorso-lateral to the ventral sucker. Vitellaria lateral, the follicles distributed between the ends of the caeca. Uterus having long descending and ascending limbs, not quite reaching to the posterior extremity and formed into slight folds.

One other species, *C. europaeus* Blaizot, 1910, is said to occur in the intestine of the edible frog, and another, *C. amphiumae* Chandler, 1923, in a Urodele in U.S.A. An unnamed species *Cephalogonimus* sp. Lühe, 1911, was discovered in the grass snake in Europe and about eight species parasitize Chelonia in widely separated localities (Burma, India, Africa, U.S.A. and Japan). Lent & Freitas (1940*a*) regarded *C. europaeus* and *C. americanus* Stafford, 1902 as valid species, though other writers have considered them to be synonyms of *C. retusus*.

Family **DICROCOELIIDAE** Odhner, 1910 (p. 95)

Syn. Brachycoeliidae Johnston, 1913, in part.

Subfamily **Brachycoeliinae** Looss, 1899 restricted

Genus *Brachycoelium* (Dujardin, 1845: subgenus) Stiles & Hassall, 1898

Brachycoelium salamandrae (Frölich, 1789) (Fig. 49C)

Syn. *Fasciola salamandrae* Frölich, 1789; *Distoma salamandrae* Zeder, 1803; *Distoma crassicolle* Rudolphi, 1809; *Distomum flavocinctum* Linstow, 1879; *Brachycoelium crassicolle* of Looss, 1899; *Lecithodendrium crassicolle* of Stossich, 1899; *Distomum hospitale* Stafford, 1900; *Brachycoelium hospitale* (Stafford, 1900) Stafford, 1903; *B. obesum* Nicoll, 1914; *B. trituri* Holl, 1928; *B. storeriae* Harwood, 1932; *B. meridionalis* Harwood, 1932; *B. daviesi* Harwood, 1932; *B. mesorchium* Byrd, 1937; *B. georgianum* Byrd, 1937; *B. ovale* Byrd, 1937; *B. dorsale* Byrd, 1937; *B. louisianae* Byrd, 1937.

HOSTS. Common frog, common toad, crested newt, common newt, salamander, slow worm.

LOCATION. Intestine, sometimes rectum.

LOCALITY. Europe and North America (various Amphibia and Reptilia).

DIAGNOSIS. *Shape and size:* elongate, 3–5 mm. long and 0·8–1·2 mm. broad. *Cuticle:* spinous, more especially in the anterior region. *Suckers:* ventral slightly smaller than the oral and situated about one-third of the distance along the body, diameters 0·20–0·26 mm. and 0·26–0·31 mm. *Gut:* prepharynx and pharynx small, oesophagus of medium length, caeca very short, scarcely reaching the level of the ventral sucker. *Excretory system:* vesicle Y-shaped. *Reproductive systems:* genital pore situated slightly in front of the ventral sucker, the small cirrus pouch mainly in front of it. Testes rounded, side by side, slightly behind the terminations of the caeca and the ventral sucker. Ovary in front of the left testis, beside the ventral sucker. Vitellaria confined to the anterior region, between the testes and the oral sucker, encroaching on the median plane. Uterus having long and much-folded descending and ascending limbs, tending to fill the posterior half of the body. Eggs light brown, measuring 0·045–0·050 × 0·032–0·036 mm.

Note. Salamandra maculosa is probably the principal host of this trematode, which is referred by some zoologists to the family Brachycoeliidae S. J. Johnston, 1912. Certainly it is a distinctive trematode on account of the very short caeca, but this character does not provide sufficient grounds for such a major change.

Rankin (1938) studied variability in the specific characters of both living and preserved American specimens of *Brachycoelium* and relegated eleven species to synonymity with the type.

Genus *Leptophallus* Lühe, 1909

Leptophallus nigrovenosus (Bellingham, 1844) Lühe, 1909 (Fig. 49B)

Syn. *Distomum nigrovenosum* Bellingham, 1844; *Distoma signatum* Dujardin, 1845; *D. nigrovenosum natricis torquatae* Diesing, 1855; *Lecithodendrium nigrovenosum* of Lühe, 1899; *Brachycoelium nigrovenosum* of Looss, 1902.

HOSTS. Grass snake and *Natrix chrysarga.*

LOCATION. Oesophagus and intestine.

LOCALITY. Europe (including Britain).

DIAGNOSIS. *Shape and size:* outline elongate oval, body only slightly flattened, rounded anteriorly, tapering posteriorly, 1·1–1·8 mm. long, 0·40–0·43 mm. broad and 0·35–0·40 mm. thick. *Cuticle:* spinous. *Suckers:* powerful, the ventral slightly smaller than the oral and just in front of the mid-body, diameters 0·18–0·22 and 0·21–0·26 mm. *Gut:* pharynx about one-third as large as the oral sucker, oesophagus very short, caeca fairly long, extending about two-thirds of the distance along the body. Bifurcation of the intestine well in front of the ventral sucker. *Reproductive systems:* genital pore almost median, not far in front of the ventral sucker, cirrus pouch containing a large seminal vesicle in its basal part. Testes spherical, symmetrically one on either side of the body. Ovary immediately in front of the right testis. Vitellaria confined to the region between the pharynx and the ventral sucker, comprising a few large follicles of pyriform shape which encroach on the median plane. Uterus with long descending and ascending limbs passing between the testes, somewhat folded in the posterior region, S-shaped near the gonads. Eggs numerous, brown, measuring 0·034–0·038 × 0·019–0·021 mm.

Subfamily **Dicrocoeliinae** Looss, 1899

Genus *Paradistomum* Kossack, 1910

Paradistomum mutabile (Molin, 1859) Nicoll, 1924

Syn. *Distomum mutabile* Molin, 1859; *Dicrocoelium mutabile* of Braun, 1901; *Anchitrema mutabile* of Rizzo, 1902; *Paradistoma mutabile* of Dollfus, 1922.

HOSTS. Various lizards (e.g. sand lizard).

LOCATION. Gall bladder.

LOCALITY. Italy and Indo-China.

The hosts and the location of the parasite in the host aid identification of this species. More than a dozen other species parasitize reptiles in various parts of the world (Ceylon, Burma, India, the Philippines, Australia, South Africa).

Genus *Orchidasma* Looss, 1900

This genus has a species, *O. amphiorchis* (Braun, 1899) Braun, 1901, which parasitizes turtles in widely separated localities (Brazil, Florida, Japan) and has been found in the loggerhead turtle off the Lancashire coast (see Baylis, 1928*b*).

Two other trematodes which are also parasites of reptiles and have been referred to this family are *Distomum arrectum* and *D. assula*, both Dujardin, 1845. Both have been found in France, the former in *Lacerta* spp., the latter in the grass snake. Lent & Freitas (1940*b*) examined a trematode collected by Travassos from the common lizard in Hamburg and identified as '*Distoma arrectum* Molin, 1859', which they name *Plagiorchis molini* and which thus belongs to the Plagiorchiidae. As mentioned above, Perkins erected the genus *Cercolecithos* for this species in 1928 (see p. 303).

Family **LECITHODENDRIIDAE** Odhner, 1910 (p. 94)

Subfamily **Lecithodendriinae** Lühe, 1901, emend. Looss, 1902

Syn. Brachycoeliinae Looss, 1899, in part.

Genus *Lecithodendrium* Looss, 1896

Two species of this genus are parasites of reptiles in Egypt, *L. hirsutum* Looss, 1899 and *L. obtusum* (Looss, 1896) Looss, 1899, both in chamaeleons.

Subfamily **Pleurogenetinae** Looss, 1899

Macy (1936*a*) provided a key to genera of this subfamily, but it will not serve for the separation of European genera.

Genus *Pleurogenes* Looss, 1896

Pleurogenes claviger (Rudolphi, 1819) (Fig. 47H)

Syn. *Distoma clavigerum* Rudolphi, 1819.

HOSTS. Edible frog, common frog, common toad, natterjack toad, common newt.

LOCATION. Intestine.

DIAGNOSIS. *Shape, size and colour:* outline elongate oval, up to 3·3 mm. long and about one-third as broad, greenish brown. *Cuticle:* spinous. *Suckers:* ventral only about half as large as the oral and about one-third of the distance along the body. *Gut:* prepharynx short, pharynx and oesophagus of medium size, caeca long, extending to the last quarter of the body, but not quite to the posterior extremity. *Reproductive systems:* genital pore near the left margin of the body at the level of the intestinal bifurcation, far in front of the ventral sucker. Cirrus pouch large, extending diagonally towards the median plane, its base close to the anterior margin of the ventral sucker: containing a well-developed pars prostatica, a cirrus and a convoluted seminal vesicle. Testes rounded, side by side in the posterior third of the body near the terminations of the caeca. Ovary globular, on the right of the ventral sucker and thus far in front of the testes. Vitellaria lateral, confined to the anterior region between the pharynx and the ventral sucker, encroaching somewhat on the median plane. Uterus with long descending and ascending limbs, very much folded and filling the posterior region of the body. Eggs very numerous, measuring 0·033 × 0·016 mm.

OTHER SPECIES:

Pleurogenes loossi Africa, 1930 occurs in the intestine of the edible frog in Germany. Pigulewski (1931) described a species, *P. minus*, which was collected from the pike in the Ukraine!

Genus *Pleurogenoides* Travassos, 1921

Pleurogenoides medians (Olsson, 1876)

Syn. *Distomum medians* Olsson, 1876; *Distoma tacapense* Sonsino, 1894; *Pleurogenes medians* of Looss, 1899, *nec* Stafford, 1905; *P. tacapensis* of Looss, 1899.

HOSTS. Edible frog, common frog, common toad, natterjack toad (chamaeleon in Tunisia).

LOCATION. Intestine.

DIAGNOSIS. *Size:* 1·5–2 mm. long and rather more than half as broad. *Colour:* cloudy reddish yellow or golden red. *Suckers:* ventral smaller than the oral in the ratio 11/14. *Gut:* caeca short, scarcely extending to the level of the ventral sucker. *Reproductive systems:* testes one on either side of the body near the terminations of the caeca at the sides of the ventral sucker. Ovary in front of the ventral sucker. Eggs, similar to those of *Pleurogenes claviger*. Mathias (1924*d*) studied the life history of this species. Balozet & Callot (1938), who reviewed the trematodes of *Rana ridibunda* in Tunisia, proposed the erection of a new genus, *Sonsinotrema*, for *Distoma tacapense*, here included in the synonyms of *Pleurogenoides medians*.

Another species of this genus, *P. tener* (Looss, 1898) Travassos, 1921, was found in *Chamaeleo basiliscus* in Egypt and occurs in Tunisia. Travassos (1930*d*) listed six other species. His *Pleurogenoides stromi* was found by Strom in 1928 in Turkestan; the others occur in widely-separated regions (India, Australia, the Philippines).

Genus *Brandesia* Stossich, 1899

Brandesia turgida (Brandes, 1888) (Fig. 47E)

Syn. *Distomum turgidum* Brandes, 1888.

HOST. Edible frog.

LOCATION. Intestine, near the pylorus.

DIAGNOSIS. *Shape and size:* outline rounded or oval, body flat ventrally and convex dorsally, 2–2·5 mm. long and nearly as broad. *Cuticle:* spinous. *Suckers:* powerful, but not very prominent, both situated on the lower surface, the ventral smaller than the oral and located behind the mid-body, diameters 0·33 and 0·55 mm. *Gut:* prepharynx absent, pharynx present, oesophagus and caeca short, the latter terminating at the sides of the ovary. *Excretory system:* vesicle V-shaped. *Reproductive systems:* genital pore lateral, near the ventral sucker. Cirrus pouch short and thick. Gonads irregular in shape, testes side by side in the anterior region near the pharynx, ovary between and behind the testes, sometimes at the same level. Vitellaria confined to the anterior region, dorso-lateral to the oral sucker, comprising two small groups of coarse follicles. Uterus with descending and ascending limbs, formed into numerous folds, completely filling postero-dorsal part of the anterior region. Eggs very numerous, light golden brown, measuring 0·038 × 0·013 mm.

Note. Fig. 47E represents a sagittal half of the fluke showing only one testis, caecum and vitellarium.

Genus *Prosotocus* Looss, 1899

Prosotocus confusus (Looss, 1894) (Fig. 47G)

Syn. *Distomum confusum* Looss, 1894; *D. clavigerum* Pagenstecher, 1897.

HOSTS. Edible frog, common frog, common toad, natterjack toad.

LOCATION. Intestine.

DIAGNOSIS. *Shape, colour and size:* ovate, cloudy reddish yellow, 1·4 mm. long and about 1 mm. broad. *Cuticle:* spinous. *Suckers:* ventral very slightly smaller than the oral and situated in the middle of the body, diameters 0·15 and 0·16 mm. *Gut:* oesophagus short, bifurcation of the intestine far in front of the ventral sucker, caeca very short and confined to the region anterior to this sucker. *Excretory system:* vesicle V-shaped, the median stem very short, the lateral canals confined to the region behind the ventral sucker. *Reproductive systems:* genital pore on the left margin of the body at the level of the oesophagus. Cirrus pouch very large, its base within the posterior half of the body. Testes rounded, one on either side of the oesophagus. Ovary pyriform or irregular in shape, posterior to the testes, between the caeca and in front of the ventral sucker. Vitellaria consisting of a few groups of follicles on either side of the anterior region, between the testes and the oral sucker. Uterus with long descending and ascending limbs, formed into large folds encircling the ventral sucker and many subsidiary folds as well. Eggs very numerous and small (0·034 × 0·013 mm.).

Note. This species is said to be commoner in the edible than in the common frog.

Prosotocus fuelleborni Travassos, 1930

This species was erected for several trematodes found in the intestine of the edible frog near Hamburg. The smallest and largest of seven specimens for which measurements were given were 0·92 and 1·31 mm. in length 0·71 and 0·84 mm. in breadth, had oral and ventral suckers of nearly equal size (though the general measurements are variable in respect of this character), and eggs showing only slight variation in size (measuring 0·028 × 0·014–0·016 mm.). There seems to be little doubt that the specimens of Travassos are identical with *Prosotocus confusus*, of which *P. fuelleborni* must be regarded as a synonym.

Family **GORGODERIDAE** Looss, 1901 (p. 93)

Subfamily **Gorgoderinae** Looss, 1899

Genus *Gorgodera* Looss, 1899

Gorgodera cygnoides (Zeder, 1800) (Fig. 47I)

Syn. *Distoma cygnoides* Zeder, 1800.

HOSTS. Edible frog, common frog.

LOCATION. Urinary bladder.

DIAGNOSIS. *Shape and size:* elongate, lancet-like (but showing lively movements when alive), up to 15 mm. long (but generally less than 10 mm.) and not more than 1 mm. broad. *Cuticle:* non-spinous. *Suckers:* ventral almost twice as large as the oral and situated about one-third of the distance along the body. *Gut:* pharynx absent, oesophagus of medium length, caeca long, bifurcation of the intestine well in front of the ventral sucker. *Reproductive systems:* genital pore near the anterior margin of the ventral sucker. Cirrus pouch absent. Testes divided into nine parts, four on one side of the median line, five on the other, in strict alinement, partition occurring in the cercaria. Ovary reniform, situated in front of testes. Vitellaria small, compact and paired, lobed (but not deeply), situated slightly in front of the ovary. Uterus having descending and ascending limbs which extend between the lateral rows of testes to the posterior extremity, not much folded. Eggs measuring about 0·031 × 0·016 mm. shortly after leaving the ootype, much larger farther along the uterus (0·047–0·048 × 0·030–0·031 mm.). *Cercariae:* 'Cystocercous' form of '*Macrocercous*' ('Gorgodera' group), *C. macrocerca* Filippi, developing in the gills of *Sphaerium* sp., later encysting in larval insects (e.g. *Epitheca*).

Gorgodera pagenstecheri Sinitsîn, 1905

HOSTS. Edible frog, common frog.

LOCATION. Urinary bladder.

DIAGNOSIS. More elongate than *G. cygnoides* and further distinguished from it by the irregularly lobed ovary, deeply lobed vitellaria (lobes longer than wide) and slightly smaller eggs which ultimately measure 0·04 × 0·028 mm. *Cercariae:* similar to *G. cygnoides*, developing in *Sphaerium corneum* (in sporocysts), later encysting in insects, generally in the fat body or on the gut of *Epitheca* sp. or *Agrion* sp.

Gorgodera varsoviensis Sinitsîn, 1905

HOSTS. Edible and common frogs.

LOCATION. Urinary bladder.

DIAGNOSIS. More elongate than *G. pagenstecheri* and resembling it in the characters of the ovary and vitellaria, but differing in the smaller size of eggs which (when ripe) measure 0·032 × 0·025 mm. The ventral sucker is large and powerful, about two and a half times as large as the oral, and oversteps the lateral margins of the body.

Two other species of *Gorgodera*, *G. microovata* and *G. asymmetrica*, both Fuhrmann, 1925, were found in the bladder of the edible frog in Switzerland.

Genus *Gorgoderina* Looss, 1902

Pereira & Cuocolo (1940) suggested the splitting of this genus into the subgenera *Gorgoderina* and *Neogorgoderina* with *G.* (*G.*) *vitelliloba* (Olsson) and *G.* (*N.*) *simplex* (Looss) as respective types.

Gorgoderina vitelliloba (Olsson, 1876)

Syn. *Distomum vitellilobum* Olsson, 1876.

HOSTS. Edible frog, common frog (common toad also in Britain; see Baylis, 1939).

LOCATION. Urinary bladder.

DIAGNOSIS. *Shape and size:* slightly elongate, lancet-like, about 6·5 mm. long and 2 mm. broad. *Cuticle:* non-spinous. *Suckers:* ventral overstepping the lateral margins of the body. *Reproductive systems:* testes paired, not further subdivided, elongate, having slightly curved borders. Ovary irregular in shape, more or less lobed, situated in front of the testes. Vitellaria paired, formed into two or three lobes. *Other characters:* similar to *Gorgodera*, but with small eggs measuring about 0·035 × 0·025 mm. *Cercariae:* similar to *Gorgodera*, but in the 'Gorgoderina' group, developing in sporocysts, later passively entering carnivorous insects, generally of the genera *Epitheca* and *Agrion*. .

Another species, *Gorgoderina capensis* Joyeux & Baer, 1934, occurs in the bladder of frogs in Southern Tunisia (Gafsa).

Another Gorgoderid trematode, *Phyllodistomum cymbiforme* (Rudolphi, 1819) Braun, 1899, has been recorded in the turtles *Caretta caretta* and *Chelonia mydas* in Italy, Egypt and Florida.

Family **HALIPEGIDAE** Poche, 1926 (p. 90)

Genus *Halipegus* Looss, 1899

Halipegus ovocaudatus (Vulpian, 1859) (Fig. 47A)

Syn. *Distomum ovocaudatum* Vulpian, 1859; *D. kessleri* Grebnitzsky, 1872; *Halipegus kessleri* of Vlassenko, 1929; *H. rossicus* Issaitschikow & Zakharov, 1929.

HOSTS. Edible frog, common frog.

LOCATION. Buccal cavity, under tongue; sometimes in Eustachian recesses.

DIAGNOSIS. *Shape, colour and size:* elongate, almost cylindrical, but constricted near the mid-body and having pointed extremities, reddish or flesh colour, up to 13 mm. long. *Cuticle:* non-spinous. *Suckers:* powerful, the ventral much larger than the oral and occupying the middle of the body, diameters 1·3 and 0·8 mm. *Gut:* pharynx small, oesophagus very short, caeca long, bifurcation of the intestine near the oral and almost half the body-length in front of the ventral sucker. *Excretory system:* vesicle Y-shaped, having a long median stem and long lateral canals which unite above the oral sucker. *Reproductive systems:* genital pore far forward near the pharynx. Cirrus pouch absent. Seminal vesicle small, pars prostatica ill-developed. Testes rounded, side by side near the middle of the posterior region. Ovary rounded, lateral, behind the testes. Vitellaria paired, but ill-developed, comprising a small group of four or five large follicles on either side of the body and slightly behind the ovary. Uterus with an ascending limb only, having numerous folds which tend to fill the space in front of the vitellaria and are as abundant in front of the ventral sucker as behind it. Eggs elongate, golden yellow, measuring 0·063 × 0·022 mm., and having a long filament nearly half as long again as the capsule at the anopercular pole. *Cercariae:* 'Cystocercous' form, *C. cystophora* Wagener, 1866, developing in rediae in *Planorbis* spp., later penetrating and encysting in the larvae of *Calopteryx virgo.*

Family **PARAMPHISTOMATIDAE** Fischoeder, 1901, emend. Goto & Matsudaira, 1918 (p. 103)

Subfamily **Diplodiscinae** Cohn, 1904

The amphistomes of Amphibia are confined to this group. All occur in the rectum.

Genus *Diplodiscus* Diesing, 1836

Syn. *Megalodiscus* Chandler, 1923.

Posterior sucker having a small central accessory sucker, one testis (two in 'Megalodiscus') and an oesophageal bulb present.

Diplodiscus subclavatus (Pallas, 1760) (Fig. 10A)

Syn. *Fasciola subclavata* Pallas, 1760; *Planaria subclavata* of Goeze, 1787; *Amphistoma subclavatum* of Rudolphi, 1819.

HOSTS. Edible frog, common frog, common toad, common newt.

LOCATION. Rectum.

DIAGNOSIS. *Shape and size:* somewhat spindle-shaped, rounded posteriorly, tapering anteriorly from a little behind the mid-body, little flattened, up to 6 mm. long and 3 mm. broad. *Cuticle:* non-spinous. *Suckers:* ventral large and prominent, posterior in position, occupying the entire breadth of the body; oral much smaller, its length one-sixth to one-tenth of the body length, and having a pair of small postero-dorsal pouches. *Gut:* pharynx absent, oesophagus long and muscular, caeca fairly long, terminating slightly in front of the posterior sucker. *Excretory system:* vesicle sac-like and having a short terminal canal leading to a slightly dorsal pore; lateral canals sinuous, extending forward to the level of the oesophagus. *Reproductive systems:* genital pore median, slightly behind the bifurcation of the intestine, about one-third of the distance along the body. Cirrus pouch absent. Seminal vesicle small, pyriform. Testes paired in the juvenile fluke, fused in the adult into a single, spherical mass situated slightly behind the mid-body. Ovary rounded, slightly lateral, located behind the testis. Vitellaria comprising a row of large follicles lateral to the caecum on each side of the body, extending slightly towards the median plane beyond the ends of the caeca and in front of the posterior sucker. Uterus with an ascending limb only (or descending limb very short), formed into slight folds. Eggs very large, but not numerous, measuring 0·128–0·137 × 0·082–0·090 mm. *Cercariae: C. diplocotylea* Filippi, developing in rediae in various species of *Planorbis*, later encysting (when water becomes shallow) on the skin of frogs, especially on the pigmented spots.

Another species of this genus, *D. conicum* Polonio, 1859, was recorded in *Natrix chrysarga* in Italy. Other species occur outside Europe.

Genus *Opisthodiscus* Cohn, 1904

Posterior sucker having a small central accessory sucker; one testis present; oesophageal bulb absent.

Opisthodiscus diplodiscoides Cohn, 1904 (a rare species)

HOST. Edible frog.

LOCATION. Rectum.

DIAGNOSIS. *Size:* up to 2·7 mm. long and 1·2 mm. broad at the mid-body. *Shape and other characters:* similar to *Diplodiscus subclavatus*, but distinguished from this species by the paired testes (which remain discrete and are situated lateral to the ovary), the vitellaria (which are restricted to about six follicles on either side of the body), the genital pore (which occupies a slightly more anterior position), the asymmetrical caeca, and an oral sucker having long pouches extending one-third of the distance along the body. The eggs are very large, measuring about 0·13 × 0·07 mm.

The genus *Catadiscus* Cohn, 1904 was founded for a trematode, *C. dolichocotyle* (Cohn, 1903) Cohn, 1904, from the rectum of *Chironius fuscus* in South America. The posterior sucker lacks the small central accessory sucker.

Subfamily **Schizamphistomatinae** Looss, 1912, emend.

Szidat (1939*a*) included six genera of reptilian trematodes in this group, all except one of which (*Stunkardia* Bhalerao, 1931) were arranged by Hughes *et al.* (1942) in the Cladorchiinae, which Szidat was inclined to regard as a group of trematodes of mammals. The genera, with the localities in which their species occur are: *Schizamphistomum* (Mediterranean, Atlantic, Australia); *Ophioxenos* Sumwalt, 1926 (U.S.A.); *Schizamphistomoides* Stunkard, 1925 (Mexico, Tanganyika, Mediterranean); *Alassostoma* Stunkard, 1917 (North America); *Alassostomoides* Stunkard, 1924 (North America); and *Stunkardia* Bhalerao, 1931 (included in the Zygocotylinae by Hughes *et al.*) (India). Other genera mentioned by Hughes *et al.* (Cladorchiinae) are *Chiorchis* Fischoeder, 1901 (Malaya); *Halltrema* Lent & Freitas, 1939 (Brazil); and *Nematophila* Travassos, 1934 (Amazon Valley). It will be noticed that only two genera occur in Europe, these being represented by *Schizamphistomum scleroporum* (Creplin, 1844) Looss, 1912 and *Schizamphistomoides spinulosum* (Looss, 1901) Stunkard, 1925. Both are parasites of turtles in the Mediterranean.

SOME TREMATODES OF CHELONIANS

Many trematodes are characteristically parasites of Chelonia. The following, species of nine genera of the Pronocephalidae, five of the Microscaphidiidae and two of the Spirorchiidae, have been found in the Mediterranean. They are unimportant for the general purpose of this book, but are listed as trematodes showing a high degree of specificity. Species marked * occur in *Chelonia mydas* those marked † in the loggerhead turtle, near Egypt, except where otherwise stated.

Family **PRONOCEPHALIDAE** Looss, 1902

Subfamily **Pronocephalinae** Looss, 1899

Genus *Adenogaster* Looss, 1901 (†*A. serialis* Looss, 1901).
Genus *Astrorchis* Poche, 1926 (*A. renicapite* (Leidy, 1856) Poche; in the luth, leathery turtle in Tunisia and U.S.A.).
Genus *Cricocephalus* Looss, 1899 (**C. megatomus* Looss, 1902; **C. resectus* Looss, 1902).
Genus *Epibathra* Looss, 1902 (†*E. crassa* (Looss, 1901) Looss, 1902).
Genus *Glyphicephalus* Looss, 1901 (**G. lobatus* Looss, 1901; **G. solidus* Looss, 1901).
Genus *Pleurogonius* Looss, 1901 (**P. bilobus*; **P. linearis*; **P. longiusculus*; and **P. minutissimus*; all Looss, 1901).
Genus *Pyelosomum* Looss, 1899 (**P. cochlear* Looss, 1899).

Subfamily **Charaxicephalinae** Price, 1931

Genus *Charaxicephalus* Looss, 1901 (**C. robustus* Looss, 1901).
Genus *Diaschistorchis* S. J. Johnston, 1913 (*D. pandus* (Braun, 1901) Johnston, 1913, in various turtles in Australia, Japan, Cuba, Bermuda and Italy).

Family **MICROSCAPHIDIIDAE** Travassos, 1922 (p. 104)

Genus *Angiodictyum* Looss, 1902 (**A. parallelum* (Looss, 1901) Looss, 1902).
Genus *Deuterobaris* Looss, 1900 (**D. proteus* (Brandes, 1891) Looss, 1901).
Genus *Microscaphidium* Looss, 1900 (**M. aberrans* Looss, 1902; **M. reticulare* (Beneden, 1859) Looss, 1901).
Genus *Octangium* Looss, 1902 (**O. hasta* Looss, 1902; **O. sagitta* (Looss, 1899) Looss, 1902).
Genus *Polyangium* Looss, 1902 (**P. linguatula* (Looss, 1899) Looss, 1902; **P. miyajimae* Kobayashi, 1915) (the latter in the South Pacific also).

Family **SPIRORCHIIDAE** MacCallum, 1921

A key to genera of which was given by Price (1934)

Genus *Hapalotrema* Looss, 1899 (†*H. loosi* Price, 1934; *H. polesianum* (Esjmont, 1927)) (the latter in *Emys orbicularis* in Poland).
Genus *Learedius* Price, 1934 (* †*L. europaeus* Price, 1934) (**L. learedi* and **L. similis* Price, 1934) in U.S.A.
Genus *Neospirorchis* Price, 1934 (**N. schistosomatoides* Price, 1934) in U.S.A.
Genus *Amphiorchis* Price, 1934 (**A. amphiorchis* Price, 1934) in U.S.A.

Some trematodes of Chelonia (and a few other reptiles) belong to families characteristically represented in fishes. The following brief records are interesting in this connexion, though they lie outside the British fauna, and may involve inaccurate diagnosis in some instances.

Family **ALLOCREADIIDAE** (subfamily **Allocreadiinae**), *Crepidostomum cooperi* Hopkins, 1931; adults in various fishes and *Trionyx mutica* (North America).

Family **ZOOGONIDAE** (? Subfamily **Zoogoninae** Odhner, 1902), ? *Zoogonoides boae* MacCallum, 1921; in the boa constrictor (Brazil). This trematode would perhaps be more appropriately placed in the Lecithodendriidae. The solitary specimen so very briefly described by MacCallum had paired, well-developed vitellaria, and thus certainly does not belong to the genus *Zoogonoides*.

Family **AZYGIIDAE**, *Proterometra sagittaria* Dickermann, 1937; cercariae reared to adults in fishes and turtles (U.S.A.).

Family **HEMIURIDAE** (subfamily **Hemiurinae**), *Hemiurus* sp. T. H. Johnston, 1912; in *Denisonia superba* and *Pseudechis porphyriacus* (Australia).

Family **HEMIURIDAE** (subfamily **Sterrhurinae**), *Lecithochirium dillanei* Nicoll, 1918; in *Distiera* sp. (Australia).

Family **ACANTHOSTOMATIDAE** Nicoll, 1935. The genus *Acanthostomum* Looss, 1899 has nine species in reptiles, mainly crocodiles, alligators and turtles, variously occurring in Brazil, North America, Sudan, Australia and the Philippines. One other genus, *Atrophecaecum* Bhalerao, 1940, was made for an Indian form previously attributed to *Acanthostomum*.

CHAPTER 10

SOME TREMATODES OF BIRDS

Despite the richness and variety of the trematode fauna of British birds—representatives of more than twenty families are mentioned in this chapter—it has been little investigated in this country, and we owe most of our knowledge to zoologists on the Continent. Fortunately, our bird fauna agrees very closely with that of western Europe, and there is some likelihood of agreement in the trematode faunae of the two regions. Hardly a single trematode which has been found in birds in Britain is peculiar to this country, though Nicoll (1923*a*) credited *Echinostephilla virgula* Lebour, 1909 from the turnstone with this characteristic. Most of our birds occur elsewhere in Europe and the temperate parts of Asia, some migrants travelling farther afield and reaching the tropical parts of Africa and Asia, even the Arctic regions and America. Such migrations cannot be without effect and doubtless have something to do with the abundance of the parasites, trematodes acquired in distant countries being introduced into Europe. At the same time, by the manner of their feeding and in the nature of their food, birds are peculiarly susceptible to infection, though this is not a universal trait.

Some birds seem to be immune from infection with trematodes, others tolerate at most one species of fluke, while some harbour as many as twenty different species. Passerines, excluding the Corvidae, represent one extreme, gulls the other. There is a similar disparity between gallinaceous birds and ducks. Some of these differences must be correlated with differences in the nature of the food, herbivores being infected to a lesser extent than carnivores, because the latter ingest free or encysted larval stages in the tissues of the animals which form their prey. Certainly fishes are important agents in the infection of such birds.

Some trematodes occur with great frequency in a large variety of avian hosts, others seem to be rarer, even specific for one or a few hosts. Monostomes like *Notocotylus attenuatus* and *Catatropis verrucosa* belong to the former category. Holostomes like *Strigea* spp. are common parasites of birds of prey, ducks and gulls. *Prosthogonimus ovatus* likewise occurs in numerous birds, more than half of which are passerines or wading birds. The species *P. cuneatus* shows an even greater preference for passerines. At the other extreme, the two commonest Echinostomes, *Echinostoma revolutum* and *Hypoderaeum conoideum*, are almost entirely confined to ducks and their close relatives.

Nearly all these common trematodes select peculiar locations in the body of the host. As might be expected, the most popular situation in which to find them is the intestine, *Notocotylus* and *Catatropis* inhabiting the caecum as well. But *Prosthogonimus* chooses the bursa Fabricii, *Leucochloridium* the cloaca, and both Holostomes and Echinostomes show a preference for the lower end of the intestine. Some common trematodes of birds seek very unusual locations, most Monostomes inhabiting the air sacs, and *Collyriclum faba* cysts under the skin. Birds also have their liver flukes which, like those of mammals, belong to the Opisthorchiidae and Dicrocoeliidae. But there are specific forms such as

Lyperosomum longicauda, a common liver fluke of crows, and *Athesmia heterolecithodes*, a liver fluke of the moorhen.

Another type of trematode* with a distinctive habitat belongs to the Clinostomatidae and occurs in the buccal cavity and the upper ends of the oesophagus and trachea. Yet another is *Renicola pinguis* (Troglotrematidae), which inhabits the kidney, as does the rarer species *Eucotyle nephritica* (Eucotylidae). Few schistosomes are likely to be found in Britain, though Nicoll found *Gigantobilharzia acotylea* Odhner, 1910 in the abdominal veins of the black-headed gull at St Andrews more than 30 years ago.

Some birds, like other vertebrates, are exceptional in harbouring two or more species of Trematoda belonging to the same genus. The common scoter is said to be parasitized by no fewer than seven species of *Gymnophallus* (Microphallidae).. It is uncertain whether or not all the forms thus found in a single host are genuine species, but both the trematodes and the hosts in such instances are worthy of special study. Skrjabin (1926) recorded the case of an exceptional bird simultaneously parasitized by seventeen species of helminths.

It is impossible to deal in a comprehensive way with the trematodes of birds, and sufficient for our purpose to examine some of the better-known species to single out a few hosts which provide what might be called a representative sample of the trematode fauna. In Table 5 the parasites of nine birds are listed, and it will be seen that they represent no fewer than twenty families of Digenea, about half of which are represented in the domestic duck alone. Diagnoses of other Digenea deemed highly characteristic parasites of birds are given, but no attempt has been made even to enumerate *all* European species, though many are named.

Family **PLAGIORCHIIDAE** Lühe, 1901 (p. 94)

Syn. Lepodermatidae Looss, 1901.

KEY TO SOME GENERA OCCURRING IN BIRDS

1. (Plagiorchiinae.) Body elongate; testes one behind the other; genital pore distant from the anterior extremity — *Plagiorchis*
2. (Prosthogoniminae.) Body broadly oval; testes side by side; genital pores separate and near the anterior extremity
   - A. Greatest breadth behind the middle of the body; folds of the uterus overreaching the caeca laterally; genital pores close together — *Prosthogonimus*
   - B. Greatest breadth at the middle of the body; folds of the uterus confined within the limits of caeca, genital pores somewhat separated — *Schistogonimus*

Subfamily **Plagiorchiinae** Lühe, 1901, emend. Pratt, 1902

Genus *Plagiorchis* Lühe, 1899

Syn. *Lepoderma* Looss, 1899.

Plagiorchis arcuatus Strom, 1924

HOST. Fowl.

LOCATION. Oviducts.

LOCALITY. Russia and Germany.

DIAGNOSIS. *Shape and size:* outline oval, pointed at both extremities, 4–4·7 mm. long and 1·2–1·5 mm. broad. *Cuticle:* spinous, spines more abundant anteriorly. *Suckers:* almost equal in size, diameters 0·46–0·48 mm. (oral) and

* E.g. *Clinostomum complanatum* (see Table 5).

Table 5. *Some Trematodes of Various Birds in Europe*

As, air sacs; Ac, abdominal cavity; B, gall bladder; Bd, bile ducts; BF, bursa Fabricii; Bv, blood vessels; C, cloaca; c, conjunctiva; Ca, caeca; E, eye; I, intestine; IO, infra-orbital sinus; K, kidney; L, liver; M, buccal cavity; N, nasal cavity; n, nictitating membrane; O, oesophagus; R, rectum; (s), subcutaneous cysts; T, trachea; Tb, trachea and bronchi; Tc, thoracic cavity.

	Carrion crow	Rook	House sparrow	Domestic duck	Common scoter	Herring gull	Black-throated diver	Coot	White stork
PLAGIORCHIIDAE:									
Prosthogonimus ovatus	BF	BF	BF	—	—	—	—	BF	—
P. cuneatus	BF	BF	BF	—	—	—	—	BF	—
P. anatinus	—	—	—	BF	—	—	—	—	—
P. rudolphii	—	—	—	BF	—	—	—	—	—
P. skrjabini	—	—	—	BF	—	—	—	—	—
Plagiorchis cirratus	I	I	—	—	—	I	—	—	—
P. elegans	—	—	I	—	—	—	—	—	—
P. marii	—	—	I	—	—	—	—	—	—
P. potanini	—	—	—	I	—	—	—	—	—
Schistogonimus rarus	—	—	—	BF	—	—	—	BF	—
DICROCOELIIDAE:									
Lyperosomum longicauda	LB	B	—	—	—	—	—	—	—
Oswaldoia skrjabini	B	B	—	—	—	—	—	—	—
MICROPHALLIDAE:									
Gymnophallus affinis	—	—	—	—	I	—	—	—	—
G. bursicola	—	—	—	—	BF	—	—	—	—
G. dapsilis	—	—	—	—	BFI	—	—	—	—
G. deliciosus	—	—	—	—	—	B	—	—	—
G. oidemiae	—	—	—	—	I	—	—	—	—
G. ovoplenus	—	—	—	—	I	—	—	—	—
G. macroporus	—	—	—	—	I	—	—	—	—
G. somateriae	—	—	—	—	ICa	—	—	—	—
Maritrema lepidum	—	—	—	—	—	I	—	—	—
M. gratiosum	—	—	—	—	I	—	—	—	—
Spelotrema pygmaeum	—	—	—	—	I	—	—	—	—
S. excellens	—	—	—	—	—	I	—	—	—
S. simile	—	—	—	—	—	I	—	—	—
Levinseniella pellucida	—	—	—	Ca	—	—	—	—	—
L. brachysoma	—	—	—	—	I	—	—	—	—
LECITHODENDRIIDAE:									
Phaneropsolus sigmoideus	—	—	I	—	—	—	—	—	—
Eumegacetes emendatus	—	—	I	—	—	—	—	—	—

OPISTHORCHIIDAE:									
*Amphimerus speciosus**	L	—	—	—	—	—	—	—	—
Metorchis xanthosomus	—	—	—	BdL	B	—	—	—	—
M. crassiusculus	—	—	—	B	—	—	—	—	—
Opisthorchis simulans	—	—	—	Bd	—	—	—	—	—
Pachytrema calculus	—	—	—	—	—	B	—	—	—
HETEROPHYIDAE:									
'Ciureana' cryptocotyloides	—	—	—	—	—	—	I	—	—
Cryptocotyle concava	—	—	—	—	I	—	—	—	—
C. lingua	—	—	—	—	—	I	—	—	—
CLINOSTOMIDAE:									
Clinostomum complanatum	—	—	—	—	—	MO	—	—	—
STRIGEIDAE:									
Strigea sphaerula	I	I	—	—	—	—	—	—	—
Parastrigea robusta	—	—	—	I	—	—	—	—	—
Apatemon gracilis	—	—	—	I	I	—	—	—	—
Cotylurus cornutus	—	—	—	I	—	—	—	—	—
C. variegatus	—	—	—	—	—	I	IBF	—	—
C. erraticus	—	—	—	—	—	I	IBF	—	—
Apharyngostrigea cornu	—	—	—	—	—	—	—	—	I
Cardiocephalus longicollis	—	—	—	—	—	I	—	—	—
DIPLOSTOMATIDAE:									
'Tylodelphys excavatum'	—	—	—	—	—	—	—	—	I
Diplostomum spathaeceum	—	—	—	—	—	I	—	—	—
'Hemistomum pileatum'	—	—	—	—	—	—	I	—	I
SCHISTOSOMATIDAE:									
Gigantobilharzia monocotylea	—	—	—	Bv	—	—	—	—	—
Bilharziella polonica	—	—	—	Bv	—	—	—	—	—
STOMYLOTREMATIDAE:									
Stomylotrema pictum	—	—	—	—	—	—	—	—	C
PHILOPHTHALMIDAE:									
Philophthalmus lucipetus	—	—	—	—	—	c	—	—	—
PSILOSTOMATIDAE:									
Apopharynx bolodes	—	—	—	—	—	—	—	BF	—
Psilostomum brevicolle	—	—	—	—	I	—	—	—	—
Psilochasmus oxyurus	—	—	—	—	I	—	—	—	—
EUCOTYLIDAE:									
Eucotyle nephritica	—	—	—	—	—	—	K	—	—
CATHAEMASIIDAE:									
Cathaemasia hians	—	—	—	—	—	—	—	—	O

Table 5 (*cont.*)

	Carrion crow	Rook	House sparrow	Domestic duck	Common scoter	Herring gull	Black-throated diver	Coot	White stork
ECHINOSTOMATIDAE:									
Echinostoma revolutum	R	—	—	R	R	—	—	—	I
E. anceps	—	—	—	—	—	—	—	I	—
E. mesotestius	—	I	—	—	—	—	—	—	—
Echinoparyphium paraulum	—	—	—	I	—	—	—	—	—
E. recurvatum	—	—	—	I	—	—	—	—	—
E. baculus	—	—	—	—	—	—	I	—	—
Echinochasmus coaxatus	—	—	—	—	—	—	—	—	I
Hypoderaeum conoideum	—	—	—	R	—	—	—	—	—
Himasthla elongata	—	—	—	—	—	I	—	—	—
H. leptosoma	—	—	—	—	—	I	—	—	—
Parorchis acanthus	—	—	—	—	—	RBF	—	—	—
Chaunocephalus ferox	—	—	—	—	—	—	—	—	I
Stephanoprora denticulata	—	—	—	—	—	I	I	—	—
S. spinosa	—	—	—	—	—	—	I	—	—
TROGLOTREMATIDAE:									
Collyriclum faba	—	—	(s)	—	—	—	—	—	—
NOTOCOTYLIDAE:									
Notocotylus attenuatus	—	—	—	CaR	—	—	—	CaI?	—
N. thienemanni	—	—	—	Ca	—	—	—	—	—
N. gibbus	—	—	—	—	—	—	—	I	—
N. seineti	—	—	—	Ca	—	—	—	—	—
Catatropis verrucosa	—	—	—	Ca	BF	—	—	I	—
Paramonostomum alveatum	—	—	—	—	I	—	—	—	—
CYCLOCOELIDAE:									
Cyclocoelum mutabile	—	—	—	—	N	—	—	AsTc	—
*C. orientale**	—	—	—	—	—	—	—	TbTc	—
'*C. pseudomicrostomum*'	—	—	—	—	—	—	—	AcTc	—
Typhlocoelum cucumerinum	—	—	—	NTb	—	—	—	—	—
Tracheophilus cymbium	—	—	—	T	—	—	—	—	—
'*Hyptiasmus laevigatus*'	—	—	—	—	NIO	—	—	—	—
'*Hyptiasmus oculeus*'	—	—	—	—	—	—	—	N	—
'*H. sigillum*'	—	—	—	—	—	—	—	N	—
BRACHYLAEMIDAE:									
Brachylaemus fuscatus	—	I	I	—	—	—	—	—	—
Leucochloridium macrostomum	BFI	—	C	—	—	—	—	—	—
*L. insignis**	—	—	—	—	—	—	—	C	—
MICROSCAPHIDIIDAE:									
Polyangium colymbi	—	—	—	—	—	—	K	—	—

* Not characteristic European species.

0·47–0·49 mm. (ventral). *Gut:* oesophagus absent. *Reproductive systems:* testes rounded, one behind the other in the posterior half of the body. Cirrus pouch enclosing a bipartite seminal vesicle which is rounded in front and elongate behind. Vitellaria of continuous extent between the pharynx and the posterior extremity, the follicles mingling in the median plane in front of the ventral sucker and behind the testes. Uterus confined to the region in front of the anterior testis. Eggs measuring 0·035–0·042 × 0·021–0·023 mm.

OTHER SPECIES:

The best-known species parasitizing birds are *Plagiorchis maculosus* (Rudolphi, 1802), *P. cirratus* (Rudolphi, 1802), *P. elegans* (Rudolphi, 1802), *P. nanus* (Rudolphi, 1802), *P. triangulare* (Diesing, 1850), *P. vitellatus* (Linstow, 1875) and *P. permixtus* Braun, 1901, the first and second of which have appeared in this country, in the swift and the black-headed gull respectively (Baylis, 1939*a*). Mention can be made of *P. notabilis* Nicoll, 1909, also found in Britain in the rock pipet and blue-headed wagtail. Foggie (1937) described an outbreak of parasitic necrosis in the turkey in Northern Ireland caused by another species, *P. laricola* Skrjabin, 1924. Two other species, *P. marii* Skrjabin, 1920 and *P. potanini* Skrjabin, 1928, are shown in Table 5, which does not pretend to completeness. Szidat (1924) described *P. fastuosus* from *Calidris alpina.* Skrjabin & Massino (1925) described another species, *P. micromaculatus*, Massino (1927) six others (*skrjabini*, *fuelleborni*, *brauni*, *uhlwormi*, *loossi* and *blumbergi*) and Semenov (1927) three more (*multiglandularis*, *micronotabilis* and *melanderi*), all in birds of Russia. The genus is well represented also in amphibians, reptiles and mammals.

Subfamily **Prosthogoniminae** Lühe, 1909

Genus *Prosthogonimus* Lühe, 1899

Syn. *Prymnoprion*, Looss, 1899 and various misprints, e.g. *Prostogonimus* Markow, 1903.

This genus was reviewed recently by Witenberg & Eckmann (1939).

Prosthogonimus ovatus (Rudolphi, 1803) Lühe, 1899 (Fig. 50A)

Syn. *Fasciola ovata* Rudolphi, 1803; *Distoma ovatum* of Rudolphi, 1809; *Cephalogonimus ovatus* of Stossich, 1892; *Prymnoprion ovatus* of Looss, 1899.

HOSTS. Fowl and goose and various wild birds (rook, jackdaw, hooded crow, magpie, starling, house sparrow, tree sparrow, buzzard, sparrow hawk, hobby, long-tailed duck, lapwing, common gull, Richardson's skua, black guillemot, spotted crake, corncrake, coot) (carrion crow in Britain).

LOCATION. Bursa Fabricii, sometimes the oviduct, rarely the intestine.

LOCALITY. Europe, Africa, Asia.

This widely distributed fluke, like other species of the genus, is pathogenic, causing inflammation of the oviduct and various abnormalities of the egg (in which it is sometimes included) and egg production, often inducing a fatal peritonitis.

DIAGNOSIS (after Braun, 1901*c*; modified). *Shape and size:* flattened, outline pyriform, 3–6 mm. long and 1–2 mm. broad. *Cuticle:* having triangular scales anteriorly and spines behind the ventral sucker; the latter about 0·015 mm. long

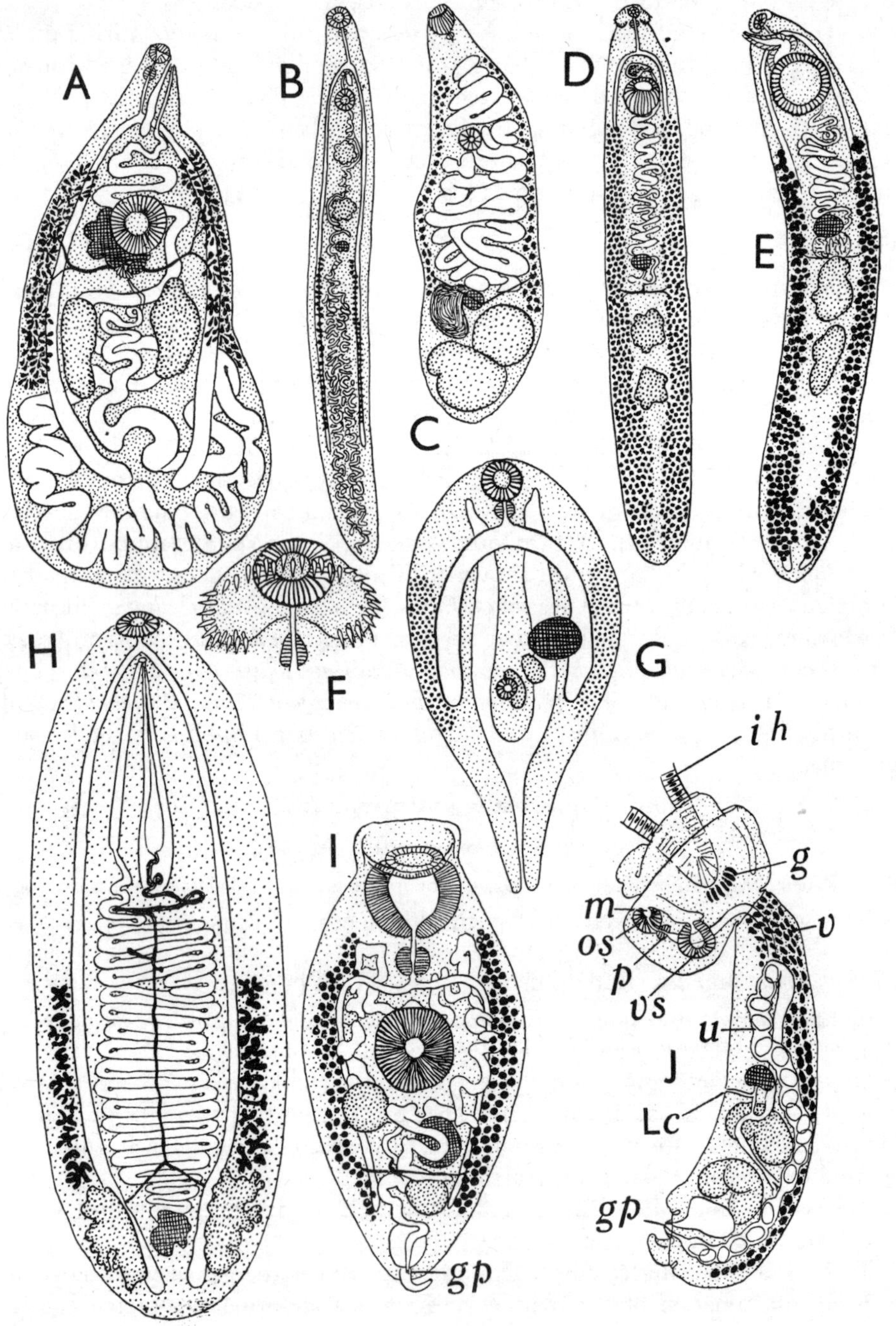

Fig. 50. Plagiorchiidae (A), Dicrocoeliidae (B), Opisthorchiidae (C), Echinostomatidae (D–F), Troglotrematidae (G), Notocotylidae (H), Brachylaemidae (I) and Strigeidae (J) from British birds. A, *Prosthogonimus ovatus*. B, *Athesmia heterolecithodes*. C, *Metorchis xanthosomus*. D, *Echinostoma revolutum*. E, *Hypoderaeum conoideum*. F, head collar of *Echinostoma revolutum*. G, *Renicola pinguis*. H, *Catatropis verrucosa*. I, *Leucochloridium macrostomum*. J, *Cotylurus erraticus* (*Strigea erratica*). (A, J, after Fuhrmann, 1928; D, E, F, after Mönning, 1934; rest after Lühe, 1909.)

and sparser posteriorly. *Suckers:* oral elongate-oval, measuring about 0·15–0·17 × 0·17–0·21 mm. Ventral about twice as large as the oral and about one-third of the length of the body behind it, diameter about 0·4 mm. *Gut:* pre-pharynx hardly developed, pharynx small, oesophagus fairly long, caeca long, ending near or in the final quarter of the body, often filled with blood. *Reproductive systems:* male and female pores close together on the left of the oral sucker. Cirrus pouch small. Testes slightly elongate, having curved borders, side by side slightly behind the middle of the body. Ovary deeply lobed, in front of and between the testes, dorsal to the ventral sucker. Vitellaria lateral, extending from slightly in front of the ventral sucker to near the posterior margin of the testes. Uterus having descending and ascending limbs extending between the testes and formed into rosette-like folds which overstep the caeca laterally in the posterior region. Folds of the uterus also occurring between the caeca in front of the ventral sucker. Eggs very small and thin-shelled, dimensions 0·022–0·024 × 0·013 mm.

Prosthogonimus cuneatus (Rudolphi, 1809) Braun, 1901

Syn. *Distoma cuneatum* Rudolphi, 1809; *Prymnoprion anceps* Looss, 1899.

HOSTS, LOCATION and LOCALITY. Similar to *P. ovatus*, but also parasitic in the swan.

DIAGNOSIS. *Shape and size:* outline almost triangular, pointed anteriorly, 5·2 mm. long and 1·7 mm. broad. *Cuticle:* spinous, spines about 0·0145 mm. long. *Suckers:* ventral twice as large as the oral, diameters 0·6–0·8 and 0·3–0·4 mm. *Other characters:* similar to *P. ovatus*, but vitellaria not extending in front of the ventral sucker, the ovary behind this haptor. Eggs measuring 0·023–0·027 × 0·013–0·016 mm.

Prosthogonimus pellucidus (Linstow, 1873) Braun, 1901

Syn. *Distoma pellucidum* Linstow, 1873; *Cephalogonimus pellucidus* of Railliet, 1890; *Mesogonimus pellucidus* of Neumann, 1892; *Prymnoprion pellucidus* of Looss, 1899.

HOSTS. Fowl, curlew.

LOCATION. Bursa Fabricii, oviducts, sometimes oesophagus (Linstow).

LOCALITY. Europe and North America.

DIAGNOSIS. *Size:* large, up to 9 mm. long and 4–5 mm. broad. *Cuticle:* spinous only near the mid-body. *Suckers:* ventral not much larger than the oral, diameters 0·8–1·3 and 0·7–0·9 mm. *Reproductive systems:* vitellaria extending only between the posterior margin of the ventral sucker and the anterior ends of the testes. Uterus less folded posteriorly than in *P. ovatus*, few folds if any occurring in front of the ventral sucker. Eggs dark brown and small (0·027–0·029 × 0·011–0·013 mm.).

OTHER SPECIES:

These are mainly Russian, viz. *P. anatinus* Markow, 1902 (in the domestic duck in European Russia), *P. putschkowskii* Skrjabin, 1913 (in Russian Turkestan), *P. dogieli* Skrjabin, 1915, *P. rudolphii* Skrjabin, 1919, *P. skrjabini* Zakharow, 1920 and *P. fuelleborni* Skrjabin & Massino, 1925. Strom (1940*b*) described three new species, *P. extremis*, *P. subbuteo* and *P. oviformis*. Several of these species occur in the domestic duck (see Table 5).

Genus *Schistogonimus* Lühe, 1909

Schistogonimus rarus (Braun, 1901) Lühe, 1909

Syn. *Prosthogonimus rarus* Braun, 1901.

HOSTS. Shoveller, coot, domestic duck.

LOCATION. Bursa Fabricii.

LOCALITY. Europe.

DIAGNOSIS. *Shape and size:* flattened, pointed anteriorly, rounded and spinous posteriorly, about 4·2 mm. long and 2 mm. broad. *Suckers:* ventral much larger than the oral, diameters 0·52 and 0·32–0·35 mm. *Gut:* oesophagus short, caeca long, diverging towards the middle of the body, converging behind the testes, extending almost to the posterior extremity. Bifurcation of the intestine far in front of the ventral sucker. *Reproductive systems:* male pore at the tip of the body near the margin of the mouth, female pore on the lateral margin of the body almost 0·3 mm. farther back and opposite the posterior margin of the oral sucker. Testes rounded or ovoid, situated immediately behind the ventral sucker, ovary dorsal to this haptor. Vitellaria lateral in the anterior region, extending slightly in front of the ventral sucker and behind the testes. Folds of the uterus confined between the caeca posteriorly, tending to fill the available space. Eggs brown and small (0·024–0·027 × 0·013 mm.).

Family **LECITHODENDRIIDAE** Odhner, 1910 (p. 94)

Subfamily **Lecithodendriinae** Lühe, 1901, emend. Looss, 1902

Syn. Brachycoeliinae Looss, 1899, in part.

KEY TO GENERA OCCURRING IN BIRDS

1. Cuticle bearing fine spines; cirrus pouch long and tubular, straight or folded into an S-shape; excretory vesicle V-shaped *Phaneropsolus*
2. Cuticle without spines; cirrus pouch short and sac-like; excretory vesicle Y-shaped, having a short stem and long branches *Eumegacetes*

The systematic positions of these genera seem to be in some doubt. Srivastava (1934) included both in the Lecithodendriinae, but Mehra (1935) placed them in new subfamilies, Phaneropsolinae and Eumegacetinae, while Macy (1936*a*) referred *Phaneropsolus* to the Pleurogenetinae.

Genus *Eumegacetes* Looss, 1900

Syn. *Megacetes* Looss, 1899, *nec* Thomas, 1859.

Eumegacetes contribulans Braun, 1901

Syn. *Eumegacetes crassus* (Siebold, 1836) Poche, 1907.

HOSTS. Swallow, house marten.

LOCATION. Cloaca and intestine.

LOCALITY. Europe.

DIAGNOSIS. *Shape and size:* little flattened, outline oval, 2·5 mm. long, 1·5 mm. broad at the level of the pharynx, tapering slightly towards the posterior extremity. *Suckers:* powerful, of nearly equal size, about 0·7 mm. diameter, the oral subterminal, the ventral near the middle of the body. *Gut:* pharynx large, oesophagus absent, caeca long, their anterior ends directed transversely outwards. *Reproductive systems:* genital pore median, immediately behind the pharynx. Testes ovoid, far forward and side by side in front of the ventral sucker. Ovary ovoid, almost median, behind the ventral sucker. Vitellaria scarcely reaching this sucker at their anterior ends. Uterus with descending and ascending limbs forming (the latter limb especially) long, sinuous folds extending on either side of the ventral sucker. Eggs numerous, having pointed poles, measuring 0·023 × 0·014 mm.

Eumegacetes emendatus Braun, 1901 (Fig. 51 A)

Syn. *Distoma meropis* Parona, 1896, *nec* Rudolphi, 1819; *Megacetes triangularis* Looss, 1899, *nec* Diesing, 1850.

HOSTS. House sparrow, nightjar, bee eater, pratincole.
LOCATION. Intestine.
LOCALITY. Europe.

Genus *Phaneropsolus* Looss, 1899

Phaneropsolus sigmoideus Looss, 1899

HOSTS. House sparrow, nightjar.
LOCATION. Intestine.
LOCALITY. Egypt.

It is unlikely that this trematode will be found in Britain, but if discovered it can be distinguished from the foregoing species, which has similar hosts, by the generic characters mentioned in the key above. Other species are more particularly parasites of mammals, but *P. micrococcus* (Rudolphi, 1819) was discovered in a pratincole.

Family **DICROCOELIIDAE** Odhner, 1910 (p. 95)

Subfamily **Dicrocoeliinae** Looss, 1899

KEY TO GENERA OCCURRING IN BIRDS

1. Body very elongate, ribbon-like	
A. Vitellaria unpaired	*Athesmia*
B. Vitellaria paired	*Lyperosomum*
2. Body less elongate, lancet-like or spindle-shaped	
A. Body broadest in the middle	*Platynosomum*
B. Body broadest near the anterior end	*Oswaldoia*

Genus *Athesmia* Looss, 1899

Athesmia heterolecithodes Braun, 1899 (Fig. 50 B)

HOST. Moorhen.
LOCATION. Liver.
LOCALITY. Europe and Africa.

DIAGNOSIS. *Shape and size:* elongate, ribbon-like, 8–9 mm. long and 1·5–2 mm. broad. *Cuticle:* non-spinous. *Suckers:* relatively close together, oral

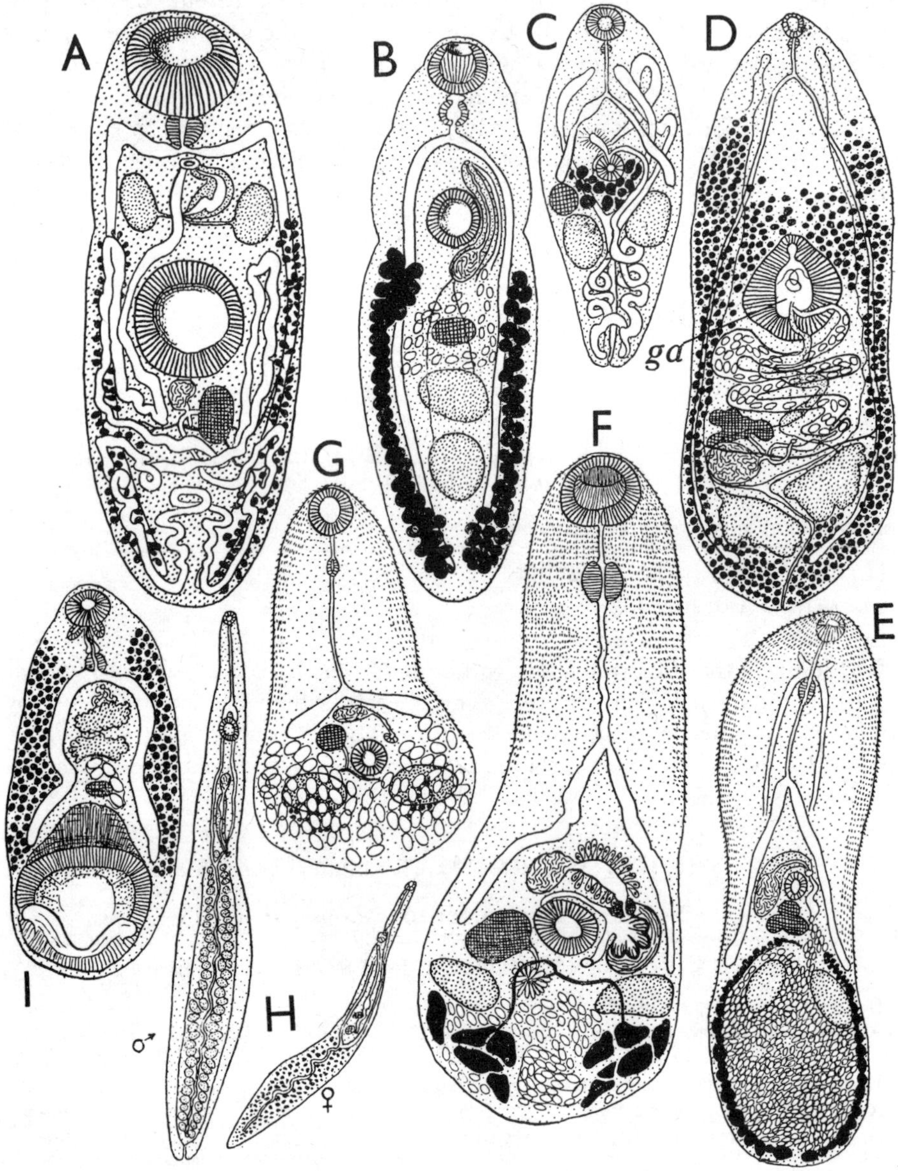

Fig. 51. Lecithodendriidae (A), Psilostomatidae (B), Microphallidae (C, E, F, G), Heterophyidae (D), Schistosomatidae (H) and Paramphistomatidae (I) of birds. A, *Eumegacetes emendatus*. B, *Psilostomum brevicolle*. C, *Gymnophallus deliciosus*. D, *Cryptocotyle lingua*. E, *Maritrema gratiosum*. F, *Levinseniella propinqua*. G, *Spelotrema claviforme*. H, *Bilharziella polonica*. I, *Zygocotyle lunata*. (A, B, F, after Fuhrmann, 1928; C, H, after Lühe, 1909; D, I, after Neveu-Lemaire, 1936; E, G, after Nicoll, 1909.)

slightly larger than the ventral, diameters 0·46 and 0·37 mm. *Gut:* pharynx small, oesophagus long, caeca extending almost to the posterior extremity. *Reproductive systems:* testes one behind the other, ovary behind the hindmost testis. Uterus having descending and ascending limbs, only slightly folded. Vitellaria fairly extensive, posterior to the gonads, sometimes asymmetrically developed. Eggs numerous and small (0·031–0·040 × 0·019–0·023 mm.).

McIntosh (1937) provided a key to five species of the genus *Athesmia*, four of them American and one of these, *A. foxi* Goldberger & Crane, 1911, a parasite of a mammal (monkey). Travassos (1941) transferred *Lyperosomum rudectum* Braun, 1901—a parasite of the ibis in Brazil—to this genus.

Genus *Lyperosomum* Looss, 1899

Lyperosomum longicauda (Rudolphi, 1809) Braun, 1901

Syn. *Distoma longicauda* Rudolphi, 1809; *Dicrocoelium longicauda* of Looss, 1899; *Distoma macrourum* Rudolphi, 1819.

HOSTS. Carrion crow, hooded crow, blackbird, song thrush, rook, magpie, starling, jay, sprosser.

LOCATION. Liver and gall bladder.

LOCALITY. Europe.

DIAGNOSIS. *Shape and size*: ribbon-like, 8–11 mm. long. *Suckers:* ventral about twice as large as the oral, diameters 0·75–0·8 and 0·40–0·42 mm. *Gut:* pharynx measuring about 0·18–0·19 × 0·23 mm. *Reproductive systems:* gonads small, vitellaria extending for about 4 mm. anteriorly to the hinder border of the anterior testis. Eggs measuring 0·023 × 0·019 mm.

Note. This is one of the most elongate of liver flukes, the body being scarcely broader than it is thick.

Numerous other species of the genus are known, and the type and those most likely to be encountered in Europe can be separated by the following key, which is based on that given by Skrjabin (1913*a*).

I. Ovary smaller than the testes
  A. Both testes situated behind the ventral sucker
    (*a*) Vitellaria extending forward to the hinder end of the anterior testis — *L. longicauda*
    (*b*) Vitellaria extending forward only as far as the ovary
      (i) Testes widely separated — *L. corrigia*
      (ii) Testes close together (overlapping slightly) — *L. olssoni*
    (*c*) Vitellaria not extending forward as far as the ovary — *L. lobatum*
  B. Anterior testis dorsal to the ventral sucker — *L. salebrosum*
II. Ovary larger than the testes
  (*a*) Testes longitudinally ovoid — *L. filiforme*
  (*b*) Testes transversely ovoid — *L. strigosum*

The full titles, hosts, locations and localities in respect of these trematodes are:

L. corrigia Braun, 1901, *Tetrao tetrix*, *Caccabis chukar* (gut); Europe, Turkestan.
L. olssoni (Railliet, 1900), *Cypselus apus* (liver); Europe.
L. lobatum (Railliet, 1900), *Accipiter nisus* (liver); Europe.
L. salebrosum Braun, 1901, *Cypselus melba* (location ?); locality ?.
L. filiforme Skrjabin, 1913, *Circus cinereus* (liver) Turkestan.
L. strigosum (Looss, 1899), *Merops apiaster* (liver); Egypt.

The characters which separate some of these species seem to be rather trivial, yet the genus has continued to grow. Some species which parasitize Russian birds are *L. donicum* Issaitschikow, 1919; *L. fringillae* Layman, 1923; *L. laniicola*, *L. collurio*, *L. alaudae* and *L. loossi*, all Layman, 1926; *L. papabejani* Skrjabin & Udinzew, 1930. Semenov (1927) described three new subspecies, Lopez-Neyra (1941 *c*) another. Strom (1940 *a*) proposed the partitioning of the genus into three subgenera, *Lyperosomum* s.s., *Brachylecithum* and *Corrigia*, respective types proposed being *longicauda*, *filum* Dujardin and *corrigia*. Price & McIntosh (1935) described an unusual American species, *Lyperosomum monenteron*, which occurs in the gall bladder and bile ducts of a robin and which, as the name implies, has only one caecum, at the same time pointing out that some species are based on material in such poor condition that it is not possible to decide if any other species show this feature, as some may.

Genus *Platynosomum* Looss, 1907

Platynosomum acuminatum Nicoll, 1915

HOST. Kestrel.

LOCATION. Liver.

LOCALITY. Scotland (rare), India.

DIAGNOSIS. *Shape and size:* acutely pointed at both extremities, spindle-like, 6·3 mm. long and 1·5 mm. in maximum breadth slightly behind the ventral sucker. *Suckers:* oral much smaller than the ventral, the latter about 2 mm. from the anterior extremity, dimensions 0·45 × 0·40 and 0·60 × 0·75 mm. *Gut:* pharynx small, oesophagus short, caeca long and narrow, their proximal ends making an acute angle with one another. *Reproductive systems:* genital pore far forward, near the pharynx. Cirrus pouch fairly large, but terminating little more than half-way between the pharynx and the ventral sucker. Testes ovoid, side by side and widely separated slightly behind the ventral sucker. Ovary transversely ovoid, situated some distance behind the left testis. Vitellaria confined to the middle third of the body, behind the ventral sucker, lateral. Uterus with descending and ascending limbs, tending to fill up all available space posteriorly. Eggs numerous and small (0·033–0·039 × 0·018–0·020 mm.).

OTHER SPECIES:

Platynosomum olectoris (Nöller & Enigk 1933); *P. semifuscum* Looss, 1907; *P. clathratum* (Deslongchamps, 1824) Looss, 1907; *P. petiolatum* (Railliet, 1900) Looss, 1907; *P. deflectens* (Rudolphi, 1819) Looss, 1907; all well-known European species. *P. clathratum* occurs in the swift (López-Neyra (1941 *c*) referred it to the genus *Lyperosomum*), *P. petiolatum* in the jay, green woodpecker and stone curlew, in Britain. Other species, making a total of fourteen in birds and five in mammals, though *P. petiolatum* is unmentioned, are named by Heidegger & Mendheim (1939), who describe a new species, *P. fallax*, from a cockatoo, *Cacatua sulfurea*, from Malaya.

Genus *Oswaldoia* Travassos, 1919

Several species of this genus occur in various birds in Russia, one being *O. skrjabini* (Solowjew, 1911) (Syn. *Dicrocoelium skrjabini*) from the gall bladder of the carrion crow and the rook (Table 5), others *O. collurionis* and *O. mosquensis*, both Skrjabin & Issaitschikow, 1927; *O. pawlowskyi* Strom, 1928; *O. alagesi* Skrjabin & Udinzew, 1930; and *O. dujardini* Strom & Sondak, 1935.

Family **MICROPHALLIDAE** Viana, 1924 (p. 96)

Subfamily **Gymnophallinae** Odhner, 1905

Genus *Gymnophallus* Odhner, 1900

Gymnophallus deliciosus (Olsson, 1893) Odhner, 1900 (Fig. 51 C)

Syn. *Distoma deliciosum* Olsson, 1893.

HOSTS. Common gull, herring gull, greater and lesser black-backed gulls, glaucous gull.

LOCATION. Gall bladder.

LOCALITY. North Europe.

DIAGNOSIS. *Shape and size:* small, thick, pyriform, broad anteriorly, 1·1–2·3 mm. long and 0·5–0·75 mm. broad. *Cuticle:* spinous. *Suckers:* ventral slightly smaller than the oral and situated slightly in front of the mid-body, diameters 0·16–0·20 and 0·20–0·25 mm. *Gut:* pharynx small, oesophagus long, caeca very short, bifurcation of the intestine far in front of the ventral sucker, termination of the caeca hardly behind it. *Reproductive systems:* genital pore near the anterior border of the ventral sucker, genital atrium tubular. Gonads rounded, testes side by side, slightly diagonally orientated in the posterior half of the body, ovary in front of the testes and lateral in position. Vitellaria comprising two small clusters of follicles between the gonads and the ventral sucker. Uterus with descending and ascending limbs, folded behind the testes. Eggs numerous and small (0·022–0·026 × 0·014 mm.).

OTHER SPECIES:

A number occur in various birds, at least four in the intestine of the common scoter, namely, *G. oidemiae*, *G. affinis* (Fig. 52 A), *G. ovoplenus* and *G. macroporus* (Fig. 52 B), all Jameson & Nicoll, 1913. Another species occurs in the caeca of this host as well, *G. somateriae* (Levinsen, 1881), and two inhabit the bursa Fabricii, *G. diapsilis* Nicoll, 1907 and *G. bursicola* Odhner, 1900, the former occurring also in the intestine. Nicoll (1923*a*) also recorded *G. choledochus* Odhner, 1900 and *G. micropharyngeus* (Lühe, 1898). Most of them are extremely small, and all bear a superficial resemblance to Heterophyids as a result of convergence (Price, 1940*a*). Possibly some of them are not valid species, and all of them should be re-examined.

Subfamily **Maritrematinae** Lal, 1939, emend. Nicoll, 1940

Genus *Maritrema* Nicoll, 1907

Syn. *Streptovitella* Swales, 1933.

Lal (1939) emended this genus and made it the type of a new subfamily, Maritrem(at)inae.

Maritrema gratiosum Nicoll, 1907 (Fig. 51 E)

HOSTS. Dunlin, ringed plover, oystercatcher, black-headed gull, common scoter, bar-tailed godwit, etc.

LOCATION. Intestine.

LOCALITY. Europe, including Britain; also Tunisia (see Balozet & Callot, 1938).

DIAGNOSIS. *Shape and size:* elongate, flat and linguiform, slightly constricted in the region of the ventral sucker, 0·45–1·1 mm. long, 0·24–0·44 mm. in maximum

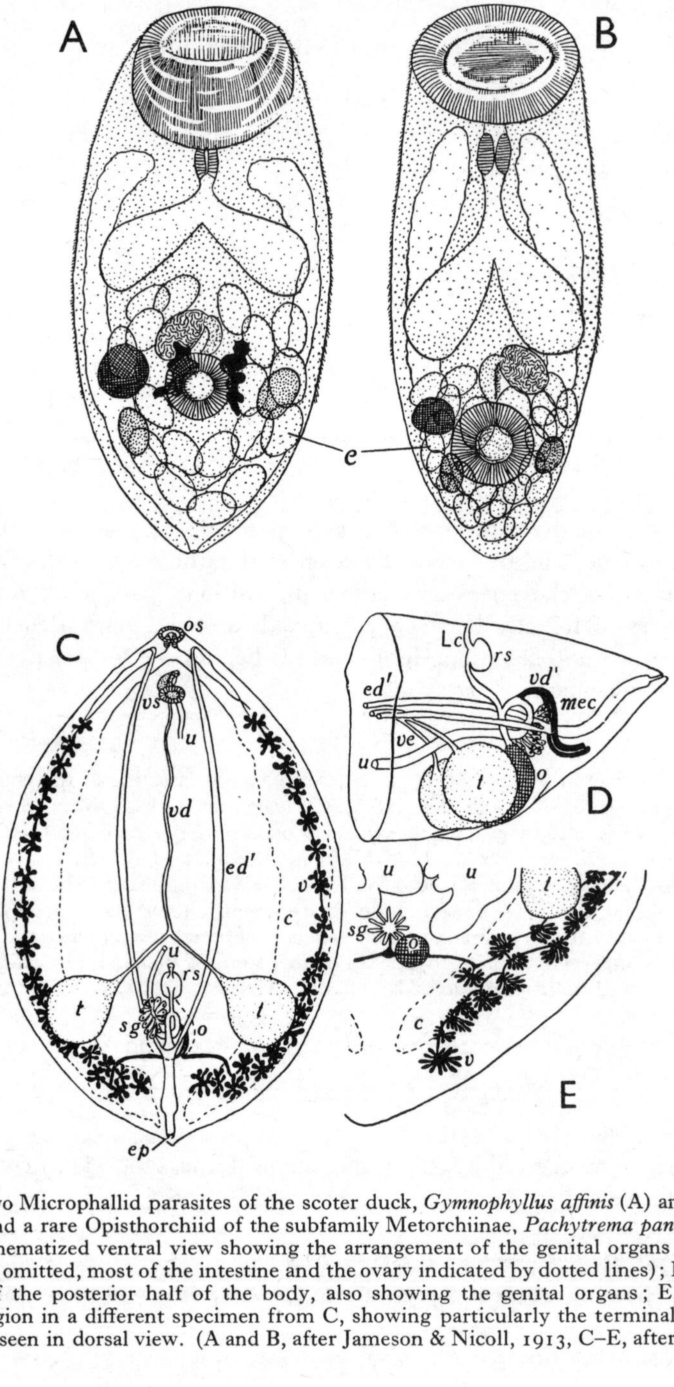

Fig. 52. Two Microphallid parasites of the scoter duck, *Gymnophyllus affinis* (A) and *G. macroporus* (B), and a rare Opisthorchiid of the subfamily Metorchiinae, *Pachytrema paniceum* (C–E). C, a half-schematized ventral view showing the arrangement of the genital organs (most of the uterus being omitted, most of the intestine and the ovary indicated by dotted lines); D, schematic side view of the posterior half of the body, also showing the genital organs; E, part of the posterior region in a different specimen from C, showing particularly the terminal parts of the vitellaria as seen in dorsal view. (A and B, after Jameson & Nicoll, 1913, C–E, after Brinkmann, 1942*c*.)

breadth. *Cuticle:* scaly. *Suckers:* small, of almost equal sizes, 0·043–0·062 mm. in diameter, oral larger than the ventral in juveniles. *Gut:* prepharynx twice as long as the pharynx, oesophagus long (15–20 % of the body length), caeca short, extending back only to the anterior ends of the testes. *Reproductive systems:* genital pore situated on the left of the ventral sucker, cirrus pouch pyriform. Testes elongate-ovoid, or spherical, side by side. Ovary slightly in front of the testes near the ventral sucker. Vitellaria forming a ring-like chain of follicles encircling the testes and most of the uterus posteriorly. Uterus having descending and ascending limbs, abundantly folded behind the gonads, but not obscuring the testes. Eggs colourless when first formed, becoming yellow later, small (0·020–0·022 × 0·010–0·012 mm.).

OTHER SPECIES:

Maritrema lepidum Nicoll, 1907.

In the intestine of the herring gull (Europe).

DIAGNOSIS. Distinguished from *M. gratiosum* by the larger suckers (ventral smaller than the oral and situated in front of the mid-body, diameters 0·068 and 0·059 mm.), shorter prepharynx and caeca and smaller eggs, which measure 0·018–0·019 × 0·009–0·010 mm.

Maritrema humile Nicoll, 1907

In the intestine of the redshank (Europe).

DIAGNOSIS. Distinguished from the foregoing species by the size relations of the suckers, the oral being smaller than the ventral, diameters 0·025–0·031 and 0·030–0·034 mm. in specimens of usual size (0·28–0·40 × 0·12–0·16 mm.), and by the smaller eggs, which measure 0·016–0·018 × 0·008–0·010 mm.

Maritrema subdolum Jägerskiöld, 1908

In the intestine of the tufted duck (Britain) and the common sandpiper (Europe).

DIAGNOSIS. Very similar to *M. humile*, 0·35–0·40 mm. long. Oral sucker smaller than the ventral, diameters 0·031–0·035 and 0·042–0·054 mm. Caeca extending to the posterior border of the ventral sucker, whereas in *M. gratiosum* they reach to the anterior borders of the testes, but in *M. lepidum* and *M. humile* only to the anterior border of the ventral sucker.

Maritrema linguilla Jägerskiöld, 1908

In the intestine of the purple sandpiper (Europe).

Maritrema sachalinicum Schumakowitsch, 1932

In the intestine of the herring gull (Sakhalin; Siberia).

M. ovata Rankin, 1939 and *M. arenaria* Hadley & Castle, 1940 occur in U.S.A., *M. eroliae* Yamaguti, 1939 in Japan. Rankin (1939 *c*) erected a new genus *Maritreminoides*, type *Maritrema nettae* Gower, to include *M. obstipum* and *M. medium*, both (van Cleave & Mueller).

Subfamily **Microphallinae** Ward, 1901

KEY TO GENERA

1. Genital atrium almost filled by a large genital papilla, but without a complicated pouch near the latter
   A. Female pore deep in the genital atrium, near the base of the papilla *Spelotrema*
   B. Female pore superficial, near the opening of the genital atrium *Spelophallus*
2. Genital atrium having small genital papilla, but a complicated pouch *Levinseniella*

Genus *Spelotrema* Jägerskiöld, 1901

Rankin (1940*a*) has emended the diagnosis of this genus and redescribed five species which, together with a new American species, *Spelotrema papillorobusta*, were regarded as the only valid ones, although *S. brevicaeca* (Africa & Garcia, 1935), a species found in the Philippines, was placed tentatively in the genus.

Spelotrema pygmaeum (Levinsen, 1881) Looss, 1902

Syn. *Distoma pygmaeum* Levinsen, 1881; *Levinsenia pygmaeum* of Stossich, 1899; *L. pygmaea* of Jägerskiöld, 1900; *Levinseniella pygmaea* of Stafford, 1903.

HOSTS. Eider-duck, king eider, common scoter, tufted duck, velvet scoter.

LOCATION. Intestine.

LOCALITY. Europe (Germany, Sweden, Britain) and North America.

DIAGNOSIS. *Shape and size:* club-shaped, 0·3–0·5 mm. long and 0·2–0·3 mm. broad. *Cuticle:* spinous. *Suckers:* oral slightly larger than the ventral, diameters about 0·4–0·5 mm. Ventral sucker one-third of the body length from the posterior extremity. *Gut:* caeca extending back only to the hinder margin of the ventral sucker. *Reproductive systems:* genital atrium about half as broad as the ventral sucker. Ejaculatory duct short and straight. Testes relatively large. Uterus voluminous, but not obscuring the testes. Eggs measuring 0·021–0·023 × 0·012 mm.

Spelotrema claviforme (Brandes, 1888) Nicoll, 1907 (Fig. 51 G)

Syn. *Distoma claviforme* Brandes, 1888; *Lecithodendrium claviforme* of Stossich, 1899. *Levinsenia claviforme* of Looss, 1899, Lühe, 1899 and Jägerskiöld, 1900.

HOSTS. Lapland dunlin, ringed plover, rock pipit, curlew, blue-headed wagtail, black-headed gull, etc.

LOCATION. Intestine.

LOCALITY. Europe (Germany, France, Sweden, Britain).

DIAGNOSIS. *Shape and size:* club-shaped, 0·2–0·4 mm. long and 0·15–0·2 mm. broad. *Cuticle:* spinous. *Suckers:* oral larger than the ventral, which is situated in the ultimate third of the body, diameters 0·035–0·04 and 0·03–0·035 mm. *Gut:* caeca short and wide apart, not extending to the level of the ventral sucker. *Reproductive systems:* genital atrium less than half as broad as the ventral sucker. Testes not prominent. Uterus voluminous, tending to obscure the testes, extending slightly in front of the ventral sucker. Dimensions of the eggs 0·020–0·024 × 0·011–0·014 mm.

Spelotrema simile (Jägerskiöld, 1900) Looss, 1902

Syn. *Levinsenia pygmaea* var. *simile* Jägerskiöld, 1900; *L. similis* of Nicoll, 1906.

HOSTS. Herring gull, lesser black-backed gull, black-headed gull.

LOCATION. Intestine.

LOCALITY. Europe (Sweden).

DIAGNOSIS. *Size:* 0·4–0·6 mm. long and 0·2 mm. broad. *Suckers:* oral slightly smaller than the ventral, diameters 0·05–0·06 mm. *Gut:* caeca extending to the hinder margin of the ventral sucker. *Reproductive systems:* genital atrium about two-thirds as broad as the ventral sucker, ejaculatory duct long and convoluted, uterus not voluminous and not obscuring the testes or vitellaria. Eggs pale in colour, measuring 0·023–0·025 × 0·010–0·013 mm.

Spelotrema excellens Nicoll, 1907

Syn. *S. feriatum* Nicoll, 1907.

HOST. Herring gull.

LOCATION. Intestine.

LOCALITY. Europe (found at St Andrews and in Sweden).

DIAGNOSIS. *Shape and size:* club-shaped, 0·7–1·4 mm. long and 0·35–0·5 mm. broad. *Suckers:* oral slightly larger than the ventral, diameters 0·6–0·85 mm. *Gut:* caeca terminating opposite the centre of the ventral sucker. *Reproductive systems:* genital atrium almost as large as the ventral sucker, ejaculatory duct short and straight, uterus voluminous, obscuring the testes, but not usually extending in front of the ventral sucker. Eggs numerous and small (0·023–0·025 × 0·010–0·013 mm.).

Cable and Hunninen (1938 *b*, 1940) described a new species, *S. nicolli*, and gave an account of its life history, in which crabs are the second intermediate and herring gulls the final hosts respectively.

Genus *Levinseniella* Stiles & Hassall, in Ward, 1901

Syn. *Levinsenia* Stossich, 1899, *nec* Mesnil, 1897, in part.

Levinseniella propinqua Jägerskiöld, 1907 (Fig. 51 F)

HOSTS. Ringed plover, oystercatcher, turnstone, tufted duck, golden-eye, redshank.

LOCATION. Intestine and caeca.

LOCALITY. Europe.

DIAGNOSIS. *Size:* 0·5–1·2 mm. long and 0·27–0·40 mm. broad. *Cuticle:* spinous. *Suckers:* oral larger than the ventral, diameters 0·075–0·90 and 0·06 mm. *Gut:* prepharynx short, pharynx large, oesophagus long, caeca short, scarcely passing behind the ventral sucker, bifurcation of the gut well in front of the latter. *Reproductive systems:* genital pore situated on one side of the ventral sucker, genital atrium larger than the latter. Gonads grouped posteriorly, the testes postero-lateral to the ventral sucker, the ovary in front of them beside it. Vitellaria situated behind the gonads, uterus filling up the space behind and between them. Eggs brown and small (0·020–0·024 × 0·009–0·012 mm.).

OTHER SPECIES:

Levinseniella brachysoma (Creplin, 1837) Stiles & Hassall, 1902

Syn. *Distoma brachysomum* Creplin, 1837; *Levinsenia brachysomum* of Stossich, 1899.

HOSTS. Oystercatcher, long-tailed duck, sheld-duck, common scoter, golden-eye, dunlin, redshank, common sandpiper, lapwing, turnstone, ringed plover, grey plover, common heron, short-eared owl.

LOCATION. Intestine.

LOCALITY. Europe, America.

DIAGNOSIS. *Size:* 0·56–0·8 mm. long and 0·24–0·32 mm. broad. *Cuticle:* spinous. *Suckers:* oral larger than the ventral, diameters 0·07–0·08 and 0·06–0·07 mm. *Gut:* prepharynx long, pharynx large, oesophagus long, caeca wide and short, scarcely extending to the ventral sucker. Eggs small (0·021–0·023 × 0·012–0·013 mm.).

Levinseniella pellucida Jägerskiöld, 1907

HOSTS. Mallard, tufted duck.

LOCATION. Intestine and caeca.

LOCALITY. Europe, America.

DIAGNOSIS. *Size:* 0·55–0·8 mm. long and 0·025–0·03 mm. broad. *Cuticle:* spinous. *Suckers:* oral much larger than the ventral, diameters 0·075–0·08 and 0·055–0·065 mm. *Gut:* prepharynx shorter than oesophagus. Eggs small (0·018–0·021 × 0·009–0·011 mm.).

Levinseniella macrophallos (Linstow, 1875)

Syn. *Distoma macrophallos* Linstow, 1875; *Levinsenia macrophallos* of Stossich, 1899.

HOST. Common sandpiper.

LOCATION. Intestine.

LOCALITY. Europe.

Genus *Spelophallus* Jägerskiöld, 1908

Spelophallus primas Jägerskiöld, 1908

HOSTS. Oystercatcher, eider duck.

LOCATION. Intestine.

LOCALITY. Europe, Japan.

DIAGNOSIS. *Size:* 0·56–0·88 mm. long and 0·25–0·4 mm. broad. *Cuticle:* spinous. *Suckers:* oral much smaller than the ventral, the latter two-thirds of the distance along the body, diameters 0·05–0·07 and 0·07–0·105 mm. *Gut:* prepharynx and pharynx short and of about equal length, oesophagus very long, caeca short, terminating at the sides of the ventral sucker. Oesophagus about one-quarter of the body length, longer than the caeca. *Reproductive systems:* testes small, ovary large. Uterus and large vitellaria obscuring the testes. Eggs small (0·022–0·024 × 0·011–0·012 mm.).

One other genus, *Gynaecotyla* Yamaguti, 1939 (Syn. *Cornucopula* Rankin, 1939) has species in U.S.A. and Japan.

Family **OPISTHORCHIIDAE** Braun, 1901 (p. 96)

Syn. Opisthorchidae Lühe, 1901.

Subfamily **Opisthorchiinae** Looss, 1899 emend.

KEY TO GENERA OCCURRING IN BIRDS

1.	Vitellaria interrupted opposite the ovary on each side	*Amphimerus*
2.	Vitellaria each comprising an uninterrupted row of follicles	*Opisthorchis*

Genus *Opisthorchis* Blanchard, 1898

Syn. *Prosthometra* Looss, 1896, in part; *Clonorchis* Looss, 1907; *Notaulus* Skrjabin, 1913; *Gomtia* Thapar, 1930 and various misprints: *Opistorchis* Railliet, 1896; *Opisthorchic* Stiles, 1901; *Opiscorcus* Woolley, 1906.

Opisthorchis simulans (Looss, 1896) Kowalewsky, 1898

Syn. *Distoma simulans* Looss, 1896; *Opisthorchis simulans* var. *poturzycensis* Kowalewsky, 1898; *O. longissimus simulans* of Erhardt, 1935.

HOSTS. Honey buzzard, marsh harrier, mallard, wigeon.

LOCATION. Bile ducts.

LOCALITY. Europe, Egypt.

DIAGNOSIS. *Shape and size:* very elongate, 7–23 mm. long and 1–1·5 mm. broad. *Suckers:* ventral much smaller than the oral and about one-quarter of the distance along the body, diameters 0·2 and 0·5 mm. *Gut:* pharynx large, oesophagus very short, caeca long, extending almost to the posterior extremity. *Reproductive systems:* testes rounded or of irregular shape, one behind the other in the posterior region. Ovary lobed, situated in front of the testes. Uterus having an ascending limb only, forming numerous folds between the ovary and the ventral sucker. Vitellaria extending from the middle of the body to the level of the ovary. Eggs measuring 0·028–0·029 × 0·016–0·018 mm.

OTHER SPECIES:

Opisthorchis geminus (Looss, 1896) Looss, 1899

Syn. *Distomum geminum* Looss, 1896; *Opisthorchis gemina* (Looss) Kowalewski, 1898; *O. geminus* var. *kirghisensis* Skrjabin, 1913.

HOSTS. Marsh harrier, kites, duck.

LOCATION. Liver.

LOCALITY. Egypt, Russian Turkestan.

DIAGNOSIS. *Size:* 7–12·5 mm. long and 1·3–2 mm. broad. *Suckers:* of equal size, diameters about 0·17 mm. (0·24–0·25 × 0·31–0·34 mm. in somewhat smaller specimens than the smallest examined by Looss, according to Skrjabin). *Eggs:* dimensions, according to Looss, 0·021–0·027 × 0·01–0·013 mm., but possibly much larger (0·032 × 0·017 mm. according to Skrjabin).

O. entzi Ratz, 1900 occurs in the bile ducts of the purple heron in Europe, *O. asiaticus* (Skrjabin, 1913) in a harrier and an eagle in Russian Turkestan and *O. skrjabini* Zhukova, 1934 in the domestic duck in Siberia.

Genus *Amphimerus* Barker, 1911

This genus is better known for its species parasitizing mammals—*Amphimerus lancea* (Diesing, 1850) occurring in the bile ducts (?) of a dolphin in Brazil and Asia—and reptiles, but *Amphimerus speciosus* (Stiles & Hassall, 1896) occurs in the liver of crows in America.

Subfamily **Metorchiinae** Lühe, 1909

KEY TO SOME GENERA OCCURRING IN BIRDS

I. Excretory pore terminal; vitellaria mainly situated in the anterior region *Holometra*

II. Excretory pore ventral; vitellaria mainly situated in the posterior region
- A. Cuticle smooth; suckers close together; receptaculum seminis absent or small *Pachytrema*
- B. Cuticle spinous; suckers fairly widely separated (one-third of the body length apart); receptaculum seminis large *Metorchis*

Genus *Metorchis* Looss, 1899

Metorchis xanthosomus (Creplin, 1846) Braun, 1902 (Fig. 50C)

Syn. *Distoma xanthosomum* Creplin, 1846; *Opisthorchis xanthosoma* Kowalewsky, 1898.

HOSTS. Common scoter, razorbill, duck, garganey, red-throated diver, carrion crow.

LOCATION. Gall bladder, intestine.

LOCALITY. Europe.

DIAGNOSIS. *Shape and size:* elongate, pointed anteriorly, rounded posteriorly, 2·8–3·2 mm. long and 0·8–0·9 mm. broad. *Cuticle:* non-spinous (?). *Suckers:* oral larger than the ventral, dimensions 0·22 × 0·18 mm. and 0·17 mm. diameter. *Gut:* oesophagus very short, caeca long. *Reproductive systems:* testes large, filling the entire breadth of the body posteriorly, diagonally orientated. Ovary situated in front of the testes. Vitellaria extending from a level slightly behind the pharynx to the anterior border of the foremost testis. Uterus having an ascending limb only, formed into wide folds. Eggs golden brown, measuring 0·027–0·032 × 0·014 mm.

Metorchis crassiusculus (Rudolphi, 1809) Looss, 1899

Syn. *Distoma crassiusculum* Rudolphi, 1809; *Opisthorchis crassiuscula* (Rudolphi) of Kowalewsky, 1898; *O. xanthosomus* var. *compascuus* Kowalewsky, 1898.

HOSTS. Buzzard, golden eagle, rough-legged buzzard, marsh harrier, white-tailed sea eagle, red-breasted merganser, garganey, mallard, duck.

LOCATION. Gall bladder.

LOCALITY. Europe.

DIAGNOSIS. *Shape and size:* spatulate, 3·7 mm. long and 1–1·5 mm. broad. *Cuticle:* spinous. *Suckers:* of approximately equal sizes, oral 0·21–0·28 mm. diameter, ventral 0·1–0·3 mm. *Gut:* pharynx relatively large, oesophagus absent.

Reproductive systems: testes not nearly filling the posterior region of the body, leaving the lateral margins free. Vitellaria extending from the level of the pharynx to that of the ovary. Eggs golden brown, measuring 0·030–0·032 × 0·016 mm.

OTHER SPECIES:

Metorchis tener (Kowalewsky, 1898)
M. zakharovi Layman, 1922

Genus *Holometra* Looss, 1899

Holometra exigua (Mühling, 1898) Looss, 1899

Syn. *Distoma exiguum* Mühling, 1898; *Opisthorchis exigua* (Mühling) of Kowalewsky, 1898.

HOST. Marsh harrier.

LOCATION. Liver.

LOCALITY. Europe, Egypt.

DIAGNOSIS. *Shape:* pointed anteriorly. *Cuticle:* spinous. *Gut:* caeca extending to a level in front of the testes. *Reproductive systems:* testes close together and diagonally orientated near the posterior extremity. Ovary situated in front of the testes. Folds of the uterus filling the available space between the testes and the ventral sucker. Vitellaria confined to the region in front of the caeca, lateral to the oesophagus.

Genus *Pachytrema* Looss, 1907

Pachytrema calculus Looss, 1907

HOSTS. Herring gull, black-headed gull, lesser black-backed gull, (a teal in Japan).

LOCATION. Gall bladder.

LOCALITY. Europe (Italy, Sweden).

DIAGNOSIS. According to Brinkmann (1942*c*), who examined Looss's specimens of this species and also specimens collected on the coast of Sweden by Odhner in 1905, the main differences from *P. paniceum* (described below) are: oesophagus much shorter (0·1 mm. long), ovary globular, vitelline reservoir present, receptaculum seminis absent, eggs rather narrower, measuring 0·11 × 0·044 mm.

Pachytrema paniceum Brinkmann, 1942*c* (Fig. 52C–E)

HOST. Scandinavian lesser black-backed gull.

LOCATION. Gall bladder.

LOCALITY. Europe (Norway; Herdla, nr. Bergen).

DIAGNOSIS (after Brinkmann). *Shape and size:* outline broad oval, pointed at both extremities, mature specimens 8·5 mm. long, 5·5 mm. broad and 3·5 mm. thick, decidedly convex ventrally, slightly convex dorsally. *Cuticle:* transversely striated, more especially dorsally and laterally. *Suckers:* ventral slightly larger than the oral and near to it (the centre 1·00 mm. behind the mouth), dimensions 0·43 × 0·36 × 0·26 mm. and 0·41 × 0·34 × 0·27 mm. *Gut:* pharynx globular (0·19 mm. long, 0·14 mm. broad, and 0·20 mm. thick), pyriform gland cells occupying the space between it and the oral sucker, oesophagus very short (0·34 mm. long), caeca long and simple, extending almost to the posterior extremity, widening posteriorly. Pharynx and oesophagus lined with cuticle, caeca having an epithelial lining. *Excretory system:* pore slightly dorsal, median canal short, lateral canals extending median to the caeca almost as far as the oral sucker, then recurring between the vitelline ducts and the caeca. *Reproductive systems:* genital pore median, slightly nearer the ventral than the oral sucker. Cirrus pouch delicate, but muscular, containing an S-shaped ejaculatory duct and surrounding prostate gland cells, also a seminal vesicle 0·30–0·34 mm. long and 0·15 mm. broad. Testes globular, situated at the same transverse level (the penultimate sixth of the body), but widely separated, ventro-lateral to the caeca, 0·7–0·8 mm. diameter and 1·75 mm. apart. Ovary reniform, 0·75 mm. long and 0·28 mm. broad, situated in the median plane between the posterior parts of the testes. Seminal receptacle ovoid, 0·3 mm. long, 0·25 mm. broad and 0·38 mm. thick, situated on the course of Laurer's canal, dorsal to the gonads. Vitellaria dorso-lateral to the caeca, extending between the ventral sucker and the terminations of the caeca, comprising 28–30 rosette-like clusters of follicles arranged in a single series on each side of the body. Transverse vitelline ducts arising near the posterior ends of the lateral ducts, extending slightly anteriorly and then directly inwards, the common duct opening into the oviduct near the ootype, which is situated dorsal to the ovary. Vitelline reservoir absent. Uterus abundantly folded, the folds filling the available space between the ovary and the ventral sucker, metraterm 0·3 mm. long, opening into the posterior wall of the genital atrium. Eggs ovoid, operculate, golden brown, measuring 0·11 × 0·055–0·06 mm., each prior to laying containing a miracidium.

OTHER SPECIES:

Pachytrema hewletti (Phadke & Gulati) Purvis, 1937 (in a house crow: India).
P. magnum Travassos, 1921 (in *Sterna maxima*: Brazil).
P. proximum Travassos, 1921 (in *Heteropygia fuscicollis*: Brazil)
P. sanguineum (Linton, 1928) Purvis, 1937 (in *Larus atricilla* L.: America).
P. tringae Layman, 1926 (in the Lapland dunlin: Russia).

The genus has been the subject of special study by Purvis (1937) and Brinkmann (1942*c*). By some writers it is regarded as the type of a separate subfamily, the Pachytrematinae Railliet, 1919, emend. Ejsmont, 1931.

Family **HETEROPHYIDAE** Odhner, 1914 (p. 97)

Subfamily **Cryptocotylinae** Lühe, 1909

Genus *Cryptocotyle* Lühe, 1899

Syn. *Tocotrema* Looss, 1899; *Hallum*, Wigdor, 1918; *Dermocystis* Stafford, 1905; *Ciureana* Skrjabin, 1923.

Cryptocotyle lingua (Creplin, 1825) Fischoeder, 1903 (Fig. 51 D)

Syn. *Distoma lingua* Creplin, 1825; *Tocotrema lingua* of Looss, 1899; *Dermocystis ctenolabri* Stafford, 1905; *Hallum caninum* Wigdor, 1918.

HOSTS. Greater black-backed gull, herring gull, lesser black-backed gull, Slavonic grebe, night heron, kittiwake, razorbill, common tern (also mammals).

LOCATION. Intestine.

LOCALITY. Europe, North America, Japan.

DIAGNOSIS. *Shape and size:* spatula-like, 0·5–2·0 mm. long and 0·2–0·9 mm. broad (large, for a heterophyid). *Cuticle:* spinous. *Suckers:* feeble, the oral larger than the ventral (which is enclosed in a genital sinus), diameters 0·07–0·11 and 0·05–0·08 mm. *Gut:* prepharynx short, pharynx 0·03–0·05 mm. diameter, oesophagus slightly longer than the pharynx, caeca long and slender. *Reproductive systems:* genital atrium situated near the mid-body, 0·12–0·25 mm. in diameter, accommodating a papillate copulatory organ. Testes slightly lobed, side by side, or diagonally orientated, in the posterior region of the body. Ovary more deeply trilobed, situated in front of the testes and on one side of the median plane. Vitellaria occupying all available space lateral to the caeca. Uterus with few wide folds. Eggs of variable size (0·032–0·050 × 0·018–0·025 mm.).

Cryptocotyle concava (Creplin, 1825) Lühe, 1899

Syn. *Distoma concavum* Creplin, 1825; *Tocotrema concavum* of Looss, 1899.

HOSTS. Long-tailed duck, tufted duck, common scoter, golden-eye, velvet scoter, scaup, great crested grebe, black-necked grebe, razorbill, goosander, red-breasted merganser, shag, heron, red-throated diver.

LOCATION. Intestine*.

LOCALITY. Europe.

DIAGNOSIS. *Shape and size:* outline oval, 0·8–1·1 mm. long, 0·6–0·9 mm. broad. *Suckers:* oral larger than in *C. lingua*, 0·06–0·09 mm. diameter. *Reproductive systems:* testes side by side, ovary irregularly lobed. Papillate copulatory organ not present in the genital atrium. Eggs dark brown, measuring 0·037 × 0·016 mm. (according to Markowski (1933*a*), 0·033 × 0·016 mm.).

OTHER SPECIES:

Cryptocotyle jejuna Nicoll, 1907 inhabits the intestine of the redshank, *C. cryptocotyloides* (Issaitschikow, 1923) the small intestine of the black-throated diver. E. W. Price (1931*b*) gave a key to these and two other species of *Cryptocotyle*, *C. quinqueangulare* (Skrjabin) and *C. echinata* (Linstow).

Subfamily **Apophallinae** Ciurea, 1924

Genus *Apophallus* Lühe, 1909

Syn. *Rossicotrema* Skrjabin & Lindtrop, 1919; *Cotylophallus* Ransom, 1920.

Apophallus mühlingi (Jägerskiöld, 1899) Lühe, 1909 (Fig. 78 G–I)

Syn. *Distoma mühlingi* Jägerskiöld, 1899; *Tocotrema mühlingi* of Looss, 1899; *Metorchis oesophagolongus* Katsurada, 1914.

HOSTS. Black-headed gull, cormorant. Larvae encysted in fishes, especially the bream, *Blicca björkna*, in the fins and muscles (see Fig. 78 A–F).

LOCATION. Intestine.

LOCALITY. Europe.

* Markowski (1933*a*) found a specimen of this species 1·5 mm. long in the coelom of the broad-nosed pipefish, *Siphonostoma typhle*, in the Polish Baltic.

DIAGNOSIS. *Shape and size:* elongate, constricted at the level of the ventral sucker, 1·2–1·6 mm. long and 0·19–0·23 mm. broad. *Cuticle:* thickly spinous, spines fine. *Suckers:* small and approximately equal in size, 0·055 mm. diameter, the ventral near the mid-body. *Gut:* prepharynx short, pharynx small, oesophagus very long, caeca long, reaching to the posterior extremity. Bifurcation of the gut well in front of the ventral sucker. *Reproductive systems:* genital pore immediately in front of the ventral sucker, gonads confined to the posterior region of the body. Genital atrium deep and flask-like. Cirrus pouch absent, pars prostatica ill-developed, seminal vesicle large. Testes rounded, one behind the other, slightly diagonal. Ovary rounded, opposite the anterior testis. Uterus short and having few folds. Vitellaria lateral, behind the ventral sucker, merging across the median plane in the region of the testes. Eggs light brown and small (0·032 × 0·018 mm.).

OTHER SPECIES:

Apophallus major Szidat, 1924

According to Witenberg (1929) this species is identical with *A. mühlingi*.

HOST. Lesser black-backed gull.

LOCATION. Small intestine.

LOCALITY. Europe (East Prussia).

Subfamily **Galactosomatinae** Ciurea, 1933 emend.

Syn. Cercarioidinae Witenberg, 1929.

Genus *Galactosomum* Looss, 1899

Syn. *Microlistrum* Braun, 1901; *Cercarioides* Witenberg, 1929; *Tubanguia* Srivastava, 1935.

Price (1932) gave good reasons for upholding the first two synonyms and in 1940 added the third to the list.

Galactosomum lacteum (Jägerskiöld, 1896) Looss, 1902

Syn. *Monostomum lacteum* Jägerskiöld, 1896.

HOSTS. Cormorant, shag, heron. Larvae encysted in the brain of the father lasher (short-spined cottus).

LOCATION. Intestine.

LOCALITY. Europe.

Galactosomum cochlear (Diesing, 1850)

Syn. *Distoma cochleariforme sternae* Rudolphi, 1819; *D. cochlear* Diesing, 1850; *D. diesingii* Cobbold, 1861; *Microlistrum cochlear* of Braun, 1902; *M. cochleariforme* of Nicoll, 1923.

HOST. Sandwich tern.

LOCATION. Intestine.

LOCALITY. Europe.

Subfamily **Centrocestinae** Looss, 1899

Genus *Phagicola* Faust, 1920

Syn. *Ascocotyle* Looss, 1899 in part

Note. Price (1936*a*) restricted *Ascocotyle* for forms having two rows of spines in the oral coronet, adopting *Phagicola* Faust, 1920 (Syn. *Parascocotyle* Stunkard & Haviland; *Metascocotyle* Ciurea, 1933) for forms having a single row of oral spines. The type species of *Ascocotyle* s.str. is *A. coleostoma* (Looss, 1896), a parasite of the pelican in Egypt.

Phagicola minuta (Looss, 1899)

Syn. *Ascocotyle minuta* Looss, 1899; *Parascocotyle minuta* of Witenberg, 1929.

HOSTS. Heron (also dog and cat).

LOCATION. Small intestine.

LOCALITY. Egypt.

DIAGNOSIS. *Shape and size:* pyriform, small and delicate, 0·5 mm. long, 0·2 mm. broad. *Cuticle:* spinous. *Suckers:* oral smaller than the ventral, which is included in the genital atrium near the mid-body, and having a simple coronet of eighteen to twenty cylindrical spines, diameters 0·04 and 0·045–0·055 mm. Buccal sac more extensive than the pharynx and 0·05–0·07 mm. long. *Gut:* prepharynx short, pharynx small, oesophagus short, caeca short, scarcely extending beyond the level of the ventral sucker. *Reproductive systems:* testes small, side by side in the posterior region. Ovary in front of the testes, on the right. Vitellaria lateral, in the hinder part of the body, extending forward towards the ventral sucker. Eggs sculptured and small (0·023–0·024 × 0·014 mm.).

Note: Price (1936*a*) gave a key to seven species of *Ascocotyle* s.str., *A. minuta* being excluded, presumably to be included in the genus *Phagicola*.

OTHER SPECIES (for the most part parasites of mammals):

Subfamily **Haplorchiinae** Looss, 1899

Genus *Haplorchis* Looss, 1899

Syn. *Monorchotrema* Nishigori, 1924; *Kasr* Khalil, 1932.

Haplorchis pumilio (Looss, 1896) Looss, 1899

Syn. *Monostoma pumilio* Looss, 1896.

HOSTS. Pelican, black kite.

LOCATION. Intestine.

LOCALITY. Egypt.

DIAGNOSIS (after Gohar, 1934*a*, modified). *Shape and size:* pyriform, 0·45–0·62 mm. long, 0·25–0·34 mm. in greatest breadth and 0·15–0·28 mm. in greatest thickness. *Cuticle:* spinous, spines about 0·0055 mm. long. Oral spines absent. *Suckers:* ventral rudimentary, oral ovoid, 0·026–0·037 mm. long and 0·056–0·060 mm. broad. *Gut:* prepharynx short, pharynx 0·037 mm. long and 0·03 mm. broad, oesophagus long, caeca extending back to the posterior border of the testis. *Reproductive systems:* genital atrium cask-shaped (0·060 × 0·037 mm.), its opening situated somewhat on the right, slightly behind the bifurcation of the intestine, seminal vesicle on the left. Testis spherical, very large (0·175–0·3 mm. long and 0·1–0·25 mm. broad and thick), situated in the middle of the hind-body. Ovary spherical, 0·06–0·11 mm. diameter, ventro-lateral to the testes and on the right. Receptaculum seminis globular, as large as or larger than the ovary and situated dorsal to it. Vitellaria comprising groups of follicles of various sizes, extending from the anterior end of the testis to the posterior extremity, confluent in the median region posteriorly. Uterus with ascending and descending limbs filling the available space in the broad posterior region. Eggs cask-shaped, thick-shelled, operculate, having a button-like process at the anopercular pole, measuring 0·026–0·028 × 0·014–0·015 mm.

Note. The genital and atrial apertures are equipped with spines.

OTHER SPECIES:

Haplorchis taichui (Nishigori, 1924) (Syn. *Monorchotrema taichui* Nishigori, 1924).
H. milvi Gohar, 1934 (this and the above species occur in the black kite in Egypt).
And several others (e.g. *H. cahirinum* (Looss, 1896), *H. microrchia* and *H. yokogawi* (Katsuta, 1932).

Family **CLINOSTOMATIDAE** Lühe, 1901, emend. Dollfus, 1932 (p. 106)

Subfamily **Clinostomatinae** Pratt, 1902

KEY TO GENERA

1.	Caeca with long lateral diverticula which are sometimes branched	*Euclinostomum*
2.	Caeca with short lateral diverticula which are never branched	*Clinostomum*

Genus *Euclinostomum* Travassos, 1928

Syn. *Clinostomum* Leidy, 1856, in part.

Euclinostomum heterostomum (Rudolphi, 1809) Baer, 1933

Syn. *Distoma heterostomum* Rudolphi, 1809; *Dicrocoelium heterostomum* of Braun, 1899; *Clinostomum heterostomum* of Braun, 1900; *Ithyoclinostomum heterostomum* of Joyeux & Houdemer, 1928.

HOSTS. Purple heron, heron, night heron (also in reptiles).

LOCATION. Buccal cavity and nasal cavities.

LOCALITY. Europe.

DIAGNOSIS. *Shape and size:* oval or elongate, narrowing in front of the ventral sucker, widening behind it, 6·5–9 mm. long, 2·5 mm. broad anteriorly, 3·5 mm. broad posteriorly. *Cuticle:* spinous. *Suckers:* well developed, close together, oral smaller than the ventral, dimensions 0·30–0·36 × 0·27–0·33 mm. and 1·4 mm. diameter. *Gut:* pharynx absent, oesophagus short, caeca long and with nine to eleven long lateral diverticula, some of which are branched. *Reproductive systems:* gonads in the posterior half of the region behind the ventral sucker. Testes one behind the other, the anterior testis having a depression in front, where the genital atrium and cirrus pouch abut on it. Ovary between the testes and on the right. Genital pore immediately in front of the anterior testis. Vitellaria extending laterally from the ventral sucker to just behind the posterior testis. Eggs very large (0·125–0·135 × 0·062–0·073 mm.).

Genus *Clinostomum* Leidy, 1856, *nec* Girard, 1856 (which is a fish)

Clinostomum complanatum (Rudolphi, 1809) Braun, 1900

Syn. *Distoma complanatum* Rudolphi, 1809; *Clinostomum marginatum* (Rudolphi, 1809).

HOSTS. Herons, herring gull.

LOCATION. Buccal cavity, pharynx, oesophagus.

LOCALITY. Europe, Asia.

DIAGNOSIS. *Size:* 3–8 mm. long, 2·6 mm. in maximum breadth. *Cuticle:* spinous. *Suckers:* close together, oral smaller than the ventral, diameters 0·28–0·30 and 0·5–0·8 mm. *Reproductive systems:* gonads midway between the ventral sucker and the posterior extremity, testes broader than long and slightly lobed, ovary smaller than the testes. Genital pore on the right, near the anterior testis, in compressed preparations sometimes in front of it. Uterus extending almost to the ventral sucker. Vitellaria enveloping the gonads, extending almost to the posterior margin of the ventral sucker. Eggs very large and variable in size (0·104–0·140 × 0·066–0·073 mm.). Yamaguti (1934) described the peculiar miracidium of this species, which is ciliated only at the extremities and contains a single germ ball. The metacercariae occur in numerous fishes, in Europe in the perch (see Ciurea, 1911).

Maccagno (1934) dealt with the subject of *Clinostomum complanatum* in Europe. Baer (1933) has given a key to the separation of the above and ten other species of *Clinostomum*, together with a list of the specific characters. They are typically parasites of herons and bitterns, but one occurs in a cormorant. Perhaps the best-known species in Europe is *C. foliiforme* Braun, 1899 (in the pharynx and oesophagus of the purple heron).

Family **ORCHIPEDIDAE** Skrjabin, 1924 (p. 98)

Genus *Orchipedum* Braun, 1902

Orchipedum tracheicola Braun, 1901 (Fig. 53B)

HOST. Velvet scoter.

LOCATION. Trachea.

LOCALITY. Europe (Vienna).

DIAGNOSIS. *Shape and size:* elongate and with pointed extremities, constricted at the level of the ventral sucker forming an anterior region 2 mm. long and a flat, broad, posterior region 5 mm. long and 1·6 mm. in maximum breadth. *Cuticle:* non-spinous. *Suckers:* not situated very close together, the oral smaller than the ventral, diameters 0·4–0·5 and 0·7 mm. *Gut:* prepharynx and oesophagus absent, pharynx small (0·23–0·24 mm. diameter), bifurcation of the intestine far in front of the ventral sucker. *Reproductive systems:* genital pore just behind the pharynx, cirrus pouch absent. Testes numerous (about fifty), almost filling the space between the caeca in the posterior region. Ovary rounded, in front of the testes, out of the median plane. Vitellaria extending throughout the posterior region, but not in front of the ventral sucker, lateral and dorsal to the caeca. Uterus short and with a few folds dorsal and anterior to the ventral sucker. Eggs few, golden brown, measuring about 0·062 × 0·050 mm.

OTHER SPECIES:

Orchipedum formosum (Sonsino, 1890) Witenberg, 1922

Syn. *Distomum formosum* Sonsino, 1890; *Polyorchis formosum* of Stossich, 1892.

HOST. Crane, heron.
LOCATION. Trachea.
LOCALITY. Europe (Italy).

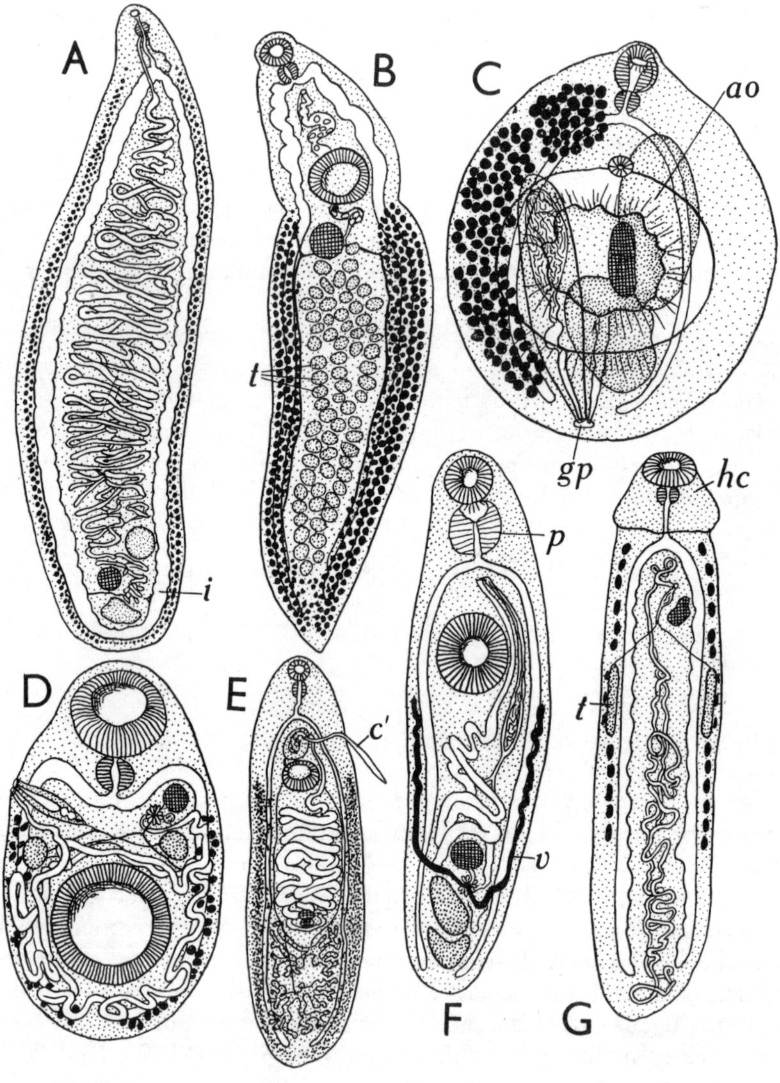

Fig. 53. More Digenea of birds. A, *Cyclocoelum orientale* (Cyclocoelidae). B, *Orchipedum tracheicola* (Orchipedidae). C, *Cyathocotyle prussica* (Cyathocotylidae). D, *Stomylotrema perpastum* (Stomylotrematidae). E, *Cathaemasia spectabilis* (Cathaemasiidae). F, *Philophthalmus palpebrarum* (Philophthalmidae). G, *Eucotyle nephritica* (Eucotylidae). (After Fuhrmann, 1928.)

Orchipedum turkestanicum Skrjabin, 1913

HOST. Spoonbill.
LOCATION. Trachea.
LOCALITY. Russian Turkestan.

This species differs from *O. tracheicola* as follows: larger size (12 × 3 mm.), less obviously divided into regions, much larger suckers (ventral measuring 2·125 mm. diameter, oral 1·02 × 1·44 mm.) which are very close together, uterus confined to the region dorsal to the ventral sucker, vitellaria extending from the pharynx to the ends of the long caeca, eggs relatively large, measuring about 0·087 × 0·049 mm.

Orchipedum centorchis Witenberg, 1922

In the bronchi of the pelican (Turkestan).

Orchipedum armeniacum Skrjabin, 1925

In the trachea of the coot (Armenia).

Family **EUCOTYLIDAE** Skrjabin, 1924 (p. 99)

KEY TO GENERA

1. Anterior region constricted off from the rest of the body	*Eucotyle*
2. Anterior region not constricted off from the rest of the body	
A. Oesophagus present; testes lobed, diagonally arranged	*Tanaisia*
B. Oesophagus absent; testes not lobed, side by side	*Tamerlania*

Genus *Eucotyle* Cohn, 1904

Price (1930) gave a key to the five species of this genus which are mentioned below.

Eucotyle nephritica (Mehlis, 1846) Cohn, 1904 (Fig. 53 G)

Syn. *Monostoma nephriticum* Mehlis, 1846.

HOST. Black-throated diver.
LOCATION. Kidney.
LOCALITY. Europe (Claustal).

DIAGNOSIS. *Shape and size:* elongate, rounded posteriorly, the anterior region triangular, about 3·5 mm. long and 0·7 mm. broad. *Suckers:* ventral absent, oral about 0·25 mm. diameter. *Gut:* pharynx small (0·09 mm. diameter), oesophagus fairly long, caeca long and having numerous small median diverticula. *Reproductive systems:* genital pore median, 0·25 mm. behind the bifurcation of the intestine, cirrus pouch absent. Testes small and situated lateral to the caeca in front of the mid-body. Ovary lateral, between the levels of the genital pore and the testes. Vitellaria extending from the bifurcation of the intestine to the testes and as far again behind them. Uterus having long descending and ascending limbs, formed into numerous folds between the caeca. Eggs small (about 0·023 × 0·012 mm.).

OTHER SPECIES:

E. zakharowi Skrjabin, 1920. In the kidney of tufted duck (Russia).

E. cohni Skrjabin, 1924. In the kidney of black-necked grebe and red-necked grebe (Turkestan).

E. hassalli Price, 1930. In the urinary tract (?) of *Cobymbus auritus* (U.S.A.).

E. wehri Price, 1930. In the urinary tract (?) of the American blue-winged teal (U.S.A.).

Genus *Tanaisia* Skrjabin, 1924

Tanaisia fedtschenkoi Skrjabin, 1924

Syn. *Prohystera rossittensis* Korkhaus, 1930.

HOSTS. Black-headed gull, common gull, common tern, water-rail, etc. (*Himantopus chettusia*, *Totanus*, *Tringa*, *Helodomas*, *Chroicephalus*, *Sterna*, *Hydrochelidon*, *Larus* and *Rallus* spp.).

LOCATION. Kidney.

LOCALITY. Europe (Russia; Don Valley), Russian Turkestan.

Genus *Tamerlania* Skrjabin, 1924

Tamerlania zarudnyi Skrjabin, 1924

HOST. Tree-sparrow (and *Hedymela atricapialla*).

LOCATION. Kidney.

LOCALITY. Russian Turkestan.

Family **PHILOPHTHALMIDAE** Looss, 1899 (p. 98)

KEY TO GENERA

I. Vitellaria lateral to the caeca; cirrus pouch extending back behind the ventral sucker; testes one behind the other *Philophthalmus*

II. Vitellaria median to the caeca; cirrus pouch not extending back behind the ventral sucker; testes diagonally one behind the other *Pygorchis*

Genus *Philophthalmus* Looss, 1899

Philophthalmus lucipetus (Rudolphi, 1819) Looss, 1899

Syn. *Distoma lucipetum* Rudolphi, 1819; *D.* (*Dicrocoelium*) *lucipetum* (Rudolphi) Dujardin, 1845.

HOSTS. Herring gull, lesser black-backed gull.

LOCATION. Nictitating membrane.

LOCALITY. Europe.

DIAGNOSIS. Location between the eyelid and the eyeball (indicating the slender line of demarcation between ecto- and endoparasitism) diagnostic for the genus. *Shape and size:* elongate oval outline, about 7 mm. long and 1·7 mm. broad. *Cuticle:* scaly. *Suckers:* transversely oval, ventral larger than the oral and about one-third of the distance along the body, dimensions 0·7–1·2 × 0·63–1·03 and 0·33–0·53 × 0·2–0·4 mm. *Gut:* prepharynx and pharynx small, oesophagus short, caeca long. *Reproductive systems:* genital pore median, slightly in front of the ventral sucker and near the bifurcation of the intestine. Cirrus pouch elongate. Testes slightly lobed, one behind the other in the posterior region. Ovary in front of the testes. Vitellaria approximately in the third quarter of the body, comprising a single row of six or seven follicles on each side, curving towards the median plane between the ovary and the anterior testis. Uterus having an ascending limb only, but the wide, transverse or backwardly directed folds may overstep caeca laterally. Eggs thin-shelled and fairly large (0·096 × 0·033 mm.), each containing a miracidium having eye-spots.

Philophthalmus palpebrarum Looss, 1899 (Fig. 53F)

HOST. Hooded crow.

LOCATION. Under the eyelid.

LOCALITY. Europe, Egypt.

DIAGNOSIS. Similar to *P. lucipetus*, but with shorter oesophagus and uterus, longer cirrus pouch and metraterm, larger pharynx and tubular instead of follicular vitellaria.

OTHER SPECIES:

Philophthalmus nocturnus Looss, 1907. In the conjunctiva of the little owl (Egypt).
P. skrjabini Efimov, 1937. In black-headed gull (Russia).
One other species, *P. lachrymosus* (and *lacrymosus*) Braun, is a parasite of a gull in Brazil.

Genus *Pygorchis* Looss, 1899

Pygorchis affixus Looss, 1899

HOST. Hooded crow, kestrel, marsh-harrier, avocet.

LOCATION. Cloaca.

LOCALITY. Egypt.

DIAGNOSIS. Similar to *Philophthalmus*, except in the characters mentioned in the key above and in the oval shape, relatively larger and more central ventral sucker, larger pharynx, shorter oesophagus, and wider uterine folds, which tend to fill the posterior half of the body.

Note. Szidat (1939*a*) drew attention to the close structural resemblances between *Philophthalmus*, *Pygorchis* and *Distomum pittacium* Braun, 1901 (a Brazilian trematode from the cloaca of *Tringa interpres*), for which he erected the new genus *Pittacium*. He extended his comparison to two genera of echinostomes, *Proctobium* Travassos, 1918 (which has a species, *gedoelsti* Skrjabin, 1924, inhabiting the cloaca of the Lapland dunlin in Russia) and *Parorchis* Nicoll, 1907.* Both these forms have a typical collar, but apart from this they are remarkably similar to *Pygorchis* and *Pittacium*, from which Szidat infers that they belong to one and the same line of evolution, and reveal the probability that the Philophthalmidae are aberrant Echinostomes.

* These genera are so similar as to be identical, and *Proctobium* is here regarded as a synonym of *Parorchis* (p. 357).

Family **CATHAEMASIIDAE** Fuhrmann, 1928, emend. Harwood, 1936 (p. 99)

Genus *Cathaemasia* Looss, 1899

Cathaemasia hians (Rudolphi, 1809) Looss, 1899

Syn. *Distoma hians* Rudolphi, 1809; *D.* (*Dicrocoelium*) *hians* (Rudolphi) Dujardin, 1845.

HOSTS. White stork, black stork, heron, night heron, purple heron.

LOCATION. Oesophagus.

LOCALITY. Europe.

DIAGNOSIS. *Shape and size:* oval or slightly elongate outline, 6–14 mm. long and 2–3 mm. broad. *Cuticle:* spinous ventrally. *Suckers:* ventral slightly the larger, diameters 0·85–1·1 and 0·8 mm. *Gut:* prepharynx and oesophagus short, pharynx present, caeca long, bifurcation of the intestine midway between the suckers. *Reproductive systems:* genital pore immediately in front of the ventral sucker. Testes lobed, one behind the other in the posterior region. Ovary small, immediately in front of the testes. Uterus having an ascending limb only, formed into wide folds. Vitellaria extending from the ventral sucker to the posterior extremity. Eggs numerous, large (about 0·1 × 0·055 mm.) (0·095–0·110 × 0·05–0·06 mm. according to Szidat (1939*a*), who recorded finding twenty-three specimens in the gullet of a single black stork at Rossitten (East Prussia)).

Cathaemasia fodicans Braun, 1901

HOST. Black stork.

LOCATION. Oesophagus.

LOCALITY. Europe (Austria).

DIAGNOSIS. Similar to *C. hians*, but oesophagus absent, vitellaria slightly less extensive, testes more deeply lobed and eggs smaller (dimensions about 0·083 × 0·042 mm.). Possibly identical with *C. hians* (see Odhner, 1926*b*).

OTHER SPECIES:

Cathaemasia spectabilis Odhner, 1926 (Fig. 53E). In '*Leptoptilus crumenifer*'.
C. famelica Odhner, 1926. In '*Tantalus ibis*'.

Szidat (*loc. cit.*) showed from consideration of the life history of *Cathaemasia* that this genus is related to the Echinostomes rather than the Fasciolids. He regarded it as a true Echinostome genus in which the typical form has been modified and the head-crown lost.

Family **PSILOSTOMATIDAE** Odhner, 1911, emend. Nicoll, 1935 (p. 99)

KEY TO GENERA

I. Body somewhat elongate
   A. Testes rounded; vitellaria extending to posterior extremity — *Psilostomum*
   B. Testes elongate; vitellaria not extending to posterior extremity — *Psilochasmus*

II. Body oval rather than elongate
   1. Ventral sucker at or behind the middle of the body — *Sphaeridiotrema*
   2. Ventral sucker anterior to the middle of the body
      A. Oral sucker present — *Psilotrema*
      B. Oral sucker absent, the pharynx taking its place — *Apopharynx*

Genus *Psilostomum* Looss, 1899

Price (1942*a*) has reviewed this genus, which is said to contain five valid species, and two forms which are probably out of place, one of them being *P. arvicolae* Schulz & Dobrova, 1933–4, a parasite of the water rat, *Arvicola amphibius*, in Russia.

Psilostomum brevicolle (Creplin, 1829) Braun, 1902 (Fig. 51 B)

Syn. *Distoma brevicolle* Creplin, 1829; *D. platyurum* Mühling, 1896; *Psilostomum platyurum* of Looss, 1899; *Distoma (Dicrocoelium) brevicolle* of Stossich, 1892.

HOSTS. Long-tailed duck, common scoter, velvet scoter, oystercatcher.

LOCATION. Intestine.

LOCALITY. Europe (in Britain: St Andrews, Lincolnshire).

DIAGNOSIS. *Shape and size:* elongate, constricted slightly at the level of the ventral sucker, about 3 mm. long and 1 mm. broad. *Cuticle:* smooth. *Suckers:* of about equal size, 0·35 mm. diameter, the oral having a transverse, slit-like opening, the ventral situated about one-third of the distance along the body. *Gut:* prepharynx short, pharynx large, oesophagus absent, caeca long. *Reproductive systems:* genital pore in front of the ventral sucker and slightly lateral. Cirrus pouch long (0·65 mm.) and bowed, curving round the ventral sucker. Gonads rounded, in line longitudinally, the ovary foremost. Vitellaria comprising coarse follicles, extending from the constriction of the body to the posterior extremity. Uterus short and little folded. Eggs few (less than fifty at any time), but large (0·112 × 0·062 mm.).

Psilostomum cygnei Southwell & Kirshner, 1937

HOST. Swan.

LOCATION. Small intestine.

LOCALITY. Europe (Sefton Park, Liverpool).

DIAGNOSIS. *Shape and size:* Y-shaped, the limbs of the Y formed by regions terminating in respective suckers, 1–2·5 mm. long. Stem of the body slender, elongate, tapering to a bluntly rounded posterior extremity. *Colour:* white. *Suckers:* ventral about twice as large as the oral, projecting almost at right angles from the body immediately behind the pharynx. *Reproductive systems:* genital pore on the left of the median plane, between the suckers. Cirrus pouch large, elongate, having a feeble musculature. Testes rounded, one behind the other, slightly behind the mid-body. Ovary rounded, almost as large as the testes and situated in front of them and on the left. Vitellaria extending from the level of the ovary almost to the posterior extremity, coalescing behind the testes. Uterus short, confined to the region between the ovary and the pharynx, containing about three (at most six) yellowish brown eggs measuring about 0·072 × 0·055 mm.

OTHER SPECIES:

Psilostomum progeneticum Wiśniewski, 1933, a metacercaria exhibiting progenesis. (In the body cavity of *Fontogammarus bosniacus* and *Rivulogammarus spinicaudatus* in the Balkans (Jugoslavia).)

P. marillae Price, 1942. This species was found by Price, sometimes in large numbers, in the small intestine of the lesser scaup duck, *Marilla affinis*, in association with *Sphaeridotrema globulus*. Fatality amongst the hosts on the Potomac River (U.S.A.) was due to the latter trematode, the species *P. marillae* seeming in no way responsible.

P. varium Linton, 1928 is the only other valid species of this genus.

Genus *Psilochasmus* Lühe, 1909

Syn. *Psilostomum* Looss, 1899, in part.

Psilochasmus oxyurus (Creplin, 1825) Lühe, 1909

Syn. *Distoma oxyurum* Creplin, 1825; *Psilostomum oxyurum* of Braun, 1902.

HOSTS. Scaup, tufted duck, common scoter, common sheld-duck, golden-eye, long-tailed duck.

LOCATION. Intestine.

LOCALITY. Europe (Dorset).

DIAGNOSIS. *Shape and size:* elongate, the extremities pointed, 6·5–7·3 mm. long and 1·0–1·8 mm. broad. *Cuticle:* smooth. *Suckers:* ventral larger than the oral and projecting prominently about one-quarter of the distance along the body, diameters 0·64 and 0·44 mm. *Gut:* pharynx small (0·135 mm. diameter), oesophagus long, bifurcation of the intestine well in front of the ventral sucker. *Reproductive systems:* cirrus pouch about 0·5 mm. long and 0·1 mm. broad. Testes elongate and lobed. Uterus containing at most thirty to forty eggs measuring 0·082–0·110 × 0·06–0·07 mm.

OTHER SPECIES:

Psilochasmus longicirratus Skrjabin, 1913. In intestine of white-eyed pochard (Russian Turkestan) (tufted duck in Japan). This species is smaller than *P. oxyurus* (3·7–5 mm. long, 1·0–1·5 mm. broad), has more dissimilar suckers (diameters 0·64 and 0·34 mm.), a much larger pharynx (0·255 × 0·204 mm.), an enormously larger cirrus pouch (1·3 mm. long, 0·24 mm. broad), only slightly lobed testes and larger eggs (dimensions 0·116–0·124 × 0·072–0·087 mm.).

Note. '*Psilochasmus lecithosus* Otte, 1926', from the intestine of the duck, was shown by Baylis (1932*a*), after examination of specimens named by Otte, to be identical with *Hypoderaeum conoideum*.

Genus *Sphaeridiotrema* Odhner, 1913

Syn. *Shaeridiotrema* Sprehn, 1933, misprint.

Sphaeridiotrema globulus (Rudolphi, 1819) Odhner, 1913

Syn. *Distoma globulus* Rudolphi, 1819.

HOSTS. Tufted duck, scaup, pintail, goosander, red-breasted merganser, long-tailed duck, whooper swan, razorbill.

LOCATION. Intestine.

LOCALITY. Europe, America.

DIAGNOSIS. *Shape and size:* compact, ovate or almost spherical, 0·5–0·75 mm. long. *Suckers:* ventral much larger than the oral and situated almost at the mid-body, diameters 0·25 and 0·105–0·125 mm. *Gut:* pharynx ovoid, relatively large, measuring 0·08 × 0·06 mm. *Reproductive systems:* genital pore not far behind the oral sucker, slightly lateral. Cirrus pouch flask-shaped. Testes transversely ovoid, one behind the other in the posterior region. Vitellaria well developed, obscuring other organs, but not extending to the posterior extremity. Eggs few (four to six), but large, measuring 0·105 × 0·075 mm. The life history has been described by Szidat (1937*a*).

Genus *Psilotrema* Odhner, 1913

Syn. *Psilostomum* Looss, 1899, in part.

Psilotrema simillimum (Mühling, 1898) Odhner, 1913

Syn. *Distoma simillimum* Mühling, 1898; *Psilostomum simillimum* of Looss, 1899.

HOST. White-eye pochard.

LOCATION. Intestine.

LOCALITY. Europe (East Prussia).

DIAGNOSIS. *Shape and size:* elongate, 1·4–1·9 mm. long and 0·8 mm. broad. *Cuticle:* spinous dorsally only at the anterior end, ventrally only behind the genital pore in the region of the ventral sucker. *Suckers:* ventral much larger than the oral and situated in the second quarter of the body, dimensions 0·11 × 0·17 mm. and 0·385 mm. diameter. *Gut:* pharynx larger than the oral sucker (about 0·2 mm. diameter), oesophagus very short or absent. *Reproductive systems:* genital pore near the pharynx, well to the left of the median plane. Cirrus extending almost to the base of the cirrus pouch. Ovary immediately behind the ventral sucker, on the right. Vitellaria extending from near the ovary to the posterior extremity, almost merging in the median plane behind the testes. Uterus very short, not folded, containing only five to seven eggs measuring about 0·09 × 0·05 mm.

Psilotrema spiculigerum (Mühling, 1898) Odhner, 1913

Syn. *Distoma spiculigerum* Mühling, 1898; *Psilostomum spiculigerum* of Looss, 1899; *P. oligoon* (Linstow, 1877) Odhner, 1913.

HOST. White-eyed pochard.

LOCATION. Intestine.

LOCALITY. Europe (East Prussia).

DIAGNOSIS. Similar to *P. simillimum*, but suckers of approximately equal size. Eggs measuring 0·097 × 0·050 mm.

Genus *Apopharynx* Lühe, 1909

Apopharynx bolodes (Braun, 1902) Lühe, 1909

Syn. *Distoma bolodes* Braun, 1902.

HOST. Coot.

LOCATION. Bursa Fabricii.

LOCALITY. Europe (East Prussia).

DIAGNOSIS. *Shape and size:* oval outline, somewhat pointed at the extremities, little flattened, 2·2 mm. long and about 1 mm. broad. *Cuticle:* spinous, as in

Psilotrema. Suckers: pharynx powerful, 0·3 mm. long and 0·35 mm. broad, assuming the function of an oral sucker (otherwise absent), ventral sucker smaller, 0·31 mm. diameter, situated in the second quarter of the body. *Gut:* oesophagus fairly long, caeca long. *Reproductive systems:* genital pore near the pharynx. Testes large, angular, one behind the other in the posterior region. Ovary in front of the testes, near the ventral sucker. Uterus short and little folded. Eggs thin-shelled, yellowish brown and large (0·093 × 0·06–0·07 mm.).

Family **ECHINOSTOMATIDAE** Looss, 1902, emend. Poche, 1926, or Stiles & Hassall, 1926 (p. 101)

Note. Dietz (1910) provided a key for the determination of twenty-two echinostome genera.

Subfamily **Echinostomatinae** Looss, 1899, emend. Stiles & Hassall, 1926

KEY TO GENERA OCCURRING IN BIRDS

I. Spines of the collar differing in size in the two rows — *Echinoparyphium*

II. Spines of the collar of equal size in the two rows
   - A. Ventral sucker in the first quarter of the body; uterus having numerous transverse folds; cuticle spinous only ventrally — *Echinostoma*
   - B. Ventral sucker in the second quarter of the body; uterus only slightly folded, cuticle spinous dorsally as well as ventrally — *Euparyphium*

In a fourth genus, *Patagifer* Dietz, 1909, the head collar has a deep incision down to the oral sucker. Rudolphi's well-known species, *P. bilobus* (1819), occurs in the spoonbill, coot and glossy ibis.

Genus *Echinostoma* Rudolphi, 1809

Echinostoma revolutum (Frölich, 1802) Looss, 1899 (Fig. 50D, F)

Syn. *Fasciola revoluta* Frölich, 1802; *Distoma echinatum* Zeder, 1803; *Echinostoma mendax* Dietz, 1909.

Hosts. Domestic duck, goose and fowl; wigeon, garganey, pochard, common scoter, common sheld-duck, mute swan, whooper swan, carrion crow, woodcock, pheasant, partridge, white stork, etc.

Location. Caeca and rectum.

Locality. Europe, North and South America, Australia, Asia.

Diagnosis. *Shape and size:* elongate, 10–22 mm. long and 2–3 mm. broad. *Cuticle:* spinous anteriorly and ventrally as far back as the ventral sucker. *Colour:* rose. *Collar:* bearing thirty-seven spines, of which five on each side are 'corner-spines'. *Suckers:* ventral very much larger than the oral, diameters 1·7–3·0 and 0·25–0·5 mm. *Reproductive systems:* gonads one behind the other, the ovary foremost, variable in form. Vitellaria very well developed, especially posteriorly, reaching to the posterior extremity, but not extending in front of the ventral sucker. Uterus with only an ascending limb, which is formed into transverse folds. Eggs large (0·097–0·126 × 0·059–0·071 mm.).

OTHER SPECIES IN EUROPEAN BIRDS (and their principal hosts). All are parasitic in the intestine:

Echinostoma academicum Skrjabin, 1915; black-tailed godwit.
E. anceps (Molin, 1859) Stossich, 1892; coot.
E. chloropodis (Zeder, 1800) Dietz, 1909; moorhen and spotted crake (water-rail in Britain).
E. echiniferum (La Valette, 1855) Stossich, 1892; avocet.
E. echinocephalum (Rudolphi, 1819) Cobbold, 1860; kite.
E. exechinatum Solowiow, 1912; cormorant.
E. mesotestius Solowiow, 1912; rook.
E. megacanthum Kotlan, 1922; great-crested grebe.
E. nephrocephalum (Diesing, 1850) Cobbold, 1860; rock thrush.
E. pungens (Linstow, 1894) Stossich, 1899; little grebe (Kent).
E. sarcinum Dietz, 1909; crane.
E. stridulae (Reich, 1801) Dietz, 1909; tawny owl.
E. uralense Skrjabin, 1915; collared praticole.
Etc.

Genus *Echinoparyphium* Dietz, 1910

Echinoparyphium recurvatum (Linstow, 1873) Lühe, 1909

Syn. *Distoma recurvatum* Linstow, 1873; *Echinostoma recurvatum* (Linstow, 1873) Stossich, 1892.

HOSTS. Domestic duck, fowl, pigeon, also scaup, tufted duck, goosander, pintail.

LOCATION. Intestine.

LOCALITY. Europe, Asia.

DIAGNOSIS. *Size:* 4·5 mm. long and 0·5–0·6 mm. broad. *Collar:* 0·35 mm. broad and bearing forty-five spines, of which four on each side are larger and form corner spines, the remaining thirty-seven being arranged in two rows, spines in the oral row smaller than those in aboral. Eggs large (0·108–0·110 × 0·081–0·084 mm.). The larval development in *Paludina vivipara* has been studied by Dinulesco (1939).

Echinoparyphium paraulum (Dietz, 1909) Sprehn, 1929

Syn. *Echinostoma paraulum* Dietz, 1909; *E. columbae* Zunker, 1925.

HOSTS. Domestic duck, goose, pigeon, also great-crested grebe, wigeon, whooper swan, mallard.

LOCATION. Intestine.

LOCALITY. Europe.

DIAGNOSIS. *Size:* 6–10·5 mm. long, 0·8–1·4 mm. broad. *Collar:* 0·4–0·6 mm. wide and bearing thirty-seven spines, of which five on each side are corner spines, the remaining twenty-seven being arranged in two rows, oral spines thicker than the aboral. Eggs large (about 0·100 × 0·070 mm.).

Note. This species is of doubtful validity; Baylis (1929) thought it a possible synonym of *Echinostoma revolutum*.

OTHER SPECIES IN EUROPEAN BIRDS (and their hosts):

Echinoparyphium aconiatum Dietz, 1909: lapwing.

E. agnatum Dietz, 1909: buzzard.

E. baculus (Diesing, 1850) Lühe, 1909: smew, scaup, golden-eye, shoveler, velvet scoter, black-throated diver.

E. clerci Skrjabin, 1915: green sandpiper.

E. mordwilkoi Skrjabin, 1915: green sandpiper.

E. politum Skrjabin, 1915: lapwing, green sandpiper.

Genus *Euparyphium* Dietz, 1909

Syn. *Isthmiophora*, Lühe, 1909.

This is a genus which chiefly infects mammals, and it will be considered in the next chapter.

Another genus, *Nephrostomum* Dietz, 1909, has an Egyptian species *ramosum* (Sonsino, 1895) in the intestine of the little egret and buff-backed heron. It can be recognized by a slight dorsal invagination of the head collar.

Subfamily **Echinochasminae** Odhner, 1910

KEY TO GENERA OCCURRING IN BIRDS

1. Cirrus pouch inconspicuous; collar-spines numbering twenty-four *Echinochasmus*
2. Cirrus pouch obviously present; collar-spines numbering twenty-two or twenty-six *Stephanoprora*

Genus *Echinochasmus* Dietz, 1909, emend. Odhner, 1910

Syn. *Episthmium* Lühe, 1909.

Echinochasmus coaxatus Dietz, 1909

HOSTS. Great-crested grebe, red-necked grebe, white stork.

LOCATION. Intestine.

LOCALITY. Europe (Gloucestershire).

DIAGNOSIS. *Size:* 2·1–2·6 mm. long. *Collar:* bearing twenty-four collar spines, three on each side being corner spines and the remaining eighteen slightly larger. *Cuticle:* spinous back to the posterior extremity. *Eggs:* not very large (0·084–0·086 × 0·057–0·058 mm.).

OTHER SPECIES (and their hosts):

Echinochasmus beleocephalus (Linstow, 1873) Dietz, 1909: heron.

DIAGNOSIS: All spines in head crown of equal size; cuticle spinous only to vicinity of ventral sucker.

E. amphibolus Kotlan, 1922: cormorant. Found in the bittern in Norfolk (Baylis, 1939).

E. botauri Baer, 1923: bittern.

E. bursicola (Creplin, 1837): heron.

E. dietzevi Issaitschikow, 1927: great-crested grebe, red-necked grebe, black-necked grebe.

E. oligacanthus Dietz, 1910: heron (Egypt).

E. liliputanus (Looss, 1896) Odhner, 1910: honey-buzzard.

Genus *Stephanoprora* Odhner, 1902

Syn. *Mesorchis* Dietz, 1909; *Monilifer*, Dietz, 1909.

Stephanoprora denticulata (Rudolphi, 1802)

Syn. *Fasciola denticulata* Rudolphi, 1802; *Distoma denticulatum* of Rudolphi, 1809; *Echinostoma denticulatum* of Cobbold, 1860; *Distoma pseudechinatum* Olsson, 1876; *Mesorchis polycestus* Dietz, 1909; *M. denticulatus* of Dietz, 1909.

HOSTS. Common tern, razorbill, black-throated diver, greater black-backed gull, lessser black-backed gull.

LOCATION. Intestine.

LOCALITY. Europe.

DIAGNOSIS. *Size:* 1·7–8 mm. long. *Collar:* equipped with twenty-two spines of equal size arranged in an unbroken row, the two corner spines on each side slightly the smaller. *Cuticle:* spinous. *Reproductive systems:* gonads one behind the other in the mid-body, the ovary foremost. Vitellaria confined to the region behind the gonads. Eggs fairly large, but of variable size (0·065–0·100 × 0·043–0·059 mm.).

Stephanoprora spinosa Odhner, 1910

Syn. *Monilifer spinulosus* Dietz, 1909, *nec* Rudolphi, 1809.

HOSTS. Great crested grebe, red-necked grebe, Slavonic grebe, black-throated diver.

LOCATION. Intestine.

LOCALITY. Europe (Norfolk, Gloucestershire), Egypt.

DIAGNOSIS. *Size:* 1–2·7 mm. long. *Collar:* bearing twenty-two spines in an unbroken row, the two corner spines on each side being slightly smaller than the rest. *Cuticle:* spinous. *Reproductive systems:* testes large, situated behind the mid-body. Eggs measuring 0·069–0·072 × 0·048–0·050 mm.

OTHER SPECIES:

S. pendula (Looss, 1899) Odhner, 1910

Syn. *Echinostoma pendulum* Looss, 1899; *Mesorchis pendulus* of Dietz, 1909. In the caeca of the avocet (Egypt) and the spotted redshank (Japan).

Subfamily **Himasthlinae** Odhner, 1910

Genus *Himasthla* Dietz, 1909, emend. Odhner, 1910

Himasthla leptosoma (Creplin, 1829) Dietz, 1909

Syn. *Distoma leptosomum* Creplin, 1829; *Echinostoma leptosomum* of Cobbold, 1860; *Distoma militare* Rudolphi, 1809; *Himasthla militaris* of Dietz, 1909; *Echinostoma secundum* Nicoll, 1906.

HOSTS. Dunlin, sanderling, curlew, turnstone, black-headed gull, herring gull, bar-tailed godwit, knot.

LOCATION. Intestine.

LOCALITY. Europe (East Lothian, South Scotland, Norfolk).

DIAGNOSIS. *Size:* 6·5–10·5 mm. long. *Shape:* elongate, broadest in the posterior region. *Collar:* bearing twenty-nine spines arranged in an unbroken row, two on each side being corner spines and slightly smaller than the remainder. *Reproductive systems:* gonads one behind the other in the posterior region of the body, the ovary foremost. Vitellaria extending from the posterior extremity to the base of the cirrus pouch. Eggs measuring 0·096 × 0·062 mm.

Himasthla elongata (Mehlis, 1831) Dietz, 1909

Syn. *Distoma elongatum* Mehlis, 1831.

HOSTS. Greater black-backed gull, herring gull, black-headed gull.

LOCATION. Intestine.

LOCALITY. Europe.

DIAGNOSIS. *Size:* 7–8·5 mm. long. *Other characters:* similar to *H. leptosoma*, but collar spines larger, corner spines relatively smaller, eggs much larger (0·122 × 0·077–0·078 mm.).

OTHER GENERA:

Acanthoparyphium, *Chloeophora* and *Pelmatostomum*, all Dietz, 1909.

KEY TO ISOLATED GENERA OF ECHINOSTOMATIDAE

I. Anterior region of the body having four spines on each side — *Sodalis*
II. Anterior region of the body having a coronet of spines
- 1. Oral sucker absent — *Pegosomum*
- 2. Oral sucker present
  - (A) Having a single coronet of head spines; testes side by side — *Parorchis*
  - (B) Having a double coronet of spines; testes one in front of the other or diagonally arranged
    - (*a*) Testes deeply lobed — *Paryphostomum*
    - (*b*) Testes not deeply lobed
      - (*a′*) Cirrus pouch extending behind the ventral sucker
        - (*aa*) Collar spines of the two rows opposite — *Echinostephilla*
        - (*bb*) Collar spines of the two rows alternating — *Hypoderaeum*
      - (*b′*) Cirrus pouch not extending behind the ventral sucker
        - (*aa*) Bifurcation of the intestine midway between the pharynx and the ventral sucker — *Parechinostomum*
        - (*bb*) Bifurcation of the intestine immediately in front of the ventral sucker
          - (i) Vitellaria in the posterior part of the spindle-like body — *Petasiger*
          - (ii) Vitellaria extending forward to the oral sucker — *Chaunocephalus*

Genus *Sodalis* Kowalewsky, 1902

Syn. *Scapanosoma* Lühe, 1909.

Sodalis spatulatus (Rudolphi, 1819) Kowalewsky, 1902

Syn. *Distoma spatulatum* Rudolphi, 1819; *Echinostoma spatulatum* of Kowalewsky, 1898; *Scapanosoma spatulatum* of Lühe, 1909.

HOST. Little bittern.

LOCATION. Intestine.

LOCALITY. Europe.

DIAGNOSIS. *Shape and size:* anterior region (back to the ventral sucker) of oval outline, 2·2–2·6 mm. long and 1·4 mm. broad, posterior region elongate,

4·3–7·4 mm. long. *Collar spines:* of varying sizes, 0·044–0·056 mm. long, smallest farthest forward, largest nearest the median plane. *Suckers:* oral smaller than the ventral, diameters 0·1–0·15 and 0·33–0·43 mm. *Cuticle:* spinous in the vicinity of the ventral sucker. *Surface of the body:* having two contractile, spinous papillae which arise in the middle line near the middle of the posterior region. *Reproductive systems:* genital pore immediately in front of the ventral sucker. Cirrus pouch present. Testes elongate oval, situated one behind the other between the two ventral papillae. Ovary immediately in front of the foremost testis. Vitellaria lateral throughout the posterior region. Eggs not very large (0·076–0·084 × 0·048–0·050 mm.).

Genus *Pegosomum* Rátz, 1903

Pegosomum saginatum (Rátz, 1898) Rátz, 1903

Syn. *Distoma saginatum* Rátz, 1898.

HOST. Great white heron.

LOCATION. Bile ducts.

LOCALITY. Europe.

DIAGNOSIS. *Shape and size:* lancet-like, 14–24 mm. long, 5–9 mm. broad and 2–3 mm. thick. *Collar:* small; of the twenty or twenty-one head spines, four large ones on each side form corner spines. *Suckers:* oral absent, ventral 1·22 mm. diameter and situated in front of the mid-body. *Cuticle:* spinous. *Reproductive systems:* gonads situated in the posterior region, one behind another, the ovary foremost. Vitellaria extending from the pharynx to the posterior extremity. Uterus short and having few folds. Eggs few, large (0·096–0·130 × 0·069–0·085 mm.).

Pegosomum spiniferum Rátz, 1903

HOST. Bittern.

LOCATION. Bile ducts.

LOCALITY. Europe.

DIAGNOSIS. *Shape and size:* lancet-like, 9–10 mm. long and 3 mm. broad. *Collar:* bearing twenty-seven collar spines, four on each side being corner spines; the remainder of varying size, smallest laterally, largest near the median plane. *Other characters:* similar to *P. saginatum*, but vitellaria tending to fill up the antero-lateral region of the body. Eggs large (0·119 × 0·085 mm.).

Genus *Parorchis* Nicoll, 1907

Syn. *Zeugorchis* Nicoll, 1906; *Proctobium* Travassos, 1918.

Parorchis acanthus (Nicoll, 1906) Nicoll, 1907 (Fig. 75A)

Syn. *Zeugorchis acanthus* Nicoll, 1906; *Parorchis avitus* Linton, 1914.

HOSTS. Herring gull, common gull, flamingo.

LOCATION. Rectum and bursa Fabricii.

LOCALITY. Europe, America.

DIAGNOSIS. *Shape and size:* oval outline, more convex dorsally than ventrally, 3–5 mm. long and 1·2–1·4 mm. in greatest breadth. *Cuticle:* spinous ventrally and laterally, spines numerous anteriorly, sparse behind the ventral sucker, deeply imbedded in the cuticle, 0·019–0·031 mm. long. *Collar:* about 0·9 mm. wide, having a single row of about sixty spines. *Suckers:* oral up to 0·5 mm. diameter, ventral more than twice as large. *Gut:* prepharynx present, oesophagus about three times as long as the pharynx, caeca long. *Reproductive systems:* genital pores borne on a papilla in front of the ventral sucker. Testes slightly lobed and side by side near the posterior extremity. Ovary rounded, median and in front of the testes. Cirrus spinous. Uterus having a short descending limb and a long ascending limb, formed into wide folds which overstep the caeca laterally. Eggs increasing in size from 0·08 × 0·044 to 0·10 × 0·06 mm. while in transit through the uterus.

Notes. Infection is never very heavy. Nicoll found about a dozen flukes in a single host and noted their resistance to salinity changes, even their ability to live for more than 24 hours in distilled water. F. G. Rees (1939*b*, 1940) has added to our knowledge of the structure of the adult, and gave a complete account of gametogenesis and early development (see pp. 496–500).

P. asiaticus Strom, 1928 occurs in the gull-billed tern in Turkestan. An Indian species, *P. snipis* Lal, 1936, occurs in the common sandpiper.

Genus *Paryphostomum* Dietz, 1909

Paryphostomum radiatum (Dujardin, 1845) Dietz, 1909

Syn. *Distoma radiatum* Dujardin, 1845; *D. echinatum* (Zeder, 1803) Wedl, 1858.

HOST. Cormorant.

LOCATION. Intestine.

LOCALITY. Europe, Japan.

DIAGNOSIS. *Size:* 3·3–6·5 mm. long. *Collar:* equipped with twenty-seven collar spines arranged in a double row, the corner spines being the largest. *Cuticle:* spinous, spines arranged in transverse rows, not extending beyond the ventral sucker posteriorly. *Suckers:* ventral much larger than the oral, and situated one-quarter of the distance along the body. *Reproductive systems:* cirrus pouch small, almost entirely confined to the region in front of the ventral sucker. Testes deeply lobed, one behind the other, the more posterior midway between the ventral sucker and the posterior extremity. Ovary midway between the ventral sucker and the testes. Uterus very short and having few folds. Eggs 0·084–0·088 × 0·054–0·061 mm.

Genus *Echinostephilla* Lebour, 1909

Echinostephilla virgula Lebour, 1909

HOST. Turnstone.

LOCATION. Mid-intestine.

LOCALITY. Northumberland.

DIAGNOSIS. *Shape and size:* elongate, rounded at the anterior, pointed at the posterior end, flattened, but concave between the suckers, 4–8 mm. long and about 0·6 mm. broad near the ventral sucker. *Cuticle:* thick, tough, spinous, the spines being sharply pointed and broader near the mid-body, arranged in rows of alternating individuals, sparser and less regular behind the ventral sucker, absent except on the lateral margins of the body posteriorly. *Collar:* bearing a coronet of blunt spines 0·008 mm. long, closely set in two rows of opposite individuals, the exact number doubtful, but about fifty-six in each row. *Suckers:* ventral larger than the oral and one-fifth of the distance along the body, diameters 0·4 and 0·12 mm. *Gut:* prepharynx short, pharynx longer than broad, oesophagus long and narrow, caeca long. *Reproductive systems:* genital pore median. Cirrus pouch elongate, slightly sinuous, extending far behind the ventral sucker. Cirrus very long, often exserted for half its length, its deeper part bearing spines with anteriorly directed points. Seminal vesicle pyriform. Testes one behind the other in the posterior region, ovary slightly in front of the anterior testis, but behind the mid-body. Vitellaria lateral, appearing as two thin bands of follicles extending from the posterior testis almost to the ventral sucker. Uterus having a very short descending limb, not passing behind the anterior testis, and a folded ascending limb, the folds occurring between the anterior testis and the ovary and in front of the latter. Eggs fairly numerous, thick-shelled, yellow, measuring about 0·10 × 0·05 mm., each containing in its posterior part a miracidium with conspicuous eye-spots. (*Note.* The miracidia may hatch *in utero.*) *Excretory system:* vesicle an elongate sac extending forward beyond the posterior testis, lateral canals extending to the oral sucker.

Genus *Hypoderaeum* Dietz, 1909

Hypoderaeum conoideum (Bloch, 1782) Dietz, 1909 (Fig. 50E)

Syn. *Cucullanus conoideus* Bloch, 1782; *Echinostoma conoideum* of Kowalewsky, 1896.

HOSTS. Domestic duck, goose and fowl, also mallard, tufted duck, garganey, scaup, common sheld-duck, shoveller, white-fronted goose, goosander, common teal.

LOCATION. Intestine.

LOCALITY. Europe (Staffordshire), Japan.

DIAGNOSIS. *Shape and size:* elongate, 6–12 mm. long and 1·3–2 mm. broad. *Collar:* small, 0·38–0·6 mm. broad, bearing forty-seven to fifty-three collar spines (as a rule, forty-nine). *Cuticle:* spinous. *Suckers:* oral much smaller than the ventral, diameters 0·16–0·3 and 0·7–1·0 mm. *Gut:* prepharynx very small, pharynx present, oesophagus short and caeca long. *Reproductive systems:* genital pore immediately in front of the ventral sucker. Cirrus pouch elongate, club-shaped, extending behind the ventral sucker. Testes ovoid or sausage-shaped, one behind the other, slightly behind the mid-body. Ovary in front of the testes. Uterus short and having a number of transverse folds. Vitellaria extending from the level of the ventral sucker to the posterior extremity. Eggs 0·095–0·108 × 0·061–0·068 mm.

Genus *Parechinostomum* Dietz, 1909

Parechinostomum cinctum (Rudolphi, 1802) Dietz, 1909

Syn. *Fasciola cincta* Rudolphi, 1802; *Distoma cinctum* of Rudolphi, 1809; *D. tringae helveticae* Rudolphi, 1819; *Echinostoma cinctum* of Cobbold, 1860.

HOSTS. Grey plover, lapwing.

LOCATION. Intestine.

LOCALITY. Europe.

DIAGNOSIS. *Shape and size:* elongate, broadening posteriorly and rounded at the extremity, 1·75–2·5 mm. long. *Collar:* small, having forty-three collar spines of equal size arranged in two rows. *Suckers:* ventral large, situated one-quarter of the distance along the body. *Gut:* bifurcation of the intestine midway between the pharynx and the ventral sucker. *Reproductive systems:* cirrus pouch small, almost confined to the region in front of the ventral sucker. Testes elongate-ovoid, one behind the other in the posterior region. Ovary in front of the testes. Uterus short. Vitellaria extending from the ventral sucker to the posterior extremity, approaching the median plane behind the testes. Eggs 0·093–0·096 × 0·055–0·057 mm.

Genus *Petasiger* Dietz, 1909

Petasiger exaeretus Dietz, 1909

HOST. Cormorant.

LOCATION. Intestine.

LOCALITY. Europe (Scilly Isles).

DIAGNOSIS. *Shape and size:* spindle-like, about 2–3 mm. long and 0·7–1 mm. broad. *Collar:* bearing twenty-seven collar spines, of which four large ones on each side form corner spines, the remainder diminishing in size towards the median plane. *Cuticle:* spinous. *Gut:* bifurcation of the intestine slightly in front of the ventral sucker. *Reproductive systems:* cirrus pouch small, almost entirely in front of the ventral sucker. Testes transversely ovoid, slightly flattened, one behind the other in the posterior region. Ovary rounded and situated between the ventral sucker and the anterior testis. Vitellaria extending from the bifurcation of the intestine to the posterior extremity. Uterus short and having few folds. Eggs not numerous, fairly large (0·091 × 0·064 mm.).

Genus *Chaunocephalus* Dietz, 1909

Chaunocephalus ferox (Rudolphi, 1795) Dietz, 1909

Syn. *Fasciola ferox* Rudolphi, 1795; *Distoma ardeae* (Gmelin, 1791) Zeder, 1803; *Echinostoma ferox* of Blainville, 1828.

HOSTS. White stork, black stork, bittern.

LOCATION. Intestine.

LOCALITY. Europe.

DIAGNOSIS. *Size:* 5·5–8 mm. long, 2·3 mm. broad in the anterior, but only 0·7–1 mm. in the posterior region, flattened ventrally, convex dorsally. *Collar:* bearing twenty-seven spines which are arranged in a double row, four on each side being corner spines, the remainder being smaller and smallest in the aboral row. *Cuticle:* finely spinous in the anterior region. *Suckers:* ventral large and situated behind the middle of the body. *Reproductive systems:* cirrus pouch small and situated almost entirely in front of the ventral sucker. Testes large and globular, diagonally orientated behind the mid-body. Ovary globular, close behind the ventral sucker on the right. Uterus short, but having many folds

Vitellaria tending to fill the anterior region behind the oral sucker, extending laterally to the posterior extremity. Eggs numerous and fairly large (0·088–0·092 × 0·053–0·057 mm.).

Family **TROGLOTREMATIDAE** Odhner, 1914, emend. Braun, 1915 (p. 99)

Syn. Troglotremidae Odhner, 1914; Collyriclidae Ward, 1917.

KEY TO GENERA OCCURRING IN BIRDS

1. Ventral sucker and cirrus pouch absent — *Collyriclum*
2. Ventral sucker present, cirrus pouch absent — *Renicola*

Genus *Collyriclum* Kossack, 1911

Collyriclum faba (Bremser, in Schmalz, 1831) Kossack, 1911

Syn. *Monostoma faba* Bremser, in Schmalz, 1831; *Monostoma bijugum* Miescher, 1838; *Wedlia faba* of Cobbold, 1860.

HOSTS. Domestic fowl and turkey, great titmouse, garden-warbler, wood-warbler, willow-warbler, wheatear, starling, jay, siskin, cirl bunting, house sparrow, chaffinch, grey wagtail, redstart, etc.

LOCATION. In cysts under the skin.

LOCALITY. Europe, America.

DIAGNOSIS. *Cysts:* about 4–6 mm. diameter, each containing two worms of unequal sizes, having a central pore through which the eggs can be passed, occurring in subcutaneous tissues, particularly on the abdomen of the host near the cloaca. *Trematode. Shape and size:* discoidal, 4–5 mm. long, 4·5–5·5 mm. broad, flat ventrally, convex dorsally. *Suckers:* oral small (0·2–0·45 mm. diameter), ventral absent. *Gut:* pharynx and oesophagus small, caeca long. *Reproductive systems:* genital pore median and in front of the mid-body. Cirrus pouch small and pyriform. Testes deeply lobed and situated side by side in the posterior region. Ovary also lobed and in front of the testes. Uterus long and much convoluted. Vitellaria sometimes asymmetrical, comprising six to nine clusters of follicles on either side of the body. Eggs numerous and very small (0·019–0·021 × 0·010–0·011 mm.).

Genus *Renicola* Cohn, 1904

Renicola pinguis (Mehlis, in Creplin, 1846) Cohn, 1904 (Fig. 50G)

Syn. *Monostoma pingue* Mehlis, 1843.

HOST. Great crested grebe.

LOCATION. Encysted in the kidney.

LOCALITY. Europe.

DIAGNOSIS. *Cysts:* the flukes pair in outgrowths of the kidney tubules about 2 mm. long and 1 mm. broad. *Trematode: shape and size:* pyriform, the anterior end the broader, 1·5 mm. long, 0·85 mm. broad, 0·35 mm. in greatest thickness. *Cuticle:* spinous. *Suckers:* oral small (0·21 mm. diameter), ventral very small (it may easily be overlooked), 0·075 mm. diameter, and situated slightly behind the mid-body. *Gut:* pharynx small, oesophagus short, caeca fairly short and wide, terminating near the level of the ventral sucker, the bifurcation of the intestine being far in front of it. *Reproductive systems:* genital pore between the ventral sucker and the bifurcation of the intestine. Cirrus pouch absent. Testes dorsal to or behind the ventral sucker. Ovary in front of the testes and on the right. Vitellaria lateral, in the middle third of the body. Uterus (excluded from the figure) having much folded descending and ascending limbs, tending to fill the available space and obscuring the other organs. Eggs brown, fairly small (0·042 × 0·019 mm.).

Timon-David (1933*b*) described a species, *Renicola lari*, from the kidneys of the herring-gull; *R. secunda* and *R. tertia*, both Skrjabin, 1924, occur respectively in the ureters of the pelican and a tern in Turkestan.

Family **NOTOCOTYLIDAE** Lühe, 1909 (p. 100)

Subfamily **Notocotylinae** Kossack, 1911

KEY TO GENERA

1. Body elongate, groups of glands present ventrally
   - A. Groups of glands arranged in three to five longitudinal rows; body narrowed anteriorly; metraterm about half as long as the cirrus pouch *Notocotylus*
   - B. Groups of glands arranged in two lateral rows and along a median, keel-like ridge; body almost equally rounded at the extremities; metraterm almost as long as the cirrus pouch *Catatropis*
2. Body oval in outline; glands absent ventrally *Paramonostomum*

Herber (1942) suggested that the most useful characters for the separation of species of the Notocotylinae are: the number of ventral glands in each row, the anterior extent of the vitellaria, the lateral lobing of the testes, and the sizes of the body and the tail, as well as the arrangement of the excretory canals in the cercaria.

Genus *Notocotylus* Diesing, 1839 (Fig. 10F)

Notocotylus attenuatus (Rudolphi, 1809) Kossack, 1911

Syn. *Monostoma attenuatum* Rudolphi, 1809; *Notocotylus triserialis* Diesing, 1839.

Hosts. Domestic duck, goose and fowl; also mallard, wigeon, common teal, pintail, long-tailed duck, common sheld-duck, whooper swan, brent goose, white-fronted goose, bean-goose, shoveller, lapwing, oystercatcher, goosander, ruff, common eider, snipe. (Doubtful records, possibly referring to *Catatropis*: garganey, tufted duck, scaup, pochard, velvet scoter, barnacle-goose, coot, moorhen, water rail, etc.)

Location. Caeca and rectum.

Locality. Europe, Asia.

Diagnosis. *Shape and size:* elongate, rounded posteriorly, slightly pointed anteriorly, 2–5 mm. long and 0·65–1·4 mm. broad. *Cuticle:* spinous ventrally, spines fine. *Glands:* arranged on the ventral surface in three rows, sixteen to seventeen in the lateral rows, fourteen to fifteen in the median row, covering the surface of the body, except antero-laterally. *Suckers:* oral, 0·11–0·2 mm. diameter, ventral absent. *Gut:* oesophagus short, caeca long, pharynx absent. *Reproductive systems:* genital pore slightly behind the bifurcation of the intestine,

Cirrus pouch elongate, extending to the second third of the body. Pars prostatica cylindrical, seminal vesicle convoluted. Testes lobed, side by side in the posterior region, lateral to the caeca. Ovary between the testes and the terminal parts of the caeca. Vitellaria extending from slightly behind the mid-body to the anterior ends of the testes. Uterus having only an ascending limb, intensely folded between the ovary and the cirrus pouch, the folds occupying all available space between the caeca. Eggs each having a straight filament at either pole, the capsule measuring 0·020–0·022 × 0·1 mm., the filaments 0·2 mm. long. Eggs each containing an embryo.

Among other species occuring in birds are:

Notocotylus seineti Fuhrmann, 1919. In the caecum of the garganey and the domestic duck (Europe). This species is 2 mm. long and 0·6–0·7 mm. broad, has glands arranged in three longitudinal rows of twelve groups, those in the median row being the largest, and eggs which are 0·021 mm. long each having filaments 0·29 mm. long.

N. aegyptiacus (Odhner, 1905) Kossack, 1911. In the caecum of the domestic duck (Egypt).

N. gibbus (Mehlis, 1846) Kossack, 1911. In the coot and the moorhen (Germany) and moorhen (Britain).

N. thienemanni Szidat & Szidat, 1933. In the duck, the fowl and the teal (Europe).

N. imbricatus U. Szidat, 1935. Reared experimentally from *Cercaria imbricata* in the duck and the fowl (Europe).

N. ralli Baylis, 1936. In the water-rail (Britain: Perthshire, Norfolk, Sussex).

Genus *Catatropis* Odhner, 1905

Catatropis verrucosa (Frölich, 1789) Odhner, 1905 (Fig. 50H)

Syn. *Fasciola verrucosa* Frölich, 1789; *F. anseris* Gmelin, 1790; *Monostoma verrucosum* Zeder of Levinsen, 1881; *Notocotyle verrucosum* of Monticelli, 1892; *Festucaria pedata* Schrank, 1786, in part; *Notocotyle triserialis* Diesing, 1839, in part.

HOSTS. Domestic duck, goose and fowl; also sheld-duck, long-tailed duck, eider-duck, red-breasted merganser, barnacle-goose, pink-footed goose, brent goose, whooper swan, golden-eye, common scoter, shoveller, grey plover.

LOCATION. Intestine and caecum, also bursa Fabricii.

LOCALITY. Europe, Arctic regions.

DIAGNOSIS. *Size:* 1–5 mm. long and 0·75–1·25 mm. broad. *Cuticle:* spinous ventrally, spines fine, becoming larger anteriorly. *Glands:* arranged in two lateral rows of eight to twelve groups, opening on non-retractile papillae, and a third row on a median, keel-like ridge. *Suckers:* oral 0·13–0·16 mm. diameter, ventral absent. *Reproductive systems:* genital pore immediately beneath the bifurcation of the intestine. Cirrus pouch extending back almost to the mid-body. Pars prostatica short and flask-like. Seminal vesicle convoluted and situated in the hinder part of the cirrus pouch. Vitellaria and gonads as in *Notocotylus*. Uterus intensely folded transversely, occupying the region between the base of the cirrus pouch and the ovary. Eggs 0·018–0·028 mm. long, each having two polar filaments 0·16 mm. long.

Genus *Paramonostomum* Lühe, 1909

Syn. *Neoparamonostomum* Lal, 1936.

Paramonostomum alveatum (Mehlis, 1846) Lühe, 1909

Syn. *Monostoma alveatum* Mehlis, 1846; *M. verrucosum* (Frölich, 1789) Wedl, 1857; *Notocotyle alveatum* of Monticelli, 1892; *Monostoma alveiforme* Cohn, 1904.

HOSTS. Domestic goose, brent goose, velvet scoter, common scaup, scoter, long-tailed duck, tufted duck, eider-duck, whooper swan.

LOCATION. Caecum and rectum.

LOCALITY. Europe.

DIAGNOSIS. *Size:* 0·6–1·0 mm. long and 0·4–0·7 mm. broad (in the swan, according to Kossack (1910), 1·165 mm. long and 0·65 mm. broad). *Cuticle:* apparently smooth. *Glands:* apparently absent. *Suckers:* oral 0·05–0·08 mm. diameter, ventral absent. *Gut:* pharynx absent, oesophagus short, caeca long. *Reproductive systems:* genital pore slightly behind the bifurcation of the intestine. Cirrus pouch short and wide (0·24 × 0·18 mm.). Seminal vesicle not convoluted (Lühe stated otherwise). Metraterm having a weak musculature and about one-third as long as the cirrus pouch. Gonads much as in *Notocotylus* and *Catatropis*, the vitellaria more extensive anteriorly and uterus formed into wide folds which embrace the terminal parts of the genital ducts anteriorly. Eggs 0·019–0·021 × 0·008–0·010 mm., polar filaments 0·055 mm. long. *Development:* the experiments of Rothschild (*J. Parasit.* **27**, 1941, 363–5) would lead us to suppose that the cercaria belongs to the '*Yenchingensis*' subgroup of Monostome cercariae.

Family CYCLOCOELIDAE Kossack, 1911 (p. 100)

Syn. Monostomida Kolenati, 1856; Monostomidae Cobbold, 1864; Monostomatidae Gamble, 1896.

Witenberg (1926) reviewed the taxonomy of this family. His scheme was severely criticized by Joyeux & Baer (1927*a*).

KEY TO GENERA

1. Caeca simple, without median diverticula (Cyclocoelinae in some schemes)	
A. Testes and ovary arranged in form of triangle	*Cyclocoelum*
B. Testes and ovary in same straight line	*Hyptiasmus*
2. Caeca with median diverticula (Typhlocoelinae in some schemes)	*Typhlocoelum*

Genus *Cyclocoelum* Brandes, 1892 (Fig. 53A)

Syn. *Monostomum* Zeder, 1800, in part; *Corpopyrum* Witenberg, 1926; and various misprints, e.g. *Cyclocoelium* and *Cyclocoeleum* Fuhrmann, 1904.

Cyclocoelum mutabile (Zeder, 1800) Stossich, 1902

Syn. *Monostoma mutabile* Zeder, 1800; *M. microstomum* Creplin, 1829; *M. attenuatum* Molin, 1859; *Cephalogonimus ovatus* (Rudolphi, 1803) Stossich, 1896; *Cyclocoelum microstomum* of Kossack, 1911; *C. halcyonis* MacCallum, 1921; *C. obliquum* Harrah, 1921; *C. pseudomicrostomum* Harrah, 1922; *C. cuneatum* Harrah, 1922; *C. macrorchis* Harrah, 1922; *C. (Antepharyngeum) mutabile* of Witenberg, 1926; *C. (Antepharyngeum) microstomum* of Witenberg, 1926; *C. (Antepharyngeum) pseudomicrostomum* (Harrah, 1922) Witenberg, 1926; and possibly other forms.

HOSTS. Common scoter, moorhen, coot, possibly turkey.

LOCATION. Air sacs and nasal cavities, also trachea and under the nictitating membrane.

LOCALITY. Europe, Japan.

DIAGNOSIS. *Shape and size:* lancet-like, rounded posteriorly, pointed anteriorly, 5–24 mm. long and 2–8 mm. broad, flat ventrally, convex dorsally. *Colour:* pale rose or yellow. *Suckers:* absent. *Gut:* prepharynx, pharynx and oesophagus all of small size, intestinal crura long and simple, uniting posteriorly to form a ring-like intestine. *Reproductive systems:* genital pore ventral to the pharynx. Cirrus pouch long (0·6–0·85 × 0·23 mm.), extending slightly over the bifurcation of the intestine. Cirrus small and cylindrical. Gonads small, situated in the posterior region between the caeca, arranged in the form of a triangle. Vitellaria lateral to the caeca, extending from the bifurcation of the intestine to the posterior extremity, where they do not merge. Uterus having only an ascending limb (plus a very short descending limb) which is formed into transverse folds confined within the boundary of the ring-like intestine. Eggs large and thick-shelled, measuring 0·112 × 0·061 mm.

Several other species of the genus occur in Europe, e.g. *C. kossacki* (Witenberg, 1926), *C. exile* Stossich, 1902, *C. fasciatum** (Stossich, 1902), *C. ovopunctatum** Stossich, 1902 and *C. vicarium** (Arnsdorff, 1908). They are chiefly parasites of Charadriidae.

Genus *Hyptiasmus* Kossack, 1911

Syn. *Transcoelum* Witenberg, 1926.

Joyeux & Baer (1927*a*) regarded this genus as synonymous with *Cyclocoelum*; Lal (1939) also included *Harrahium* Witenberg and *Prohyptiasmus* Witenberg in the synonymy.

Hyptiasmus arcuatus (Brandes, 1892) Kossack, 1911

Syn. *Monostoma arcuatum* Brandes, 1892; *Cyclocoelum arcuatum* of Stossich, 1902; *Hyptiasmus laevigatus* Kossack, 1911; *H. tumidus* Kossack, 1911; *H. oculeus* Kossack, 1911; *H. magnus* Johnston, 1917; *H. coelonodus* Witenberg, 1926; *Transcoelum sigillum* Witenberg, 1926; *T. oculeum* of Witenberg, 1926, etc.

HOSTS. Golden-eye, smew, tufted duck, velvet scoter, common scoter, eider-duck, long-tailed duck, coot.

LOCATION. Infraorbital sinus, nasal cavities, suborbital sinus.

LOCALITY. Europe, Japan.

DIAGNOSIS. *Shape and size:* pyriform, more rounded posteriorly than anteriorly. 7–20 mm. long and 2–5 mm. in greatest breadth. *Other characters:* similar to *Cyclocoelum*, but gonads arranged in a straight line, uterus having wide folds overstepping the caeca laterally and eggs measuring 0·075–0·115 × 0·055–0·060 mm.

At least a dozen other species occur in Europe, two in the pochard and two others in the coot.

* Joyeux & Baer (1927*a*) regard these as synonyms of *C. obscurum* (Leidy, 1887).

Genus *Typhlocoelum* Stossich, 1902

Syn. *Tracheophilus* Skrjabin, 1913; *Typhlultimum* Witenberg, 1924.

Typhlocoelum cucumerinum (Rudolphi, 1809) Kossack, 1911

Syn. *Distoma cucumerinum* Rudolphi, 1809; *Monostoma flavum* Mehlis, 1831; *M. sarcidiornicola* Megnin, 1890; *M. cymbium* Monticelli, 1892; *M. cucumerinum* of Braun, 1899; *Typhlocoelum flavum* of Stossich, 1902; *T. sarcidiornicola* of Stossich, 1902; *T. obovale* Neumann, 1909; *T. reticulare* Johnston, 1913; *Typhlultimum sarcidiornicola* of Witenberg, 1926.

HOSTS. Domestic duck, velvet scoter, tufted duck, scaup, eider-duck, long-tailed duck, smew, red-breasted merganser, etc.

LOCATION. Bronchi, trachea, oesophagus.

LOCALITY. Europe (Germany), North and South America.

DIAGNOSIS. *Size:* 6–12 mm. long and 2–5 mm. broad. *Shape:* outline oval, blunter anteriorly than posteriorly. *Gut:* caeca widely removed from the lateral margins of the body and having short median diverticula. *Reproductive systems:* vitellaria lateral, median and ventral to the caeca, underlying the diverticula and extending from the bifurcation of the intestine to the posterior extremity. Uterus having wide folds which are confined within the limits of the intestine. Testes deeply lobed, diagonally one behind the other, the more posterior filling the space between the caeca posteriorly. Ovary situated in front of the testes and on the same side as the more posterior. Eggs measuring about 0·156 × 0·085 mm.

Typhlocoelum cymbium (Diesing, 1850) Kossack, 1911

Syn. *Monostoma cymbium* Diesing, 1850; *Haematotrephus cymbius* of Stossich, 1902; *Tracheophilus cymbium* of Skrjabin, 1913; *Tracheophilus sisowi* Skrjabin, 1913.

HOSTS. Domestic duck, pintail.

LOCATION. Trachea.

LOCALITY. Europe (France), Asia (Turkestan, Formosa), America (Mexico).

DIAGNOSIS. *Size:* 6–11·5 mm. long and 3–6 mm. broad, of greatest breadth in the mid-body. *Gut:* each limb of the ring-like intestine bearing nine to thirteen median diverticula. *Reproductive systems:* testes rounded, not lobed, diagonally orientated posteriorly. Ovary rounded, larger than the testes, beside or a little in advance of the anterior testis. Uterus confined within the limits of the intestine. Eggs very large (0·122–0·154 × 0·063–0·081 mm.). Miracidia sometimes liberated *in utero*.

Szidat (1933) outlined the development and mode of infection of the final host of '*T. sisowi*'.

Family **BRACHYLAEMIDAE** Joyeux & Foley, 1930 (p. 101)

Syn. Harmostomidae Odhner, 1912.

KEY TO SOME GENERA OCCURRING IN BIRDS

1. Posterior limits of the vitellaria in the region of the anterior testis — *Brachylaemus*
2. Posterior limits of the vitellaria behind the anterior testis
   A. Suckers feeble; vitellaria terminating near the ovary — *Urotocus*
   B. Suckers powerful; vitellaria extending beyond the ovary — *Leucochloridium*

Notes. *Urotocus* and *Leucochloridium* were included by Dollfus (1935 *a*) in a new family, Leucochloridiidae (Syn. Urogoniminae Looss, 1899; Leucochloridiinae Poche, 1907). For a recent review of the Brachylaemidae see Allison (*Trans. Amer. Micr. Soc.*, **62**, No. 2, 1943, pp. 127–68).

Subfamily **Brachylaeminae** Joyeux & Foley, 1930

Syn. Harmostominae Looss, 1900; Heterolopinae Looss, 1899.

Genus *Brachylaemus* Dujardin, 1843, emend. E. Blanchard, 1847

Syn. *Brachylaima* Dujardin, 1843; *Brachylaimus* Dujardin, 1845; *Harmostomum* Braun, 1899.

The species of this genus are notoriously difficult to separate. Dollfus (1934, 1935 *a*) made an attempt and outlined a scheme in which the hosts and zoogeographic regions are utilized. The best-known species occurring in European birds are *B. commutatus* and *B. fuscatus*. Less common forms are *B. columbae* sp. inq. (Mazzanti, 1899) (in Columbiformes), *B. mesostomus* (Rudolphi, 1803) and *B. arcuatus* (Dujardin, 1845), the first of which is possibly identical with *B. fuscatus*, though occurring in different hosts (redwing, blackbird, missel thrush, finches). The other two species occur in *Turdus* spp. *Brachylaemus commutatus* is well known as a parasite of the fowl in various parts of Europe and has been reported under other names in Tunis, Turkestan, Indo-China and Japan. *B. fuscatus* also occurs in Galliformes, e.g. the quail, but has appeared more often in pigeons.

Brachylaemus fuscatus (Rudolphi, 1819)

Syn. *Distoma fuscatum* Rudolphi, 1819; *Distomum heteroclitum* Molin, 1858; *Harmostomum nicolli* Witenberg, 1925; *H.* (*Harmostomum*) *fuscatum* of Witenberg, 1925; *H. pellucidum* Werby, 1928.

Hosts. Rock dove, wood pigeon, pheasant, rook, mistle-thrush, starling, jay, etc.

Location. Intestine.

Locality. Europe.

Diagnosis. *Shape and size:* elongate, cylindrical, 2–5 mm. long and 0·5–0·75 mm. broad. *Cuticle:* finely spinous anteriorly. *Suckers:* of nearly equal size, oral 0·25–0·3 mm. diameter, ventral 0·25–0·28 mm. diameter and situated about one-third of the distance along the body. *Gut:* pharynx large, oesophagus absent, caeca long and sinuous. *Reproductive systems:* testes ovoid, posterior in position. Ovary between the testes in extended specimens. Genital pore ventral to the anterior testis, or farther forward. Uterus having an ascending limb which extends to the bifurcation of the intestine, and a descending limb which is formed into irregular folds. Vitellaria extending laterally from the posterior margin of the ventral sucker to the anterior testis. Eggs small (0·023–0·028 × 0·014–0·018 mm.).

Brachylaemus commutatus (Diesing, 1858)

Syn. *Distoma dimorphum* Wagener, 1852, *nec* Diesing, 1858; *D. commutatum* Diesing, 1858; *Clinostomum commutatum* of Looss, 1899; *Mesogonimus commutatus* of Sonsino (1891, 1899), Barbagallo (1906), Galli-Valerio (1891, 1901); *Harmostomum commutatum* Joyeux, 1923 and other authors; *H.* (*Postharmostomum*) *commutatum* Witenberg, 1925; *Postharmostomum gallinum* Witenberg, 1923; *Harmostomum* (*Postharmostomum*) *gallinum* Witenberg, 1925; *Postharmostomum commutatum* of MacIntosh, 1934; and probably several others, e.g. *Harmostomum annamense* Railliet, 1925; *H. horizawai* Ozaki, 1925; *H.* (*Postharmostomum*) *hawaiiensis* Guberlet, 1928, etc.

HOSTS. Domestic fowl, turkey, pigeon, pheasant.

LOCATION. Intestine.

LOCALITY. Europe, Asia (Turkestan, Indo-China), Africa, probably in Annam, Japan and Hawaii.

DIAGNOSIS. *Shape and size:* elongate, the extremities rounded, 3·5–7·5 mm. long and 1–2 mm. broad. *Suckers:* of about equal size, the oral sometimes larger than the ventral and vice versa, diameters about 0·39–0·75 and 0·23–0·69 mm.; ventral situated about one-third of the distance along the body. *Reproductive systems:* similar to *B. fuscatus*, but eggs larger (0·027–0·032 × 0·013–0·018 mm.).

Subfamily **Leucochloridiinae** Poche, 1907

Syn. Urogonimidae Looss, 1899; Leucochloridiidae Dollfus, 1935.

Genus *Urotocus* Looss, 1899

Urotocus rossittensis (Mühling, 1898) Looss, 1899

Syn. *Urogonimus rossittensis* Mühling, 1898.

HOST. Fieldfare.

LOCATION. Bursa Fabricii.

LOCALITY. Rossitten (E. Prussia).

Genus *Leucochloridium* Carus, 1835 (Fig. 72 A′, B′)

Syn. *Urogonimus* Monticelli, 1888 and misprints like *Leukochloridium* Siebold, 1853.

Carus erected this genus for a larval trematode found in a snail belonging to the genus *Succinea*. The corresponding adult was described in 1803 by Rudolphi as *Fasciola macrostomum*, though it was not recognized as such until Zeller (1874) followed the course of development of the larva. Heckert (1889) gave a full historical account of the genus. Another species, *Leucochloridium insignis*, was described by Looss (1899), being obtained from the coot in Egypt, a third, *L. turanicum* Solowiow, 1912, was found in '*Totanus glareola*' in Turkestan. Monticelli (1893) described a fourth species, *L. cercatum*, from an unknown host. Since these species were discovered, others have been found in various parts of the world (America, Asia and Australia). McIntosh (1927) described five new American species, and gave comparative data regarding European species. In some schemes of classification, the genus is the type of the family Leucochloridiidae Dollfus, 1934.

Leucochloridium macrostomum (Rudolphi, 1803) Poche, 1907 (Fig. 50 I)

Syn. *Fasciola macrostoma*, Rudolphi, 1803; *Distoma macrostomum* of Rudolphi, 1809; *Leucochloridium paradoxum* Carux, 1833; *Urogonimus macrostomus* of Monticelli, 1892; *Distomum holostomum* of Braun, 1902; *D. caudale* Müller, 1897, *nec* Rudolphi, 1809.

HOSTS. Carrion crow, jay, house sparrow, tree sparrow, chaffinch, goldfinch, northern bullfinch, great grey shrike, white-throat, barred warbler, nightingale, roller, water-rail, spotted crake, moorhen, mealy redpole.

LOCATION. Rectum.

LOCALITY. Europe.

DIAGNOSIS. *Shape and size:* oval in outline and section, extremities rounded, about 1·8 mm. long, 0·8 mm. broad and 0·45 mm. thick. *Cuticle:* finely spinous. *Suckers:* powerful, the oral longitudinally ovoid (0·35 × 0·30 mm.), the ventral transversely ovoid, slightly larger (0·3 × 0·4 mm.) and situated in the mid-body. *Gut:* prepharynx small, pharynx large, oesophagus very short, caeca long and narrow. *Reproductive systems:* genital pore posterior, terminal. Testes diagonally one behind the other in the posterior region. Ovary between the testes, in front of the more posterior. Cirrus pouch enclosing only the cirrus and ejaculatory duct; pars prostatica free in the parenchyma. Vitellaria lateral to the caeca, extending almost to the level of the oral sucker anteriorly. Uterus with ascending and descending limbs intensely folded, the folds encircling the ventral sucker. *Excretory pore:* slightly dorsal. Eggs numerous, thick-shelled, small (0·023 × 0·016 mm.).

Note. According to Lühe (1909) the suckers were 0·7–0·9 mm. diameter in an exceptionally large specimen 3 mm. long.

Family **CYATHOCOTYLIDAE** Poche, 1926 (p. 105)

Subfamily **Cyathocotylinae** Mühling, 1898

Genus *Cyathocotyle* Mühling, 1896

Cyathocotyle prussica Mühling, 1896 (Fig. 53 C)

HOSTS. Long-tailed duck, shag.

LOCATION. Intestine.

LOCALITY. Europe (East Prussia).

DIAGNOSIS. This trematode combines the form of a Distome with the structure of a Strigeid. *Shape and size:* oval (almost circular) outline, about 1 mm. long and 0·65 mm. broad, somewhat flattened. *Suckers:* close together (about 0·15 mm. apart), the oral larger than the ventral, diameters 0·12–0·13 and 0·06–0·08 mm. *Main* (*accessory*) *adhesive organ:* situated immediately behind the ventral sucker, of enormous size (about two-thirds as broad as the body) and partially obscuring the true sucker, having numerous associated unicellular glands, but no compact glandular mass. *Reproductive systems:* genital pore situated near the posterior extremity and slightly dorsal. Cirrus pouch enormous. Testes large and ovoid, one behind the other and somewhat lateral. Ovary small and ovoid, median to the anterior testis. Vitellaria well developed laterally along the entire margin of the body. Eggs large (0·097–0·103 × 0·068 mm.).

OTHER SPECIES AND GENERA:

Cyathocotyle fraterna Odhner, 1902 (in *Harelda glacialis*; also a parasite in Reptilia); *C. oviformis* Szidat, 1936 (in terns). Szidat (1936*a*) erected three new genera: *Cyathocotyloides*, for *curonensis* n.sp., obtained by experimental infection of the domestic duck, and *dubius* n.sp. in *Sterna hirundo* and *S. paradisea*; *Duboisia*, having as its type *Prohemistomum syriacum* Dubois, 1934; and *Holostephanus* for *lühei* n.sp. in the terns mentioned. Other genera are known in mammals (see Chapter 11).

Family **STRIGEIDAE** Railliet, 1919 (p. 105)

Subfamily **Strigeinae** Railliet, 1919

KEY TO GENERA

1. Vitellaria occupying both the anterior and the posterior regions of the body	
A. Vitelline follicles about equal in number in the anterior and posterior regions.	
(*a*) Anterior region of the body not thicker than the posterior	
(i) Pharynx present	*Strigea*
(ii) Pharynx absent	*Apharyngostrigea*
(*b*) Anterior region having well-developed lateral expansions	*Parastrigea*
B. Vitelline follicles few in the anterior region of the body	*Ophiosoma*
2. Vitellaria confined to the posterior region of the body	
A. Anterior region small and heart-shaped	*Cardiocephalus*
B. Anterior region taking up about one-third of the body length	
(*a*) Bursa a simple invagination lacking a muscular bulb	*Apatemon*
(*b*) Bursa containing a muscular, sucker-like bulb	*Cotylurus*

Genus *Strigea* Abildgaard, 1793 (Fig. 10H)

Strigea strigis (Schrank, 1788) Abildgaard, 1793

Syn. *Festucaria strigis* Schrank, 1788; *Fasciola strigis* of Gmelin, 1790; *Holostomum variabile* Nitzsch, 1819; *H. macrocephalum* (Rudolphi, 1803) Creplin, 1839; *H. excisum* Linstow, 1906; *H. cornucopia* of Molin, 1859 and Diesing, 1859; *Tetracotyle colubri* Linstow, 1877; and others.

HOSTS. Buzzard, long-eared owl, short-eared owl, barn owl, tawny owl, snowy owl, eagle owl, marsh harrier, sparrow hawk, peregrine falcon, merlin, kestrel, osprey, etc.

LOCATION. Intestine.

LOCALITY. Europe.

DIAGNOSIS. *Shape and size:* anterior region of the body short and rounded, posterior oviform, but having a short terminal process, total length up to 6 mm. *Suckers:* oral smaller than the ventral, diameters 0·15–0·17 and 0·35 mm. *Reproductive systems:* genital papilla well developed, genital pore only slightly dorsal. Eggs brown and very large (0·12–0·14 × 0·07–0·08 mm.).

OTHER SPECIES:

More than twenty known. *S. falconis* Szidat, 1928 occurs in the gut of a number of Aquilidae (Iceland falcon and Montagu's harrier in Britain) and *S. sphaerula* (Rudolphi, 1803) in the intestine of the carrion crow, rook and other Corvidae.

Genus *Apharyngostrigea* Ciurea, 1927

Apharyngostrigea cornu (Goeze, 1800)

Syn. *Distoma cornu* Goeze, 1800 *Amphistoma cornu* of Rudolphi, 1809; *Monostoma cornu* of Rudolphi, 1819; *Holostomum cornu* of Dujardin, 1845; *H. variabile* (Nitzsch, 1819) Wedl, 1857.

HOSTS. Heron, purple heron, night heron, white stork.
LOCATION. Intestine.
LOCALITY. Europe, U.S.A.
DIAGNOSIS. *Size and shape:* 4–8 mm. long, anterior region rounded, 1–1·7 mm. long and 1 mm. broad, posterior region cylindrical, 0·4–0·65 mm. broad and thick. *Suckers:* oral 0·18–0·20 mm. diameter, ventral 0·3 mm. *Other characters:* Similar to *Strigea* spp., but pharynx absent, testes branched, genital pore terminal. Eggs about 0·1 mm. long.

Note. The finer details of the anatomy were described and illustrated (Figs. 1–3) by Byrd & Ward (*J. Parasit.*, **29**, 1943, pp. 270–4).

Genus *Parastrigea* Szidat, 1927

Parastrigea robusta Szidat, 1927

HOST. Domestic duck.
LOCATION. Intestine.
LOCALITY. Europe (East Prussia).
DIAGNOSIS. *Size:* 2–2·5 mm. long and broader in the anterior than in the posterior region, respective breadths 1·5 and 1 mm. *Suckers:* oral smaller than the ventral, diameters 0·1 and 0·15 mm. *Reproductive systems:* genital papilla large and oviform. Testes fairly compact and only slightly lobed. Vitellaria particularly well developed in the adhesive organ, but extending into the lateral expansions of the anterior region. Eggs large (0·09–0·100 × 0·055 mm.).

Genus *Ophiosoma* Szidat, 1928

Ophiosoma wedlii Szidat, 1928

Syn. *Holostomum longicolle* (Rudolphi, 1819) Brandes, 1888; *Strigea longicollis* of Lühe, 1909.

HOSTS. Lesser black-backed gull, bittern.
LOCATION. Intestine.
LOCALITY. Europe.
DIAGNOSIS. *Shape:* very elongate, cylindrical. *Size:* total length 10–18 mm., anterior region 0·7–1·5 mm. long and having slight lateral outgrowths. *Suckers:* very small. *Reproductive systems:* vitellaria mainly distributed in the posterior region of the body, extending back to the ovary, a few follicles to the bursa and some (in the anterior region) into the adhesive organs. Eggs 0·092–0·110 mm. long.

Genus *Cardiocephalus* Szidat, 1928

Cardiocephalus longicollis (Rudolphi, 1819) Szidat, 1928

Syn. *Amphistoma longicolle* Rudolphi, 1819; *Holostomum longicolle* of Dujardin, 1845; *H. bursigerum* Brandes, 1888; *Strigea bursigera* of Lühe, 1909.

HOSTS. Herring gull, black-headed gull.

LOCATION. Intestine.

LOCALITY. Europe.

DIAGNOSIS. *Shape and size:* very elongate, the anterior region pyriform or heart-shaped, the posterior broadening in the middle and about 12 mm. long. *Suckers:* feeble. *Other characters:* vitellaria confined to the posterior region.

Genus *Apatemon* Szidat, 1928

Apatemon gracilis (Rudolphi, 1819) Szidat, 1928

Syn. *Amphistoma gracile* Rudolphi, 1819; *Holostomum gracile* of Dujardin, 1845; *Strigea gracilis* of Lühe, 1909.

HOSTS. Goosander, smew, red-breasted merganser, golden-eye, common scoter, velvet scoter, domestic duck, white-fronted goose, rock dove, wigeon.

LOCATION. Intestine.

LOCALITY. Europe, Asia.

DIAGNOSIS. *Shape and size:* anterior region cup-like and taking up about one-third of the total length, posterior region cylindrical; entire fluke 1·5–2·5 mm. long, 0·4 mm. broad and thick. *Suckers:* fairly well developed, the oral about 0·1 mm., the ventral 0·15–0·2 mm. diameter. *Reproductive systems:* gonads ovoid or globular and arranged one behind another, the ovary being foremost. Eggs large (0·100–0·110 × 0·067–0·075 mm.).

Genus *Cotylurus* Szidat, 1928 (Fig. 50J)

Cotylurus cornutus (Rudolphi, 1809) Szidat, 1928

Syn. *Amphistoma cornutum* Rudolphi, 1809; *Holostomum cornutum* of Dujardin, 1845; *H. multilobum* Cobbold, 1860; *H. erraticum* Ercolani, 1881; *Strigea tarda* (Steenstrup, 1842); *Tetracotyle typica* Diesing, 1858.

HOSTS. Domestic duck, grey lag goose, rock dove, golden plover, curlew, scaup, mute swan.

LOCATION. Intestine.

LOCALITY. Europe.

DIAGNOSIS. *Shape and size:* anterior region rounded, posterior ovoid, total length 1·2–1·4 mm., breadth about 0·5 mm. *Suckers:* oral smaller than the ventral, diameters 0·1 and 0·15 mm. *Reproductive systems:* gonads arranged somewhat as in *Apatemon*. Eggs measuring 0·09–0·10 × 0·06 mm.

OTHER SPECIES:

Cotylurus erraticus (Rudolphi, 1908) Szidat, 1928; *C. variegatus* (Creplin, 1825) Szidat, 1929; *C. platycephalus* (Creplin, 1825) Szidat, 1928. The last-named has been found in the razorbill in Britain.

Family **DIPLOSTOMATIDAE** Poirier, 1886 emend. (p. 105)

Subfamily **Diplostomatinae** Monticelli, 1888

Syn. Polycotylinae Monticelli, 1892.

KEY TO GENERA

1.	Ear-like processes or groups of glandular areas present laterally	*Diplostomum*
2.	Ear-like processes or glandular areas absent	*Neodiplostomum*

Genus *Diplostomum* Nordmann, 1832

Syn. *Proalaria* La Rue, 1926.

Diplostomum spathaceum (Rudolphi, 1819) Olsson, 1876, emend.

Syn. *Distoma spathaceum* Rudolphi, 1819; *Diplostomum volvens* Nordmann, 1832; *Holostomum spathaceum* of Dujardin, 1845; *Hemistomum spathaceum* of Diesing, 1850; *Diplostoma spathaceum* of Olsson, 1876; *Conchosoma spathaceum* of Stossich, 1898; *Proalaria spathaceum* of Semenov, 1927.

HOSTS. Herring gull, common gull, greater black-backed gull, kittiwake, black-headed gull, goosander. Larvae in rainbow trout.

LOCATION. Adults in intestine.

LOCALITY. Europe, Egypt.

DIAGNOSIS. *Shape and size:* anterior region shorter and broader than the posterior, which is narrowest in its anterior part, respective breadths 1 and 0·8 mm., total length 2·3–4 mm. *Suckers:* very small, the ventral incorporated in the accessory adhesive organ, which takes up one-third of the entire breadth of the anterior region. *Reproductive systems:* vitellaria occupying most of the posterior region ventrally, extending forward at the sides of the adhesive organ. Eggs large (about 0·100 × 0·060 mm.).

Genus *Neodiplostomum* Railliet, 1919

Neodiplostomum attenuatum (Linstow, 1906) Railliet, 1919

Syn. *Holostomum spathula* Creplin, 1829, in part; *Hemistomum attenuatum* Linstow, 1906.

HOSTS. Buzzard, kite, rough-legged buzzard, sparrow hawk, tawny owl and other owls.

LOCATION. Intestine.

LOCALITY. Europe.

DIAGNOSIS. *Shape and size:* anterior and posterior regions of about equal length and breadth, total length about 2 mm. Lateral margins of the anterior region turned ventralwards and fused posteriorly. *Suckers:* ventral separate from the adhesive organ, which is about half as long and one-third as broad as the anterior region of the body. Eggs yellow and large.

OTHER SPECIES:

Neodiplostomum cuticola (Nordmann, 1832). In the buzzard, sparrow hawk and long-eared owl (larvae in the rudd).
N. pseudattenuatum (Dubois, 1927).
N. spathulaeforme (Brandes, 1888). In the snowy owl.
N. cochleare Krause, 1915.
N. morchelloides Semenow, 1927.
N. fungiloides Semenow, 1927.
Etc.

Family **SCHISTOSOMATIDAE** Looss, 1899, emend. Poche, 1907 (p. 106)

Syn. Schistosomidae Looss, 1899; Bilharziidae Odhner, 1912.

Subfamily **Bilharziellinae** Price, 1929

Gynaecophoric canal absent or incomplete.

KEY TO GENERA

I. Both sexes flat and lancet-like, female shorter than the male *Bilharziella*

II. Male only slightly flat, female cylindrical; vitellaria extending through nine-tenths of body *Gigantobilharzia*

III. Body slender and divided into a larger anterior and a filiform posterior part, separated by a slight swelling; testes numerous and showing an opposite arrangement in the posterior region *Trichobilharzia*

IV. Body thread-like, somewhat broader in the anterior region, which has a short gynaecophoric canal in the male *Pseudobilharziella*

Genus *Bilharziella* Looss, 1899

Bilharziella polonica (Kowalewsky, 1895) (Fig. 51H)

Syn. *Bilharzia polonica* Kowalewsky, 1895; *Schistosoma polonicum* of Railliet, 1898.

HOSTS. Domestic duck, mallard, common teal, tufted duck.

LOCATION. Blood-vessels; adults in the abdominal veins, especially the portal vein, juveniles in the venules of the intestine.

LOCALITY. Europe.

DIAGNOSIS. *Shape and size:* lancet-like; male 4 mm. long and 0·5 mm. broad; female 2·1 × 0·25 mm. *Suckers:* in the male, oral 0·10 mm. diameter ventral 0·14 mm., the latter situated 0·76 mm. behind the former; in the female, oral 0·05 mm. diameter, ventral 0·07 mm., the latter 0·37 mm. behind the former. *Reproductive systems: male:* genital pore 0·8 mm. behind the ventral sucker, testes numerous and situated in the posterior region along the sides of the single posterior caecum, cirrus-pouch absent, seminal vesicle long and free in the parenchyma; *female:* genital pore slightly behind the ventral sucker, ovary elongate and folded, somewhat in front of the posterior union of the caeca. Eggs rounded, very large (0·385–0·400 × 0·100 mm.), each having a small terminal spine at one end, and a long tapering process at the other.

Genus *Gigantobilharzia* Odhner, 1910

Gigantobilharzia acotylea Odhner, 1910

In the abdominal veins of the black-headed gull, lesser black-backed gull, Mediterranean black-headed gull in Europe (west Sweden).

G. monocotylea Szidat, 1930

In the veins of the gut of the black-headed gull, mallard, great crested grebe in Europe (west Sweden). [May be a stage of some species of *Pseudobilharziella* (Szidat, 1939).]

G. egreta Lal, 1937 occurs in India, *G. gyrauli* (Brackett, 1940) and *G. lawayi* Brackett, 1942 in U.S.A.

Genus *Trichobilharzia* Skrjabin & Zakharov, 1920

Trichobilharzia ocellata (La Val., 1854) Brumpt, 1931

Syn. *Trichobilharzia kossarewi* Skrjabin & Zakharov, 1920.

In the garganey.

Genus *Pseudobilharziella* Ejsmont, 1929

Pseudobilharziella kowalewskii Ejsmont, 1929

Only the male of this species is known. It is identical with a male Schistosome from *Anas crecca* which Kowalewski (1896*b*) thought to be a juvenile *Bilharziella polonica*.

Pseudobilharziella filiformis Szidat, 1939

In a young swan.

Both sexes of this Schistosome were described. The male differs from *P. kowalewskii* (and the Canadian species, *P. querquedulae* McLeod, 1937) in the size and arrangement of the testes, which are very large.

Brackett (*J. Parasit.*, **28**, 1942, pp. 25–42) gave a key for the separation of eight species of *Pseudobilharziella*, including the three species mentioned above, and also *P. yokogawai* (Oiso, 1927), as well as four new American species, *P. burnetti*, *P. kegonensis*, *P. waubensis*, and *P. horiconensis*.

Subfamily **Schistosomatinae** Stiles & Hassall, 1889, emend.

Genus *Ornithobilharzia* Odhner, 1912

Having a well-developed gynaecophoric canal.

Ornithobilharzia intermedia (Odhner, 1910)

In the veins of the gut of the lesser black-backed gull in Europe (west Sweden).

Family **PARAMPHISTOMATIDAE** Fischoeder, 1901, emend. Goto & Matsudaira, 1918 (p. 103)

Subfamily **Zygocotylinae** Stunkard, 1916

This is the only subfamily of Amphistomes represented in birds. It contains a single genus.

Genus *Zygocotyle* Stunkard, 1916

Zygocotyle lunata (Diesing, 1836) Stunkard, 1916 (Fig. 51 I)

Syn. *Amphistoma lunatum* Diesing, 1836; *Chiorchis lunatus* Travassos, 1921; *Zygocotyle ceratosa* Stunkard, 1916.

HOSTS. Domestic duck and goose, curlew; also ox.

LOCATION. Caecum of mammals, caecum and intestine of birds.

LOCALITY. America.

DIAGNOSIS. *Size:* 3–9 mm. long and 1·5–3 mm. broad. *Suckers:* oral slightly elongate (0·5 × 0·4 mm.) and having a pair of small pouches 0·3–0·35 mm. long and 0·1–0·12 mm. broad; ventral situated at the posterior extremity and directed ventrally, transversely ovoid, 1–1·2 × 0·8–1 mm., the posterior border having a thickening. *Gut:* pharynx absent, oesophagus very long and having a muscular bulb at its base, caeca fairly long, but terminating in front of the posterior sucker. *Reproductive systems:* genital pore slightly behind the bifurcation of the intestine. Cirrus pouch absent. Testes lobed and situated one behind the other in the middle of the body. Ovary behind the testes. Uterus having a short descending limb and an intensely folded ascending limb. Vitellaria lateral and composed of numerous large follicles, extending almost from the oral to the posterior sucker. Eggs large (0·130–0·150 × 0·072–0·090 mm.), yellow and operculate.

Family **MICROSCAPHIDIIDAE** Travassos, 1922 (p. 104)

Syn. *Angiodictyidae* Looss, 1902

Genus *Polyangium* Looss, 1902

Syn. *Nephrobius* Poche, 1926.

Price (1937*d*) has shown that *Nephrobius colymbi* as described (from two specimens) by Poche is without doubt congeneric with species of *Polyangium*, so that Poche's genus *Nephrobius* becomes a synonym of *Polyangium*. Incidentally, Price thought that some error had been made, deeming it incredible that a trematode bearing so close a resemblance to the intestinal parasites of turtles should be found in the kidneys of a bird. This record may be taken, therefore, as a case of accidental parasitism, for Poche is entitled to the benefit of the doubt.

Note. *Polyangium* and four other genera of Microscaphidiidae were mentioned in the previous chapter as having species in Chelonia. Price (*loc. cit.*) erected two new genera, *Octangioides* (for *O. skrjabini* Price, 1937, a parasite of the turtle *Dermatemys mawii* in Mexico) and *Hexangitrema* (for *H. pomacanthi*, a parasite of the black angel fish in the New York Aquarium), and gave a key to these seven genera and two others, *Parabaris* Travassos and *Hexangium* Goto & Ozaki.

Polyangium colymbi (Poche, 1926) Price, 1937

Syn. *Nephrobius colymbi* Poche, 1926.

HOST. Black-throated diver.
LOCATION. Kidney.
LOCALITY. Europe, Asia.
DIAGNOSIS. *Shape and size:* elongate, about 10 mm. long and 1·75 mm. broad. *Cuticle:* non-spinous. *Suckers:* oral very small (0·23 × 0·28 mm.) and lacking pouches, ventral absent. *Gut:* oesophagus very long and narrow (2 × 0·4 mm.), caeca long and wide. *Reproductive systems:* genital pore slightly behind the oral sucker. Testes globular, situated one behind the other in the posterior half of the body, the ovary being still farther back. Vitellaria lateral to the caeca in the posterior two-thirds of the body, merging behind the ovary. Uterus having only an ascending limb, which is formed into slight folds between the caeca. Eggs fairly large (0·081 × 0·052 mm.), the shell thickened at the poles.

Family **STOMYLOTREMATIDAE** Poche, 1926 (p. 98)

Genus *Stomylotrema* Looss, 1900 (Fig. 53 D)

Syn. *Stomylus* Looss, 1899

Stomylotrema pictum (Creplin, 1837) Braun, 1901

Syn. *Distoma pictum* Creplin, 1837; *D. singulare* Molin, 1858, nec Looss, 1899.

HOSTS. White stork, glossy ibis.
LOCATION. Cloaca, intestine.
DIAGNOSIS. *Shape and size:* oval, about 4 mm. long and 2 mm. broad. *Cuticle:* smooth. *Suckers:* very large, oral, 1·0 mm. diameter, ventral transversely ovoid and well in the posterior half of the body, dimensions 1·17 × 1·03 mm. *Gut:* pharynx large, oesophagus absent, caeca long. *Reproductive systems:* cirrus pouch spindle-shaped and directed laterally, genital pore lateral and on the right opposite the pharynx. Testes side by side in front of the ventral sucker, ovary to the left of the pharynx. Vitellaria comprising only about 7–9 follicles on each side. Uterus conspicuously folded around the ventral sucker. Eggs dark brown, measuring about 0·027 × 0·019 mm.

OTHER SPECIES (and their locations and hosts):

Stomylotrema perpastum Braun, 1902 (cloaca of collared pratincole); *S. fastosum* Braun, 1901 (intestine of grey plover); *S. bijugum* Braun, 1901 (intestine of black-winged stilt); *S. gratiosus* Travassos, 1922 (cloaca of *Turdus* sp.); *S. rotunda* Tubangui, 1928 (intestine of *Hypotaenidia philippensis*).

CHAPTER 11

SOME TREMATODES OF MAMMALS

Many of the trematode parasites of mammals infect man and domestic animals, thereby acquiring importance in medical and veterinary science as a consequence of which they have received ample consideration by various writers on helminthology. Many pests, perhaps less common now than hitherto, have become notorious during the growth of helminthology and are dealt with in text-books on parasitology and some similar works on general zoology.

Other trematode parasites of mammals have remained in obscurity, chiefly because they do not seriously interfere with the activities of man or domestic animals, and have received little or no attention from zoologists in general and British zoologists in particular. Nevertheless, they constitute the main part of the trematode fauna and, even if regarded as of mainly academic interest, can be used to widen our knowledge of Trematoda. Most of our knowledge of the trematodes of mammals has been due to the work of continental zoologists. Fortunately, the mammal fauna of this country has few characteristics which distinguish it from that of western Europe, and most trematodes which occur in mammals on the continent can reasonably be expected here also, more especially because possible intermediate hosts in the life histories are common to both regions. For this reason, as well as the paucity of British records, this chapter is devoted to diagnoses of a number of the best-known trematodes of mammals whether or not they are actually known to occur in these islands.

In Table 6 the trematode parasites of eleven species of mammals are shown and provide a representative list of flukes recorded in Europe. Thirteen families of Digenea are represented in these animals, more than half of them in wild and domestic cats.

Family **PLAGIORCHIIDAE** Lühe, 1901, emend. Ward, 1917 (p. 94)

Syn. Lepodermatidae Looss, 1901.

Subfamily **Plagiorchiinae** Lühe, 1901, emend. Pratt, 1902

Olsen (1937*a*) made a systematic study of this subfamily, giving a key to fourteen genera, and also keys to their species.

Genus *Plagiorchis* Lühe, 1899

Syn. Lepoderma, 1899.

Plagiorchis vespertilionis (Müller, 1784) Braun, 1900

Syn. *Fasciola vespertilionis* Müller, 1784; *Distoma vespertilionis* of Zeder, 1803; *D. lima* Rudolphi, 1809; *Lepoderma vespertilionis* of Looss, 1899.

Hosts. Greater and lesser horse-shoe bats, long-eared bat, Natterer's bat, whiskered bat, Daubenton's bat, serotine, noctule, pipistrelle and other Chiroptera.

LOCATION. Intestine.

LOCALITY. Europe, America.

DIAGNOSIS. *Shape and size:* elongate-oval outline, tapering at both extremities, but more gradually posteriorly, 4–9 mm. long and about 0·9 mm. broad. *Cuticle:* spinous, least so laterally and posteriorly. *Suckers:* oral slightly larger than the ventral, diameters 0·23–0·25 and 0·21–0·22 mm., the ventral situated about two-fifths of the distance along the body. *Gut:* pharynx large (0·145 × 0·125 mm.), oesophagus short, caeca long. *Reproductive systems:* genital pore slightly in front of the ventral sucker, not exactly median. Cirrus pouch well developed, about 1 mm. long. Testes rounded, diagonally one behind the other in the third quarter of the body. Ovary in front of the testes, relatively far forward and dorsal to the cirrus pouch. Vitellaria well developed, lateral and restricted to the region behind the ventral sucker. Uterus having descending and ascending limbs, but only slightly folded. Eggs small (0·033 × 0·018 mm.).

OTHER SPECIES OCCURRING IN THE INTESTINE OF MAMMALS (and their hosts):

Plagiorchis muris (Tanabe, 1922): brown rat, long-tailed field mouse.
P. asper Stossich, 1904: long-eared bat.
P. massino Petrow & Tichonow, 1927: cat and dog (Kasakstan and Armenia).
P. arvicolae Schulz & Skwortzow, 1931: *Arvicola terrestris* and *A. amphibius* (Russia).
P. eutamiatis and *P. obensis*, both Schulz, 1932: *Eutamias asiaticus* and *Cricetus* sp. respectively (Russia).

Family **MESOTRETIDAE** Poche, 1926 (p. 102)

Genus *Mesotretes* Braun, 1900

Mesotretes peregrinus (Braun, 1900) Braun, 1900 (Fig. 54A)

Syn. *Distoma peregrinum* Braun, 1900.

HOST. Greater horse-shoe bat.

LOCATION. Intestine.

LOCALITY. Europe.

DIAGNOSIS. *Shape and size:* elongate, pointed at the extremities, constricted slightly at the level of the ventral sucker, about 8 mm. long and one-fifth as broad. *Cuticle:* spinous in the anterior third of the body. *Suckers:* of medium size, the ventral larger than the oral and situated in the second quarter of the body. *Gut:* pharynx present, oesophagus short, caeca long. *Reproductive systems:* genital pore median, immediately behind the ventral sucker. Cirrus pouch elongate. Testes elongate-oval in outline, diagonally one behind the other in the posterior half of the body. Ovary behind the testes, near the posterior extremity. Uterus having a short descending and a long ascending limb and little folded. Vitellaria well developed laterally in the posterior two-thirds of the body. Eggs numerous, small and thin-shelled.

Table 6. *Some Trematodes of Certain Mammals in Europe*

Left out of account are numerous trematodes which occur outside Europe, the ones here mentioned having been recorded in this continent, but not necessarily in all the hosts stated.

B, gall bladder; Bd, bile ducts; FS, frontal sinuses; I, intestine; L, liver; S, stomach; (s), subcutaneous cysts; ( ), encystment.

	Hedgehog	Mole	Noctule	Pipistrelle	Common porpoise	Common seal	Grey seal	Blue and silver foxes	Pig (wild and domestic)	Cat (wild and domestic)	Dog (wild and domestic)
PLAGIORCHIIDAE:											
Plagiorchis vespertilionis	—	—	I	I	—	—	—	—	—	—	—
P. massino	—	—	—	—	—	—	—	—	—	I	I
LECITHODENDRIIDAE:											
Lecithodendrium chilostomum	—	—	I	I	—	—	—	—	—	—	—
L. lagena	—	—	—	I	—	—	—	—	—	—	—
Parabascus semisquamosus	—	—	I	—	—	—	—	—	—	—	—
Pycnoporus macrolaimus	—	—	—	I	—	—	—	—	—	—	—
P. heteroporus	—	—	—	I	—	—	—	—	—	—	—
DICROCOELIIDAE:											
Dicrocoelium dendriticum	—	—	—	—	—	—	—	—	BdB	BdB	BdB
FASCIOLIDAE:											
Fasciola hepatica	—	—	—	—	—	—	—	—	L	L	—
CAMPULIDAE:											
Campula oblonga	—	—	—	—	Bd	—	—	—	—	—	—
ALLOCREADIIDAE:											
Crepidostomum metoecus	—	—	I	—	—	—	—	—	—	—	—
OPISTHORCHIIDAE:											
Opisthorchis felineus	—	—	—	—	—	—	—	B	Bd	B	B
O. tenuicollis	—	—	—	—	L	—	L	—	—	—	—
Pseudamphistomum truncatum	—	—	—	—	—	L	L	B	—	B	B
P. danubiense	—	—	—	—	—	—	—	—	—	L	—
Metorchis albidus	—	—	—	—	—	—	L	LB	—	LB	LB
Omphalometra flexuosa	—	SI	—	—	—	—	—	—	—	—	—

HETEROPHYIDAE:											
Cryptocotyle lingua	—	—	—	—	—	I	—	I	—	—	I
Apophallus mühlingi	—	—	—	—	—	—	—	—	—	I*	—
A. donicum	—	—	—	—	—	—	—	—	—	—	I
Phagicola italica	—	—	—	—	—	—	—	—	—	—	I
'*Metagonimus romanica*' [=*M. yokogawai*]	—	—	—	—	—	—	—	—	I	I	I
Euryhelmis squamula	—	—	—	—	—	—	—	I	—	—	—
'*Ciureana quinqueangularis*'	—	—	—	—	—	—	—	—	—	I	—
DIPLOSTOMATIDAE:											
Alaria alata	—	—	—	—	—	—	—	I	—	I	I
CYATHOCOTYLIDAE:											
Mesostephanus appendiculatum	—	—	—	—	—	—	—	—	—	I*	I*
Pharyngostomum cordatum	—	—	—	—	—	—	—	—	—	I	—
ECHINOSTOMATIDAE:											
Echinostoma acanthoides	—	—	—	—	—	I	—	—	—	—	—
Euparyphium melis	I	—	—	—	—	—	—	I	—	I	—
E. suinum	—	—	—	—	—	—	—	—	I	—	—
Echinochasmus perfoliatus	—	—	—	—	—	—	—	I	I	I	I
TROGLOTREMATIDAE:											
Troglotrema acutum	—	—	—	—	—	—	—	FS	—	—	—
Pholeter gastrophilus	—	—	—	—	(s)	—	—	—	—	—	—
BRACHYLAEMIDAE:											
Brachylaemus '*helicis*' [=*erinacei*]	I	—	—	—	—	—	—	—	—	—	—
B. '*spinulosum*' [=*erinacei*]	I	—	—	—	—	—	—	—	—	—	—
Itygonimus ocreatus	—	I	—	—	—	—	—	—	—	—	—
I. lorum	—	I	—	—	—	—	—	—	—	—	—

* Experimental infections.

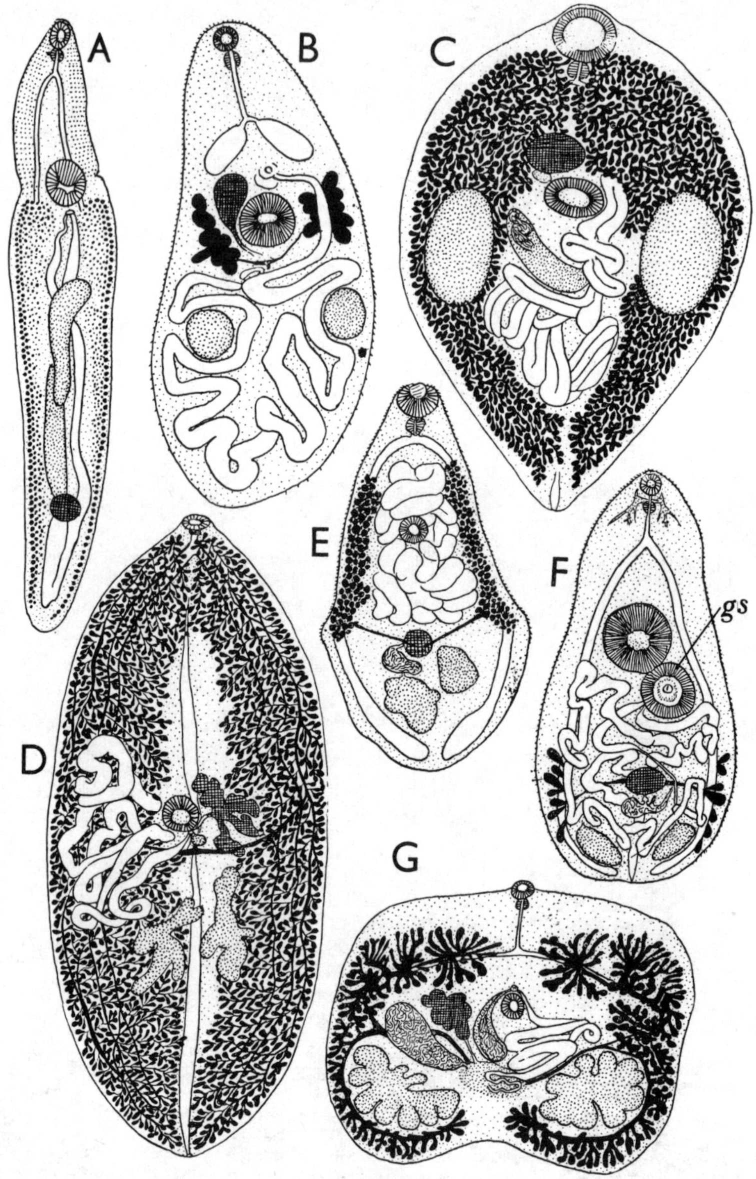

Fig. 54. Mesotretidae (A), Lecithodendriidae (B), Troglotrematidae (C, D), Opisthorchiidae (E) and Heterophyidae (F, G) from mammals. A, *Mesotretes peregrinus*. B, *Pycnoporus acetabulatus*. C, *Troglotrema acutum*. D, *Paragonimus westermanii*. E, *Metorchis albidus*. F, *Heterophyes heterophyes*. G, *Euryhelmis squamula*. (A, B, after Fuhrmann, 1928; C, D, G, after Mönnig, 1934; E, F, after Neveu-Lemaire, 1936.)

Family **LECITHODENDRIIDAE** Odhner, 1910 (p. 94)

Subfamily **Lecithodendriinae** Lühe, 1901, emend. Looss, 1902

Syn. Brachycoeliinae Looss, 1899, in part.

KEY TO GENERA OCCURRING IN MAMMALS

1. Cirrus pouch absent	
A. Cuticle smooth; testes in front of the folds of the uterus	*Lecithodendrium*
B. Cuticle spinous; testes enveloped in the folds of the uterus	*Pycnoporus*
2. Cirrus pouch present	
A. Genital pore in the vicinity of the pharynx	*Phaneropsolus*
B. Genital pore near the ventral sucker	*Parabascus*

Genus *Lecithodendrium* Looss, 1899

Syn. *Mesodendrium* Faust, 1919, and misprints like *Lecithodendrum* Pratt, 1902; *Levcithodendrium* Jameson, 1902; *Leucithodendrium*, Shipley & Hornell, 1904.

According to Dollfus (1931) and Macy (1936*b*) this genus of parasites of bats should be subdivided. The genus *Prosthodendrium* Dollfus, 1931 is supposed to contain species in which the vitellaria are situated in front of the testes, namely, *ascidia* (Beneden, 1873), *chilostomum* (Mehlis, 1831), *cordiforme* (Braun, 1900), *dinanatum* (Bhalerao, 1926) (type), *liputianum* (Travassos, 1928), *longiforme* (Bhalerao, 1926), *luzonicum* (Tubangui, 1928), *naviculum* Macy, 1936, *orospinosa* (Bhalerao, 1926), *pyramidum* (Looss, 1896), *swansoni* Macy, 1936 and *urna* (Looss, 1907). Other species belonging to the subgenus *Paralecithodendrium* Odhner, 1911 (or to a genus of this name) are, *anticum* (Stafford, 1905), *glandulosum* (Looss, 1896), *lucifugi* Macy, 1936, *nokonis* Macy, 1936, *obtusum* (Looss, 1896) and *ovimagnosum* (Bhalerao, 1926). The genus *Lecithodendrium* s.s. was restricted to species in which the vitellaria are situated behind the testes, namely, *attia* (Bhalerao, 1926), *granulosum* (Looss, 1907), *hirsutum* (Looss, 1896), *linstowi* Dollfus, 1931, *macrostomum* (Ozaki, 1929) and *spathulatum* (Ozaki, 1929).

Lecithodendrium lagena (Brandes, 1888) Looss, 1899

Syn. *Distoma ascidia* Beneden, 1872, *nec* Rudolphi, 1819; *D. lagena* Brandes, 1888; *Lecithodendrium ascidia* Lühe, 1909.

HOSTS. Greater horse-shoe bat, lesser horse-shoe bat, long-eared bat, noctule, serotine, Daubenton's bat, and other Chiroptera (pipistrelle, Cambridgeshire).

LOCATION. Intestine.

LOCALITY. Europe, North Africa.

DIAGNOSIS. *Shape and size:* spindle-shaped, 1·2 mm. long and 0·33 mm. broad. *Cuticle:* smooth, or bearing exceedingly fine spinelets dorsally. *Suckers:* oral very slightly larger than the ventral, diameters 0·08 and 0·075 mm. *Reproductive systems:* genital pore in front of the ventral sucker. Cirrus pouch absent. (*Note.* The long, tubular ejaculatory duct forms a tangled mass enclosed in parenchyma and simulates a cirrus pouch, but never contains muscular fibrils.)

Testes not far removed from the ventral sucker. Uterus forming a number of folds behind the testes. Vitellaria lateral and in the posterior region, behind the testes and beneath the limbs of the V-shaped excretory vesicle. Eggs small (about 0·020 × 0·11 mm.).

Lecithodendrium chilostomum (Mehlis, 1831) Braun, 1900 (Fig. 77E)

Syn. *Distoma chilostomum* Mehlis, 1831; *D. ascidioides* Beneden, 1873.

HOSTS. Long-eared bat, Daubenton's bat, Natterer's bat, whiskered bat, Leisler's bat, noctule, serotine, and other Chiroptera (pipistrelle in Cambridgeshire).

LOCATION. Intestine.

LOCALITY. Europe.

DIAGNOSIS. *Shape and size:* oval outline, 0·9–1·5 mm. long and 0·7 mm. broad. *Suckers:* oral much larger than the ventral, diameters 0·25 and 0·16 mm. *Reproductive systems:* vitellaria disposed laterally in the anterior region. Eggs small (0·031–0·033 × 0·013–0·015 mm.).

OTHER SPECIES IN BATS:

Lecithodendrium pyramidum (Looss, 1899). In the intestine of the lesser horse-shoe bat in Egypt. (*Note.* The development of this trematode from a Xiphidiocercaria living in *Melania tuberculata* has been outlined by Azim (1936).)

Genus *Phaneropsolus* Looss, 1899

The type-species, *P. sigmoideus*, has been mentioned as a parasite of birds. Of the species infecting mammals *P. longipenis* Looss, 1899 was discovered in an ape in the Zoological Gardens at Giza, *P. orbicularis* in *Cebus trivirgatus* in Brazil, and *P. oviforme* in *Nycticebus javanicus*.

Genus *Pycnoporus* Looss, 1899 (Fig. 54B)

Pycnoporus heteroporus (Dujardin, 1845) Looss, 1899

Syn. *Distoma heteroporum* Dujardin, 1845; *Lecithodendrium heteroporum* of Stossich, 1899.

This species occurs in the intestine of the pipistrelle in Europe and has been found in Cambridgeshire.

OTHER SPECIES:

Pycnoporus macrolaimus (Linstow, 1894) Looss, 1899. In the intestine of the pipistrelle (also found in Cambridgeshire).

P. acetabulatus Looss, 1899. In the intestine of '*Vesperugo kuhli*' in Egypt.

Genus *Parabascus* Looss, 1907

Several species of this genus also inhabit the intestine of bats. *Parabascus semisquamosus* occurs in Europe, but the type species, *P. lepidotus* Looss, 1907, in Egypt.

Subfamily **Pleurogenetinae** Looss, 1896

Genus *Prosotocus* Looss, 1899

This genus is better known as parasites in frogs and toads, but several species have been described by Mödlinger (1930), viz. *Prosotocus vespertilionis* in the mouse-eared bat, *P. trigonostomum* in the serotine and *P. amphoraeformis* in *Myotis oxygnathus*. All occur in Hungary, where *Pycnoporus heteroporus* is found also in the pipistrelle.

Family **DICROCOELIIDAE** Odhner, 1910 (p. 95)

Subfamily **Dicrocoeliinae** Looss, 1899

KEY TO GENERA IN MAMMALS

1. Body oval in outline and relatively wide. Testes side by side, somewhat widely separated *Eurytrema*
2. Body elongate and narrow. Testes directly or diagonally one behind the other
   A. Body lancet-like, widest behind the middle. Testes diagonally arranged *Dicrocoelium*
   B. Body ribbon-like, testes one behind the other *Lyperosomum*

Genus *Dicrocoelium* Dujardin, 1845

Dicrocoelium dendriticum (Rudolphi, 1819) Looss, 1899 (Fig. 57A, 81G)

Syn. *Fasciola lanceolata* Rudolphi, 1803, *nec* Schrank, 1790; *Distoma dendriticum* Rudolphi, 1819; *D. lanceolatum* of Mehlis, 1825; *D.* (*Dicrocoelium*) *lanceolatum* (Rudolphi) of Dujardin, 1845; *Dicrocoelium lanceatum* of Stiles & Hassall, 1896; *D. macaci* Kobayashi, 1915.

HOSTS. Sheep, goat, ox, pig, cat, rabbit, dog, hare and other mammals.

LOCATION. Bile ducts and gall bladder.

LOCALITY. Europe, Africa, Asia, America.

Note. The correct name of this species is that given above. Odhner (1910*a*, footnote, p. 88) re-examined Rudolphi's types of *dendriticum* (hitherto supposed to have come from *Xiphias*) and found them identical with '*lanceolatum*'.

DIAGNOSIS. *Shape, texture and size:* lancet-like, more acutely tapering anteriorly than posteriorly, delicate and translucent, 4–12 mm. long and 1·5–2·5 mm. broad. *Glands:* numerous and scattered throughout in the anterior region, opening on the anterior border of the oral sucker. *Cuticle:* smooth. *Suckers:* about one-fifth of the body length apart, oral slightly smaller than the ventral, diameters c 3–0·4 mm. and 0·6 mm. *Reproductive systems:* genital pore far in front of the ventral sucker, near the bifurcation of the intestine. Cirrus pouch elongate and small (0·5–0·6 mm. long). Testes very slightly lobed, one in front of the other slightly behind the ventral sucker. Ovary globular, close behind the posterior testis. Uterus having descending and ascending limbs, formed into numerous folds which tend to fill the posterior region of the body. Vitellaria confined to a short region behind the testes laterally. Eggs thick-shelled, sculptured, measuring 0·038–0·045 × 0·022 0·023 mm., each containing an oval miracidium with anterior ciliation.

Conklin & Baker (*J. Parasit.* **17**, 1930, pp. 18–19) recorded the occurrence of this species in sheep in Canada, Price (1943*c*) in cattle in the U.S.A. I have outlined the life history on pp. 477–8 and the growth of the trematode on p. 532 (see Fig. 81 A–G).

Genus *Eurytrema* Looss, 1907

This genus of trematodes inhabiting the pancreatic ducts, sometimes the bile ducts, of cattle, sheep and goats in Asia, India and America is typified by *Eurytrema pancreaticum* (Janson, 1889) Looss, 1907 (Syn. *Distoma pancreaticum* Janson, 1889; *Dicrocoelium pancreaticum* (Janson, 1889) Railliet, 1897), which is described in several text-books on helminthology, along with a number of other species, and we need not deal with it here.

Genus *Lyperosomum* Looss, 1899

This genus of liver flukes, occasionally found in the intestine, is typified by *Lyperosomum longicauda*, a parasite of crows, blackbirds, song thrushes and other birds, and is represented by other species in birds and also the species *L. vitta* (Dujardin, 1845) Baylis, 1927, which inhabits the intestine of the long-tailed field mouse in Britain (see Baylis, 1939*a*).

Family **FASCIOLIDAE** Railliet, 1895 (p. 102)

Syn. Fasciolopsidae Odhner, 1926.

Genus *Fasciola* Linnaeus, 1758

Fasciola hepatica Linnaeus, 1758 (Fig. 55 A, B)

Syn. *Distoma hepaticum* Abildgaard; *Planaria latiuscula* Goeze, 1782, in part; *Distoma caviae* Sonsino, 1890; *Cladocoelium hepaticum* of Stossich, 1892.

Hosts. Man, sheep, goat, ox, horse, ass, pig, rabbit, cat, and other land mammals, also killer whale (grampus) and piked whale (lesser rorqual).

Location. Bile ducts.

Locality. Cosmopolitan.

Diagnosis. *Shape and size:* oval outline, flattened, the anterior end more rounded than the posterior, but having a conical process bearing the mouth at its tip. Total length 18–51 mm., greatest breadth (near the mid-body) 4–13 mm. *Cuticle:* spinous. *Suckers:* relatively small, the ventral larger than the oral and 3–4 mm. behind it, diameters about 1·6 and 1 mm. *Gut:* small pharynx, oesophagus short, caeca long and having diverticula with primary, secondary and tertiary branches laterally. Bifurcation of intestine located within the conical anterior process. *Reproductive systems:* genital pore median and midway between the ventral sucker and the bifurcation of the intestine. Cirrus pouch absent. Testes much branched, one behind the other in the posterior half of the body. Ovary also much branched, situated in front of testes and on the right. Vitelline follicles filling up all available space between the diverticula of caeca laterally. Uterus having an ascending limb only, this forming a rosette-like cluster of folds between the anterior testis and the ventral sucker. Eggs large (0·130–0·145 × 0·070–0·090 mm.), each containing an ovum and a cluster of vitelline cells.

Fasciola gigantica Cobbold, 1856

Syn. *Distomum giganteum* Diesing, 1858; *Cladocoelium giganteum* of Stossich, 1892; *Fasciola hepatica* var. *angusta* Railliet, 1895; *Distomum hepatica* var. *aegyptiaca* Looss, 1896.

HOSTS. Man, goat, sheep, ox, etc.

LOCATION. Bile ducts and liver.

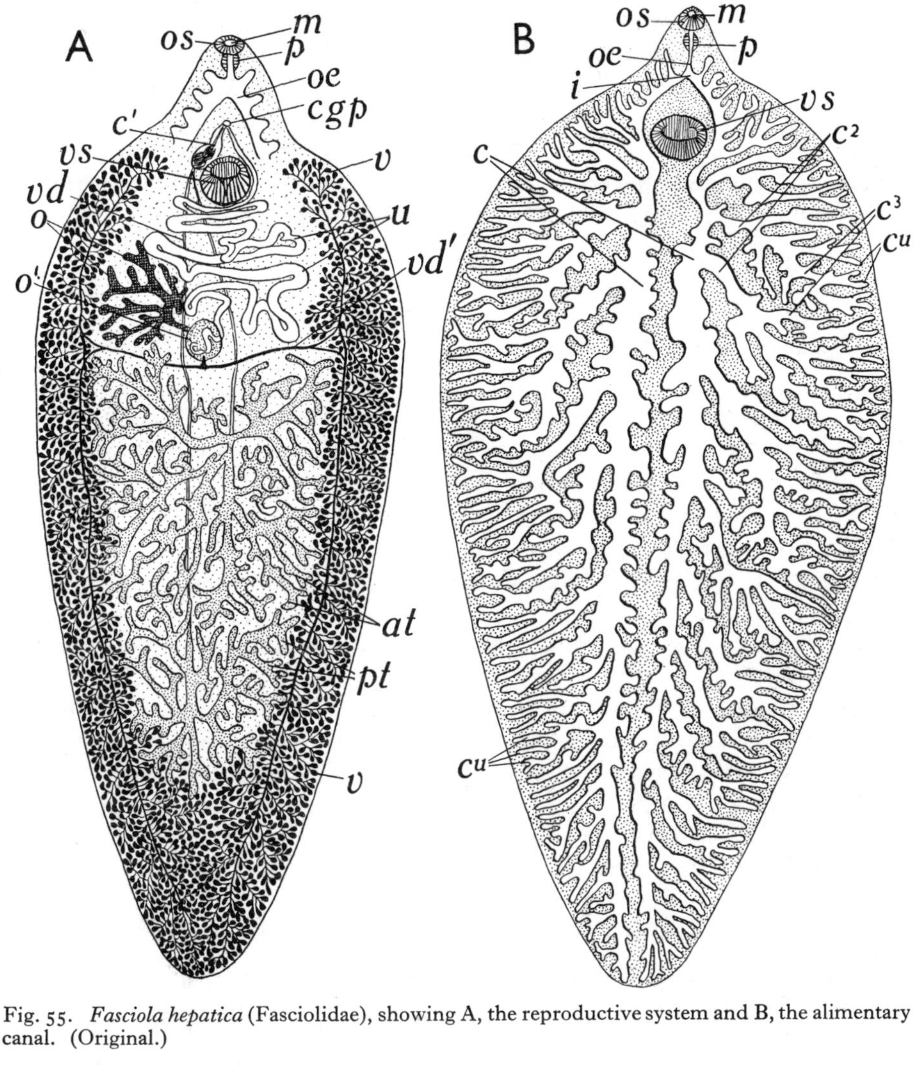

Fig. 55. *Fasciola hepatica* (Fasciolidae), showing A, the reproductive system and B, the alimentary canal. (Original.)

LOCALITY. America, Africa, Asia, Europe. (*Note.* Three cases of human infection in Russia were described by Pigulewsky (1927, 1928): the occurrence in the sheep in Spain was recorded by Almarza (1935); other records by Tabunščikova (1932*a*, *b*).

DIAGNOSIS. *Size:* 25–75 mm. long and 3–12 mm. broad. *Other characters:* similar to *Fasciola hepatica*, but more elongate, extension of the body mainly affecting the posterior region.

Genus *Fascioloides* Ward, 1917

Fascioloides magna (Bassi, 1875) Ward, 1917

Syn. *Distoma magnum* Bassi, 1875; *Fasciola magna* of Stiles, 1894; *Distoma grande* Perroncito, 1882, *nec* Rudolphi, 1819; *Fasciola carnosa* Hassall, 1891; *Distoma texanicum* Francis, 1891.

HOSTS. Man, ox, sheep, horse, etc.

LOCATION. Bile ducts.

LOCALITY. America, Europe (the occurrence in deer in Germany was recorded by Salomon (1932*a*, *b*); in sheep in Spain by Almarza (1935)).

DIAGNOSIS. Similar to *Fasciola hepatica*, but oval outline, uniformly tapering anteriorly, rounded posteriorly, 23–100 mm. long, 11–26 mm. broad and 2–4·5 mm. thick. Oesophagus relatively long, caeca even more extensively branched, eggs very large and variable in size (0·109–0·168 × 0·075–0·096 mm.).

Genus *Fasciolopsis* Looss, 1899

Fasciolopsis buski (Lankester, 1857) Looss, 1899

Syn. *Distomum buski* Lankester, 1857; *D. crassum* Busk, 1859; *D. rathouisi* Poirier, 1887; *Opisthorchis buski* of Blanchard, 1895; *Fasciolopsis rathouisi* of Ward, 1903; *F. fülleborni* Rodenwaldt, 1909; *F. goddardi* Ward, 1910; *F. spinifera* Brown, 1917.

HOSTS. Man and pig; dog and rabbit experimental hosts (discovered in a Lascar who died in London).

LOCATION. Intestine.

LOCALITY. Far East (China, Formosa, Indo-China, Siam, Java, Sumatra, India, etc.).

DIAGNOSIS. *Shape and size:* elongate-oval outline, somewhat flattened, tapering anteriorly, rounded posteriorly, 20–75 mm. long and 8–20 mm. broad. *Cuticle:* spinous. *Suckers:* close together, the ventral much larger than the oral, but both small, diameters 2–3 mm. and about 0·5 mm. *Gut:* pharynx small, oesophagus short, caeca long and unbranched. *Other characters:* in general similar to *Fasciola hepatica*, but gonads confined to the posterior half of the body, the anterior half being more extended and lacking 'shoulders'. Eggs very large (0·130–0·140 × 0·080–0·085 mm.).

Family **CAMPULIDAE** Odhner, 1926 (p. 103)

Syn. Sub-family Campulinae Stunkard & Alvey, 1930 (in the Fasciolidae).

KEY TO GENERA OCCURRING IN MARINE MAMMALS

(After Price, 1932, modified)

1. Cirrus not armed with spines
   - A. Body very elongate (60–80 mm. long); ovary deeply lobed; follicles of the vitellaria arranged in rectangular masses — *Lecithodesmus*
   - B. Body less elongate (not 20 mm. long); ovary not lobed; follicles of the vitellaria not arranged in rectangular masses
     - (*a*) Oral sucker much larger than the ventral; eggs circular in transverse section — *Zalophotrema*
     - (*b*) Oral and ventral suckers of about equal size; eggs triangular in transverse section — *Campula*

2. Cirrus armed with spines
   A. Metraterm armed with spines; follicles of the vitellaria uniformly scattered
      (*a*) Caeca simple, without diverticula; testes not deeply lobed *Orthosplanchnus*
      (*b*) Caeca with median and lateral diverticula; testes deeply lobed *Synthesium*
   B. Metraterm not armed with spines; follicles of the vitellaria arranged in distinct masses *Odhneriella*

Genus *Campula* Cobbold, 1858

Syn. *Brachycladium* Looss, 1899.

KEY TO SPECIES

(After Price, 1932, modified)

1. Testes lobed
   A. Cirrus pouch not extending behind the ventral sucker; caeca lacking anal openings *palliata*
   B. Cirrus pouch extending behind the ventral sucker; caeca having anal openings *oblonga*
2. Testes not lobed
   A. Suckers close together; testes in the anterior half of the body *rochebruni*
   B. Suckers widely separated; testes in the final third of the body *delphini*

Campula oblonga (Cobbold, 1858) of Braun, 1900, *nec* Cobbold, 1876 (Fig. 56A)

Syn. *Distomum oblongum* (Cobbold, 1858) Braun, 1891; *Distoma* (*Brachylaimus*) *oblongum* of Stossich, 1892; *Distomum tenuicolle*, Rudolphi, 1819 of Olsson, 1893; *Brachycladium oblongum* of Looss, 1902; *Opisthorchis oblonga* of Kowalewski, 1898.

HOST. Common porpoise.

LOCATION. Bile ducts.

LOCALITY. Europe, North America.

DIAGNOSIS. *Shape and size:* elongate, the anterior end bluntly pointed, the posterior end rounded, 4–7 mm. long and 1–3 mm. broad. *Cuticle:* spinous, spines arranged in alternating longitudinal rows. *Suckers:* oral 0·31–0·36 mm. diameter, ventral 0·43–0·46 mm. diameter and situated in the first quarter of the body. *Gut:* prepharynx short and wide, pharynx ovoid, oesophagus very short, crura long and sinuous, each of them having short median and lateral diverticula and an anterior prolongation extending to the oral sucker. Posterior ends of the crura opening into the tubular excretory vesicle. *Reproductive systems:* genital pore situated in front of the ventral sucker. Cirrus pouch slightly curved, containing a sinuous seminal vesicle and an unarmed cirrus. Testes deeply lobed, one behind the other in the middle third of the body. Ovary transversely ovoid, immediately in front of anterior testis and on the right. Vitellaria distributed in the dorsal region, extending from the oesophagus to the posterior extremity, overlapping the caeca ventrally. Uterus short and having a few loops between the ovary and the ventral sucker. Metraterm about half as long as the cirrus pouch, muscular and unarmed. Eggs yellow, each having a knob-like projection at the anopercular pole, flat at the opercular pole, triangular in transverse section, measuring 0·090–0·097 × 0·045 mm.

Campula palliata (Looss, 1885) Looss, 1901 (Fig. 56B)

Syn. *Distomum palliatum* Looss, 1885; *Cladocoelium palliatum* of Stossich, 1892; *Brachycladium palliatum* of Looss, 1899.

HOST. Common dolphin.

LOCATION. Liver and bile ducts.

LOCALITY. Europe.

DIAGNOSIS. *Shape and size:* elongate, more rounded anteriorly than posteriorly, 9–10 mm. long, 1·5–2 mm. broad and 0·75–1 mm. thick, constricted at the level of the ventral sucker. *Cuticle:* spinous, the spines being closely set. *Suckers:* of about equal sizes and 2·5–3·5 mm. apart. In other respects similar to *C. oblonga*, but cirrus pouch not extending posterior to the ventral sucker, caeca not opening into the excretory vesicle, eggs much smaller and blunter at the opercular than at the anopercular pole, dimensions about 0·059 × 0·043 mm.

Campula delphini (Poirier, 1886) Bittner & Sprehn, 1928 (Fig. 56C, D)

Syn. *Distomum delphini* Poirier, 1886; *Cladocoelium delphini* of Stossich, 1892; *Brachycladium delphini* of Looss, 1899.

HOST. Common dolphin.

LOCATION. Liver.

LOCALITY. Europe.

DIAGNOSIS. *Shape and size:* elongate, slightly attenuated at both extremities, 14 mm. long and 2 mm. broad. *Cuticle:* spinous. *Suckers:* widely separated (7 mm. apart), of almost equal sizes, the ventral slightly the larger and 0·7 mm. diameter. *Other characters:* similar to other species, but testes ovoid, not lobed, and situated in the posterior third of the body; anterior extensions of caeca each having three branches; posterior ends of the caeca ending in the parenchyma; eggs measuring 0·060 × 0·045 mm. and slightly pointed at the anopercular pole.

Campula rochebruni (Poirier, 1886) Bittner & Sprehn, 1928 (Fig. 56E)

Syn. *Distomum rochebruni* Poirier, 1886; *Cladocoelium rochebruni* of Stossich, 1892; *Brachycladium rochebruni* of Looss, 1899.

HOST. Common dolphin.

LOCATION. Liver.

LOCALITY. Europe.

DIAGNOSIS. *Shape and size:* elongate and narrow, flat ventrally and slightly convex dorsally, 10 mm. long and 1 mm. broad. *Cuticle:* spinous, the spines more abundant and more closely set anteriorly. *Suckers:* close together (0·7 mm. apart), of equal sizes and about 0·38 mm. diameter. *Other characters:* similar to other species, but testes ovoid, not lobed, and situated in front of the middle of the body, eggs larger than in *C. palliata* and *C. delphini* (0·082 × 0·045 mm.).

Genus *Lecithodesmus* Braun, 1902

Lecithodesmus goliath (Beneden, 1858) Odhner, 1905 (Fig. 56F)

Syn. *Distomum goliath* Beneden, 1858.

HOSTS. Piked whale (lesser rorqual), sei whale, Greenland right whale.

LOCATION. Liver.

LOCALITY. Europe.

DIAGNOSIS. *Shape and size:* very elongate, bluntly rounded anteriorly, slightly attenuated posteriorly, 60–80 mm. long, 8 mm. broad and 1·6–1·8 mm. thick. *Cuticle:* spinous anteriorly (possibly, more extensively). *Suckers:* ventral smaller than the oral and one-third of the body-length from it (28 mm.), diameters 1·3–1·8 and 2·3 mm. *Gut:* prepharynx short, pharynx present, oesophagus very short, caeca very long, having branched median and lateral diverticula and anterior extensions which reach to the level of the pharynx, each having four lateral outgrowths. *Excretory system:* vesicle tubular, extending to the level of the ovary. *Reproductive systems:* genital pore in front of the ventral sucker. Cirrus pouch club-shaped, containing a large seminal vesicle and an unarmed cirrus 3–4 mm. long. Testes branched, one behind the other in the posterior half of the body. Ovary lobed, just in front of the testes and on the right. Vitellaria extending from the pharynx to the posterior extremity, the follicles arranged in dorsal, ventral and lateral cuboidal masses. Uterus convoluted, formed into rosette-like folds dorsal to the ventral sucker. Eggs triangular in transverse section and very large (0·120 × 0·075 mm.).

Genus *Zalophotrema* Stunkard & Alvey, 1929

This genus of parasites of Pinnipedia is typified by *Zalophotrema hepaticum* Stunkard & Alvey, 1929, which occurs in the bile ducts of *Zalophus californianus* in North America.

Genus *Orthosplanchnus* Odhner, 1905

Orthosplanchnus arcticus Odhner, 1905 (Fig. 56G)

HOSTS. Bearded seal, ringed seal.

LOCATION. Gall bladder and liver.

LOCALITY. Europe, Greenland, Spitzbergen, North America (Canada).

DIAGNOSIS. *Shape and size:* elongate-oval outline, both ends tapering, the posterior extremity more attenuated, slightly flattened, 3·5–7 mm. long (more often 4·5–6 mm.) and 0·85–1·15 mm. broad. *Cuticle:* spinous, spines slightly smaller at the extremities of the body. *Suckers:* oral 0·48–0·6 mm. diameter, ventral 0·43–0·53 mm. diameter and about one-quarter of the distance along the body. *Gut:* prepharynx fairly long, pharynx present, oesophagus short, caeca long, simple, each caecum having a forward extension reaching to the level of the prepharynx. *Excretory system:* vesicle tubular, pore terminal. *Reproductive systems:* genital pore in front of the ventral sucker and median. Cirrus pouch club shaped, extending back well beyond the limits of the ventral sucker, containing a spherical seminal vesicle, a cylindrical pars prostatica and a spinous cirrus. Testes of elongate oval outline with indented margins, one in front of the other behind the mid-body. Ovary globular, in front of the testes and on the right. Vitellaria above and beneath the caeca from the level of the pharynx to the posterior extremity, approaching the median plane dorsally in front of the ventral sucker and ventrally behind it, merging in the median plane behind the testes. Uterus having a few folds between the gonads and the ventral sucker. Metraterm having a spinous cuticular lining. Eggs thickened at the anopercular pole, large (0·091–0·100 × 0·054–0·058 mm.).

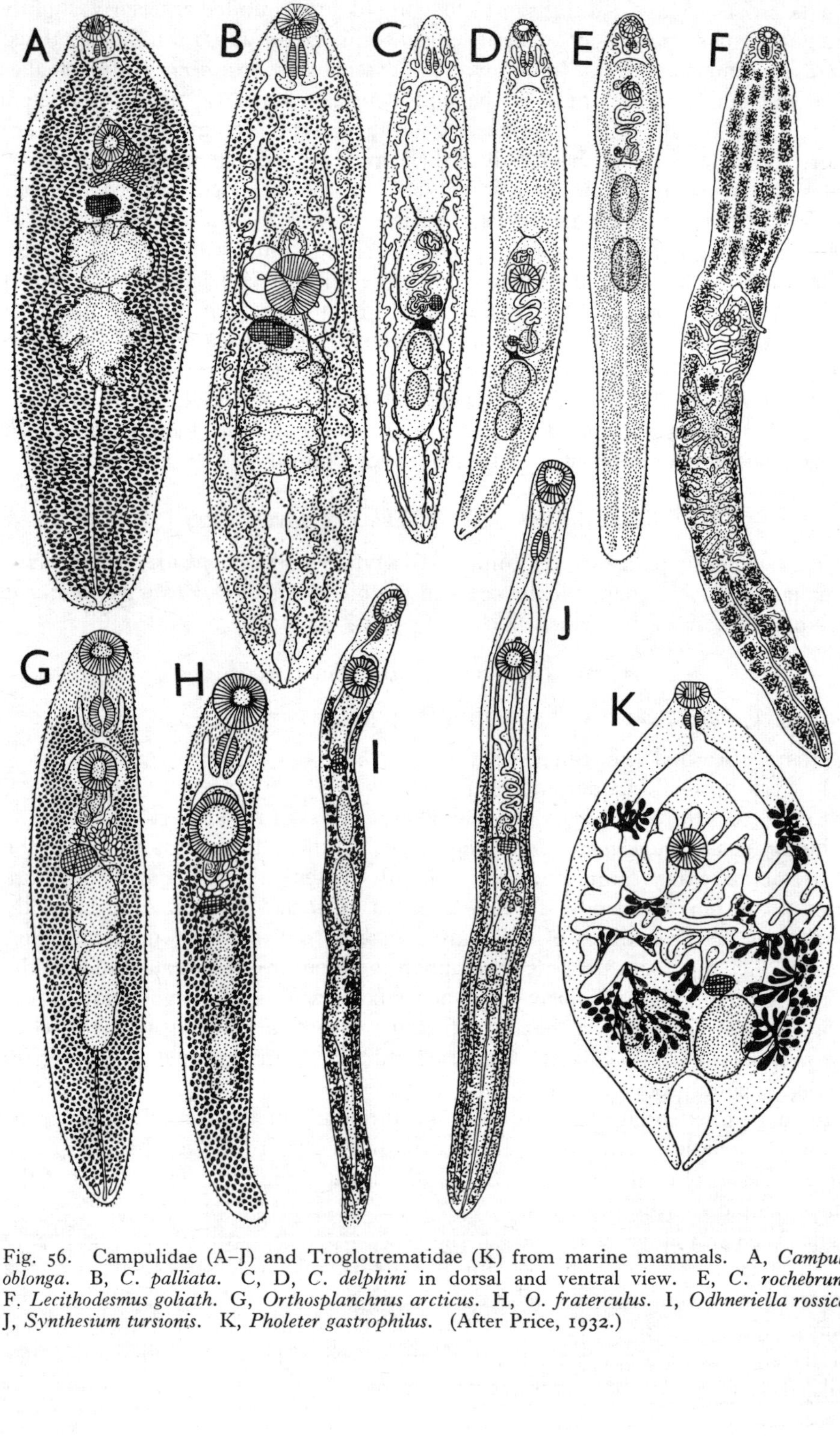

Fig. 56. Campulidae (A–J) and Troglotrematidae (K) from marine mammals. A, *Campula oblonga*. B, *C. palliata*. C, D, *C. delphini* in dorsal and ventral view. E, *C. rochebruni*. F. *Lecithodesmus goliath*. G, *Orthosplanchnus arcticus*. H, *O. fraterculus*. I, *Odhneriella rossica*. J, *Synthesium tursionis*. K, *Pholeter gastrophilus*. (After Price, 1932.)

Orthosplanchnus fraterculus Odhner, 1905 (Fig. 56H)

HOSTS. Walrus, bearded seal.
LOCATION. Gall bladder.
LOCALITY. Europe (Spitzbergen).
DIAGNOSIS. *Shape and size:* almost cylindrical, 3–4 mm. long and 0·5–0·6 mm. broad. *Cuticle:* spinous to a greater extent than in *O. arcticus*. *Suckers:* ventral slightly larger than the oral, diameters 0·4–0·5 and 0·37–0·44 mm. *Other characters:* as in *O. arcticus*, but testes deeply indented (not evident in Price's copy of Odhner's figure), vitellaria not approaching the median plane dorsally, except behind the ventral sucker in the region of the gonads. These differences are so slight as to suggest that the two forms are identical, '*O. fraterculus*' probably representing small specimens of *O. arcticus* in which the ratio of the sizes of the suckers has not attained a definitive value.

Genus *Synthesium* Stunkard & Alvey, 1930

Synthesium tursionis (Marchi, 1873) Stunkard & Alvey, 1930 (Fig. 56J)

Syn. *Distomum tursionis* Marchi, 1873; *D. longissimum* Poirier, 1886, *nec* Linstow, 1896; *D.* (*Dicrocoelium*) *tursionis* of Parona, 1896; *Fasciolopsis* (?) *tursionis* of Nicoll, 1923.

HOST. Bottle-nosed dolphin.
LOCATION. Intestine.
LOCALITY. Europe.
DIAGNOSIS. *Shape, size and colour:* greatly elongate and flattened, 20 mm. long and 1·5 mm. broad, whitish. *Cuticle:* spinous. *Suckers:* of equal size (0·8 mm. diameter), the ventral 3–4 mm. behind the oral. *Gut:* prepharynx well developed, pharynx long and narrow (0·72 × 0·3 mm.), caeca long and simple. *Reproductive systems:* genital pore near the anterior margin of the ventral sucker. Cirrus pouch long and tubular, containing a seminal vesicle, a long pars prostatica and a spinous cirrus. Testes lobed, one in front of the other in the posterior third of the body. Ovary small and globular, situated near the middle of the body. Vitellaria rather delicate, but well developed, extending laterally from slightly behind the ovary to the posterior extremity. Uterus having slight sinuosities between the ovary and the ventral sucker. Metraterm spinous. Eggs measuring 0·056 × 0·033 mm., the anopercular pole thickened and pointed.

Genus *Odhneriella* Skrjabin, 1915

Odhneriella rossica Skrjabin, 1915

HOST. Walrus.
LOCATION. Bile ducts.
LOCALITY. Europe (Russia).
DIAGNOSIS. *Shape and size:* very elongate (ribbon-like), the lateral margins of the body almost parallel, flattened, 9 mm. long and 0·76 mm. broad. *Cuticle:* spinous in front of the ventral sucker. *Suckers:* oral directed ventrally, 0·5 mm.

long and 0·48–0·53 mm. broad; ventral raised from the surface of body and close behind the oral, 0·5 mm. long and 0·68 mm. broad. *Gut:* prepharynx, pharynx, oesophagus present, caeca long, simple and without anterior extensions. *Reproductive systems:* genital pore near the anterior margin of the ventral sucker. Cirrus pouch sac-like, its base behind the ventral sucker. Cirrus spinous. Testes ovoid, one behind the other in the second quarter of the body. Ovary spherical and in front of the testes. Vitellaria extending in the lateral regions from a level between the ovary and the ventral sucker to the posterior extremity. Uterus short and having few folds. Metraterm not spinous. Eggs measuring 0·100 × 0·060 mm., triangular in transverse section, the shell thickened at the anopercular pole.

Genus *Hadwenius* Price, 1932

This genus was erected to accommodate *Hadwenius seymouri* Price, 1932, an intestinal parasite of the white whale or beluga in Alaska, and Price considered this species to be more closely related to *Synthesium tursionis* than to any other trematode.

Family **ALLOCREADIIDAE** Stossich, 1904 (p. 86)

Genus *Crepidostomum* Braun, 1900

Syn. *Acrodactyla* Stafford, 1904; *Stephanophiala* Nicoll, 1909; *Acrolichanus* Ward, 1917; *Crepidastomum* Pratt, 1902 (a misprint).

This genus typically occurs in fresh-water fishes, but one species occurs (somewhat rarely) in bats, namely:

Crepidostomum metoecus (Braun, 1900)

Syn. *Distomum metoecus* Braun, 1900.

HOST. Noctule, *Vespertilis lasiopterus*
LOCATION. Intestine.
LOCALITY. Europe.

C. suecicum (Nybelin, 1932), a parasite of fresh-water fishes in Sweden, may be identical with this species. According to Hopkins (1934) bats may be only 'accidental' hosts.

Family **OPISTHORCHIIDAE** Lühe, 1901, emend. Braun, 1901 (p. 96)

Syn. Opisthorchidae Lühe, 1901.

Subfamily **Opisthorchiinae** Looss, 1899, emend.

Genus *Opisthorchis* Blanchard, 1895

Syn. *Prosthometra* Looss, 1896, in part; *Notaulus* Skrjabin, 1913; *Clonorchis* Looss, 1907; *Gomtia* Thapar, 1930, etc.

Opisthorchis felineus (Rivolta, 1884) Blanchard, 1895 (Fig. 57B)

Syn. *Distoma conus* Gurlt, 1831, *nec* Creplin, 1825; *D. lanceolatum felis cati* Siebold, 1836; *D. felineum* Rivolta, 1884; *D. lanceolatum canis familiaris* Tright, 1889; *D. sibiricum* Winogradoff, 1892; *Dicrocoelium felineus* of Moniez, 1896; *Distoma winogradoffi* Jaksch, 1897; *Campula felinea* of Cholodkowsky, 1898; *Opisthorchis wardi* Wharton, 1921; *O. tenuicollis felineus* of Erhardt, 1935.

HOSTS. Dog, cat, fox, pig, man.

LOCATION. Bile ducts, sometimes the pancreatic ducts.

LOCALITY. Europe, Asia.

DIAGNOSIS. *Shape, size and colour:* lanceolate and flattened, pointed anteriorly and rounded posteriorly, 7–12 mm. long and 1·5–2·5 mm. broad, golden red. *Cuticle:* smooth. *Suckers:* close together (1·5–2 mm. apart), of about equal size (0·17–0·28 mm. diameter). *Gut:* pharynx present, oesophagus short, caeca long; bifurcation of the intestine far in front of the ventral sucker. *Excretory system:* pore terminal, vesicle S-shaped, meeting the lateral canals slightly in front of the testes. *Reproductive systems:* genital pore near the anterior margin of the ventral sucker. Testes diagonally one behind the other in the posterior quarter, the anterior having four lobes, the posterior five. Ovary transversely ovoid and situated in front of the testes. Receptaculum seminis large and disposed between the ovary and the anterior testis. Uterus having an ascending limb only and formed into numerous transverse folds between the ovary and the ventral sucker. Vitellaria lateral and confined to the middle third of the body, extending from a short distance behind the ventral sucker to the level of the ovary, the follicles being arranged in seven to eight groups on either side of the body. Eggs small (0·028–0·030 × 0·011–0·015 mm.).

Opisthorchis tenuicollis (Rudolphi, 1819) Stiles & Hassall, 1896

Syn. *Distoma tenuicollis* Rudolphi, 1819; *D. viverrini* Poirier, 1886; *Opisthorchis viverrini* Stiles & Hassall, 1896; *O. tenuicollis-felineus* Looss, 1899, etc.

HOSTS. Dog, cat, civet cat, man; also bearded seal, grey seal, common porpoise, etc.

LOCATION. Bile ducts.

LOCALITY. Europe, Asia.

DIAGNOSIS. Very similar to *Opisthorchis felineus* and practically impossible to distinguish from it. Several writers have maintained that the two forms are synonymous (Muhling, 1896; Looss, 1905; Barker, 1911; Morgan, 1924; Price, 1940*a*). Ejsmont (1937) also concurred, pointing out that the marine mammals which serve as the definite hosts enter estuaries, there to feed on cyprinid fishes which serve as the intermediate hosts of *O. felineus*; he affirmed that even the distinction of different races or subspecies is fictitious. Some forms which bear close resemblance to *O. felineus* have been placed in different genera, e.g. *O. novocerca* (Braun, 1902) in *Amphimerus* Barker, 1911 and *O. caninus* Barker, 1911 in *Paropisthorchis* Stephens, 1912. The latter genus has been regarded as synonymous with the former (Sprehn, 1932), and it is possible that both are synonymous with *Opisthorchis*. It is now generally admitted that *Clonorchis* should be included in the synonyms, the species *Clonorchis sinensis* (Cobbold, 1875) Looss, 1907 having been retained by some writers (e.g. Price, 1940*a*) on account of long-standing usage of the name.

Subfamily **Metorchiinae** Lühe, 1909

KEY TO SOME GENERA IN MAMMALS

A. Posterior extremity truncated, simulating a sucker	*Pseudamphistomum*
B. Posterior extremity rounded, not simulating a sucker	*Metorchis*

Note. Another genus, *Parametorchis* Skrjabin, 1913, contains several species which occur in the gall bladder of various fur-bearing mammals (cat, silver fox, mink) in North America (see Price, 1929*b*).

Genus *Metorchis* Looss, 1899

Metorchis albidus (Braun, 1893) Looss, 1899 (Fig. 54E)

Syn. *Distoma albidum* Braun, 1893; *Opisthorchis albidus* of Railliet, 1896.

Hosts. Cat, dog, fox; also grey seal.

Location. Gall bladder and bile ducts.

Locality. Europe, North America.

Diagnosis. *Shape and size:* spatulate, pointed anteriorly, rounded and flat posteriorly, 1·6–4·6 mm. long and 0·8–2 mm. broad. *Cuticle:* spinous, spines small. *Suckers:* of equal size, oral 0·2–0·32 mm. diameter, ventral 0·2–0·3 mm., the latter about one-third of the distance along the body. *Gut:* pharynx very small, oesophagus very short, caeca long. *Reproductive systems:* genital pore in front of the ventral sucker. Testes lobed, situated diagonally one behind the other in the posterior region. Ovary rounded, median and a little in front of the anterior testis. Receptaculum seminis large and sac-like, situated between the ovary and the testes. Vitellaria lateral, extending from the level of the genital pore to that of the ovary. Uterus having an ascending limb only, formed into wide folds that sometimes overstep the caeca laterally and extend in front of the ventral sucker. Eggs small (0·024–0·034 × 0·013–0·016 mm.). Freeman & Ackert (1937) first recorded this species in North America, in the bile duct of an Eskimo husky dog imported into U.S.A. from Alaska.

Other species of the genus *Metorchis* infect birds, *M. xanthosomus* and *M. crassiusculus* having been mentioned in Chapter 10.

Genus *Pseudamphistomum* Lühe, 1909

Pseudamphistomum truncatum (Rudolphi, 1819) Lühe, 1909 (Fig. 57C)

Syn. *Amphistoma truncatum* Rudolphi, 1819; *Distoma conus* Creplin, 1825, *nec* Gurlt, 1831; *D. lanceolatum* Mehlis of Diesing, 1858; *D. campanulatum* Ercolani, 1875; *D. truncatum* of Railliet, 1886; *Opisthorchis truncatus* of Railliet, 1896; *Metorchis truncatus* of Looss, 1899.

Hosts. Dog, cat, fox, silver fox; also common seal, greenland seal, ringed seal, grey seal.

Location. Bile ducts.

Locality. Europe (Germany, Hungary, Italy, France, Holland, Russia), east coast of Greenland.

DIAGNOSIS. *Shape and size:* conical, pointed anteriorly and truncated posteriorly, 2–2·25 mm. long and 0·6–0·8 mm. broad. *Cuticle:* thickly and regularly spinous. *Suckers:* of about equal size (0·13–0·14 mm. diameter), the ventral very slightly in front of the mid-body. *Excretory system:* pore situated in a depression at the posterior extremity. *Gut:* pharynx of medium size (about 0·09 mm. long), oesophagus very short, caeca long and slightly sinuous. *Reproductive systems:* genital pore slightly in front of ventral sucker and far behind the bifurcation of the intestine. Seminal vesicle convoluted and lying free in the

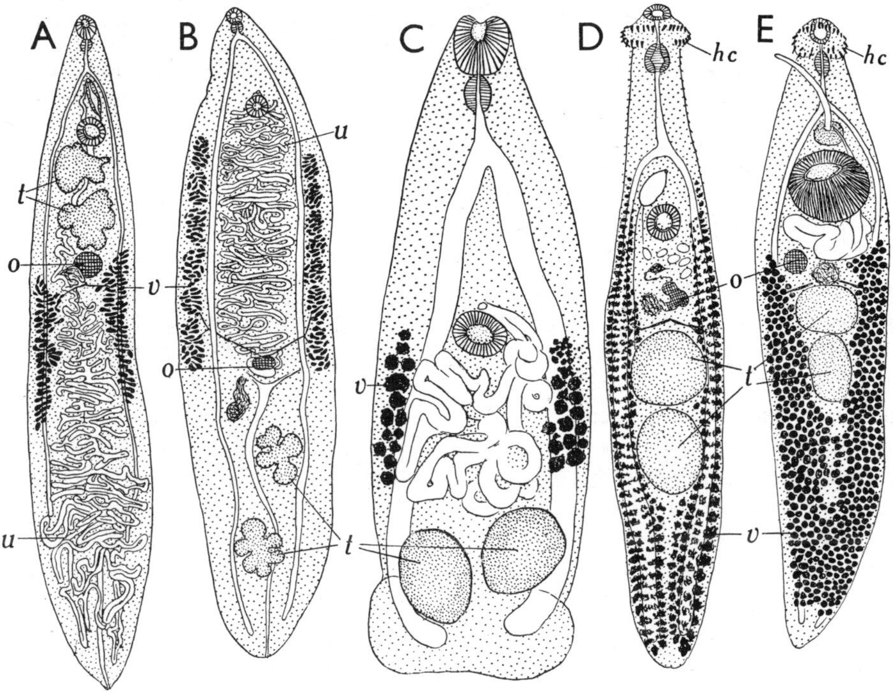

Fig. 57. Other Digenea of mammals. A, *Dicrocoelium dendriticum* (Dicrocoeliidae). B, *Opisthorchis felineus* (Opisthorchiidae). C, *Pseudamphistomum truncatum* (as B). D, *Echinochasmus perfoliatus* and E, *Euparyphium melis* (Echinostomatidae). (A, B, original; C, E, after Mönnig, 1934; D, after Baylis, 1929.)

parenchyma. Testes rounded, almost side by side near the posterior extremity. Ovary rounded, midway between the testes and the ventral sucker. Uterus with a short descending and a long ascending limb, formed into numerous folds which occupy most of the space between the testes and the genital pore, sometimes overreaching the ventral sucker anteriorly and overstepping the caeca laterally. Vitellaria lateral, confined to the middle third of the body, extending from the level of the genital pore to that of the ovary. Eggs small (0·029 × 0·011 mm.).

Pseudamphistomum danubiense Ciurea, 1913 occurs in the domestic cat in Roumania.

Family **HETEROPHYIDAE** Odhner, 1914 (p. 97)

Subfamily **Heterophyinae** Ciurea, 1924

Genus *Heterophyes* Cobbold, 1866

This genus contains a number of species which occur in Egypt and Asia. The best known of them is:

Heterophyes heterophyes (Siebold, 1852) Stiles & Hassall, 1900 (Fig. 54F)

Syn. *Distoma heterophyes* Siebold, 1852; *Dicrocoelium heterophyes* of Weinland, 1858; *Fasciola heterophyes* of Moquin-Tandon, 1860; *Heterophyes aegyptiaca* Cobbold, 1866; *Mesogonimus heterophyes* of Railliet, 1890; *Coenogonimus heterophyes* of Looss, 1899; *Cotylogonimus heterophyes* of Lühe, 1899; *Heterophyes nocens* Onji & Nishio, 1915; *H. pallidus* Looss, 1902; *H. fraternus* Looss, 1902; *H. heterophyes sentus* Looss, 1902.

HOSTS. Man, dog, cat, fox, rat (experimentally); also birds, e.g. black kite.

LOCATION. Small intestine and caecum.

LOCALITY. Egypt, East Asia.

DIAGNOSIS. *Size:* small, about 2 mm. long and 0·4 mm. broad. *Cuticle:* covered with rectangular scales, especially anteriorly. *Suckers:* oral much smaller than the ventral, diameters 0·08–0·09 and 0·2–0·23 mm. *Gonotyle:* 0·1–0·15 mm. diameter, having sixty to eighty radially arranged cuticular rodlets, slightly lateral behind the ventral sucker, the genital pore at its centre. *Glands:* numerous in the anterior region, opening on the anterior border of the oral sucker. *Gut:* prepharynx small, pharynx present, oesophagus fairly long, caeca long, bifurcation of the intestine far in front of the ventral sucker. *Reproductive systems:* testes ovoid and side by side in the posterior region, near the ends of the caeca. Ovary situated in front of the testes and median. Receptaculum seminis between the ovary and the testes. Vitellaria lateral, opposite the gonads, comprising about fourteen groups of follicles on each side of the body. Uterus much folded, the folds occupying all available space in the posterior half of the body. Eggs thin-shelled, brown and small (0·026–0·030 × 0·015–0·017 mm.).

Subfamily **Metagoniminae** Ciurea, 1924

Genus *Metagonimus* Katsurada, 1912

Syn. *Loxotrema* Kobayashi, 1912, *nec* Gabb, 1868; *Yokogawa* Leiper, 1913; *Loossia* Ciurea, 1915; *Dexiogonimus* Witenberg, 1929.

Metagonimus yokogawai (Katsurada, 1912)

Syn. *Heterophyes yokogawai* Katsurada, 1912; *Loxotrema ovatum* Kobayashi, 1912; *Tocotrema yokogawai* Katsurada, 1912; *Metagonimus ovatus* Yokogawa, 1913; *Yokogawa yokogawai* Leiper, 1913; *Paragonimus yokogawai* Montel, 1914; *Loossia romanica* Ciurea, 1915; *L. parva* Ciurea, 1915; *L. dobrogiensis* Ciurea, 1915; *Metagonimus romanica* (Ciurea, 1915).

HOSTS. Cat, dog, pig, man; also various birds, including the pelican.

LOCATION. Small intestine.

LOCALITY. Far East (Japan, China, Korea, Formosa, Dutch East Indies), Europe (Roumania).

DIAGNOSIS. *Shape and size:* pyriform, bluntly pointed anteriorly, 1–2·5 mm. long and 0·42–0·75 mm. broad. *Cuticle:* scaly, the scales more abundant anteriorly. *Suckers:* oral smaller than the ventral, dimensions 0·048–0·11 mm. diameter and 0·065–0·165 × 0·055–0·115 mm., the ventral lateral and enclosed in the genital sinus. *Gonotyl:* inconspicuous. *Gut:* prepharynx short, pharynx and oesophagus fairly long, caeca long. *Reproductive systems:* testes diagonally one behind the other or side by side in the posterior region. Receptaculum seminis in front of the testes and the ovary in front of these. Vitellaria lateral in the posterior region, beside the gonads (not extending in front of the ovary), comprising about ten large follicles on either side of body. Uterus with many folds between the genital sinus and testes, extending behind the ovary and the receptaculum seminis. Eggs small (0·027–0·030 × 0·015–0·017 mm.).

Another species of *Metagonimus* is *M. ciureanus* (Witenberg, 1929) (Syn. *Dexiogonimus ciureanus* Witenberg).

The genus *Metagonimoides* Price, 1931 differs from *Metagonimus* in the anterior extent of the vitellaria to the pharynx and the absence of uterine folds behind the ovary and receptaculum seminis. The type species, *M. oregonensis*, is parasitic in the small intestine of the raccoon in U.S.A.

Subfamily **Cryptocotylinae** Lühe, 1909

Genus *Cryptocotyle* Lühe, 1899

The type species of this genus, *C. concava*, and the species, *C. lingua*, have already been described as parasites of numerous birds. The latter occurs also in the dog and fox, as well as in the common seal. Other species mainly parasitize birds.

Subfamily **Apophallinae** Ciurea, 1924

Genus *Euryhelmis* Poche, 1926

Syn. *Eurysoma* Dujardin, 1845, *nec* Gistl, 1829, *nec* Koch, 1839.

Euryhelmis squamula (Rudolphi, 1819) Poche, 1926 (Fig. 54G)

Syn. *Distoma squamula* Rudolphi, 1819; *Monostoma squamula* of Diesing, 1850; *Eurysoma squamula* of Dujardin, 1845.

HOSTS. Polecat, fox, weasel, mink. Larvae encysted in the skin of frogs.

LOCATION. Intestine.

LOCALITY. Europe.

DIAGNOSIS. *Shape and size:* quadrilateral, 0·6 mm. long and 1·45 mm. broad. *Suckers:* very small, the oral 0·07 mm. diameter, the ventral even smaller and situated approximately at the mid-body. *Gut:* prepharynx very short, pharynx much smaller than the oral sucker, oesophagus fairly long and caeca long and extending more or less parallel to the anterior and lateral margins of the body, terminating near the median plane posteriorly. *Reproductive systems:* testes lobed and side by side, but widely separated in the posterior region, the main excretory canals extending between them. Ovary in front of the right testis. Cirrus pouch long and slender (somewhat club-shaped), half encircling the ventral sucker on the right side. Vitellaria very well developed laterally filling the margins of the body lateral to the caeca, reaching towards the oesophagus anteriorly, almost completely encircling the reproductive organs. Uterus forming a number of wide transverse folds between the testes and the ventral sucker. Eggs brown and small (0·028–0·032 × 0·012–0·014 mm.).

Genus *Apophallus* Lühe, 1909

Syn. *Rossicotrema* Skrjabin & Lindtrop, 1919; *Cotylophallus* Ransom, 1920.

The type-species of this genus, *Apophallus mühlingi*, and *A. major* are parasites of sea birds and have been mentioned. The following species and *A. zalophi* Price, 1932 (a parasite of *Zalophus californianus* in North America) infect mammals:

Apophallus donicus (Skrjabin & Lindtrop, 1919) Price, 1931

Syn. *Rossicotrema donicum* Skrjabin & Lindtrop, 1919; *R. simile* (Random, 1920) Ciurea, 1924; *Cotylophallus venustus* Ransom, 1920; *C. similis* Ransom, 1920.

Hosts. Dog, cat, fox and common seal.

Location. Small intestine.

Locality. Europe and North America.

Diagnosis. *Shape and size:* oval outline, but bluntly pointed anteriorly and rounded posteriorly, 0·5–1·15 mm. long and 0·2–0·4 mm. broad. *Cuticle:* spinous, the spines scale-like. *Suckers:* ventral smaller than the oral and in front of the mid-body, dimensions 0·04–0·06 × 0·05–0·06 and 0·65–0·86 mm. diameter. *Gut:* prepharynx very short, pharynx small, oesophagus long and slender, caeca long. Bifurcation of the intestine midway between the suckers. *Reproductive systems:* testes large and rounded, obliquely one beside the other in the posterior region. Ovary between the testes and the ventral sucker and lateral. Vitellaria extending from the posterior end of the oesophagus to the posterior extremity. Uterus having a few folds between the testes and genital pore, which is anterior to the ventral sucker. Eggs not very small (0·049–0·050 × 0·018–0·025 mm.).

Subfamily **Galactosomatinae** Ciurea, 1933 n.nom.

Syn. Cercarioidinae Witenberg, 1929.

Genus *Galactosomum* Looss, 1899

Syn. *Microlistrum* Braun, 1901; *Astia* Looss, 1899, in part; *Astiotrema* Looss, 1900, in part; *Cercarioides* Witenberg, 1929; *Tubanguia* Srivastava, 1935.

Two species of this genus have been mentioned already (p. 340) as parasites of birds. Another species occurs in the intestine of the common dolphin in Europe. It is:

Galactosomum erinaceus (Poirier, 1886) Bittner & Sprehn, 1928

Syn. *Distoma erinaceum* Poirier, 1886; *Astiotrema erinacea* of Stossich, 1904.

Diagnosis. *Shape and size:* elongate, broader anteriorly than posteriorly, 3 mm. long and 0·8 mm. broad. *Cuticle:* spinous, spines arranged in regular transverse rows, sparser in the posterior region. *Suckers:* oral 0·3 mm. diameter, ventral very much smaller and situated near the mid-body. *Gut:* prepharynx longer than the oesophagus, pharynx small (0·017 mm. long), caeca long and simple. *Reproductive systems:* genital pore median and near the ventral sucker, seminal vesicle long, slender and muscular. Testes globular, diagonally arranged in the posterior third of the body, ovary smaller and globular, situated between the anterior testis and the ventral sucker. Receptaculum seminis large. Uterus having long descending and ascending limbs. Eggs?

All other members of the genus occur in birds, and this species is possibly an accidental parasite of the dolphin, being known as a juvenile form enclosed in a cyst about 1 mm. diameter and occurring free in the intestine. Jägerskiöld (1908) showed that it is very similar to the encysted form of *G. lacteum*, which occurs in the brain of the father lasher, and had little doubt about the real host, which he regarded as being a fish-eating sea bird.

Subfamily **Centrocestinae** Looss, 1899

Genus *Phagicola* Faust, 1920

Syn. *Ascocotyle* Looss, 1899, in part; *Parascocotyle* Stunkard & Haviland, 1924; *Metascocotyle* Ciurea, 1933.

In addition to *Phagicola minuta* (Looss, 1899), which has been mentioned as a parasite of the heron and of mammals, this genus contains a number of species which occur only in mammals, notably:

Phagicola longa (Ransom, 1920) Price, 1935

Syn. *Ascocotyle longa* Ransom, 1920; *Parascocotyle longa* of Witenberg, 1929.

HOSTS. Dog, cat, fox and black kite.

LOCATION. Intestine.

LOCALITY. North America and Palestine.

DIAGNOSIS. Can be distinguished from *P. minuta* by the length of the caeca, which extend far beyond the ventral sucker and almost to the posterior extremity, and by the short buccal sac, which is much shorter than the pharynx.

Phagicola italica (Alessandrini, 1906)

Syn. *Ascocotyle italica* Alessandrini, 1906; *Echinostoma piriforme* Blanc & Hedin; *Parascocotyle italica* (Alessandrini, 1906) Witenberg, 1929.

HOST. Dog.

LOCATION. Intestine.

LOCALITY. Italy, Palestine.

DIAGNOSIS. Can be distinguished from *P. minuta* by the same characters which separate *P. longa*, and from the latter by the lack of a beak-like dorsal lip to the oral sucker.

Genus *Pygidiopsis* Looss, 1907

This genus can be distinguished from *Ascocotyle* by the limited extent of the vitellaria, which reach the level of the ovary, but never that of the ventral sucker.

Pygidiopsis genata Looss, 1907

Syn. *Ascocotyle plana* Linton, 1928.

HOSTS. Dog, wolf and various birds, e.g. black kite. Also cat and rabbit (experimentally).

LOCATION. Intestine.

LOCALITY. Europe, Asia, Africa, North America.

DIAGNOSIS. *Size:* very small, recurved anteriorly, 0·4–0·7 mm. long and 0·2–0·4 mm. broad. *Cuticle:* spinous. *Suckers:* ventral near the mid-body, larger than the oral, diameters 0·4–0·6 and 0·3–0·5 mm. *Gut:* prepharynx long, pharynx present, oesophagus short, caeca fairly long, extending almost to the posterior extremity. *Reproductive systems:* testes side by side near the posterior extremity. Receptaculum seminis large, median in position between and in front of the testes. Ovary ventral to the receptaculum seminis. Vitellaria lateral to the testes, barely extending forward as far as the level of the ovary. Uterus forming wide folds between the gonads and the ventral sucker. Eggs small (0·018–0·022 × 0·009–0·012 mm.).

Subfamily **Adleriellinae** Witenberg, 1930

Syn. Adleriinae Witenberg, 1929.

Genus *Adleriella* Witenberg, 1930

Syn. *Adleria* Witenberg, 1929, *nec* Rohwer & Fagan, 1917.

Adleriella minutissima (Witenberg, 1929) Witenberg, 1930

Syn. *Adleria minutissima* Witenberg, 1929.

HOSTS. Dog, cat.

LOCATION. First part of small intestine.

LOCALITY. Palestine.

DIAGNOSIS. *Shape and size:* pyriform, 0·27–0·47 mm. long and 0·09–0·15 mm. broad. *Cuticle:* spinous. *Suckers:* oral 0·025–0·035 mm. in diameter, ventral absent (but gonotyl prominent in the anterior region of the body). *Gut:* prepharynx long, pharynx relatively large (0·025–0·03 mm. long), oesophagus short, caeca very short, extending only to the mid-body. *Reproductive systems:* testis single and ovoid, slightly behind the gonotyl, ovary and receptaculum seminis farther back. Vitellaria dorsal in the posterior region. Uterus having descending and ascending limbs, but little folded. Eggs delicate and small (0·024 × 0·012–0·014 mm.).

Family **ECHINOSTOMATIDAE** Looss, 1902, emend. Poche, 1926 (or Stiles & Hassall, 1926)

Subfamily **Echinostomatinae** Looss, 1899, emend. Stiles & Hassall, 1926

Genus *Echinostoma* Rudolphi, 1809

This genus is represented in mammals by one or two species, but is almost exclusively confined to birds. *Echinostoma spiculator* (Dujardin, 1845) Cobbold,

1860 has been recorded in the small intestine of the brown rat in Europe and *Echinostoma acanthoides* (Rudolphi, 1819) Cobbold, 1860 in the intestine of the common seal. According to Price (1932) it is not certain that the latter species really belongs to the genus *Echinostoma*, because descriptions of it are too incomplete to warrant definitive generic assignment.

Genus *Euparyphium* Dietz, 1909

Syn. *Isthmiophora* Lühe, 1909.

Euparyphium melis (Schrank, 1788) Dietz, 1909 (Fig. 57E)

Syn. *Planaria teres duplici poro* Goeze, 1782, in part; *Fasciola putorii* Gmelin, 1790; *F. melis* Schrank, 1788; *F. armata* Rudolphi, 1793; *F. trigonocephala* Rudolphi, 1802; *Distoma armatum* of Zeder, 1803; *D. trigonocephalum* of Rudolphi, 1809; *Echinostoma mehlis* Dietz, 1909; *E. trigonocephalum* of Cobbold, 1860; *Isthmiophora melis* Lühe, 1909.

HOSTS. Polecat, pine-marten, badger, otter, hedgehog, fox, cat.

LOCATION. Intestine.

LOCALITY. Europe.

DIAGNOSIS. *Shape and size:* lancet-like, 3·5–11·2 mm. long and 1·3–1·6 mm. broad. *Collar:* about 0·5 mm. broad and equipped with twenty-seven collar spines, of which four larger ones on each side form corner spines, the remaining nineteen being arranged in a double row, those of the oral row larger than those in the aboral. *Cuticle:* spinous anteriorly. *Suckers:* oral very much smaller than the ventral, which is situated 1·1–1·25 mm. behind it, diameters, 0·2–0·3 and 0·7–0·75 mm. *Gut:* prepharynx short, pharynx large, oesophagus and caeca long, bifurcation of the intestine not far in front of the ventral sucker. *Reproductive systems:* cirrus pouch not muscular, 0·5–0·75 mm. long and 0·35–0·5 mm. broad, extending beyond the anterior border of the ventral sucker. Cirrus spinous. Testes ovoid and one behind the other, the anterior near the mid-body. Ovary globular, in front of testes and on the right. Ootype and shell gland opposite the ovary. Eggs very large (0·120–0·125 × 0·091–0·094 mm.).

The development of this species was studied recently by Beaver (1941).

Euparyphium jassyense Léon & Ciurea, 1922

HOST. Man.

LOCATION. Intestine.

LOCALITY. Europe (Roumania).

DIAGNOSIS. *Size:* 5·4–7·6 mm. long and 1·05–1·35 mm. broad. *Collar:* as in *E. melis*. Eggs exceptionally large, e.g. more than 0·13 mm. long (0·132–0·154 × 0·079–0·085 mm.). According to Hsü (1940) this form is a synonym of *E. melis*.

Euparyphium suinum Ciurea, 1921

HOST. Pig.

LOCATION. Intestine.

LOCALITY. Europe (Roumania).

DIAGNOSIS. Similar to the foregoing species, but as a rule less than 3·5 mm. long (dimensions: 2·7–3·6 × 0·8–1·2 mm.). Eggs very large, but smaller than those of *E. melis* and *E. jassyense* (0·117–0·127 × 0·078–0·088 mm.).

Another species is *E. ilocanum* (Garrison, 1908), a human intestinal fluke, the life history of which was studied by Tubangui & Pasco (1933). *E. murinum* Tubangui, 1931 is yet another species (see Tubangui (1932) for an account of the life cycle).

Subfamily **Echinochasminae** Odhner, 1910

Genus *Echinochasmus* Dietz, 1909

Echinochasmus perfoliatus (Rátz, 1908) Dietz, 1909 (Fig. 57D)

Syn. *Echinostoma perfoliatum* Rátz, 1908; *E. gregale* Railliet & Henry, 1909.

HOSTS. Man, dog, cat, fox, pig (also night heron).

LOCATION. Intestine.

LOCALITY. Europe (Hungary, Roumania, Russia, Italy), Japan.

DIAGNOSIS. *Size:* 2–4 mm. long and 0·4–0·8 mm. broad. *Collar:* 0·22–0·3 mm. broad and equipped with twenty-four collar spines, three small ones on either side forming corner spines, the remainder arranged in a single row, interrupted dorsally. *Suckers:* ventral nearly twice as large as the oral and about one-third of the distance along the body, diameters 0·17–0·22 and 0·085–0·136 mm. *Gut:* prepharynx short, pharynx of medium size, oesophagus fairly long, caeca long. *Reproductive systems:* testes large, transversely ovoid, one behind the other slightly behind the mid-body. Ovary small and in front of testes on the right. Vitellaria distributed along the courses of the caeca, confluent behind the testes. Eggs of very variable size (0·092–0·135 × 0·057–0·094 mm.).

Family **TROGLOTREMATIDAE** Odhner, 1914, emend. Braun, 1915 (p. 99)

Syn. Troglotremidae Odhner, 1914; Collyriclidae Ward, 1917.

KEY TO GENERA OCCURRING IN MAMMALS

(*Paragonimus* is included because of its importance in helminthology)

1. Genital pore on the anterior margin of the ventral sucker; uterus formed into large, transverse folds — *Pholeter*
2. Genital pore near the posterior margin of the ventral sucker; uterus formed into a rosette-like mass of folds
   - A. Gonads rounded or only slightly lobed; cirrus pouch present — *Troglotrema*
   - B. Gonads deeply lobed; cirrus pouch absent — *Paragonimus*

Genus *Pholeter* Odhner, 1914

Pholeter gastrophilus (Kossack, 1910) Odhner, 1914 (Fig. 56K)

Syn. *Distomum gastrophilum* Kossack, 1910.

HOST. Common porpoise.

LOCATION. Encysted in the mucosa of the stomach near the pylorus.

LOCALITY. Europe.

DIAGNOSIS. *Shape and size:* broad spindle-shaped: according to Odhner (1914*b*), 1·5–3·3 mm. long and 1·7–2·1 mm. broad; according to Kossack (1910), 3·15–3·66 mm. long and 1·8–2·25 mm. broad. About one-third as thick as broad. *Cuticle:* spinous, spines small and pointed, not arranged in groups. *Suckers:* fairly small, the ventral the larger and about one-third of the distance along the body, diameters 0·17–0·2 and 0·25–0·3 mm. *Gut:* pharynx 0·15–0·17 mm. diameter, oesophagus short and wide, caeca fairly long, extending to the level of the testes, bifurcation of the intestine approximately midway between the ventral sucker and the anterior extremity. *Reproductive systems:* genital pore at the anterior margin of the ventral sucker. Cirrus pouch absent. Pars prostatica short, dorso-ventrally orientated; seminal vesicle undivided and tubular. Testes longitudinally ovoid and side by side in the posterior region. Ovary deeply lobed, slightly in front of the testes and slightly lateral. Receptaculum seminis fairly large, dorsal to the ovary. Vitellaria dorso-lateral, extending from a level midway between the bifurcation of the intestine and the ventral sucker to the ends of the caeca. Uterus long and much folded, occupying almost the entire breadth of body between the ovary and the genital pore. Eggs small (0·023–0·025 × 0·014 mm.).

Genus *Troglotrema* Odhner, 1914

Troglotrema acutum (Leuckart, 1842) Odhner, 1914 (Fig. 54C)

Syn. *Distoma acutum* Leuckart, 1842.

HOSTS. Polecat, mink, fox. Larvae in frogs.

LOCATION. Frontal and ethmoidal sinuses.

LOCALITY. Europe (Germany).

DIAGNOSIS. *Shape and size:* broad pyriform or oval, generally rounded anteriorly and pointed posteriorly, about 3·3 mm. long and 2·25 mm. broad. *Habit:* sometimes occurring in pairs in cysts, though in the fox freely attached to the mucosa. *Suckers:* of about equal sizes, the oral slightly the larger, the ventral a little in front of the middle of the body. *Gut:* pharynx small, oesophagus very short, caeca long. *Reproductive systems:* testes ovoid, side by side but widely separated behind the ventral sucker. Cirrus pouch median, behind the ventral sucker and between the testes. Ovary ovoid, in front of the testes and above the ventral sucker, slightly lateral. Uterus having descending and ascending limbs and much folded between and behind the testes. Vitellaria dorso-lateral, abundantly developed, extending from the oral sucker to the posterior extremity. Eggs measuring about 0·08 × 0·95 mm.

Genus *Paragonimus* Braun, 1899

Syn. *Polysarcus* Looss, 1899 and misprints like *Paragominus* Daniels & Stanton, 1907.

Paragonimus westermanii (Kerbert, 1878) Stiles & Hassall, 1900 (Fig. 54D)

Syn. *Distoma westermani* Kerbert, 1878; *D. ringeri* Cobbold, 1880; *D. pulmonale* Baelz, 1883; *D. pulmonis* of Suga, 1883; *Mesogonimus westermani* of Railliet, 1890; *Polysarcus westermani* of Looss, 1899; *Mesogonimus ringeri* of Railliet, 1890; *Paragonimus ringeri* of Ward & Hirsch, 1915.

HOST. Man.

LOCALITY. Lungs, the flukes generally paired in cysts, rarely occurring in other organs.

LOCALITY. Asia (China and Japan).

DIAGNOSIS. *Shape and size:* oval outline, 8–16 mm. long and 4–8 mm. broad. *Cuticle:* spinous, the spines scale-like, variable in different individuals and in different situations in the same individual. *Colour:* reddish brown. *Suckers:* small and of about equal size, diameters 1·0–1·4 mm. ventral situated near the middle of the body. *Gut:* as in *Troglotrema*. *Excretory system:* flame-cell formula 2[(3 + 3 + 3 + 3 + 3 + 3 + 3) + (3 + 3 + 3 + 3 + 3 + 3 + 3 + 3 + 3)] (generic: see Byrd, 1941). *Reproductive systems:* genital pore just behind the ventral sucker, slightly on the right, cirrus pouch and true cirrus absent. Testes lobed, side by side in the posterior region. Ovary lobed and situated in front of the testes, on the left and slightly behind the ventral sucker. Vitellaria extending along the entire length of the body in the dorso-lateral regions. Uterus having folds bunched into a rosette-like mass on the right of the ventral sucker. Eggs golden brown, variable in size (0·075–0·100 × 0·042–0·067 mm.) [according to Craig & Faust (1940) 0·080–0·118 × 0·048–0·060 mm.]. *Development:* cercaria microcercous, developing in a redia. For a brief account of the life cycle see Chapter 13, p. 480.

Family **NOTOCOTYLIDAE** Lühe, 1909 (p. 100)

Subfamily **Ogmogasterinae** Kossack, 1911

Genus *Ogmogaster* Jägerskiöld, 1891

Ogmogaster plicata (Creplin, 1829) Jägerskiöld, 1891, emend.

Syn. *Monostomum plicatum* Creplin, 1829.

HOSTS. Piked whale (lesser rorqual), sei whale, fin whale (common rorqual), also some seals (*Leptonychotes weddellii* and *Lobodon carcinophaga*).

LOCATION. Intestine.

LOCALITY. Europe (Norway) and the Antarctic regions.

DIAGNOSIS. *Shape and size:* oval outline, flattened, 6–14 mm. long and about 4 mm. in mean breadth, the margins fluted, the ventral surface having fifteen to seventeen longitudinal ribs or rugae. *Suckers:* oral about 0·5 mm. diameter, ventral absent. *Gut:* pharynx absent, oesophagus short, caeca long and sinuous. *Excretory system:* pore postero-dorsal, and about 0·7 mm. from the posterior extremity, vesicle Y-shaped and ventral to the gonads. *Reproductive systems:* genital pores median, slightly behind the oral sucker, close together at the base of a short genital atrium. Cirrus pouch cylindrical, 3 mm. long and 0·3 mm. broad, containing an elongate seminal vesicle and the ejaculatory duct. Cirrus lined with a membrane bearing papillae. Testes deeply lobed and side by side in the posterior region. Ovary deeply lobed and between the testes. Vitellaria comprising twelve to sixteen isolated follicles, below the caeca on each side, extending from the anterior ends of the testes to the base of the cirrus pouch. Uterus much folded in the posterior half of the body, the folds overstepping the caeca laterally. Metraterm muscular and lined with spinelets. Eggs 0·025 mm. long, each having two polar filaments.

Subfamily **Notocotylinae** Kossack, 1911

Genus *Notocotylus* Diesing, 1839

This genus contains a number of characteristic trematodes of birds, but *N. noyeri* Joyeux, 1922 occurs in the caecum of the water vole. In North America *N. quinqueserialis* Barker & Laughlin, 1911 is very common in the caecum of the muskrat. This species was recently transferred by Harwood (1939) to the genus *Quinqueserialis* Skwortzow, 1935 (Syn. *Barkeria* V. Szidat, 1936). Two species of *Paramonostomum*, *P. echinum* Harrah, 1922 and *P. pseudalveatum* Price, 1931, also occur in the muskrat, although the best-known species is a parasite of geese and ducks (see p. 362). Skrjabin & Schulz (1933) described a new genus and species of Notocotylid, *Ogmocotyle pygardi*, from a roe deer, and erected for it the new subfamily Ogmocotylinae.

Family **CYATHOCOTYLIDAE** Poche, 1926 (p. 105)

Subfamily **Prohemistomatinae** Lutz, 1935, emend. Hughes, Higginbotham & Clary, 1942

Genus *Mesostephanus* Lutz, 1935

Syn. *Prohemistomum* Odhner, 1913, in part; *Paracoenogonimus* Katsurada, 1914.

Mesostephanus appendiculatum (Ciurea, 1916) Lutz, 1935 (Fig. 58A)

Syn. *Prohemistomum appendiculatum* Ciurea, 1916.

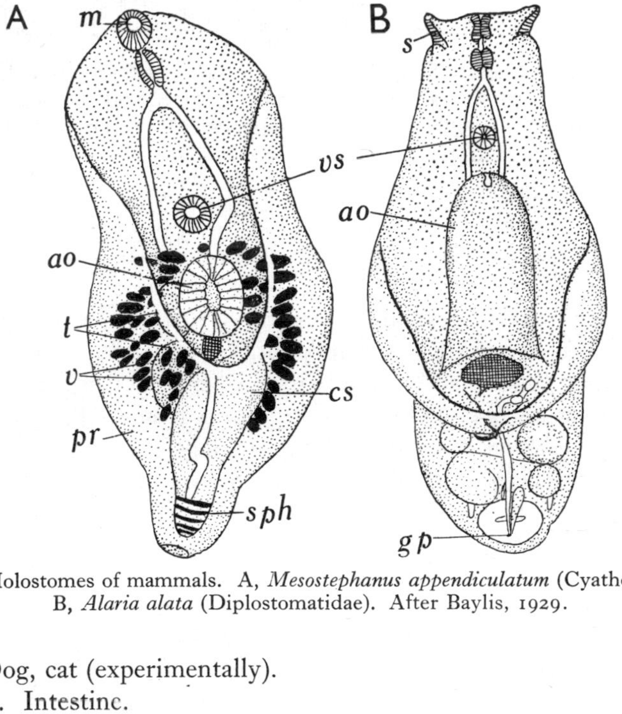

Fig. 58. Holostomes of mammals. A, *Mesostephanus appendiculatum* (Cyathocotylidae). B, *Alaria alata* (Diplostomatidae). After Baylis, 1929.

Hosts. Dog, cat (experimentally).
Location. Intestine.
Locality. Europe (Roumania).

DIAGNOSIS. *Shape and size:* slightly elongate, the anterior and posterior regions not sharply differentiated, but the former tongue-shaped, flattened and having ventrally inturned margins, the latter conical; total length 0·9–1·75 mm., breadth 0·4–0·6 mm. *Cuticle:* scaly. *Suckers:* ventral slightly in front of the mid-body, sometimes transversely oval, its dimensions 0·05–0·09 × 0·065–0·075 mm.; oral 0·055–0·09 mm. diameter. *Adhesive organ:* relatively small, slightly behind the ventral sucker, elliptical, 0·15–0·25 mm. long and 0·125–0·2 mm. broad, lacking a compact glandular mass, but having numerous isolated gland cells. *Gut:* prepharynx short, pharynx elongate, oesophagus very short, caeca long, terminating slightly in front of the conical posterior extremity. Bifurcation of the intestine far in front of the ventral sucker. *Reproductive systems:* genital pore posterior, cirrus pouch well developed and sac-like, extending diagonally from right to left, containing a tubular seminal vesicle, a straight pars prostatica and a muscular cirrus. Testes ovoid, diagonally one behind the other above and behind the adhesive organ. Ovary rounded, in front and on the left of the anterior testis. Vitellaria lateral, merging between the ventral sucker and the adhesive organ, large elliptical follicles radiating about the adhesive organ. Uterus short, extending forward only as far as the adhesive organ, containing four or five large, golden eggs measuring 0·100–0·117 × 0·063–0·068 mm.

Subfamily **Pharyngostomatinae** Szidat, 1936, emend.

Genus *Pharyngostomum* Ciurea, 1922

Pharyngostomum cordatum (Diesing, 1850) Ciurea, 1922

Syn. *Hemistomum cordatum* Diesing, 1850; *Alaria cordata* of Railliet, 1919; *Hemistomum kordatum* Schneidemühl, 1898; *Holostomum linguaeformis* Dubois, 1938; *Diplostomulum mutadomum* Wallace, 1937; *Diplostomum putorii* Linstow, 1877; etc.

HOSTS. Domestic cat, wild cat.

LOCATION. Intestine.

LOCALITY. Europe, Asia (China).

DIAGNOSIS. *Shape and size:* oval outline, anterior and posterior regions only faintly indicated, 2·6–3·8 mm. long and 1·6–2·0 mm. broad. *Suckers:* concealed by the adhesive organ, the oral larger than the ventral, diameters about 0·2 and 0·06 mm. *Adhesive organ:* broad, somewhat heart-shaped, connected with the body only in the median plane and by a narrow bridge of tissue. *Gut:* pharynx slightly larger than the oral sucker (0·22–0·24 mm. diameter), oesophagus short, caeca long. *Reproductive systems:* vitellaria mainly contained in the adhesive organ. Gonads posterior in position. Testes very large and deeply lobed, side by side. Ovary small and situated in front of the testes.

Wallace (1939) described the life cycle of this species at Canton. The cercaria is of Strigeid type, with two pairs of penetration glands and a flame-cell formula $2[(1+1)+(1+1)+(1)]$. It develops to a metacercaria of diplostomulum type in tadpoles of frogs and toads. Diplostomula occur naturally in frogs, also in snakes (accumulator hosts) which feed on frogs and thereby accumulate hundreds of parasites. The trematodes mature in about one month in the intestine of the cat, which becomes as heavily infected after eating one snake host as by living on frogs for weeks. The white rat, chicks and ducklings act as experimental final hosts, and the short-eared owl also acquires the trematode from the same sources.

Harkema (1942) described a new species and genus, *Pharyngostomoides procyonis*, a parasite of the racoon in U.S.A.

Family **DIPLOSTOMATIDAE** Poirier, 1886 emend. (p. 105)

Subfamily **Alariinae** Hall & Wigdor, 1918

Genus *Alaria* Schrank, 1788

Syn. *Hemistomum* Diesing, 1850.

Alaria alata (Goeze, 1782) Hall & Wigdor, 1918 (Fig. 58B)

Syn. *Planaria alata* Goeze, 1782; *Alaria vulpis* Schrank, 1788; *Festucaria alata* of Schrank, 1790; *Fasciola alata* of Rudolphi, 1793; *Distoma alatum* of Zeder, 1800; *Holostomum alatum* of Nitzsch, 1819; *Hemistomum alatum* of Diesing, 1850; *Diplostomum alatum* of Parona, 1894; *Conchosomum alatum* of Railliet, 1896.

HOSTS. Dog, cat, fox.

LOCATION. Intestine.

LOCALITY. Europe, Australia.

DIAGNOSIS. *Shape and size:* flat, expanded anterior much longer than the cylindrical posterior region and having two lateral tentacle-like processes bearing the openings of glands, total length 3–6 mm. *Suckers:* very small, the oral slightly larger than the ventral. *Adhesive organ:* two long folds situated behind the ventral sucker, curving forward and having distinct lateral margins. *Gut:* pharynx small, oesophagus short, caeca long. *Reproductive systems:* gonads in the posterior region, vitellaria in the anterior and in the adhesive organ. Testes bipartite and in the posterior region. Ovary in front of the testes, at the base of the adhesive organ. Uterus having an ascending limb which extends into the adhesive organ and a descending loop which is little folded. Male and female ducts opening to the exterior on a small genital papilla contained in the genital atrium, which may be inconspicuous. Cirrus and cirrus pouch absent.

Family **SCHISTOSOMATIDAE** Looss, 1899, emend Poche, 1907 (p. 106)

Genus *Schistosoma* Weinland, 1858

Syn. *Gynaecophorus* Diesing, 1858; *Thecosoma* Moquin-Tandon, 1860; *Bilharzia* Cobbold, 1879, etc. (See Stiles & Hassall, 1926, p. 96.)

Schistosoma haematobium (Bilharz, 1852) Weinland, 1858

Syn. *Distoma haematobium* Bilharz, 1852; *Gynaecophorus haematobius* of Diesing, 1858; *Bilharzia haematobia* of Cobbold, 1859; *Thecosoma haematobium* of Moquin-Tandon, 1860; *Distoma capense* Harley, 1864.

HOSTS. Man (rat, mouse, hedgehog and other mammals experimentally).

LOCATION. Adults inhabit portal, vesicle, mesenteric and splenic veins, sometimes the vena cava.

LOCALITY. Southern Europe, Africa, Asia, Australia, America.

DIAGNOSIS. *Male. Size:* 12–14 mm. long (but mature when 4 mm. long) and about 1 mm. broad, but having the lateral regions rolled up to form a kind of tube 0·4–0·5 mm. diameter. Body widening sharply about 0·6 mm. from the anterior extremity. *Cuticle:* finely tuberculated dorsally behind the ventral sucker, fine spinelets also occurring in the suckers, larger ones in the gynaecophoric canal. *Suckers:* close together, the ventral slightly smaller than the oral, about 0·28 mm. diameter. *Gut:* pharynx absent, oesophagus having two dilatations (and equipped with numerous gland cells), caeca long and united behind the testes to form a single median caecum which continues to the posterior extremity. Bifurcation of the intestine just in front of the ventral sucker. *Reproductive systems:* genital pore behind the ventral sucker and near the anterior end of the gynaecophoric canal. Testes four or five in number, situated between the caeca a little behind the ventral sucker. *Female. Size:* very elongate (and much longer than the male), cylindrical, 16–20 mm. long and 0·25 mm. broad and thick. *Cuticle:* lacking spines, except in the cavities of suckers and near the posterior extremity. *Suckers:* small, the oral slightly smaller than the ventral, which is about 0·2–0·3 mm. from the anterior extremity, diameters 0·06 and 0·07 mm. *Gut:* pharynx absent, oesophagus as in the male, caeca long and separate back to the level of the ovary, behind this level uniting to form a single median caecum. Bifurcation of the intestine in front of the ventral sucker. *Reproductive systems:* genital pore slightly behind the ventral sucker. Ovary ovoid, about two-thirds of the distance along the body. Vitellaria unpaired and situated behind the ovary more or less filling the final quarter of the body. Uterus having a long, little folded ascending loop. Eggs large (0·12–0·19 × 0·05–0·07 mm.), spindle-like, but pointed at one pole and rounded at other, each containing a developing miracidium.

OTHER SPECIES:

Schistosoma bovis (Sonsino, 1876) (Fig. 59A). In the abdominal veins of the ox, sheep and other mammals in Egypt, Sicily, Sardinia, France, Africa, India, Malaya and Indo-China.

S. japonicum Katsurada, 1904. In the hepatic portal vein and mesenteric arteries of man, dog, cat, cattle, horses and other mammals in Japan.

S. mansoni Sambon, 1907. In the mesenteric veins of man in Africa.

S. indicum (Montgomery, 1906). In the portal and mesenteric veins of horse, ass, sheep, zebra, etc., in India and Africa (Rhodesia).

And others. Most of the Schistosomes are dealt with in the standard text-books of helminthology.

Genus *Ornithobilharzia* Odhner, 1912

Members of this genus are similar to *Schistosoma*, but the caeca tend to form anastomoses in front of the median posterior caecum, the testes are numerous (sixty or more) and the ovary is formed into a spiral and lies in the anterior region of the body. They are parasites of both mammals and birds.

Ornithobilharzia bomfordi (Montgomery, 1906). In the mesenteric veins of the ox in France, and the same location in the zebu in India.

O. turkestanica (Skrjabin, 1913). In the portal veins of cattle in Russian Turkestan, Iraq and Europe (France).

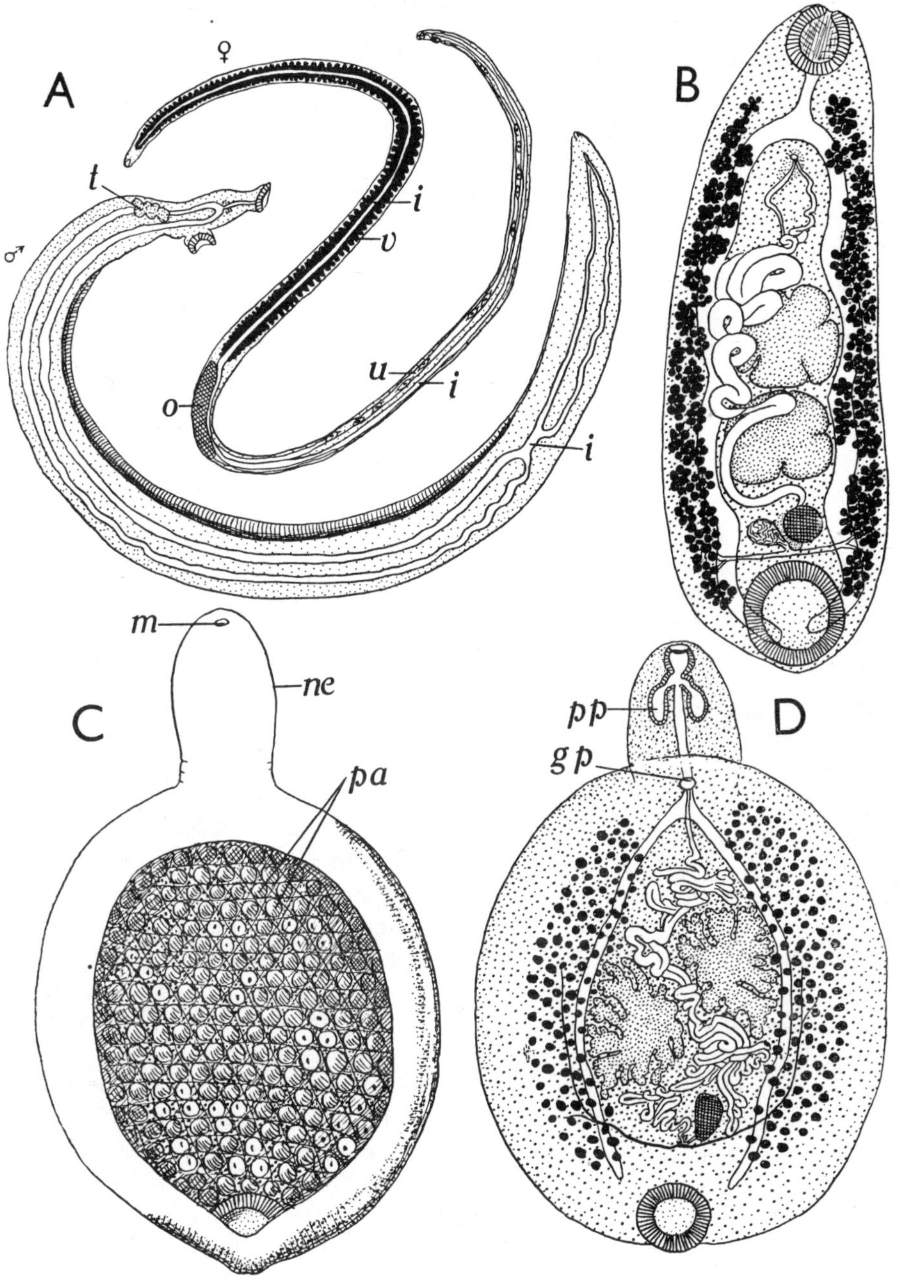

Fig. 59. Schistosomatidae (A) and Paramphistomatidae (B–D) of mammals. A, *Schistosoma bovis*, male and female. B, *Paramphistomum cotylophorum*. C, D, *Gastrodiscus*, showing external characters in ventral view (C) and internal organs (D). (A, B, D, after Mönnig, 1934; C, after Fuhrmann, 1928.)

Family **BRACHYLAEMIDAE** Joyeux & Foley, 1930 (p. 101)

Syn. Harmostomidae Odhner, 1912.

Subfamily **Brachylaeminae** Joyeux & Foley, 1930

Syn. Harmostominae Looss, 1900.

KEY TO GENERA OCCURRING IN MAMMALS

1. Cuticle smooth; uterus not extending in front of the ventral sucker; parasites of the mole *Itygonimus*
2. Cuticle sometimes spinous; uterus extending in front of the ventral sucker; parasites of the hedgehog, badger and other mammals, but not the mole *Brachylaemus*

Genus *Brachylaemus* Dujardin, 1843, emend. E. Blanchard, 1847

Syn. *Harmostomum* Braun, 1899; *Heterolope*, Looss, 1899.

Brachylaemus erinacei E. Blanchard, 1847

Syn. *Distoma linguaeforme* Diesing, 1850; *D. caudatum* Linstow, 1873; *D. leptostomum* Olsson, 1876; *Mesogonimus linguaeformis* of Stossich, 1896; *Heterolope leptostoma* of Looss, 1899; *Harmostomum leptostomum* of Braun, 1899; *Distoma spinulosum* Hofmann, 1899; *Harmostomum* (*Harmostomum*) *helicis* (Meckel, 1846) Witenberg, 1925; *H.* (*H.*) *spinulosum* of Witenberg, 1925; *Brachylaemus helicis* (Meckel, 1846) Baer, 1932.

Hosts. Badger, hedgehog.

Location. Intestine.

Locality. Europe.

This species is sometimes referred to under two names, *B. spinulosus* and *B. helicis*, supposedly spinous and non-spinous species, but Dollfus (1935 *a*) found it impossible to decide whether or not this is justified, because few writers have noted the nature of the cuticle. That such separation is not justified is suggested by the occurrence of spinous and non-spinous forms in the same locality. Dollfus also pointed out that Meckel did not propose the name '*helicis*' for this species, Braun (1891) being the first to introduce it, so that the name based on *Cercariaeum helicis* Meckel, 1846 is an error and has not got priority, which thus goes to *Brachylaemus erinacei*.

Diagnosis (after Dollfus). *Size:* specimens 1·3 mm. long may be mature, though larger specimens (1·7 mm. long) may remain immature. *Suckers:* of about equal sizes, the oral 0·25–0·26 mm. diameter, the ventral 0·26 mm. in a specimen 3·34 mm. long and 0·55 mm. broad. *Genital pore:* near and generally slightly in front of the anterior margin of the foremost testis (0·9 mm. from the posterior extremity). *Eggs:* measuring 0·030–0·040 × 0·022–0·025 mm. (according to Baer, 1928, 0·027–0·034 × 0·015–0·023 mm.)

Other Species:

B. advena Dujardin, 1843 and *B. fulvus* Dujardin, 1843 occur in shrews of the genera *Crocidura* and *Sorex* in France, *Brachylaemus dujardini* (Baer, 1928) and *B. corrugatus* also in the common shrew, *Sorex araneus*. *Brachylaemus advena* has also been recorded in rodents, including the black rat, but the best-known form in such mammals is *B. recurvatus* Dujardin, 1845, which occurs in the long-tailed fieldmouse in France and Britain (Baylis, 1928*b*) and the housemouse in Vienna, as well as in other Rodentia in Algeria and Egypt. At least three species are

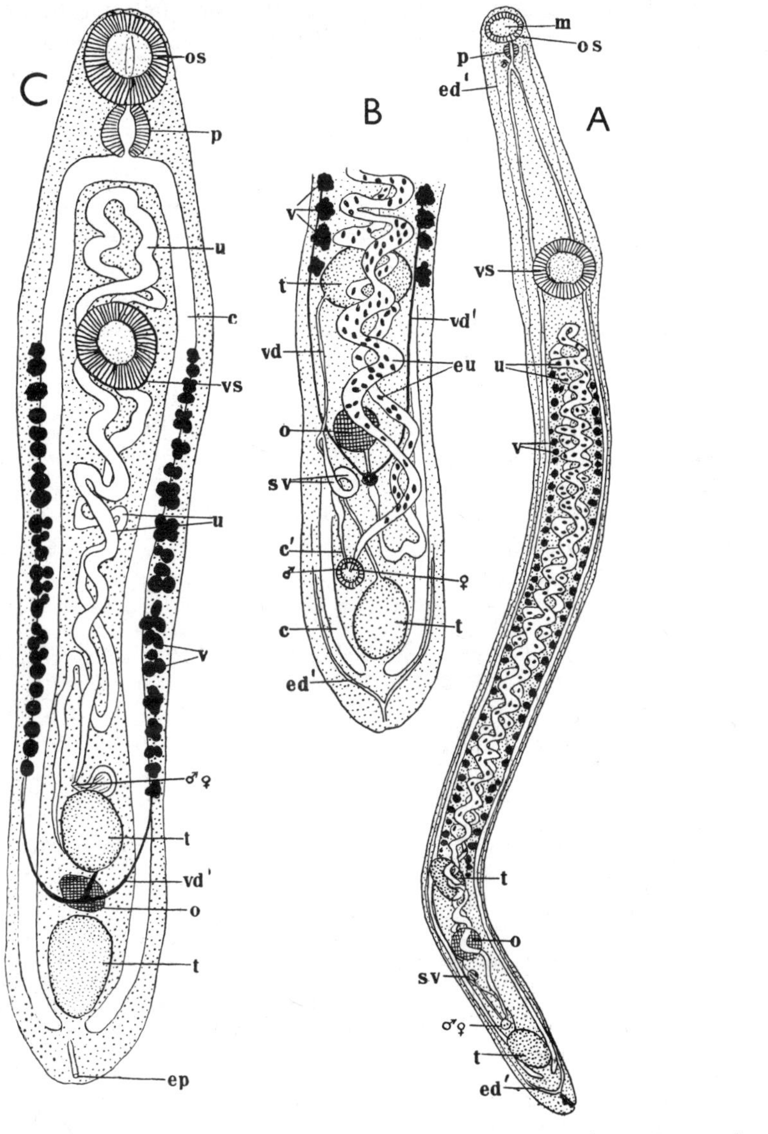

Fig. 60. Brachylaemidae from the hedgehog (C) and mole (A, B). A, *Itygonimus lorum*. B, posterior end of A, magnified. C, *Brachylaemus* sp. (A, B, after Davies, 1932; C, after Fuhrmann, 1928.)

known to occur respectively in the genera of marsupials, *Dasyurus*, *Parameles* and *Didelphys*, a fourth also in the last-named. Dollfus (1934, 1935 *a*) has reviewed the members of the genus *Brachylaemus* contributing to the fauna of France. Allison (1943) reviewed both the genus and the family.

Genus *Itygonimus* Lühe, 1899

Two species of this genus seem to occur in the intestine of moles, but there has been much confusion about their names. Davies (1932), who seems to have followed Witenberg (1925) into error as regards nomenclature, claimed to have identified both species in moles found in Wales, and the following diagnoses are based on his conclusions. I am indebted to Dr H. A. Baylis and his assistant, Mr Prudhoe, for the lists of synonyms here given, those marked with an asterisk having been taken on trust, references for checking them not now being available. Dr Baylis has pointed out to me (*in litt.*) that *Itygonimus talpae* has no *locus standi* at all. Dujardin (1845) gave '*Fasciola talpae* Goeze (1782), p. 182, Pl. 15, figs. 6–7' as a synonym of '*Monostoma ocreatum* Zeder', but Goeze never seems to have used the name *Fasciola talpae*, and Witenberg and later writers have been in error in following Dujardin. The '*Itygonimus*' figured by Goeze is called by him 'Der Stiefelwurm des Maulwurfs', *Fasciola ocreata* (p. 182). The species *ocreata* 'Zeder' was made the type of *Itygonimus* by Lühe (1899), but this is really Goeze's species. Incidentally, Dujardin described *Distoma lorum* as a new species, but added after the description that it is 'identique' with the *Monostoma ocreatum* of the German helminthologists, despite the fact that his two descriptions show a distinct difference in the ratio of size of the oral and ventral suckers.

Itygonimus ocreatus (Goeze, 1782) Braun, 1902

Syn. *Fasciola ocreata* Goeze, 1782; **Cucullanus talpae* Müller, 1787; **C. ocreatus* Schrank, 1788; *Monostoma ocreatum* Zeder, 1800; '*Fasciola talpae* Goeze' Dujardin, 1845; *Monostomum ocreatum* Küchenmeister, 1855; **Distoma lorum* of Melnikow, 1865; *'*D. ocreatum* Zeder' Lühe, 1899; **Dolichosomum lorum* of Looss, 1899; **Itygonimus lorum* of Looss, 1907; **I. lorum* of Gonder, 1910; '*Ithygonimus talpae* (Goeze, 1782)' Witenberg, 1925; '*Itygonimus talpae* (Goeze)' Baer, 1932 (also Davies, 1932; Sprehn, 1932).

DIAGNOSIS. *Shape and size:* very elongate, cylindrical, 8–54 mm. long and 0·25–1·0 mm. broad (according to Gonder, 8–15 × 0·25–0·7 mm; to Dujardin, 15–54 × 0·03–0·05 of the body length; to Davies, 19–32 × 0·3–1·0 mm.). *Suckers:* breadth of the ventral about one-third of the diameter of the oral. *Vitellaria:* extending in front of the ventral sucker. *Receptaculum seminis:* present (cp. *I. lorum*).

Itygonimus lorum (Dujardin, 1845)

Syn. *Distoma lorum* Dujardin, 1845; **Mesogonimus lorum* Monticelli, 1893; **Itygonimus filum* Looss, 1907; *Ithygonimus lorum* of Witenberg, 1925.

DIAGNOSIS. *Shape and size:* very elongate and cylindrical, the greatest breadth occurring in the vicinity of the ventral sucker, 3·8–10·5 mm. long (mean 5·9 mm.), greatest breadth 0·42–1·00 (mean 0·66 mm.), thickness 0·2–0·25 at the anterior end in specimens 5 mm. long. *Suckers:* ventral larger than the oral and about one-third or one-quarter of the length of the body behind it, diameters 0·32–0·67 and 0·22–0·47 mm. *Vitellaria:* lateral, approaching the median plane posteriorly, but not extending forward in front of the ventral sucker. *Eggs:* measuring 0·03–0·35 × 0·015–0·017 mm. *Receptaculum seminis:* absent. *Locality:* both species are fairly widely distributed throughout Europe (Denmark, France,

Germany, Austria, Wales). Out of a total of 201 moles in the Aberystwyth district, Davies (1932) found thirty-seven parasitized by *I. lorum* and forty-eight by *I. ocreatus*, one to eleven trematodes per host in the former and two to six in the latter species.

Family **PARAMPHISTOMATIDAE** Fischoeder, 1901, emend. Goto & Matsudaira, 1918 (p. 103)

About six subfamilies of Amphistomes parasitize mammals, according to Szidat (1939*a*) the Cladorchiinae, Gastrodiscinae, Chiorchiinae, Balanorchiinae, Gastrothylacinae and Paramphistomatinae. Although only an inconspicuous element in the trematode fauna of mammals in Europe, Amphistomes are important helminthological types, and we must examine the best-known ones briefly. In the case of many genera it will be unnecessary to specify morphological characters, the hosts being more or less distinctive.

Subfamily **Cladorchiinae** Fischoeder, 1901, emend. Hughes, Higginbotham & Clary, 1942

Nine genera are well known in various parts of the world, one of them being of regular occurrence in Europe. *Watsonius* Stiles & Goldberger, 1910 occurs in the duodenum and small intestine of man and monkeys in Africa, possibly in Asia. *W. watsoni* (Conyngham, 1904) was obtained at the autopsy of a negro. *Wardius* Barker, 1915 occurs in the caecum of rodents and *Stichorchis* Fischoeder, 1901 in the caecum and large intestine of these hosts and Suidae as well. *Cladorchis* is found in the caecum of the tapir, *Hawkesius* and *Pfenderius*, both Stiles & Goldberger, 1910, in the colon of the Indian elephant, *Brumptia* Travassos, 1921 in the African elephant and the rhinoceros, and *Pseudodiscus* Sonsino, 1895 in the colon of the horse.

Genus *Stichorchis* Fischoeder, 1901

Stichorchis subtriquetrus (Rudolphi, 1814) Fischoeder, 1901

Syn. *Amphistoma subtriquetrum* Rudolphi, 1814; *Distoma amphistomoides* Bojanus, 1817.

Hosts. Beaver, muskrat, voles.

Location. Small intestine, caecum, colon.

Locality. Europe, America.

Diagnosis. *Size:* 4–12 mm. long, generally 6–10 mm. *Suckers:* posterior large, 2·6 mm. in diameter and 0·6 mm. deep, situated slightly in front of the posterior extremity, its hinder margin about 1 mm. farther forward; oral absent, the pharynx taking its place. *Gut:* pharynx large (1–1·5 mm. in diameter) and having a pair of small pouches confined within the musculature of its posterior wall, caeca long. *Reproductive systems:* genital pore far forward, near the posterior end of the pharynx. Genital atrium having a feeble, sucker-like musculature and not sharply set off from the parenchyma. Cirrus pouch ovoid, 0·5–0·6 mm. long and 0·3–0·4 mm. broad. Testes much branched, one behind the other and situated between the bifurcation of the intestine and the posterior sucker, the ovary behind them. Vitellaria slightly dorsal and ventral (but not lateral) to the caeca in the posterior region, extending forward only slightly into the anterior half of the body. Eggs very large (0·156–0·166 × 0·090–0·095 mm.).

Subfamily **Gastrodiscinae** Monticelli, 1892

Szidat (1939*a*) recognized three genera in this subfamily, one of them being *Gastrodiscoides* 'Lane, 1924', but it was Leiper (1913) who proposed this genus, and it still seems doubtful whether his action was justified, even if the hosts are utilized as aids to classification. The genus is said to occur in the pig and man, being represented by the species better known as *Gastrodiscus hominis* (Lewis & McConnell, 1876) Sonsino, 1896. The remaining two genera are *Gastrodiscus* Leuckart, 1877, with species in the large intestine of Suidae, rhinoceros and horse, and *Homalogaster* Poirier, 1882, which occurs in the large and small intestine of cattle. In *Gastrodiscus aegyptiacus* (Cobbold, 1876) Looss, 1896 (Fig. 59C, D), which occurs in Africa and India, there is a small, conical anterior and a wide posterior region, the ventral surface of which bears numerous wart-like papillae set in regular transverse rows. The posterior sucker is small and inconspicuous, but diagnostic, and the internal organs show no significant difference from the typical Amphistome arrangement. In *Homalogaster poloniae* Poirier, 1883, a parasite of sheep and oxen, the body is also divided into anterior and posterior regions, but the former is large and the latter small and cylindrical.

Subfamily **Chiorchiinae** Szidat, 1939, emend.

This group contains the solitary genus *Chiorchis* Fischoeder, 1901, which occurs in the caecum of the manatee, but also in a turtle in Malaya.

Subfamily **Balanorchiinae** Fuhrmann, 1928, emend.

This group also contains a solitary genus, *Balanorchis* Fischoeder, 1901, in which the pharynx has two well-developed pouches and the mouth is surrounded by a number of spinous papillae. The type-species, *B. anastrophus* Fischoeder, 1901 was obtained from the stomach of a deer in Brazil.

Subfamily **Gastrothylacinae** Stiles & Goldberger, 1910

Stiles & Goldberger (1910) claimed recognition for four distinct genera. *Gastrothylax*, *Fischoederius*, *Carmyerius* and *Wellmanius*, the last three being new, Maplestone (1923) showed *Wellmanius* to be invalid, its only species, *W. wellmani*, being identical with *Carmyerius spatiosus*. I showed (Dawes, 1936*b*) that the three remaining genera must be reassembled into the original one, *Gastrothylax* Poirier, 1883, because the characters upon which generic distinction is claimed (slight differences in the positions of the testes and the slightly variable course of the uterus) are insufficient for the purpose. *Gastrothylax* (Fig. 10B) differs from all other Digenea in having an enormous ventral pouch, probably a very enlarged genital atrium, which occupies most of the ventral region of the body and part of the lateral regions, and into which the genital pore opens. The type-species, *G. crumenifer* (Creplin, 1847) occurs in the rumen of the buffalo in the Far East. Other species are *G. elongatus* Poirier, 1883, discovered in the stomach of a Javanese gayal and now well known in cattle in Ceylon, Siam, Java, Celebes, Japan, Formosa and the Philippines; *G. cobboldi* Poirier, 1883, another genuine Asiatic species, *G. spatiosus* Brandes, 1898, *G. synethes* Fischoeder, 1901, etc.

Subfamily **Paramphistomatinae** Fischoeder, 1901, emend. Dollfus, 1932

This group contains two well-marked genera, *Paramphistomum* Fischoeder, 1901 and *Stephanopharynx* Fischoeder, 1901. A third genus, *Cotylophoron* Stiles & Goldberger, 1910, differs from the first only in the presence of a rudimentary sucker around the genital pore (i.e. a gonotyl), a character which hardly justifies generic distinction so that *Cotylophoron* can be regarded as a synonym of *Paramphistomum*. *Stephanopharynx* can be distinguished from *Paramphistomum* by the single diverticulum which arises from the pharynx. Both genera are found in ruminants.

Genus *Paramphistomum* Fischoeder, 1901

Paramphistomum cervi (Schrank, 1790)

Syn. *Fasciola cervi* Schrank, 1790; *Festucaria cervi* Zeder, 1790; *Fasciola elaphi* Gmelin, 1791; *Monostoma conicum* Zeder, 1803; *Amphistoma conicum* Rudolphi, 1809; *Paramphistomum gracile* Fischoeder, 1901; *P. microbothrium* Fischoeder, 1901; *P. bathycotyle* Fischoeder, 1901; *P. liorchis* Fischoeder, 1901; *P. epiclitum* Fischoeder, 1904; *P. papillosum* Stiles & Goldberger, 1910; *P. papilligerum* Stiles & Goldberger, 1910; *P. pisum* Leiper, 1910; *P. ichikawai* Fukui, 1922.

P. explanatum (Creplin, 1849)

Syn. *Paramphistomum microon* Railliet, 1924; *P. birmiense* Railliet, 1924; *P. calicophorum* Fischoeder, 1901; *P. crassum* Stiles & Goldberger, 1910; *P. caliorchis* Stiles & Goldberger, 1910; *P. fraternum* Stiles & Goldberger, 1910; *P. gigantocotyle* (Brandes, 1896); *P. ijimai* Fukui, 1922; *P. formosanum* Fukui, 1929.

HOSTS. Cattle, zebu, Asiatic buffalo, sheep, goat, deer, antelopes, etc.

LOCATION. Rumen.

LOCALITY. Europe, Asia, Africa, North America. (*Note.* A fatal outbreak of parasitism due to *P. cervi* in Cheshire was recorded by Craig & Davies, 1937.)

DIAGNOSIS. *P. cervi* and *P. explanatum* cannot be separated satisfactorily on the morphological characters of adults and the latter is probably a synonym of the former (see Dawes, 1936*b*). The latter is regarded as having a larger posterior sucker than the former, and the testes are diagonally arranged instead of regularly one behind the other. In both forms, however, the sucker varies enormously in relative size, its size-relation with the body being represented by the ratios 1/8–1/3·5 for '*P. cervi*' and 1/1·9–1/1·6 for '*P. explanatum*' (Maplestone, 1923), and the position of the testes depends on the size of the sucker (Dawes, 1936*b*).

DIAGNOSIS. *Size:* 5–13 mm. long and 2–5·5 mm. in maximum breadth. *Gut:* pharynx large, 0·8–1·2 × 0·5–1·0 mm., oesophagus short, caeca long, terminating above the posterior sucker. *Excretory system:* pore situated on the postero-dorsal surface of the body, in front of the opening of Laurer's canal, vesicle globular. *Reproductive systems:* genital pore median, ventral and about one-third of the distance along the body. Testes rounded or slightly lobed, directly or diagonally one behind the other. Ovary behind the testes and dorsal to the sucker. Vitellaria lateral and coarsely follicular, extending from the pharynx to the posterior extremity. Uterus having slight folds. Eggs very large (0·145–0·156 × 0·075–0·082 mm.).

OTHER SPECIES:

Paramphistomum cotylophorum Fischoeder, 1901, *P. orthocoelium* Fischoeder, 1901, *P. siamense* Stiles & Goldberger, 1910, *P. gotoi* Fukui, 1926, etc. For synonyms of these Amphistomes, and also of species of *Gastrothylax*, see Dawes, 1936*b*.

CHAPTER 12

THE LARVAE OF THE DIGENEA

Some Digenea develop from larvae closely resembling those of *Fasciola hepatica*, both in the structure of corresponding forms and the sequence in which these appear, but the majority have somewhat different larval forms which may appear in a different sequence, the redia or sporocyst stage being omitted. Sporocysts vary between a simple, sac-like and a much-branched form and in some extreme instances are scarcely distinguishable from rediae, which can generally be recognized by the saccular gut. Miracidia of different species of the Digenea also vary and here, as well, differences may cast some light on the difficult problems of phylogenetic relationship. The larval form which shows the greatest variety of structure and is most helpful in this respect, however, is the cercaria. For this reason, cercariae will receive more detailed consideration than other larval forms in this chapter, although we shall pay some attention to the sequence in which larvae appear during the life cycle.

Various attempts have been made to arrange cercariae in groups according to their structure and habits, but this so-called classification has not necessarily anything to do with real taxonomy and is of somewhat artificial and academic interest, except in so far as it can be correlated with the phylogeny of adults. Another noteworthy point is that cercariae and the corresponding adults may be discovered and are frequently described and named separately, which explains how a species of trematode has two totally different names, e.g. *Cercaria diplocotylea* and *Diplodiscus subclavatus*. It is important to realize that '*Cercaria*' is *not* a generic name, but merely a group name. In spite of this, specific names combined with it have nomenclatural validity, and may take priority over later names given to adults of the same species.

Lühe (1909)* made the first attempt to classify cercariae in a comprehensive manner, his scheme lending itself to the following summary (with examples in parentheses):

1. **LOPHOCERCARIAE** (Fig. 61 A): having a dorsal, longitudinal fin-fold along the body (*Cercaria cristata* La Valette St George, 1855).
2. **GASTEROSTOME CERCARIAE** (Fig. 61 B): tail having two symmetrical processes; intestine simple and sac-like; mouth centrally situated on the ventral surface of the body (the cercaria of *Bucephalus polymorphus*).
3. **MONOSTOME CERCARIAE** (Fig. 61 C): ventral sucker absent (*Cercaria ephemera* Nitzsch, 1807; *C. monostomi* Linstow, 1896; *C. imbricata* Looss, 1893; *C. lophocerca* Filippi, 1857).
4. **AMPHISTOME CERCARIAE** (Fig. 61 D): ventral sucker at or very near the posterior extremity (*Cercaria pigmentata* Sonsino, 1892; *C. diplocotylea* Pagenstecher, 1857).
5. **DISTOME CERCARIAE** (Fig. 61 E–M): ventral sucker distant from the posterior extremity:
   A. '**Cystocercous**' **cercariae**: base of the tail containing a cavity into which the body can be retracted (*Cercaria macrocerca* Filippi, 1854; Fig. 61 E).

* In many subsequent references to the work of Lühe, the paper published in 1909 is implied although unspecified by date.

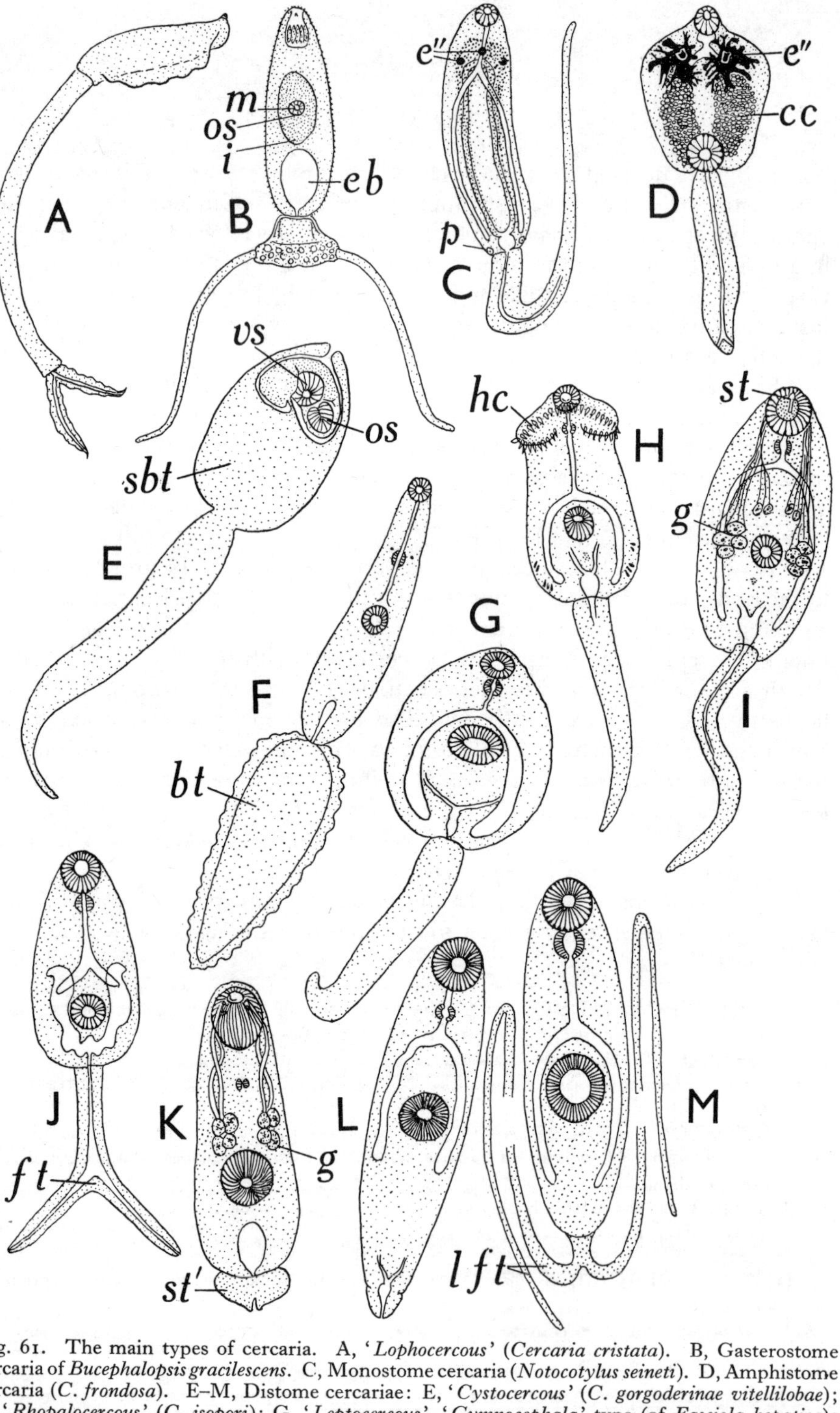

Fig. 61. The main types of cercaria. A, '*Lophocercous*' (*Cercaria cristata*). B, Gasterostome cercaria of *Bucephalopsis gracilescens*. C, Monostome cercaria (*Notocotylus seineti*). D, Amphistome cercaria (*C. frondosa*). E–M, Distome cercariae: E, '*Cystocercous*' (*C. gorgoderinae vitellilobae*); F, '*Rhopalocercous*' (*C. isopori*); G, '*Leptocercous*', '*Gymnocephala*' type (of *Fasciola hepatica*); H, Echinostome cercaria (*Echinoparyphium recurvatum*); I, *Xiphidiocercaria* (*Cercaria* XI Harper, 1929); J, '*Furcocercous*' (*C. dichotoma*); K, '*Microcercous*' (*C. brachyura*); L, *Cercariaeum* (*C. politae nitidulae*); M, cercaria from *Nucula*. (A, F, after Lühe, 1909; C, D, G, after Faust, 1930; E, after Vickers, 1940; B, J, K, after Lebour, 1911; G, H, I, L, after Harper, 1929; M, after Idris Jones & Rothschild, 1932.)

B. '**Rhopalocercous' cercariae**: tail as wide as or wider than the body (*Cercaria isopori* Looss, 1894; Fig. 61 F).
C. '**Leptocercous' cercariae**: tail straight, slender and narrower than the body:
(*a*) '**Gymnocephalous' cercariae**: anterior extremity rounded and lacking a piercing spine or *stylet* (the cercaria of *Fasciola hepatica*; Fig. 61 G).
(*b*) '**Echinostome' cercariae**: anterior extremity provided with a 'head collar' and a coronet of stout 'head spines' or 'collar spines' (*Cercaria echinata* Siebold, 1837; the cercariae of other Echinostomes; Fig. 61 H).
(*c*) '**Xiphidiocercariae**': anterior extremity provided with a stylet (*Cercaria ornata* La Valette, 1855; *Cercaria* X 1 Harper, 1929; Fig. 61 I).
D. '**Trichocercous' cercariae**: tail provided with spines or 'bristles' (the cercaria of *Pharyngora bacillaris*; Fig. 62 F).
E. '**Furcocercous' cercariae**: tail forked distally (*Cercaria furcata* Diesing; *C. dichotoma* Lebour, 1911; Fig. 61 J).
F. '**Microcercous' cercariae**: tail short and stumpy (*Cercaria brachyura* Lespés, 1857; Fig. 61 K).
G. '**Cercariaea**': tail undeveloped (the cercaria of *Asymphylodora tincae*; *Cercaria politae nitidulae* Harper, 1929; Fig. 61 L).
H. '**Rat-King' cercariae** (marine): cercariae arranged in groups having the tips of the tails united to form a kind of colony.

Numerous attempts have been made to modify this scheme and improve our conception of the relationships between the various types of cercariae, but it is partially true, at any rate, to say that Lühe's classification is still the basis of any scheme which can be put forward at the present time. Lebour (1911) formulated an alternative scheme to guard against misconceptions that might arise by rigid insistence on the taxonomic importance of the tail. She drew attention to the separation of forms which in other respects show close relationship because of the loss of the tail by abbreviation of the life history. We now know that the tails of cercariae show adaptive characters which illustrate, not affinity, but convergence in evolution. It is unlikely that Lebour's scheme could be amplified to include all known cercariae, but it is noteworthy because it attempted a break with tradition which even yet has not come about. The scheme, which lays greater emphasis on the mode of origin of cercariae than did Lühe's, may be represented (with examples in parentheses) as follows:

1. **GASTEROSTOMATA** Cercariae developing in sporocysts. Mouth situated in the middle of the ventral surface. Tail having a broad base and two very contractile lateral filaments (the cercariae of Bucephalidae).
2. **PROSTOMATA** Cercariae in which the mouth is situated anteriorly:
A. **Monostomum group**: having one sucker only (*Cercaria ephemera* Nitzsch, 1807).
B. **Distomum group**: having two suckers:
I. Cercariae developing in sporocysts:
(*a*) '**Gymnophallus' group**: cercaria tailless (*Cercaria glandosa* Lebour, 1911).
(*b*) '**Fork-tail' cercariae**: tail forked; excretory canals extending down the furcal rami, to open at their tips (*Cercaria dichotoma* Muller in La Valette, 1855).
(*c*) '**Spelotrema' group**: stylet and penetration glands present (the cercaria of '*Spelotrema excellens*').
(*d*) '**Stumpy-tail' cercariae**: tail broad and stump-like (*Cercaria brachyura* Lespés, 1857).
(*e*) '**Lepodora**': (the cercaria of *Lepidapedon rachion*).

II. Cercariae developing in rediae:

(*a*) '**Cercariae neptunae**': tail large and thick; wall of the excretory vesicle thick; having two eye-spots (*Cercaria neptunae* Lebour, 1911).

(*b*) '**Acanthopsolus**' **group**: caeca long; tail thin, cast off before full growth is attained; having two eye-spots (cercaria of *Acanthopsolus lageniformis*).

(*c*) '**Echinostomum**' **group**: a 'head collar' and a coronet of 'collar spines' present (the cercaria of '*Echinostomum leptosomum*').

Sewell (1922) modified the scheme of Lühe in several particulars, not invariably satisfying subsequent workers in this field. He made great use of the characters to which Lühe attached importance, extending them and forming groups where Lühe, for want of information, had to be content with solitary, but characteristic species. This was in keeping with advance in knowledge, to which Sewell himself contributed substantially. We shall see how some of Sewell's ideas proved untenable in the light of later work.

Mention may be made also of the conception of the dual origin of the Digenea, which is attributable to Dubois (1929), following the lead of Sewell. The latter writer recognized in his group of Monostome cercariae two types of excretory system which receive the names '*Stenostoma*' and '*Mesostoma*'. In the former type, the lateral excretory canals extend forward to the pharynx and turn about before dividing into anterior and posterior branches; in the latter type, the lateral excretory canals extend forward only to the vicinity of the ventral sucker, there dividing into anterior and posterior vessels. Dubois showed other characteristics to be correlated with the '*Stenostoma*' and '*Mesostoma*' types of excretory system. Thus, the cercariae develop respectively in rediae and sporocysts, and the miracidium has two pairs or only one pair of flame cells respectively. Xiphidiocercariae and Furcocercariae show the '*Mesostoma*', Amphistome, Monostome, Echinostome and Gymnocephalous cercariae the '*Stenostoma*' type.

From such correlations Dubois inferred that some Digenea followed one or another path in evolution consequent upon this initial cleavage. Groups other than those mentioned show combinations of characters in one or the other series, e.g. '*Lophocercariae*' (Sanguinicolidae), '*Cystocercous*' cercariae (*Gorgodera cygnoides*), '*Microcercous*' cercariae (*Sphaerostoma bramae*), '*Rhopalocercous*' cercariae of the '*Isopori*' and '*Parapleurolophocerca*' groups, cercariaea of the '*Helicis*' and '*Leucochloridium*' groups and, perhaps, the Monostome cercariae belonging to the Notocotylidae. Cercariaea form an unnatural group, however, and many of them have the attributes of the 'Stenostoma' series. Excepting only '*Lophocercariae*', most others mentioned which have been studied sufficiently seem to have followed one suggested line of evolution or the other.

We can now examine the main characters of the principal kinds of cercariae, noting also the preceding stage in development. Some types which are not likely to be found in this country are mentioned, and many which occur here are left out, but the scheme followed will enable one to refer most cercariae to their respective groups. Precise diagnosis would entail substantial excursions into the original literature, which is already both deep and wide.

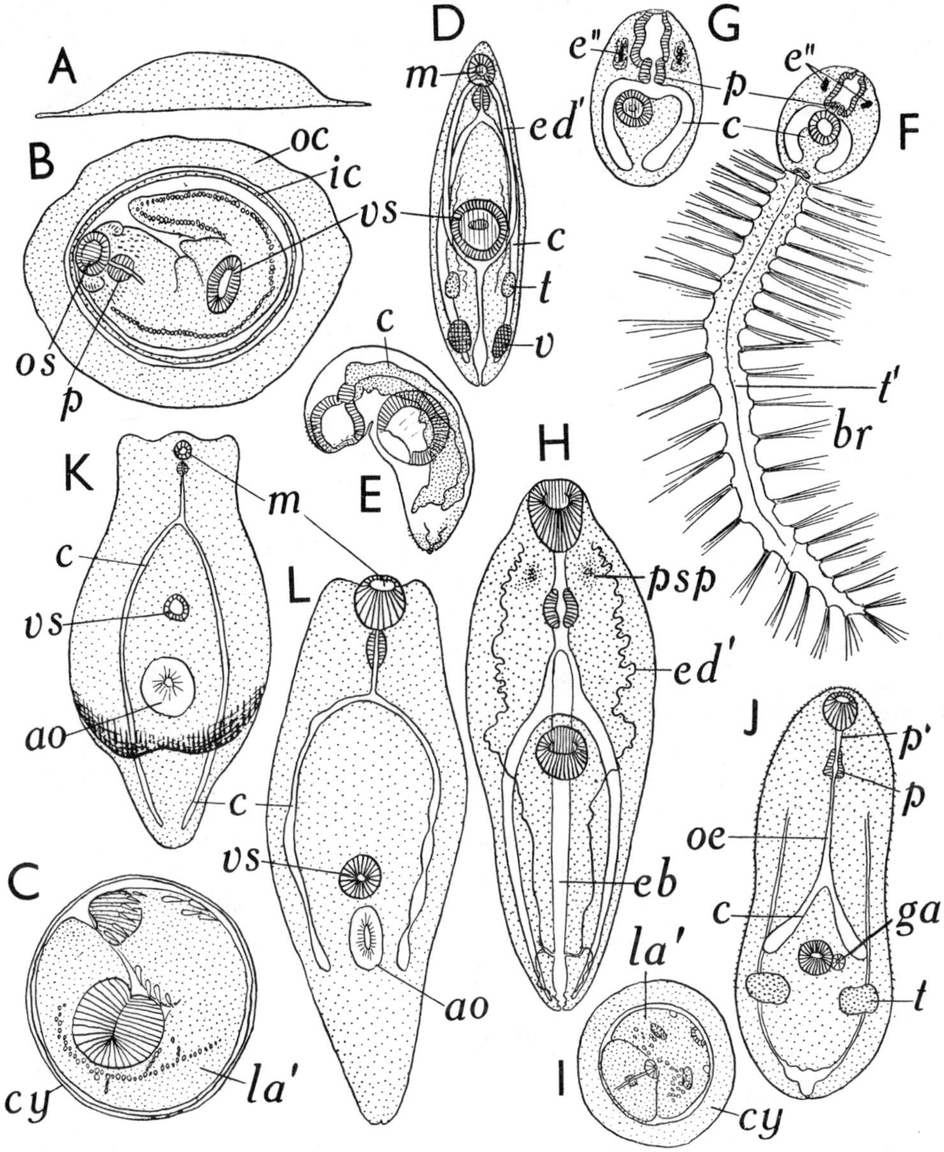

Fig. 62. Various larval and juvenile Digenea. A, B, cysts of *Parorchis acanthus* (*C. purpurae*) from the mantle of *Cardium edule* and *Mytilus edulis*. A, in side, B in surface view. C, encysted metacercaria of *Echinostoma secundum* (0·23 mm. diameter). D, E, *Derogenes varicus* from *Sagitta*. F, free-swimming cercaria of *Opechona bacillaris*. G, same species as F, but tailless, from a medusa. H, as G. I, J, *Spelotrema excellens*: I, encysted metacercaria (*C. ubiquita*) from *Carcinus maenas*; J, larva pressed out of cyst (0·8 mm. long). K, '*Diplostomum volvens*'. L, '*Tylodelphis rhachiaea*'. (A, B, after Lebour & Elmhirst, 1922; C, I, J, after Lebour, 1911; H, after Lebour, 1916–1918*a*; D, E, F, G, after Lebour, 1916–1918*b*; K, L, after Lühe, 1909.)

(1) AMPHISTOME CERCARIAE

These are amongst the largest known cercariae. They have a prominent body, a small tail, a distinctive posterior sucker, and globular masses of highly refractive material which fill the main excretory canals. They are feeble swimmers having only a short free life and inhabit the deeper parts of ponds and pools. They develop in rediae, are born at a relatively early stage of development, and come to full development as free parasites in the tissues of snails.

Cort (1915*a*) referred five Amphistome cercariae known at that time to the well-known subfamilies Paramphistominae and Diplodiscinae. Beaver (1929) has pointed out that this cannot be done on the basis of larval characters, and suggested that Sewell's classification into '*Pigmentata*' and '*Diplocotylea*' types is appropriate.

(*a*) '**Pigmentata**' **type**, e.g. *Cercaria pigmentata* Sonsino, 1892 (the larva of *Paramphistomum cervi*)

Cercariae of this type are uncommon in Europe, perhaps because large mammals are scarce, and belong to tropical or subtropical regions. Briefly, their main characters are: *Size:* well over 1 mm. long, including the tail. *Colour:* body deeply pigmented by stellate melanophores which merge to produce dendritic patches and, in some instances, dorsal and lateral lines of pigment. *Sense organs:* eye-spots anteriorly, provided with spherical lens-like bodies. *Cystogenous cells:* numerous, containing oval or rod-like masses. *Suckers:* oral smaller than the ventral. *Gut:* pharyngeal pouches absent, caeca long. *Excretory system:* vesicle opening by a small pore at its anterior end; main canals (connected by transverse anastomoses) extending to the oral sucker, turning about, and terminating in a network of capillaries near the posterior sucker. *Development:* cercariae develop in sausage-shaped rediae lacking locomotor processes, but having a birth-pore and a short, saccular gut, as well as several gland cells and three (according to Looss, five) flame cells.

Sewell (1922) described three cercariae of this type, *C. indica* XXVI, XXIX and XXXII. Another example is the cercaria of *Cotylophoron* [=*Paramphistomum*] *cotylophorum* (see Bennett, 1936).

(*b*) '**Diplocotylea**' **type**

Cercaria diplocotylea Pagenstecher, 1857 (Fig. 63A) (the larva of *Diplodiscus subclavatus*)

Host. *Planorbis umbilicatus.*

Diagnosis. *Size:* 0·4–0·9 × 0·3–0·8 mm., the tail generally longer than the body. *Shape:* variable, elongate spindle-shape to broad oval and having a pointed tip. *Colour:* abundantly pigmented, especially in the anterior region and around the eyes, sometimes along the courses of the main nerves. *Sense organs:* eye-spots having spherical lens-like bodies. *Cystogenous glands:* numerous rounded or pyriform cells containing oval or rod-like granules, dorsal and ventral in position, filling the available space between the suckers. *Suckers:* ventral much larger than the oral, taking up one-third of the ventral surface, the oral having lateral diverticula. *Gut:* oesophagus long, thin and transversely striated, caeca extending laterally, then posteriorly almost to the posterior sucker. *Excretory system:* vesicle triangular, giving off two antero-lateral canals, unconnected by anastomoses, and a median posterior canal extending along the tail and opening by two short lateral canals near its tip. Main canals generally as

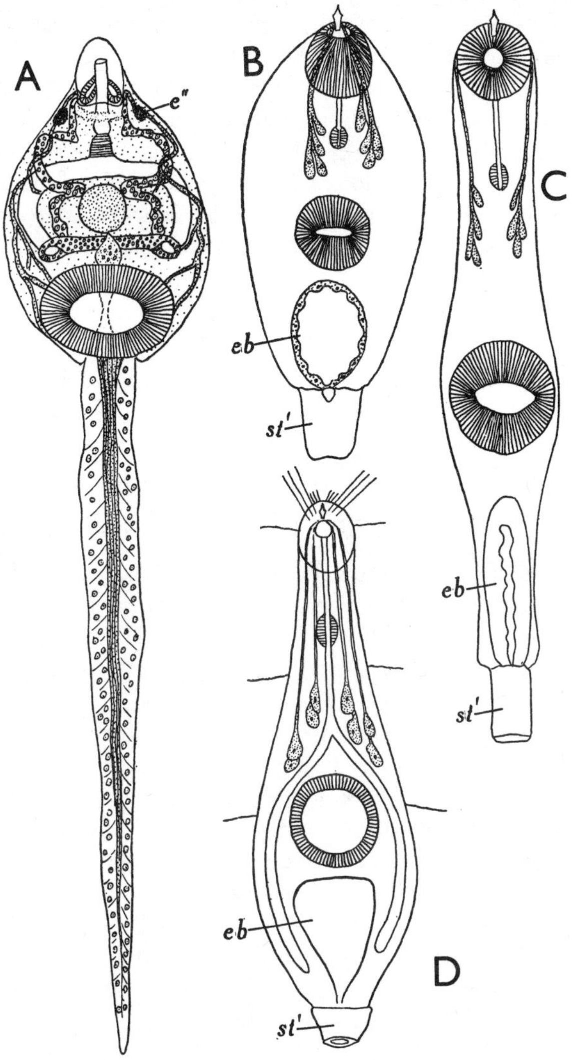

Fig. 63. An Amphistome and '*Microcercous*' cercariae. A, *Cercaria diplocotylea*. B, C, *C. micrura* in retracted and extended conditions. D, *C. micrura*. (A, B, C, after Wesenberg-Lund, 1934; D, after Lühe, 1909.)

in cercariae of '*Pigmentata*' group, similarly filled with highly refractive material. *Development:* in long, cylindrical, active rediae lacking a collar and procruscula, but having a birth pore, a large globular pharynx and an intestine, germ balls and three pairs of flame cells. The cercariae leave the redia at a very early stage and disperse in the tissues of the snail.

Notes. Lang (1892) showed that encystment occurs on the skin of frogs, infection following the ingestion of the cast stratum corneum at ecdysis, but it is possible that cysts on the bottom of ponds are also ingested. Krull & Price (1932) showed that cercariae of the American form, *Diplodiscus temperatus*, encyst when evaporation makes the water shallow, so that the larva are exposed to air, and will not encyst in aquarium tanks even when stones, leaves and other suitable objects are provided. As soon as frogs enter the aquarium, however, the cercariae encyst, showing a preference for the pigmented regions of the skin, and the final host is infected as Lang believed it to be in the case of *D. subclavatus*. Wesenberg-Lund (1934) observed that tadpoles sometimes draw cercariae into the buccal cavity along with the water which forms the respiratory stream, encystment occurring in this location. Herber (1939) found that the tadpoles of *Rana clamitans* can be infected with cercariae of *Diplodiscus temperatus* when 10 days old, or as soon as the mouth becomes open.

SOME OTHER CERCARIAE OF DIPLOCOTYLEA TYPE:

The cercaria of *Gastrodiscus aegyptiacus* (Cobbold, 1876): *Cercaria diastropha* Cort, 1915; *C. inhabilis* Cort, 1915; *C. missouriensis* McCoy, 1929. The cercaria of *Allassostoma parvum* Beaver, 1929; *Cercaria frondosa* Cawston, 1918; *C. corti* O'Roke, 1917; *C. convoluta* Faust, 1919.

(2) MONOSTOME CERCARIAE (Fig. 61 C)

This type of cercaria, as limited by Dubois (1929) and Wesenberg-Lund (1934) to the '*Urbanensis*' and '*Ephemera*' groups of Sewell, invariably develops in a redia, which in turn may arise from a miracidium without the intervention of a sporocyst stage. Development is completed only after the escape of the cercaria from the redia into the tissues of the snail. Later on, the emergent cercaria encysts on stones, water plants and the like in the open, no other intermediate host being required. The free life of the cercaria is short, possibly no more than a few minutes. Strong pigmentation of the cercaria may be correlated with the acute phototactic responses which it shows, and both light and a high atmospheric pressure seem to determine the moment of emergence from the snail, or at any rate to play an important part in its determination.

The '*Urbanensis*' and '*Ephemera*' groups of Sewell correspond roughly to the '*Binoculate*' and '*Trioculate*' groups of Faust (1917), characterized by having two or three eye-spots respectively, but it must be made clear that *Cercaria urbanensis* itself is a species with three eyes.

(*a*) **'Ephemera' type**, e.g. *Cercaria ephemera* Nitzsch, 1807

HOST. *Planorbis corneus*.

DIAGNOSIS. *Size:* body large (about 0·5 × 0·15–0·2 mm.) and capable of considerable extension, but much smaller when contracted and ready to encyst. *Colour:* anterior end heavily pigmented, especially around the eyes. Pigment bands extending back along the body to its posterior end. *Sensory organs:* young cercariae having two eye-spots, older ones also a third median eye-spot which is situated between the other two. *Adhesive pockets:* two postero-lateral pits, strengthened by infolded cuticle. *Suckers:* ventral absent, oral small (0·05 mm. diameter). *Gut:* pharynx absent, oesophagus short, caeca long. *Excretory system:* vesicle rounded, opening at the base of the tail; the main antero-lateral canals following the courses of the caeca and uniting in the median plane slightly behind the eyes; the posterior canal extending along the entire length of the tail. *Tail:* extremely contractile and ductile, equipped with circular and longitudinal muscles underneath the cuticle. *Development:* rediae very variable in shape, having a powerful pharynx, a large intestine and, as a rule, two to four cercariae ready for liberation, as well as a posterior mass of germ cells.

Notes. The redia is very mobile, but lacks definite locomotor organs. The cercariae encyst within a short time of emergence from the snail, sometimes on the shells of molluscs, but generally on water plants.

Some other Species:

Cercaria monostomi Linstow, 1896 (the larva of *Notocotylus seineti* Fuhrmann; see p. 363); *C. imbricata* Looss, 1893; *C. (glenocercaria) lucania* Leidy, 1877; *C. hyalocauda* Haldemann, 1842; *C. zostera* Sinitsîn, 1911; *C. pellucida* Faust, 1917; *C. spatula* Faust, 1919; *C. plana* Faust, 1922; *C. indica* XI Sewell, 1922; *C. trabeculata* Faust, 1924; *C. helvetica* I Dubois, 1929; *C. infracaudata* Horsfall, 1930; *C. triophthalmia* Faust, 1930; *C. lebouri* Stunkard, 1932.

(*b*) '**Urbanensis**' type

Cercariae of this type closely resemble those of the '*Ephemera*' group in the non-spinous nature of the cuticle, the heavy pigmentation, the absence of a ventral sucker and a pharynx, the presence of adhesive pockets and a spherical excretory vesicle, etc., but they are only of medium size (0·3–0·46 × 0·1–0·16 mm. long in the body), and they lack the median eye-spot (with the exception noted above). The group is represented by a number of species, including *Cercaria urbanensis* Cort, 1914; *C. konadensis* Faust, 1917; *C. fulvoculata* Cawston, 1911; *C. robusta* Faust, 1918; *C. aurita* Faust, 1918; *C. hemispheroides* Faust, 1924; and *C. yenchingensis* Faust, 1930. Some of these form the types of subgroups, e.g. the last-named of the '*Yenchingensis*' group, some members of which develop into flukes of the genus *Paramonostomum* in the caeca of ducks (Rothschild, 1941*b*).

(3) GYMNOCEPHALOUS CERCARIAE (Fig. 61 G)

Some cercariae belonging to this group are poor swimmers, live near the bottom of ponds, and are either unresponsive to light or show no special behaviour in response to light stimulation. The best-known example is the cercaria of *Fasciola hepatica*, which can subsist with very little water, perhaps needing little more than morning dew. The free life of the cercaria is generally short, normally a few hours, but possibly only a matter of minutes. Most Gymnocephalous cercariae develop in prosobranchiate snails, *Bithynia* and *Paludina* being the commonest genera of hosts. Encystment occurs on the herbage in the open, and the cysts are transferred passively to the final host, no second intermediate host being required. The metacercaria is protected by the cyst under conditions of fairly severe desiccation. In some instances encystment seems to occur in snails, and it can be induced experimentally by allowing the fluids of snails to come into contact with water containing the cercariae.

Included in this group are the '*Pleurolophocerca*', '*Parapleurolophocerca*', '*Isopori*', '*Agilis*' and '*Reflexae*' groups of Sewell. Both Dubois (1929) and Wesenberg-Lund (1934) removed the first-named from Sewell's group of Monostome cercariae, and Rothschild (1938*f*) and others have shown that there are insufficient grounds for separating it from the second group. Wesenberg-Lund indicated that Sewell's groups are not of any great value in the separation of cercariae which occur in Europe. Cercariae of the '*Pleurolophocerca*' and '*Parapleurolophocerca*' groups develop into trematodes belonging to the Heterophyidae and Opisthorchiidae, those of the latter to members of the Haplorchiinae in the former family. A number of them parasitize *Peringia ulvae* and some have been described by Lebour, others by Rothschild, the latter writer having provided a modified list of characters including the following:

Cuticle: entirely or only anteriorly beset with backwardly directed spines. *Sense organs:* a pair of pigmented eye-spots generally present, but sometimes absent. *Suckers:* anterior modified to form a protrusible organ of penetration, ventral absent, or feebly developed. *Gut:* pharynx invariably present, the rest of the gut absent, or rudimentary. *Glands:* penetration glands present, their ducts opening anteriorly. *Excretory system:* vesicle reniform, or roughly globular, and having thick walls. Main canals entering the vesicle antero-laterally. *Tail:* longer than the body and powerful, provided with a cuticular fin-fold or a cuticular sheath. *Reactions:* phototropic, swimming by short dashes, each followed by a pause, during which the body hangs below the tail. *Development:* in rediae having an intestine and a birth pore, but lacking locomotor appendages. The cercariae leave the redia at an early stage, continuing their development within the tissues of the host, which is a fresh-water, brackish-water or sea-water gasteropod. Second intermediate host generally a fish, but exceptionally an amphibian. Encystment may occur in the first intermediate host.

It must be mentioned that Rothschild (1938*f*) experienced considerable difficulty in distinguishing between related species. Measurements cannot be relied on and the excretory system is too complex to serve in a practical way. According to her, the most useful characters are:

(*a*) The shape and size of the body under a coverslip at death.
(*b*) The precise extent, position and shape of the caudal fin-folds.
(*c*) The pigmentation of the body.
(*d*) The behaviour and length of life of the cercaria.

What is referred to as the 'death attitude' is determined immediately bubbles begin to break through the cuticle, the camera lucida being used as an aid to drawing. The manner of swimming is common to the whole group, differences observed falling under a number of headings, e.g.

(*a*) The length of free-swimming life (42 hours in *Cryptocotyle lingua*: 8 hours in *C. jejuna*).
(*b*) The relative duration of passive and active periods of swimming.
(*c*) The nature and degree of the phototropic response.
(*d*) The types of intermediate host or hosts selected.
(*e*) The site of encystment.
(*f*) The behaviour of the tail after it has been cast off.
(*g*) The stage of development reached when the cercaria leaves the redia.

The inference is plain. Morphological characters, and even the swimming activities of cercariae may be insufficient, as here, for purposes of diagnosis, which may call for characteristics which seem at first sight to have little to do with taxonomy. For further details, and a list of the forms which parasitize *Peringia ulvae*, the reader is referred to Rothschild's paper.

(4) CYSTOCERCOUS CERCARIAE (Figs. 61E, 64, 65)

A number of cercariae which are able to retract the body into a large chamber hollowed out of the basal part of the tail have been referred to this group. The first of such cercariae to be described was *C. macrocerca*, Filippi, 1854, and Sinitsîn (1907) showed that the original description covers at least four species. Lühe (1909) retained the original name, *C. macrocerca*, for a form which Sinitsîn regarded as the larva of *Gorgoderina vitelliloba*, and Sewell (1922) pointed out that the real '*macrocerca*' is the larva not of this trematode, but of *Gorgodera cygnoides*.

Lühe separated the *Cystocercous* cercariae, which because of their large size are sometimes called '*Macrocercariae*', by the following characters:

I. Having a chamber containing the retracted body of the young Distome; development in sporocysts
  A. Tail flat, ribbon-like and forked distally; piercing spines apparently absent; development in sporocysts in *Limnaea* — *C. mirabilis* Braun, 1891
  B. Tail cylindrical, not forked; piercing spines present; development in sporocysts in *Sphaerium* — *Macrocercous cercariae of Gorgoderids*

II. Having a chamber containing the body of the young Distome, and also a peculiar 'tail'; development in rediae in *Planorbis* — *C. cystophora* Wagener, 1866

E. L. Miller (1936) recognized three groups which are based roughly on such characters as these, calling them '*Cystophorous*', '*Cystocercous*' and '*Macrocercous*' (*Gorgoderine*) cercariae. Sewell (1922) removed *C. mirabilis* and certain allied forms from the group '*Cystocerke*' of Lühe, arranging them with *Furcocercariae* in a group called '*Mirabilis*', with which Szidat (1932*b*) has dealt in a special publication. Wesenberg-Lund (1934) seemed to concur, and it may ultimately be proved that the *Cystocercous* cercariae are true *Furcocercariae* (i.e. forked tail), although there seems to be sufficient practical grounds for retaining them at present where Miller would have them. The cystocercous character of the tail may provide here another example of convergence, but it is thought advisable here to allow Miller's grouping to stand, more particularly because the *Furcocercariae* form a large and somewhat unwieldy group.

(*a*) '**Cystophorous**' **cercariae**

The first cercaria of this type to be described was *C. cystophora* Wagener, 1866, which was discovered in *Planorbis marginatus*. E. L. Miller (1936) listed fourteen other species, some of which occur in marine hosts, the best-known species being those described by Pelseneer (1906), namely, *C. appendiculata* (in *Natica alderi* at Boulogne-sur-Mer) and *C. vaullegeardii* (in *Trochus cinerarius* at Wimereux), which Sewell referred to a separate group called '*Appendiculata*'. One species to be added to Miller's list is *C. sinitsîni* Rothschild, 1938, which was found in *Peringia ulvae* at Plymouth. Several other species are known.

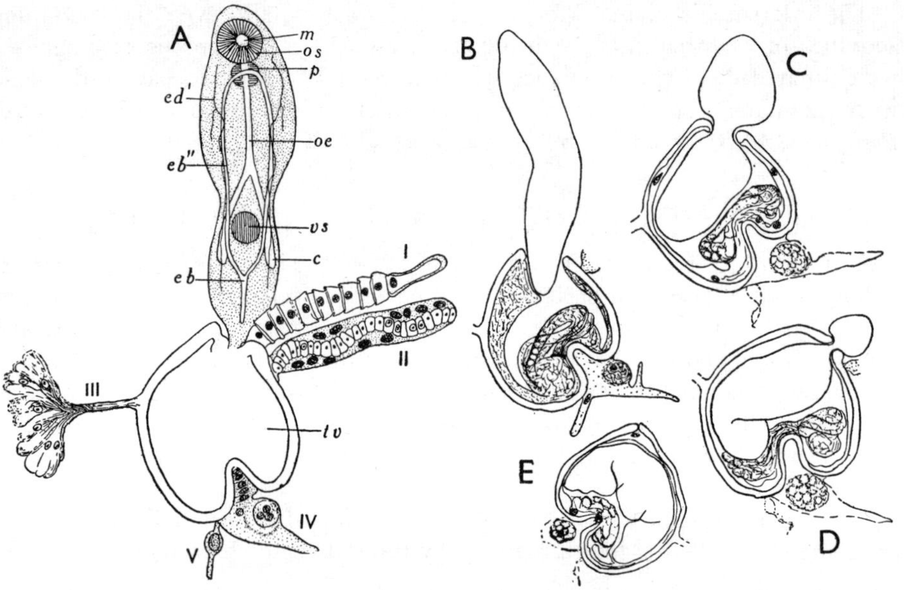

Fig. 64. A '*Cystophorous*' cercaria, *C. sinitsîni*. A, body in ventral, tail in lateral view (appendage I telescoped). B–D, stages in encystment of the cercaria (appendages II and III omitted, their bases faintly indicated). E, encysted cercaria. (After Rothschild, 1938*c*.)

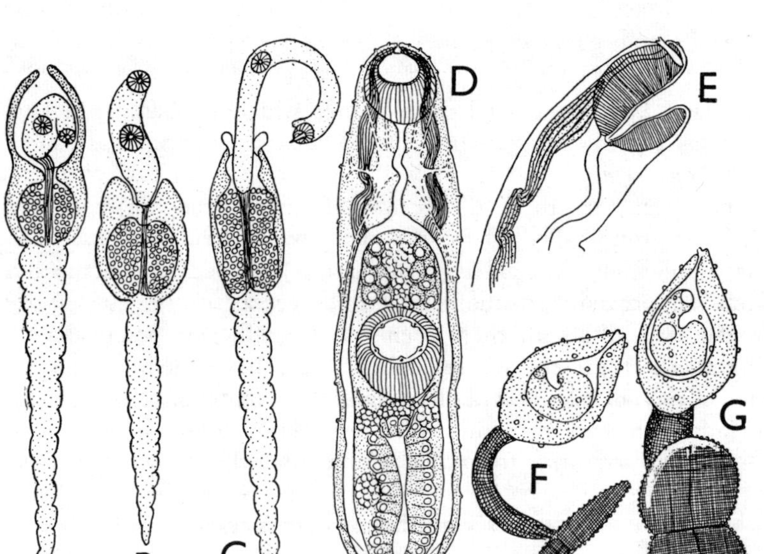

Fig. 65. '*Macrocercous*' and '*Cystocercous*' cercariae. A, B, C, *Cercaria gorgoderinae vitellilobae* in retracted, half-extended and fully extended conditions. D, E, *C. macrocerca*. F, G, *C. splendens*. In E the anterior extremity is seen in profile, the stylet shown in its normal position. (D, E, after Vickers, 1940; rest after Wesenberg-Lund, 1934.)

Only the baldest details are available about *C. cystophora*, which, according to Sinitsîn (1905), is the larva of *Halipegus ovocaudata* (Vulpian, 1859). According to Lühe, the cercaria itself is 0·15 mm. long, has an oral sucker 0·02 mm. in diameter, and inhabits a chamber 0·28 mm. in diameter. The whole cyst, together with the protracted body of the cercaria, is 0·27 mm. long, and the free 'tail' projecting from the chamber slightly longer (0·32 mm.). The cercaria develops in a redia in one or another species of *Planorbis* (e.g. *P. marginatus*), and after emergence it penetrates and encysts in the body of the larval dragon-fly, *Calopteryx virgo*.

Krull, W. H. (1935), studied the life history of *Halipegus occidualis* Stafford, 1905 and Thomas (1939) that of another American species, *H. eccentricus*, which he named. From the work of these writers it is clear that our interpretation of the structure of *Cercaria cystophora* is both incomplete and faulty. The structure of the tail is evidently much more complicated than Lühe's description indicated, calling for further study. It is interesting to find that Thomas saw in the miracidium of his species large, retractile anterior spines, because Creutzburg (1890) recorded a spiny surface for the European form at this stage. It seems that there is very probably a sporocyst stage in the life history, elongate and very mobile larvae of this kind having been found in *Halipegus eccentricus*. Thomas also showed that the metacercaria is similar to the young fluke, but has a number of finger-like posterior processes. He could not infect dragon fly nymphs with the cercariae, but found that large *Cyclops* would gorge themselves with the larvae, some of which were digested, while others made their way through the gut into the body cavity. Tadpoles in this instance became infected by ingesting the copepods and their parasites.

Cercaria sinitsîni Rothschild, 1938* (Fig. 64A)

This is an unusual larva in many respects, not least in that it is best studied in stained preparations, only the excretory system being visible in the living state (Rothschild). It develops in a redia inhabiting the gonad of *Peringia ulvae*, the diagnostic characters of the two stages being as follows:

Redia. Shape, size and colour: elongate and cylindrical, 1·00 mm. long and 0·20 mm. broad and thick pearly white, but having a conspicuous yellow intestine. *Collar and procruscula:* present. *Birth pore:* situated slightly behind the pharynx. *Gut:* length of the intestine about one-eighth of the body length. *Glands:* surrounding the pharynx. *Excretory system:* well developed, more than fourteen flame cells on either side of the body, apparently arranged in groups of four. *Movements:* very active, accompanied by violent local contractions of the body. *Contained cercariae:* numerous (more than 100) displaying all stages of development.

Cercaria. Body: shape and size: elongate, 0·06–0·10 mm. long and 0·012–0·025 mm. broad. *Cuticle:* non-spinous. *Suckers:* ventral only half as large as the oral and situated about two-thirds of the distance along the body, dimensions about 0·007 × 0·009 and 0·14 mm. diameter. *Gut:* pharynx fairly large (0·006 mm. diameter), oesophagus long, caeca short, but extending behind the ventral sucker. *Excretory system:* the main canals together Y-shaped, the lateral canals uniting above and behind the pharynx. *Cystogenous glands:* not observed.

* The original spelling was '*sinitzini*'.

(*Note.* The body is invested by an epithelium in which the nuclei are arranged in pairs; it is inserted by a short peduncle in an anterior aperture of the tail.) *Tail:* main part jug-like, having a wide mouth when viewed from the side, oviform in other views. *Size:* main part of the tail (apart from the appendages), 0·043–0·052 mm. long and 0·035–0·042 mm. broad. *Appendages:* five in number (see Fig. 64A): I, variously known as the 'Arrow', 'Excretory projection', 'Delivery tube' by various writers (in allied forms); II, the 'Ribbon' of Sinitsîn; III, the 'Sultan's plume' of Sinitsîn; IV, the 'Phrygian cap' of Sinitsîn; V, a thin outgrowth of IV. *Encystment of the cercaria:* by retraction of the body into the tail (Fig. 64B–E). *Cyst:* when fully formed measuring 0·039 × 0·039–0·043 × 0·043 mm., the aperture closed, but its position indicated by a slight thickening of the cyst wall. *Appendages:* persistent, but altered or reduced: I is folded up within the cyst; II shows degenerate nuclei; III has its stalk elongated and the free ends frayed into a number of separate strands; IV and V become mere strands of tissue without nuclei. *Excystment of cercaria:* light pressure on the cyst results in swift eversion of appendage I, immediately following which the cercaria is drawn rapidly down its tube and projected outside. *Note on the delivery mechanism:* Rothschild believed that the 'delivery tube' (appendage I) serves to project the cercarial body through the intestine wall into the body cavity of some copepod which serves as the intermediate host. Other appendages are believed to attract the copepod, which brings about delivery of the larval body from the cyst by interfering with the latter during attempts to eat it. It is scarcely possible that delivery can be as forcible as Rothschild supposes, but the mechanism seems to be an alternative to release through the agency of digestive enzymes, which seem to have no effect on the cysts of some related forms. Willey (1930), who described an American species, has reviewed the literature and discussed the mechanism.

SOME OTHER 'CYSTOPHOROUS' CERCARIAE:

Four species have been mentioned; some additional ones are (* indicates marine forms): *Cercaria capsularia* Sonsino, 1892; *C. sagittarius** and *C. laqueator** Sinitsîn, 1911; *C. yoshidae* Yoshida, 1917; *C. californiensis* Cort & Nicols, 1920; *C. syringicauda* Faust, 1922; *C. indica* XXXV Sewell, 1922; *C. calliostomae** Dollfus, 1923; *C. invaginata* Faust, 1924; *C. macrocercoides* Faust, 1926 (Syn. *C. macrura* Faust, 1921, preoccupied); *C. biflagellata* Faust, 1926; *C. projecta* Willey, 1930; the cercaria of *Lecithaster confusus** Odhner (see Hunninen & Cable, 1941*b*).

(*b*) '**Cysticercaria**' (anchor-tailed cercariae)

The first cercaria of this group to be described was *C. mirabilis* Braun, 1891. E. L. Miller (1936) listed ten species, two from Europe, *C. mirabilis* and *C. splendens* Szidat, 1932. The former is a giant among cercariae, being 6–7 mm. long, and both develop into species of the genus *Azygia*, the latter without the intervention of an intermediate host (Szidat, 1932*b*).

Cercaria splendens Szidat, 1932 (Fig. 65F, G)

Wesenberg-Lund (1934) extended Szidat's description of this species, which is peculiar in that it consists of a pyriform, hyaline chamber or 'house' containing the cercarial body and a broad, flat stalk of a brown colour which terminates in two broad plates. The latter, which may be regarded as furcal rami, are set perpendicular to the stalk and can be closed down on one another and reopened, which is roughly what happens when the cercaria moves with a sudden leaping motion from a suspended poise in the water.

DIAGNOSIS. *Chamber: shape and colour:* about 0·76 mm. long and 0·5 mm. broad, hyaline, bluish, consisting of loose parenchyma, on its surface six or seven circlets of knob-like papillae, and at its apex the opening through which the cercarial body can be withdrawn. *Peduncle:* measuring about 0·77 × 0·10 mm., ribbon-like, brown, containing powerful muscles, invariably curved. *Furcal rami:* extremely broad, almost circular, more hyaline than the peduncle, bordered with fine spines (Szidat) and/or covered with short papillae (Wesenberg-Lund), having conspicuous longitudinal and transverse muscles. *Cercarial body:* yellow in colour. Excretory canal extending along the peduncle, dividing to supply a branch to each ramus, each branch opening by a pore on one distal margin. *Development:* in a redia which resembles a sporocyst and has a pharynx [which also serves as a birth-pore (Szidat)], but no intestine, a conspicuous excretory system and germinal layer posteriorly. The cercariae leave the redia before their development is complete, nourishing themselves in the tissues of the snail. The chamber develops after liberation of the cercaria from the snail.

SOME OTHER 'CYSTICERCARIA':

Cercaria wrighti Ward, 1916; *C. anchoroides* Ward, 1916; *C. macrostoma* Faust, 1918; *C. brookoveri* Faust, 1918; *C. fusca* Pratt, 1919; *C. pekinensis* Faust, 1921; *C. stephanocauda* Faust, 1921; *C. melanophora* Smith, 1932; *C. hodgesiana* Smith, 1932.

(*c*) '**Macrocercous**' (**Gorgoderine**) **Cercariae**

(i) '**Gorgoderina**' **group** (Fig. 65 A–E)

In this group, the chamber into which the larval body can be withdrawn takes up about one-third of the total length of the tail; behind it there is a broad region containing numerous spherical cells.

Cercaria macrocerca Filippi, 1854

HOST. *Sphaerium corneum.*

LOCATION. Between the gill lamellae (together with the developing embryos of the snail). Vickers (1940) reinvestigated the structure of this species, his specimens conforming to the characters of Sewell's '*Gorgoderina*' group, the type of which is *Cercaria gorgoderinae vitellilobae* (Sinitsîn, 1905).

DIAGNOSIS (after Vickers). *Cercaria* (Fig. 65 A, B, C). *Tail: size:* the proximal, swollen region measuring 0·42 × 0·25 mm., the elongate distal region 0·8 × 0·12 mm. *Cuticle:* forming a complete covering, raised into a circular fold around the cercarial body, completely enclosing it in a chamber having a minute anterior pore (the only other connexion between body and tail being a socket formed by extension of the ventral lip of the excretory pore and a small knob at the proximal end of the tail, together with muscle fibres radiating from the knob to the posterior wall of the body). *Gland cells:* about twelve very large cells which have clear cytoplasm and spherical nuclei aggregated at the proximal end of the tail globe, and a group of small greenish cells which project beyond this chamber into the tail, continuing as a central strand to its tip. *Distal end of tail:* covered with minute spines and formed by the action of muscles into a small 'sucker', by which the larva periodically attaches itself to the substratum. *Cercarial body* (Fig. 65 D, E). *Shape and size:* when removed from the chamber, elongate oval outline, measuring 0·54 × 0·12 mm. when quiescent, but up to 0·78 mm. long when extended and as short as 0·31 mm. when contracted. *Suckers:* oral smaller than the ventral and lined by finely spinous cuticle, while the latter has a fringe of longer spines; dimensions 0·085 × 0·08 and 0·096 × 0·092 mm. *Stylet:* set dorso-ventrally in the median plane near the dorsal lip of the oral sucker, measuring 0·017 × 0·006 mm., and having three backwardly directed points. *Penetration glands:* twelve large cells and twelve ducts, two sets of six each. *Cuticle of the body:* non-spinous, but bearing an irregular double row of sensory papillae, which are more numerous anteriorly, six of them occurring round the opening of the ventral sucker. *Gut:* pharynx absent, oesophagus long and sinuous, caëca long. *Nervous system:* comprising two lateral ganglia, a wide commissure, two lateral, two posterior and four anterior nerves. *Excretory system:* vesicle wide, having a wall composed of columnar cells, packed with granules which are discharged into the lumen. Eighteen flame cells on each side of the body are arranged in groups of three. *Genital rudiments:* rudiments of the ovary and the testes situated behind the ventral sucker, rudiments of the terminal parts of the ducts in front of it and between the right and left series of penetration glands. *Behaviour of the cercaria:* after their liberation (which occurs at any time during the day or night, the daily output of a single snail averaging up to 145), the cercariae do not swim, but remain close to the snail, the body retracted into the chamber. Only moribund specimens emerge from the chamber. The cercariae used in laboratory experiments will live half as long again in normal saline as in water, and twice as long in ox serum as in saline, the maximum duration of life under these conditions being at least 60 hours. Low temperatures increase the duration of life, temperatures above 20° C. decrease it considerably. *Development:* in white, club-shaped sporocysts (measuring 2 × 0·7 mm. when fully developed) having a wrinkled cuticle and an excretory system comprising on each side a twisted excretory canal together with the anterior and posterior factors, and a total of twenty flame cells of peculiar pattern, the cilia being fused into a membrane. Birth pore absent.

Note. According to Sinitsîn (1907) the cercariae passively enter a second intermediate host, being ingested by carnivorous larval insects belonging to the genera *Epitheca* and *Agrion*. On one occasion Vickers obtained cysts from the larvae of *Chironomus pedellus*, two or three in each larva occurring amongst the anterior muscles. Further attempts to infect this host experimentally were unsuccessful.

(ii) '**Gorgodera**' **group**

The chamber is much smaller in this than in the '*Gorgoderina*' group, occupying not more than one-sixth, frequently not more than one-tenth, of the total length of the tail. E. L. Miller (1936) expressed the opinion that Sewell's separation of the two groups lacks experimental justification and, accordingly, he declined to adopt Sewell's scheme.

Cercaria gorgoderae pagenstecheri Sinitsîn, 1905

Syn. *C. macrocerca* Thiry, 1859.

The most obvious character of this species, albeit a negative one, is the absence of a swelling of the tail behind the chamber. The cercariae range between 2 and 5 mm. in length. In a cercaria 4·45 mm. long, the length of the chamber is 0·50 mm., the breadth 0·45 mm., and the tail measures 3·55 × 0·5 mm. There is no constriction between the chamber and the more distal part of the tail, of which the part nearer to the chamber is filled with large hyaline cells having small nuclei. *Cercarial body. Suckers:* in cercariae extracted from the chamber, oral slightly smaller than the ventral, diameters 0·06 and 0·07 mm.; in cercariae without a tail, oral larger than the ventral, diameters 0·12 and 0·10 mm. *Stylet:* having three points. *Penetration glands:* comprising six cells on either side of the ventral sucker. *Excretory system:* the vesicle tubular, extending almost to the ventral sucker, the lateral canals having anterior and posterior factors, the median canal extending from the vesicle into the tail; situated at the sides of the vesicle there are seven cells, diminishing in size in a posterior direction and having granular contents. *Development:* in sporocysts, which never contain fully developed cercariae. While the cercaria remains in the sporocyst, or in the mollusc for that matter, the chamber in the tail is undeveloped, developing later when the larva has left tissues of the mussel, but remains suspended between the gape of the valves.

SOME OTHER 'MACROCERCOUS' CERCARIAE:

Cercaria gorgoderinae vitellilobae Sinitsîn, 1905 (= *C. macrocerca* Filippi, 1854 of Lühe); *C. gorgodera cygnoides* Kowalewski, 1904; *C. g. loossi* Sinitsîn, 1905; *C. g. varsoviensis* Sinitsîn, 1905; *C. sphaerocerca* Miller, 1935; *C. mitocerca* Miller, 1935; *C. conica* Goodchild, 1939.

(5) TRICHOCERCOUS CERCARIAE (Figs. 62F, 66A–J)

Lühe erected this distinctive group, but mentioned only *Cercaria major* Nitzsch, 1817, which is a translucent, milky white larva about 2–2·5 mm. long having a slender tail which is about as long as the body, is marked out into annuli, and bears numerous bristles near the posterior end. Our knowledge of this type of cercaria is still fragmentary, but at least one species occurs in the plankton off our coasts, the cercaria of *Opechona* [= *Pharyngora*] *bacillaris*. Lebour (1916–18*b*), who found the cercaria, stated that the tail is several times as long as the body, the bunches of long bristles which occur at regular intervals along it giving the larva the appearance of an annelid. The tail is an efficient swimming organ, the bristles no doubt assisting flotation. Other characters to some extent indicate the genus to which the adult trematode belongs, the oral sucker having the typical *Opechona* form, resembling a pharynx, although a true pharynx is also present. The entire larval body is spinous, the gut is well developed, and there is a pair of large reniform eye-spots (Fig. 62F, *e''*).

The first intermediate host of this trematode has not been discovered, the cercaria having been found free in the plankton, but it is probably a mollusc. Late cercariae have been found in *Sagitta* and clinging to the manubrium or the wall of the 'stomach' in medusae like *Cosmetira pilosella*, *Turris pileata* and *Obelia* sp. The first of these hosts was the commonest one in early summer when *Opechona* was most abundant, but in late summer *Obelia* frequently contained

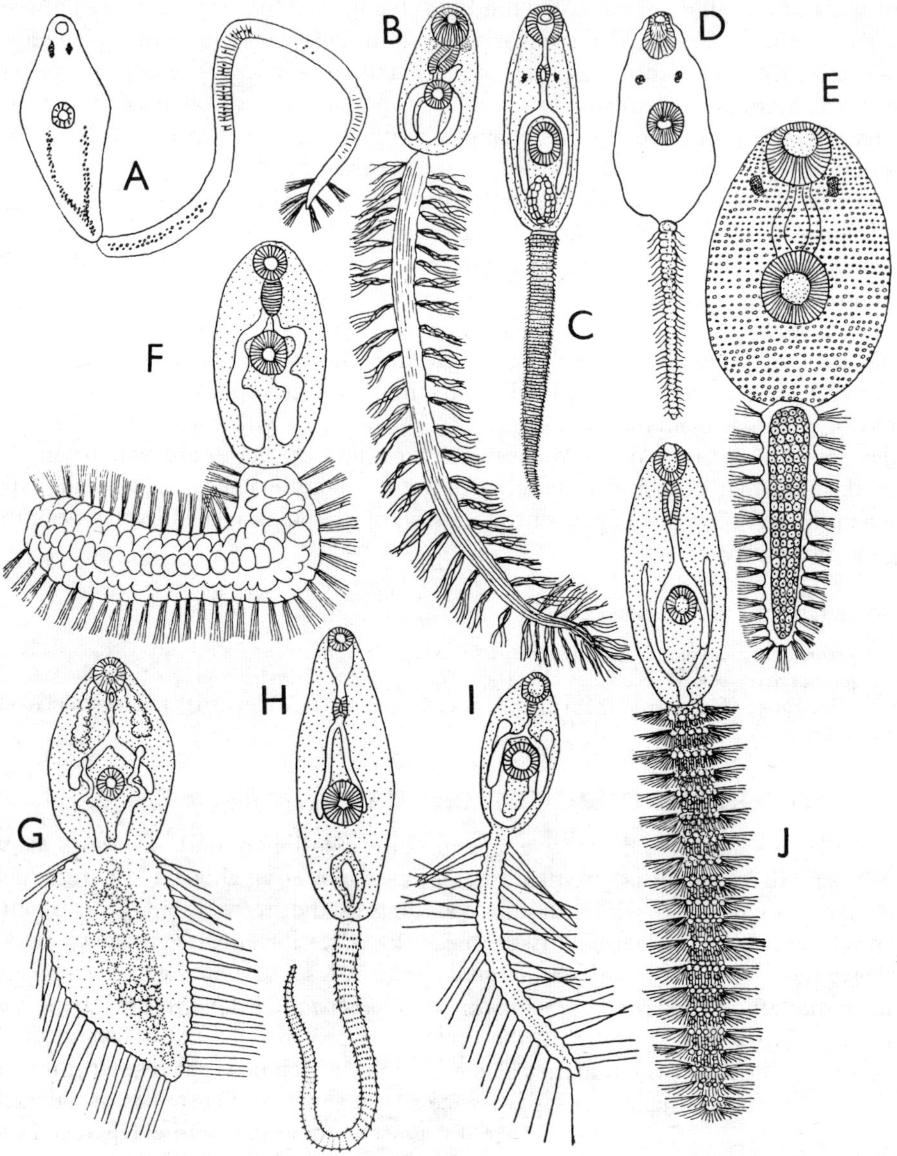

Fig. 66. *Trichocercous* cercariae. A, *Cercaria elegans* Müller. B, cercaria of *Lepocreadium album*. C, *Cercaria myocerca* Villot. D, *C. quintareti* Dollfus. E, *C, fascicularis* Villot. F, *C. setifera* Müller. G, *C. pelseneeri* Monticelli. H, *C. myocercoides* Pelseneer. I, *C. chiltoni* Dollfus. J, *C. villoti* Monticelli. (After Dollfus, 1925*a*.)

the cercariae. *Phialidium hemisphericum* also served as a host in later summer and autumn, even in December, and at times late larvae have been found, devoid of a tail (Fig. 62 G), in the 'stomach' of the ctenophore *Pleurobrachia pileus*. The frequency with which such pelagic organisms were infected (sometimes every specimen in a haul) convinced Lebour that this is a case of true parasitism, an opinion which was strengthened by the absence of the cercariae in other planktonic animals caught in the same hauls.

Dollfus (1925*a*) divided marine Trichocercariae into two groups, according to the presence or absence of eye-spots. Our knowledge is scanty, but the following forms can be listed along with the fragmentary information which is available.

(*a*) Cercariae having eye-spots

Cercaria elegans Müller (in La Valette St George, 1855) (Fig. 66A). (Tail has twelve bundles of spines on each side; not three, as figured.) Found in the sea near Marseilles.

The cercaria of *Lepocreadium album* (Stossich, 1890) (Fig. 66B). (Tail up to four times as long as the body, having twenty-seven to twenty-eight bundles of five bristles on each side.) Like the cercaria of *Opechona bacillaris*, it becomes attached to various pelagic animals, but has been found free in the plankton at Naples.

Cercaria echinocerca Filippi, 1855. (Tail not as long as the body and having about twenty annuli.)

C. thaumantiatis Graeffe, 1860. (Tail long and having about twenty-five annuli each bearing a pair of bristles.) Sometimes occurring in ctenophores; found at Nice and Naples.

Macrurochaeta acalepharum Costa, 1864. (Tail about half as long again as the body and having forty-nine bundles of bristles along each side; ventral sucker much larger than the oral.) Found at Naples.

Cercaria echinocerca Lankester, 1873. Found at Naples.

Histrionella setocauda Daday, 1888. (Tail about three times as long as the body and having twenty-five setigerous annuli each bearing four bundles of bristles.) Found at Naples.

Cercaria of *Opechona bacillaris* (mentioned above).

Cercaria claparedei sp.inq. Dollfus, 1925. (Tail four or five times as long as the body and having nineteen pairs of bundles of bristles.) Found free in the plankton in the English Channel.

C. fascicularis Villot, 1875 (Fig. 66E). (Tail about as long as the body and having rigid spines arranged in fifteen pairs of bundles.) Found at Roscoff in *Nassa reticulata*.

C. myocerca Villot, 1878 (Fig. 66C). (Tail narrow and having about sixty-nine annuli bearing very short bristles, resembling the tail of a rat.) Found at Roscoff.

C. quintareti sp.inq. Dollfus, 1925 (Fig. 66D). (Tail similar to that of *C. myocerca*, but having fewer annuli and longer bristles.) Found at Banyuls.

(*b*) Cercariae lacking eye-spots

Cercaria setifera Müller (in La Valette St George, 1855) (Fig. 66F). (Tail half as long again as the body, having spines arranged in up to thirty pairs of bundles, four to five in a bundle.) Found at Trieste.

C. villoti Monticelli, 1888 (Fig. 66J). (Tail very extensile, when extended half as long again as the body, having 14–21 circlets of bristles arranged radially in the transverse plane.) Found at Roscoff.

C. pelseneeri Monticelli, 1914 (Fig. 66G). (Tail large and flat, and having twenty-six to twenty-eight bundles of long bristles on each side.) Found at Boulogne.

C. myocercoides Pelseneer, 1906 (Fig. 66H). (Tail half as long again as the body and slender, having numerous (sixty-seven?) annuli and as many circlets of very short bristles.) Found at Boulogne.

C. chiltoni Dollfus, 1925 (Fig. 66I). (Tail twice as long as the body and slender, having twenty-one to twenty-four bundles of long bristles on each side.) Found in New Zealand.

C. pectinata Huet, 1891. (Tail twice as long as the body and having twenty-seven annuli, each bearing a pair of bundles of six long bristles.) Found at Wimereux and Boulogne.

In addition, Dollfus mentioned two species of Trichocercariae which are found in fresh water. One of them is *Cercaria major* Nitzsch, 1817, found in *Limnaea* and *Planorbis* in Germany; the other an unnamed species in South Africa, *Cercaria* sp. Gilchrist, 1911.

(6) ECHINOSTOME CERCARIAE (Figs. 61H, 67)

In this group are placed cercariae which are easily recognized by the 'head collar' and 'collar spines' and are further characterized by a complete gut, including long caeca. They are vigorous swimmers and move swiftly by the action of a long and powerful tail, but they frequently live near the bottom of ponds and spend the last stages of their free life creeping on it. They are common in pulmonate snails, and seem to occur most frequently in species of *Limnaea*. Wesenberg-Lund found them much less frequently in species of *Planorbis*, and not at all in *Bithynia*. Some species seem to be phototactic, yet not to a marked degree. Propagation is mainly by rediae and takes place all the year round. Many generations of rediae may follow one another, each snail containing thousands of them, yet the numbers of cercariae produced are relatively small, each redia giving rise to only a few, and these generally in summer. The rediae are amongst the most highly differentiated of their kind; the well-developed collar and procruscula give them good powers of locomotion, and the large pharynx is an indication of their voracity. The cercariae generally encyst in a second intermediate host, which is frequently a mollusc, sometimes the same species as the first host.

Several writers have expressed the opinion that Echinostome cercariae cannot be separated in small natural groups (Cort, 1915*a*; Dubois, 1929; Wesenberg-Lund, 1934; E. L. Miller, 1936). Sewell recognized four groups called '*Echinatoides*', '*Coronata*', '*Echinata*' and '*Megalura*'. The last-named was erected by Cort (1915*a*), and E. L. Miller does not concur with Sewell as to its inclusion here. The earliest known forms of the first three groups were divided into four kinds by Lühe on the following basis:

I. Redia having a long, sinuous gut
   1. Tail of the cercaria having a fin-fold — *C. echinatoides* Filippi, 1854
   2. Tail of the cercaria lacking a fin-fold — *C. coronata* Filippi, 1855

II. Redia having a short, club-shaped gut
   1. Tail of the cercaria lacking a fin-fold — *C. echinata* Siebold, 1837
   2. Tail of the cercaria having a fin-fold — *C. spinifera* La Valette, 1855

Sewell apparently did not recognize these distinctions between the form of the gut in the rediae, because he included *C. spinifera* and *C. echinatoides* in the same group, '*Echinatoides*'. Dubois (1929) found some evidence of two groups in the details of the excretory system; in one type of cercaria, e.g. that of *Echinostoma revolutum*, the flame-cell formula was $2[(3+3+3)+(3+3)+(3)+(3+3)] = 48$; in another, e.g. the cercaria of *Echinoparyphium aconiatum*, $2[(4+4+4)+(4+4)+(4)+(4+4)] = 64$. These types would seem to represent the groups '*Echinata*' and '*Coronata*'. We can examine all three groups of Sewell.

(*a*) '**Echinata**' **group**

Cercaria echinata Siebold, 1837 (Fig. 67B) (the larva of *Echinostoma revolutum*).

HOST. *Limnaea stagnalis.*

LOCATION. Digestive gland.

Sewell placed this cercaria not in this group, but in '*Coronata*'. I have followed Wesenberg-Lund in letting it stand here, where it represents the type in Lühe's scheme.

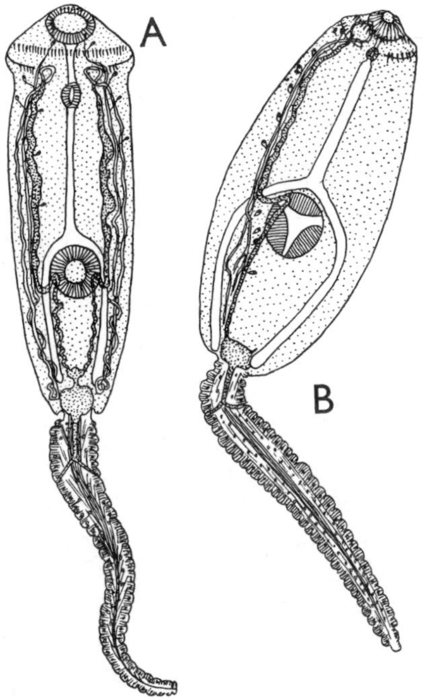

Fig. 67. Three Echinostome cercariae. A, the cercaria of *Hypoderaeum conoideum.* B, *C. echinata.* C, the cercaria of *Echinoparyphium recurvatum.* (After Wesenberg-Lund, 1934.)

DIAGNOSIS (after Brown, 1926). *Cercaria: Size:* body comparatively large (0·58–0·70 mm. long), tail slender, slightly longer than the body (0·76 mm. in length and 0·105 mm. in breadth where it meets the body). *Collar:* having thirty-seven collar spines arranged in a double row, which is continuous dorsally, four spines on each side being larger than the remainder and forming corner spines. *Cuticle:* spinous anteriorly. *Suckers:* ventral larger than the oral and well behind the middle of the body, dimensions 0·10 × 0·095 and 0·065 mm. diameter. *Gut:* prepharynx short, pharynx present, oesophagus long, caeca extending to the posterior end of the body. *Glands:* cystogenous gland cells distributed throughout the body, partially obscuring other gland cells in the anterior region, the latter connected by twelve ducts opening separately by pores around the mouth. *Excretory system:* vesicle large and rounded, main lateral canals uniting near the vesicle, then extending in three loops between the ends of the body, the median posterior duct extending along the proximal sixth of the tail, then opening by a pair of transverse ducts with lateral pores, the flame cells numbering forty-three on each side of the body. *Genital rudiments:* comprising three masses of small, rounded cells, one in front of and lateral to the margin of the ventral sucker, the others in line between this sucker and the excretory vesicle. *Development:* in elongate, orange-coloured rediae 0·71–3·5 mm. long, having a collar and procruscula and a short, sac-like gut. *Encystment of the cercaria:* after leaving the snail the cercariae swim for a time, then penetrate a snail of the same or another species, there to encyst near the salivary gland and oesophagus, generally on the left side of the body. *Limnaea stagnalis* and *L. pereger* serve as second intermediate hosts, but the cysts occur also in the mantle of *Sphaerium corneum*, sometimes in *Paludina vivipara*. They are thick-walled, 0·24 mm. diameter and each contains a juvenile fluke so curled up that the anterior and posterior regions of the ventral surface are in contact. A fluke extracted from the cyst measures 0·71 × 0·16 mm. and has larger collar spines than are found in the cercariae prior to encystment, also a ventral sucker situated in the anterior half of the body and approximately twice as large as the oral (0·12 mm. diameter as against 0·064 mm.). Already it shows close resemblance to the adult *Echinostoma revolutum*.

OTHER SPECIES:

The cercaria of *Hypoderaeum conoideum* Mathias, 1925 (Fig. 67A)

HOSTS. *Limnaea ovata*, *L. stagnalis*, *L. pereger*; apparently not *Planorbis*.

LOCATION. Sporocyst containing rediae in the 'lung'.

DIAGNOSIS. Collar spines relatively numerous (50–54); a slight constriction occurring behind the collar; ventral sucker slightly larger than the oral and situated behind the middle of the body. (*Notes*. Mathias (1925) showed that ducks which had been fed with snails containing cysts harbour *Hypoderaeum conoideum* in the intestine. He infected species of *Limnaea* with miracidia, which seemed to show a preference for *L. ovata*.)

Nöller & Wagner (1923) believed that encystment may occur in tadpoles. F. G. Rees (1932) found this and another species of Echinostome cercaria encysted in the mantle cavity and digestive gland of *Lymnaea pereger* in South Wales, in no instance finding free-swimming cercariae.

(*b*) '**Coronata**' **group**

Spines of the collar of about equal size; cuticle spinous; tail lacking a fin-fold; cystogenous cells abundant; main excretory canals extending to the pharynx, recurring, and dividing opposite the ventral sucker.)

The cercaria of *Echinoparyphium recurvatum* Mathias, 1926 (Fig. 67C)

HOSTS. *Planorbis umbilicans*. According to Harper (1929) *Valvata piscinalis*.

LOCATION. Digestive gland.

DIAGNOSIS. *Cercaria: Size:* body small and ovate, measuring 0·25 × 0·12 mm. *Collar:* bearing forty-three to forty-five collar spines arranged in a double row, the oral row containing the smaller spines. *Cuticle:* spinous anteriorly, the spines becoming progressively smaller towards the ventral sucker. *Suckers:* ventral larger than the oral and situated in the final third of the body. *Tail:* blunt, stout and muscular, about as long as the body and having a terminal caudal pocket. *Gut:* oesophagus long, caeca extending almost to the posterior extremity. *Excretory system:* vesicle ovoid or rectangular, the lateral canals slender to the level of the ventral sucker, wider, bending outwards and filled with granules farther forward, recurring behind the collar, and branching opposite the ventral sucker, the median canal passing along the tail and bifurcating in its anterior third. *Development:* in inert, yellow rediae 1·0–1·5 mm. long and 0·18–0·22 mm. broad and thick, having a collar and a birth pore, a prominent pharynx and a short intestine, numerous constrictions occurring in the cuticle between the anterior extremity and the collar. Rarely more than six cercariae are contained in a single redia, but a snail can produce 50–100 cercaria per day. (*Note.* The young redia is colourless and mobile.)

According to Harper (1929), whose observations have been drawn upon, the cercariae encyst in the same species of host, fifty or more cysts occurring in the digestive gland and tissues of the mantle in *Valvata piscinalis.* Fewer cysts (3–10) were found in *Planorbis albus.* The cysts are transparent, spherical, 0·11–0·13 mm. in diameter; the contained metacercariae practically quiescent, having forty-five collar spines and displaying the rudiments of the gonads. When the cysts are fed to ducklings the larvae emerge and develop in the intestine of the final host into the adult *Echinoparyphium recurvatum*, the eggs of which are recoverable from the faeces 18 days afterwards.

(*c*) '**Echinatoides**' group

(Having a fin-fold on the tail)

Cercaria echinatoides Filippi, 1854

Syn. *C. echinifera* La Valette, 1855.

HOST. *Paludina vivipara.*

LOCATION. On gonads and heart(?)

DIAGNOSIS (after Lühe, 1909). *Cercaria: Size:* 0·19–0·58 mm. long and 0·05–0·26 mm. broad. *Collar:* bearing more than forty collar spines, of which the corner spines are slightly larger than the remainder, respective lengths being 0·016 and 0·013 mm. *Suckers:* ventral much larger than the oral and situated behind the mid-body, diameters 0·059 and 0·26–0·033 mm. *Tail:* 0·52 mm. long and 0·065 mm. broad, tapering posteriorly and having both a dorsal and a ventral fin-fold. *Redia: Size:* 0·3–1·5 mm. long and 0·07–0·23 mm. thick. *Gut:* pharynx well developed but small, intestine extending almost to the posterior extremity. *Procruscula:* present.

SOME ECHINOSTOME CERCARIAE OF FRESH WATER:

Cercaria echinata Siebold, 1837; *C. echinatoides* Filippi, 1854; *C. coronata* Filippi, 1855; *C. spinifera* La Valette, 1855; *Cercaria* 7 Nakagawa, 1915; *C. catenata* Cawson, 1917; *C. trisolenata* Faust, 1917; *C. chisolenata* Faust, 1918; *C. acanthostoma* Faust, 1918; *C. constricta* Faust, 1919; *C. echinostomi xenopi* Porter, 1920; *C. cucumeriformis* Faust,

1921; *C. indica* XII, XX, XXIII, XLVIII, Sewell, 1922; *C. serpens* Faust, 1922; *C. cristacantha* Faust, 1922; *C. chekiensis* Faust, 1924; *C. limnicola* Faust, 1924; *Echinostoma* C, B and A Tsuchimochi, 1926; cercaria of *Hypoderaeum conoideum* Mathias, 1925; *C. isodorae* Faust, 1926; *C. granulosa* Brown, 1926; cercarium of *Echinoparyphium flexum* McCoy, 1928; *C. rebstocki* McCoy, 1929; *C. mehrai* Farugui, 1930; cercaria of *Euparyphium murinum* Tubangui, 1932; *C. palustris* Chatterji, 1933; *C. equispinosa* Brown, 1926; *C. helvetica* II, XX–XXVI, XXXII, Dubois, 1929; *C. oscillatoria* Brown, 1931; *C. limbifera* Seifert.

SOME MARINE ECHINOSTOME CERCARIAE:

Cercaria leptosoma Villot, 1878; *C. purpurae* Lebour, 1907; *C. patellae* Lebour, 1907; cercaria of *Echinostoma secundum* Lebour, 1912; *C. littorinae obtusata* Lebour, 1912; *C. quissetensis* Miller & Northup, 1926.

SOME ECHINOSTOME CERCARIAE WITHOUT COLLAR SPINES:

Cercaria agilis Filippi, 1857; cercaria of *Himasthla militaris* Beneden, 1861; *C. reflexae* Cort, 1914; *C. fusiformis* O'Roke, 1917; *C. arcuata* Cawson, 1818; *C. complexa* Faust, 1919; *C. penthesilla* Faust, 1921; *C. indica* XLI Sewell, 1922; *C. semirobusta* Faust, 1924; *C. pseudoechinostoma* Faust, 1924; *C. redicystica* Tubangui, 1928; *C. chitinostoma* Faust, 1930.

(7) **MICROCERCOUS CERCARIAE** (stumpy tail) (Figs. 61 K, 63 B–D)

This is one of the lesser-known groups of cercariae, though, in recent times, a number of both marine and fresh-water species have been described. Lühe separated the three best-known forms of his day in the following way:

I. In Prosobranchiata; cercariae creep forth. Pharynx and boring-spine present
   A. Development in sporocysts in *Bithynia tentaculata* — *Cercaria micrura*, the larva of *Sphaerostoma bramae*
   B. Development in rediae in *Neritina fluviatilis* — *Cercaria myzura*

II. In *Dreissenia polymorpha*; cercariae encyst without emerging from the mother sporocyst. Pharynx and boring-spine absent — *C. catoptroidis macrocotylis*

Cercaria myzura Pagenstecher, 1881 was found only once prior to this time (by Braun, 1893?), and Wesenberg-Lund failed to find it in about 200 individuals of the host genus.

Dollfus (1914) separated a number of short-tailed cercariae subsequently discovered from other Microcercous forms, establishing the '*Cotylocercous*' group of cercariae, which develop in marine gasteropods within simple sporocysts and have an elongate body and a very short cup-like tail which has thick walls and large cells and which acts as an organ of adhesion. Other significant characters are the presence of a stylet, penetration glands and a large non-bifurcate excretory vesicle which almost fills the region between the posterior extremity and the ventral sucker and has cellular walls of glandular appearance. Stunkard (1932) thought there was sufficient evidence for regarding these forms as distinct. In this group Dollfus placed *Cercaria pachycerca* Diesing, 1858* and other marine

* This form is *C. brachyura* Lespés, 1857, renamed by Diesing because he used the name earlier (1850) for a different larva. As Stunkard has pointed out (1932), the term '*Cercaria*' is not a generic name, so that the correction is invalid and the name *C. brachyura* Lespés still stands.

forms, but he excluded *C. micrura* Filippi, 1857 because it is a fresh-water form. The first members of the group were described by Lespés in 1857 as *C. brachyura* and *C. linearis*. Sewell (1922) separated the latter (along with *C. pachycerca* Diesing, 1858, *C. buccini* Lebour, 1912 and *C. indica* XXXVIII), placing them in a separate group called '*Linearis*', referring *C. micrura* to the alternative group. Palombi (1938*a*) described three new '*Cotylocercous*' cercariae, and listed twelve others, which are named at the end of this section. Hunninen & Cable (1941 *a*) have pointed out that *Podocotyle atomon* has a '*Cotylomicrocercous*', not a *Trichocercous*, cercaria as suggested by Palombi (1938*b*), so that this must be added to the list. It is said to be similar to the cercaria of *Opecoeloides manteri*, both larvae having a double-pointed stylet, three pairs of penetration glands and a flame-cell formula 2[(2 + 2) + (2 + 2)], the chief differences being the size of the sucker and of the tail. It might be mentioned that recent descriptions of two cercariae which were referred provisionally to *C. linearis* and *C. brachyura* Lespés are readily available to British readers (see Stunkard, 1932), and we can now examine some of the remaining forms briefly. McCoy (1928, 1929*c*, 1930) suggested that '*Cotylocercous*' cercariae are the larvae of the Allocreadiidae. Some undoubtedly are, but others belong to other families, e.g. the Zoogonidae and Gorgoderidae. Nor is this type alone found in the Allocreadiidae, which, as has been mentioned, is a somewhat unnatural assemblage of trematodes.

Cercaria micrura Filippi, 1857 (the larva of *Sphaerostoma bramae*) (Fig. 63 B–D)

HOST. *Bithynia tentaculata*.

LOCATION. Digestive gland.

DIAGNOSIS (after Lühe). *Cercaria: Shape and size:* elongate, tapering anteriorly, about 0·1 mm. long and 0·04 mm. broad. *Cuticle:* bearing six spines about 0·009 mm. long, at anterior extremity, and alongside them three long bristles, and three others laterally, respectively opposite the oral and ventral suckers and the mid-point between them. *Suckers:* ventral larger than the oral and behind the mid-body, diameters 0·06 and 0·04 mm. *Stylet:* present. *Gut:* prepharynx, oesophagus and caeca relatively long, pharynx well marked. *Tail:* very short and stump-like, about 0·03 mm. long and 0·04 mm. broad. *Excretory system:* vesicle large. *Sporocysts:* tubular, very contractile, up to 2 mm. long.

Note. The emergent cercariae creep under the skin of the snail and aggregate there, notably in the vicinity of the tentacles. When the snail approaches a leech of the genus *Herpobdella* the cercariae emerge, penetrate this second intermediate host and encyst beneath the skin. The subcutaneous cysts are ovoid, measuring 0·2 × 0·1 mm. and each contains a juvenile fluke about 0·42 mm. long and 0·15 mm. broad.

Wesenberg-Lund (1934) examined this cercaria without finding the spines or bristles in the cuticle. He noted that the enormous excretory vesicle is elliptical or triangular, having thick, glandular walls, and that the tail has a small posterior separated from a larger anterior region. McCoy (1929*a*) remarked that *C. micrura* differs from most '*Cotylocercous*' cercariae in having a well-developed alimentary canal.

Cercaria catoptroidis macrocotylis Lühe, 1909 (the larva of *Catoptroides* [= *Phyllodistomum*] *macrocotyle* (Gorgoderidae)

HOST. *Dreissenia polymorpha.*

LOCATION. Gills.

Cercariae of this species develop, twelve to fourteen at a time, in large sporocysts 4–8 mm. long. The tail is about one-tenth as long as the body. The cercariae encyst within the sporocyst, which leaves the gills of the snail, rises in the surrounding water, becomes a free-swimming organism, and is ultimately ingested by a cyprinid fish, in which the juvenile flukes emerge, to wander about in the gut for 24 hours before passing into the urinary passages.

The cercaria of *Zoogonoides viviparus* (Zoogonidae) (Fig. 68 A–F)

HOST. *Buccinum undatum.*

LOCATION. Digestive gland.

DIAGNOSIS (after Lebour, 1916–18*c*). *Size, shape and colour:* when full grown 0·33–0·48 mm. long, oval or elongate in outline according to state of contraction or extension (B and C), the anterior end the more rounded, the greatest breadth in front of the ventral sucker, colourless and translucent. *Cuticle:* spinous, the spines being larger towards the posterior extremity, much larger at this point. *Tail:* replaced by a peculiar sucker-like disk (F) (which may be inconspicuous owing to change in shape (B)), in the centre of which the excretory pore opens (F). *Suckers:* ventral larger than the oral and situated behind the middle of the body, diameters 0·10 and 0·06 mm. *Stylet:* 0·015 mm. long and having a long central and smaller lateral points (E). *Penetration glands:* grouped between the oral and ventral suckers, the openings of the ducts occurring on either side of the stylet. *Gut:* prepharynx short, pharynx 0·03 mm. diameter, oesophagus short and narrow, caeca reaching only to about the level of the centre of the ventral sucker, the walls of the caeca in section comprising very few cells (sometimes only two) having large nuclei. *Gonads:* rudiments of the testes well developed, but compact, ovoid and situated side by side behind the ventral sucker. Ovary and vitellaria undifferentiated, but represented by masses of nuclei. *Excretory system:* vesicle an oblong sac having thick walls, its opening posteriorly situated on a small papilla. *Development:* in sporocysts which are 0·5–1·0 mm. long and one-quarter or one-half as broad, faintly yellow, and contain germinal balls and one to eight cercariae in various stages of development.

Note. The infected digestive gland of the mollusc has a characteristic sickly grey colour, which can be seen to best advantage through a small aperture cut in the spire of the shell. The further history of the cercaria remains unknown, but its structure suggests that penetration into a second intermediate host occurs.

OTHER '*COTYLOCERCOUS*' CERCARIAE (listed by Palombi (1938*a*)) (Marine):

Cercariae A and B Miller, 1925*b* (respectively the larvae of *Hamacreadium mutabile* and *H. gulella* both Linton; see McCoy, 1929*c*, 1930); Cercariae M and I Miller, 1925; *C. searlesiae* Miller, 1925; *Cercariae pisaniae, stunkardi, ruvida* and *tridenata*, all Palombi.

Note. Dobrovolny (1939*a*) included *Cercaria cotylura* Pagenstecher 1862 and *C. pachycerca* Pelseneer, 1906, and the cercaria of *Anisoporus manteri* Hunninen & Cable, 1940 must be added.

'Microcercous' Cercariae of Fresh Water:

Cercaria micrura Filippi, 1857; *C. myzura* Pagenstecher, 1881; *C. columbellae* Pagenstecher, 1862; *C. limacis* Moulinié, 1856; *C. trigonura* Cort, 1915; certain eastern forms (e.g. the cercaria of *Paragonimus westermanii*; *Cercaria* H Yeoheda and *Cercaria* A Kobayashi).

Note. Dobrovolny (*loc. cit.*) included in a shorter list the cercariae of three species of *Plagioporus* (*siliculis, virens, sinitsîni*), the cercaria of *Allocreadium angusticolle*, and two new cercariae, *C. dioctorenalis* and *C. trioctorenalis*. Cable (*Trans. Amer. Micr. Soc.* **58**, 1939, 62–2) described two new species, *Cercaria trichocephala* and *C. abbrevistyla*, having previously (*J. Parasit.* **21**, 1935, 436) described *C. trichoderma*.

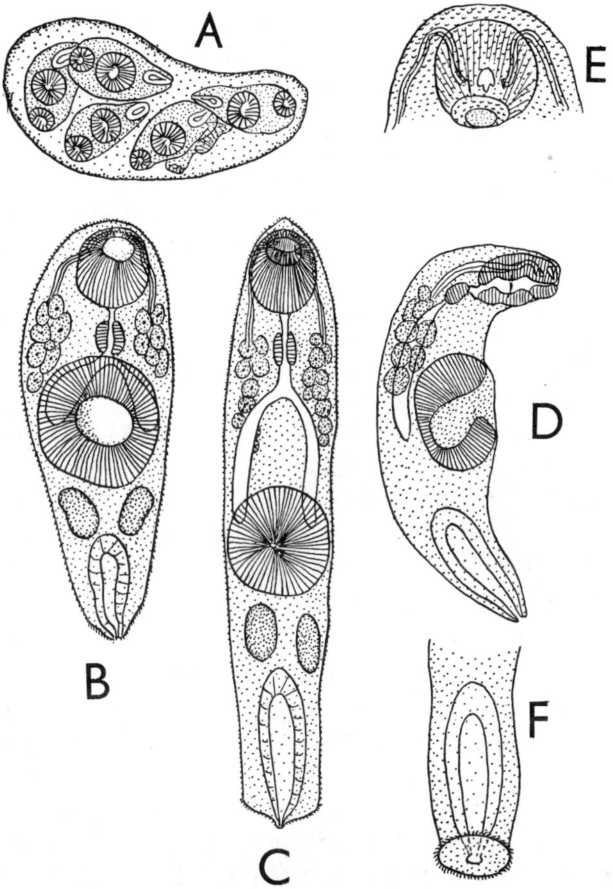

Fig. 68. The larval stages of *Zoogonoides viviparus*. A, sporocyst and contained cercariae; from the digestive gland of *Buccinum undatum*. B, D, contracted cercariae in ventral and side views. C, extended cercaria. E, anterior end of cercaria showing stylet and ducts of penetration glands. F, posterior end of cercaria showing the disk-like 'sucker', bordered by spines. (After Lebour, 1916–1918*c*.)

(8) XIPHIDIOCERCARIAE (stylet cercariae) (Figs. 61 I, 69 A–I)

Xiphidiocercariae, which are characterized by the stylet and associated penetration glands, are poor swimmers, rarely pelagic or with powers of flotation. When the larva must swim the body is strongly contracted and the tail lashes in all directions, but these movements soon give way to a caterpillar-like creeping motion. Free life is short and the whole development occurs in the sporocyst, separate life in the tissues of the snail occurring in exceptional instances. The sporocysts, which multiply by budding and partition, liberate huge swarms of cercariae, generally in bright sunshine and when the barometer stands high, the water around the snail becoming turbid with the larvae. Encystment in the open does not occur, the cercariae entering the body of an invertebrate such as an insect, or more rarely a fish. When conditions are unfavourable, the cercariae may encyst in the snail in which they developed, moribund individuals actually encysting in excised fragments of the digestive gland. The miracidia of some species seem to seek out particular hosts (snails), for Wesenberg-Lund never found *C. virgulae* or *C. vesiculosa* in snails of any other genus than *Bithynia*, and *Planorbis* and *Limnaea* spp. caught with the same sweep of the net as well infected *Bithynia* sp. were invariably free of them.

The chief characters by which different Xiphidiocercariae are distinguished from one another are the relative sizes of the suckers, the position of the ventral sucker, the nature of the gut, the number and position of the penetration glands, and the presence or absence of a fin-fold along the margins of the tail. Lühe recognized four groups, called *Cercaria Microcotylae*, *C. Virgulae*, *C. Ornatae* and *C. Armatae*. Lebour (1911) placed a number of forms in another group called '*Spelotrema*', and Cort (1915*a*) added another called '*Polyadenous*'. Sewell (1922) subdivided the four groups of Lühe: *Microcotylae* into '*Cellulosa*', '*Pusilla*', '*Parapusilla*' and '*Vesiculosa*'; *Virgulae* into '*Virgula*' and '*Paravirgula*'; Ornatae into '*Prima*' and *Cercaria ornata* La Val. (the larva of *Pleurogenes claviger*); and *Armatae* into '*Polyadena*' (of Cort) and '*Daswan*'.

McMullen (1937*b*) proposed the emendation of the superfamily Plagiorchioidea Dollfus, 1930 to include all trematodes which develop from this kind of cercaria, claiming that members of the family Plagiorchiidae develop from cercariae belonging to the '*Polyadena*' group.

(*a*) **Cercariae Microcotylae**

Lühe erected this group for small cercariae, less than 0·2 mm. long, in which the ventral sucker is smaller than the oral and situated behind the mid-body, the tail is not forked, not provided with a fin-fold and not of very different length from the body. Sewell regarded the group as representing the most primitive Xiphidiocercariae. There are two to four penetration glands and a simple bicornuate excretory vesicle. The flame-cell formula seems to vary, for Sewell gave it as $2[(1+1+1)+(1+1+1)]$, Dubois as $2[(2+2+2)+(2+2+2)]=24$.

(i) '**Cellulosa**' **group** (*having a very simple excretory system and only two pairs of penetration glands*)

Sewell (*loc. cit.*) included in this group a species which Looss (1896) found in Egypt and he himself found in India. Wesenberg-Lund (1934) also recorded this species, *Cercaria cellulosa*, in Denmark, so that there is a possibility of its occurrence in Britain, and it is worth while considering the diagnostic characters.

Cercaria cellulosa Looss, 1896 (Fig. 69A)

HOST. *Paludina vivipara.*

LOCATION. Sporocysts around the kidney, but not in it.

DIAGNOSIS. *Cercaria: Size:* body, 0·175 mm. long and 0·095 mm. broad, tail 0·145 mm. long and 0·025 mm. broad. *Cuticle:* spinous, the spines very small and more prominent in the anterior region. *Stylet:* having a small and acute lateral enlargement. *Suckers:* ventral smaller than the oral and situated a little behind the mid-body, diameters 0·025 and 0·030 mm. *Penetration glands:* numbering two on each side and located in front of ventral sucker. *Gut:* pharynx present, oesophagus and caeca apparently absent. *Excretory system:* vesicle having short cornua, the lateral canals dividing into anterior and posterior factors slightly in front of the ventral sucker, one posterior canal extending into the tail. *Other characters:* the interior of body filled with greyish droplets which render it opaque. (*Note.* When the larva is swimming the form of the body varies, but it may be broad and heart-shaped, and strongly-curved in profile.) *Development:* in irregularly ovoid sporocysts, which are half as long again as broad, grey in colour and scaly in appearance, and contain ten to fifteen germ balls in all stages of development.

(ii) '**Vesiculosa**' group

Sewell* erected this group for small *Microcotylous* cercariae with an anterior pair of penetration glands and three more pairs in a postero-lateral position. Wesenberg-Lund* found one such form with the following characters:

Cercaria vesiculosa Diesing, 1855 (Fig. 69B)

HOST. *Bithynia tentaculata.*

LOCATION. Digestive gland.

DIAGNOSIS. *Cercaria: Size:* body 0·265 mm. long and 0·07 mm. broad, tail 0·255 × 0·015 mm. *Colour:* brown. *Cuticle:* spinous. *Stylet:* having a wing-like projection one-third of the length of the tip. *Penetration glands:* four pairs; one pair in front of the ventral sucker having relatively dark and granular cytoplasm; the three pairs lateral to this sucker being of rather bright appearance and having finely granular contents; the ducts of the posterior group of cells opening near the stylet, the remaining ducts at the base of the sucker. *Suckers:* ventral much smaller than the oral and situated behind the middle of the body, diameters 0·030 and 0·050 mm. *Gut:* comprising pharynx, oesophagus and caeca. *Excretory system:* vesicle bicornuate in the extended larva, almost flat in the state of contraction, the lateral canals bifurcating in front of the ventral sucker, the posterior canal extending into the tail. *Development:* in sporocysts which are small and sausage-shaped (0·335 × 0·19 mm.), and contain three or four mature and a number of immature cercariae.

Notes. The digestive gland of the host contains cysts (enclosing metacercariae) as well as sporocysts and cercariae. Cysts may occur within the sporocysts. Infected may be twice as large as uninfected snails.

* Frequent references are made to these zoologists in this chapter and, unless it is otherwise stated, the particular papers referred to are Sewell (1922) and Wesenberg-Lund (1934).

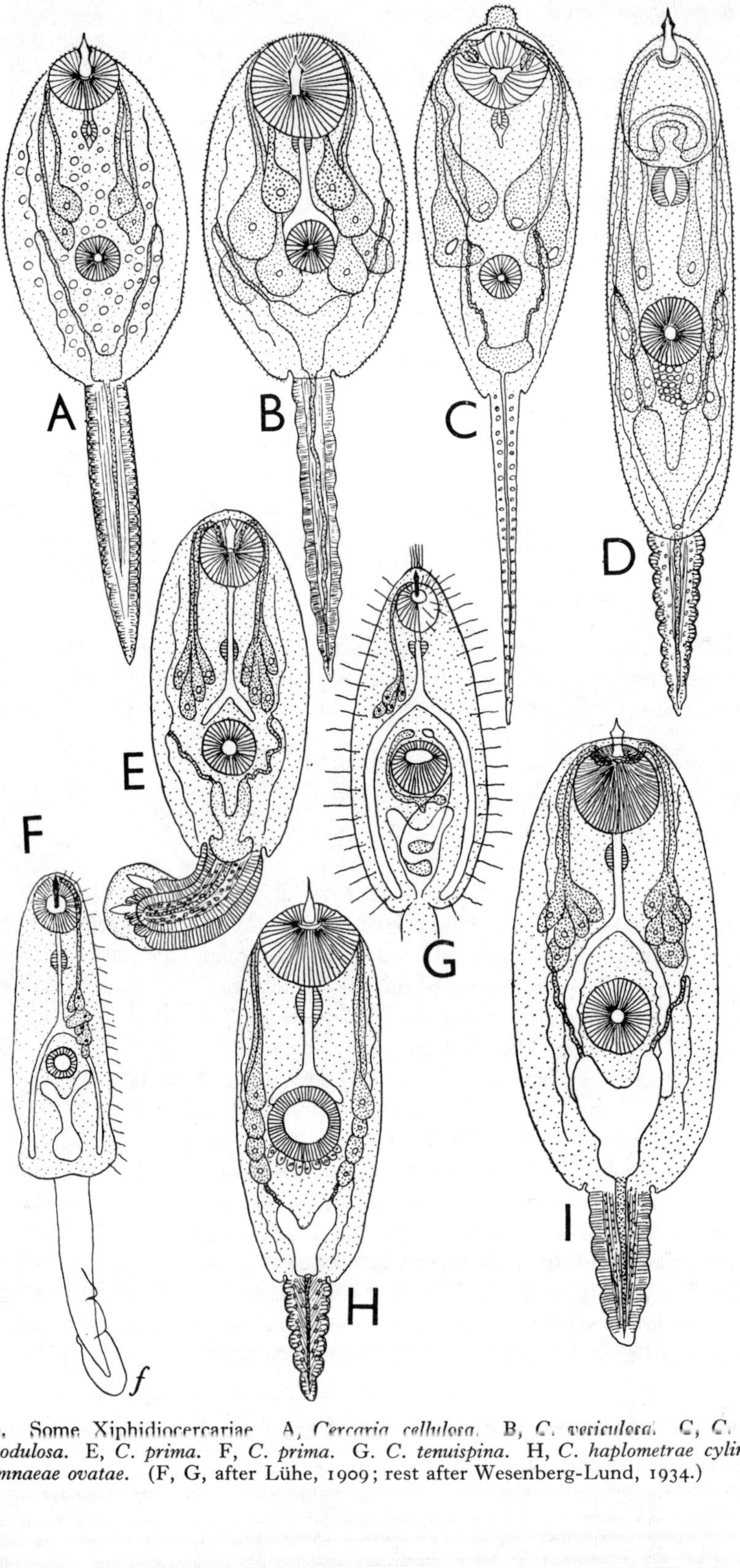

Fig. 69. Some Xiphidiocercariae. A, *Cercaria cellulosa*. B, *C. vesiculosa*. C, *C. pusilla*. D, *C. nodulosa*. E, *C. prima*. F, *C. prima*. G. *C. tenuispina*. H, *C. haplometrae cylindracea*. I, *C. limnaeae ovatae*. (F, G, after Lühe, 1909; rest after Wesenberg-Lund, 1934.)

(iii) '**Pusilla**' group

This was erected by Sewell for small Xiphidiocercariae having a slender tail, three or four pairs of penetration glands, an undeveloped gut, a small ventral sucker (smaller than the oral), a bicornuate or reniform excretory vesicle, which develop in small ovoid sporocysts. The characters here given for the type, *Cercaria pusilla*, are based on a description by Wesenberg-Lund.

Cercaria pusilla Looss, 1896 (Fig. 69C)

HOST. *Bithynia tentaculata.*

LOCATION. Digestive gland.

DIAGNOSIS. *Cercaria: Size:* body small (0·17 mm. long and 0·075 mm. broad), tail not quite as long as the body (0·15 × 0·015 mm.). *Cuticle:* spinous, the spines very delicate. *Stylet:* lacking conspicuous lateral thickenings. *Penetration glands:* two pairs of large size, those of the anterior pair granular, their ducts having peculiar median outgrowths. *Suckers:* ventral much smaller than the oral and situated in the final third of the body, diameters 0·02 and 0·035 mm. *Gut:* pharynx small, oesophagus rudimentary, caeca absent. *Excretory system:* vesicle reniform, the anterior canals bifurcating into anterior and posterior factors slightly in front of the ventral sucker, the posterior canal extending throughout the length of the tail. *Sporocysts:* very small (0·16 × 0·08 mm.), sometimes globular, containing two or three mature cercariae and the same number of germ balls, sometimes as many as five. (*Note.* The sporocysts may be exceedingly numerous, the entire liver being permeated with them.)

SOME OTHER CERCARIAE MICROCOTYLAE:

Cercaria chlorotica Diesing, 1850; *C. brunnea* Ercolani, 1850; *C. microcotyla* Filippi, 1854; *C. pugnax* La Valette, 1855; *C. subulo* Pagenstecher, 1861; *C. exigua* Looss, 1896; *C. pseudomata* Lühe, 1909; *C. brevicaeca* Cort, 1915; *C. leptacantha* Cort, 1915; *C. haskelli* O'Roke, 1917; *Cercariae indicae* V, XVI, XVIII, XIX, XL, XLIV, XLVI, LI and LIX Sewell, 1922; *Cercariae helveticae* XI, XII and XXVIII Dubois, 1929; *C. cordiformis* Wesenberg-Lund, 1934; *C. cystorhysa* E. L. Miller, 1935; *C. meniscadena* E. L. Miller, 1935; *C. cyclica* E. L. Miller, 1936.

(*b*) **Cercariae Virgulae**

Lühe erected this group for spinous Xiphidiocercariae in which the ventral sucker is smaller than the oral, the tail lacks a fin-fold, the excretory vesicle almost V-shaped, and a highly characteristic '*virgula*' organ present, consisting of two pyriform sacs which are fused in the median line and have their pointed ends directed forward and situated near the oral sucker. He mentioned five forms, the best known of which are *Cercaria virgula* Filippi, 1837 and *C. vesiculosa* Diesing, 1855. Wesenberg-Lund regarded the former as perhaps indistinguishable from *C. nodulosa*, which Lühe separated from it only by the size of the sporocysts and the numbers of contained cercariae. Both forms develop in *Bithynia tentaculata*, *Cercaria vesiculosa* in *Paludina vivipara*. According to Dubois (1929) the flame-cell formula in this group is $2[(2+2+2)+(2+2+2)]=24$, i.e. the same as in *Cercariae microcotylae*.

Cercaria nodulosa Linstow, 1873 (Fig. 69D)

DIAGNOSIS. *Shape and size:* body of variable shape, about 0·28 mm. long and 0·18 mm. broad, tail 0·24 × 0·04 mm.* *Cuticle:* spinous anteriorly. *Penetration glands:* four pairs, the anterior pair club-shaped and situated in front of the ventral sucker. *Suckers:* ventral smaller than the oral, diameters 0·045 and 0·065 mm. *Excretory system:* vesicle with two short cornua, lateral canals extending to or slightly in front of the ventral sucker, there dividing into anterior and posterior factors, the posterior canal extending along the whole length of the tail. *Gut:* pharynx present, oesophagus and caeca apparently absent. *Sporocysts:* small, almost globular, containing only one or two cercariae each, never more than four to six.

Note. Wesenberg-Lund found this cercaria at all seasons of the year, and he studied its movements closely. It is an active swimmer and can remain suspended in mid-water with apparent ease. When about to ascend, the cercaria arches its body, keeps the concave side uppermost, suddenly turns from side to side, moving the tail with great force. When the larva is pausing in mid-water the tail is extended fully and the body remains quiescent. On the bottom, the cercaria rests on its back.

(*c*) **Cercariae Ornatae**

Lühe placed in this group Xiphidiocercariae having a tail provided with a fin-fold. He separated two species known in his time as follows:

I. Fin-fold extending along the entire tail; ventral sucker larger than the oral — *C. ornata*

II. Fin-fold extending only along the distal half of the tail; ventral sucker smaller than the oral — *C. prima*

Sewell separated the latter species and certain related forms from the former and others in a subgroup called '*Prima*'. Other groups have been made, but need not be considered here.

Cercaria ornata La Valette, 1855 (according to Linstow, the larva of *Pleurogenes claviger*)

HOST. *Planorbis corneus.*

LOCATION. Digestive gland.

DIAGNOSIS (after Lühe). *Cercaria: Size:* body 0·23–0·5 mm. long when contracted, tail very contractile, cylindrical, but tapering posteriorly and 0·2–0·6 mm. long, also having an unmistakable fin-fold. *Suckers:* ventral much larger than the oral and located somewhat behind the mid-body, diameters 0·07 and 0·03 mm. *Gut:* pharynx and oesophagus present, caeca extending only slightly behind the ventral sucker, the bifurcation of the intestine being well in front of it. *Development:* in cylindrical, yellow sporocysts containing numerous cercariae.

* According to Lühe the tail is only half as long as the body, but Wesenberg-Lund's figures clearly show this to be due to change in form, which is extreme.

'Prima' group

Cercaria prima Sinitsîn, 1905 (Fig. 69E, F)

HOSTS. *Planorbis vortex*, *P. albus* and *Aplexa hypnorum*.

DIAGNOSIS (after Lühe). *Cercaria: Size:* body 0·31–0·35 × 0·04–0·1 mm., tail 0·22–0·24 mm. long, but capable of contracting to half this length, tapering sharply in its posterior half. *Tail-fin:* median, dorsal and ventral, confined to the posterior half of the tail. *Cuticle:* having numerous bristles characteristically arranged; two longitudinal rows of fourteen to sixteen each, two arch-like rows on the ventral surface and behind the mouth, and two pairs of isolated bristles between the suckers. *Stylet:* 0·025 mm. long, slightly thickened behind the point. *Penetration glands:* five pairs situated slightly in front of the ventral sucker. *Suckers:* ventral slightly behind the mid-body, somewhat smaller than the oral. *Gut:* prepharynx and pharynx present, oesophagus long, caeca extending almost to the posterior extremity. *Excretory system:* vesicle Y-shaped. *Development:* in small sac-like sporocysts 0·4 mm. long, the cercariae being liberated in spring, particularly on sunny days, duration of free life only about 15 hours, encystment occurring in ephemerid larvae, *Corethra* and *Ilybius*, cysts yellow and spherical, 0·1 mm. diameter.

Note. This diagnosis can be supplemented by observations made by Wesenberg-Lund, who examined immature cercariae. When the body is 0·29 mm. long, the diameters of the oral and ventral suckers are 0·055 and 0·050 mm., the stylet is rather short and has marked lateral projections, the penetration glands number four on each side, the fin-fold is very broad at the rounded end of the tail, and is there folded in a characteristic manner (Fig. 69E). Sporocysts each contain not more than two cercariae.

SOME OTHER CERCARIAE ORNATAE:

Cercaria hemilophura Cort, 1914; *C. racemosa* Faust, 1917; *C. trifurcata* Faust, 1919; *C. indicae* XXIV and XXVIII, Sewell, 1922; *C. helveticae* VII Dubois, 1928; *C. longistyla* McCoy, 1929; *C. mesotyphla* E. L. Miller, 1935; *C. laticauda* Wesenberg-Lund, 1934.

(*d*) Cercariae Armatae

Lühe defined this group as Xiphidiocercariae having a tail and a body of about equal lengths, lacking a fin-fold and a 'virgula' organ, having oral and ventral suckers of almost equal size (the latter perhaps being slightly the larger and situated just behind the middle of the body), and a Y-shaped excretory vesicle. He mentioned ten species, most of which are imperfectly known. Cort erected a group called '*Polyadena*' for the American species *C. polyadena* Cort, 1914 and *C. isocotylea* Cort, 1924, but also *C. limnaeae ovatae* Linstow, 1884 and *C. secunda* Sinitsîn, 1905. Sewell erected a group called '*Daswan*' for two Indian forms, tentatively suggesting the future inclusion into the '*Daswan*' group, or one near it, of *C. tenuispina* and *C. limnaeae ovatae*, which have been regarded as the larvae of species of *Opisthioglyphe*. The subgroups of the *Cercariae Armatae* are thus rather ill-defined, and both Wesenberg-Lund and E. L. Miller (1935) were unable to concur with Sewell as regards his proposed grouping, both neglecting it when dealing with this type of cercariae. We can be content to examine typical members of the major group.

Cercaria tenuispina Lühe, 1909 (the larva of *Opisthioglyphe ranae*) (Fig. 69G)

Syn. *C. gibba* Sinitsîn, 1906, *nec* Filippi, 1854.

HOSTS. *Limnaea stagnalis* and *L. palustris: L. ovata* and *L. palustris* in the Elbe; see Komiya (1939).

DIAGNOSIS (after Lühe, but modified by the findings of Komiya which are shown in parentheses). *Cercaria:* body ovoid or spindle-shaped, 0·4 mm. long and 0·06 mm. broad, but when contracted only half this length and twice the breadth (0·31–0·35 × 0·12–0·13 mm.); tail longer than the body, 0·45 mm. long (shorter than the body, 0·22–0·25 × 0·04–0·05 mm.). *Cuticle:* spinous, spines very fine and 0·0016 mm. long, arranged in transverse rows which become more widely separated posteriorly, leaving the extremity smooth. *Stylet:* uniformly slender, 0·03 × 0·003 mm. (0·018 × 0·03 mm. and provided with a ring-like thickening at the base of the point). *Penetration glands:* four pairs, situated in front of the ventral sucker (six pairs opening at the margin of the mouth near the point of the stylet; six more pairs opening above the oral sucker). '*Sensory hairs*'*:* characteristically arranged; a triple coronet around the mouth comprising six short bristles in the inner and ten longer in each of the outer rows, two additional rows at the anterior tip of the body, around the stylet, and two lateral rows of ten to twelve each along each side of the body. *Suckers:* ventral larger than the oral and situated behind the mid-body, diameters 0·07 and 0·06 mm. (oral measuring 0·05–0·07 × 0·065–0·08 mm.; ventral close behind the mid-body and 0·06–0·07 mm. diameter). *Gut:* prepharynx short, pharynx small and 0·036 mm. diameter (0·026–0·030 mm. long), oesophagus present, caeca long (fairly long). *Excretory system:* vesicle Y-shaped, the stem about 0·07 mm. long, the cornua 0·05 mm. (flame-cell formula in young cercariae $2[(3)+(3)]$, in mature cercariae $2[(3+3+3)+(3+3+3)]=36$). *Genitalia:* evident as rudiments, that of the ovary on the right close behind the ventral sucker, those of the testes one behind the other posteriorly, genital pore indicated slightly in front of the ventral sucker. *Development:* in sac-like sporocysts 0·2–4·6 mm. long, the emergent cercariae penetrating tadpoles.

Note. According to Komiya the cercariae encyst in the snail in which they developed, or in a nearly related species, ripe cysts from *Limnaea ovata* measuring 0·17–0·27 mm. diameter and containing metacercariae 0·3–0·4 mm. long, which (when 14 days old) will develop experimentally in frogs, but not in goldfish, ducks or mice.

Cercaria limnaeae ovatae Linstow, 1884 (Fig. 69I) (the reputed larva of '*Opisthioglyphe hystrix*' = *Dolichosaccus rastellus*)

HOST. *Limnaea ovata.*

DIAGNOSIS (after Lühe). *Cercaria:* body 0·31 × 0·25 mm., tail 0·16 × 0·06 mm. *Cuticle:* having fine parallel striations, but lacking spines. (*Note.* Lühe is ambiguous about the nature of the cuticle (referring to superficial intersecting lines in which no spines are recognizable), but Wesenberg-Lund was more categoric and is quoted here. *Stylet:* 0·029 mm. long, having a thickening behind the tip. *Penetration glands:* Lühe did not mention them and Wesenberg-Lund stated that 'five' exist, but his figure indicates five pairs situated at the sides of the bifurcation of the gut. *Suckers:* ventral smaller than the oral, diameters about 0·07 and 0·1 mm. (smaller and more nearly equal in size according to Wesenberg-Lund). *Gut:* pharynx present, oesophagus fairly long, caeça long. *Excretory system:* vesicle small and having two short cornua. *Development:* in elongate sporocysts, 3·2 mm. long, 0·48 mm. broad and thick, the emergent cercariae penetrating various larval insects (*Limnophilus* spp., *Anabolia* spp., *Ephemera* spp.) there to encyst, cysts 0·18–0·30 mm. long.

Note. Wesenberg-Lund described interesting experiments with Xiphidiocercariae from *Planorbis corneus*, *Limnaea ovata* and *L. stagnalis*. *Cercaria limnaeae ovatae* immediately attacked *Corethra* larvae when these insects were introduced into the vessel containing the cercariae, and in an incredibly short time the cercariae had penetrated the skin and could be seen creeping about (minus the tail, which is thrown off during penetration) inside the body. Within a few hours the insect larvae were packed with cysts numbering several hundreds and had succumbed.

THE CERCARIA OF *HAPLOMETRA CYLINDRACEA* (Fig. 69H)

HOSTS. *Limnaea ovata* and *L. stagnalis*.

LOCATION. Digestive gland.

DIAGNOSIS. *Cercaria*: the body measuring 0·33 × 0·12 mm., the tail 0·31 mm. long, 0·05 mm. broad at the base. *Stylet:* 0·04–0·046 mm. long and slender, sharply pointed and lacking lateral thickenings. *Penetration glands:* four or five on each side of the body (Wesenberg-Lund). *Suckers:* oral the larger, diameters 0·077 and 0·051 mm. *Gut:* prepharynx short, pharynx well developed, oesophagus long, caeca short. *Excretory system:* cornua of vesicle extending only half-way to ventral sucker. (*Note.* Body is opaque because filled with cystogenous cells (Wesenberg-Lund)). *Development:* in yellowish sporocysts measuring 2·35 × 0·3 mm., the emergent cercariae penetrating larval insects (*Ilybius fuliginosus*) there to encyst in the body cavity. Cysts thick-walled, spherical, 0·34 mm. in diameter, the contained Distome being 0·66 mm. long and 0·13 mm. broad anteriorly, 0·098 mm. posteriorly.

OTHER CERCARIAE ARMATAE:

This large group contains well over fifty species occurring in various parts of the world, at least a dozen being American. European forms include those described and also *C. secunda* Sinitsîn, 1905; *Cercariae helveticae* IV, V (=VII), VI, XXVII and XXX Dubois, 1929; *C. gracilis* Wesenberg-Lund, 1934; *C. elbensis* Komiya, 1939, etc.

(9) FURCOCERCOUS CERCARIAE (Figs. 61J, 70, 71A–H)

Lühe defined this group as Distome cercariae generally developing in sporocysts and having a long, forked tail into which the body cannot be retracted. He separated nine forms according to the presence or absence of eye-spots, the degree of definition between tail stem and forks, the nature of the preceding stage in development (redia or sporocyst) and the hosts. One of his other major groups, *Lophocercariae*, characterized by a long fin-fold on the body, was included by Sewell (1922) in the group of Monostome cercariae, but is more appropriately referred to this group, which has grown considerably since Lühe's time. Another group of Sewell's, '*Lophoides*', also belongs here, and strigeid cercariae of the '*Proalaria*' and '*Strigea*' types as well.

It is impossible to deal with this formidable group in any detail, but the scheme of classification adopted by H. M. Miller (1926*b*) must be mentioned. This writer reviewed the Furcocercariae known at that time, of which there were about 126 and eight additional marine forms, but excluded Gasterostome cercariae from consideration. He separated furcocercous cercariae into those which lack a pharynx (*apharyngeal*) and those which possess one (*pharyngeal*). Each category was further divided according to the relative lengths of the tail stem and the furcal rami into two groups; in one of them (*brevifurcate*) the rami are less than half as long as the tail stem, in the other (*longifurcate*) they are quite as long and may be longer than the tail stem. The four groups thus formed are further subdivided, as far as they can be, into Distome and Monostome cercariae, and these into small groups denoted by letters (which really refer to characteristics of the excretory systems). The scheme has much to commend it, especially where comprehensiveness is aimed at. A clear impression of three important characters is gained when we think of the larvae of *Schistosoma japonicum*, *S. haematobium* and *S. mansoni* as apharyngeal brevifurcate Distome cercariae. We shall find it sufficient, however, to refer to several groups based on the terminology of Lühe, including his *Lophocercariae* and Gasterostome cercariae of the *Bucephalus* groups, as well as Strigeid forms. One group, '*Cysticercaria*', which Wesenberg-Lund included along with the Furcocercariae, has been considered (as *Cercaria splendens*) with other '*Cystocercous*' forms.

(*a*) **Bucephalus group**

Lühe included under the heading of Gasterostome cercariae forms having two symmetrical furcal rami arising from a basal cushion, a centrally situated mouth and a simple, sac-like gut. He mentioned only one form, *Bucephalus polymorphus*, but other Bucephalids must be included in the group.

Bucephalus polymorphus Baer, 1827.

HOST. Species of *Anodonta* and *Unio*.

This well-known cercaria has been thoroughly studied, the anatomy by Ziegler (1883), the biology by Wunder (1924) and Wesenberg-Lund (1934), and the life history and germ-cell cycle by Woodhead (1929, 1930; see 1931). It is unmistakably characterized by the fresh-water hosts and may be regarded as a Furcocercaria having a very short tail stem and very well-developed furcal rami. One of the most interesting of planktonic organisms, it lives mainly in mid-water and, by peculiar vertical movements, attracts the attention of cyprinid fishes, especially the rudd, beneath whose skin it encysts. It has powers of contraction and elongation unequalled in any other cercaria and, when poised in mid-water, carries the tail upturned and the body projecting downwards. For brief periods of rest it remains motionless on the bottom, right side up, with the furcal rami folded over the body in a characteristic manner (Fig. 70A), but generally it is engaged in making vertical ascents and descents in the water, during which the furcal rami show considerable changes of form. When at the surface, the furcae

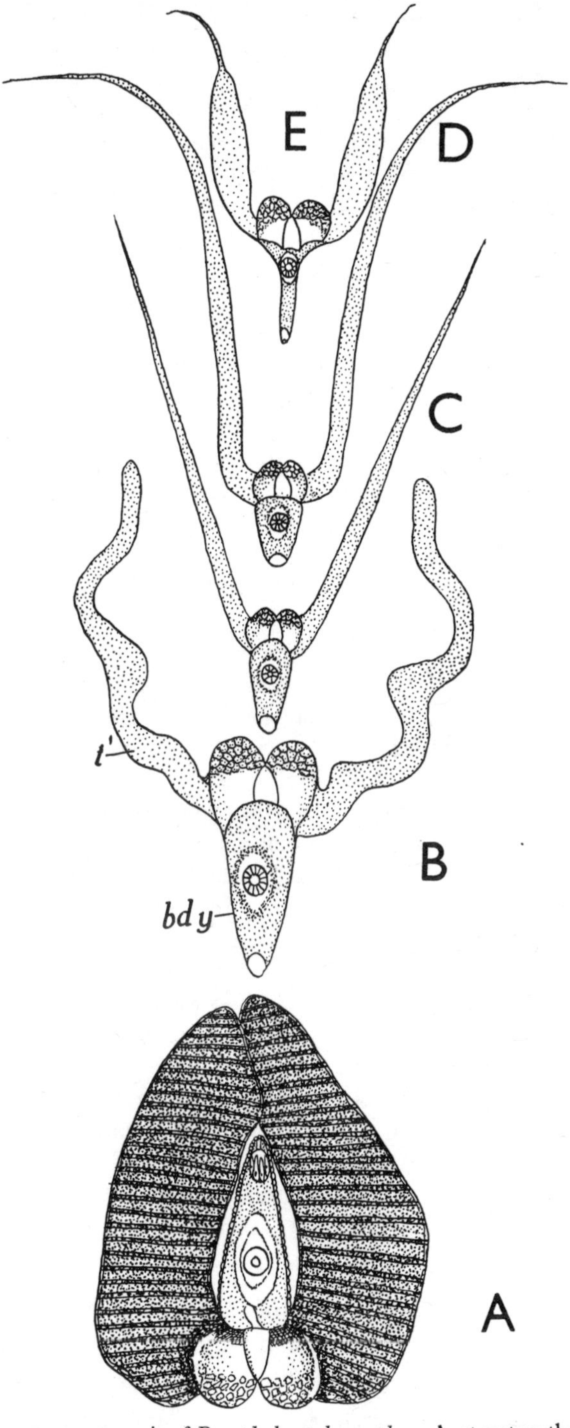

Fig. 70. The Furcocercous cercaria of *Bucephalus polymorphus*. A, at rest on the bottom. B, C, D, at successive stages during ascent in the water. E, as the larva is beginning slowly to sink again. (After Wesenberg-Lund, 1934.)

are fully extended and laterally curved (Fig. 70 D), the larva literally resting on the tips, buoyancy being restored as the larva tends to sink slowly by slight contractions of the rami. Primary ascent from the bottom is accompanied by and consequent upon gradual extension of the furcae, descent by their contraction. Wesenberg-Lund likens the whole process to constant treading of water, which is scarcely disturbed. He points out that most Furcocercariae are endowed with powers of locomotion with which to counteract the tendency to sink, the furcae being exclusively organs of flotation, but in *Bucephalus* both purposes are served by one organ, the furcal rami. Other Furcocercous cercariae depend on the furcae only for counteracting descent, using the median stem of the tail for powerful swimming motion. *Bucephalus* seems to be incapable of horizontal motion, but is able to carry out an uninterrupted series of ascents and descents which provide good chances of meeting the second intermediate host.* According to Wunder, the rudimentary tail secretes viscid material which serves to fasten the cercaria to the body of the fish, and sometimes also as a bait which the fish takes. He described how the cercariae ascend to the surface and remain suspended there, with the tail fully extended, for hours at a time. The swimming attitudes are shown in Fig. 70 B–E. The trematode finally comes to maturity in the intestine of fishes like the pike and perch, which prey upon cyprinids infected with metacercariae.

Bucephalopsis haimeanus Lacaze-Duthiers, 1854.

HOSTS. Oyster, also cockles in the vicinity of oyster-beds.

LOCATION. Sporocysts all over the body, closely packed, but not in the foot.

DIAGNOSIS (after Lebour, 1911). *Cercaria:* body 0·26 mm. long, tail extremely contractile and when extended several times as long as the body. *Cuticle:* spinous, the spines minute and arranged in transverse rows. *Tail:* median part at the hinder end of the body triangular, bearing on each side a thin, elastic process, which alternately contracts and expands as the cercaria swims. *Gut:* mouth situated near the mid-body, communicating with a simple sac-like intestine. *Suckers:* oral (? pharynx) surrounding the mouth, the additional holdfast occurring near the anterior extremity. *Excretory system:* vesicle large and pyriform. *Development:* in very long, irregularly branched sporocysts having much-tangled processes and containing cercariae in all stages of development; the emergent cercariae encysting in the nerve cord and the nerves of various gadoid fishes, commonly in the haddock. Cysts occurring particularly in the region of the auditory and the spinal nerves, sometimes beneath the skin, ovoid and thin-walled, 0·6 mm. long, the contained Bucephalid (2·5 mm. long) already having the main characters of *Bucephalopsis gracilescens.*

(*b*) '**Lophocerca**' **group** (apharyngeal brevifurcate Monostome cercariae)

Lühe mentioned two cercariae belonging to this group, *C. cristata* and *C. microcristata.* Sewell placed both of them in his '*Lophocerca*' group along with Monostomes, but expressed the opinion that they are closely related to certain

* In experiments of my own the cercariae soon settled on the bottom when the water in the tank was allowed to become quiescent by withdrawal of the circulation siphon, but immediately the latter was re-introduced the larvae became very active, maintaining steady pulsations of the furcal rami (10–12 per minute at room temperature in spring).

Furcocercariae. H. M. Miller (1926*b*) grouped them with the apharyngeate brevifurcate monostome Furcocercariae. According to Odhner (1911*b*) *Cercaria cristata* develops into a species of *Sanguinicola*, a genus of parasites in the blood-vascular system of cyprinid fishes, and others (Scheuring, 1922; Ejsmont, 1925) have supported this contention. It is unlikely, therefore, that Lophocercous cercariae can legitimately be included with Monostomes. They are unique among Furcocercariae in having no flame cells in the tail stem. Incidentally, the group seems to be confined to the Old World, occurring in India, Africa and Asia, as well as Europe. Dubois (1929) remarked on many analogies between these and other cercariae which develop into blood parasites, e.g. the non-operculate eggs, the mode of development in sporocysts and the lack of a stage of encystment, penetration into the final host being direct. Lophocercous cercariae differ, however, in having a fin-fold on the body, in lacking a ventral sucker, and in having a very different excretory system with two canals in the tail. The body, tail stem and furcal rami of the cercaria all show great powers of contractility, the sporocyst lacks a birth pore, and the miracidium has one pair of 'frontal glands' and one pair of flame cells.

Cercaria cristata La Valette, 1855 (Fig. 71 A)

HOST. *Limnaea stagnalis.*

LOCATION. Digestive gland.

DIAGNOSIS. According to Lühe the body of the cercaria is 0·13–0·19 mm. long and 0·033 mm. broad, the tail 0·039 mm. long, the cylindrical tail stem measures 0·325 × 0·018 mm. and the furcal rami 0·065 × 0·006 mm. The fin-fold is about 0·002 mm. wide. Wesenberg-Lund has examined this species in greater detail, and the following diagnosis is based on his determinations. *Cercaria:* body 0·095 mm. long and 0·03 mm. broad, bearing a membraneous crest 0·025 mm. tall; tail stem measuring 0·24 × 0·025 mm.; furcal rami 0·115 × 0·02 mm. The dorsal crest extends from close behind the anterior organ of penetration to a point near the attachment of the tail, and its height is somewhat variable. *Anterior organ:* protrusible, bluntly conical, delimited from the body by a conspicuous constriction, the extreme tip bearing a pair of conical spines which are probably hollow and act as piercing organs, delicate bristles or papillae sometimes being apparent in regular circlets around the cone. (*Note.* The whole organ is capable of great telescopic movements.) *Penetration glands:* two fine, inconspicuous threads. *Suckers:* absent. *Gut:* sac-like and lacking diverticula. *Eye-spots:* not observed, but present in related Indian species (Sewell). *Excretory system:* vesicle of variable form, sometimes bicornuate, having long lateral canals extending as far forward as the anterior penetrating organ, flame-cell formula 2 × 3 × 1 = 6, the two posterior canals extending into the tail, one entering each ramus. *Parenchyma:* containing large, rounded cells having large nuclei and numerous yellow granules. *Development:* in yellowish white, thread-like sporocysts 1–2 mm. long, which are numerous and occur chiefly on the surface of the digestive gland, to which each is attached by a muscular suctorial disk. The mother sporocyst contains several broods of sporocysts: (*a*) the long forms mentioned, which are packed with germ balls, (*b*) shorter forms (0·112–0·15 mm. long and half as broad) containing sporocysts, (*c*) germ balls showing a cleft

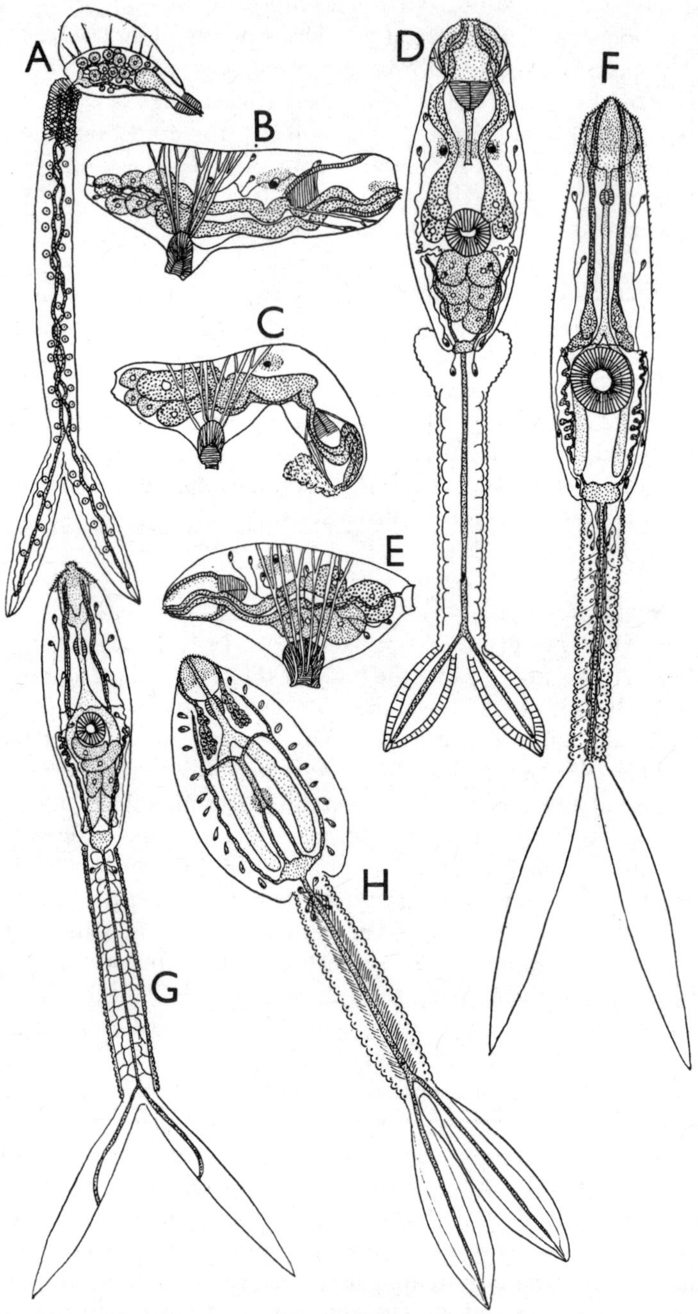

Fig. 71. Some Furcocercariae. A, *C. cristata*. B, C, *C. ocellata*. D, E, the cercaria of *Bilharziella polonica*. F, *Cercaria A* Szidat. G, *Cercaria C* Szidat. H, *C. vivax*. In B, the body is seen at rest, in C when an attempt at penetration is to be made. The aspect is lateral, as in E. (After Wesenberg-Lund, 1934.)

tail rudiment. The emergent cercariae are planktonic, penetrating cyprinid fishes, especially of the genus *Carassius*, members of which Ejsmont (1926) at Warsaw infected in fifty-five cases out of 250, though Wesenberg-Lund failed to infect any.

SOME OTHER CERCARIAE OF THE 'LOPHOCERCA' GROUP:

Cercaria microcristata Ercolani, 1881; the cercaria of *Sanguinicola inermis* Scheuring, 1920; *Cercaria bombayensis* No. 8 Soparkar, 1921; *Cercariae indicae* IX, XIII, XXXIX and LV Sewell, 1922; *C. helvetica* XVI Dubois, 1929.

(*c*) '**Ocellata**' **group** (apharyngeal brevifurcate Distome cercariae)

This is a group of Sewell's second series, typical characters being the presence of an anterior penetrating organ, paired penetration glands (divisible into anterior, coarsely granular and posterior finely granular clusters), a sac-like intestine (but not a pharynx), a long tail (which is attached to the ventral side of the body, making an angle with it, and has short furcal rami), eye-spots, and a small excretory vesicle having paired anterior canals and a main median posterior canal. Development occurs in either sporocysts or rediae. The type, *Cercaria ocellata* has a flame-cell formula $2[2 \times 3(+1)] = 12(+2)$, the figures in round brackets representing the numbers of flame cells in the tail stem. Members of the group differ from nearly all related forms in that the ducts of the penetration glands open on papillae at the tip of the anterior organ of penetration, not on hollow, conical spines as in Schistosome cercariae. At least four species occur in Europe, some of them in this country.

Cercaria ocellata La Valette, 1855 (according to Brumpt (1931) the larva of *Trichobilharzia ocellata*) (Fig. 71 B, C)

HOSTS. *Limnaea stagnalis* and *L. ovata*.

DIAGNOSIS (after Wesenberg-Lund, 1934). *Cercaria:* the body tapering at both ends and measuring 0·35 × 0·09 mm., the tail 0·365 × 0·045 mm. (*Note.* When fully extended the tail and the body may be of about equal length. The furcal rami are short, but can be extended to more than two-thirds of the length of the tail, although generally they are not more than half as long.) *Ventral sucker:* greatly protrusible (telescope fashion), but capable of retraction into the body, the region in front of it then being flexed, two groups of thread-like muscles being inserted in the folds of its cuticle. *Cuticle:* smooth, the anterior region having fine striations separated by a fold from the posterior region, concentric circlets of very fine spines also occurring on the ventral sucker. *Anterior organ:* differentiated into thin-walled anterior and posterior parts. *Head gland:* pyriform and situated above the anterior organ. *Penetration glands:* five pairs, those of the two anterior pairs having coarse granules, those of the three posterior finer granules and smaller nuclei, the sinuous ducts merging gradually with the glands. *Gut:* oesophagus extending to a point midway between the anterior organ and the ventral sucker, dividing into two short protuberances representing the caeca, pharynx absent. *Eye-spots:* numerous brownish red granules and a well-developed lens-like body situated at the sides of the oesophagus. *Excretory system:* vesicle very small, the lateral canals extending to the ventral sucker, each there dividing into anterior and posterior factors with three and four flame cells respectively, the posterior canal extending into the tail, bifurcating in front of the furcal rami, the branches penetrating to their tips. *Development:* in very long sporocysts of variable form, having grey and narrow alternating with very broad and transparent regions. Old sporocysts consisting mainly of grey regions and containing a few fully developed cercariae.

Note. Taylor & Baylis (1930) gave an account of this cercaria accompanied by some measurements made by La Valette, Dubois, Mathias, Miller and themselves.

The Cercaria of *Bilharziella polonica* (Kowalewsky, 1895) (Fig. 71 D, E)

HOST. *Planorbis corneus.*

DIAGNOSIS. *Cercaria:* body 0·24–0·30 mm. long, the tail 0·24–0·31 mm. and having relatively short furcal rami. *Other characters:* as shown in the figure.

Notes. No sooner have the cercariae emerged from the snail than they swim to the surface, attach themselves by the ventral sucker to the surface film, and hang there tail downwards. They are poor swimmers and, apart from this special trait, feebly pelagic. The tail sometimes breaks away, but whether or not this happens the body is soon invested by a gelatinous secretion binding numerous cercariae together. Swimming birds, especially ducks, find their feathers coated with this slime and are soon infested with the cercariae which penetrate the skin, young flukes being found in the mesenteric veins after 10–12 days (Szidat, 1929).

Regarding the previous species, *C. ocellata*, different behaviour is seen. In the laboratory the cercaria shows strong phototaxis, and after swimming about for some time attaches itself by the ventral sucker to the vessel containing it. In nature, vegetation may be used for anchorage, and although the cercaria generally remains quiescent after attachment, it may seek a fresh anchorage from time to time, according to the prevailing conditions of illumination. In contact with human skin the cercaria penetrates beneath it, contributing towards the dermatitis which develops, but proceeding no further with its development. Infected ducks, the natural hosts, harbour the corresponding adult, *Trichobilharzia ocellata*, in the mesenteric veins. Both of these species occur in Europe, *Cercaria ocellata* having been recorded in Britain.

(*d*) '**Strigea**' and '**Proalaria**' groups (pharyngeal, longifurcate Distome cercariae)

Certain Furcocercariae which are not bucephalids, Cystocercariae or Lophocercariae have a complicated mode of development involving three, sometimes possibly four, different hosts. A characteristic feature of development is the intercalation between the cercaria and the mature fluke of a larva of a kind not previously considered. This larval form differs to some extent in even closely related Trematoda, so that it has come to be known by several names such as *Diplostomulum* (Fig. 62 K), *Tylodelphis* (Fig. 62 L), *Tetracotyle*, *Codonocephalus* and *Neascus*. Species of *Tetracotyle* abound in certain leeches, molluscs and vertebrates, and are characterized by a definite cyst wall. Species of *Diplostomulum* are found in the eyes, brain and spinal cord of fishes and Amphibia;

they lack a cyst wall and move freely in the tissues of the host. Ashworth & Bannerman (1927) found them in the brain of the minnow, Braun (1894) in the Ammocoete, Rushton (1937, 1938) in the eye of the trout, Baylis (1939*b*) in the lens of the trout. The literature bearing on such forms is enormous, and we can at most touch on one or two of them.

Early investigators were misled or puzzled by the occurrence of *Tetracotyle* within rediae, some referring them to the form in whose larvae they occurred (Steenstrup, 1842; Filippi, 1854, 1855, 1859), others declaring their presence to be accidental (Siebold, 1843; Moulinié, 1856). The adult trematodes corresponding to the *Tetracotyle* stages were quite unknown till Ercolani (1881) showed that ducks which have been fed with the latter develop adult Holostomes in the gut. Linstow (1877) ingeniously supposed as much, after studying the miracidia which developed from the eggs of a Holostome, but believed that the *Tetracotyle* developed directly from the miracidium. Lutz (1921) first showed that it actually develops by metamorphosis of a Furcocercaria developing in thread-like sporocysts. Ruszkowsky (1922) next saw the development of the eggs of *Alaria alata* into sporocysts which produced Furcocercariae. Marked progress came soon afterwards when Mathias (1922) and Szidat (1924) traced the development of single species from egg to adult.

Cercaria A Szidat, 1924 (the larva of *Tetracotyle typica* (= *Cotylurus cornutus*) (Fig. 71F).

HOSTS. *Limnaea palustris* on the Continent; *L. pereger* in this country (Harper).

DIAGNOSIS (after Harper (1931); observations by Wesenberg-Lund (1934) shown in parentheses*). *Cercaria: Size:* body 0·14–0·16 mm. long (0·34 × 0·085 mm.), flat and translucent; tail stout, the stem 0·18 mm. long (0·225 × 0·04 mm.), the furcal rami 0·18–0·20 mm. long (0·28 mm.). *Cuticle:* spinous, the spines fine and arranged in transverse rows on the anterior region, minute tubercles also occurring on the ventral sucker. *Gut:* prepharynx short, pharynx spherical, oesophagus tubular, caeca long, bifurcation of intestine slightly in front of the ventral sucker. *Anterior organ* (large, globular or pyriform). *Penetration glands:* two pairs, situated immediately in front of the ventral sucker, their ducts sinuous and dilated in passing through the anterior organ. *Excretory system:* vesicle bluntly bilobed (globular or rectangular), the two main lateral canals extending forward slightly in front of the ventral sucker (having a transverse canal here), the posterior canal sending a branch into each furcal ramus to open by a pore 0·06 mm. from its tip. Flame cells arranged as seven on each side. *Note:* According to Mathias (1925) the flame-cell formula varies as follows: in the cercaria: $2[(1 + 1 + 1) + (1 + 1 + [2])]$; in very young metacercariae: $2[(1 + 1 + 1) + (1 + 1)]$ or $2[(3 + 3 + 3) + (3 + 3)]$; in the ripe Tetracotyle: $2[(3^n + 3^n + 3^n) + (3^n + 3^n)]$. *Development:* in active white, thread-like sporocysts, numerous and involved in the digestive gland and gonad of the host (the anterior extremity often projecting), individual sporocysts up to 4 mm. long and 0·15 mm. broad having a birth pore 0·1 mm. from the anterior end.

* These two sets of observations are so different as to suggest that the writers were dealing with different species.

Harper (1931) showed that emergent cercariae migrate from the first host, *Limnaea pereger*, and seek a second of the same species, generally an individual which is free of sporocysts, in accordance with the findings of Szidat (1924). They swim tail foremost and can survive 24–30 hours in pond water. Transition from the cercarial to the encysted *Tetracotyle* stage takes place experimentally in 25–30 days, and there are generally five to ten individuals in one host, which means that only a small fraction of the individuals which enter actually encyst. Penetration by the cercaria is rapid, taking only 15–20 seconds (Szidat), and generally takes place near the respiratory opening. By the time the cercaria has reached the hermaphrodite gland, which is the principal location, the penetration glands have disappeared and growth has increased the length of the body from 0·15 to 0·23 mm. Next the gut disappears, and the entire body comes to have a curious reticular appearance. When the metacercaria is 0·4 mm. long, the ventral sucker has almost disappeared. After 15 days, when the body measures 0·66 × 0·3 mm., growth ceases and a process sets in whereby the internal organs are regenerated, the body shrinking slightly meanwhile. The suckers, the accessory organ of adhesion and other structures soon appear, and the form of *Tetracotyle typica* is gradually assumed. These facts fully support Szidat's opinion that development from the Furcocercaria to the *Tetracotyle* stage represents a true holometabolic metamorphosis. As the degeneration of the internal organs ensues, nutriment must reach the deeper parts of the larva by endosmotic means through the integument, a process which is correlated with much degeneration of the snail's tissues and their gradual transformation into a slimy mass.

'Proalaria' ('Hemistomum') group

In 1924 Szidat described another Furcocercaria, naming it *Cercaria C* (Fig. 71 G). This develops not into a *Tetracotyle*, but into a *Diplostomulum*. The chief difference between the two forms has been mentioned already, and we may briefly consider the development of certain forms which regularly spend part of their lives in the eyes of fishes. In this situation the presence of worms produces a milkiness and causes the protrusion of the lens through the pupil. In some instances the lens may completely disappear. The larvae responsible for these dire results arrive at their location by migration through the blood vessels and lymph channels, ultimately piercing the capsule of the lens. Sometimes reaction of the host's tissues kills the parasites, lens epithelium growing around them, but the host generally suffers severely, becoming emaciated to a fatal extent. One cercaria which produces such results is Szidat's *Cercaria C*, which gives rise to '*Diplostomum volvens*' Nordmann, 1832 and ultimately to *Diplostomum spathaceum* (Rudolphi, 1819). Emerging from the snail (*Limnaea stagnalis* or *L. auricularia*) the cercaria swims tail uppermost with the body bent over the base of the tail stem. Both Szidat and Wesenberg-Lund devoted much attention to its movements. The former writer proved that the cercariae do not seek out their piscine hosts, but find them by chance. They must penetrate with remarkable rapidity, because the fishes are never covered with creeping or adherent cercariae. Wesenberg-Lund thought that some may enter by way of the mouth, but undoubtedly most of them penetrate the skin. He described how Crucian carp when exposed to the cercariae become restless, jumping high out of the water, and how in some instances the fishes died within 15 minutes, their bodies crowded with tailless cercariae in every conceivable location, on the gills, beneath the scales, in the stomach and heart, even already in the lens of the eye, and still of characteristic shape apart from the lack of a tail. Dissection proved that they reach the eye by way of the blood circulatory system. We see here not only a great difference from the mode of development of *Tetracotyle*, the cercariae of which return to snails, but also a mode of entry into a vertebrate which reminds one of Schistosomes, except that in the case of the *Diplostomulum* the vertebrate is the second intermediate, not the definitive, host. It might be emphasized that *Diplostomulum* stages are not found in invertebrates and that another peculiarity is the gradual and progressive nature of development, which lacks the metamorphic and holometabolic changes which characterize members of the '*Strigea*' group (*Tetracotyle*). Vertebrates other than fishes are affected. Cort & Brackett (1938) found a new cercaria which they named *C. ranae* in several species of *Stagnicola*. It penetrated ultimately into a tadpole and after 10–16 days had developed into a *Diplostomulum*, heavy infections causing a characteristic 'bloat-disease'.

(*e*) **'Vivax' group** (pharyngeal longifurcate Monostome cercariae)

This group gets its name from the type species, *Cercaria vivax*, and, according to Sewell, the chief characteristics are a small and rudimentary ventral sucker, fin-folds which extend round the furcal rami, and an excretory system which comprises twelve pairs of flame cells in the body and three pairs in the tail. Dubois (1929) regarded the group as transitional between Distome and Monostome Furcocercariae. Looss (1900*a*) gave a full description of specimens of *C. vivax* which he obtained from *Cleopatra bulimoides* in Egypt. The species was found in Tunis by Sonsino and has been discovered in Denmark by Wesenberg-Lund. The corresponding adult is *Prohemistomum vivax* (Sonsino, 1892) (Syn. *P. spinulosum* Odhner, 1913), which belongs to the Cyathocotylidae. The following points are based on the observations of Wesenberg-Lund (1934):

Cercaria vivax Sonsino, 1892 (Fig. 71H)

HOST. In Denmark, *Bithynia tentaculata*.

DIAGNOSIS. *Cercaria*: body 0·26 mm. long and 0·145 mm. broad, the tail only slightly shorter (0·23 × 0·05 mm.), the furcal rami 0·22 mm. long. When fully extended, the body and the tail are of about equal length. *Cuticle:* non-spinous. *Anterior organ:* rounded or pyriform, very protrusible, having circlets of spines. *Head glands:* one pair, situated immediately behind the anterior organ. *Ventral sucker:* absent, its place being taken by a circular group of parenchymatous cells. *Penetration glands:* absent. *Gut:* pharynx globular, oesophagus short, caeca long, bifurcation of the intestine situated in the first third of the body. *Excretory system:* vesicle broad, sometimes three-lobed, the main anterior canals four in number, the two median ones between the caeca uniting to form a median canal, which is connected with the lateral canals by lateral vessels at the level of the oesophagus, the posterior canal dividing into two branches extending one along each furcal ramus to open at the tip, the flame cells ten to twelve in number on each side, but not present in the tail.

Furcal rami: having fin-folds. *Development:* in sporocysts 3–4 mm. long, each having a terminal birth pore, sometimes constrictions separating what have been called 'brood chambers', and numerous germ balls (except in older individuals, which rarely have more than fifty). The sporocysts probably propagate by division. According to Looss (1900*a*) the emergent cercariae may live in freedom for more than two days, remaining suspended at the surface of the water. Wesenberg-Lund observed that the cercariae stand near the bottom of their container, rising to the surface if this is agitated and remaining there with the body flexed and the furcal rami extended and upturned. Abdel Azim (1933) showed that under experimental conditions in Cairo *Cercaria vivax* later encysts in *Gambusia affinis* and *Tilapia nilotica* which die when heavily infected. Seven days after infected fishes were fed to cats and dogs, eggs were recovered from the faeces, subsequent autopsy of the dog revealing numerous mature Holostomes of the species *Prohemistomum vivax*, a parasite originally found to be common in the Egyptian kite, *Milvus milvus aegyptiacus*.

(10) CERCARIAEA

Lühe reserved this group for cercariae in which the tail is not developed, recognizing two types, *Cercariaeum sensu stricto*, which develop in rediae or unbranched sporocysts, and *Leucochloridium*, which develop in very much branched sporocysts. Correlated with the absence of the tail, of course, is the lack of any free-swimming stage in the life history which, as Sewell pointed out, might arise as an adaptation in any group of the Digenea. Sewell divided Lühe's *Cercariaeum* group into two groups called '*Mutabile*' and '*Helicis*'. In the former he included those forms which develop in rediae, in the latter those which develop in sporocysts. Dubois (1929) proposed two other groups called '*Helveticum*' and '*Squamosum*', which were split off from the '*Mutabile*' group. *Cercaria helveticum* I, the type of the former group, has a stylet and an excretory system which differs considerably from typical members of the '*Mutabile*' group.

(*a*) '**Mutabile**' group

Cercaria paludinae impurae (Filippi, 1854) (Fig. 73A) (the larva of *Asymphylodora tincae*)

Host. *Bithynia tentaculata.*

Location. Digestive gland.

Diagnosis. *Cercaria:* body ovoid or spindle-shaped, more pointed posteriorly than anteriorly, 0·91 mm. long and 0·375 mm. broad. *Cuticle:* spinous, the spines being more numerous anteriorly and present also on the suckers. *Suckers:* of equal size, 0·155 mm. diameter, the ventral situated slightly behind the mid-body. *Gut:* prepharynx very small, pharynx large and broader than long, oesophagus wide, caeca not very long, but extending behind the ventral sucker and rudiments of the testes, bifurcation of the intestine immediately in front of the ventral sucker. *Cystogenous cells:* masses of cells having small nuclei and granular cytoplasm, some at least of which are cystogenous, filling the lateral regions of the body. *Excretory system:* vesicle tubular, almost reaching the ventral sucker and opening by a pore at the posterior extremity, the two main canals extending forward to the oral sucker, the portions in front of the ventral sucker forming branches which divide repeatedly. *Reproductive systems:* rudiments of testes present, occurring side by side slightly behind the ventral sucker. *Development:* in sac-like rediae 2·5 mm. long and 1·0 mm. broad, lacking a collar and procruscula, but having a small pharynx and a fairly long intestine (0·8 mm. long), also containing four to six cercariae as well as several immature forms. The emergent cercariae may encyst in the same snail, near the rectum or round the heart (encysted individuals having a well-developed ovary and uterus and a spinous cirrus) or may creep about and can sometimes be found on the bottom of a pool. Wesenberg-Lund watched them creep out of the pulmonary opening, over the surface of the head on to the tentacles, till fifty or more had aggregated there. If the tentacles come into contact with another snail, the cercariae pass over, penetrate the second individual and encyst within its body.

(*b*) '**Helicis**' group

Cercaria limnaeae auriculariae (Filippi, 1854) (Fig. 73C)

HOST. *Limnaea auricularia.*

LOCATION. Digestive gland, sometimes tumour-like swellings bordering the mantle.

DIAGNOSIS. *Cercarium:* 0·58 mm. long and 0·23 mm. broad. *Cuticle:* spinous, the spines being more numerous anteriorly. *Suckers:* ventral larger than the oral and situated near the middle of the body, diameters 0·13 and 0·095 mm. *Gut:* prepharynx short, pharynx globular, oesophagus long and sinuous, caeca not long, but extending well behind the ventral sucker, the bifurcation of intestine being immediately in front of it. *Excretory system:* vesicle small and sac-like, the main lateral canals extending in small loops almost to the level of the oral sucker, there recurring and, dividing into anterior and posterior branches close to the ventral sucker, branches of the third order terminating in flame cells. *Reproductive systems:* the rudiment of a single testis is situated in front of the excretory vesicle, that of the ovary farther forward, rudiments of the uterus, a spinous cirrus and the cirrus-pouch also present. *Development:* in rediae of reddish colour which are 4·2 mm. long and 0·9 mm. broad and thick, and which protrude from the digestive gland, each redia having an extremely small pharynx and oesophagus, but not procruscula, and containing twenty to thirty cercariae.

Note. Wesenberg-Lund's observation on the development of this species in rediae seems to necessitate its removal from the '*Helicis*' group in which Sewell placed it.

(*c*) '**Leucochloridium**' **group** (Fig. 72)

The cercariaeum of *Leucochloridium paradoxum* (= *macrostomum*) develops in a peculiar much-branched sporocyst, with one or a number (2, 4, 6–8) of large saccular outgrowths, in snails of the genus *Succinea*. The sacs are of two kinds, which, because of the banded blue-green or brown colour, have become known as 'green' or 'brown' sacs. The older the sacs, the more numerous are their bands of pigment. The green sacs are the more slender and are capped by a red spot (Fig. 72B). While the root-like branches of the sporocyst lie inextricably mixed with the visera in the visceral hump, the sacs project into the body cavity of the snail, frequently into the antennae, which become enormously distended (Fig. 72C). The relative size of the sporocyst, especially the sacs, is shown in Fig. 72D, which represents a partially dissected snail; frequently its weight approaches one-fifth of the total weight of the snail. If the sporocyst has only a single sac, this lies on the left side (perhaps due to the crowding of certain viscera of the snail on the opposite side), but as a rule both antennae are

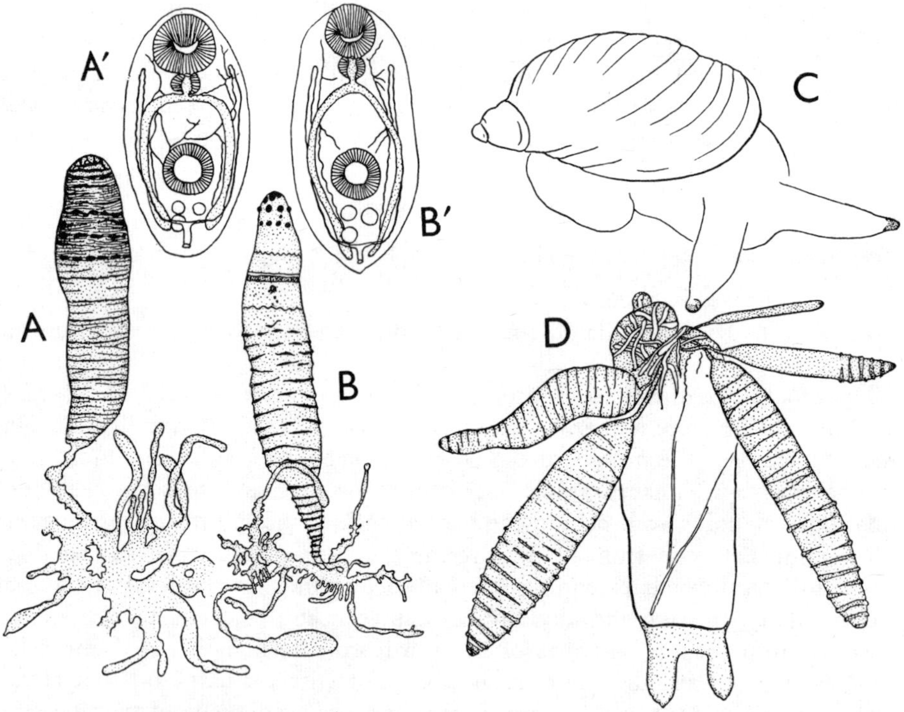

Fig. 72. *Leucochloridium paradoxum*. A, B, sporocysts with brown and green sacs respectively, A′, B′, the cercariae from A and B. C, a snail (*Succinea*) with distended tentacles containing pulsatile sacs. D, a snail with three large and two small sacs revealed by dissection, showing their relative size. (After Wesenberg-Lund, 1931.)

affected. Infected snails rest on the edges of leaves, seeming to seek the light, their distended antennae showing the obvious pulsations of the sacs, which die down in darkness and vary in rate with temperature, increasing from forty to fifty a minute at 12–14° C., to about 70 at 20–22° C. and to 120–125 at 30° C. These rhythmically pulsating, bright-coloured antennae attract birds which prey upon the snails, and thus become infected with cercariae which develop in the sacs. These show certain slight differences, according to their place of origin in the green or brown sacs (Fig. 72B′, A′). Sometimes a snail ejects the sacs, the antennae being so distended that the slightest contact with an external object causing the taut skin to rupture. According to Wesenberg-Lund (1931), who studied the biology of this trematode in detail, each sporocyst produces a few thousand cercariaea. Mönnig (1922) wrote a monograph on this species. McIntosh (1932) described six new species and devised a key for the separation of all known forms. Other writers in this field subsequently include Woodhead (1935, 1936), Gower (1936), Yamaguti (1935*a*) and T. H. Johnston (1938). We cannot hope to do justice here to such a considerable body of work.

(*d*) '**Gymnophallus**' group

Cercaria strigata Lebour, 1908 (Fig. 73B)

HOSTS. First intermediate hosts, *Cardium edule* and *Tapes decussata*. Second intermediate hosts, *Tellina tenuis* and *Donax vittatus*.

LOCATION. Sporocysts deep in the digestive gland; metacercariae between the mantle and the shell, in groups of one to five individuals.

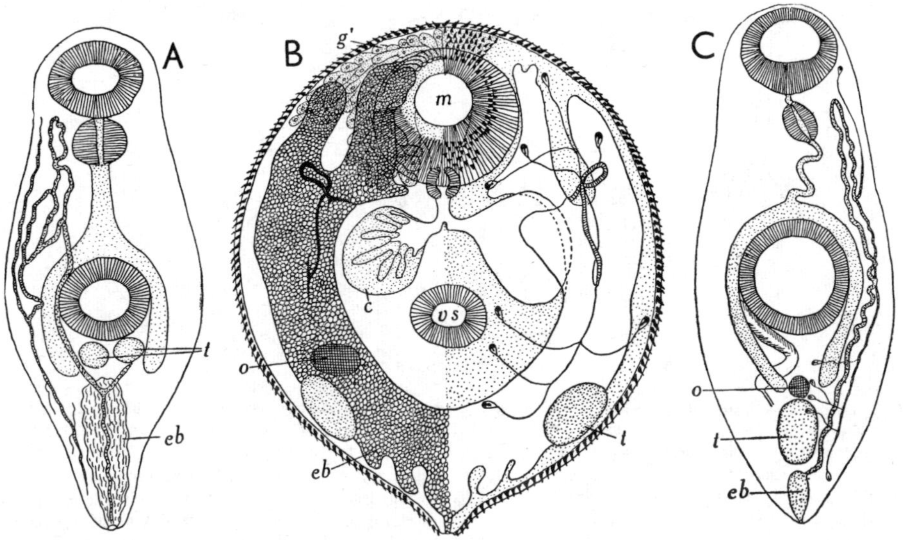

Fig. 73. Three *Cercariaeae*. A, *C. paludinae impurae*. B, metacercaria of *C. strigata*. C, *C. limnaeae auriculariae*. (B, after F. G. Rees, 1939*a*; A, C, after Wesenberg-Lund, 1934.)

DIAGNOSIS (after F. G. Rees, 1939*a*). *Metacercaria: Shape:* broadly ovoid, tapering posteriorly, 0·40 × 0·095 mm. when extended, 0·13 × 0·30 mm. when contracted. *Cuticle:* spinous, the spines being arranged in transverse rows, diminishing in size posteriorly, small area around the mouth being devoid of them. *Penetration glands:* numerous, situated at the anterior end of the body, around the oral sucker, their ducts opening by a series of pores at the rim of the spineless circumoral area. *Suckers:* ventral smaller than the oral in the ratio of 5:9 and situated slightly behind the mid-body, diameters 0·05 and 0·09 mm. *Gut:* prepharynx absent, pharynx small (0·025 × 0·021 mm.), oesophagus very short, caeca large and round, generally empty and with folded walls, but capable of great distention, the bifurcation of intestine far in front of the ventral sucker, the caeca not reaching its level. *Excretory system:* vesicle consistently lyre-shaped and with two lobes on either side of the pore (at *eb*) and a ventral lobe beneath each caecum, situated dorsally, filled with granules, having conspicuous long limbs extending to the oral sucker, flame-cell formula $2[(2+2)+(2+2)]=16$, the four anterior and four posterior flame-cell capillaries having ducts which merge to form a pair of main canals with internal tufts of cilia that enter a ventral lobe of the vesicle. *Reproductive systems:* rudiments of the testes located midway between the ventral sucker and the posterior extremity, lateral to the excretory vesicle, the rudiment of the ovary slightly in front of the rudiment of the right testis. *Development:* from a cercaria having the same structure as the meta-cercaria, the excretory vesicle being diagnostic. This larva in turn develops in

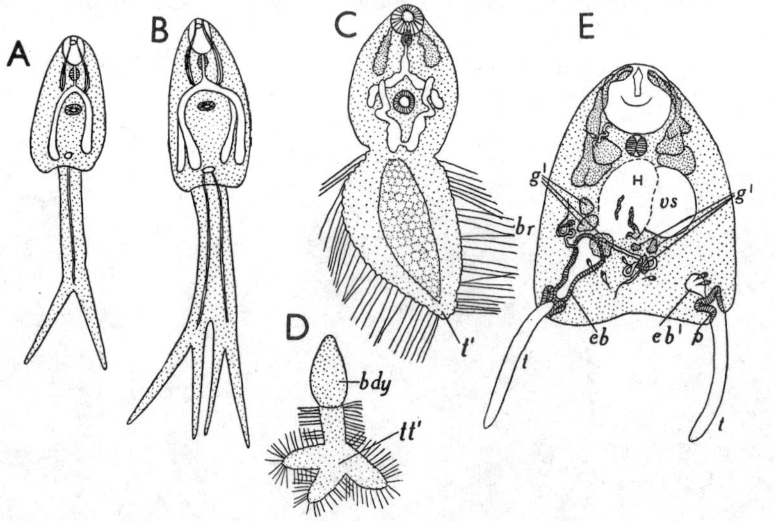

Fig. 74. Malformations of larval Trematoda. A, normal Furcocercous cercaria from *Bithynia tentaculata*. B, 'double' cercaria with bifid tail. C, normal specimen of *Cercaria setifera*; D, cercaria of same species with a trifid tail. E, partial twinning in a Xiphidiocercaria. One twin is well developed, having normal suckers, stylet, penetration glands, excretory vesicle (*eb*) (with collateral canals), caudal pockets and tail (on the left); the other twin is represented by a head process (*H*) with associated rudimentary penetration glands (*g'*), a thin-walled excretory vesicle, but no canals, caudal pockets (*p*) and a normal tail. (A–D, after Brumpt, 1936; E, after Hussey, 1941 *a*.)

a thin-walled, ovoid and colourless sporocyst, which is filled with cercariae in various stages of development. The emergent cercariae apparently leave the cockle and enter the second intermediate host, remaining stationary and contracted, attached by both suckers midway along the body and one-third of the distance between the umbo and the gape. It does not encyst, but epithelial cells of the mantle seal it off (Lebour, 1911). Adult a species of *Gymnophallus* infecting a bird, probably *G. deliciosus*, perhaps *G. macroporus*.

Abnormalities of larval development

Mention must be made, in conclusion, of various abnormalities which have appeared in larval Digenea. Mathias (1930) found a Furcocercous cercaria (in *Bithynia tentaculata*) in which the distal end of the tail stem was bifurcate, each branch terminating in a pair of furcal rami (Fig. 74B; the normal cercaria is shown in A). The same writer (1934) reported a Trichocercous cercaria, *C. setifera*, with a trifurcate tail, each branch having an apparently normal complement of bristles (Fig. 74C, D). More recently, Hussey (1941*a*) recorded partial twinning in the Xiphidiocercaria of *Alloglossidium corti*. In this instance the posterior region of the body was almost fully duplicated (Fig. 74E). The abnormal larva had two complete tails (*t*), each with a pair of caudal pockets (*p*) and two excretory vesicles, one (*eb*) thick-walled and apparently normal, having two lateral canals, the other (*eb'*) with indistinct walls and without lateral canals. Near the ventral sucker of this individual there was a partially developed anterior region of dwarfed proportions, no larger than the sucker itself (*H*). Other characters are shown in the figure and explained in the caption.

Twinning in miracidia has also been recorded, by Janer (1941) for *Schistosoma mansoni*. Eggs of normal size which were collected from human faeces each contained a pair of miracidia, formed of individuals which were fused anteriorly for about one-third of the body length. Each individual in the twins had the normal set of four flame cells, which at the time of observation were functional. Another instance of miracidial twinning was recorded by Hoffmann & Janer (1936).

CHAPTER 13

THE LIFE HISTORIES OF THE TREMATODA

The contrast between direct and indirect modes of development in Trematoda, without or with change of host, was emphasized in an earlier chapter, and we have seen that *Polystoma* develops from a ciliated larva (Fig. 3A) somewhat dissimilar to the parent, yet possessing the obvious characters of a monogenetic fluke, *Fasciola* through a sequence of larval forms (Fig. 4A–K) which differ markedly in structure and mode of life from any adult trematode. We have examined diverse larval Digenea, especially cercariae, and we can now turn to the correspondingly varied life histories in which these larvae play a part.

The distinction between Monogenea and Digenea is based upon evidence of the most meagre kind. A substantial fund of knowledge about digenetic life histories has been accumulated, and it is unlikely that we shall be called upon in the future to modify significantly our conception of this mode of development. But we have so little knowledge about monogenetic life histories, not more than half a dozen having been worked out in any detail, that future discoveries may necessitate substantial modification of our conceptions about this mode of development. It is very likely that we have credited the Monogenea with an altogether too simple life history. Of the few monogenetic life cycles which are known, one includes viviparity, another the union of two individuals in permanent copula, and a third embryonic development. In some instances development is larval and in all it is monogenetic, but our knowledge is too fragmentary to permit us to generalize about the mode of development of Monogenea as a whole. We must be content for the present to consider briefly what little is known before proceeding to examine the principal differences between the modes of development in various Digenea. The arrangement of digenetic life histories according to the zoological status of the intermediate hosts is admittedly an unscientific procedure, but it is the most that can be achieved in the present state of knowledge. It illustrates the fact that the main steps of the life history bear little, if any, relation to phylogenetic characters, but also indicates how trematodes attain final success as a result of the feeding habits of many definitive and some intermediate hosts.

A. Monogenea

Members of the Gyrodactylidae may differ from other Monogenea in being viviparous. The egg of *Gyrodactylus elegans* does not become encapsulated, but develops into an embryo within the body of the parent (Fig. 14A, *e′*). Before the embryo is born, even before it has taken on the general characters of the parent, it has given rise to a smaller embryo within itself, and this in turn to a still smaller embryo which contains a fourth and smallest embryo. The relation between these embryos, which simulate the elements of a chinese puzzle box, has not been satisfactorily explained. Kathariner (1904), who investigated the life history, regarded the phenomenon as natural merogony, which we might call true polyembryony, i.e. the formation of a number of individuals by partition of the ovum, in this instance unequally developed individuals. Fuhrmann (1928) saw in it a form of paedogenesis with very early origin and development of ova, and drew a parallel between this case and that of the digenetic fluke *Parorchis*, in which the eggs still contained in the uterus of the parent each contain a miracidium, and this a fully formed redia (Fig. 75B, *re*).

When the largest, first-formed embryo of *Gyrodactylus* has assumed the general characters of the parent, and the opisthaptor with its two large hooks and sixteen marginal hooklets has appeared, it is born and at once attaches itself to the same host as the parent. Unlike the newly liberated larva of *Polystoma*, it is non-ciliate and could not be mistaken for a trematode of any other genus than that to which it belongs. The two or three embryos still contained within its body develop and are born before this individual itself attains its full size.

Oviparous Gyrodactyloidea, on the other hand, have larvae which are more similar to that of *Polystoma*. *Dactylogyrus crassus*, which is found on the gills of the common carp in Poland, has a free-swimming larva with anterior, middle and posterior bands of cilia. Kulwiéc (1927), who discovered this larva, observed that of *D. anchoratus* also and formulated what she regarded as the typical life history of members of the genus *Dactylogyrus*. After the eggs are laid the ciliated larvae hatch. They have eye-spots such as exist in the parent, also a typical pharynx, and they seek out an appropriate fish host and cling to its gills by means of the fourteen larval hooklets, which may be set in pairs on the naked ventral surface of the body. Once attachment has occurred the cilia are lost, the haptorial disk and large hooks develop, the hooklets are dispersed and, as the fluke increases in size, maturity is gradually attained. Wilde (1937) studied the development of '*D. macracanthus*', which also has a free-swimming, ciliated larva.

According to Siwàk (1931) the larva of '*Ancyrocephalus vistulensis*',* also a Dactylogyrid, but a member of the Tetraonchinae, has three bands of cilia and twelve larval hooklets, and develops in much the same way as *Dactylogyrus*. The more specialized fluke studied under the name of *Epibdella melleni* (now *Benedenia melleni*) by Jahn & Kuhn (1932) also has a free-swimming larva with three ciliated areas, but it is better developed in the sense that it already possesses a definite opisthaptor with three pairs of hooks and fourteen larval hooklets, as well as two pairs of eye-spots and simple digestive and excretory systems. After attachment to the host, a fish, the cilia are lost and the worm gradually takes on the adult form.

Microcotyle spinicirrus, a parasite on the gills of *Aplodinotus grunniens*, has a somewhat similar larva with three bands of cilia, an eye-spot and rudimentary digestive and excretory systems, but there is more marked dissimilarity from the parent, which has an opisthaptor comprising a paired series of about 200 small clamps. The haptor in the liberated larva is without the clamps, having two pairs of comparatively large anchor-like hooks and six pairs of marginal hooklets. The larvae attach themselves to the same host as the parent, and by the time this happens six pairs of clamps have developed from a ventral syncytium in the posterior region. Later on others are added behind, and by the time the genital organs have developed there are about thirteen pairs. Up till this time the larval haptor persists at the extreme posterior end of the body, but at this time or

* This form is probably identical with *Haplocleidus siluri*.

subsequently it is lost. According to Remley (1942), on whose study this brief account is based, the adult opisthaptor grows continually throughout the life of the fluke, as is shown by the production of new clamps in the hindmost region.

On the other hand, *Sphyranura oligorchis*, a parasite of *Necturus maculosus* and a closer relative of that other parasite of Amphibia, *Polystoma* (belonging to the same family, but a different subfamily), has a somewhat protracted embryonic development as distinct from a larval one. After the eggs are laid, the embryo develops for about four weeks without showing marked signs of activity. On the 27th–29th days it becomes active in the egg, and at this stage is similar to the juvenile fluke which emerges finally and creeps on the bottom in search of a host, dying if the quest fails. It is non-ciliate, has a well-developed opisthaptor bearing hooks and hooklets, and after a few days of attachment to the host has grown considerably and resembles the adult more closely. The large hooks are said to grow like the rest of the larva (see Alvey, 1936).

In *Diplozoon paradoxum* we have a monogenetic fluke which at the time of hatching is uniformly ciliated, but also has a specialized opisthaptor comprising one pair of clamps, two eye-spots, a well-developed pharynx and a sac-like intestine. After a short period of freedom the larva attaches itself to the gills of the minnow or another suitable cyprinid fish, loses the cilia and eye-spots, becomes more elongate and broader posteriorly, and develops a branched intestine. A small sucker develops in the middle of the ventral surface and a small conical process in a corresponding position dorsally. At this stage the larva is the well-known '*diporpa*' of Dujardin (1845). Two individuals next unite in such a way that the midventral sucker of one embraces the dorsal cone of the other. This is followed by a twisting of both bodies, which also brings the free midventral sucker and dorsal cone into contact. The two worms become even more intimately related by fusion not only at the points of contact mentioned, but also at the external openings of the genital ducts, the male pore of one individual becoming closely apposed to the vaginal pore of the other, and vice versa. Firmly united, they represent the *Diplozoon* of Nordmann (1832), which remains thus, as Zeller (1872*b*) showed, two individuals in permanent copula. Not until the two individuals have become united do the remaining three pairs of clamps which characterize the adult opisthaptor develop (Fig. 21 A, *b*).

In general, therefore, the development of Monogenea is monogenetic, but the free-swimming stage may be omitted (*Sphyranura*), and the juvenile fluke may be born at a comparatively late stage in development (*Gyrodactylus*). In *Diplozoon* we encounter an example of precocious sexual behaviour which is unequalled in complexity even in the most highly specialized cyst-dwellers or blood flukes amongst the Digenea.

B. Digenea

I. *One intermediate host only*

(i) *Cercariae not encysting, but actively penetrating the definitive host*

The best known examples of this type of life history occur in blood flukes of the family Schistosomatidae. As they are fully dealt with in several standard works on helminthology, it will not be necessary to deal with them in detail here. Briefly, adults *in copula* migrate against the current flowing in the venules which form the habitat till size makes further progress impossible. In *Schistosoma haematobium* migration is mainly in the direction of the bladder, in *S. japonicum* and *S. mansoni* towards the rectum and large intestine. In the small venules the female lays her eggs singly, while both adults retire with the blood stream, so that the eggs are spaced out. Elastic recoil of the wall of the blood vessel, together with the fluid pressure of the blood itself, forces the spines of the capsules into the wall, through which the eggs gradually work their way into the urinary bladder, or the rectum, to be cast out of the body of the host with the urine, or faeces. At this stage considerable damage is done to the tissues of the final host.

If the eggs fall into water, the contained miracidia hatch within a few hours and attack the snail intermediate host forthwith, penetrating it and developing within 48 hours into saccular mother sporocysts. Various fresh-water snails act as the intermediate hosts, according to the species of fluke and the geographical locality. The germ balls of the mother sporocyst soon develop into daughter sporocysts, which escape into the tissues of the snail by rupture of the wall of the mother sporocyst, make their way to the digestive gland, grow and give rise to the next stage in the life cycle, the cercariae. These are Furcocercariae of the apharyngeal brevifurcate Distome group and they emerge from the snail and swim about. The definitive host, man, may be infected by drinking water containing the cercariae, but generally acquires the parasites as a result of allowing the naked skin to come in contact with the larvae. By the use of secretions of the penetration glands and the activity of the anterior organ of penetration, cercariae pass swiftly through the skin, as Looss first demonstrated on himself, enter the lymphatics or blood vessels and ultimately congregate in the venules of the hepatic portal or other appropriate system. In pregnancy they may pass from the maternal to the foetal blood vessels in the placenta, so that the newly born infant is already in the first stages of schistosomiasis.

Cattle and sheep may be infected in the same way with *Schistosoma bovis* (Sonsino, 1876) (Fig. 59A) in Africa, southern Europe and southern Asia, or with *S. spindale* (Montgomery, 1906) in India and Sumatra; and musk-rats and other rodents with *S. douthitti* (Cort, 1914) in North America. Wild and domesticated ducks in Europe and America may similarly be infected with *Bilharziella polonica* (Kowalewski, 1896) (Fig. 51H), whose cercariae we have also examined (Fig. 71D, E). In this instance the cercariae, on liberation from the snail, rise to the surface of the water and fasten themselves by means of the ventral sucker to the surface film. A slimy exudate from the body causes the cercariae to aggregate in small swarms and also serves to fasten them to the feathers or skin of swimming birds which come in contact with them, after which penetration is effected as in the case of human Schistosomes. The principal snail host of this species is *Bulinus truncatus* (Syn. *B. contortus*; see Baylis, 1931), snails like *Planorbis corneus* and species of *Limnaea* being immune from attack by the miracidia (see Praetorius, *Z. Urol.* Sonderband, 1930, p. 392).

Many instances of the infection of man by non-human Schistosomes are on record, in which penetration is incomplete and gives rise to a dermatitis which has been called 'swimmer's itch'. Matheson (1930) and Taylor & Baylis (1930) have studied an outbreak of this kind in this country, and similar outbreaks have been studied elsewhere by Cort (1928) and others.* Taylor & Baylis found *Cercaria ocellata* La Valette (Fig. 71 B, C) to be the cause, and within five minutes of allowing the cercariae to penetrate the skin of the forearm, Taylor experienced a tingling sensation, later suffering an eruption of the skin which increased in intensity for several days, only gradually subsided and had not disappeared seven weeks later. In a similar outbreak in Germany, Vogel (1930*a*, *b*) allowed himself to be attacked by the cercariae responsible and had a small area of the inflamed region of skin excised and sectioned, finding not only the cercariae but also the canals which they had constructed in the tissues of the skin. Vogel also identified the cercariae as *C. ocellata*. Various helminthologists attempted to trace the larva to its adult stage, and Brumpt (1931) identified it with *Trichobilharzia kossarewi* Skrjabin & Zakharow, 1920, the name of this species being changed to *T. ocellata* (La Valette, 1854) Brumpt, 1931, in accordance with the rule that larval specific names have nomenclatural validity and take priority over names given later to adults of the same species. According to Brackett (1940) *Cercaria stagnicola*, *C. elvae* and *C. physella* are the agents responsible for 'swimmer's itch' in the U.S.A. The cercariae are destroyed in the superficial (epidermal) layer of the skin, but Penner (1941) found that one non-human Schistosome cercaria which produces dermatitis, that of *Schistosomatium douthitti*, penetrated deeply into a rhesus monkey, migrating flukes being found in the lungs at a post-mortem examination several ($5\frac{1}{2}$) days after the time of infection.

Other trematodes which have the same kind of life history as Schistosomes belong to the family Sanguinicolidae and are also parasites of the blood-vascular system. They have Furcocercous cercariae of the '*Lophocerca*' type, which are pelagic and attack cyprinid fishes that constitute the definitive hosts. We have dealt with the general characters of the best-known form, *Cercaria cristata* (Fig. 71 A). Odhner (1924) placed snails (*Limnaea auricularia*) infested with this cercaria in an aquarium containing goldfish and carp. The fishes were fatally attacked by the larvae. Young worms were found in hundreds on the mucous membrane of the buccal cavity and on the gills, adults which had laid their eggs occurring in the heart. Odhner was satisfied that they belonged to the genus *Sanguinicola*.

Spirorchiidae have this type of life history also, according to Wall (*Trans. Amer. Micr. Soc.*, **60**, 1941, pp. 221–60), who described the first to be elucidated, that of *Spirorchis parvus* (Stunkard). Adults of this species inhabit arterioles in the wall of the gut of a turtle (*Chrysemys picta*). The eggs pass into the lumen of the gut and out with the faeces, hatching in four to six days. The miracidia swim freely for a short time and then penetrate into snails of the species *Helisoma trivolvis* and *H. campanulata* and develop into mother sporocysts, which give rise to daughter sporocysts. These in turn produce Furcocercous cercariae of apharyngeal brevifurcate distome type, which leave the snail and swim about in the surrounding water. When the cercaria meet

* Johnston wrote a recent account (*Trans. Roy. Soc. S. Australia*, **65**, 1941, pp. 276–84) of schistosome dermatitis in bathers in the Antipodes.

a turtle belonging to any one of ten species they penetrate the soft tissues and migrate to the final location, where in *Chrysemys picta* (but not other species) they reach maturity in about 3½ months.

(ii) *Cercariae encysting on herbage after emergence from the snail host*

Fasciola hepatica shows this type of life history, already described. In *Fasciolopsis buski*, a closely related fluke which inhabits, not the liver, but the small intestine and stomach of the final host (pig and man), the main differences from what has been mentioned for the liver fluke are: (*a*) the delay in hatching of the miracidium, which may take three weeks after the laying of the eggs, (*b*) the particular kind of snail host, a species of *Planorbis* or *Segmentina*, and (*c*) the site of encystment of the cercariae, which is generally the tubers of the water caltrop, *Trapa natans*, or other plant on which the snails feed. These tubers are sold in Chinese markets and eaten raw there and then, with consequent infection if they happen to bear the encysted metacercariae. Barlow (1925) obtained as many as 200 cysts on a single water caltrop, so that the risk of infection may be heavy. This is explained by the practice of fertilizing the crops with human faeces, incidentally containing the eggs of the fluke.

Certain Amphistomes also have life histories belonging to this category. The cercariae of *Paramphistomum cervi* develop in snails of several genera, *Bulinus*, *Pseudosuccinea* and *Galba*, and leave the intermediate host, forthwith encysting upon vegetation. Metacercariae thus enter the body of the definitive host passively, along with herbage.

(iii) *Cercariae encysting upon the shell of the snail intermediate host*

The cercariae of the Monostome, *Notocotylus seineti* Fuhrmann, develop in rediae in the digestive gland and gonads of *Limnaea pereger* and *Physa fontinalis*. According to Harper (1929), who studied the development of this species in Scotland, the rediae migrate towards the surface of these organs as they develop, and when at the surface have acquired a characteristic dark brown colour and cylindrical form. Each contains eight to twelve dark brown Monostome cercariae closely similar to *C. monostomi* Linstow (Fig. 61 C). The emergent cercariae leave the snail, but encyst at once, generally on the outer surface of the opening of the shell, in hemispherical cysts formed with the flat base towards the shell. In nature, cysts occur on the shells of molluscs other than those which serve as intermediate hosts, especially those of *Limnaea truncatula*, *Valvata piscinalis* and *Planorbis albus*, and were found once on the ostracod, *Eurycypris pubera*. They are abundant during the summer, but also occur in winter. Under experimental conditions cercariae selected *Limnaea pereger* from a number of other snails at the time of encystment and showed a preference for young individuals not infected with the earlier stages. Ducklings which were fed on the snails bearing cysts subsequently harboured various stages of *Notocotylus seineti* from the juveniles to the adults. Wunder (1923) found that *Cercaria monostomi* encysted on the leaves of *Elodea canadensis* and *Myriophyllum*, as well as on the shells of *Limnaea ovata* (the host involved) and *Planorbis marginalis*. The selection of the shells of molluscs in preference to water plants is not rigidly exercised, but would seem to increase the possibility of infection of the final host, many water birds showing a preference for an animal instead of a vegetable diet. But many Monostomes, it must be admitted, encyst in the open, not necessarily on the shells of snails.

(iv) *Cercariae encysting in the tissues of the snail intermediate host*

The cercariae of the Echinostome, *Echinoparyphium recurvatum*, develop in pale yellow rediae with a birth pore, within the digestive gland of *Valvaia piscinalis*. They can be recognized by their small size, being about 0·25 mm. long, and by the head collar with its double circlet of forty-three to forty-five collar spines, those of the oral being slightly smaller than those of the aboral row. On leaving the body of the parent redia the mature cercariae immediately encyst in the digestive gland or mantle of the snail. Harper (1929) found more than fifty transparent, spherical cysts 0·11–0·13 mm. diameter in these situations. Cysts also occur in *Planorbis albus*, and are sometimes found in snails which are not infected with rediae, showing that the cercariae can and do migrate from the body of the intermediate host under certain circumstances. Certainly, they can swim by lashing the tail, but generally there seems to be no reason for them to do so. Harper fed snails containing the encysted metacercariae to ducklings and obtained adult flukes in the intestine. Presumably, the natural hosts such as the scaup and tufted duck also become infected by ingesting molluscs infested with metacercariae.

Echinostoma revolutum (Fig. 50D), a common parasite in the rectum and caeca of wild and domestic ducks and other aquatic birds, has a similar life history. According to Johnson (1920) the miracidia escape from the eggs about three weeks after these are laid and penetrate the snail *Physa occidentalis*, in which each develops into a mother redia. This gives rise to a generation of daughter rediae in which the cercariae develop. The typical Echinostome cercaria, *C. echinata* Siebold (Fig. 67B), may encyst within the tissues of the snail in which the rediae developed, or may leave this host. In the latter event it may re-enter the same or enter another individual in order to encyst, and may even select another species of snail. A brief excursion outside the body of the intermediate host seems to be undertaken sometimes, though this is not the general procedure. Infection of birds follows their ingestion of infested snails, and man may be infected as a result of eating raw or insufficiently cooked freshwater mussels of the species *Corbicula producta*. Human consumption of *Corbicula lindoensis* in the central Celebes is the cause of a widespread echinostomiasis due to *Echinostoma lindoensis* (see Bonne, 1941*a*, *b*).

Euparyphium ilocanum, an Echinostome parasite of man in the Philippines, produces cercariae which will encyst in almost any snail available. Human infection generally results from eating infected *Pila luzonica*, a mollusc which labourers of Luzon regard as a delicacy and which they eat straight from the shell or with a light sprinkling of salt or a dash of vinegar or lemon insufficient to injure the encysted metacercariae. Tubangui & Pasco (1933) made a special study of this fluke. Trematodes other than Echinostomes may encyst in the snail intermediate host, e.g. *Brachylaemus fuscatus*, a parasite of the sparrow in France (Joyeux, Baer & Timon-David, 1934) and Monostomes of the genus *Tracheophilus* (see Szidat, 1933).

The trematode *Monorcheides cumingiae* (Martin, 1938) has a life history which bears a close resemblance to those included in this section, but the intermediate host is a clam (*Cumingia tellinoides*), not a snail. According to Martin (*Biol. Bull. Woods Hole*, **79**, 1940, pp. 131–44), the cercariae develop in sporocysts in the visceral mass of the mollusc, later emerging with the exhalent current to swim freely for a short time in the surrounding water. Eventually some come in contact with the mantle or foot of the mollusc (sometimes being drawn into the mantle cavity by the inhalent current), penetrate into the tissues and encyst as tailless metacercariae in various situations (mantle, siphons, foot, gills, etc.). Puffers were observed to bite off portions of the siphons and these and other fishes (flounders and eels) were infected experimentally by feeding them on the flesh of clams containing encysted larvae.

(v) *Cercariae encysting in the snail, but cysts deposited in 'slime-balls' on herbage*

Dicrocoelium dendriticum (Figs. 57A, 81A–G), a familiar parasite in the bile ducts and gall bladder of sheep and some other herbivorous or omnivorous animals, occasionally man, has a life history which is unique in several particulars. The eggs, when freshly laid, each contain a miracidium and this is hatched, not in water, but only when the eggs have been ingested and acted upon by the digestive secretion of land snails such as *Torquilla frumentum*, *Helicella candidula* and *Zebrina detrita* on the Continent, and *Helicella itala* in Scotland. The miracidium for a time swims by means of its cilia in the gut of the snail, eventually entering the digestive gland by one of the ducts. It promptly penetrates the wall of the gland, after which the cilia are lost and the modified larva wanders about for a time in the connective tissue of the gland before settling down. Subsequently, two generations of sporocysts are formed, those of the second generation forming Xiphidiocercariae known as *Cercaria vitrina* (Fig. 81A), which leave the sporocyst through a terminal birth pore.

Now the presence of a stylet and an equipment of penetration glands has been taken to indicate that the cercariae should penetrate a second intermediate host in order to encyst, and Brown (1933) suggested that in all probability this is a larval insect, perhaps the larva of one of the Diptera, the imago of which is accidentally eaten by sheep. He pointed out that grass harbours many winged insects, and that soil and roots form suitable habitats for larval insects of various kinds. On the other hand, evidence has accumulated which shows this view to be fallacious. Cameron (1931) succeeded in infecting sheep by feeding them with infected snails, and Mattes (1934) showed that the cercariae encyst in the snail and are deposited in 'slime balls', and he infected rabbits and a sheep directly by feeding them with these, thus showing that a second intermediate host is unnecessary. In 1936 he gave a full account of the development in the snails, and experimentally infected *Helicella ericetorum*, *H. candidula* and *Zebrina detrita*. A year later he reviewed the whole subject and gave a considerable list of snails which serve as intermediate hosts. He expressed the view that hundreds of cercariae leave the snails in 'slime balls' when rain follows drought. Neuhaus (1936) also infected a sheep by feeding it with the 'slime balls' and showed (1939) that each cercaria becomes invested in its own cystogenous secretion before becoming enveloped, along with others, in slime. Shortly before encystment the cercariae emerge from the sporocysts, leave the digestive gland with the venous blood stream and enter the pulmonary chamber, through the wall of which they penetrate. Encystment follows immediately.

Neuhaus (1939) also revealed several unique features during the later stages of this life cycle, when the larvae have entered the definitive host and are migrating to the final location. The cysts are ingested with the food of the sheep or other mammal, pass through the stomach with little or no change, and are dissolved in the intestine. The liberated cercariae, still possessing both the stylet and the tail, penetrate the wall of the gut and enter a vein. They travel with the blood stream into the portal vein, where they remain for several days, after which they leave the blood vessels and enter the tissue of the liver, eventually the smaller bile ducts. At first only the smaller ducts are inhabited, but the growing fluke later seeks the larger ones and eventually the gall bladder. Not until the cercariae have reached the liver is the stylet cast out and the tail lost, so that we may presume that both organs function during the migration. Some interesting growth changes in the juvenile flukes will receive consideration later, but it is opportune here to note that the liver is first reached 8 days after the time of infection, the portal vein after 24 hours, and that within 2 months the young flukes are about 6 mm. long and have produced innumerable eggs. Neuhaus suggested that the life-span of the fluke is not less than 1–1½ and may be as much as 3–5 years.

II. *Two intermediate hosts*

(i) *Both intermediate hosts molluscs*

(*a*) *Gastropod and lamellibranch* (*adult flukes in birds*). The adult *Parorchis acanthus* is a parasite in the bursa Fabricii or rectum of the herring gull, sometimes the common gull. As the eggs travel through the first part of the uterus they increase in size from about 0·08 × 0·044 to 0·10 × 0·06 mm. and contain each a developing miracidium. In the distal part of the uterus the miracidia are ripe, and hatching generally occurs shortly after the eggs are laid, though it may occur *in utero*. Another feature, which is almost unique, is the presence inside the miracidium at the time of hatching of a single mother redia* (Fig. 75 B, *re*).

When at rest, the miracidium shows uniform ciliation (as in Fig. 75 B), but, when swimming, two sets of anterior cilia (anterior and posterior cephalic) are elevated in a highly characteristic manner, only partially shown in Fig. 75 C. The cilia arise from tiers of cells, successive tiers from the anterior extremity containing six, seven, four and two cells respectively, and the cephalic cilia spring from the cells of the anterior tier. The miracidium also has a pair of eye-spots, two lateral sensory processes (Fig. 75 B and C, *lp*), a circlet of twenty-four small papillae (*ap*) in alternate ones and threes between the cells of the first and second tiers, and a circlet of isolated papillae (*ppa*) in the interval between the cells of the third and fourth tiers. It also has a vestigial gut (or apical gland), a brain, an excretory system consisting of two flame cells and two short canals

* *Tracheophilus sisowi* Skrjabin, 1913 [= *Typhlocoelum cymbium*] (Cyclocoelidae) and *Stichorchis subtriquetus* (Paramphistomatidae) are other Digenea which show this feature in development.

with lateral openings (Fig. 75 B, *fc*, *ep*), and the contained redia already mentioned. The latter, though unborn, has a well-developed pharynx, a sac-like intestine, and numerous germ balls each of which is destined later to give rise to a daughter redia. Early development in this fluke is thus somewhat accelerated.

After a brief period of freedom the miracidia attack and penetrate the marine snail, *Nucella* [= *Purpura*] *lapillus*. The mother redia is liberated in the tissues of this mollusc, after which the miracidium probably degenerates, no trace of it having been found. The germinal balls of the mother redia develop into at least twenty daughter rediae, each ultimately larger than the parent, and these emerge

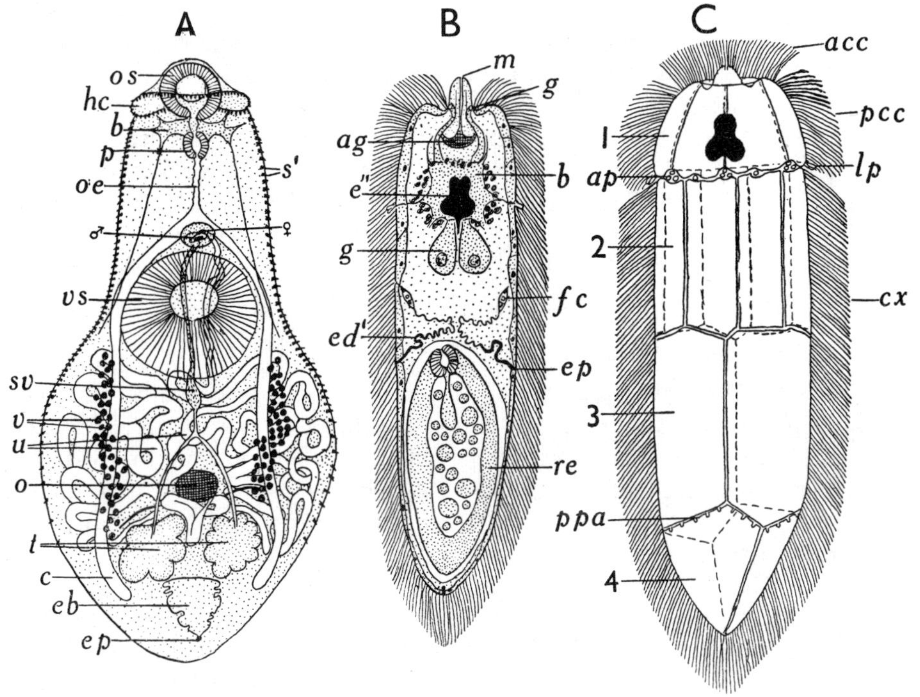

Fig. 75. *Parorchis acanthus*. A, ventral view of adult. B, general anatomy of the miracidium, showing the first generation redia. C, external characters of miracidium during the condition of rest. (After F. G. Rees, 1939*b*; 1940.)

by way of the birth pore and pass into the tissues of the snail, migrating gradually to the surface of the digestive gland. As they feed and grow, the germinal cells multiply and arrange themselves into germinal balls, complete germinal continuity being maintained as in the earlier stages of development with the original propagatory cell of the embryo miracidium. The germinal balls eventually develop into Echinostome cercariae of the '*Megalura*' group, *C. purpurae* Lebour, 1911, which leave the snail. As Lebour & Elmhirst (1922) showed, the cercariae penetrate a second intermediate host, either *Cardium edule* or *Mytilus edulis*, forming characteristic cysts (Fig. 62 A, B) in the mantle and foot. According to F. G. Rees (1937), who studied the early stages of this trematode in great detail, cercariae in aquarium tanks will encyst on the outsides of *Purpura* shells and on other objects. In nature the definitive host is presumed to be infected by ingesting mussels or cockles containing encysted metacercariae, which already possess well-developed rudiments of the genital organs.

In America, another Echinostome parasite of the herring gull, *Himasthla quissetensis* (Miller & Northrup, 1926), utilizes similar hosts. According to Stunkard (1938 *a*), the cercariae develop in *Nassa obsoleta* at Woods Hole, but encystment can occur in numerous lamellibranchs, species of the genera *Mya*, *Modiolus*, *Mytilus*, *Cumingia*, *Pecten*, *Ensis* and *Crepidula*.

(*b*) *Gastropod and nudibranch* (adult flukes in fishes). *Lepocreadium album*, a parasite of the gattorugine in some parts of Europe, shows this kind of life history. According to Palombi (1937*b*) the miracidium of this trematode penetrates the gastropods *Nassa mutabilis* and *N. corniculum*, which later come to contain rediae and then Trichocercous cercariae closely similar to *C. setifera* (Fig. 66F). The latter escape and penetrate nudibranchs of the species *Rizzolia peregrina* and *Polycerca quadrilineata*, in which they encyst, the definitive host being infected by devouring infested nudibranchs. *Opechona bacillaris* has a similar cercaria (Fig. 62G) and may have a similar life history. But late cercariae have been found free in the plankton, as well as in or on various coelenterates, e.g. *Obelia* sp., *Cosmetira pilodella*, *Turris pileata*, *Phialidium hemisphericum* and *Pleurobrachia pileus*, and in *Sagitta* (Lebour, 1916–1918*a*, *b*).

(ii) *First intermediate host a mollusc; second a crustacean*

(*a*) *Second intermediate host a fresh-water crustacean; definitive host man.* *Paragonimus westermanii* (Fig. 54D), the lung fluke of man in India, the Far East and South America, produces eggs which are voided in sputum. Miracidia of somewhat retarded development are hatched 16–60 days later, according to the season, winter temperatures lengthening the process, and attack any one of six or seven species of snails belonging to the genus *Melania*. Within the body of the snail the miracidium is transformed into a sporocyst and this generally produces a mother redia which in turn yields daughter rediae. From the rediae of the second generation Xiphidiocercariae with a very short tail (*Microcercous*) arise, and these can penetrate any one of a number of fresh-water crustaceans belonging to the genera *Potamon*, *Eriocheir*, *Astacus*, *Cambarus*, *Sesarma* and *Pseudotelphusa* (the last in South America, the others in the Orient) and encyst in the gills, or liver, or muscles. When the cysts are taken in with food by the definitive host, the metacercariae emerge in the duodenum, penetrate the wall of the gut, pass from the coelom into the pleural cavity through the diaphragm, enter the lungs and eventually settle down in the bronchioles, where they grow and develop during several weeks into adults, encapsulated in tissue which the host lays down.

(*b*) *Second intermediate host a marine copepod, definitive host a marine fish.* Further research is called for before this type of life history can be regarded as established. In 1923, and again in 1935, Lebour reported abundant copepods of the species *Acartia clausi* containing larval trematodes in the tow-nettings at Plymouth, the parasites being identified as *Hemiurus communis*. In some instances comparatively large young flukes were caught in the act of emerging, hind-end foremost, one *Acartia clausi* 1·7 mm. long thus liberating a parasite 0·63 mm.

long. Pratt (1898) also saw a young *Hemiurus* emerging from a copepod, but such rare 'accidents' seem to have nothing to do with the transference of the trematode from one host to another.

In 1923 Dollfus reviewed our knowledge of trematodes which occur in copepods and described a peculiar cercariae, *C. calliostomae*, now included in the group of '*Cystophorous*' cercariae, which was found frequently in *Calliostoma conuloides*. In attempting to decide between two apparent possibilities (*a*) that the miracidium penetrates directly into the copepod and (*b*) that it requires an earlier host, Dollfus inclined to believe that the miracidium enters some mollusc, there giving rise to cercariae which leave this host and then enter the copepods. Going further, he expressed the opinion that *C. calliostomae* is the larva of a species of *Hemiurus*. Next to nothing is known about Hemiurid life histories, but Dollfus's contention was probably correct, '*Cystophorous*' cercariae having come to be recognized as the larvae of Hemiuridae or Halipegidae. The existence of parasitized *Acartia clausi* in Plymouth Sound is circumstantial evidence of the release of swarms of '*Cystophorous*' cercariae from some mollusc or molluscs in its depths, and of the probable ingestion of these by copepods. One such cercaria, *C. sinitsîni* Rothschild, 1938 (Fig. 54), has been found in the district. Several post-larval fishes in the Plymouth district are known to feed upon these crustaceans, gadoids amongst them, and *Acartia clausi* itself forms part of the diet of post-larval whitings. It is not surprising that one of our commonest copepods perhaps plays the part of second intermediate host to one of our commonest Hemiurids, and that it should be eaten by some of the fishes most consistently parasitized by the corresponding adult trematode.

In the Mediterranean, juvenile Hemiurids have been found, not in copepods, but in *Sagitta*, or free in the sea. Young fishes may devour young flukes which leave either of these hosts. Juvenile *Derogenes varicus* frequently occur in *Sagitta bipunctata*, the local species at Plymouth, generally singly and near the ovary. According to Hall (1929), young flukes of this species may also occur in *Acartia*. The life history of this species no doubt closely resembles that of Hemiurus sp., although its larva has been found in Polychaete worms.

Hunninen & Cable (1941*b*) showed that the Hemiurid *Lecithaster confusus* has minute, simplified '*Cystophorous*' cercariae, each with an inverted delivery tube and a simple, non-motile filament. They develop in marine snails of the genus *Odostomia* and are eaten by copepods, 20 % of *Acartia* sp. from the American locality in which the snails occur being infected naturally. The metacercariae develop rapidly in the body of the copepod, but do not encyst. In another paper (1941*a*) the same writers claim that *Podocotyle atomon*, a member of the Allocreadiidae, has '*Cotylomicrocercous*' cercariae, which develop in *Littorina rudis* and later encyst in various amphipods, chiefly in species of *Gammarus*, but also in *Carinogammarus mucronatus* and *Amphithoe longimana*. In *Bacciger bacciger*, the first, second and final hosts are lamellibranchs (*Tapes*, *Pholus*, *Donax*), an amphipod (*Erichthonius difformis*) and an atherine (see Palombi, 1934*b*).

(*c*) *Second intermediate host a fresh-water copepod, definitive host a frog.* *Halipegus ovocaudatus* (Fig. 47A) lives under the tongue and elsewhere in the buccal cavity of frogs. According to Lühe (1909) its larva, *Cercaria cystophora*

Wagener, the type of the '*Cystophorous*' group, develops in rediae in species of *Planorbis* and undergoes a period of further development in the dragonfly, *Calopteryx virgo*. That a fresh-water copepod may take the place of an insect larva as second intermediate host is shown by the development of the American species *Halipegus eccentricus* Thomas, 1939, which inhabits the Eustachian recesses of frogs. According to this writer, the eggs contain embryos when they are laid. They leave the body of the definitive host with the faeces and are eaten by snails belonging to the genera *Physa* and *Helisoma*. The miracidium, which hatches within the body of the snail, is not ciliated, but has a spinous cuticle such as Creutzburg (1890) found in the European species *H. ovocaudatus*. The miracidium penetrates the gut of the snail, after which it sheds its cuticle and anterior spines,* and in the connective tissue becomes transformed into a sporocyst containing up to eight rediae having conspicuous orange bands. Within a month of the time the eggs were ingested by the snail, the rediae have each produced more than fifty '*Cystophorous*' cercariae, which leave the snail in small swarms and are devoured by copepods of the species *Cyclops vulgaris* and *Mesocyclops obsoletus*. Some are digested by the crustacean, but others are unharmed and these migrate through the wall of the gut into the haemocoele. The metacercariae at this stage resemble juvenile flukes, but each has a peculiar tuft of posterior papillae, said to represent an everted excretory vesicle. The definitive host acquires the parasites when a tadpole by ingesting infected copepods, the metacercariae released by the digestive processes remaining in the cardiac portion of the stomach until the tadpole begins to metamorphose, when they migrate to the final location, the Eustachian recesses, to grow and mature.

(iii) *First intermediate host a mollusc; second a larval insect*

(*a*) *Definitive host a fish. Crepidostomum farionis* (Fig. 76H) is an Allocreadiid inhabiting the gall bladder, intestine and pyloric caeca of the trout and grayling. Brown (1927) found the early larval stages of this parasite in Yorkshire rivers, inhabiting the snails *Pisidium amnicum* and *Sphaerium corneum*, more frequently the former. Two generations of identical rediae (Fig. 76A, B) develop and are found attached to the gills of the mollusc. They are 2–2·5 mm. long and bear some resemblance to sausage-shaped sporocysts, partly because procruscula are lacking, but each has a sac-like gut, pharynx and birth pore. The cercariae (C) which develop in the daughter rediae are almost 1 mm. in total length, with a tail as long as the body proper. They swim vigorously with the tail flexed into an S-shape. Having both a stylet and two groups of three penetration glands, they provide fairly certain indications that at the appropriate time they will encyst in a second intermediate host. This happens to be a larval mayfly, *Ephemera danica*, in the fat-bodies or muscles of which light brown, pear-shaped cysts may subsequently be seen through the translucent cuticle. As many as twenty-six cysts, each 0·25–0·3 mm. long, may occur in an individual host, a favourite situation being

* Manter (1926) remarked on five radiating areas near the anterior extremity of the miracidium of *Otodistomum cestoides* from which bristles or spines project. Thomas (*loc. cit.*) noticed the protraction and retraction of a similar coronet of spines in the miracidium of *Halipegus eccentricus* as a result of movements of the anterior region of the body. Such records are of interest on account of their rarity.

the ventral region of the body near the second to the fifth gills. The obvious cysts (D) are produced by the tissues of the insect, but a thinner, true cyst lies in this capsule, enveloping the folded metacercaria which produced it (E). Neither cysts nor metacercariae interfere with the development of the larval insect, and both may later be found in the adult mayfly. At this time (F), the eye-spots

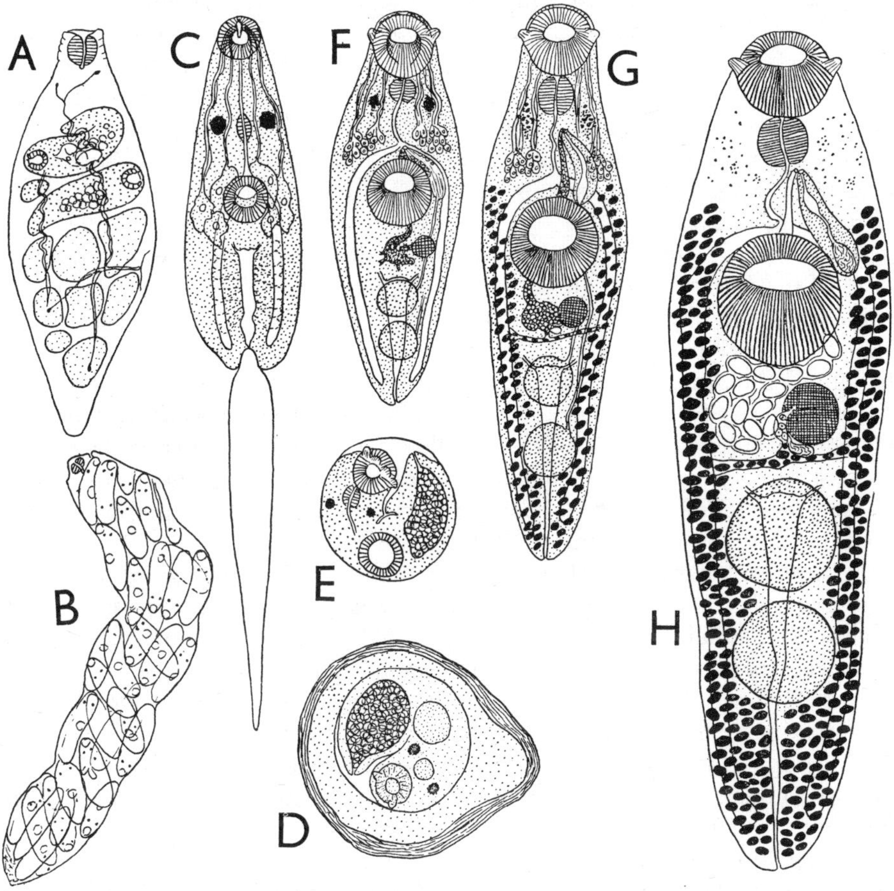

Fig. 76. Stages in the life cycle of *Crepidostomum farionis* (Allocreadiidae). A, parent redia. B, daughter redia with cercariae. C, cercaria. D, cyst from the mayfly larva. E, true cyst. F, larva removed from the cyst. G, juvenile fluke from the trout. H, adult. (After Brown, 1927.)

are less conspicuous than they were in the free cercaria and show signs of degeneration, and the larva has some of the characters of the adult, notably the distinctive processes of the oral sucker. In the pyloric caeca of the trout and grayling slightly older but still juvenile flukes (G) may be found, now released from their cysts, and the rational explanation is that they have been ingested along with the mayflies or their larvae which served as food. Brown examined trout and grayling from the River Wharfe for these parasites, finding all the former and 84 % of the latter infected. Only 3–10 % of *Pisidium* were infected with the early larval stages, and young molluscs were uninfected. This testifies to the intensive infection of the second intermediate host and the efficacy of the insect in disseminating parasites among the piscine hosts.

Allocreadium isoporum (Fig. 29A), another trematode parasite of fresh-water fishes and a member of the same family as *Crepidostomum*, has a similar life history. The cercaria, *C. isopori* Looss (Fig. 61F), belongs to the '*Parapleurolophocerca*' group of Sewell and develops in a redia in *Sphaerium corneum* or *S. rivicola*. The rediae are elongate, colourless, pinched at intervals into 'brood chambers', with a mouth and an oral sucker, but without a functional gut. The cercariae continue their development, according to Lühe, in various larval insects, especially the mayfly, *Ephemera vulgata*, and the caddis-flies, *Anabolia nervosa* and *Chaetopteryx villosa*, within thin-walled cysts which measure 0·36 × 0·34 mm. Fresh-water fishes are infected by devouring insects containing the encysted metacercariae.

(*b*) *Definitive host a frog.* Several well-known trematodes of frogs have this type of life history. In various species of *Gorgodera* (Fig. 47I), which when adult live in the urinary bladder, the cercariae are of the distinctive '*Macrocercous*' type (Fig. 65A, B, C), develop in sporocysts in the gills of snails belonging to the genus *Sphaerium*, and are born before attaining full development. They have very feeble powers of locomotion, but gain access into various larval insects which are attracted by the chamber containing the cercarial body when seeking food. These, the second intermediate hosts, are generally dragonflies of the genera *Epitheca* and *Agrion*, which in turn are devoured by the definitive host and thus convey the encysted metacercariae to it.

Haplometra cylindracea (Fig. 47D), a parasite in the lungs of frogs, and *Opisthioglyphe ranae* and *Dolichosaccus rastellus*, parasites in the intestine of frogs, have Xiphidiocercariae belonging to the '*Armatae*' group which develop in sporocysts in snails of the genus *Limnaea*. The cercariae of the first of them (Fig. 69H) leave the first intermediate host, *Limnaea ovata* or *L. stagnalis*, and penetrate into larvae of the beetle, *Ilybius fuliginosus*, the metacercariae being found in thick, spherical cysts within the haemocoele of either the larval or the imaginal insect. *Cercaria tenuispina* (the larva of *Opisthioglyphe ranae*) leaves the intermediate host (*Limnaea palustris* or *L. stagnalis*) and, according to Wesenberg-Lund (1934), penetrates into tadpoles. This would seem to be a very unusual procedure for a Xiphidiocercaria, and Hall (1929) listed a number of second intermediate hosts that are insects, including mayflies and caddis-flies, and also an amphipod. *Cercaria limnaeae ovatae* (Fig. 69I), the larva of *Dolichosaccus rastellus* (also known by the incorrect name *Opisthioglyphe hystrix*), develops in *Limnaea ovata*. The cercariae leave this host, later encysting in larval Trichoptera (*Limnophilus* spp. and *Anabolia* sp.) and Ephemerida (*Ephemera vulgata* and *Chloëon dipterum*). Final hosts ingest these insects and are infected when they contain metacercariae. Trematodes of the genus *Haematoloechus* (Syn. *Pneumonoeces*) (Fig. 47C), which is also included in the Plagiorchiidae, similarly use larval insects as second intermediate hosts. Thus, the cercariae of *H. variegatus* and *H. similis* encyst in dragonflies of the species *Calopteryx virgo*, the adults inhabiting the lungs of frogs and toads. The same type of life history is seen, however, in members of other families. The cercariae of *Pleurogenes claviger* encyst in water-beetles, those of *P. medians* and *Prosotocus confusus* also. These species, which belong to the family Lecithodendriidae, inhabit the intestine of the definitive host.

(*c*) *Definitive host a bat.* The distinctive trematode fauna of bats includes several species of the genus *Lecithodendrium*, the type of the Lecithodendriidae. According to Braun (1879–93), the cercariae of *L. lagena* are the Xiphidiocercariae known as *Cercaria armata*, which develop in sporocysts in the digestive gland

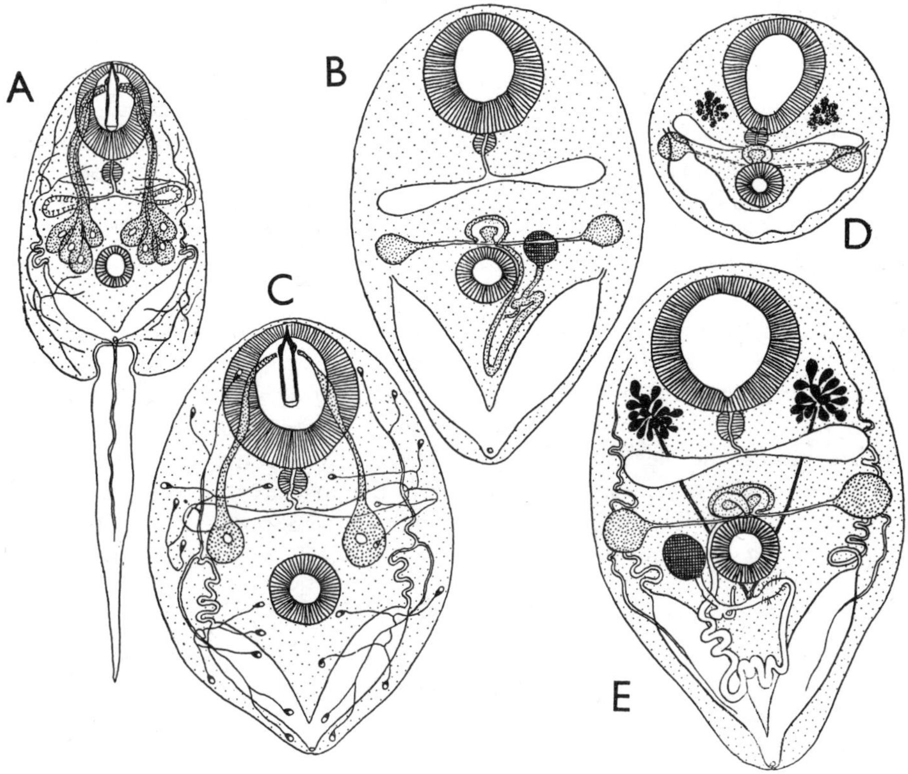

Fig. 77. *Lecithodendrium chilostomum* (Lecithodendriidae). A, cercaria. B, early, post-cercarial stage. C, metacercaria. D, cyst. E, juvenile fluke from cyst. A, B and C occur in the body cavity, D and E in the thoracic muscles of *Phryganea grandis*. (After Brown, 1933.)

of *Limnaea stagnalis*. Brown (1926) described a similar cercaria, *C. pseudarmata*, which he at first regarded as the larva of *Lecithodendrium lagena* but later (1933) as that of *Plagiorchis vespertilionis*. In the same paper he described larval stages of another bat trematode, *Lecithodendrium chilostomum* (Fig. 77E), the cercariae of which show rather unusual behaviour which we may consider briefly. It is believed that *Cercaria armata* leaves the first intermediate host and encysts as a metacercaria in 'Ephemera' and 'Perla' larvae (Siebold) or in the larva of *Chironomus plumosus* (Linstow). On the other hand, the cercaria of *Lecithodendrium chilostomum* (Fig. 77A) penetrates the larva of the caddis-fly, *Phryganea grandis*, the unencysted metacercaria (Fig. 77B, C) inhabiting the haemocoele, sometimes attaching itself to the pale green fat-body and feeding upon it. In this condition it remains throughout a winter and until the following spring, growing meanwhile to about thrice its former length and breadth, and developing the reproductive organs, but not to the functional state. When the caddis larva pupates, the metacercariae migrate from the abdominal region into the thorax and promptly encyst (in condition D, Fig. 77) in the thoracic muscles. The cysts persist in the imago, and within a very short time of the ingestion of the caddis-fly by the definitive host, the trematode commences to lay its eggs.

(*d*) *Definitive host a bird.* From what has gone before it would seem likely that many birds become infected with trematodes which make use of insects as the second intermediate host. Few instances are known at present, but more are likely to appear as our knowledge of life histories becomes extended. The genus *Prosthogonimus* contains several species which infest the bursa Fabricii of birds, including domestic poultry (Fig. 50A). One species which is known to cause a fatal inflammation of the oviduct in the fowl, or, at least, the production of defective eggs, is *P. intercalandus* Hieronymi & Szidat, 1921 (according to Szidat (1927)* = *pellucidus* (Linstow, 1873)). Szidat identified metacercariae which he found in the body cavity of the dragonfly *Libellula quadrimaculata* with this species. An American species, *Prosthogonimus macrorchis* Macy, 1934,† develops from cercariae which occur in the snail *Amnicola limosa porata*. During a short spell of freedom, the cercariae search out various dragonfly nymphs, particularly species of the genera *Leurorrhinia*, *Tetragoneuria* and *Epicordulis*. When successful they enter the rectal respiratory chamber of the insect and migrate into the muscles as the nymph grows, later encysting in the haemocoele. The definitive hosts, birds, are infected as in like instances by eating either the nymphs or the imagines, migration of the metacercariae to the cloaca and bursa Fabricii taking about one week, at the end of which the parasite is fully mature.

(iv) *First intermediate host a mollusc, second a fish*

(*a*) *Definitive host also a fish.* The Bucephalidae form an outstanding example of this type of life history, which is exemplified by *Bucephalus polymorphus* (Fig. 24), a well-known parasite of fresh-water fishes such as the pike, perch, burbot and gudgeon. The cercariae, which are of the characteristic Furcocercous type (Fig. 70), develop in sporocysts in fresh-water mussels, mainly *Anodonta*. They are pelagic and make periodic ascents and descents in the water till they chance to encounter a cyprinid, when they penetrate the skin and encyst in the subcutaneous tissues. The rudd is perhaps the principal second intermediate host, but the definitive hosts are infected after devouring any cyprinids which happen to constitute a reservoir of metacercariae.

Marine Bucephalids have a similar life history, though in most instances the details are still unknown. '*Bucephalus haimeanus*' is a well-known larval form which occurs in oysters and cockles and develops in very long and much-branched sporocysts. The encysted metacercariae occur in gadoid fishes and are common in the cod and whiting. The cysts are found in local swellings of the nerves, mainly the cranial nerves and especially the auditory, but also in the

* *Arch. Geflügelkunde Berlin*, 1, Heft 5.

† This is supposed to be the species mentioned by Baylis (1929) as having been found by Kotlan & Chandler as metacercariae encysted in both larval and imaginal dragonflies.

spinal nerves near the tail, and sometimes in the brain. Worms pressed out of the cysts (Fig. 25B) have most of the characters of *Bucephalopsis gracilescens*, but they become mature (Fig. 25A) only in the stomach and intestine or caeca of the angler which devours infected gadoids. Prosostomatous Digenea may have this type of life history also, e.g. *Stephanostomum* (Acanthocolpidae) (see Martin, 1939*b*) and *Azygia lucii* (Azygiidae) (see Szidat, 1932*b*).

(*b*) *Definitive host a bird or mammal.* Many Digenea, some of them of economic importance, have this kind of life history. One of the best known is *Metagonimus yokogawai*, the commonest Heterophyid trematode in the Far East and an intestinal fluke of man and the dog. The eggs when laid contain miracidia, but these develop only after ingestion by snails belonging to the genus *Melania*, giving rise to a sporocyst, two generations of rediae and finally to typical distome cercariae of the '*Pleurolophocerca*' group with eye-spots and a well-developed penetration apparatus. After alternately swimming and floating for some time, the cercariae penetrate the skin of fresh-water fishes, some of which are important food fishes, notably the trout, *Plecoglossus altivelis*, and encyst under the scales or in the subcutaneous tissues of the gills, fins and tail. Man and dogs are infected by eating raw or insufficiently cooked fishes containing the metacercariae, and fish-eating birds such as the pelican are not immune from infection.

Another trematode belonging to the same family, *Heterophyes heterophyes* (Fig. 54F), which is a common parasite of man, dogs and cats in Egypt and Palestine as well as the Far East, resembles *Metagonimus* closely in both its life cycle and its structure. In this instance the eggs are eaten (in Egypt) by fresh-water snails of the species *Pirenella conica* and the Lophocercariae, which eventually develop, penetrate and encyst in the mullet, *Mugil cephalus*, or some other fresh-water fishes. The encysted metacercariae are able to withstand treatment for 7 days in brine, a period in excess of that allowed in some forms of commercial enterprise, so that both raw and salted fish are sources of infection.

Another Heterophyid, *Apophallus mühlingi* (Fig. 78I), which is a parasite in the intestine of birds such as the black-headed gull and cormorant, encysts as a metacercaria (Fig. 78G) in the musculature or fins of various cyprinids, mainly the bream, *Blicca björnka*, on which these birds feed. The locations in which the cysts occur and the appearance of juvenile and adult forms are shown in Fig 78A–I. M. J. Miller (1941*b*) has outlined a similar life cycle in a related species in America. Perhaps the best known life history of this kind is that of *Cryptocotyle lingua* (Fig. 51D), which was studied experimentally and in great detail by Stunkard & Willey (1929) and Stunkard (1930*a*). The cercariae develop in rediae in *Littorina littorea* and penetrate the cunner, large numbers of them entering the fish by way of the fins. Immediately after penetration into the subcutaneous tissues has been effected, the metacercariae encyst, excystment occurring in the gut of birds and mammals which serve as definitive hosts.

Trematodes of the family Opisthorchiidae also encyst in fishes. *Opisthorchis felineus* (Fig. 57B), a liver fluke of cats and other mammals, including man, in Europe as well as India and the Far East, lays eggs containing characteristic asymmetrical miracidia. These do not hatch in the open, but when the eggs have been eaten by snails of the species *Bythinia leachii*, the miracidia then liberated give rise to a generation of sporocysts and of rediae. The '*Pleuro-*

lophocercous' cercariae which emerge from the rediae attack a wide variety of freshwater fishes, e.g. common carp, barbel, bream, silver bream, roach and rudd, but, according to Ciurea (1917), most commonly *Idus melanothus* and tench. Metacercariae are found encysted in the subcutaneous tissues, infection of the final host following their ingestion. In '*Clonorchis*' *sinensis*, the Asiatic liver-fluke of man, the life cycle is similar, but the snail hosts belong to species of the

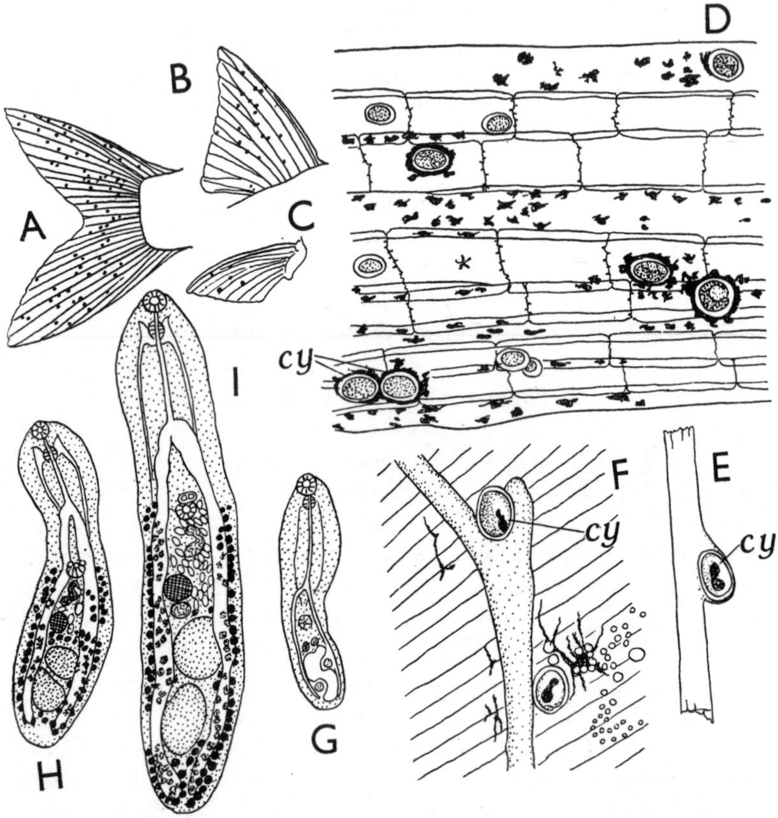

Fig. 78. *Apophallus mühlingi* (Heterophyidae) A, B, C, caudal, dorsal and pectoral fins of the bream, *Blicca björnka*, infested with encysted metacercariae of this fluke. D, portion of caudal fin showing cysts. E, muscle fibre and F, muscular tissue containing metacercariae. G, metacercaria extracted from a cyst. H, young fluke when 5 days old. I, young fluke 7 days old taken from the dog. (After Ciurea, 1924.)

genera *Parafossarulus* and *Bythinia*, and more than forty species of fishes, belonging to the Cyprinidae, Gobiidae, Anabantidae and Salmonidae, serve as second intermediate hosts. It is only rarely that encysted metacercariae of the Opisthorchiidae are found in this country, but Lebour (1914) mentioned three instances in which they were found in the skin of fishes, in lythe from Millport, haddock from the Lancashire coast and cod from Millport. She named the species provisionally *Cercaria chromatophila* on account of the pigment which partially covered the oval cyst, regarding it as an *Opisthorchis* which when adult probably infects some fish-eating bird.

Mention may be made of the similar kind of life history met with in a member of the Troglotrematidae. *Troglotrema salmincola*,* a frequent concomitant of 'salmon poisoning' in dogs in North America, has a cercaria belonging to the '*Microcotylous*' class of Xiphidiocercariae, which penetrates and encysts in Salmonidae. Finally, Cable & Hunninen (1941*a*) have reported that *Siphodera vinaldwardsii*, which is referred to the Cryptogonimidae, has '*Pleurolophocercous*' cercariae which encyst in flounders near Woods Hole.

(v) *First intermediate host a mollusc, second an amphibian*

(*a*) *Definitive host an amphibian.* The older view of the life cycle of the Amphistome, *Diplodiscus subclavatus* (Fig. 10A), stated that no second intermediate host is necessary, infection of the final host resulting from the ingestion of cysts containing metacercariae along with mud from the floor of ponds (Lühe). Krull & Price (1932) followed the development of the American species, *D. temperatus*, and their observations indicate that our conception of the mode of development in the European form may be incomplete. The eggs of *D. temperatus* contain miracidia which develop in the snail, *Helisoma trivolvis*, giving rise to rediae and ultimately to Amphistome cercariae belonging to the '*Diplocotylea*' group (Fig. 63A). These encyst on various objects, but make great use of the skin of frogs, showing a marked preference for the dark spots. Adults are infected with the trematode after eating the cast stratum corneum and contained cysts after ecdysis. Tadpoles may become infected by taking cercariae into the buccal cavity along with the respiratory stream of water. When this occurs the cercariae encyst during their passage along the oesophagus, and excystment takes place in the intestine.

Olsen (1937*b*) described a similar life history in a very different kind of trematode, *Haplometrana utahensis*, which belongs to the Plagiorchiidae. In this instance the cercariae (Xiphidiocercariae of the '*Ornatae*' group) develop in the gastropod *Physella utahensis*, and after emerging from the snail penetrate the epidermis of frogs, tunnelling there and finally encysting. As in the case of *Diplodiscus temperatus*, infection of the definitive host follows the ingestion of the moulted skin containing cysts, the adult flukes inhabiting the intestine of frogs of the species *Rana pretiosa*. *Dolichosaccus rastellus*, which belongs to the same family, has a similar life history, but the Xiphidiocercariae penetrate tadpoles, and frogs are said to be infected by eating these (Joyeux & Baer, 1927*b*).

(*b*) *Definite host a bird.* L. Szidat (1939*a*) obtained a number of specimens of *Cathaemasia hians* from the oesophagus of the white stork in East Prussia, and reared miracidia by experimental infection of snails, finally obtaining peculiar cercariae previously described under the name *C. choanophila* by U. Szidat (1936*b*), who found the larvae in *Planorbis planorbis*, *P. septemgyratus* and *Limnaea palustris*. Out of eleven different species of snail used by L. Szidat in his experiments, the cercariae develop only in these species and in *Planorbis contortus*. Ten to twelve days after the entry of miracidia into the snails, mother rediae appear in the sporocysts which first develop. These give rise to daughter rediae which, like the parent, are colourless and have shoulder-like processes, as well as procruscula,

* This species (=*Nanophyetus salmincola*) has generally been regarded as a Heterophyid, though Witenberg considers it to be a species of *Troglotrema*.

and a tail-like appendage. The first cercariae emerge 37 days after penetration of the miracidia into the snail. As U. Szidat discovered, a second intermediate host is required for completion of the life cycle, tadpoles of Anura such as *Rana esculenta* or larvae of salamanders serving equally well. The cercariae are Echinostome-like, having a feebly developed head crown with forty-seven small, uniform spines. They penetrate into the larval amphibian and encyst in the nares in peculiar elongate, bun-shaped cysts, the metacercaria retaining the vestigial head crown. Attempts to infect the hedgehog, rat, fowl, duck and plover experimentally failed, but the natural definitive hosts (storks and herons) may be presumed to infect themselves by feeding on amphibians which contain the encysted metacercariae. The juvenile flukes lack the 'head crown' and have suckers of about equal size, being thus less like young Echinostomes than are the cercariae.

(*c*) *Definitive host a mammal.* It has long been known that the Heterophyid trematode, *Euryhelmis squamula* (Fig. 54G), which is a parasite of polecats in some parts of Europe, encysts as a metacercaria in the skin of various frogs. Zeller (1867) found these larvae in two-thirds of the *Rana temporaria* taken near Tübingen, Germany, and Joyeux, Baer & Carrère (1934) in all tadpoles and young of *Rana esculenta* in the vicinity of Marseilles. The amphibian second intermediate hosts are devoured by polecats, which thus acquire their parasites. The life cycle in this instance is only incompletely known.

Ameel (1938) investigated the life cycle of a new American species, *Euryhelmis monorchis*, which had been collected regularly for years from mink without attracting attention until identical forms were obtained from rats which had been fed on metacercariae developed experimentally in the green frog, *Rana clamitans*. The adults closely resemble *Euryhelmis squamula*, but have only one testis, which functions only for a short period and then degenerates. The cercaria develops in the snail, *Pomatiopsis lapidaria*, and is not a typical heterophyid cercaria, i.e. '*Pleurolophocercous*', lacking eye-spots and anterior spines near the openings of the penetration glands. It can penetrate frogs and tadpoles equally readily, the skin blistering at the points of entry and the cysts later being clearly visible to the naked eye in the swollen areas. All regions of the body are thus blistered, but more particularly the lateral and ventral regions in the tadpoles and the limbs of the adults. Two-thirds of the green frogs (*Rana clamitans*) in the locality inhabited by infected snails were themselves infected with the metacercariae.

Tadpoles serve as intermediate host in the life cycle of the Echinostome *Euparyphium melis*, adult flukes occurring in the mink, otter and (experimentally) ferret (see Beaver, 1941).

The main features of the life histories which have been discussed are tabulated on p. 491.

This list of intermediate hosts is by no means exhaustive, though the principal groups are represented, and it suggests that all cercariae develop in snails. This is the general rule, but an exception is *Cercaria loossi* Stunkard, 1929, which develops in the polychaete *Hydroides dianthus*. Cercariae have been found in animals which lie outside the groups mentioned in the table. Lebour (1916–18*a*) found late cercariae of *Opechona* [= *Pharyngora*] *bacillaris* (Fig. 62G) on or in a number of coelenterates, clinging to the manubrium or to the wall of the enteron or under the manubrium of the medusae of *Obelia* sp. and in *Cosmetira pilosella*, *Turris pileata*, *Phialidium hemisphericum* and *Pleurobrachia pileus* at

Table 7. *The Life Histories of some Trematodes*

Early larvae in snails; cercariae leave this host and	Penetrate the definitive host	*Schistosoma* spp. (mammals, man)
		Bilharziella polonica (birds)
		Sanguinicola (fishes)
	Encyst on herbage	*Fasciola hepatica* (sheep)
		Fasciolopsis buski (man)
		Paramphistomum cervi (ruminants)
	Encyst on shells of snails	*Notocotylus* spp. (birds)
	Are deposited in slime balls on herbage	*Dicrocoelium dendriticum* (mammals)
	Penetrate the same or another snail to encyst	*Echinostoma* spp. (birds, mammals)
		Echinoparyphium spp. (birds)
		Euparyphium ilocanum (man)
	Encyst in other Mollusca	*Lepocreadium album* (fishes)
		Parorchis acanthus (birds)
	Encyst in copepods or other small Crustacea	*Hemiurus* spp. (fishes)
		Halipegus eccentricus (frogs)
	Encyst in Decapoda	*Paragonimus westermanii* (man)
	Encyst in larval or imaginal Insecta	*Crepidostomum farionis* (fishes)
		Allocreadium isoporum (fishes)
		Gorgodera spp. (frogs, bladder)
		Haplometra spp. (frogs, lungs)
		Pleurogenes spp. (frogs, lungs)
		Opisthioglyphe spp. (frogs, intestine)
		Haematoloechus spp. (frogs, intestine)
		Prosotocus sp. (frogs, intestine)
		Prosthogonimus spp. (birds)
		Lecithodendrium spp. (bats)
	Encyst in fresh-water fishes	*Bucephalus polymorphus* (fishes)
		Apophallus mühlingi (birds)
		Opisthorchis felineus (cat)
		Metagonimus yokogawai (man, dog)
		Heterophyes heterophyes (man, dog, cat)
	Encyst in marine fishes	*Bucephalopsis gracilescens* (fishes)
	Encyst in or on amphibians	*Dolichosaccus rastellus* (frogs)
		Diplodiscus sp. (frogs)
		Cathaemasia hians (storks, herons)
		Euryhelmis squamula (polecats)
		Euparyphium melis (mink, otter)

various times of the year. *Cosmetira* was the commonest 'host' in early summer, when *Opechona* was most abundant, but in late summer *Obelia* and in autumn, and even in December, *Phialidium*. As has been mentioned, this cercaria occurs also in *Sagitta*. Poche (1911) recorded the occurrence of an unidentified trematode in a siphonophore. Levinsen (1881) recorded the larva of *Derogenes varicus* in the polychaete, *Harmothoë imbricata*, fragments of which he found in the stomach of *Cottus*. This unusual kind of life history, in which the first intermediate host is a mollusc and the second a worm, may become more familiar as our knowledge grows. The penetration of the modified Trichocercous cercaria of *Deropristis inflata* into *Nereis virens* and its encystment in the axial region and parapodia was demonstrated by Cable & Hunninen (1942*a*) and has been mentioned in an earlier chapter (p. 220), likewise (p. 249) the use of this intermediate host by the cercariae of *Zoogonus lasius* in America. The latter example is all the more interesting because in Europe the cercariae of *Zoogonus* spp. cannot be induced to utilize worms as intermediate hosts, just as *Z. lasius* will not utilize echinoderms, which seem to be the normal second intermediate hosts of the European species. To turn to fresh-water forms, H. E. Wallace (1941), who reared *Cercariaeum mutabile* Cort, obtained from the snail *Helisoma* spp., to the adult, found that the free cercariae are eaten by the oligochaete, *Chaetogaster limnaei*, which lives in commensal relationship with the snails, and also by *Planaria* sp. The metacercariae penetrate the wall of the gut and encyst in the body cavity or in the parenchyma. Adult flukes were obtained from *Erimyzon sucetta*, which feeds on these hosts and thus acquires the metacercariae, which belong to the species *Triganodistomum mutabile*. Macy (1942) showed that *Planaria* serves as the second intermediate host of a species of *Echinostoma* from the musk-rat. It would seem that no invertebrate is beyond suspicion as a possible intermediate host of digenetic trematodes.

Echinoderms may serve as the intermediate and even the final hosts of Digenea (for references see Timon-David, 1938). About fifty mature specimens of the unusual Holostome, *Cleistogamia holothuriana* Faust, 1924, were found in the gut of the holothurian, *Actinopyga mauritiana*, in the Andamans. Cuénot (1892) found the metacercariae of the Echinostome, *Himasthla leptosoma*, encysted in the circumoral tentacles of the related echinoderm, *Leptosynapta inhoerens*, at Arcachon, what seems to have been the same larva having been found by Villot (1879)* in the foot of the lamellibranch, *Scrobicularia tenuis*, which is perhaps a more usual host. In this country adults of this species occur in various Charadriiformes (see p. 355), which may thus acquire the parasites by devouring holothurians as well as molluscs. Two other metacercariae have been found in holothurians, one by Schneider (1858) in *Holothuria tubulosa* at Naples. The other, '*Metacercaria capricosa* (Cuénot, 1893)', occurs in the circumoral tentacles of *Leptosynapta inhoerens*, but also in the gonads of the ophiuroids *Ophiothrix fragilis* and *Ophiura albida* at Roscoff. Metacercariae may also be found in echinoids. Timon-David found one to sixty cysts containing the metacercariae of *Zoogonus mirus* in 50–60 % of individuals of the species *Paracentrotus lividus* in the Bays of Marseilles and Banyuls, and fewer cysts in *Sphaerechinus lividus* and *Arbacia aequituberculata*. The cysts, which are about 0·25 mm. in diameter, occur in the muscles of the dental apparatus. A second metacercaria was described by Timon-David (1934) as '*Metacercaria psammechinus*' and occurs frequently in *Psammechinus microtuberculatus*, sometimes in *Sphaerechinus granularis* at Marseilles and Banyuls, encysted in muscles near the mouth. The systematic position of the adult remains in doubt, tentative allocation to a genus of the Fellodistomatidae being rescinded by Timon-David in 1938.

* *C. R. Acad. Sci. Paris*, 91, 938–40.

Nor does the above list include all types of life history. Thus, when the second intermediate host is an amphibian, two more hosts may be required. Olivier (1940) showed that the Strigeid metacercariae of *Diplostomum micradenum* and of *Apharyngostrigea pipientis* develop from cercariae in tadpoles of *Rana pipiens*, the second intermediate host, and that the domestic pigeon can serve as a definitive host, though perhaps not the natural one. Odlaug (1940) claimed that another Strigeid, *Alaria intermedia*, requires three intermediate hosts for the completion of its life cycle. Early larvae (miracidia, two generations of sporocysts, cercariae) develop in the snails *Planorbula armigera* or *Helisoma trivolvis*, later larvae (mesocercariae) occur in the tadpoles or adults of the species *Rana pipiens* and then (metacercariae) in mice or rats, the definitive hosts being cats and dogs. Bosma (1934) had already proved that four hosts are required for the completion of the life history of the closely related species, *A. mustelae* Bosma, 1931. In this instance the cercariae produced in snails of the species *Planorbula armigera* emerge and penetrate frogs or tadpoles, in which they develop into mesocercariae in three weeks. Metacercariae mature in nine weeks in the muscles and lungs of mice which feed on the infected amphibians, and ten days after devouring infected mice, mink or weasels have sexually mature individuals in the intestine.

Strigeid life histories are unusual as compared with others, but not necessarily as complicated as these three examples suggest. Linstow (1877), Brandes (1888, 1891) and others believed that the metacercariae long known as '*Tetracotyle*' and '*Diplostomum*' (Fig. 62K) developed directly from the egg, sporocyst and cercarial stages being eliminated from development. Lutz (1921) first showed them to be linked in development with Furcocercous cercariae. Then Ruszkowski (1922) traced the development of such a cercaria obtained from *Planorbis vortex* to the trematode called *Hemistomum alatum* [now *Alaria alata* (Goeze, 1782)]. Szidat (1924) next traced the development of his *Cercaria A*, obtained from *Limnaea palustris*, to '*Tetracotyle typica*', the snail-infesting metacercaria of *Cotylurus cornutus* (Syn. *Strigea tarda*), a parasite of the domestic duck and other birds. He also followed the development of his *Cercaria C* (Fig. 71G), obtained from *Limnaea ovata* and *L. stagnalis*, to '*Diplostomum volvens*' (Fig. 62K), the fish-infesting metacercaria of *Hemistomum*, now *Diplostomum spathaceum*, a parasite of gulls and other birds. In a later work (1929)* Szidat demonstrated that *Tetracotyle typica* can develop in the leeches *Herpobdella atomaria* and *Haemopis sanguisuga*, and by feeding experiments similar to these which produced the adult *Cotylurus cornutus* showed that another new '*Tetracotyle*' from the leech, *Herpobdella atomaria*, develops in birds into the well-known Strigeid, *Apatemon gracilis*. The cercariae of Strigeid trematodes may thus encyst in a variety of second intermediate hosts, viz. the same or another snail, leeches or fishes, sometimes amphibians, even bats and mice, and two or three such hosts may be required. In some works the terms 'mesocercaria' and 'metacercaria' are used to denote the successive post-cercarial stages.

* *Zool. Anz.* **86**, 133–49.

CHAPTER 14

REPRODUCTION, GEOGRAPHICAL DISTRIBUTION AND PHYLOGENY

Reproduction in the Trematoda

I. *Fecundity*

Monogenea and some of the Digenea produce few offspring, but in most Digenea fecundity is the outstanding feature of the reproductive process. Invariably it is notoriously difficult to assess the numbers of eggs laid by such trematodes, and the virtual impossibility of counting them in the uterus has led to wild guesses in some instances. Thus, Ishii (1935*b*) recorded the existence of 'millions' of capsules in the uterus of individual Didymozoidae. Such exaggerated claims can scarcely be accepted, although other helminths have been credited with similar enormous numbers of eggs, a single female, *Ascaris lumbricoides*, with 27,000,000 (Cram, 1925) and a productive capacity of 200,000 per day. We have to admit for trematodes that if far fewer eggs are produced they certainly number many thousands and that the number is far greater than it would seem because many eggs are laid regularly, others being formed to take their place in the uterus. *Fasciolopsis buski* is said to lay as many as 25,000 eggs per day (see Stoll, Cort & Kwei, 1927).

This is only part of the whole story of fecundity, and it does not make clear one of the major differences between Monogenea and Digenea. In the former, a single egg represents one potential adult, but in the latter the egg shelters not one, but a host of potential individuals. The rate of multiplication probably varies greatly in different trematodes. Faust & Hoffmann (1934) estimated that a single miracidium may give rise during several months to 100,000–250,000 cercariae. McCoy (1935) provided a more modest estimate of 10,000 cercariae. Krull (1941) exposed twenty-one snails (*Pseudosuccinea columella*) individually to infection with single miracidia of *Fasciola hepatica*: eight of them became infected and sporocysts emitted cercariae subsequently after 68–69 days, continuing to do so for 10–77 days. The total numbers of cercariae thus developing from single miracidia varied between fourteen and 629. Perhaps under natural conditions the larger number would be exceeded.

This multiplicity of individuals is brought about largely by the activity of sporocysts or rediae of a second generation. In this, as in other matters concerning trematodes, no general rule can be laid down, because the number of progeny produced seems to depend on their type. A single infected mollusc may emit as many as 100,000 Xiphidiocercariae per day, while another produces a mere 250 Echinostome cercariae in the same period (Dubois, 1929). The relative fecundity is doubtless a measure of the comparative risks involved in continuing the life cycle and bringing it to completion. Many Echinostome cercariae encyst in the host in which they developed or in some other similar host which is close at hand. Xiphidiocercariae are pelagic creatures which must roam far and wide in quest of a suitable host.

The experiment of Meyerhof & Rothschild (1940) admirably enumerated the numbers of cercariae that can be emitted by one snail. These writers isolated a single *Littorina littorea* already infected with the miracidia of *Cryptocotyle lingua*, presumably by only a small fraction of the total progeny of this fluke, and fed it on the sea lettuce, *Ulva*, for five years. During the first week of captivity the periwinkle emitted 3300 cercariae per day, and at the end of the first year it had a total of 1,300,000 cercariae to its credit. Cercariae continued to emerge during suceeding years, the daily emission falling off to about 830. This captive mollusc not only produced several million cercariae, many of which emerged at least five years after the time of infection, but while so doing increased the size of its shell by 3·5 mm. We need only consider further the high degree of infection of periwinkles in some localities (40 % of the population) to realize the astronomical numbers of cercariae which come into existence and are largely overlooked in the records of planktonic life.

II. *Gametogenesis and development*

Gametogenesis has been studied in only a few trematodes, and in consequence is imperfectly understood. Apart from two or three recent studies inharmonious results have accrued, and issues like the precise stage at which reduction occurs have not been clearly elucidated. Goldschmidt (1902) studied oogenesis in *Polystoma integerrimum* and credited this species with eight diploid chromosomes. Janicki (1903) reported the same number tentatively in *Gyrodactylus elegans*, and Kathariner (1904) agreed, although Gille (1914) held that the diploid number of chromosomes in this species is twelve.

In regard to Digenea, Goldschmidt (1905) claimed that *Zoogonus mirus* has ten chromosomes, but Schreiner & Schreiner (1908), Grégoire (1909) and Wassermann (1913) disputed the claim and were unanimous in affirming the number to be twelve. This number of chromosomes also characterizes *Fasciola hepatica* (Schellenberg, 1911) and *Cryptocotyle lingua* (Cable, 1931). In other Digenea the chromosomes are more numerous, sixteen in *Paragonimus kellicotti* (Chen [Pin-Dji], 1937), twenty in *Brachycoelium salamandrae* (Kemnitz, 1913) and twenty-two in *Haematoloechus* [= *Pneumonoeces*] *medioplexus* (Pennypacker, 1936, 1940) and *Parorchis acanthus* (F. G. Rees, 1939*b*, 1940). (The names in parentheses are those of authorities providing these figures.)*

Schistosomes have received the special consideration they merit as dioecious forms. Lindner (1914) discovered when studying spermatogenesis in *Schistosoma haematobium* that the male of this species has six bivalent and two univalent chromosomes. The latter join one or the other set of six univalents at the reduction division, so that one-half of the spermatozoa formed have eight chromosomes, the other half only six. The ovum after maturation has eight chromosomes. Faust & Meleney (1924) identified two kinds of spermatozoa also in *S. japonicum*, claiming that one kind has seven and the other eight chromosomes. Severinghaus (1928) corrected the numbers to six and eight. Females of this species have sixteen diploid chromosomes and are homozygous, producing one kind of ovum only. Sex is determined by the spermatozoa at fertilization, two sex-chromosomes

* Markell (1943) specified twelve diploid chomosomes in *Probilotrema californiense* Stunkard, 1935, a member of the Gorgoderidae.

that pass over into the nucleus of the male-producing sperm during spermatogenesis being the deciding factors. As would be expected from these considerations, sex is already determined in the miracidium, which produces sporocysts and then cercariae which develop into adults of one sex or the other, but not both.

Perhaps the best account of gametogenesis in the literature on trematodes is that provided by F. G. Rees (1939*b*, 1940) for *Parorchis acanthus*, and the following details are attributable to her. She stated that this trematode provides an excellent subject for the study of maturation, fertilization, and cleavage of the ovum, all these processes taking place in the proximal part of the uterus of a single individual. Sperms from another individual are assembled in the receptaculum seminis *uterinum* after passage through the entire uterus prior to the liberation of ova from the ovary, Laurer's canal being functionally unconnected with the process of insemination. Fertilization occurs immediately after maturation of the ovum, and by the time the eggs have reached the middle region of the uterus each of them contains a fully formed miracidium with a conspicuous eye-spot. In the terminal part of the uterus the miracidia are hatched, and by this time each contains a fully formed redia. Much information can thus be gained from the study of a solitary mature fluke. It might be mentioned here that both in spermatogenesis and in oogenesis, in this species, reduction occurs during the first meiotic division in the primary spermatocytes and oocytes respectively.

A. *Spermatogenesis* (Fig. 79A–R)

Attached to the peripheral wall of each testis in *Parorchis acanthus* there is a layer of rounded cells 0·005 mm. diameter with large, spherical nuclei and little cytoplasm (A). These are the primary spermatogonia, each of which divides into two secondary spermatogonia, which are hemispherical at first, but become spherical after separating (B–H). They pass into the cavity of the testis and by division give rise to a group of four tertiary spermatogonia (I) which grow and move towards the centre of the testis. One further mitotic division results in the production of eight pyriform cells joined by their bases; these are the primary spermatocytes (J). At the next division sixteen secondary spermatocytes of smaller size come into being (K–L), and these possess the haploid number of chromosomes, eleven, so that reduction has occurred. One further mitotic division occurs, giving rise to thirty-two spermatids (M) which, without further division, are modified into spermatozoa. First, the chromatin mass of each spermatid becomes irregular (N), then elongate, and pyriform (O), its pointed tip protruding through a projection at the free end of its cell. The extruded portion develops, as the chromatin mass further elongates (P), into the tail of the spermatozoon, which is thus forced out of the cytoplasm, tail first, till eventually it is free (Q), leaving behind a residual mass of cytoplasm (R). The cytoplasm of the spermatid, therefore, plays no direct part in the formation of the spermatozoon, giving rise only to one of the numerous cytoplasmic bodies which are dispersed throughout the spent testes of older flukes.*

* According to Anderson (1935) the entire spermatozoon of *Proterometra macrostoma* Horsfall, 1933 is formed from nuclear materials, although Dingler (1910) specified that the spermatozoa of *Dicrocoelium lanceolatum* [=*dendriticum*] originate from both the nuclear and the cytoplasmic components of the spermatids.

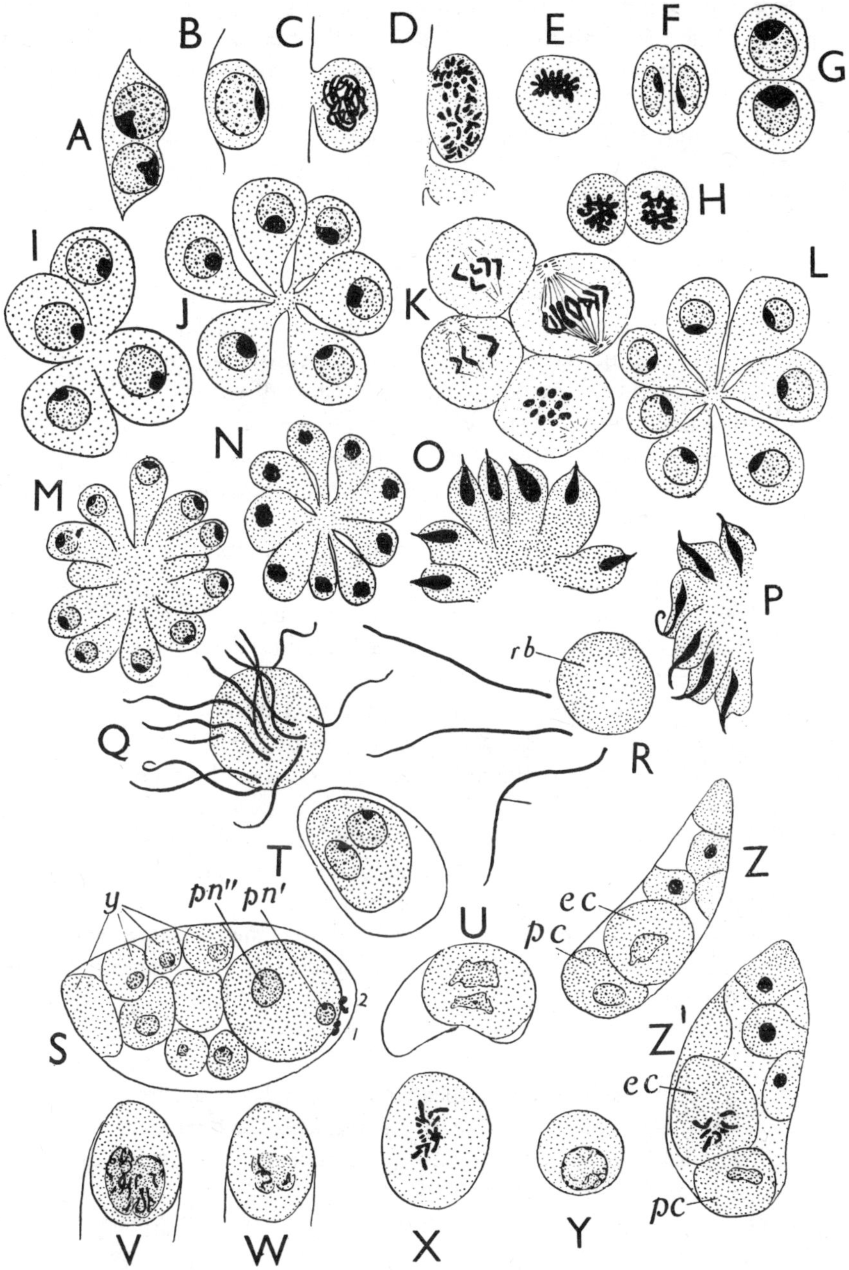

Fig. 79. Stages in spermatogenesis (A–R) and oogenesis (S–Z′) in *Parorchis acanthus*. A, primary spermatogonia on wall of testis. B, primary spermatogonium just prior to division. C, prophase to first mitotic division of primary spermatogonium. D, anophase. E, telophase of same. F, two secondary spermatogonia. G, the same, prior to division. H, late prophase in division of primary spermatogonium. I, four tertiary spermatogonia. J, 'eight' primary spermatocytes. K, anaphase of first meiotic division of the primary spermatocyte. L, 'sixteen' secondary spermatocytes, resting. M, 'thirty-two' spermatids. N, O, P, Q, R, stages in formation of spermatozoa. S, male and female pronuclei, and both polar bodies extruded. T, male and female pronuclei fully formed. U, disappearance of nuclear membrane of the pronuclei. V, W, section of same egg as U with eleven chromosomes appearing from each pronucleus. X, fertilization. Y, zygote nucleus, resting. Z, two-cell stage. Z′, prophase in the 'ectodermal' cell, and undivided 'propagatory' cell. (After F. G. Rees, 1939*b*.)

The spermatozoon, when fully formed, is extremely long and slender, the tail tapering gradually from the body. After swimming about for a time in the seminal fluid, it passes with its fellows along the vas deferens and into the seminal vesicle, to await the opportunity afforded by copulation of passing into the female pore of another fluke and down to the receptaculum seminis uterinum.

B. *Oogenesis* (Figs. 79 S–Z′, 80 A–D)

(*a*) *Maturation and fertilization.* The wall of the ovary is lined internally by a layer of spherical cells about 0·0045 mm. diameter, each of which has a rounded nucleus with a prominent nucleolus, and relatively little cytoplasm. These cells, the resting oogonia, divide mitotically to form oocytes, which grow rapidly, leave the wall of the ovary and move towards its centre till, centrally situated near the origin of the oviduct, they attain full size. Closely packed as they are in this situation, the oocytes appear polyhedral, but in the vicinity of the ovicapt, where they are unrestrained by pressure, they are ovoid or spherical and about 0·028 mm. diameter, i.e. several times as large as the oogonia from which they were derived.

After being liberated singly through the ovicapt, the oocytes pass along the oviduct, through the ootype, and into the receptaculum seminis *uterinum*, where they encounter the spermatozoa from the same or another fluke. At the same time, vitelline cells enter the oviduct from the vitelloducts and pass close behind the advancing oocytes, becoming applied to them in small clumps in the first loop of the uterus, prior to fertilization. Fertilization then occurs. As the sperm enters the oocyte, it shortens and takes up a position beside the nucleus of the oocyte, which has suffered no change as yet. In spite of the fact that countless spermatozoa envelop the oocyte at this stage, no single instance of polyspermy has been witnessed, so that it is safe to assume that it does not occur. Many spermatozoa may attempt to penetrate the surface membrane of the oocyte but, in *Parorchis acanthus*, only one succeeds in doing so.*

Passing swiftly through the first part of the uterus, the egg develops its shell, which is thin at first but gradually thickens to a maximum as development proceeds. Immediately following the first appearance of the shell, maturation division of the oocyte nucleus occurs. The prophase of this division, like that of the spermatocytes, is prolonged. Then the nuclear membrane disintegrates and twenty-two oval chromosomes appear in the cytoplasm and become arranged in pairs, yielding eleven stout, bivalent chromosomes which occupy a central position. The spindle then appears, and the chromosomes become attached to it. As the division proceeds, the chromosomes separate, eleven passing towards each pole of the spindle. Of the two groups of chromosomes thus formed, one is the precursor of the nucleus of the ovum, while the other moves into a superficial position in the oocyte and is soon extruded from the cytoplasm as the first polar body. Immediately afterwards, another division takes place, without the formation of a resting nucleus meanwhile, in which eleven chromosomes, already split longitudinally, divide and separate. One group passes towards the surface and is extruded as the second polar body, the other forms the nucleus of the

* The entry of the entire spermatozoon into the ovum at fertilization was also observed by Anderson (1935) in *Proterometra macrostoma* and by Markell (1943) in *Probilotrema californiense.*

mature ovum, i.e. the female pronucleus. For a time, this nucleus and the closely similar male pronucleus remain in a resting condition, side by side and discrete (T). The egg in the meantime has entered the uterus proper. Its contents at this stage are represented in Fig. 79 S. Fertilization takes place in one of the first loops of the uterus, lying above the left testis. Stages in the formation of the zygote nucleus are shown in Fig. 79 U–Y.

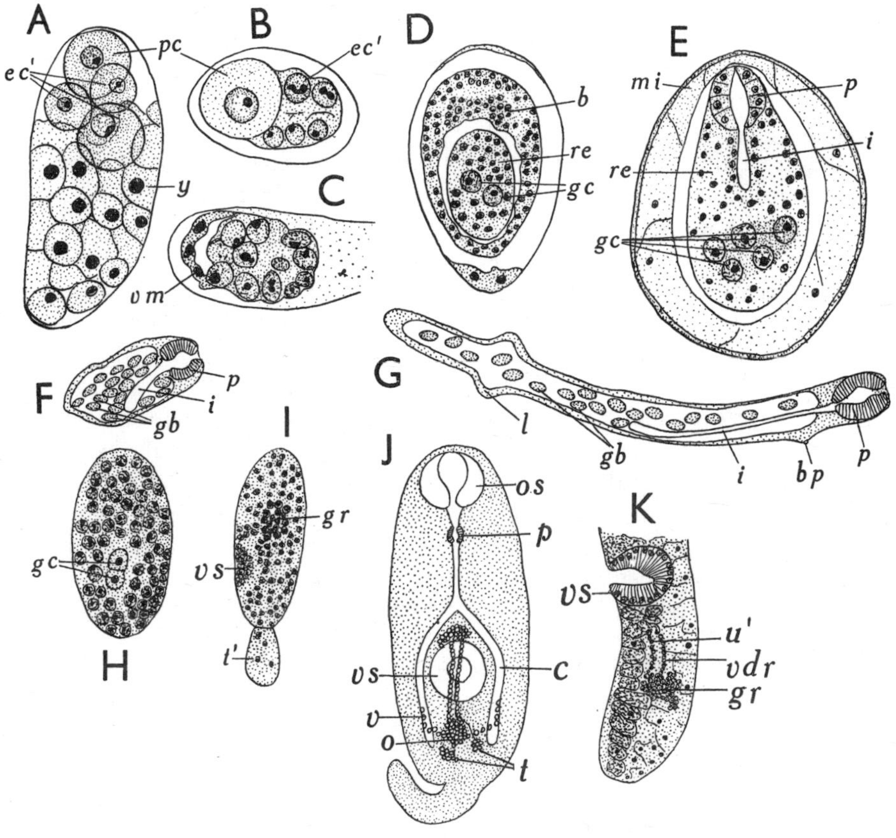

Fig. 80. A–C, early development of fertilized ovum of *Parorchis acanthus*. A, third cleavage (four-cell stage) with 'propagatory' cell unchanged. B, young embryo. C, the same and the developing vitelline membrane. D, differentiation of miracidium. E, transverse section of miracidium showing first generation redia with germ cells. F, first generation redia contracted. G, the same specimen as in F, extended. H, embryo cercaria showing division of 'propagatory' cell into two germ cells. I, longitudinal section of young cercaria, showing development of ventral sucker and genital primordium. J, cercaria in ventral view. K, longitudinal section of cercaria showing development of genitalia. (After F. G. Rees, 1940.)

(*b*) *Segmentation.* After fertilization the ovum begins almost at once to segment. After a mitotic division it gives rise to two cells of unequal size. The larger one is known as the '*ectoderm cell*', the smaller one as the '*propagatory cell*' (Z). Only the former is involved in the second cleavage (Z′), which thus leads to a three-cell stage. One of the daughter ectoderm cells thus formed divides again, so that in the four-cell stage (Fig. 80 A) one large cell (situated nearest the pole of the egg) is the unchanged propagatory cell (*pc*), the remaining three (*ec'*) being derivatives of the original ectoderm cell following two divisions. The latter cells are destined to form most of the body of the miracidium. They divide repeatedly to form, first of all, an ovoid mass of cells with indistinct boundaries, the sizes diminishing with each division. Some of the small resultant cells next draw away from the rest (B, C) and become arranged into a cellular layer of vitelline membrane (C, *vm*), which has been observed in the development of other digenetic trematodes.

When these changes have taken place, the propagatory cell divides for the first time. Of the resultant daughter cells, one contributes to the formation of the miracidial body and the other retains the characters of the parent cell. Further divisions of the same kind occur several times, with only slight amounts of growth intervening, so that the resultant propagatory cell or germ cell is smaller than the original one but is not altered as regards cytological characters. Meanwhile, the whole embryo has enlarged by division of the somatic cells and presses into the surrounding mass of vitelline cells, which gradually break down and disappear as their materials are absorbed by the developing embryo.

A small cavity next appears in the posterior region of the embryo, and the germ cell moves into it. This remarkable cell now behaves exactly as did the fertilized ovum at an earlier stage. It divides into two cells of unequal size, again '*propagatory*' and '*ectoderm*' cells, the latter of which soon divides again and again. In this way there comes into being *within* the body of the still incompletely developed miracidium a second embryo, also possessed of germinal and somatic cells, the former in direct descent from the original propagatory cell of the fertilized ovum. This second embryo represents the next larval stage in the life history of the fluke, a solitary *redia*, thus early segregated from the miracidial body (Fig. 80 D).

C. *The miracidium and mother redia*

As the embryo miracidium enlarges, a superficial layer of ciliated cells arranged in four tiers becomes differentiated. Then an anterior mass of cells is segregated and forms the rudiments of the brain and eye-spots. The contained redia tends at first to lag slightly as regards its differentiation, but it is fully formed when the miracidium is ready for hatching, by which time the second propagatory cell has produced a small mass of redial germ cells (Fig. 80 E, *gc*). Some structural details of the miracidium have been noted already (p. 478), and for others the reader is referred to the latter of the two excellent papers by Miss Rees.

D. *Rediae of the second generation and cercariae*

These larvae originate within the body of the mother redia from the germinal balls, which develop exactly as did those of the miracidium during the formation of the parent, after initial formation of '*ectoderm*' and '*propagatory*' cells. As in the previous stage, there is no reduction and no formation of polar bodies. Upwards of twenty daughter rediae (each ultimately larger than the parent) develop thus within the mother redia, their germinal cells being in direct lineal descent from the original propagatory cell which was formed at the first cleavage of the ovum. Germinal cells do not originate from the body wall of the embryos, as Thomas stated to be the case in *Fasciola hepatica*. By the time each daughter redia is ready to leave the mother redia, the germinal cells have segmented to give rise to germinal balls, the precursors of the future cercariae. Once again, '*ectoderm*' and '*propagatory*' cells are first formed, the latter giving rise to the rudiments of the reproductive organs, the former to the rest of the cercarial body. By the time the tail and rudiments of the suckers have developed, the genital primordium has become elongate, then hourglass-shaped. The anterior half represents the rudiment of the cirrus and its pouch, the metraterm and part of the uterus; the constricted region is the precursor of the vas deferens and much of the uterus; and the posterior half will give rise to the ovary, testes and vitellaria. These genital rudiments (see Fig. 80 I–K) are derivatives of the original propagatory cell of the segmenting ovum, traced thus through three generations of larvae (miracidium, mother and daughter rediae) to a fourth generation, the cercaria. But the germinal lineage thus established seems not to apply to trematodes in general, and we may now consider various theories which apply or have been thought to apply to the germ-cell cycle.

III. *The germ-cell cycle in Trematoda*

Although the past century has brought many efforts at explaining the nature and significance of the distinctive life cycle of Digenea, it has not seen our understanding of it placed on a firm basis. Controversy has centred in the main on the origin and subsequent development of the germinal cells in successive larval stages. Apart from the difficulties involved in pioneering in the field of cytology with animals whose life is a disjointed series of stages, it would seem that there has been more insistence on universality in the cycle of changes which contribute to the development of Digenea than can now be justified. It seems likely that the complicated life histories of various Digenea involve not a single phenomenon so much as a group of phenomena. Some ideas on the subject have outlived their usefulness and retain only historical interest, but we may review the hypotheses which have been put forward briefly. They have been classified under five headings by Brooks (1930):

(1) *Metagenesis*, essentially the theory of an alternation of sexual and asexual stages, first propounded by Steenstrup (1842; translated ed., Ray Soc., 1845) and applied to coelenterates and tunicates as well as trematodes. Steenstrup regarded cercariae as '*pupae*' and the '*king's yellow worms*' of Bojanus (rediae) as '*nurses*' which arise from '*parent nurses*' (presumably, first generation rediae), these in turn developing from ova. This idea received the support of other zoologists, notably Moulinié (1856), Pagenstecher (1857), Wagener (1866), Balfour (1880) and Biehringer (1884). When we consider that the first digenetic life history to be worked out in its entirety was that of *Fasciola hepatica* in 1883, it is evident that the hypothesis of metagenesis is based largely upon superficial homologies between larval and adult stages.

(2) *Heterogeny*, namely, the alternation of a parthenogenetic with a sexual generation. Grobben (*Arb. Zool. Inst. Univ. Wien*, 1882, 4) first formulated the opinion that cercariae develop from parthenogenetic ova, and since his time close scrutiny of the germinal cells of sporocysts and rediae for signs of maturation

phenomena, the formation of polar bodies and such like, has yielded only inconclusive and contradictory statements. One line of progress may be represented by the observations of Tennent (1906) that the germinal cells of '*Bucephalus haimeanus*' originate from the wall of the sporocyst, in which this writer saw also what he took to be polar bodies. Looss (1892) committed himself to the view that any cell of the wall can produce germinal balls. Some other writers claim to have seen polar bodies in larval trematodes, but many have searched in vain for these and other signs of maturation. Dollfus (1919) examined a number of Monostomes and more Distomes and was forced to conclude that later larvae originate not from the body wall of earlier larvae, but from the germinal line which is established at cleavage of the ovum. In the view of Dollfus and those who concur with his findings, perhaps the majority of helminthologists and others competent to judge, the larval bodies are, so to speak, only superimposed on the germinal line, which is uninterrupted from the ovum to the adult. On the other hand, competent persons have encountered conflicting evidence. Thus, Faust (1917) affirmed the origin of *Cercaria flabelliformis* from cells of the base of the gut and *C. diaphana* and *C. micropharynx* from any cell of the body wall, and Dubois (1929) insisted that germinal discontinuity and parthogenesis occur during the development of *Cercaria helvetica* V. An impartial interpretation can only suggest that both sets of opinions may be justified by the existence of different modes of germ-cell formation in Digenea.

(3) *Extended metamorphosis*, briefly, the concept of a very complicated metamorphosis during the course of which several forms arise to take up intermediate positions. Balfour (1880) seems to have originated this idea, which received the support of distinguished helminthologists such as Leuckart (1886) and Looss (1892), and which has been modified to include not only larval budding to produce fresh rediae and cercariae and asexual reproduction, but also a true alternation of generations or metagenesis. Looss (1892) stressed the common general structure of sporocyst, redia and cercaria, tending to regard them as homologous. Sewell (1922) wondered if the distinction generally made between sporocysts and rediae is valid, indicating that on several occasions he obtained sporocysts corresponding to the Furcocercous *Cercaria indica* XV which produced miracidia, not daughter sporocysts or cercariae, and that daughter sporocysts still within the parent sporocyst may contain not cercariae, but miracidia (this in a single instance, miracidia taking the place of Furcocercariae related to the '*Vivax*' group). Sewell seemed to uphold the idea of periods of sexual reproduction alternating with a succession of parthenogenetic stages, coupled with paedogenesis.

(4) *Germinal lineage*, essentially the concept of germinal continuity, of which we have seen an example in *Parorchis acanthus*, with or without multiplication of progeny (polyembryony). Leuckart (1886) first expressed the opinion that the germinal cells of Digenea remain quiescent in the bodies of sporocysts and rediae during the somatic development of these larvae. Many writers, Dollfus among them, have upheld slightly modified versions. Thus, Brooks (1930) maintained that the primordial germ cells of the miracidium segment to form germinal clusters which dissociate, their components forming by subsequent division secondary germinal clusters, or developing into germinal balls which

represent the primordia of larval forms. Tertiary germinal clusters, it is suggested, may arise in the same manner. The process of dissociation, in the interpretation of Brooks, is typical polyembryony, sporocysts and rediae being regarded as structurally homologous. Pointing out that polyembryony may occur as a result of lowered oxygen tension (such as might conceivably be experienced by an animal embarking on a parasitic mode of life) or of altered temperature, Brooks suggested that the cause in Digenea may lie in the external medium, and the manifestation might be an adaptive response on the part of the organism.

(5) *Paedogenesis*, i.e. the relative retardation of the general structure of the body as compared with the organs of reproduction. This hypothesis alone cannot explain the successions of forms which arise during the life cycle, and we have indicated how it has been linked with other hypotheses.

Perhaps the most unorthodox opinion is that expressed by Woodhead (1931, 1932), who extended the idea of polymorphism to include Bucephalids, regarding both sporocysts and rediae as adults rather than larvae, multiplying by sexual means, and he postulated a life cycle comprising not one, but three sexually mature generations. Of these, the sporocyst is regarded as a dendritic colony. It is extremely doubtful if this hypothesis can be applied to most trematodes, but if such surmises prove correct in regard to Bucephalidae they might possibly apply also to Strigeidae, Diplostomatidae and Schistosomatidae, which, according to La Rue and others, are amongst the closest relatives of the Gasterostomes. Woodhead (1932) himself believed that *Leucochloridium* (Brachylaemidae) has a similar polymorphic life cycle.

These views provide a striking contrast with Brooks's hypothesis of sudden adaptive change to polyembryony as a result of the combined action of altered conditions in the environment and the organic response. Cable (1934), who provided an exhaustive review of the literature which has been drawn on liberally, also opposed the theory of Brooks, affirming that germinal lineage supplemented by intercalated polyembryony is insufficient to explain the germ-cell cycle in trematodes in general, and suggesting that the changes brought about as a result of the parasitic habit appeared not suddenly, but gradually. In his view, the primitive trematode became sexually mature in the molluscan host and here completed its life cycle at the time when vertebrates had not yet appeared. Refinements of parasitism in the subsequent evolution of the Digenea included the postponement or loss of sexual reproduction, parthenogenesis taking its place in the intermediate stages which came into being along with the acquisition of fresh hosts.

The Geographical Distribution of the Trematoda

The various classes of vertebrates tolerate trematode parasites which are peculiar to them in greater or lesser degree, and the class of fishes is especially important because it is the most ancient and widespread, its members being the hosts of one-third of all known families of Digenea, as well as almost all the Monogenea. In surveying the distribution of trematodes we are at once struck with the well-marked examples of compatibility and incompatibility between various classes of vertebrate host and whole families of trematodes. We find accordingly all over the world broadly distinctive types of Trematoda which in general are distributed according to the distribution of their hosts. Within this framework occur minor discrepancies, like the comparative absence of Digenea in Selachii and the occurrence of *Polystoma* in Amphibia.

Monogenea are subject to limitations which do not seem to affect Digenea. They attack only aquatic vertebrates, obviously incapable of overriding the threat of desiccation to the extent of making a conquest of land vertebrates, condemned to live in superficial locations in aquatic vertebrates and in consequence much more dependent upon and modified by changes in the external environment, which for the majority of Digenea might not exist. Unlike Digenea they are independent of invertebrates, being able to complete their life cycle on the vertebrate host. We might reasonably expect them to be distributed in accordance with the distribution of the hosts.

Digenetic trematodes are largely dependent upon invertebrates for the completion of their life cycle, and in view of the degree of host-specificity which their larvae show, we might expect their distribution to be confined to those localities in which the ranges of their intermediate and definitive hosts overlap. Thus *Schistosoma japonicum* reaches maturity only in localities where molluscs of the family Rissoidae occur (Craig & Faust, 1940). Where the life cycle involves two intermediate hosts we might expect further limitations in the distribution of the flukes. But in many instances both intermediate hosts are widespread and little or no host-specificity is shown by the cercariae in the selection of the second intermediate host, e.g. in '*Clonorchis*' *sinensis*, which utilizes over forty species of fresh-water fishes.

Because infection with trematodes often arises in consequence of the selection on the part of the definitive host of food of a specific kind, we may expect distribution to be related to diet. In this respect we see notable instances where zoologically distinct hosts acquire closely related parasites, e.g. some fresh-water fishes and also bats thus acquire species of *Crepidostomum*. Some human infections similarly are determined by the consumption of particular plants, molluscs and fishes, and the incidence is increased by herding conditions in regard to the definitive hosts. But for peculiarities of diet and characteristic feeding habits the trematodes responsible for such infections could not have survived. In other words, the mere overlap of distribution as between intermediate and definitive hosts is insufficient to account for the distribution of adult parasites; success for the trematode involves co-operation of one kind or another between these hosts.

In establishing themselves in fresh localities trematodes may show in their larval stages some loss of host-specificity. The miracidia of *Fasciola hepatica* prefer species of *Limnaea* as hosts, but can thrive in their absence provided that species of other genera (*Succinea*, *Fossaria*, *Praticolella*, *Bulinus*, etc.) are available. We cannot regard the distribution of the common liver fluke of sheep as characteristic of trematodes in general, because it has been modified by man; but it illustrates how parasites may continue to thrive after their definitive hosts have made extensive migrations. Quite apart from man's activities, the migration of animals has been an important factor in the distribution of parasites. Johnston (1912) commented on the tendency of trematodes to occur in faunal groups, i.e. to be found not so much in particular zoogeographical regions as in zoologically related hosts. Thus, in Europe, Asia, North America and Australia, closely related amphibian hosts are parasitized by trematodes which likewise show close affinities, e.g. the intestinal Distomes of the genera *Opisthioglyphe* (Europe), *Glyphthelmins* (North America) and *Dolichosaccus* (Australia) and the even more closely related rectal Amphistomes, which in these widely separated regions are represented by various species of *Diplodiscus*. Such faunal groups tend to show, in Johnston's view, that trematodes are animals of great antiquity, the distant ancestors of present-day forms like those mentioned having parasitized ancestral frogs whose distribution was much less extensive than that of their descendants of the present time. The migrating hosts carried their parasites with them and both evolved. The closer affinity which exists between some widely distributed parasites of to-day than between their hosts is explained as due to the more uniform environment in which they have lived, one which favours slower evolution. Johnston left larval forms out of consideration in formulating his opinions, which lose some of their force in consequence. The larvae of trematodes are highly susceptible to factors in the outside world which do not affect adults. But we may presume that molluscs and arthropods able to serve as intermediate hosts were available in new territories which were explored. Perhaps some slight loss of host-specificity assured the larvae of success.

Many evidences of spreading during the evolution of animals have been adduced, some of them relating to parasitism. Metcalf (1929) drew attention to the restricted distribution of the Opalinid protozoon *Zelleriella*, which occurs in the rectum of 'Southern Frogs' (Leptodactylidae) in South America and Australia, but is not represented in other frogs of either the Old or the New World. Claiming that in this instance parallel evolution can scarcely account for very close resemblances between the Protozoa in question and between their hosts in the two localities, Metcalf concluded that radiation occurred by way of a land bridge in the South Pacific. Zschokke (1904) showed also that the cestodes of the two regions show close resemblances. In the case of vertebrates which have made the most extensive migrations, birds and mammals, the problem of tracing out such relationships as can be shown for frogs becomes much more difficult. In many widely separated countries of the world the trematode faunae of these classes of vertebrates show much uniformity. It is in fishes, which have been much studied in recent years by Manter and others, that we find some of the most clear-cut examples of the limitation of spreading by environmental conditions.

Isolation has long been regarded as a cause of differentiation, and Darwin and many after him have used the fauna of the Galapagos Archipelago as an example for special study. Manter (1939) found thirty-nine species of Digenea in the marine fishes of the Archipelago, and of these twenty-one were new to science. Of the new species, eight had their closest relatives in the region of Tortugas (Florida), four along the Pacific coast of Mexico, three in Japan and two in the Northern American Pacific (the relatives of the remaining four not being evident). From these facts Manter concluded that the trematode species endemic to the Galapagos are more closely related to West Indian and Gulf of Mexico forms than to any others. Of the eighteen species previously known, ten occur in fishes of Tortugas (five here only) or the tropical American Atlantic, six occur on the Pacific coast of Mexico (four here only) and three on the Pacific coast of South America (one here only). One of the species occurs elsewhere only in Europe.

The Galapagos Archipelago is rather more than 500 miles west of Ecuador, and according to Hesse, Allee & Schmidt (1937) the fauna has been influenced by two ocean currents, one from the coast of Peru and the other from the Gulf of Panama; even the dispersal of the land snails may have been favoured by oceanic currents, for they are little specialized and most of the genera are found elsewhere in Central or South America. So much the more likely is it that marine snails which constitute the intermediate hosts of trematodes of marine fishes have been dispersed by such currents to produce this notable effect.

Even more remarkable are Manter's discoveries (1934) concerning the digenetic trematodes of deep-water fishes of Tortugas. Ninety species of fishes were collected from depths of 40–582 fathoms and forty-nine species of trematodes found. Almost 80 % of the species of fishes and 30 % of the individuals were infected, these figures being comparable with findings concerning shallow-water fishes. There was no indication of comparative rarity of Digenea at increasing depths to the limit stated. Linstow (1888), dealing with the Entozoa of the *Challenger* Expedition, assumed that trematodes are absent from the depths of the ocean, being of the opinion that their larvae are too delicate to withstand great pressures such as prevail there. We now know that this view involves a fallacy, for the pressures inside and outside their tissues would be equal and no crushing effect would be experienced. I have personally exposed tadpoles to pressures of 200 atmospheres and have observed through the quartz window of the bomb-calorimeter containing them that they show no signs of discomfort at such high pressures. Witenberg & Yofe (1938) submitted the cercariae of *Schistosoma* to exactly this pressure for a period of 6 hours, without impairing the vitality of the larvae. Linstow's other supposition, that the eggs are likely to be too widely scattered in enormous volumes of water to permit larvae to make contact with their hosts, is probably equally fallacious. Linstow himself recorded two immature trematodes from invertebrates under such circumstances, *Distomum filiferum* Sars and *D. glauci* Bergh, the latter a larval Hemiurid. Bell (1887)* also described *D. halosauri* from the ureters of *Halosaurus macrochir*, collected from a depth of three-quarters of a mile (1090 m.) by H.M.S. *Challenger*. It is a Gorgoderid near *Phyllodistomum* but cannot be assigned to a genus because the description is incomplete. The helminth from the greatest depth of ocean yet recorded is the Acanthocephalan, *Echinorhynchus abyssicola* Dollfus, 1931, which came from a depth of more than 3 miles (4783 m.). The deepest water trematode of Manter's collection belongs to the Fellodistomatidae and came from the comparatively shallow depth of 582 fathoms. Monogenea related to *Diplectanum* came from the stomach (!) of a fish at 190–280 fathoms, and a species of *Microcotyle* was found on the gills of the same host-species from 197 fathoms.

We might expect trematodes of deep-water fishes to differ from those of shallow-water and surface forms, not only because the definitive hosts differ but also because the deep-water Crustacea, Mollusca and other invertebrates are distinctive. The outstanding result of Manter's work was the demonstration of close similarity between the trematode faunae of fishes of the deep sea at Tortugas and the fishes of shallow northern waters, and the differences between the trematode faunae of deep- and shallow-water fishes of Tortugas. No fewer than seven species of trematode found only in the deep sea at Tortugas occur also in shallow waters farther north, five of them being well known in Europe (*Derogenes*

* *Ann. N.H.* (5) **19**, p. 116.

varicus, *Lepidapedon rachion*, *L. elongatum*, *Helicometra fasciata* and *Sterrhurus laevis*). The deep-water trematodes of Tortugas are more closely related to the forms in British surface waters than to shallow-water forms found only a few miles away.

Manter is no doubt correct in his surmise that the future will show the distribution of trematodes of marine fishes to be continuous from Arctic to Antarctic, through deep-water hosts of tropical seas. Hesse, Allee & Schmidt (1937) stated that this is true of many snails and bivalves of the north Atlantic, which inhabit the Arctic littoral (to a depth of 50 m.) and may be traced to the Canaries and St Helena (at 2000 m.) and to the West Indies and Pernambuco (at 800 m.). Examples are the Lamellibranchs *Yoldia*, *Nucula*, *Lima* and *Abra*. The same is true of shrimps of the northern genera *Crangon* and *Pandalus* and crabs like *Maia*. I have singled out such invertebrates as these because they belong to groups which are important as the intermediate hosts of trematodes, and to emphasize the fact that the factors which influence the distribution of definitive hosts tend to affect that of possible intermediate hosts also.

Manter has recently (1941) compared collections of digenetic trematodes from marine fishes of Tortugas (175 species), the Tropical American Pacific (82 species), Bermuda (26 species) and Woods Hole (about 70 species). Assuming that sampling has been adequate, as certainly seems to be the case, he found it safe to conclude that the Digenea of Tortugas fishes are extremely similar to those of Bermudan fishes on the one hand and fishes of the Tropical American Pacific on the other, with more than 28 % identity in the latter comparison. They show only 12 % identity with the trematodes of Woods Hole fishes, which show pronounced similarity with those of North Carolina (Beaufort) fishes and with Arctic fishes. Manter is of the opinion, no doubt a very reliable one, that the distribution of these trematodes can be explained largely by oceanic currents and temperatures of past and present times.

One of the most successful trematodes of marine fishes is *Derogenes varicus*, which is known to parasitize more than fifty species of fishes, well over forty species in British waters alone. This remarkable fluke has been recorded from two hosts in Japan (Yamaguti, 1934*a*), from five hosts in the Russian Arctic (Issaitschikow, 1928), from six hosts on the coast of Maine (Manter, 1926), and from five hosts in deep water in the Gulf of Mexico (Manter, 1934). It has not been recorded on the coast of North Carolina and is absent from surface fishes of Tortugas, circumstantial evidence that there is a definite limiting factor, perhaps temperature, operating through adult or larval stages or both, which would explain the geographical distribution.

The Phylogeny of the Trematoda

Numerous efforts to define the affinities of Trematoda with certain other animals have lacked concordance. Early zoologists, marking superficial resemblances between flukes and leeches (Hirudinea), classified these animals together, and Schmarda (1871)* was prepared to exclude Cestoda and Turbellaria from the group thus formed. Looss discerned in the apparent origin of some larval trematodes from the body wall of others a resemblance to the mode of formation of gametes in Annelida, which suggested affinity. Sinitsîn (1911) singled out

* *Zoologie* (I). Wien, 372 pp.

the non-ciliated, cuticularized integument of trematodes as a character which suggests relationship between the Arthropoda and the Trematoda. Stunkard (1937) suggested that all Platyhelminthes were derived from planula-like ancestors, thus implying affinity with the Coelenterata.

As Baylis (1938*b*) has pointed out, very little is known or can be conjectured with any probability of accuracy about the phylogenetic origin of any helminths. But there would seem to be good reasons for grouping together the classes now included in the Platyhelminthes and for upholding the opinion that parasitic members of the phylum have been derived from non-parasitic forms which bore some resemblance to the Turbellaria. Apart from the obvious fact that they are all 'flat-worms', Turbellaria, Trematoda, Temnocephalida and Cestoda show fundamental analogy and homology in the structure of the important excretory, nervous and reproductive organs. Most of them have free-swimming, turbellarian-like larvae, a fact which in itself is sufficient to account for the common origin which has been propounded. Leuckart (1889) made the comparison, by no means as obvious in his time as now, between the sporocysts of Digenea and the Acoela, a group of gut-less Turbellaria. If we accept this basis, the discovery by Thomas (1939) of the anterior spines in the cuticularized miracidia of *Halipegus eccentricus*, and a corresponding discovery made some thirteen years earlier by Manter in regard to the miracidium of *Otodistomum cestoides*, may be more than merely accidental reminders of affinity between trematodes and other helminths, especially the Cestoda.

Little can be added to the account by Baylis (*loc. cit.*) of the opposing opinions of Meixner and Fuhrmann, who claimed that Monogenea, Digenea and Cestoda had evolved separately from Rhabdocoelida stock, and of Lönnberg and Wilhelmi, who believed that the Digenea had originated in somewhat specialized Turbellarians belonging to the Tricladida. Baylis has also provided a plausible sketch of the possible manner in which a turbellarian-like creature came to adopt the parasitic mode of life, perhaps by change in feeding habits and by wandering into the interior of some animal on whose surface it was wont to feed. He has outlined also the possible physiological limitations experienced by Monogenea and probably having to do with their superficial parasitism, and has discussed the possible significance of the larval stages of Digenea in the evolution of the group. Beyond this, it is scarcely possible to go without more facts to guide our judgements, and, as emphasized in Chapter 6, we know all too little about existing interrelationships between various families of Digenea, and are perhaps too prone to lay emphasis on certain larval stages in attempting to elucidate problems of phylogeny. Thus, Poche (1926) represented the Strigeidae, Schistosomatidae and Bucephalidae as terminal twigs of long and distinct branches of the trematode 'family tree', whereas La Rue and others, pointing to the furcocercous nature of the cercariae of these three families, affirmed that close relationship exists between them. In the face of such taxonomic changes as have been proposed recently on the basis of larval and especially cercarial characters, it is positively dangerous to make any assertions.

Many errors of judgement have resulted from the misuse of developmental features in attempting to solve the difficult problems of phylogeny, not the least of them being the misinterpretation of Baer's law to the extent of affirming that ontogeny recapitulates phylogeny. Thanks in large measure to De Beer (1930, 1940) we no longer accept the idea that an animal climbs up its own ancestral tree. In regard to trematodes, Faust (1917) put forward an antithetic view, that the adult stage is more closely related to the ancestral group than are larval stages, which have been simplified secondarily. According to this view, larval trematodes have sacrificed complexity of structure in order to preserve plasticity and thus qualify for parasitic life by producing numerous 'ova' which are physiologically 'young' and develop not into the specialized adult, but into other intermediate stages.

While the phylogeny of Trematoda is a problem which cannot be solved in the present state of our knowledge, certain considerations must be borne in mind if the concepts of modern embryology are to be applied to it. De Beer has stressed the interplay in ontogeny between external factors and internal factors which reside in the ovum and are transmitted from it to the ovum of the next generation. The latter, represented fundamentally by genes or gene complexes, can vary, and by modifying the rates at which structures develop can alter their times of appearance. By such changes, organs may reverse the order of their appearance in successive ontogenies, adult structures may become vestigial or larval, and embryonic structures may persist in the adult stage. *Heterochrony* is the name given to a principle according to which structure may appear sooner, at the same time or later than in previous ontogenies, and it presents a number of possibilities which De Beer has discussed. Of these, some would seem to be directly applicable to trematode developmental stages. Perhaps of greatest importance are *caenogenetic* characters, i.e. those which appear only in young stages. In this category are characters like the nature of the tail in many cercariae, the presence of penetration glands, stylets and the like. These are adaptive characters which rightly belong to larval life and may have little if any bearing on phylogeny. We have seen that the modern tendency in helminthology is the postulation of affinity for trematodes which have cercariae of similar type, e.g. Furcocercous cercariae. Yet this type of tail may be regarded as an adaptation to pelagic life. Baylis pointed out that it is possible to conceive of a stage resembling the cercaria at which maturity was attained. The tail may be regarded as an organ which has been developed to further the chances of reaching a second host when this became practicable and essential. The various kinds of cercarial tail which came into being need have no phylogenetic significance and may merely represent various adaptive structures by which the end was achieved. One type of larva, the cercariaeum, does not develop a tail, and manages to find the final host in spite of its handicap.

Paedogenesis or neoteny, the outcome of retardation of development of the bodily as compared with reproductive structures, has been demonstrated for *Polystoma* by Gallien, whose work I have outlined. Progenesis such as has been described in various digenetic trematodes may be included in this category of phenomena, to which may also belong in a restricted sense events during the intermediate stages of the life cycle in Digenea. *Reduction*, i.e. the occurrence in young stages only of characters which occurred in young and adults of the ancestors, is perhaps the best explanation which can be put forward of the greatly reduced ventral sucker which appears in the ontogeny of some Monostomes but does not persist in the adult. The peculiar metamorphosis of Strigeid trematodes may be said to constitute *hypermorphosis*, i.e. an addition to the ontogeny which was unrepresented in ancestral development. *Acceleration*, i.e.

the precocious appearance of ancestral characters, may be held to account for the development of some Cercariaea which at the time of 'birth' already possess many of the distinctive characters of the adult. *Adult variations*, i.e. characters which appear only in the adult, and give rise to the distinctions which are utilized in their taxonomy, is perhaps widespread in Trematoda, as is evinced by the difficulty of assigning a trematode to its systematic position till the larval stages have been passed through and some at least of the adult characters taken on.

These speculations may seem like piling Ossa upon Pelion, but the need for making such conjectures, on the other hand, emphasizes our abysmal ignorance of fundamental processes which are linked with the problem of trematode phylogeny. Not until we know more about the developmental physiology of Trematoda shall we be able to survey their interrelationships and elucidate their origin. We need to know more also about the relation between the parasite and its host, which may have undergone a kind of parallel evolution.

Stunkard (1929*b*) pointed out that parasites have probably existed since the earliest times, the original species or their descendants having survived to a large extent, in some instances at least being transformed as the hosts evolved. Later (1930*a*) he explained the presence of related parasites in marine and fresh-water hosts by assuming that the primitive hosts were parasitized by the ancestors of present species, and that both hosts and parasites became modified as the hosts differentiated according to the habitat finally utilized. Sandground (1929) concluded that specificity in most parasites is the product of an evolution and a reciprocal relationship between parasite and host. The closer the phylogenetic relationship between two hosts, the more similar the conditions of parasitism in those hosts when young, i.e. before they acquire specific characteristics unfavourable to certain parasites. Considerations of specificity thus parallel the anatomical changes which may be explained by the biogenetic law of Baer.

Szidat (1939*a*) had more faith in specificity, claiming that it is not so much the non-specific tissues and juices of young hosts as constitutional weakness, hunger and sickness which permit a parasite to attack a 'wrong' or unsuitable host, as is suggested by the comparative ease with which hosts that have been fed on unnatural food, e.g. food poor in vitamins, can be infected with parasites which would never attack them in nature. Further, he held that the lists which have been compiled of the parasites in various hosts are frequently incorrect and in sad need of revision. He remarked that '*Notocotylus attenuatus*' was once regarded as a trematode showing very little host-specificity, although study of the life history and better diagnosis has revealed an entire line of new species which show decided specificity in the selection of their final hosts. Fasciolids and Campulids, which also show a high degree of host-specificity, are claimed to be derived from Echinostomes, which typically infect insectivorous and carnivorous vertebrates. Such trematodes as followed the evolution of their hosts in this way have adapted their life cycle to the ways of life of their hosts. Study of the adjustments between parasites and their hosts thus appears to be necessary before a reliable taxonomic scheme can be worked out, and the evolution of the Trematoda elucidated. The latter aim can be achieved only after a much more complete survey has been carried out of the morphological relationships of larval and adult trematodes to the zoological status and ways of life of their hosts.

CHAPTER 15

THE BIOLOGY OF THE TREMATODA

A. Larvae

Larval trematodes conform to a number of types, the behaviour of which varies considerably. We have examined in passing some of the outstanding types of larval behaviour and can be content here to summarize differences which are worthy of note in regard to various processes like the hatching of miracidia, the effects of early larval stages (especially sporocysts and rediae) on their hosts, the emergence and activities of cercariae and their encystment. Most of the information is meagre, practically nothing being known of fundamental physiological processes in trematode larvae, but it represents the basis on which future research will perhaps build up a considerable body of knowledge.

(a) *The hatching of miracidia*

The miracidia of *Parorchis acanthus* hatch while the eggs are passing through the uterus of the parent, those of *Fasciola hepatica* only after a relatively long period of incubation in the world outside the parent. We may take these as extremes. The miracidia of Schistosomes are ready to hatch when the eggs are laid, but do not do so until the latter come into contact with water. In other trematodes the eggs may also contain fully formed miracidia, but hatching occurs not in water in the open, but only after the eggs have been ingested by the first intermediate host. This occurs in '*Clonorchis*', *Metagonimus* and other genera. In *Fasciola* the eggs are only beginning to segment at the time of laying, and the miracidium is not ready for hatching until 9–15 days have elapsed at temperatures of 22–25° C., much longer periods being required for development at lower temperatures. Willey (1941) showed that miracidia of *Zygocotyle lunata* hatch in 19–22 days in summer (June–Aug.) but in 30–40 days in winter (Nov.–Dec.). Mathias (1925) found that the miracidia of *Strigea tarda* hatch in 8 days at 27° C., 12 days at 24° C., 20–21 days at 20° C. and 45 days at 16° C. Dubois (1929) discovered that hatching follows the law of van't Hoff and Arrhenius, the temperature coefficients (Q_{10}) calculable from the data being 4 at 17–27° C. and 5 at 16–26° C., indicating that lowering the temperature by 10° C. within these limits quadruples or quintuples the period required for hatching. It would be interesting to ascertain for wider ranges of temperature whether there is a close correspondence between this process and physiological processes in animals generally, as seems likely. At lower temperatures hatching is inhibited, and it is important to determine for some trematodes what periods of inhibition can be tolerated by the eggs without loss of viability.

(b) *The emergence of cercariae*

Dubois (1929) remarked on differences in the mode of emergence of cercariae from sporocysts or rediae, especially in the relative ease or difficulty as indicated by the time required. Furcocercariae emerge swiftly by comparison with other cercarial types, *Cercaria helvetica* XV taking less than 30 seconds, *Cercaria* A Szidat two to six times as long. Further illustration of the comparative ease of emergence is given by the fact that several cercariae may leave simultaneously after one sustained traction on the parent larva. Echinostome cercariae are slower to emerge, requiring 1–5 minutes according to size, which in some instances seems to be inversely proportional to speed. Xiphidiocercariae emerge more slowly still and with greater difficulty, birth calling for many minutes of violent effort in some instances (*C. helvetica* XXX—a Xiphidiocercaria of the '*Armata*' group) and yet in others requiring only a few seconds even for larger cercariae such as *C. helvetica* V (=VII).

Various factors influence the liberation of cercariae from the snail host, some of them no doubt by direct action on the developmental processes. On two successive days when respective temperatures were 18 and 31° C., Dubois obtained 28,000 and 100,000 *C. helvetica* XXX from the same *Limnaea stagnalis*. Low temperatures limit the production of cercariae, which may cease at about 10° C. (Mathias, 1925) or slightly higher temperatures, e.g. 12–13° C. for *Cercaria strigeae tardae* and 16° C. for *C. hypoderaei conoidei*. The cercariae of *Bucephalus elegans* cease to emerge from the mussel, *Eurynia iris*, at 15° C. (Bevelander, 1933*a*). Light and darkness may also exert an influence, *Cercaria limbifera* Seifert and *Cercaria* No. 2, Rees, 1931 emerging from *Limnaea palustris* only by day. This is not invariably true, but in many instances more cercariae emerge from the host during some part of the day or night than during others. Cable (1938) claimed that *Cercaria vogeli* emerges from the snail almost exclusively during late afternoon and early evening. Dubois found that *Limnaea stagnalis* emitted 14,000 cercariae per hour between 8 and 10 p.m., but only 1800 per hour between 10 p.m. and 8 a.m. The swarms of cercariae steadily become smaller as the vitality of the host becomes impaired, but some idea of the total size of the brood may be gained from the fact that for some months, and until the very day on which the host succumbs, thousands of cercariae may be emitted daily. These are not necessarily the progeny of a single miracidium, of course, but it seems safe to say that they do not fully represent the powers of a single trematode, which is capable of heavily infecting many molluscs simultaneously.

(*c*) *The movements of cercariae*

(i) *Swimming*. The tail of the cercaria is commonly regarded as the organ of locomotion, where it is not an organ of flotation or of anchorage. Strictly, natation depends on the co-ordinated interaction of the muscles of both body and tail. It may continue almost uninterruptedly for some hours, or may be of only a few minutes' total duration. In many instances (Monostomes, Echinostomes and Xiphidiocercariae) the tail when at rest is flexed and laid along the ventral surface of the body, projecting beyond the anterior extremity. Often, the tail goes in advance of the body during natation, drawing the rest of the animal along, not pushing it along, essentially the same difference as between rowing a boat and sculling it with a single oar from the stern.

Dubois (1929) discussed the natatory movements of the very mobile Xiphidiocercariae and Echinostome cercariae, during which the body is partially contracted and arched dorsally in the sagittal plane and the tail lashes vigorously, its undulations being confined to one plane. As a result of such dispositions the body oscillates round a nodal point situated near the ventral sucker and the larva is drawn rapidly along. Only in forms which are sensitive and responsive to light is the body orientated and, consequently, rates of movement are difficult to determine except where such forced movements (*taxes*) prevail. But Dubois found that in four cercariae of progressively larger body size, movement was correspondingly more rapid:

	Length (extended)	Moved 1 cm. per	Moved body-length per
C. hypoderaei conoidei	0·5 cm.	6–12 sec.	3–6 sec.
C. helvetica XXXII	0·6 cm.	5–8 sec.	3–4·8 sec.
C. helvetica XXII	0·9 cm.	4–5 sec.	3·6–4·5 sec.
C. echinostoma sp. from *Fulica*	0·9–1·2 cm.	2–3 sec.	2·7–2·4 sec.*

The figures which I have added (the last column) indicate that the time taken for the cercariae to move through their own length is approximately constant, the slight advantage possessed by larger cercariae scarcely being significant in the rather crude estimates.

In swimming Monostome cercariae the body is contracted, and in this condition is drawn along by vibratory movements of the tail. Even the isolated tail, cast off during encystment, moves with the same liveliness, shows the same or similar vibrations and always advances with the posterior tip foremost.

Mathias (1925) gave good descriptions of natation in the Furcocercous *Cercaria strigeae tardae*, which progresses tail foremost with the furcal rami extended and closely apposed, thus continuing the longitudinal axis of the tail stem. In this instance the entire body undulates, the static impression gained by an observer being of a figure of 8 which displaces itself. All Furcocercariae studied by Dubois execute similar movements, making frequent short and swift ascents, and slow descents during which the mobile body is suspended as efficaciously as possible by the now divaricated furcal rami. During flotation and gradual descent, the body is contracted into an oblique position, owing to the flexure of the tail into an obtuse angle, this no doubt increasing buoyancy. In *C. lophocerca* movement is highly characteristic: there is a brusque displacement due to the vibrations of a tail which is invisible, so rapid are its movements, followed by instantaneous cessation of movement, the cercaria remaining poised and motionless: then, equally suddenly, the larva awakens to violent motion and darts through the water as an arrow moves through the air, only to stop in its course as before. Rothschild (1939) found the active and passive phases to last 2¼ and 15¼ seconds in a 'Pleurolophocercous' cercaria believed to be *C. lophocerca* Lebour *nec* Fil.

(ii) *Creeping*. This, the only form of locomotion in a cercariaeum and in most Microcercous cercariae, described in some detail by Faust (1917), Dubois (1929) and others, results from the interaction of the general muscles of the body and those of the suckers. With the ventral sucker firmly affixed to the substratum, the fluke extends the anterior part of the body to its fullest extent by contraction of the circular and relaxation of the longitudinal muscles. Stretched out as far

* The smaller cercariae are assumed to take longer in covering the distance of 1 cm., and vice versa.

in front of the ventral sucker as possible, the oral sucker is now applied to the substratum. The ventral sucker then releases its hold and is drawn forward by contraction of the longitudinal muscles until it is situated close behind the oral, where it again becomes attached to the substratum. Oft repeated, such movements result in a looping progression which varies in amplitude in different cercariae and is equal to rather more than the distance between the two suckers, generally one-third to one-half the length of the extended body at each step, as the following figures of Dubois indicate:

	Length of body extended (mm.)	Amplitude of movement (mm.)
Cercaria helvetica XXX (*armatae*)	0·27–0·35	0·09–0·14
,, IV (*armatae*)	0·36–0·45	0·12–0·18
,, X (*virgulae*)	0·12–0·16	0·075
,, XI (*microcotylea*)	0·18–0·225	0·06–0·075
,, XX (*echinostome*)	0·54–0·63	0·30
,, II (*echinostome*)	0·45–0·60	0·20–0·25

From what has been said it will be clear that conspicuous change of form occurs during reptation, the fully extended body often exceeding three times its mean length when at rest. It might be mentioned that Echinostome cercariae and Xiphidiocercariae are seldom immobile, their movements alternating between natation and reptation. The cercaria of *Echinostoma revolutum* swims spasmodically, taking periodic rests, till it comes into contact with some object, when it stops, attaches itself by the suckers and creeps over the surface, resuming swimming if the object is an unsuitable site for encystment. In Monostome cercariae, the posterior locomotor pockets serve in place of a ventral sucker during creeping. In other cercariae there may be characteristics which are worth noting, e.g. in *C. lophocerca* reptation is combined with slow unilateral contraction of the body, which is thus freshly orientated at each movement. After a dozen such movements the cercaria may have returned to the point of departure after having marked out a circular or polygonal path. Furcocercariae, expert in floating, do not creep during free life, but do so after penetrating into a molluscan second intermediate host, continuing their movements till they reach the digestive gland or the gonad.

(iii) *Rotation*. Dubois regarded this type of motion as of rare occurrence, having been observed only in *Cercaria A* Szidat and *C. lophocerca* Fil. The body is laterally flexed and vibrations of the tail bring about spiral motion, movement being round a fixed point roughly corresponding to the position of the oral sucker. In the latter of these two cercariae the high frequency of the tail vibrations results in especially rapid motion of this kind.

(iv) *Forced Movements* (*Taxes*). It is not difficult to see in the behaviour of larval trematodes activities in the nature of forced movements or *taxes*, which are notoriously difficult to study critically in any lower animal and have not been investigated more than superficially in Trematoda. But miracidia show a pronounced *chemotaxis*, the mucous secretions of snails having the power to attract the larvae whenever they happen to come within a few millimetres of the mollusc (Faust & Meleney, 1924). Blood seems to have a similar attraction for the Furcocercous cercariae of Schistosomes, so that these larvae eventually arrive at the blood circulatory system of the definitive host, though other cercariae do not as a rule show marked chemotactic responses towards their hosts. Generally, however, they are influenced greatly by light, *Cercaria lophocerca* and *C. ocellata* showing positive phototaxis, *Cercaria A* Szidat the converse. According to Willey (1941) the cercariae of *Zygocotyle lunata* will 'follow' a beam of light back and forth.

Some cercariae are especially responsive to shadow stimuli. Thus, *C. floridensis*, which encysts in small fishes, swims vertically upwards whenever some object casts a shadow on it (McCoy, 1935). In some instances the change of light intensity must be rapid, so as to invoke a shock reaction, e.g. in the cercariae of *Bucephalus elegans* (see Bevelander, 1933*a*). The latter writer found that the reaction time of the cercaria in this respect varies inversely as both the temperature and the intensity of the stimulus. In other instances there is a dual mechanism relating shadow and touch stimuli, as in *Cercaria hamata*, which is kept in a state of almost constant movement by repeated touch stimuli, whereas a relatively long interval must be interposed between consecutive shadow stimuli in order to invoke a regular response (Miller & Mahaffey, 1930). Various writers have recognized forced movements of other kinds, some of which are not beyond suspicion. Thus, Dubois claimed that certain cercariae show *thermotaxis* because between certain temperature limits they do not leave the snail, and others were said to show *aerotaxis* because they aggregate in a zone at the surface of a column of water having uniform temperature, some *thigmotaxis* because they adhere to a substratum with which they happen to come in contact. The study of forced movements cannot be carried out in as elementary a way as this, and some conclusions so far reached are of little value.

At the same time, we must realize the fundamental importance of studying the behaviour of larval trematodes. Wheeler (1939) emphasized the well-known difficulty of identifying certain closely related cercariae on purely morphological grounds, pointing out that, where structural differences are so slight as to be valueless, differences in behaviour may be very marked, especially in regard to the frequency and duration of spontaneous natatory movements and responses to shadow stimulation. Thus, of two almost identical Pleurolophocercariae, *C. opacocorpa* and *C. semicarinatae*, the former swims intermittently and for periods which average less than 1 second, the latter almost continuously for the first few hours after emergence from the snail. It may prove a valuable aid in identifying closely similar cercariae to ascertain such characters as the number of spontaneous swims per given period, the nature and duration of each swim and the comparative effects of shadow stimulation.

(*d*) *Encystment*

Apart from blood flukes of the families Schistosomatidae and Aporocotylidae, trematodes at one stage in the life cycle enter a resting phase when no nourishment is taken and the larva remains sealed up in a cyst. Encystment may be regarded in most instances as a larval adaptation to new conditions which suddenly prevail when the larva emerges from the tissues of the snail. Strictly, it is scarcely a resting condition, for the young fluke is able to move, though only in a restricted sense, and much development occurs while it lies thus ensconced;

most important, the reproductive organs gradually attain their full development. And the process can scarcely be adaptive in instances when encystment occurs relatively early, e.g. in *Asymphylodora tincae* in Japan, where the cercariaeum may encyst in the tissues of the intermediate host, or may emerge and encyst in another individual, or may even encyst within the redia in which it developed while this remains in the tissues of the snail (Nagano, 1930).

In a few instances progenesis occurs, though doubted by some writers. Sinitsîn (1905) found eggs in the uterus of encysted metacercariae of *Pleurogenes medians* within the body of *Agrion* sp., and Fuhrmann (1928) mentioned that the corresponding stage of *Ratzia parva*, encysted in the muscles of frogs, may have eggs containing fully developed miracidia. Metacercariae of *Anisoporus manteri* containing numerous eggs inhabit cysts in the amphipods, *Carinogammarus mucronatus* and *Amphithoë longimana* (Hunninen & Cable, 1940). *Coitocaecum anaspides* Hickman, 1934 encysts as a metacercaria in the Tasmanian mountain shrimp, *Anaspidis tasmaniae* (in New Zealand, *Paracalliope fluviatilis*; see MacFarlane, 1939), and in this condition lays eggs from which miracidia hatch, to swim about in the liquid contents of the cyst. Similarly, *Psilostomum progeneticum* occurs in gammarids in the Balkans (Wiśniewski, 1933*a*). Serkova & Bychowsky (1940) claimed that *Asymphylodora progenetica* has alternative life cycles; it may mature in the intestine of fresh-water fishes of the genera *Rutilus* and *Carassius*, or in the snail, *Bithynia tentaculata*, which serves as the intermediate host in the former case. Another instance of progenesis, in a species of *Phyllodistomum*, was reported by Wu (1938).

We have already seen that various types of cercariae encyst, and that the process occurs in various situations, on herbage or on the shells of snails, in the body of the intermediate host or in an accessory host. Handling or irritation will sometimes bring about spontaneous encystment, even in a pipette used to suck up cercariae (Willey, 1941). Consequently, we can scarcely regard encystment as a standardized process. Indeed, it differs in many particulars in different cercariae, and we cannot review all the means whereby it is achieved. Thomas (1883) described the process for *Fasciola hepatica* and many others have done so for other flukes. We might select from many accounts that of F. G. Rees (1937) for *Parorchis acanthus*, already mentioned in other connexions. The cercariae of this trematode, *C. purpurae*, will encyst on the outsides of *Purpura* shells, on the sides of glass vessels or on objects in an aquarium tank. If the cercariae are placed in a watch-glass, the whole process may be watched under the microscope. The cyst is white, circular in outline, flat on the attached side and convex on the free surface, measuring about 0·3 mm. diameter. It consists of a thin, outer layer (about 0·045 mm. thick) and a thicker inner layer (0·113 mm. thick). The outer wall becomes opaque, but the inner remains translucent and closely envelops the metacercaria curled up inside.

The cercaria which is preparing to encyst attaches itself at an appropriate spot by means of the ventral sucker, after which mucus secreted by conspicuous ventral glands pours out through pores in the cuticle and brings about firm adherence along the anterior median-ventral line. The body of the cercaria now contracts, the oral sucker becomes attached, and mucus from anterior glands pours over it, firmly cementing it to the substratum. Mucus next flows through pores in the cuticle dorsally, being derived from dorsal gland cells, and it covers the larva with a thin film. These secretions form the mould in which cystogenous materials are cast into the form of the cyst proper. Granules stream out of the ventral cystogenous cells and are passed by the body against the mucus film, to form the outer margin of the cyst. As the cercaria withdraws slightly, further cystogenous secretion is extruded and forced upwards, to form the dome of the cyst. Opposite the oral sucker the cyst is incomplete, but this defect is remedied by retraction of this sucker into the body, which brings the paired cystogenous glands into a terminal position, from which they operate to seal the opening. Meanwhile, the tail is being gradually forced away from the body and eventually becomes severed from it, a minute indentation of the cyst wall marking the spot where the cystogenous secretion hardened whilst the tail was still attached to the body.

When the cystogenous secretion has been completely emitted, the metacercaria awakens to activity, moving round the cyst, smoothing down the still plastic materials from the inside. Such movements continue for several hours, after which the larva becomes quiescent. Later, it secretes a membrane, probably akin to the original mucous secretion, which forms the thin, elastic inner wall. Throughout the entire process the cuticle remains complete, as its spiny nature makes obvious, thus playing no part in cyst formation.

According to Wunder (1924, 1932) and Bovien (1931)* the cuticle of the cercaria may in some instances swell, cystogenous materials being deposited within it. Rothschild (1936*b*) described an unusual mode of cyst formation in a Gymnocephalous cercaria, in which cystogenous materials are forced round the larva between outer and inner layers of the cuticle, the former layer eventually being ruptured by the pressure generated. The finished cyst, therefore, consists of secreted materials *plus* the outermost layer of the cuticle, which is itself peculiar in containing pigment granules and a 'primitive epithelium'. The cercariae in this instance were encysting in a species of *Limnaea* from Iraq.

Abnormal pigmentation in the host. In some instances, the host's tissues react towards the invading metacercariae as they might towards other foreign bodies, forming about each a capsule of connective tissue. Cyst formation parallels similar inflammatory or local immunity reactions in higher vertebrates (Hunter & Hamilton, 1941). This is notably the case where fishes are the hosts, and in many instances the capsules are surrounded by masses of pigment, especially of melanophores containing the black pigment melanin. Smith's experiments (1931, 1932)† on the effects of mechanical injury (the eruption of corial melanophores and general cutaneous melanosis) show the reaction to be a defensive one. The question whether the melanophores migrate to the site of infection or are formed there *de novo* is at present unanswerable, though the numbers of melanophores may be so far in excess of that normally present as to overrule the practical effects of migration alone. Hsiao (1941) described a cod (*Gadus callarias*) which was heavily infested with Heterophyid metacercariae and which was very deeply pigmented in the affected regions of the integument, though not in unaffected regions. In this instance (which is only one of many which have been recorded)

* *Vidensk. Medd. Naturh. Foren. Københ.* **92**, 223–6.

† *Biol. Bull. Woods Hole* **61**, 73–84; **63**, 484–91.

the cysts and melanophores were abundant even in the cornea, so that the fish was blind, its eye scarcely distinguishable from the blackened skin around the orbit.

It is well known to physiologists that melanin is formed from various substances, characteristically the amino-acid tyrosine, by the action of a number of enzymes collectively known as tyrosinase. The substrate which is acted upon is sometimes spoken of as the 'chromogen'. Hunter (1941) believed that where infection is due to *Cryptocotyle lingua*, and possibly elsewhere, certain cells of the fish contain the necessary enzyme but lack the chromogen, which the parasite itself supplies at or after penetration. He has to admit, however, that the dearth of pigment cells around some cysts (e.g. those of *Clinostomum*) may be due to the absence of the chromogen in the parasites. It seems to me that these assumptions are both unwarranted and unnecessary. In reviewing my own work on additive and subtractive processes which modify the pigmentation of frogs under certain conditions and also that of others on reversible changes affecting melanin, I emphasized (1941 *c*) that the enzyme tyrosinase is regulated by oxidation-reduction potentials and that natural melanin, like the pigment produced by the autoxidation of dihydroxyphenylalanine ('dopa', a precursor of melanin), is capable of reversible oxidation and reduction, with consequent change in the intensity of blackness in the pigment. These features strongly suggest that the formation of melanin in the vicinity of encysted metacercariae in fishes is an effect produced, not by the exchange of chromogen or enzyme as between parasite and host, but by alteration of pre-existing conditions of oxidation reduction. The parasite does not require fresh nutrients in order to continue its development, but must effect oxidation of nutrients it already possesses in order to gain the necessary energy for further development. How it does this we do not know, but it seems likely that the processes by which it is achieved are such as tend to promote the development of melanin in the surrounding host tissues. Since melanin results from an oxidative process, the effect must be due to heightened oxidation in the tissues around the parasites. Such a change would lead to the production of apparently greater amounts of melanin, apart from considerations relating to quantities of interacting enzyme and substrate.

(*e*) *The effect of the parasite on the host*

(i) *General.* The effects of larval trematodes on their molluscan hosts are twofold injuries, mechanical and chemical. We saw in the case of *Fasciola hepatica* that rediae literally eat their way through the digestive gland, and in like instances the tunnelling larvae may completely riddle this and other glands. The total mass of larvae may greatly exceed that of the infected organ, whose capsule may be distended almost to bursting point, a prick with a needle causing it to burst and discharge a copious stream of larvae. In addition, the parasites secrete toxins, which tend further to upset the physiological equilibrium of the host.

Faust (1917) found no infection to be so light that the host was unharmed, and amongst the cytological injuries marked the accumulation of fatty bodies in some cells, the occurrence of large vacuoles in the vicinity of the nucleus in others, cytolysis, karyolysis and the sloughing of tissue. Agersborg (1924) recognized four infective stages in the fresh-water snails *Physa gyrina* and *Planorbis trivolvis*: (1) the parasite invades the host, whose tissues shrink and become friable and refractory to cytological treatment; (2) the tissues secrete a granular substance which is discharged into the intercellular spaces and becomes apparent everywhere in the tissues of the host, but does not occur in the parasites; (3) during prolonged heavy infection the tissues degenerate, epithelia becoming transformed from the columnar to the squamous type, and the host may die; (4) if resistance is still offered to the activity of the parasites, the host's tissues gradually return to normal as the parasites decrease in number. Dubois (1929) and F. G. Rees (1931) confirmed these findings in the main. The latter writer regarded physiological injury done to *Patella vulgata* by *Cercaria patellae* of greater importance than the mechanical injury. The invaded organ ceases to function properly, being occupied in ridding itself of the toxic products of its parasites, a process which leads to atrophy of the tissues concerned.

W. J. Rees (1936*b*) studied the differential effects of various larval trematode parasites of *Littorina littorea*, recording only mechanical injuries in organs such as the gills and kidneys, but physiological injury in others such as the digestive gland and gonads. He found differences in the degree of damage to the latter organs explicable according to three interdependent factors: (*a*) the nature of the larvae, whether sporocysts or rediae; (*b*) the size of the larvae; and (*c*) the hindrance afforded by certain inactive sporocysts, forming what he called a 'blocking layer'.

When *Littorina* is attacked by rediae the first organ to suffer is the gonad, which is eaten by the larvae (whose gut is frequently filled with eggs and yolk) and disappears. These effects are produced by the rediae from which *Cercaria himasthla secunda* and *C. lophocerca* develop. The digestive gland is only indirectly damaged, by pressure exerted by the rediae, by the loss of nutrient substances, and by the enormous production of waste materials.

Specimens infected with sporocysts instead of rediae preserve the gonad much longer, and its eventual disappearance is due to other causes, which are similar to those affecting the digestive gland in redial infections. Such effects are seen when *Littorina* is parasitized by larval stages corresponding to *Cercaria emasculans* and *C. littorinae*.

Of the larvae considered by W. J. Rees, those corresponding to *C. himasthla secunda* (rediae) and *C. littorinae* (sporocysts) were of very large size, and parasitism by these forms resulted in a precocious degeneration of the tubules of the digestive gland locally, tubules in other (unparasitized) regions being unaffected. Rediae corresponding to *C. lophocerca* are much smaller and produce more even degeneration, which Rees attributed to a not so heavy local output of waste products as in the case of larger rediae. The sporocysts in which *C. ubiquitoides* develop are the smallest of the larvae considered, and the effects of parasitism are mildest.

The characteristic spoken of as a 'blocking layer' is seen where the larvae are inactive and not migratory. The sporocyst of *C. emasculans* belongs to this category, multiplying enormously and remaining in a circumscribed location, so that a section of the digestive gland (the distal portion) is isolated from the rest by a layer of tightly packed sporocysts (the 'blocking layer'). The isolated region suffers starvation through stoppage of lymph channels and its tissue disappears.

(ii) *The problem of sex change.* Pelseneer (1906, 1928) first observed that specimens of *Littorina littorea* infected with the larval trematode *Cercaria emasculans* and its progenitors have a penis of greatly reduced size as compared with that of uninfected individuals. The name which he gave to the cercaria indicates his opinion that in this instance parasitism resulted in castration. Periwinkles are dioecious molluscs, and the sexes are easily distinguishable, even out of the breeding season, by the prominent accessory organs of reproduction, the penis and the oviduct. The size of the organs depends on the activity of the gonads, and if the latter are partially or wholly destroyed, both penis and oviduct are reduced in size and the sexes consequently difficult to determine. W. J. Rees (1936*b*) discovered that where such reduction occurs in this mollusc as a result of parasitism, the active agents are rediae of *C. himasthla secunda* or *C. lophocerca*, which devour the gonads. He found that larvae corresponding to *C. emasculans* rarely achieve this result in full, the penis and oviduct nearly always remaining clearly visible.

Wesenberg-Lund (1931) first suggested that changes in sex might occur in such instances. He found that in the hermaphrodite snail *Succinea putris*, parasitized by *Leucochloridium paradoxum*, spermatozoa continue to develop after the ovary has been destroyed, although regeneration following severe atrophy restores the balance. The only real evidence suggestive of change of sex in this instance is a slight preponderance of functional females among the older members of the uninfected snail population. Rees found no transitional stages in *Littorina littorea*, and in his view effective sex change in this species is unlikely of achievement because of the low intensity of infection, a mere 2–10 %.

Krull (1935) found infected specimens of *Peringia ulvae* which, though functionally females, possessed a small penis and which he regarded as individuals that had suffered sex reversal. Rothschild (1938*e*) observed that such infected individuals invariably have an abnormal penis, so that it is almost always possible to judge of infection by casual inspection of this organ. She thought it highly improbable, however, that all snails with a small penis were individuals which had undergone sex reversal, as Krull believed, deeming it likely that some specimens devoid of the organ were not females, but males which had suffered castration. Taking into consideration the sex ratio in populations of uninfected snails, she arrived at the conclusion that males are more frequently infected than females, suggesting several possible causes, the preference of miracidia for individuals of this sex, or their inhibited development in the opposite sex. Another possibility is higher mortality in infected females than in infected males. Yet she is unable to deny her support to the opinion of Krull that *some* snails with a small penis are really females which have undergone sex reversal. So, at present, the position stands.

(iii) *Gigantism and abnormal growth in the host.* Many writers have accepted as a rule the more frequent infection of large than of small snails with larval trematodes, and this had been interpreted as meaning that the miracidium shows intraspecific selectiveness, preferring larger hosts. In some instances this is undoubtedly true. Ameel (1934), however, failed to infect old specimens of *Pomatiopsis lapidaria* with the miracidia of *Paragonimus* sp., but attained 100 %

success in attempts to infect young snails of the same species with similar larvae. Wesenberg-Lund (1934) noticed that infected snails are sometimes abnormally large, and he was the first to suggest that the phenomenon is attributable not to greater age of the snails, but to excessive growth induced by the parasites. In attempting to explain such examples of gigantism he formed the opinion that the probable cause is excessive food consumption designed to meet the demands of the parasites. Rothschild (1936*a*) confirmed the effect for *Peringia ulvae*, regarding the cause not as excessive nutrition so much as the destruction of the gonads and other organs, implying that reserves of nutrients become available for general growth by the loss of some organ which normally makes use of them.

Like other problems of its kind, that of gigantism in the snail hosts of larval trematodes is difficult to solve. In this case, as Rothschild (1938*e*) showed, molluscs collected from apparently uniform habitats only a few feet apart possess entirely different growth characteristics. Snails kept in bowls in the laboratory grow much faster than snails kept in tubes (Rothschild, 1939). But Spooner's analysis of Lysaght's data (1941) concerning *Littorina neritoides* parasitized by *Cercaria B* (see Rothschild, 1941*a*) revealed apparent stimulation of growth as a result of parasitism. Rothschild's later experiments (1941*c*) clinched the matter by showing that *Peringia ulvae* attains a size of 9–10 mm. when parasitized by larval trematodes, uninfected specimens being no more than two-thirds as large. We may take it, therefore, that some factor or factors at present unknown are at work during parasitism by larvae to enhance the growth rate of the host.

The more frequent infection of larger than of smaller snails may be due to causes other than growth stimulation, however. Young snails may be less attractive to miracidia than old ones, thus being immune to attack. If this is so, then it is not a universal trait, as Ameel's experiments quoted above show. Again, lethal consequences of infection may be more frequent in young than in old snails. Many writers have produced evidence to this effect, showing that infected snails are less resistant to changes in environmental conditions than uninfected snails, and more liable to die as a result of a sudden change in one of these conditions. Or growth may be inhibited in snails after a certain size has been reached, the time factor accounting for the phenomenon. All animals suffer retardation of growth in the later stages of development, and mollusca are no exception to this general rule. Since young individuals grow faster than old ones, their size relative to that of older specimens is not a reliable criterion of the *period* during which they have been open to risk of infection. They have been exposed to infection for a shorter period than mere size indicates. Other explanations of the more frequent infection of large snails are no doubt available. In some instances small snails are segregated from larger and older ones, sometimes under conditions in which the risk of infection seems to be smaller. Possibly definitive hosts prefer large snails as food and, coming to the localities in which large snails occur, leave behind in their excrements larvae for the more intensive infection of the large snails that survive their depredations. The probability is that a number of factors are operative, and these can only be ascertained by further research.

Other abnormalities of infected molluscs relate to the shape and colour of the shells. Wesenberg-Lund (1934) noted that shells may be thinned or corroded and that certain whorls may show 'ballooning'. Rothschild (1936*a*) suggested that these and other abnormalities, such as asymmetrical spires, are probably due to pressure generated by the swarms of larvae in the body of the host. Later (1938*e*, 1941*a*) she discovered that defects in the shell may differ markedly in specimens from similar but geographically distinct habitats, showing environmental conditions to be a contributory cause. Snails from different localities which differ in size, the shape, colour and texture of the shell, etc. show no such differences when reared in the laboratory (Rothschild, 1939). Rankin (1939*a*) found that infected snails frequently have gnarled and warty shells, or shells with broken spires, and sometimes the lip of the shell is flared and forms a prominent shelf. Finding that uninfected snails are not invariably without such defects, he arrived at the same conclusion. Brumpt (1941) found that erosion of the shell in *Planorbis glabratus* is caused by the browsing of other snails. As regards colour in the shells of infected molluscs, Wesenberg-Lund (1934) noted the development of black pigmentation in shells which normally are bright yellow, e.g. in *Limnaea auricularia*. He regarded such pigment as 'excreta', but it would seem to be akin to the pigmentation of the skin in some fishes which harbour metacercariae, which we have discussed already. These observations suggest that in collecting infested molluscs it may be worth while to pay special attention to any with deformed or dark-coloured shells.

(*f*) *Host-specificity*

Most larval trematodes show marked preference for certain hosts and will not attack others proffered to them, though refusal to do so means death. This remarkable degree of specificity to some extent indicates a primitive nature and old-established relationships, i.e. is indicative of stability in evolution. Wesenberg-Lund found that all Danish Gymnocephalous cercariae occur only in *Bithynia tentaculata* (Dubois, all Swiss forms also), and that Xiphidiocercariae of the '*Virgula*' group and Furcocercariae of the '*Vivax*' group also confine themselves to this host. Gorgoderinid cercariae occur only in *Sphaerium corneum*, the Monostome *Cercaria ephemera* only in *Planorbis corneus* and the Amphistome *Cercaria diplocotylea* generally in the latter host. *Fasciola hepatica* is a trematode of world-wide distribution, but its miracidia shows a clear preference for *Limnaea truncatula* where this snail occurs, though in its absence other snails, chiefly species of *Limnaea*, can serve as hosts.

In some instances this precise specificity is relaxed, cercariae being found in a number of species of a genus, or in species of more or less related genera, the latter being rare. Thus, the Furcocercous *Cercaria ocellata* occurs with equal frequency in *Limnaea stagnalis* and *L. ovata*, and the Cercariaeum *C. strigata* in *Tellina tenuis* and *Donax vittatus*. These exceptions do not in their rarity detract from the general rule of host-specificity. How remarkable this may be is shown by the observations of Faust (1924*b*) and Faust & Meleney (1924) to the effect that three species of blood fluke utilize different species of molluscs, and by Manson's predictions that rectal bilharziosis could be expected where *Planorbis* (the host of *Schistosoma mansoni*) was to be found, and that where *Bulinus* (the host of *Schistosoma haematobium*) took the place of *Planorbis* the disease would necessarily be of the urinary type.

This specificity relates to miracidia, which probably represent a more primitive and ancient type of organism than cercariae, the only other larvae called upon to choose between a number of available hosts. Cercariae are much less selective. The cercariae of '*Clonorchis*' *sinensis* will penetrate and encyst in more than forty fresh-water fishes belonging to several families, and any gadoid (as well as a number of other fishes) will serve equally well as the second intermediate host of *Bucephalopsis gracilescens*.

Apparent specificity in relation to herbage utilized during encystment, e.g. the preference shown by the cercariae of *Fasciolopsis buskii* for the water caltrop and other plants, probably results not from choice by the cercariae but from the dietetic preferences of the snail, together with the promptitude with which encystment occurs once the cercariae have been freed.

(*g*) *Ecological*

After an exhaustive study of the larval trematodes of fresh-water molluscs, Wesenberg-Lund (1934) concluded that there does not seem to be a limitation of special trematode faunae to lakes, smaller lakes and ponds, nor to ponds of a dystrophic or entrophic nature. At the same time, he pointed out, Prosobranchiate snails preponderate over Pulmonate snails at greater depths than 5 m. in lakes and, as each type has its special attraction for different trematode larvae, we might expect that trematodes utilizing Prosobranchiate hosts would be more abundant than those using Pulmonates. One type of cercaria which shows a marked preference for lakes is the Furcocercous type, which is indicative of its essentially planktonic nature. At the same time, numbers of such cercariae emerge from *Planorbis corneus* and *Limnaea stagnalis*, which occur mainly in ponds.

As might be expected from the possible conditions of crowding or of dispersal, the intensity of infection of a snail population varies enormously, 3–4 % infections being rare in lakes, 50–100 % infections not uncommon in ponds. Neither the degree of infection nor the nature of the fauna of larval trematodes is predictable from one year to the next, because the stock of snails may be so reduced one year as to constitute a poor reservoir for the next. On the other hand, heavy infection in some localities favours excessive growth, and the regenerative capacity in snails is so good as sometimes to permit of the carrying over from one year to another of stocks of large snails for which miracidia show a preference.

Some marine habitats are ideal nurseries for larval trematodes, though they have been little studied. Rothschild (1941*c*) provided an interesting study of *Peringia ulvae* in pools and on mud-flats of the River Tamar. There is a high infection rate which shows seasonal fluctuations apparently having to do with the migrations of wading birds, the chief definitive hosts of trematodes belonging to the Heterophyidae, Echinostomatidae, Microphallidae and Notocotylidae, which utilize this intermediate host. According to Rothschild, this mollusc is for such reasons unique among marine and brackish water species. So varied and abundant are its larval trematode parasites that life-history studies incriminate all the main groups of animals found in the vicinity, showing the importance of larval trematodes in the local ecology.

Cort recently discussed (1942) the human factors in parasite ecology. He pointed out that while man has made great cultural advances in the past century, he is still subject to parasitism to the same extent as most animals. By conscious efforts to control disease man has created conditions unfavourable to parasites (drugs and preventive measures of many kinds), but his activities in many ways favour these enemies. Man's migrations, herding instincts and personal habits tend to create fresh areas of infection, and overcrowding, bad sanitation and careless feeding habits enable the parasites to establish themselves. War may create favourable conditions for parasites, which then take their toll of soldiers and civilians alike. Infection with *Schistosoma haematobium* cost the British Government millions of pounds in disability pensions after the Boer war, and there is evidence to show that the Japanese invaders of China were infected with helminths, as well as parasitic Protozoa, which hindered their progress and added to their discomfiture.

While immigrants may convey parasitic disease from one country to another, as hookworm disease was taken from Africa to America by the slave trade, in other instances they take up the parasites of the new country. Cort pointed out that the American Government was greatly concerned during the recent war about the dangers to which troops and administrative officers were exposed by their work in the countries of the Orient. The same might be said of our Government. Of diseases caused by trematodes, schistosomiasis is perhaps the most important.

This disease generally occurs in small districts which are widely scattered, where the incidence of infection may be low. But in some regions, notably in the Yangtse Valley in China and the Delta and Valley of the Nile, the disease is at once intense and widespread, millions of persons being infected and mortality rates being high. In the Far East, schistosomiasis is an occupational disease of rice farmers, as Faust & Meleney (1924) showed. Ancient agricultural practices lead to the infection of tens of millions and the death of hundreds of thousands each year. Human excrement, often containing the eggs of Schistosomes, is spread in fertilizing the fields, and the bare-legged farmer in rice-paddies is an easy prey for cercariae which develop in high concentrations there. In Egypt drinking water is purified (Witenberg & Yofe (1938) reviewed the efficacy of various methods and the failure of a few), but mortality is high. J. A. Scott (1937) estimated that seven million out of a population of twelve million people in rural districts are infected with one or both species of Schistosome prevalent there. Methods of irrigation, using the waters of the Nile in extensive canal systems, favour the parasite. Slowly flowing water becomes clogged with vegetation on which myriads of snails feed, most of them suitable hosts for the larval trematodes. Insanitary habits provide a fountain of larvae, pouring into such reservoirs of potential flukes, and human activities bring definitive hosts into contact with parasites whose major problem has been solved for them. Children play in the canals, women wash clothes there, and men expose themselves in other ways, all to the advantage of the parasites. Plans to bring still greater areas into the irrigation system promise greater abundance not only of wealth, but of sickness, suffering and death. Education about parasites and parasitism as well as rigid methods of control are still more urgently needed than they were in the past.

B. Adults

(*a*) *Effect of the parasite on the host*

(i) *Monogenea.* Many monogenetic trematodes seem to do little if any harm to their hosts, gaining sustenance from the natural secretion of mucus, and the detritus mixed with it, on organs like the gills. *Polystoma* lives for several years in the bladder of frogs, how is still something of a mystery, but these animals show no signs of ill-effects, and the bladder wall remains, as far as has been ascertained, undamaged.

In heavy infections, however, the host may suffer intensely and finally succumb to its parasites. MacCallum (1927) observed that certain fishes (the puffer-fish, spade-fish and various angel-fish) in the New York Aquarium now and again become blind, investigation revealing the causal agent to be *Benedenia* [=*Epibdella*] *melleni*. The puffer-fish has eyelids which close over the eye, and the parasites were found to lodge in the conjunctival sac thus formed, to pierce and finally lay open the eyeball, completely destroying it within the space of three weeks. The parasites occur on the skin, one remarkable fish bearing about 2000 of them (Nigrelli, 1937), and in the gills and nasal cavities, the last two being important nurseries for the young. Jahn & Kuhn (1932) prepared lists of fishes which are and others which are not susceptible to the attacks of this parasite, and these have been extended by Nigrelli & Breder (1934). Nigrelli, who investigated the control of *Benedenia melleni* (1935*b*, 1937), examined (1940) 1600 deaths among fishes, reptiles and invertebrates at the New York Aquarium during 1939, and showed that the most virulent parasitic diseases of aquatic forms are caused by external parasites (copepods as well as flukes) which live attached to the skin, eyes and gills. Three genera of Monogenea besides *Benedenia* are included by Mellen (1928) amongst organisms causing disease and sickness in fishes. Various species of *Ancyrocephalus*, *Gyrodactylus* and *Microcotyle* produce these effects. *Gyrodactylus* which, among other effects, produces shrivelling of the gills, has been investigated by a number of zoologists, methods of control having been proposed by a few of them (Hess, 1930; Wilde, 1937). MacCallum also found fishes fresh from the sea abundantly infected with *Diplectanum*, which in some instances caused injuries that proved fatal.

Microcotyle sometimes occurs on the gills of certain fishes in enormous swarms of adults and tangled masses of eggs. No sooner have the larvae hatched than they fasten on the gills by their clamps and add to the ravages done by their progenitors. Presumably, great irritation is set up by the flukes, which move briskly and continually over the surfaces of the gills, and to it the host responds by producing abnormally copious streams of mucus, which clog the gills and cause the suffocation of the host. It is claimed by some that the worms also feed on blood sucked from the capillaries of the gills, but this does not seem at all likely, although the anaemia induced in the fish is indisputable. At one time no less than 90 % of angel-fish and butterfly-fish in the New York Aquarium died as a result of infection with *Microcotyle*. E. M. Brown (1929) investigated such an epidemic of flukes of this genus at the Zoological Society's Aquarium in London.

Other monogenetic flukes cause the suffocation of their hosts by inducing hypersecretion of mucus. Southwell & Kirshner (1937*c*) found upwards of 100 *Discocotyle sagittata* on the gills of individual trout from a reservoir in North Wales, death resulting under these conditions. In other instances there is evidence of hypersecretion, but the appetite of the fluke is sufficient to obviate fatal results. This seems to be the case with *Hexostoma*, a parasite on the tunny which excavates the tissue by the grip of the opisthaptor in maintaining a hold on the gills, causes other histological changes in the gill tissues, perhaps as a result of stasis of the blood stream, and causes considerable erosion of the tissues at the upper ends of the gills, to which the mouth is closely applied during feeding. Goblet cells which have outlived their function are replaced by differentiating embryonic cells in the deeper part of the ectoderm, this contributing to the erosion. The effects do not seem to be baneful, but the possibility of perforation of important blood-vessels and of ulceration is not ruled out (Dawes, 1940*a*).

(ii) *Digenea.* Like most Monogenea, the vast majority of Digenea are ideal parasites in that they rarely adversely affect the host. Intestinal parasites generally produce no lesions of the gut, which is what the absence of organs more formidable than suckers would lead us to expect. But in other instances, sometimes in this, the effect on the host depends on the intensity of the infection and the organ or tissues attacked. Even in man and other mammals, large numbers of small flukes like *Heterophyes* and *Metagonimus* attached to the intestine produce only mild symptoms, though some slight damage may be done. Thus, Ciurea (1933*b*) found, two days after experimental infection of cats with members of these genera, flukes with their anterior extremities sunk deeply in the crypts of Lieberkühn, where the epithelium showed histological signs of erosion and atrophy. This is true of other flukes like *Apophallus mühlingi* and *A. donicus* which, histological examination of tissues apart, seem harmless enough. In other instances, e.g. *Cryptocotyle concava*, *C. jejuna* and *Pygidiopsis genata*, like effects are seen at the bases of the intestinal villi. On the other hand, the presence of a few flukes of the genus *Fasciolopsis* in the intestine may produce a profound toxaemia.

Perhaps the most injurious flukes are cyst dwellers such as *Didymocystis*, which causes extensive lesions in the gill tissues of fishes like the tunny, but the most notorious are lung and liver flukes, especially the latter. *Opisthorchis* in fair numbers may produce marked local tissue reactions, proliferation and desquamation of epithelium, hyperplasia of the connective tissue, and the development of fibrous enclosures around eggs, without serious symptoms developing in the host. The effects of parasitism in such instances are progressive, however, and the final effect depends on the intensity of infection and on reinfection. Extreme cases involve cirrhosis.

Damage is also done to the host's tissues by deep-seated parasites during the migration of the metacercariae to the final location, and this is intelligible when we consider that both the gut and the capsule of the liver are penetrated. Damage takes the form of ulcer or abscess formation and the subsequent development of fibrous tissue with consequent injury of a functional kind. In addition, toxins produced by the parasites are absorbed by the host and leucocytosis and other deleterious effects may result.

In the case of blood flukes, damage to the host is brought about by the eggs on the way to the exterior. Capillaries and other tissues are pierced and there may be considerable loss of blood. According to one opinion, the unhatched miracidia produce substances which assist penetration and are thus injurious to the host. Again, when Schistosome cercariae penetrate the final host there results a dermatitis, the intensity of which depends on the amount of exposure and on individual sensitivity, and further symptoms due to toxaemia.

Strigeid trematodes do little damage to their normal hosts, but may seriously damage 'wrong' hosts, which have no natural immunity to them. Blindness may be caused by '*Diplostomula*' which enter the eyes of fishes that serve as intermediate hosts, and damage may be done by the adults to the intestine of their final hosts, especially when the latter are not the normal hosts. Fatal results of the infection of pigeons by *Cotylurus cornutus* have been recorded (Bittner, 1927; Baudet, 1929). Baylis (in Lowe & Baylis, 1934) attributed the deaths of fifteen razorbills found on Littleton Reservoir to the effects of enormous numbers of *C. platycephalus* in the intestine, which was much inflamed and full of blood. The birds were presumed to have fed on an unusual diet of fresh-water fishes, from which they acquired metacercariae that normally use other avian hosts, and thus accidentally to have acquired parasites to which they were unaccustomed. Price (1934*b*) recorded a fatal disease in the lesser scaup, *Marila affinis*, in North America, involving ulceration of the intestine, but in this case the trematode, *Sphaeridiotrema globulus*, was thought previously to be innocuous. It caused deaths of lesser scaup in three successive years (1928–1930) along the Potomac River (Price). Foggie (1937) reported an outbreak of necrotic enteritis due to *Plagiorchis laricola* in turkeys in Northern Ireland.. Conditions favoured the visits to the feeding grounds of terns and gulls, in which Skrjabin first found the flukes.

(*b*) *Immunity*

It is well known that hosts often respond to the presence of helminth parasites by the production of antibodies, and that these may be identified by serological tests such as complement-fixation (a test akin to the Wassermann reaction for syphilis). The simpler diagnosis of helminth infections and diseases like schistosomiasis, fascioliasis and paragonimiasis by means of samples of eggs which are easy to obtain, however, usually renders the method unnecessary. Accounts of immunity to helminths refer mostly to nematodes, but also to some cestodes and a few trematodes of economic or medical importance, and are given in some standard works (e.g. Craig & Faust, 1940). An experiment not mentioned in such works but of some interest is that of Wu (1937), who tested the susceptibility or immunity of twelve species of mammal to infection with *Fasciolopsis buski*, which normally attacks man and the pig. The animals used were fed cysts containing metacercariae, practically all of which developed to maturity in pigs. In dogs the flukes failed to attain maturity and in rabbits some showed a similar failure to develop. Monkeys, cats, sheep, goats, oxen, guinea-pigs, rats and mice all proved to be resistant to infection.

Immunity to infection with monogenetic flukes may also exist or be developed by certain hosts. Nigrelli & Breder (1934) demonstrated that the hosts of

Benedenia melleni show varying degrees of susceptibility and resistance to attack. Some fishes have natural immunity, others soon acquire complete immunity which persists for long periods, and many are always susceptible and are invariably parasitized to some extent. The black angel-fish is susceptible for one or two weeks early in life, but individuals acquire immunity which lasts till they are becoming old, when they again become susceptible. In the moon-fish immunity is localized in areas of the skin which have been parasitized previously, the trematodes of subsequent infections seeming to avoid such regions.

Nigrelli (1937) tried to produce resistance in the host by various means, periodic injections of extracts prepared from fresh *Benedenia melleni*, or ground dried trematodes, and treatment with serum from immunized fishes. Such attempts largely failed, but some experiments suggested the possibility of protection afforded by mucus. Living *Benedenia* were placed in mucus from various fishes immune from attack (grouper, dog-fish, ray), and one fish which is very susceptible, as well as in sea water as a control. Parasites died in the mucus of immune fishes in periods of less than 8 hours, living in that of susceptible fishes and in sea water for 3 days, indicating not only that mucus affords protection, but that some kinds of mucus are much more efficacious than others. It seems possible that hypersecretion of mucus may be a defensive mechanism, but it is too soon to form reliable conclusions. In moderate amounts the mucus seems to be a food supply for the parasites, and the parasites may stimulate its production to their own advantage. But over-production of mucus seems at least as bad for the host as for the parasites, causing suffocation.

Willey (1941) also found that an established infection of the Amphistome *Zygocotyle lunata* in ducks and rats forbids reinfection. In this instance, primary infection was established by feeding metacercariae to experimental animals, up to 144 flukes thus becoming established in the host. Attempts then made to reinfect the hosts by feeding 50–150 metacercariae to them 6–261 days after the time of the primary infection failed. Metacercariae ingested under these conditions may remain in the body of the host for 4 days, but usually either die or are extruded within this period. Apparently, infection of the host with trematodes in this instance results in the development of an immunity which prevents superinfection.

Taliaferro (1940), reviewing immunity to helminths generally, expressed the opinion that it rests primarily on endocrine responses in the host. He mentioned that two types of immunity have been demonstrated in regard to cestode infections, one acquired soon after infection, the other later. In each case, antibodies are produced, but those of the 'late' immunity are not absorbed by fresh-ground helminth material. Ackert (1942) believed that the degree of natural resistance of a host to its helminth parasites is to a large extent dependent on the dietary, genetic constitution and age of the host. Resistance is lowered if vitamins are lacking in the diet, balanced rations containing sufficient quantities of vitamins A, B and D, together with minerals such as occur in cereals, meat and milk, generally enabling the host to maintain a high degree of resistance. Where intestinal parasites are concerned, he believed that this is due, in the first place, to the production by the goblet cells of a thermostable substance or substances which inhibit the development of the parasites.

(*c*) *Hyperparasitism*

Nematodes are well known to have their parasites, but trematodes seem to be less liable to hyperparasitism. Nevertheless, several interesting records exist in the literature, though auxiliary parasites are rare. Cort (1915 *c*) found larvae of *Gordius* in the trematode *Brachycoelium hospitale* Stafford, 1903, itself a parasite in the intestine of the green newt, *Diemictylus viridescens*. This was apparently not a case of accidental parasitism, because eight out of sixteen trematodes from several hosts were infected, two with two larvae each and others with one apiece. Fischthal (1942 *a*) found a single larva of *Paragordius* free in the parenchyma of a trematode closely related to *Plagiorchis sinitsîni* Mueller, 1934 (Plagiorchiidae).

Lebour (1908 *b*) found a living copepod closely related to *Ergasilus* attached by a pair of large anterior hooks to *Derogenes varicus* inhabiting the buccal cavity of the long rough dab. The unlikely occurrence of a living copepod in deeper situations like the stomach or intestine led her to believe that the buccal cavity of the fish is the normal location of the fluke. Another interesting yet somewhat different case is the occurrence of *Udonella caligorum* on the ovisacs and posterior region of species of *Caligus*, itself a parasite of the cod and other gadoid fishes. In these two examples members of the same two zoological groups play the reciprocal parts of parasite and host.

Most of the records of hyperparasitism in Trematoda concern Protozoa, which are well known to parasitize nematodes, but less often found as parasites of flukes. Dollfus gave the name *Nosema legeri* to a microsporidian discovered by Léger in 1897 as a parasite of marine metacercariae. Brumpt (1936) mentioned another species, *Nosema echinostomi*, which frequently parasitizes the cercariae and rediae of fresh-water Echinostomes and prevents the metacercariae from reaching maturity. The hyperparasites in this instance showed some host-specificity, for they did not attack other species of cercariae inhabiting the same mollusc. This record is especially interesting because several species of the genus *Nosema* are well-known parasites: one, *N. bombycis*, is the causal agent of the silkworm disease known as pébrine; another, *N. apis*, that of Nosema sickness in honey bees; and other species occur in various insects, especially mosquitoes and larvae of the black fly, *Simulium*, where they may exert an effect of economic advantage to man. Brumpt mentioned one other microsporidian, *Glugea danilewskyi*, which occurs in a trematode parasitic in the stomach of the grass snake, and Caullery & Chappelier reported a haplosporidian parasitic in the sporocysts of a marine trematode. More recently, Martin (1936) found larval trematodes to harbour a sporozoan in such numbers as to interfere with the completion of their life cycle. Hunninen & Wichterman (1938) found a flagellate of the genus *Hexamita* to be parasitic in the eggs and uterus of the trematode *Deropristis inflata*, itself a parasite of the marine eel, *Anguilla chrysypa*. These hyperparasites occurred also in the oviduct, receptaculum seminis, vitellaria and testes. In several instances counts revealed the fact that all the eggs were infected, and it was found that heavily infected eggs, some of which contained twenty or more flagellates, did not develop miracidia.

Another example of the infection of the eggs of Trematoda, this time with fungi, was recorded by Brumpt. Large numbers of eggs of *Fasciola hepatica* and

Fasciolopsis buski may be destroyed by '*Chytridiacea*', and Butler & Humphries (1932) found that the eggs of the former, together with gelated white of egg, make a perfect culture medium for the fungus *Catenaria anguillulae*.

Without doubt the most interesting example of hyperparasitism is that discovered by Cort, Olivier & Brackett (1941), in which the cercariae of the Strigeid *Cotylurus flabelliformis* develop inside the rediae and sporocysts of other trematodes in snails which are antagonistic to the Strigeid larvae. This is true of Planorbid and Physid snails, the normal hosts being species of the genus *Limnaea*. Very remarkably, under such conditions of hyperparasitism, the tetracotyle stage is reached more rapidly than in the normal host. Cort and his colleagues suggest the reason to be the protection afforded against the immunizing reactions of the abnormal host, and the utilization of nutrient substances which the other trematodes have absorbed for the nourishment of their own progeny.

(*d*) *Nutrition*

The depredations of trematode parasites are largely due to their nutritional requirements. Early zoologists were inclined to believe that flukes like *Fasciola hepatica* live chiefly if not entirely on blood sucked from the host. Careful observers such as Looss were always on their guard against overstatement in such matters, but Looss himself believed that *Haematoloechus variegatus* feeds on blood. Müller (1923) examined the gut contents of the liver fluke microscopically, but found neither blood corpuscles nor haemoglobin amongst them, though he did find fragments of biliary epithelium and leucocytes. Believing the gut epithelium to be secretory rather than phagocytic, Müller expressed the opinion that digestion in *Fasciola* must be extracellular, not intracellular as earlier writers believed.

Weinland & Brand (1926) classified more than 12,000 *Fasciola hepatica* from the livers of 148 sheep into four groups according to the full or empty condition of the caeca and the particular location in the host, some being found in the gall bladder and large bile ducts, others only in the smaller ducts. From consideration of the glycogen, fat and nitrogen content of the four sets of flukes, they reached the conclusion that regular feeding migrations occur. Hungry flukes migrate into the smaller bile ducts and capillaries to feed, while satiated flukes pass into the larger ducts and gall bladder, later to return to their former feeding ground when inanition prevails. It is difficult to think of good reasons for such a migration, but it may well be that the tissues of the smaller bile ducts can be attacked more efficaciously. These workers confirmed Müller's finding, namely, that flukes feed on tissue elements and inflammatory exudates, but not on blood. Many helminths cause necrosis and liquefaction of the host's tissue in their vicinity, though most of those studied are nematodes. Such effects have been investigated by Hoeppli (1927, 1929), who reviewed the literature on external digestion by parasites (1933). Nematodes commonly liquefy the mucosa of the host by secretions which they extrude (external digestion), the nutriment thus rendered available being sucked into the gut. Szidat (1929*a*) observed similar necrosis and liquefaction in tissues adjacent to the adhesive organ of the Strigeid *Cotylurus cornutus*, fluid nutriment similarly being ingested by the parasite. Other baneful effects on hosts that have been mentioned are probably of a similar nature.

It would be useless, however, to try to formulate the means by which trematodes in general derive sustenance from their hosts, because investigations have been so few. From what has been stated it must be evident that Monogenea and Digenea vary greatly amongst themselves in their manner of feeding, and we may expect corresponding differences to occur in their digestive processes. Hsü (1940) made histological examinations of the gut contents in a number of parasites, and found that some trematodes feed on blood (*Azygia*), others on inflammatory exudates and mucosa cells (*Pleurogenes* sp. and *Plagiorchis* sp.), and some on ingesta from the gut of the host (*Diplodiscus*). It must be emphasized that great care is necessary in interpreting the histological appearance of the gut contents in trematodes. Having formed the tentative opinion because of the microscopic appearance of the food in the caeca that *Hexostoma* was a blood feeder, I scrutinized the gut contents for irrefragable proof, finding many bodies which might have passed for blood corpuscles denuded by digestion, but none which were indubitably corpuscles. Then patience was rewarded, a single perfect erythrocyte showing up in one of the caeca. My final conclusion, however, was that the fluke is not a blood feeder, and my explanation of the solitary corpuscle in the caecum that it had been swept there by the microtome knife during the preparation of the specimen, which was sectioned *in situ* on a fragment of the host's gill.

It might be mentioned that adult flukes and cercariae do not exhibit the cannibalistic tendencies which Wesenberg-Lund (1934) observed in the daughter rediae corresponding to *Cercaria echinata*, which will devour other rediae and cercariae. It is interesting to note, however, that *Diplodiscus temperatus* will feed on Opalinids which also live in the gut of the green frog, *Rana clamitans* (see Hazard, 1941).

(*e*) *Respiration*

Many Monogenea and some Digenea which live in superficial locations in the host have access to abundant supplies of oxygen, and might be presumed to live after the fashion of other aerobic organisms. Little evidence is available about respiratory processes in such forms, or in those living in deeper locations where oxygen is available only in small amount, if at all. The ultimate oxidation of substances to produce energy is now well known to depend on respiratory enzymes like cytochrome and other oxidases. Both haemoglobin and cytochrome have been discovered in nematodes (Keilin, 1925), only cytochrome in cestodes (Friedheim & Baer, 1933). About these important respiratory enzymes and pigments little is known for Trematoda, though *Telorchis robustus* possesses haemoglobin which is different from that of the host, having a distinctive dissociation curve, and *Allassostoma magnum* has cytochrome (Wharton, 1941).

Trematodes which live in the deeper parts of the host's body largely metabolize the carbohydrate glycogen. Weinland & Brand (1926) concluded that *Fasciola hepatica*, like *Ascaris*, gains energy by an anaerobic fermentation process, but that in this case glycogen is split into carbon dioxide and certain fatty acids of a higher, non-volatile kind than those produced in Nematoda. The duration of life and the rate of carbon dioxide production in flukes is unaltered if the external medium is saturated with hydrogen, nitrogen or air freed of carbon dioxide. From such results Weinland & Brand concluded that oxygen is of no special significance in the life processes of Trematoda. Harnish (1932) confirmed the uniform output of carbon dioxide by *Fasciola hepatica* under aerobic and anaerobic conditions. When oxygen is available the flukes can take it up, but there is no fixed relation between the uptake of oxygen and the output of carbon dioxide. For partial pressures of oxygen ranging from that in air down to that of one-hundredth of an atmosphere, oxygen consumption varies directly with the oxygen tension of the external medium. The conclusion reached is that energy is produced by means of anaerobic processes, and that oxygen consumption is an unrelated process which serves perhaps merely to raise the degree of oxidation of certain organic substances present in the fluke. Similar general conclusions hold for cestodes and nematodes, and analyses of the oxygen content of bile and the liquids of organs such as the gut support them.

(*f*) *Growth*

The study of growth in Trematoda is a difficult undertaking. The shape of the living fluke is continually changing, defying the best efforts of an investigator to measure the sizes of organs and parts, and age cannot be ascertained unless facilities are available for making experimental infections. Much can be achieved, however, by the use of carefully 'fixed' specimens, and the time factor in growth is not necessarily the most important, as Huxley (1932) has shown. In general, the organs and parts which make up the body of an animal do not all grow at the same rate, or at a rate characteristic of the body as a whole. The larval and juvenile stages of *Dicrocoelium dendriticum* (Fig. 81 A–F) indicate as much when merely inspected. The net results of such *allometric* growth is alteration of the proportions of the body as growth proceeds. Where the growth of a particular organ is especially intensive, the result may be the limitation of the total size reached by the animal. It is possible that many errors of judgement have occurred in the erection of taxonomic units in the Trematoda because we lack vital information concerning the amounts of variability, in total size and in the proportions, which may be expected to arise in consequence of growth. Scrutiny of size data sometimes reveals overlapping variability in forms regarded as distinct species, which may indicate taxonomic errors of this kind (Dawes, 1942*b*). Measurements of the organs in a mere hundred individuals may obviate such errors and reveal great changes in the proportions during growth. In *Styphlodora elegans* the ventral sucker and the gonads become progressively *relatively* smaller during growth, the amount of space between the suckers and also behind the gonads *relatively* greater, the latter space coming to accommodate a uterus showing an increased number of folds (Dawes, 1941*a*). Such differences as exist between individuals of various sizes might, in some circumstances, give rise to the erroneous notion of specific distinctions. In revealing such alteration of the proportions in *Styphlodora*, I indicated a method whereby the varying proportions can be shown graphically, by plotting the projected positions of the organs on the sagittal plane, and also their sizes in relation to the size of the body as a whole. Logarithmic plotting, from which growth coefficients can be determined, is highly desirable also. Ultimately it may well be that specific characters will have to include growth characteristics of one kind or another which have proved their worth in allometry, thus giving a measure of variability in the proportions of trematodes in regard to total size. For Monogenea also a start has been made, Mizelle (1940*b*) having re-described certain well-known species, incorporating in the descriptions data regarding the dimensions of the body and of certain critical internal structures.

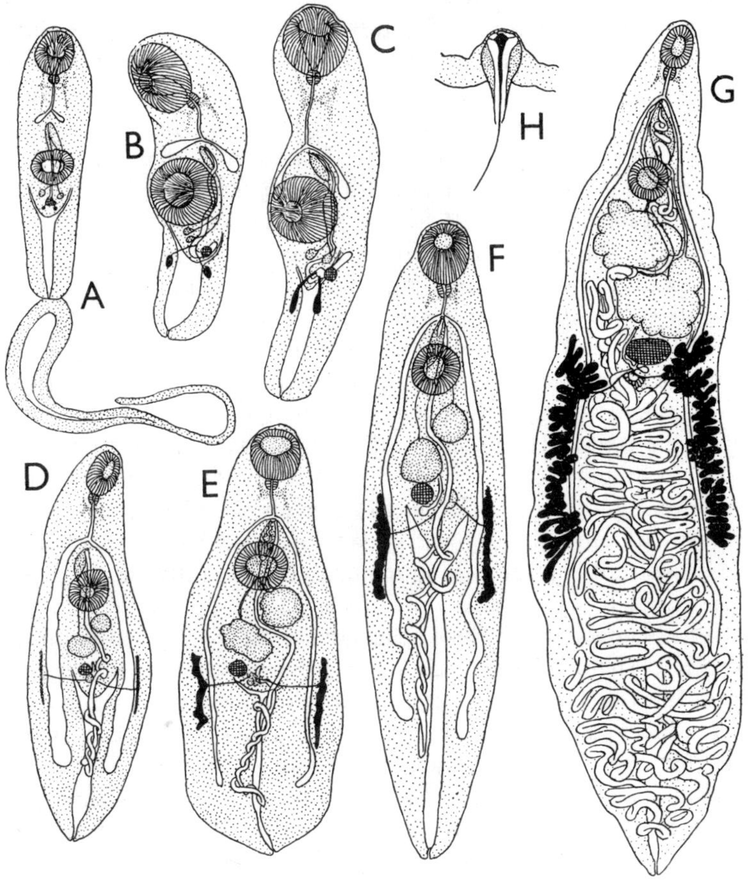

Fig. 81. Stages of *Dicrocoelium dendriticum* developing in the definitive host. A, *Cercaria vitrina*, removed from cyst. B–F, juvenile flukes collected 8–16 days after infection: B, 8; C, 9; D, 12: E, 14; and F, 16 days old. G, full-grown fluke. H, a touch papilla (on the suckers of the adult). (After Neuhaus, 1939.) (The magnifications of ABC/DEF/G stand in the relation 5·1/2·4/1.)

Willey (1941) has provided by far the most comprehensive account of growth in a trematode to date, dealing with the Amphistome *Zygocotyle lunata*, and taking the time scale into consideration, but not allometry. Flukes of this species and belonging to the same age group vary in size according to the numbers present in the host. In one infection 92 days old the mean length of the twenty-eight flukes which had established themselves in the host was 5 mm.; in another infection of the same duration the mean length of two flukes which represented the total infection was 7 mm. Lyubinski & Kulakovskaya (1940) found a similar but even greater variation in the size of *Opisthorchis felineus* according to the intensity of infection, one cat containing sixty-three flukes whose mean length was 10·6 mm., another 1187 flukes the mean length of which was only 5·5 mm. The two infections were of the same age. Rankin (1937*a*) recorded a similar effect in trematodes of the genera *Brachycoelium*, *Plagitura* and *Megalodiscus*. Whether or not such conditions of crowding lower growth rates by finer partition of available nutrients remains to be ascertained, but seems likely.

Willey also demonstrated that *Zygocotyle lunata* continues to grow long after attaining sexual maturity, and in fact as long as it lives, or at least for two years,* the limiting factor in some experiments being the life span of the host (the rat). The smallest mature specimen obtained was 3·1 mm. long and 1·4 mm. broad, the largest 9·2 mm. long and 4·7 mm. broad. Assuming that thickness increased proportionately with length and breadth, these figures indicate a thirty-fold increase in volume *after the attainment of sexual maturity*, which is reached 45–711 days after the time of infection with metacercariae.

A further point of interest regarding this trematode is that it reached maturity more speedily in the duck than in the rat, the requisite periods after infection being 41–44 days and 46–61 days respectively. Willey interprets this result partly in terms of the higher body temperature of the avian host, partly in terms of greater or smaller hazards involved in reaching the final location in the hosts, the caeca of the bird and the caecum of the mammal. The latter factor operates by limiting the number of parasites which find the final location (the only one used) to a lower figure in the bird than in the mammal. Conditions in the host are thus less crowded, other things being equal. The effect has another manifestation, the larger size of flukes just reaching maturity in ducks than in rats, comparative figures for length being 6·1 and 7·9 mm. (bird) and 5·4 and 4·1 mm. (mammal). The growth advantage which flukes parasitizing birds enjoy early during growth and shortly after maturity disappears in older worms, the size differences between flukes from the two kinds of host gradually diminishing after 200 days.

Further effects seen during the growth of *Zygocotyle lunata* parallel those mentioned for *Styphlodora elegans*. In sexually mature specimens, the relative size of organs like the ventral sucker and the gonads diminishes, so that the proportions of the mature body are inconstant. Prior to the attainment of maturity the growth of the body and its parts are claimed as being proportionate by Willey, but his figures show distinctly that the gonads increase in relative size up to the time of maturity, and that this probably holds good for organs like the ventral sucker. Willey emphasized the need for caution in attaching importance to size as a specific character, and we might extend this warning to include also the proportions of the body in relation to total size. But even logarithmic plotting of growth data must be cautiously undertaken, for the growth factors calculable by such means depend on the *slope* of the curve, which can be grossly misleading, because sampling is almost invariably poorer towards the extreme ends of the curves, where greatest accuracy is needed.

J. N. Miller (1939) also studied the time scale of growth in a trematode, feeding the cercariaeum of *Postharmostomum laruei* to mice and obtaining mature

* Longer life-spans than this are on record, e.g. *Fasciola hepatica* may live for at least 4 years 9 months (see J. W. G. Leiper, 1938).

trematodes 3·07 mm. long and 1·09 mm. broad 400 hours later. During a further period of 700 hours, length and breadth were increased, but by less than one-third, to the final proportions 3·99 × 1·32 mm. In this species, we are told, there is a marked decrease in size of the juvenile fluke beginning shortly after ingestion by the definitive host and pronounced 30 hours afterwards. Miller attributed this to a failure to feed, which seems improbable considering that encysted metacercariae regularly do not feed but are not known to diminish appreciably in size and may even grow. He also thinks it possible that loss of water has something to do with shrinkage, which seems more likely, suggesting that other trematodes may show similar loss on entry into the definitive host. But this is not always the case, as is shown for *Zygocotyle lunata* by Willey. Miller is correct, however, in maintaining that the entry into the final host represents one more hazard which the young fluke must overcome. The short period during which there is a more or less abrupt change from a poikilothermal to an homiothermal host and the sudden presentation of a different set of chemical factors in the environment may well constitute a critical period in the lives of many Trematoda.

The Maintenance of Living Trematodes *in vitro*

Several attempts have been made to keep trematodes alive outside the bodies of their hosts, not always with encouraging results, and we might review briefly what little progress has been made in the difficult process of isolating the living parasite, which might prove useful one day in analysing physiological processes that are otherwise out of reach.

Ortner-Schönbach (1913) kept *Lecithochirium rufoviride* and *Haplometra cylindracea* alive for 6–7 days in a saline medium, changed every 8 hours, finding no decrease in glycogen in the flukes (when the Best method of carmine staining was used) though decrease occurred in *Fasciola hepatica* and *Dicrocoelium dendriticum* maintained at the body temperature of their hosts. Meier (1913) kept *Azygia lucii* from the pike alive for 21 days at room temperature in Locke's solution to which a small amount of peptone had been added, transferring the worms to fresh medium every two days. Vogel (1934) maintained the encysted cercariae of *Opisthorchis felineus* for 60 days in 0·9 % salt solution, and Kollath & Erhardt (1936) the adult of this species for 40 days in Ringer's solution to which a little potassium indigotrisulphonate (1/50,000) had been added.

Van Cleave & Williams (1943) experimented on the maintenance of *Aspidogaster conchicola* outside the body of its natural hosts (pond mussels), using various media such as physiological salt solution, tissue culture medium and mussel blood. When immersed in salt solution kept in a refrigerator, all the trematodes used remained alive for 20 days and became active at room temperature, but died one by one during the next 19 days. When immersed in Hedon-Fleig's solution, a medium recommended by Gatenby (*Biological Laboratory Technique*; London, 1937) for the culture of the tissues of snails, the trematodes lived for 29–38 days, or for a maximum of 20 days at room temperature. Better results were obtained with mussel blood, and even a crude mixture of body fluid and the watery liquid from the viscera and mantle cavities of freshly-opened mussels kept the animals alive for 32 days at temperatures of 2–9° C. Pure mussel blood proved to be the most satisfactory medium, and by merely adding small quantities of fresh blood to the medium at intervals of a fortnight and keeping the culture in a refrigerator, *Aspidogaster conchicola* was kept alive in it for a period of at least 75 days.

Rather more refined methods of culture were described by Lee & Chu (1935) who kept *Schistosoma japonicum* from the portal vein of the rabbit alive for 2½ months at 37° C. in horse or rabbit serum contained within modified Carrel flasks closed by rubber caps. By a slight modification of the same technique Hoeppli & Chu (1937) kept '*Clonorchis*' *sinensis in vitro* for a maximum period of 5 months. The Carrel flask they used held 3 c.c. of horse serum, and to this some or all the following substances were added: 2 drops glycogen, 2 drops 10 % cystine, 1 drop 2 % lecithin, 2 drops haemoglobin and 1 drop bile. They did not find it necessary to change the medium until the end of the third week, and additions of mercurochrome-220 or prontosil enabled them to lengthen this period to 53–65 days. Hoeppli, Feng & Chu (1938) cultivated '*Clonorchis*' *sinensis* in similar flasks, using serum diluted with Tyrode solution and also heparinized rabbit plasma, which gave the best results. They found the serum of horse, rabbit, cat and man to be equally serviceable, and better diluted than undiluted. Changing the medium once a week, and with diluted serum, they kept this adult trematode alive for as long as 5 months, and maintained fragments alive for as much as one month. The latter continued to show movements and wounds quickly healed, but there was no regeneration of tissue, as Meier found also with *Azygia lucii*. Without changing the medium Hoeppli *et al.* kept the flukes alive for 3 weeks in diluted horse serum, and almost trebled this period by adding mercurochrome-220 or prontosil to the medium. They also kept *Schistosoma japonicum* alive for a maximum of 5 months, changing the medium once or twice a week, and in this instance found that the addition of prontosil killed the worms in a few days. Using a similar technique they kept the living metacercariae of *Clonorchis sinensis* (freed by treating the cysts for 2 hours with artificial gastric juice, and then for 6–8 minutes with trypsin solution) for a maximum of 2 weeks. Immature flukes used for experimental infection of cats lived under these conditions for the same period as adults (1 month), but did not grow or develop. Ferguson (1940*a*, *b*) also described methods whereby Strigeid metacercariae can be removed from their piscine host, excysted, separated from tissue debris and freed from micro-organisms. Sterile metacercariae remained alive and healthy for a month when stored in a refrigerator in dilute Tyrode solution, and developed into adults when incubated at 39° C. for several days in a similar medium to which chicken serum and yeast extract had been added. Thus, the future holds promise for experiments which utilize and improve on this and related methods of culture *in vitro*.

CHAPTER 16

A SHORT HISTORICAL ACCOUNT OF THE TREMATODA

Tapeworms and roundworms were known to man for centuries before the advent of the Christian era, but trematodes remained much longer in obscurity. Infection of man is known to have occurred in ancient Egypt, however, Ruffer having discovered the calcified eggs of *Schistosoma* in mummies dating back beyond 1000 B.C.

It seems likely that the shepherd, Jehan de Brie in 1379,* was the first man consciously to set eyes on a trematode, the now well-known liver fluke, *Fasciola*. The fluke then lapsed into obscurity till Gabucinus saw it a century and a half later (1547, 1549). Acute observers as these men were, they probably took for granted the existence of worms inside the bodies of other animals, never questioning their mode of entry. Not till the middle of the seventeenth century was the false doctrine of spontaneous generation seriously combated, one of those having the honour of first refuting it being Francesco Redi (1629–1697), after whom the 'Redia' stage in the digenetic life history is named.

Before the end of the seventeenth century fresh observations were made. Leeuwenhoek wrote several times about the liver fluke between 1695 and 1702, and saw in the herring at the earlier date the parasite we know as *Hemiurus ocreatus*. Not till the eighteenth century opened were definite records made, however, Swammerdam observing and to some extent describing in his *Biblia naturae*... (1737) a worm from the lungs of the frog. This illustrious microscopist also recorded living worms, evidently cercariae, in *Paludina vivipara*.

Goeze, in 1782, published a work of 471 pages on the natural history of parasitic worms. He was familiar with flukes from birds and moles which we now call *Alaria alata* and *Itygonimus ocreatus*. More than twenty years earlier (1760) Pallas saw and named the parasite of the bladder of frogs which we know as *Diplodiscus subclavatus*, but used the generic title *Fasciola*, which Linnaeus had invented some years before and which still stands in a restricted sense. In 1758 came the discovery by Roesel von Rosenhof of a 'leech' in the same location and host as *Diplodiscus*—the trematode we now know as *Polystoma integerrimum*. Two years earlier (1756) a short account of 'worms in animal bodies' by Frank Nicholls was published in the *Philosophical Transactions of the Royal Society*.

Before the beginning of the nineteenth century many more flukes had been discovered, described and figured. A very accurate observer, O. F. Müller, wrote a series of memoirs between 1773 and 1786, and laid the foundation of our knowledge of trematodes such as *Derogenes varicus*, *Azygia lucii*, *Sphaerostoma bramae*, *Bunodera luciopercae* and *Crepidostomum farionis*, which were at that time all referred to the genus *Fasciola*. Müller also observed and described a number of larval forms. It is remarkable also that between this time and the end of the eighteenth century Schrank discovered the Amphistome *Paramphistomum cervi*, as well as several other trematodes, including *Dicrocoelium dendriticum*, Abildgaard founded the genus *Strigea*, Frölich described the species

* See Huber (1890).

Opisthioglyphe ranae, *Macrodera longicollis*, *Bunodera nodulosa* and *Brachycoelium salamandrae*, and Carlisle revealed the course of the excretory canals in *Fasciola* by a method of injection, believing he had injected the alimentary canal (1794, Figs. 18, 19).

The new century opened with the publication by Zeder in 1800 of a treatise of 320 pages in which he attempted for the first time a systematic classification of parasitic worms of all kinds, dividing them into five 'classes' which he named (i) 'roundworms', (ii) 'hookworms', i.e. Acanthocephala, (iii) 'sucking worms', i.e. Trematoda, (iv) 'tapeworms' and (v) 'bladderworms'. It goes without saying that he knew many trematodes; certainly the species we know as *Haplometra cylindracea*, *Gorgodera cygnoides* and *Cyclocoelum mutabile*, because he founded them. He also erected a species, *Monostoma bombynae*, for what may have been the form we call *Haematoloechus variegatus*, but if so he overlooked the ventral sucker and mistook the worm for a Monostome.

Rudolphi (1771–1832) broadened the foundations of our knowledge of parasitic worms in two monumental works published in 1808–1810 (1370 pp.) and 1819 (811 pp.). He invented the name 'trematode' to replace Zeder's term 'sucking worm' and adopted the term 'order' instead of Zeder's 'class'. A large amount of the knowledge of what we now regard as the class Trematoda, as well as of other kinds of parasitic worms, dates from his time. In 1802 he founded many species of trematodes from fishes and Amphibia and described the larva of *Euryhelmis squamula* and species of '*Tetracotyle*'. In 1819 he established many other species. He described about forty of the species of Trematoda that may be found in British birds, and he provided the basis of our knowledge of Gasterostomata by describing *Prosorhynchus crucibulum*. His work fully justifies the title of 'Father of Helminthology' that has been conferred upon him.

During the early part of the nineteenth century the study of trematodes had gained such impetus that a volume would be needed to do justice to the labours of those who strove to advance knowledge in this field. Names deserving of mention include Otto, Nitzsch, Bosc, Olfers, Bremser, Creplin, Bojanus, Mehlis, Baer, Schmalz, Leuckart, Blainville, Kuhn, Eysenhardt, Laurer, Nordmann, Wagener, Siebold, Owen, Diesing, Carus, Steenstrup, Goodsir, Dujardin, Blanchard and Müller, to proceed merely half-way through the century. Nitzsch in 1807 gave the first account of the encystment of a cercaria, nine years later describing several new cercariae, including *C. ephemera*. Bojanus (1818) witnessed the emergence of cercariae from rediae. Creplin (1825) recorded observations on numerous trematodes, erecting several new species. Bojanus (1821) had already contributed much to a knowledge of the anatomy of flukes, and Otto (1816) and Schmalz (1827) studied especially that of the nervous system. Baer (1826–1827) founded the genera *Aspidogaster*, *Bucephalus* and *Nitzschia*, and gave a detailed account of *Polystoma integerrimum*. In laying the basis of our knowledge concerning Monogenea, Baer was much helped by contemporaries like Nordmann, Nitzsch, Diesing, Leuckart, Kuhn and Creplin, most early workers having neglected this group. Nordmann (1832) erected the genera *Gyrodactylus* and *Diplozoon*. Creplin, who wrote on Digenea as well as Monogenea, published a number of important works and described more than twenty species of Trematoda belonging to the fauna of Europe.

Outstanding works of a comprehensive kind appearing at this time were Dujardin's *Histoire naturelle*... (1845; 654 pp.), Blanchard's *Recherches* (1847) and Diesing's *Systema helminthum* (1850, 679 pp.). A notable event was the observation by Carus (1849) of rediae developing within rediae. During the next ten years our knowledge of larval Digenea was much advanced by the publication of Filippi's three *Mémoires* (1854, 1855, 1859) and other works (1856, 1857), by Müller's discovery of Trichocercous cercariae (1850), Diesing's *Revision der Cercarieen* (1855), the *Symbolae ad trematodum evolutionis historia* of La Valette de St George (1855), and by the works of Pagenstecher (1857), Moulinié (1856), Wagener (1857) and Lespès (1857). There was no equivalent work in this country till 1881–3, when Thomas worked out the life history of *Fasciola hepatica*.

Early in the second half of the century (1851) Bilharz discovered the first known dioecious trematode, *Schistosoma haematobium*, in the blood of a native of Cairo. He also discovered *Heterophyes heterophyes*, and both of these discoveries were published by Siebold in the following year, though the name *Schistosoma* was first applied to the former parasite by Weinland in 1858.

British writers on Trematoda were few at this time, but one rose to some eminence, Cobbold, who published his *Observations on Entozoa* in 1858, having already described *Fasciola gigantica* (1856) and later producing his *Synopsis of the Distomidae* (1861). Still later he described flukes from the horse, elephant and ox (1875) and from dolphins (1876), and in 1879 published his *Parasites; a Treatise on the Entozoa of Man and Animals* (510 pp.).

Other notable events occurred during the latter half of the nineteenth century. Leidy's *Contributions to Helminthology* and other works appeared (1852, 1856–1857), Diesing discovered many new species (1855, 1858), Lankester (1857) described *Fasciolopsis buski*, a parasite of man first found by Busk 14 years earlier in the intestine of a Lascar in London. Wedl (1858), Stieda (1870, 1871) and others advanced our knowledge of the structure of trematodes, Metschnikoff (1870), Zeller (1872, 1874) and Willemoes-Suhm (1871, 1872, 1873) that of trematode development, especially in regard to Monogenea. Molin (1861) produced his *Prodromus faunae helminthologicae Venetae*, and described many trematodes that are now well known. Olsson (1869) investigated the Entozoa of Scandinavian fishes, founding many new species of Digenea. Later (1876) he described others from frogs. His work was not confined to Digenea, however, and he founded the species now known as *Rajonchocotyloides emarginata* and *Dactycotyle denticulata*, as well as the genus *Dactylodiscus*.

In 1856 Leuckart proposed the division of trematodes into two 'families', 'Distomea' and 'Polystomea', to accommodate endo- and ectoparasites respectively. Seven years later Carus adopted for these groups the names Digenea and Monogenea, which we use to-day. In 1856 also Burmeister proposed to separate *Aspidogaster* from the remainder in a threefold division of trematodes into 'Malacobothrii' (Distomids), 'Pectobothrii' (Polystomids) and 'Aspidobothrii' (*Aspidogaster*). At this time, however, the publications of P. J. van Beneden were beginning to appear, and in 1858 his memoir on intestinal worms (376 pp.) appeared, to be followed by an account, for which Hesse prepared the figures, of researches on marine trematodes (1863–4) and a memoir on the parasites of fishes of the Belgian coast (1870, 100 pp.). These researches led to the modern conception of direct (monogenetic) and indirect (digenetic) modes of development, and they completely outweighed the proposals of Burmeister, with the result that a twofold division of the Trematoda remained in vogue till Monticelli, in 1892, revived the three groups of Burmeister under new names, Heterocotylea (monogenetic forms), Aspidocotylea (*Aspidogaster*) and Malacocotylea (digenetic forms). This alternative scheme prevailed till the early part of the present century, when it gave way to van Beneden's conception of the Monogenea and Digenea, *Aspidogaster* being included in the latter group. Faust & Tang (1936) appear to favour the revival of Burmeister's scheme once again, reminding us that Aspidogastrids are not qualified for admission into either group.

The final quarter of the nineteenth century opened with McConnell's study (1875) of the structure and pathological relations of the asiatic liver fluke of man, '*Clonorchis*' *sinensis*, and the discovery by Lewis & McConnell (1876) of another fluke of man, the Amphistome *Gastrodiscus hominis*. In the latter year Sonsino discovered *Schistosoma bovis*, and two years later Kerbert found and described the lung fluke *Paragonimus westermanii*. Taschenberg was at this time preparing his authoritative work on the systematics of Monogenea, which appeared in 1879, to be followed by papers on *Didymozoon* and other trematodes. Lang (1880) was engaged in a detailed comparative study of the nervous system in Platyhelminthes. Ercolani (1881, 1882) was proving by feeding experiments that the larval forms known as 'Tetracotyle' and 'Diplostomum' develop in suitable hosts into Holostomes, and Leuckart (1881, 1882) and Thomas (1881, 1882, 1883) were independently making their classical discoveries concerning the life history of *Fasciola hepatica*. At this period also Linstow was involved in extensive study of trematodes and other helminths, publishing many papers between 1873 and 1906. One of his most notable contributions to knowledge is the *Compendium der Helminthologie* (1878, 1889), a forerunner of American bibliographical works, and one which may still be consulted with much profit.

Proceeding slightly beyond this period, several names come into prominence. Brandes distinguished himself by his studies (1888–1898) on detailed structure of Trematoda, and also by his systematic studies on Holostomes and other groups. Fischoeder (1901) wrote an important monograph on Amphistomes. Stossich compiled many helminthological studies between 1880 and 1905. Braun made detailed studies of the trematodes of bats and other mammals, of Chelonia and of birds, and during the years 1879–1900 published his classical articles on 'Vermes' in Bronn's comprehensive work. These are distinguished names in helminthology, but they are partially eclipsed by that of Looss, one of the most prolific and accurately observant helminthologists of any period. His works on nematodes are as superlatively excellent as those on trematodes, with which we are alone concerned. The appearance of his *Beiträge zur Kenntnis der Trematoden* in 1885 marked the beginning of a new era in helminthology. This famous work was followed by others, now equally famous, on the structure and development of *Diplodiscus subclavatus* (1892); on the *Trematodes of Fishes and Frogs* (1894)—the figures of which are an inspiration to students; on *The Parasitic Fauna of Egypt* (1896, 1899); on *Haploporidae and Monorchiidae* (1901–1902); on *Old and New Trematodes of Chelonians* (1902); on *Schistosomes* (1905); and on *Hemiuridae* (1907–8).

Looss did not devote himself particularly to the trematode parasites of birds, and a name which tends to fill this gap is that of Jägerskiöld. The influence of a long series of papers which this helminthologist published during the period 1896–1908 will be evident in the chapter of this book which is devoted to the trematodes of birds. Other valuable contributions to knowledge in this field were an important paper on Echinostomes by Dietz (1910), one on Monostomes by Kossack (*Zool. Jb.*, *Syst.* **31**, 1911, 491–590) and several by Skrjabin, that of 1913 being the best known. At this period one other name stands out boldly, that of Odhner. This author published a long series of important papers on trematodes between 1900 and 1928, and did much towards establishing a natural system of classification of the Digenea.

In passing thus into the twentieth century, many illustrious names have been overlooked. Among other notable publications regarding trematodes during the early part of the present century must be mentioned Stiles & Hassall's *Index-Catalogue of Medical and Veterinary Zoology—Authors* and *Trematoda and Trematode Diseases*, and Lühe's article on 'Trematoda' in *Die Süsswasserfauna Deutschlands*, which is still a most useful work of reference. In the development of our knowledge of Monogenea after Taschenberg, Monticelli was the dominant figure, his researches extending throughout the period 1889–1912. Lesser figures in this field were Cerfontaine, Goto, Parona and Perugia, St Remy, Kathariner and Wegener. To do justice to their work we should need to range through most of what is known about Monogenea. They did not leave us with very clear ideas about the classification of the group; but they made valuable contributions concerning the structure and life history of monogenetic flukes. In America, Pratt (1900) published an account of New World forms, the first of which had been described as an insect by La Martinière in 1787 and shown to be a fluke by Bosc in 1811. Others who have continued the work in America include Guberlet, Linton, MacCallum, Manter, Mizelle, Mueller, Price, Stunkard, Ward and others, in Australia S. J. and T. H. Johnston, Murray, Tiegs, and Woolcock, in India Bhalerao, Dayal, Lal, Mehra, Moghe, Sinha, Soparkar, Srivastava, Thapar, Verma and others, in the Philippines Tubangui and in Japan Yamaguti. In Europe the Monogenea have received comparatively little attention, but the work of Gallien on *Polystoma* is outstanding. In Britain Johnstone, Lebour, Little, A. and T. Scott, and others have made contributions to our knowledge.

Too numerous to mention here are many of those who have worked during the current century on Digenea, and the bibliography of this book must serve to show what has been done in this country and, to some extent, in other countries. Lebour has made outstanding contributions to our knowledge of the trematode parasites of British fishes, while Nicoll has compiled lists of such parasites for all classes of Chordata, as well as performing much original work. Baylis, Idris Jones, Little and others have added something to our conceptions of the trematode fauna of British fishes. In America Manter, in Japan Yamaguti, have studied the faunae of fishes in great detail. The Russian trematode fauna, which is more important in relation to the British than is the American or Japanese, has been studied by many helminthologists, Bychowsky, Issaitschikow, Skrjabin, Skwortzoff, Strom, Vlasenko, Zhakarov and others, with much improvement of our knowledge. López-Neyra (1941 *a*, *b*) has outlined the history of helminthology from 1550 onwards, especially in Spain and Portugal, and has reviewed the Trematoda of the Iberian peninsula, where Gasterostomata are so far unknown and only three monogenetic flukes have been found, *Gyrodactylus elegans*, *Hexostoma thynni* and *Diplozoon paradoxum*. Thapar (1938) has outlined the progress of helminthology in India, where several zoologists have attempted to improve the taxonomy of the Trematoda.

Larval trematodes also have received comparatively little attention in this country during the present century. Lebour has described a number of larvae which may be found in marine molluscs and crustaceans, and Brown, Cole, Harper, F. G. Rees, W. J. Rees and Rothschild have worked with some success in this field. Abroad, numerous helminthologists have followed this line of research intensively and names like Ciurea, Cort, Dollfus, Dubois, Ejsmont, Faust, Hughes, Katsudara, Kobayashi, La Rue, Mathias, E. L. Miller, H. M. Miller, Palombi, Porter, Sewell, Stunkard, Szidat, Timon-David, Wesenberg-Lund and many others stand out prominently in connexion with studies on the life histories of digenetic trematodes.

APPENDIX

NOTES ON THE COLLECTION AND PRESERVATION OF TREMATODA

Information on collecting and preserving parasitic worms is not widespread, but fortunately the most appropriate methods of dealing with adults have been described in sufficient detail by Baylis (1922) and Baylis & Monro (1941). We have seen that trematodes occur in various locations in the host. Monogenea are found on the skin and gills, in the gill chamber and the cloaca of fishes and in the bladder of Amphibia. In examining fishes and amphibians for Monogenea, these are the first locations to search. Digenea can be sought afterwards. They penetrate to varying depths into the host and occur in practically all organs and tissues. Most of them are parasites of the alimentary canal, and the first situation in which to seek them is the buccal cavity. Other locations should be called to mind before beginning to dissect the host. Larvae may be encysted beneath the scales or under the integument of fishes, and in the skin, muscles and connective tissues of amphibians. Cyst-dwelling Didymozoidae may occur under the skin of the head in fishes, though a more usual location is near the gills or pharynx, sometimes in the mucosa of the gut. Other locations are the nasal cavities; eustachian recesses (in Amphibia); orbit and eyeball; trachea, bronchi and lungs; liver, bile ducts and gall bladder; kidneys, ureters, urethra and bladder; heart and larger blood vessels, especially the hepatic-portal system; and the thoracic and abdominal cavities. In fishes, the swim bladder should not be overlooked, and the cloaca and the skin near its aperture should be examined before setting out to dissect a bird. Such locations can be examined one by one as the dissection proceeds.

To find trematodes of the alimentary canal, slit open the gut with blunt-ended scissors, remove the contents and extract any obvious parasites without delay. If the host is small the entire gut may be removed and laid open on a tile or sheet of glass, on to which the fine contents can be carefully scraped for scrutiny with a binocular or other low-power microscope. Most trematodes of the gut lie free in its lumen, but a few adhere to the wall and may be revealed after the gut has been washed with salt solution or water and is re-examined under the microscope. The possibility of regurgitation of the intestinal contents into the stomach must be allowed for, because the discovery of an intestinal parasite in the stomach might add to the difficulties of diagnosis. It is a good plan to sever definite regions of the gut, taking care not to mutilate large parasites in the process. The different portions may then be shaken separately with saline solution in a jar or tube, to detach any trematodes which may be adhering to the mucosa. Layman (1933*b*) noted that the intestinal Digenea of *Rana esculenta* at Jakovlevskii (Russia) inhabit distinct zones: first *Prosotocus* in the duodenum, then *Opisthioglyphe* and *Pleurogenes* in succession in the intestine, and *Diplodiscus* in the rectum. Similar considerations should be made whenever possible.

Trematodes should be examined alive whenever possible, though this is sometimes impracticable when collecting is intensive. Worms should not be fixed, however, till they have been thoroughly cleansed of the debris which generally clings to them. The best method of achieving this is to shake them vigorously in a tube or bottle containing 1 % NaCl. Cold or lukewarm tap water may be used instead of a saline solution if care is taken to avoid adverse osmotic effects by restricting the time factor as much as possible. Shaking does not damage specimens, but when accidents happen it is as well to remember that a better impression of trematode structure may sometimes be gained from a mutilated specimen than from a beautifully stained and permanently mounted preparation. Structures like the cirrus pouch and its contents, and especially the flame cells of the excretory system, can sometimes be seen with great clarity in a fragment of a living worm. Mutilated specimens also afford an opportunity to collect eggs, which may be important for diagnosis, though here great care is called for to avoid the mixing of samples and other forms of contamination.

Fixation aims at preserving the animal in a condition as nearly like that of life as possible. All fixatives cause some contraction of the body, and care is needed in choosing appropriate reagents. If the worms are required for sectioning, some suitable histological fixative such as Bouin's fluid can be used. Corrosive sublimate is sometimes used, and after being fixed in this fluid specimens must be washed in running tap water for 12–24 hours. A good plan is to transfer the fixed specimens to a tube of water, and to tie muslin over the mouth. The whole tube of specimens may then be immersed in running water with less risk of loss. A quicker and more certain method is to transfer the specimens to 70 % alcohol with sufficient iodine in solution to give it the colour of sherry, and to change the solution as frequently as it loses its colour, finally storing the specimens in 70 % alcohol. Other good histological fixatives such as Müller's fluid or Helly's fluid can be used, and preliminary treatment with 10 % formalin followed by storage in 3 % solution is satisfactory. For ordinary purposes, such as preparing whole mounts, 70 % alcohol provides suitable fixation.

Various methods of applying the fixative are available. For whole mounts specimens can be fixed on a slide, and it may be advantageous to apply some slight pressure, that of another slide on top of the first in the case of a large worm, or of a cover-slip in that of a small one. The amount of pressure can be regulated to some extent by adjusting the amount of liquid between the sheets of glass. In applying pressure either to the living or to the fixed worm it is as well to bear in mind that drying of the fluid will increase the pressure, perhaps with dire results. The use of props of glass or wax may obviate such undue compression. But, at best, this method has its drawbacks, the chief of which is unavoidable distortion. A better method of fixation was used by Looss and is described by Baylis & Monro (1941) as follows: 'After vigorously shaking up the worms in 1 % salt solution in a tube, pour off the liquid and replace it by a small quantity of clean solution. Continue the shaking, and add quickly an equal quantity of 10 % formalin. Resume the shaking at once.' By this method most trematodes can be fixed in an extended condition, after which they can be stored in 3–5 % formalin. According to Mendheim (1939) the shaking method of Looss may have two drawbacks. Echinostomes are sometimes not extended, and small living flukes may be lost in the decantation of saline at the first step. This writer suggests that Echinostomes can be extended by the use of a fine paint brush, and that the tendency for the anterior end to roll up may be counteracted by treatment with plain water. He also recommends that the decanted saline be examined in flat glass dishes on a dark background, the floating slime sometimes containing small forms such as Heterophyids which would otherwise be lost.

How to find time for all matters requiring attention during the collection of trematodes is left to the individual. But it is important at such times to note any special features of parasitism, such as the appearance of cysts, signs of the extravasation of blood, erosion of tissues, and any special characters of the living parasites, e.g. their movements and the nature of their gut contents. The association of other kinds of parasites, cestodes and nematodes, with the flukes found in a host is also noteworthy. The precise identity of the host, and the locality and circumstances in which it was found, are other points to record. In collecting trematodes of the gut the nature of food organisms in the stomach or intestine of the host should be noted, for they may throw some light on the life history of the worms, perhaps indicating what animals might serve as intermediate hosts. The feeding habits of the same species of host may vary with age and size. Young fishes do not generally feed on the same organisms as the adults. When the gut is empty of food it may be important to note the size and age of the fish, because independent studies may have determined the characteristic food of fishes of particular ages. In teleosts, age may be determined by 'reading' the scales, i.e. counting the annual rings of growth, though determinations of this kind may be misleading, because a setback in growth may produce a ring and simulate a seasonal phenomenon. The identification of food organisms is a highly specialized matter, but, if any doubts prevail, the obvious policy is to preserve specimens and seek help from a specialist.

The staining and mounting of trematodes after fixation follows well-known rules of technique, which need not be reviewed here. Paracarmine is perhaps a more suitable stain than borax carmine, and excellent results may also be obtained with Delafield's haematoxylin if the flukes are small. The effect is enhanced by 'blueing', which is best carried out by the use of a trace of ammonia in the 90 % or absolute alcohol.

Larval trematodes call for special methods, which have been reviewed by Stunkard (1930*c*) and others, and it is a good principle to regard fixed and stained larvae such as cercariae as greatly inferior to living ones. There is something to be said for the method of making a smear preparation from the gonad or digestive gland of a snail in collecting cercariae, but the larvae thus obtained may be incompletely developed. A better plan is to maintain the snails alive and to collect the cercariae as they emerge. Different helminthologists use various methods to achieve this end. Miss F. G. Rees (1931) advocated the segregation of snails in groups of about six in suitable glass vessels. If any group shows emergent cercariae, the snails of that group are segregated, so that individuals containing the mature larvae can be identified. Single snails may be kept in 3 × 1 in. tubes filled with water containing a sprig of *Elodea*, and covered with a piece of muslin. When counts of the free cercariae have to be made, the snails can be transferred to identical tubes of water, leaving the originals containing the cercariae free for manipulation. Care should be taken to regulate the temperature and control other factors likely to influence the snails when such a transfer is made.

Different cercariae may emerge from snails under different conditions of illumination, sometimes during different periods of the day or night, and collecting can be carried out accordingly. The difficulty of identifying certain cercariae on purely morphological grounds has already been emphasized. Peculiarities in the behaviour of such larvae can be noted, their modes of swimming and creeping, their responsiveness to shadow stimuli, etc., as valuable aids to identification and observations of some biological value. Rediae and sporocysts can be obtained by crushing snails in normal saline solution. The powers of contraction and extension of the larvae should be noted. When examining living larvae of all kinds the distorting effect of pressure due to the weight of the cover-slip should be avoided. In all cases mounts of living larvae last much longer if the edges of the cover-slip are sealed with vaseline. Intra-vitam staining with neutral red and other suitable stains may be advantageous. When a specimen stained in this way loses its stain it is becoming moribund.

Methods of fixing larval trematodes are not essentially different from those used for adults. Harper (1929) obtained good results with 5 % formalin for larvae and 10 % formalin or corrosive sublimate for the snail hosts. Staining methods also are those generally employed, counterstaining with an 'acidic' dye after using a 'basic' stain like haematoxylin sometimes being advantageous. Some workers have maintained that the unicellular glands of certain cercariae which show differences in staining reaction have different functions. This is perhaps extending staining technique beyond useful limits. Gray (1931) has shown how grossly misleading differential staining results may be, the effect sometimes depending on the order in which the stains are used.

Not infrequently only a solitary trematode is collected from the host. It is natural, under such circumstances, to be pessimistic about what can be achieved by work on the 'collection'. Stunkard's experience (1940*a*) disproved the idea that such scanty materials are only of slight value. While examining a grebe he found a single Cyclocoelid of the species *Typhocoelum cymbium*, previously found only twice in America. After examining the fluke he placed it in a saline solution, in which it soon began to lay eggs. The worm was then fixed and mounted whole. The eggs were put in tap water and miracidia soon hatched. Laboratory snails were available and the miracidia were given access to them, infection resulting in some instances. Development continued in the snails and later larvae were obtained, with the result that the entire life cycle was worked out, from a *single* adult specimen and its carefully maintained progeny. Zoologists of lesser experience, skill and knowledge could scarcely hope for such good fortune, but the moral is obvious; living material is precious and may provide the means of extensive study if handled skilfully. Methods for the culture of adults have been described (p. 535). Schumacher (1939) gave details of methods used in dealing with the larvae of *Fasciola hepatica*, and in some institutions conditions for the culture of larval helminths are available, like other amenities of study.

The Use of the Camera Lucida

Though not usually recommended for use by students, who must develop skill in draughtsmanship, this device is indispensable for some parts of the study of Trematoda. The principle according to which it works (the superimposition of an image of a pencil point and drawing paper on the image of a microscopic object by means of a perforated prism and a mirror) can be mastered with the help of a teacher in an hour. The main use of the instrument is for measurement of the eggs, or of various organs and parts. Knowing the magnification of the microscope, absolute sizes can easily be determined from the drawings. Perhaps the best method is to draw part of the scale of a micrometer slide beside the drawing of the object. Choose a number of divisions, say several tenths of a millimetre, divide into hundredths, and transfer them to the paper. Carefully measure the *apparent* size of a given number of tenths of a millimetre, and divide this by the *absolute* size of the part of the scale chosen, thus obtaining the required magnification. The true sizes of the organs or parts or eggs may then be found readily from the drawings, by dividing the apparent sizes by the magnification. Care must be taken to note which objectives and eyepieces are used in determining specific magnifications, and to maintain the draw-tube of the microscope in a constant position.

The camera lucida can be modified for making larger drawings in ways discussed by Harding (1941). It can also be used for transcribing drawings from journals in compiling records of the works of others as described by Dawes (1942*a*). Used apart from the microscope in this way it is a useful device for bringing figures of unequal size to the same absolute size in the copy, thus indicating differences in relative size of organs and parts.

THE HOSTS REFERRED TO IN THE TEXT BY THEIR COMMON ENGLISH NAMES

The scientific names (* excepted) are taken from the *List of British Vertebrates* (Hinton *et al.* 1935), where some synonyms are to be found.

A. Fishes

Allis shad	*Alosa alosa* (L.)
Anchovy	*Engraulis encrasicholus* (L.)
Angel-fish (monk-fish)	*Squatina squatina* (L.)
Angler (fishing frog)	*Lophius piscatorius* L.
Argentine	*Argentina sphyraena* L.
Atherine (sand smelt)	*Atherina presbyter* Cuvier
Axillary bream	*Pagellus owenii* Günther
Ballan wrasse (bergylt)	*Labrus bergylta* Ascanius
Barbel	*Barbus barbus* (L.)
Bass	*Morone labrax* (L.)
Bib (pout)	*Gadus luscus* L.
Black-mouthed dogfish	*Pristiurus melastomus* (Rafinesque)
Black sea bream (Ray's bream)	*Brama raii* (Bloch)
Bleak	*Alburnus alburnus* (L.)
Blue shark	*Carcharinus glaucus* (L.)
Boar-fish	*Capros aper* (L.)
Bogue	*Box boops* (L.)
Bordered ray (Burton skate)	*Raja marginata* Lacepède
Bream (carp bream)	*Abramis brama* (L.)
Brill	*Scophthalmus rhombus* (L.)
Brook lamprey (Planer's lamprey)	*Lampetra planeri* (Bloch)
Burbot	*Lota lota* (L.)
Butter-fish (gunnel)	*Pholis gunnellus* (L.)
Butterfly blenny	*Blennius ocellaris* L.
Char	*Salvelinus* sp.
Chub	*Squalius cephalus* (L.)
Coal-fish (saithe)	*Gadus virens* L.
Cod	*Gadus callarias* L.
Comber (wrasse)	*Serranus cabrilla* (L.)
Common carp	*Cyprinus carpio* L.
Common eel (eel)	*Anguilla anguilla* (L.)
Common (freckled) goby	*Gobius minutus* Gmelin
Common sea bream	*Pagellus centrodontus* De la Roche
Common topknot (Bloch's topknot)	*Zeugopterus punctatus* (Bloch)
Conger	*Conger conger* (L.)
Corkwing (Baillon's wrasse)	*Crenilabrus melops* (L.) ('Gilt-head' here reserved for *Sparus aurata*)
Cornish sucker	*Lepadogaster gouani* Lacepède
Couch's sea bream	*Pagrus pagrus* (L.)
Crucian carp	*Carassius carassius* (L.)
Cuckoo ray	*Raja naevus* Müller & Henle
Cuckoo wrasse (striped wrasse)	*Labrus mixtus* L.
Dab	*Limanda limanda* (L.)
Dace	*Leuciscus leuciscus* (L.)
'Darkie Charlie'	*Scymnorhinus licha* (Bonnaterre)
Dentex	*Dentex dentex* (Gmelin)
Dover sole (common sole)	*Solea solea* (L.)
Dragonet	*Callionymus lyra* L.

Eagle ray	*Myliobatis aquila* (L.)
Fifteen-spined stickleback	*Spinachia spinachia* (L.)
Five-bearded rockling	*Onos mustelus* (L.)
Flapper skate (skate)	*Raja batis* L.
Flounder (fluke)	*Platichthys flesus* (L.)
Gar-fish (green-bone)	*Belone belone* (L.)
Gattorugine (tompot blenny)	*Blennius gattorugine* Bloch
Gilt-head	*Sparus aurata* L.
Golden grey mullet	*Mugil auratus* Risso
Goldfish	*Carassius auratus* (L.)
Gold-sinny	*Ctenolabrus rupestris* (L.)
Grayling	*Thymallus thymallus* (L.)
Greater fork-beard	*Urophycis blennoides* (Brünnich)
Greater pipe-fish	*Syngnathus acus* L.
Greater sand eel	*Ammodytes lanceolatus* Lesauvage
Greater weever	*Trachinus draco* L.
Grey gurnard	*Trigla gurnardus* L.
Grey mullet	*Mugil* sp.
Gudgeon	*Gobio gobio* (L.)
Haddock	*Gadus aeglifinus* L.
Hake	*Merluccius merluccius* (L.)
Halibut	*Hippoglossus hippoglossus* (L.)
Hammerhead shark	*Sphyrna zygaena* (L.)
Herring	*Clupea harengus* L.
Horse mackerel (scad)	*Trachurus trachurus* (L.)
Houting	*Coregonus oxyrhynchus* (L.)
John Dory	*Zeus faber* L.
Lampern (river lamprey)	*Lampetra fluviatilis* (L.)
Lemon sole (smear dab)	*Microstomus kitt* (Walbaum)
Lesser fork-beard	*Raniceps raninus* (L.)
Lesser sand eel	*Ammodytes tobianus* L.
Lesser weever (sting-fish)	*Trachinus vipera* Cuvier & Valenciennes
Ling	*Molva molva* (L.)
Loach (stone loach)	*Cobitis barbatula* (L.)
Long-nosed skate	*Raja oxyrhynchus* L.
Long rough dab	*Hippoglossoides platessoides* (Fabricius)
Long-spined cottus	*Cottus bubalis* Euphrasen
Lump-sucker (lump-fish)	*Cyclopterus lumpus* L.
Mackerel	*Scomber scombrus* L.
Megrin (sail-fluke)	*Lepidorhombus whiff-iagonis* (Walbaum)
Miller's thumb (bull-head)	*Cottus gobio* L.
Minnow	*Phoxinus phoxinus* (L.)
Montagu's sea-snail	*Liparis montagui* (Donovan)
Murry	*Muraena helena* L.
Norwegian topknot	*Phrynorhombus norvegicus* (Günther)
Nurse hound (greater-spotted dogfish)	*Scyliorhinus stellaris* (L.)
Ocean pipe-fish	*Entelurus aequoreus* (L.)
Painted ray	*Raja microcellata* Montagu
Pandora	*Pagellus erythrinus* (L.)
Pelamid (short-finned tunny)	*Sarda sarda* (Bloch)
Perch	*Perca fluviatilis* L.
Pike	*Esox lucius* L.
Piked dogfish (spur dog)	*Squalus acanthias* L.
Pilchard (sardine)	*Sardina pilchardus* (Walbaum)
Plaice	*Pleuronectes platessa* L.
Pollack	*Gadus pollachius* L.
Poor cod (power)	*Gadus minutus* L.
Poutassou	*Gadus poutassou* (Risso)

Rabbit fish (chimaera)	*Chimaera monstrosa* L.
Red band-fish	*Cepola rubescens* L.
Red gurnard (elleck)	*Trigla cuculus* L.
Red mullet	*Mullus surmuletus* L.
Roach	*Rutilus rutilus* (L.)
Rock cook	*Centrolabrus exoletus* (L.)
Rock goby	*Gobius paganellus* Gmelin
Rough hound (lesser-spotted dogfish)	*Scyliorhinus caniculus* (L.)
Rudd	*Scardinius erythrophthalmus* (L.)
Ruffe (pope)	*Acerina cernua* (L.)
Salmon	*Salmo salar* L.
Sand sole (French sole)	*Pegusa lascaris* (Risso)
Sandy ray	*Raja circularis* Couch
Sea lamprey	*Petromyzon marinus* L.
Sea snail	*Liparis liparis* (L.)
Shadow fish (meagre)	*Sciaena aquilla* Risso
Shagreen ray (Fuller's ray)	*Raja fullonica* L.
Shanny	*Blennius pholis* L.
Short-spined cottus (father lasher)	*Cottus scorpius* L.
Silver (white) bream	*Blicca bjoernka* (L.)
Six-gilled shark	*Hexanchus griseus* (Gmelin)
Smelt (sparling)	*Osmerus eperlanus* L.
Smooth hound	*Mustelus mustelus* (L.)
Spanish mackerel	*Pneumatophorus colias* (Gmelin)
Spinax	*Spinax spinax* (L.)
Spined loach	*Cobitis taenia* L.
Spotted (homelyn) ray	*Raja montagui* Fowler
Sprat	*Clupea sprattus* L.
Starry ray	*Raja radiata* Donovan
Sting ray	*Trygon pastinaca* (L.)
Streaked gurnard	*Trigla lineata* Gmelin
Sturgeon	*Acipenser sturio* L.
Sun-fish	*Mola mola* (L.)
Sword-fish	*Xiphias gladius* L.
Tench	*Tinca tinca* (L.)
Ten-spined stickleback (tinker)	*Pygosteus pungitius* (L.)
Thickback sole	*Microchirus variegata* (Donovan)
Thick-lipped grey mullet	*Mugil chelo* Cuvier
Thin-lipped grey mullet	*Mugil capito* Cuvier
Thornback ray	*Raja clavata* L.
Three-bearded rockling	*Onos tricirratus* (Bloch)
Three-spined stickleback (tittlebat)	*Gasterosteus aculeatus* L.
Tope	*Eugaleus galeus* (L.)
Torpedo (electric ray)	*Torpedo nobiliana* Bonaparte
Torsk (tusk)	*Brosme brosme* (Müller)
Trout (sea: brown)	*Salmo trutta* L.
Tunny	*Thunnus thynnus* (L.)
Turbot	*Scophthalmus maximus* (L.)
Twaite shad	*Alosa finta* (Cuvier)
Two-spotted goby	*Gobius ruthensparri* Euphrasen
Viviparous blenny (eel pout)	*Zoarces viviparus* (L.)
Whiting	*Gadus merlangus* L.
Witch (pole flounder)	*Glyptocephalus cynoglossus* (L.)
Wolf-fish (cat-fish)	*Anarhichas lupus* L.
Yellow gurnard (tub-fish)	*Trigla lucerna* L.

B. Amphibia and Reptilia

Common frog	*Rana temporaria temporaria* L.
Common lizard	*Lacerta vivipara* Jacquin
Common toad	*Bufo bufo bufo* (L.)
Crested newt	*Triturus palustris palustris* (L.)
Edible frog	*Rana esculenta* L.
Grass snake	*Natrix natrix natrix* (L.)
Loggerhead turtle	*Caretta caretta* (L.)
Luth (leathery turtle)	*Dermochelys coriacea* (L.)
Natterjack toad	*Bufo calamita* Laurenti
*Salamander	*Salamandra maculosa*
Sand lizard	*Lacerta agilis agilis* L.
Slow worm	*Anguis fragilis* L.
Smooth newt	*Triturus vulgaris vulgaris* (L.)
*Tree frog	*Hyla arborea* L.
Viper	*Vipera berus berus* (L.)

C. Birds

Avocet	*Recurvirostra avocetta* L.
Barnacle-goose	*Branta leucopsis* (Bechstein)
Barn-owl	*Tyto* Billberg
Barred warbler	*Sylvia nisoria nisoria* (Bechstein)
Bar-tailed godwit	*Limosa lapponica lapponica* (L.)
Bean-goose	*Anser fabalis fabalis* (Latham)
Bee-eater	*Merops apiaster* L.
Bittern	*Botaurus stellaris stellaris* (L.)
Blackbird	*Turdus merula merula* L.
Black guillemot	*Uria grylle grylle* (L.)
Black-headed gull	*Larus ridibundus ridibundus* L.
Black kite	*Milvus migrans migrans* (Boddaert)
Black-necked grebe	*Podiceps nigricollis nigricollis* Brehm
Black stork	*Ciconia nigra* (L.)
Black-tailed godwit	*Limosa limosa limosa* (L.)
Black-throated diver	*Colymbus arcticus arcticus* L.
Blue-headed wagtail	*Motacilla flava flava* L.
Brent (dark-breasted) goose	*Branta bernicla bernicla* (L.)
Buff-backed heron	*Ardeola ibis ibis* (L.)
Buzzard (common)	*Buteo buteo buteo* (L.)
Carrion crow	*Corvus corone corone* L.
Chaffinch	*Fringilla coelebs coelebs* L.
Cirl-bunting	*Emberiza cirlus cirlus* L.
Common eider	*Somateria mollissima mollissima* (L.)
Common gull	*Larus canus canus* L.
Common heron	*Ardea cinerea cinerea* L.
Common partridge	*Perdix perdix perdix* (L.)
Common sandpiper	*Tringa hypoleucos* L.
Common scoter	*Melanitta nigra nigra* (L.)
Common sheld-duck	*Tadorna tadorna* (L.)
Common snipe	*Capella gallinago gallinago* (L.)
Common teal	*Querquedula crecca crecca* (L.)
Common tern	*Sterna hirundo hirundo* L.
Coot	*Fulica atra atra* L.
Cormorant	*Phalacrocorax carbo carbo* (L.)
Corncrake (land-rail)	*Crex crex* (L.)
Crane	*Grus grus grus* (L.)
Curlew	*Numenius arquata arquata* (L.)

Domestic duck	*Anas*
Eagle-owl	*Bubo bubo bubo* (L.)
Fieldfare	*Turdus pilaris* L.
Fowl	*Gallus*
Garden-warbler	*Sylvia borin* (Boddaert)
Garganey	*Querquedula querquedula* (L.)
Glaucous gull	*Larus hyperboreus* Gunnerus
Glossy ibis	*Plegadis falcinellus falcinellus* (L.)
Golden eagle	*Aquila chrysaëtus chrysaëtus* (L.)
Golden-eye	*Glaucionetta clangula clangula* (L.)
Golden oriole	*Oriolus oriolus oriolus* (L.)
Golden (southern) plover	*Pluvialis apricarius apricarius* (L.)
Goldfinch (continental)	*Carduelis carduelis carduelis* (L.)
Goosander	*Mergus merganser merganser* L.
Great crested grebe	*Podiceps cristatus cristatus* (L.)
Great grey shrike	*Lanius excubitor excubitor* L.
Great titmouse (continental)	*Parus major major* L.
Great white heron	*Egretta alba alba* (L.)
Greater black-backed gull	*Larus marinus* L.
Greenfinch	*Chloris chloris chloris* (L.)
Green sandpiper	*Tringa ochropus* L.
Green woodpecker	*Picus viridis virescens* (Brehm)
Grey lag-goose	*Anser anser* (L.)
Grey plover	*Squatarola squatarola squatarola* (L.)
Grey wagtail	*Motacilla cinerea cinerea* Tunstall
Gull-billed tern	*Gelochelidon nilotica nilotica* (Gmelin)
Hawfinch	*Coccothraustes coccothraustes coccothraustes* (L.)
Herring gull	*Larus argentatus argentatus* Pontoppidan
Hobby	*Falco subbuteo subbuteo* L.
Honey-buzzard	*Pernis apivorus apivorus* L.
Hooded crow	*Corvus cornix cornix* L.
House-martin	*Delichon urbica urbica* (L.)
House-sparrow	*Passer domesticus domesticus* (L.)
Jackdaw	*Corvus monedula spermologus* Vieillot
Jay (continental)	*Garrulus glandarius glandarius* (L.)
Kestrel	*Falco tinnunculus tinnunculus* L.
King-eider	*Somateria spectabilis* (L.)
Kite	*Milvus milvus milvus* (L.)
Kittiwake	*Rissa tridactyla tridactyla* (L.)
Knot	*Calidris canutus canutus* (L.)
Lapland dunlin	*Erolia alpina alpina* (L.)
Lapwing	*Vanellus vanellus* (L.)
Lesser black-backed gull (Scandinavian)	*Larus fuscus fuscus* L.
Little bittern	*Ixobrychus minutus minutus* (L.)
Little egret	*Egretta garzetta garzetta* (L.)
Little grebe (dabchick)	*Podiceps ruficollis ruficollis* (Pallas)
Little owl	*Carine noctua vidalii* (A. E. Brehm)
Long-eared owl	*Asio otus otus* (L.)
Long-tailed duck	*Clangula hymalis* (L.)
Magpie	*Pica pica pica* (L.)
Mallard	*Anas platyrhyncha platyrhyncha* L.
Marsh-harrier	*Circus aeruginosus aeruginosus* (L.)
Mealy redpole	*Acanthis flammea flammea* (L.)
Mediterranean black-headed gull	*Larus melanocephalus* Temminck
Merlin	*Falco columbarius aesalon* Tunstall
Mistle-thrush	*Turdus viscivorus viscivorus* L.
Moorhen	*Gallinula chloropus chloropus* (L.)
Mute swan	*Cygnus olor* (Gmelin)

Night-heron	*Nycticorax nycticorax nycticorax* (L.)
Nightingale	*Luscinia megarhyncha megarhyncha* Brehm
Nightjar	*Caprimulgus europaeus europaeus* L.
Northern bullfinch	*Pyrrhula pyrrhula pyrrhula* (L.)
Osprey	*Pandion haliaetus haliaetus* (L.)
Oystercatcher (British)	*Haematopus ostralegus occidentalis* Neumann
*Pelican	*Pelecanus onocrotalus* L.
Peregrine falcon	*Falco peregrinus peregrinus* Tunstall
Pheasant	*Phasianus colchicus* L.
Pink-footed goose	*Anser brachyrhynchus* Baillon
Pintail	*Dafila acuta acuta* (L.)
Pochard	*Nyroca ferina ferina* (L.)
Pratincole	*Glareola pratincola pratincola* (L.)
Purple heron	*Ardea purpurea purpurea* L.
Quail	*Coturnix coturnix coturnix* (L.)
Razorbill	*Alca torda* L.
Red-breasted merganser	*Mergus serrator* L.
Red-necked grebe	*Podiceps griseigena griseigena* (Boddaert)
Redshank (common)	*Tringa totanus totanus* (L.)
Redstart	*Phoenicurus phoenicurus phoenicurus* (L.)
Red-throated diver	*Colymbus stellatus* Pontoppidan
Redwing	*Turdus musicus musicus* L.
Richardson's (Arctic) skua	*Stercorarius parasiticus* (L.)
Ringed plover	*Charadrius hiaticula hiaticula* L.
Rock-dove	*Columba livia livia* Gmelin
Rock-pipit	*Anthus spinoletta petrosus* (Montagu)
Rock thrush	*Monticola saxatilis* (L.)
Roller	*Coracias garrulus garrulus* L.
Rook	*Corvus frugilegus frugilegus* L.
Rough-legged buzzard	*Buteo lagopus lagopus* (Brünnich)
Ruff	*Philomachus pugnax* (L.)
Sanderling	*Crocethia alba* (Pallas)
Sandwich tern	*Sterna sandvicensis sandvicensis* Latham
Scaup-duck (scaup)	*Nyroca marila marila* (L.)
Shag	*Phalacrocorax aristotelis aristotelis* (L.)
Sheld-duck	*Tadorna tadorna*
Short-eared owl	*Asio flammeus flammeus* (Pontoppidan)
Shoveler	*Spatula clypeata* (L.)
Siskin	*Spinus spinus* (L.)
Slavonic grebe	*Podiceps auritus* (L.)
Smew	*Mergellus albellus* (L.)
Snowy owl	*Nyctea scandiaca* (L.)
Song-thrush (continental)	*Turdus ericetorum philomelus* Brehm
Sparrow-hawk	*Accipiter nisus nisus* (L.)
Spoonbill	*Platalea leucorodia leucorodia* L.
Spotted crake	*Porzana porzana* (L.)
Spotted redshank	*Tringa erythropus* (Pallas)
Sprosser (thrush-nightingale)	*Luscinia luscinia* (L.)
Starling	*Sturnus vulgaris vulgaris* L.
Stone-curlew	*Oedicnemus oedicnemus oedicnemus* (L.)
Swallow	*Hirundo rustica rustica* L.
Swan	*Cygnus* Bechstein
Swift	*Micropus apus apus* (L.)
Tawny owl	*Strix aluco sylvatica* Shaw
Tree-sparrow	*Passer montanus montanus* (L.)
Tufted duck	*Nyroca fuligula* (L.)
Turkey	*Meleagris*
Turnstone	*Arenaria interpres interpres* (L.)

Velvet scoter	*Melanitta fusca fusca* (L.)
Water-rail	*Rallus aquaticus aquaticus* L.
Wheatear	*Oenanthe oenanthe oenanthe* (L.)
White-eyed pochard (ferruginous duck)	*Nyroca nyroca nyroca* Güldenstadt
White-fronted goose	*Anser albifrons* (Scopoli)
White stork	*Ciconia ciconia ciconia* (L.)
White-tailed sea eagle	*Haliaetus albicilla albicilla* (L.)
Whitethroat	*Sylvia communis communis* Latham
Whooper swan	*Cygnus cygnus cygnus* (L.)
Wigeon	*Mareca penelope* (L.)
Willow-warbler	*Phylloscopus trochilus trochilus* (L.)
Woodcock	*Scolopax rusticola rusticola* L.
Wood-pigeon	*Columba palumbus palumbus* L.
Wood-warbler	*Phylloscopus sibilatrix sibilatrix* (Bechstein)

D. Mammals

Ass	*Equus asinus asinus* Pocock
Badger	*Meles meles meles* (L.)
Bats:	
Daubenton's	*Myotis daubentonii* (Kuhl)
Greater horseshoe	*Rhinolophus ferrum-equinum insulanus* Barrett Hamilton
Leisler's	*Nyctalus leisleri* (Kuhl)
Lesser horseshoe	*Rhinolophus hipposideros minutus* (Montagu)
Long-eared	*Plecotus auritus* (L.)
Mouse-eared	*Myotis myotis* (Borkhausen)
Natterer's	*Myotis nattereri* (Kuhl)
Noctule (great bat)	*Nyctalus noctula* (Schreber)
Pipistrelle	*Pipistrellus pipistrellus* (Schreber)
Serotine	*Eptesicus serotinus* (Schreber)
Whiskered	*Myotis mystacinus* (Kuhl)
Bearded seal	*Erignathus barbatus* Fabr. (*Phoca barbata*)
Black rat	*Rattus rattus* (L.)
Bottle-nosed dolphin	*Tursiops truncatus* (Montagu)
*Civet cat	*Viverra civetta*
Common dolphin	*Delphinus delphis* L.
Common porpoise	*Phocaena phocaena* (L.)
Common seal	*Phoca vitulina* L.
Common shrew	*Sorex araneus castaneus* Jenyns
Deer	*Cervus elaphus* L. and *C. dama* (L.)
*Dog	*Canis lupus* L.
*Domestic cat	*Felis cattus* L.
Fin whale (common rorqual)	*Balaenoptera physalus* (L.)
Fox	*Vulpes vulpes crucigera* (Bechstein)
*Goat	*Capra hircus* L.
Greenland (harp) seal	*Phoca groenlandica* Fabricius
*Greenland right whale	*Balaena mysticetus* L.
Grey seal	*Halichoerus grypus* (Fabricius)
Hare (brown)	*Lepus europaeus occidentalis* De Winton
Hedgehog	*Erinaceus europaeus* L.
*Horse	*Equus caballus caballus* Lydekker
Killer whale (grampus)	*Orcinus orca* (L.)
Long-tailed field mouse	*Apodemus sylvaticus sylvaticus* (L.)
*Mink	*Mustela putorius vison* Schreber
Mole	*Talpa europaea* L.
Mouse (house)	*Mus musculus* L.
Musk-rat	*Ondatra zibethica* (L.)

Otter	*Lutra lutra* (L.)
*Ox	*Bos taurus* L.
*Pig	*Sus scrofa* L.
Piked whale (lesser rorqual)	*Balaenoptera acuto-rostrata* Lacepède
Pine marten	*Martes martes martes* (L.)
Polecat	*Mustela putorius putorius* L.
Rabbit	*Oryctolagus cuniculus* (L.)
Rat (brown)	*Rattus norvegicus* (Erxleben)
Ringed seal	*Phoca hispida* Schreber
Sei whale	*Balaenoptera borealis* Lesson
*Sheep	*Ovis aries* L.
Silver fox	*Mustela erminea erminea* L.
Voles	*Microtus* spp.
Walrus	*Odobenus rosmarus* (L.)
Water voles	*Arvicola* spp.
Weasel	*Mustela nivalis nivalis* L.
White whale (beluga)	*Delphinapterus leucas* (Pallas)

LIST OF LITERATURE*

A. *Some Notable Publications prior to 1800 (chronological order)*

(1379) Jehan de Brie. *Le bon Berger* (original not discovered, but extracts published in *Bon Berger* by Paul Lacroix: Isid Liseux, Paris, 1879). (First mention of liver fluke and liver rot in sheep.)

(1547) Gabucinus, H. F.. *De lumbricis alvum occupantibus ac de ratione eurandi eos, qui ab illis infestantur* (Venetiis).

(1549) Gabucinus, H. F. *Commentarius de lumbricis alvum occupantibus et eorum cura* (Ludg. Batav.). (Mention of liver fluke.)

(1551) Gesnerus, C. *Historiae animalium liber; de quadripedibus viviparis* (Tiguri). (Statement about *duvae*, liver-fluke sickness.)

(1684) Redi, F. *Opusculorum pars tertia, sive de animalculis vivis, quae in corporibus animalium vivorum reperiuntur, observationes* (Lugd. Batav. 1729; earlier edition in Italian). (Parasites arranged according to their hosts: birds, mammals, etc.)

(1686) Redi, F. *Opusculorum pars prima, sive experimenta circa generationem insectorum* (Amstelod.). (Refers to liver fluke as 'vermis vervecini hepatis', and gives a tolerably good picture of it.)

(1695) Leeuwenhoek, A. *Arcana naturae detacta* (Delph. p. 147). (Liver fluke.)

(1697) Leeuwenhoek, A. *Continuatio Arcanorum naturae* (Delph. Epist. 97). (Reference to trematode we know as *Hemiurus ocreatus*, from the herring.)

(1702) Bidloo, G. Brief over de dieren, die man in't Lever der Schaapen vind. *Philos. Trans.* **22**, for 1700–1, p. 571.

Leeuwenhoek, A. Letter concerning the worms in sheep's livers, gnats and animalcula in the excrements of frogs. *Philos. Trans.* **22**, 509. (First paper in a British journal.)

(1706) Leeuwenhoek, A. Letter concerning worms observed in sheep's livers and pasture ground. *Philos. Trans.* **24**, for 1704–5, p. 1522.

(1737) Swammerdam, J. *Biblia naturae s. historia animalium*...(Leid.). (First mention of a worm from the lung of the frog (*Haplometra*) and one of the earliest records of larvae (cercariae).)

(1753) Schäffer, J. C. *Die Egelschnecken in den Lebern der Schafe und die von diesen Würmern entstehende Schafkrankheit* (Regensb.). (Description and figure of *F. hepatica*, and an opinion that it develops from a free-living worm which the sheep ingests.)

(1756) Nicolls, F. An account of worms in animal bodies. *Philos. Trans.* **49**, for 1757, pp. 246–8. (Mention of *F. hepatica*.)

(1758) Linnaeus, C. *Systema naturae*, ed. x (Holmiae). (*Fasciola hepatica* named.)

Rosenhof, A. J. R. von. *Historia naturalis ranarum nostratium....* (First mention of the fluke we call *Polystoma integerrimum*.)

(1759) Baster, J. *Opuscula subseciva, observationes miscellaneas de animalculis et plantis quibusdam marinis earumque ovariis et seminibus continentia* (Harlem). (Diagram of erstwhile *Epibdella hippoglossi*.)

(1760) Pallas, I. S. *De infestis viventibus intra viventia* (Diss. in Lugd. Bat.) (First mention of fluke we call *Diplodiscus subclavatus*.)

(1773) Müller, O. F. *Vermium terrestrium et fluviatilium, seu animalium infusorum, helminthicorum et testaceorum, non marinorum succincta historia* (Havn. and Lips.) (Recognition of cercariae, some as Infusoria.)

* The exact dates of some of the references given in this list are difficult to determine because of the old practice of issuing the component parts of a journal separately, without giving particulars about the dates of issue in the complete volume, which may bear the misleading date of the final part. Happily, this practice has given way to a system which is more acceptable to the bibliographer, so that the recent works in the list are the least liable to bear the erroneous dates. For a more complete list of the early literature see Braun (1879–93).

(**1776**) Retzius, A. J. *Lectiones publicae de vermibus intestinalibus, imprimis humanis* (Holm.) (Name *Distoma* proposed for trematodes.)

Müller, O. F. *Zoologicae danicae prodromus seu animalium Daniae et Norvegiae indigenarum characteres, nomina et synonyma imprimis popularium* (Havn.). (Planarians and trematodes separated as '*Planaria*' and '*Fasciola*'.)

(**1779**) Müller, O. F. *Zoologia danica seu animalium Daniae et Norvegiae rariorum et minus notorum descriptiones et historia* (first ed. 1777, last 1788–1806). (Descriptions of trematodes from fishes, bats, etc.)

(**1782**) Hermann, J. Helminthologische Bemerkungen, I. *Naturforscher*, **17**, Halle. (Genus *Mazocraes* erected.)

(**1786**) Müller, O. F. *Animalcula infusoria, fluviatilia et marina, quae detexit, systematice descripsit et ad vivum delineari curavit; op. posthum. cura O. Fabricii* (Havn.).

(**1787**) Müller, O. F. Verzeichniss der bisher entdeckten Eingeweidewürmer, der Thiere, in welchen sie gefunden werden und der besten Schriften, die derselben erwähnen. *Naturforscher St.* **22**, 33–86, Halle.

Goeze, J. A. E. *Versuch einer Naturgeschichte der Eingeweidewürmer thierischer Körper* (Leipzig).

(**1788**) Schrank, F. v. P. *Verzeichniss der bisher hinlänglich bekannten Eingeweidewürmer nebst einer Abhandlung über ihre Anverwandschaften* (Munich). (Erection of genus *Alaria*.)

Braun, M. Beytrag zur Geschichte der Eingeweidewürmer. *Schr. Berl. Ges. naturf. Frde*, **8**, 236–8. (*Lecithodendrium lagena* founded.)

(**1789**) Frölich, J. M. Beschreibungen einiger neuer Eingeweidewürmer. *Naturforscher*, Part 24, pp. 101–62. (*Brachycoelium salamandrae* and *Catatropis verrucosa* founded.)

(**1790**) Abildgaard, P. C. Almindelige Betragtninger over indvoldeorme, bemaerkninger ved Hundsteilens Baendelorum. *Skr. af naturh. selsk.* **1**, 26–64, Copenhagen. (Genus *Strigea* erected.)

Schrank, F. v. P. Frotekning på nagra hittils obeskrigene intestinalkrak. *Kgl. svensk. Vetensk. Acad. nye for LL Stockholm*, pp. 118–26. (Founding of several species, *Dicrocoelium* '*lanceatum*', *Paramphistomum cervi*, etc.)

Modeer, A. 'Tillägningar.' *Kgl. svensk. Vetensk. Acad. nye for LL Stockholm*, pp. 125–30. (*Asymphylodora tincae*.)

(**1792**) Frölich, J. M. Beyträge zur Naturgeschichte der Eingeweidewürmer. *Naturforscher St.* **25**, 52–113. (Founding of *Opisthioglyphe ranae*, *Macrodera longicollis*, *Bunodera nodulosa*.)

(**1794**) Carlisle, A. Observations upon the structure and oeconomy of those intestinal worms called Taeniae. *Trans. Linn. Soc. Lond.* **2**, 247–62.

B. **1800–1850** (*chronological order*)

(**1800**) Zeder, F. G. H. *Erster Nachtrag zur Naturgeschichte der Eingeweidewürmer von J. A. C. Goeze* (Leipzig).

(**1801**) Rudolphi, C. A. Beobachtungen über die Eingeweidewürmer. Wiedemann's *Arch. Zool. Zoot.* **2**, 1–65.

(**1802**) Rudolphi, C. A. Fortsetzung der Beobachtungen. Wiedemann's *Arch. Zool. Zoot.* **2**, 2, 1–67.

Rudolphi, C. A. Zweite Fortsetzung. Wiedemann's *Arch. Zool. Zoot.* **3**, 1, 67–125.

Frölich, J. A. Beiträge zur Naturgeschichte der Eingeweidewürmer. *Naturforscher St.* **29**, 5–96.

(**1803**) Rudolphi, C. A. Neue Beobachtungen über die Eingeweidewürmer. Wiedemann's *Arch. Zool. Zoot.* **3**, 2, 1–32.

(**1807**) Nitzsch, C. L. Seltame Lebens- und Todesart eines bisher unbekannten Wasserthierchens. *Georgia*, pp. 33–6. (Encystment of a cercaria.)

(**1809**) Rudolphi, C. A. *Entozoorum sive vermium intestinalium historia naturalis*, **2** (Amstelod.).

(1810) Risso, A. *Ichthyologie de Nice, ou histoire naturelle des poissons du département des Alpes maritimes* (Paris).

(1811) Bosc, L. A. G. Sur deux nouveaux genres de vers. *Nouv. Bull. soc. Philom. Paris*, An. 4 (51), **2**, 384–5.

Delaroche, F. Sur deux animaux vivants sur les branchies des poissons. *Nouv. Bull. soc. Philom. Paris*, pp. 270–3.

(1814) Rudolphi, C. A. Erster Nachtrag zu meiner Naturgeschichte der Eingeweidewürmer. *Ges. naturf. Frde Berlin Mag. neuest. Entdeck.* **6**, 83–113.

(1815) Oken, L. *Lehrbuch der Naturgeschichte*, 3. Theil: Zoologie. 1. Abt.: Fleischlose Thiere, **1**, pt. 1, Leipzig, 842 pp., 40 Pl.

(1815–22) Lamarck, J. B. P. *Histoire naturelle des animaux sans vertèbres* (Paris).

(1816) Otto, A. Ueber das Nervensystem der Eingeweidewürmer. *Ges. naturf. Frde Berlin Mag. neuest Entdeck.* **7**, 223–33.

Nitzsch, C. L. Beitrag zur Infusorienkunde oder Naturbeschreibung der Zerkarien und Bazillarien. *Neue Schrift d. nat. Ges. Halle*, **3**, 1.

(1817) Olfers, J. F. M. de. *De vegetativis et animatis corporibus in corporibus animalium reperiundis* (Berol.).

Cuvier, G. *Le règne animal distribué d'après son organisation, etc.*

(1818) Bojanus, L. Kurze Nachricht über die Zerkarien und ihren Fundort. Oken's *Isis*, pp. 729–30.

(1819) Nitzsch, C. L. Artikel 'Amphistoma'. Ersch und Gruber's *Allgemeine Encyclop. wiss. Künste*, Leipzig.

Rudolphi, C. A. *Entozoorum synopsis* (Berol. 811 pp.).

Bremser, J. G. *Ueber lebende Würmer im lebenden Menschen* (Wien).

(1821) Bojanus, L. Enthelminthica. *Isis*, **2**, 162–90.

(1824) Bremser, J. G. *Icones helminthum systema Rudolphi entozooligicum illustrantes* (Vienna, I–XIII).

(1825) Creplin, F. C. H. *Observationes de entozois* (Gryphisw.).

Mehlis, E. *Observationes anatomicae de Distomate hepatico et lanceolato* (Gotting.).

(1827) Schmalz, E. *De entozoorum systemate nervoso* (Diss. in Lipsiae).

Baer, K. E. v. Beiträge zur Kenntniss der niederen Thiere. *Nova Acta Leop. Carol.* **13**, II, Bonnae, 524–762. (Abstracted in 1826.)

Nitzsch, C. L. Cercaria. In Ersch und Grüber, Leipzig, **1**, Sect. 16, 66–9.

(1828) Leuckart, F. S. *Brevis animalium quorundam maxima ex parte marinorum descriptio* (Heidelb.).

Blainville. In *Dictionnaire des sciences naturelles*, p. 57 (Paris).

(1829) Kuhn, J. Description d'un nouveau genre de l'ordre des douves et de deux espèces de Strongyles. *Mém. Mus. hist. nat.* **18**, 357–68.

Creplin, F. C. H. *Novae observationes de entozois* (Berol.).

Creplin, F. C. H. Filariae et Monostomi speciem novam in *Balaena rostrata* repertam describit. *Nova Acta Leop. Carol.* **14**, II, Bonnae, 871–82.

Eysenhardt, C. G. Einiges über Eingeweidewürmer. *Verg. Ges. naturf. Frde Berlin*, **1**, 144–52.

(1830) Laurer, F. F. *Disquistiones anatomicae de* Amphistomo conico (Diss. in Gryphiae).

(1831) Schmarz, E. *XIX Tabulae anatomiam entozoorum illustrantes* (Dresden).

Mehlis, E. Anzeige von Creplin's Novae observationes de entozois. *Isis*, pp. 68–99, 166–99.

(1832) Nordmann, A. *Mikrographische Beiträge zur Naturgeschichte der wirbellosen Thiere*, First Part (Berlin, 118 pp.).

Wagner, R. Beobachtungen über den Bau und die Entwicklung der Infusorien mit besonderer Berücksichtigung von Ehrenberg's Arbeit. *Isis*, pp. 343–98.

(1833) Nordmann, A. Du *Diplozoon paradoxum*. *Ann. sci. nat.* **30**, 372–98.

(1835) Siebold, C. T. v. Helminthologische Beiträge. *Arch. Naturgesch.* **1**, 45–83.

Johnston, G. Illustrations in British zoology, 44, *Udonella caligorum*. *Mag. Nat. Hist. Lond.* **8**, 496–8.

Owen, R. On the anatomy of *Distoma clavatum*. *Trans. Zool. Soc. Lond.* **1**, 381–4.

Diesing, C. M. *Aspidogaster limacoides*, eine neue Art Binnerwurm. *Isis*, p. 1231, and *Med. J. Österr. Kaiserstaates*, N.F. **7**, 420–30.

DIESING, C. M. Monographie der Gattungen *Amphistoma* und *Diplodiscus*. *Ann. Wiener Mus. Naturg.* **1**, 235–60.

CARUS, C. G. Beobachtungen über einen merkwürdigen schöngefärbten Eingeweidewürm, *Leucochloridium paradoxum* mihi und dessen parasitische Erzeugung in einer Landschnecke, *Succinea amphibia* Drap., *Helix putris* L. *Nova Acta Leop. Carol.* **17**, 1, 85–100.

(1836) DIESING, C. M. Monographie der Gattung *Tristoma*. *Nova Acta Leop. Carol.* **18**, 1, 1–16.

(1837) CREPLIN, F. C. H. Distoma. In Ersch und Grüber, Leipzig, Sect. 1, Part 29, 309–29.

(1838) JOHNSTON, G. Miscellanea zoologica. *Ann. Nat. Hist.* or *Mag. Zool. Bot. Geol. Lond.* **1**, 471–7.

(1839) DIESING, C. M. Neue Gattungen von Binnenwürmern nebst einem Nachtrag zur Monographie der Amphistomen. *Ann. Wiener Mus. Naturg.* **2**, 219–42.

CREPLIN, F. C. H. Artikel 'Eingeweidewürmer'. Ersch und Grüber's *Allg. Encyclop. wiss. Künste*, pp. 277–302.

(1840) NORDMANN, A. Les vers (Vermes). In Lamarck's *Histoire naturelle des animaux sans vertèbres*..., Paris, 3rd ed., pp. 542–686.

GULLIVER, G. Note on the ova of *Distoma hepaticum*. *Proc. Zool. Soc. Lond.* **8**, 30–1.

(1842) LEUCKART, F. S. Helminthologische Beiträge. *Zool. Bruckstücke*, **3**, Freiburg, 60 pp.

CREPLIN, F. C. H. Endozoologische Beiträge. *Arch. Naturg.* **1**, 315–39.

STEENSTRUP, J. J. S. *Ueber den Generationswechsel oder die Fortpflanzung und Entwicklung durch abwechselnde Generationen, eine eigenthümliche Form der Brutpflege in den niederen Thierklassen.* (Copenhagen, pp. 50–110.)

(1843) DUJARDIN, F. Mémoire sur les helminthes des musaraignes et en particulier sur les Trichosomes, les Distomes, et les Ténias, sur leurs métamorphoses et leurs transmigrations. *Ann. Sci. nat.* (*Zool.*), (2), **20**, 329–49.

(1844) BELLINGHAM, O'B. Catalogue of Irish Entozoa with observations. *Ann. Mag. Nat. Hist.* **13**, 335–40, 422–30.

GOODSIR, H. D. S. On the development, structure and economy of the acephalocysts of authors, with an account of the natural analogies of the entozoa in general. *Trans. Roy. Soc. Edinb.* **15**, 561–71; *Proc. Roy. Soc. Edinb.* **1**, 466–8; *Ann. Mag. Nat. Hist.* **14**, 481–4.

(1845) GOODSIR, H. D. S. On the structure and development of the cystic entozoa. *Rep. Brit. Ass. Adv. Sci.* 14th Meeting, 1844; in *Transactions*, pp. 67–8.

DUJARDIN, F. *Histoire naturelle des Helminthes ou vers intestinaux* (Paris, 654 pp. atlas, Pl. 1–12).

THOMPSON, W. Additions to the fauna of Ireland with descriptions of some new species of Invertebrata. *Ann. Mag. Nat. Hist.* **15**, 308–22.

(1847) BLANCHARD, E. Recherches sur l'organisation des vers. *Ann. Sci. Nat.* Sér. 3, Zool. pp. 271–341.

(1849) CARUS, J. V. *Zur näheren Kenntniss des Generationswechsels* (Leipzig).

C. 1850–1900 (*chronological order*)

(1850) MÜLLER, J. Ueber eine eigenthümliche Wurmlarva aus der Classe der Turbellarien und aus der Familie der Planarien. Müller's *Arch. Anat. Physiol.* pp. 485–500.

DIESING, C. M. *Systema helminthum*, **1** (Vindobonae, 679 pp.).

(1851) THAER, A. Ueber *Polystomum appendiculatum* (*Onchocotyle appendiculata* Diesing). *Arch. Anat. Physiol. wiss. Med.* Year 1850, pp. 602–32.

THAER, A. *De polystomo appendiculato* (Diss. Berolini, 32 pp.).

CREPLIN, F. C. H. Nachträge von Creplin zu Gurlt's Verzeichniss.... *Arch. Naturg.* **17**, 269–310.

(1852) LEIDY, J. Contributions to helminthology. *Proc. Acad. Nat. Sci. Philad.* **6**, 1850–1, 205–9, 224–7, 239–44, 284–90.

(1853) Baird, W. *Catalogue of Entozoa* (London, pp. 1–132).

Baird, W. Description of some new species of Entozoa from the collection of the British Museum. *Proc. Zool. Soc. Lond.* **21**, 18–25.

Bilharz, T. Ein Beitrag zur Helminthographia humana aus brieflichen Mittheilungen, nebst Bemerkungen von C. T. v. Siebold. *Z. wiss. Zool.* **4**, 53–76.

Bilharz, T. Fernere Mittheilungen über *Distomum haematobium*. *Z. wiss. Zool.* **4**, 454–6.

Siebold, C. T. v. Ueber *Leucochloridium paradoxum*. *Z. wiss. Zool.* **4**, 425–37.

Beneden, P. J. van. Espèce nouvelle du genre *Onchocotyle* vivant sur les branchies du *Scimnus glacialis*. *Bull. Acad. Belg. Cl. Sci.* **20**, Part 3 (9), 59–68.

Dalyell, J. G. *The powers of the Creator displayed in creation*.... London.

(1854) Lacaze-Duthiers. Mémoire sur le Bucephale Haime, helminthe parasite des huitres et des bucardes. *Ann. sci. nat.* 4th Ser. **1**, 294–302.

Filippi, F. de. Mémoire pour servir à l'histoire génétique des Trématodes. *Mém. R. Accad. Torino*, 2nd Ser. **15**, 331–58.

(1855) Filippi, F. de. Deuxième Mémoire pour servir..., etc. *Mém. R. Accad. Torino*, 2nd Ser. **16**, 419–42; also in *Ann. sci. .nat.* 4th Ser. Zool. **2**, 1854, 255–85; **3**, 1855, 111–13.

Beneden, P. J. v. Sur les vers parasites du poissons lune (*Orthagoriscus mola*)..., etc. *Bull. Acad. Belg. Cl. Sci.* **2**, 22, Part 2, 520–7.

Aubert, A. Ueber das Wassergefässsystem, die Geschlechtsverhältnisse, die Eibildung und die Entwicklung des *Aspidogaster conchicola*..., etc. *Z. wiss. Zool.* **6**, 349–76.

Diesing, K. M. Revision der Cercarieen. *S.B. Akad. Wiss. Wien*, **15**, 377–400.

Küchenmeister, F. *Die in und an dem Körper des lebenden Menschen vorkommenden Parasiten* (Abt. 2, Leipzig).

La Valette de St George, A. de. *Symbolae ad trematodum evolutionis historiam* (Berolini, 38 pp.).

Diesing, K. M. Sechzehn Gattungen von Binnenwürmern und ihre Arten. *Denkschr. Akad. Wiss. Wien*, **9**, 171–85.

Diesing, K. M. Neunzehn Arten von Trematoden. *Denkschr. Akad. Wiss. Wien*, **10**, 59–70.

Wedl, K. Helminthologische Notizen. *S.B. Akad. Wiss. Wien*, **16**, 371–95.

(1856) Cobbold, T. S. Description of a new species of trematode worm. *Rep. 25th Meeting Brit. Ass. Adv. Sci.*; also *Edinb. Phil. J.* N.S. 2, 262–7.

Burmeister, H. *Zoonomische Briefe. Allgemeine Darstellung der thierischen Organisationen* (Leipzig).

Filippi, F. de. Quelques nouvelles observations sur les larves des Trématodes. *Ann. sci. nat.* 4th Ser. **6**, 83–6.

Moulinié, J. J. De la reproduction chez les Trématodes endoparasites. *Mém. Inst. nat. génév.* **3**, 7–279.

(1856–7) Leidy, J. A synopsis of Entozoa and some of their ectocongeners observed by the authors. *Proc. Acad. Nat. Sci. Philad.* **8**, 42–58.

(1857) Filippi, F. de. *Ann. Mag. Nat. Hist.* 2nd Ser. pp. 127–32.

Wagener, G. Beiträge zur Entwicklungsgeschichte der Eingeweidewürmer, etc. *Naturk. Verh. holland. maatsch. wet. Haarlem, 2 versamel.* **13**, 1–112.

Wagener, G. Helminthologische Bemerkungen aus einem Senschreiber an C. Th. von Siebold. *Z. wiss. Zool.* Leipzig, **9** (1), 73–90.

Claparède, E. Ueber die Kalkkörperchen der Trematoden und die Gattung *Tetracotyle*. *Z. wiss. Zool.* Leipzig, **9**, 99–105.

Lespès, C. Observations sur quelques cercaires parasites de mollusques marins. *Ann. sci. nat.* 4th Ser. Zool. **7**, 113–17.

Lankester, E. On the occurrence of species of Distome in the human body. In Küchenmeister, F., *On Animal and Vegetable Parasites of the Human Body*, **1**, 433–7.

Pagenstecher, H. A. *Trematodenlarven und Trematoden, helminthologischer Beitrag* (Heidelberg, 56 pp.).

(1858) WEDL, C. Anatomische Beobachtungen über Trematoden. *S.B. Akad. Wiss. Wien*, **26**, 241–78.

DIESING, K. Revision der Myzhelminthen, Abtheilung Trematoden. *S.B. Akad. Wiss. Wien*, **32**, 307–90.

WAGENER, G. R. Enthelminthica No. VI. Ueber *Distoma campanula* (*Gasterostoma fimbriatum* Siebold) Duj. und *Monostoma bipartitum* Wedl. *Arch. Naturgesch.* 24. Jahrg. **2**, 250–6.

DIESING, K. Berichtigungen und Zusätze zur Revision der Cercarien. *S.B. Akad. Wiss. Wien*, **31**, 239–90.

DIESING, K. Vierzehn Arten von Bdelliden. *Denkschr. Akad. Wiss. Wien.* **14**, 63–80.

COBBOLD, T. S. Observations on Entozoa with descriptions of several new species. *Trans. Linn. Soc. Lond.* **22**, 3, 155–72, 363–70.

BENEDEN, P. J. VAN. Mémoire sur les vers intestinaux. Suppl. *C.R. Acad. Sci., Paris*, **2**, 376 pp.

(1859) VULPIAN. Note sur un nouveau Distome de la grenouille. *C.R. Soc. Biol., Paris*, Year 1858, pp. 150–2.

FILIPPI, F. DE. Troisième mémoire, etc. *Mem. R. Accad. Torino*, 2nd Ser. **18**, 201–32.

WEINLAND, D. F. Systematischer Katalog aller Helminthen die im Menschen gefunden worden. *Arch. Naturg.* 25. Jahrg. **1**, 276–85.

MOLIN, R. Nuovi myzhelminta raccolti ed esaminati. *S.B. Akad. Wiss. Wien*, **37**, 818–54.

(1860) POLONIO, A. F. *Prospectus helminthi faunae Venetae* (Pavia).

WAGENER, G. R. Ueber *Gyrodactylus elegans* von Nordmann. *Arch. Anat. Phys. wiss. Med.* pp. 768–97.

(1861) COBBOLD, T. S. Synopsis of the Distomidae. *Proc. Linn. Soc. Lond.* **5**, 1–56, 117–27.

COBBOLD, T. S. Further observations on entozoa with experiments. *Trans. Linn. Soc. Lond.* **23**, 349–58.

BRADLEY, C. S. On the occurrence of *Gyrodactylus elegans* on sticklebacks in the Hampstead ponds. *Proc. Linn. Soc. Lond.* (*Zool.*), **5**, 209–10. Also *G. anchoratus*. *Proc. Linn. Soc. Lond.* (*Zool.*), **5**, 257.

MOLIN, R. Prodromus faunae helminthologicae Venetae adjectis disquisitionibus anatomicis et criticis. *Denkschr. Akad. Wiss. Wien*, **19**, 189–338.

(1862) HOUGHTON, W. On the occurrence of *Gyrodactylus elegans* in Shropshire. *Ann. Mag. Nat. Hist.* (3) 10, p. 77.

PAULSON, O. Zur Anatomie von *Diplozoon paradoxum*. *Mém. Acad. Sci. St-Pétersb.*, 7th Ser. **4**, no. 5, 16 pp.

LEARED, A. Description of a new parasite found in the heart of the edible turtle. *Quart. J. Micr. Sci.* N.S. **2**, 168–70.

(1863) PAGENSTECHER, H. A. Untersuchungen über niedere Seethiere aus Cette. *Z. wiss. Zool.* **12**, 263–311.

LEUCKART, R. *Die menschlichen Parasiten und die von ihnen herrührender Krankheiten*, **1** (Leipzig and Heidelberg, pp. 448–634, 765–6).

CLAPARÈDE, E. *Beobachtung über Anatomie und Entwicklung wirbelloser Thiere, an den Küste der Normandie angestellt* (Leipzig).

BENEDEN, P. J. VAN & HESSE, C. E. Recherches sur les bdellodes ou hirudinées et les trématodes marins.* *Mém. Acad. R. Belg. Cl. Sci.* **34**, 1–142; also *Mém. Acad. R. Belg. Cl. Sci.* for 4 appendices, vols. **34**, **35**.

CARUS, J. V. Räderthiere, Würmer, Echinodermen, Coelenteraten und Protozoen. In Peters, Carus & Gerstaecker's *Handb. d. Zool.* **2**, 422–600 (Leipzig).

(1864) MCINTOSH, W. C. Notes on the food and parasites of *Salmo salar* of the Tay. *Proc. Linn. Soc. Lond.* (*Zool.*), **7**, 145–54.

COBBOLD, T. S. *Entozoa, an introduction to the study of helminthology, with reference more particularly to the internal parasites of man* (London, 480 pp.).

* This work was first published in 1863, but the better known and more accessible edition dated 1864 has been consulted and is quoted in the text.

(1865) Melnikow, N. Ueber das *Distomum lorum*. *Arch. Naturgesch.* **1**, 49–55.

(1866) Cobbold, T. S. *Catalogue of the specimens of entozoa in the museum of the Royal College of Surgeons of England* (London, 24 pp.).

Wagener, R. Ueber Redien und Sporocysten Filippi. *Arch. Anat. Physiol.* pp. 145–50.

(1867) Maddox, R. S. Some remarks on the parasites found in the nerves of the common haddock, *Morrhua aeglifinus*. *Trans. Roy. Micr. Soc. Lond.* N.S. **15**, 87–99; also in *Quart. J. Micr. Soc.* N.S. **7**, p. 87, 1867.

Zeller, E. Ueber das encystierte Vorkommen von *Distomum squamula* Rud. im braunen Grasfrosch. *Z. wiss. Zool.* **17**, 215–20.

(1868) Beneden, P. J. van. Sur le cigogne blanche et ses parasites. *Bull. Acad. Belg.* 2nd Ser. **25**, 294–303.

Olsson, P. Entozoa iakttagna hos skandinaviska hafsfiskar. *Lunds Univ. Årsskrift*, **4**, 64 pp.

(1869) Cobbold, T. S. *Entozoa, being a supplement...etc.* (London, 124 pp.).

Olsson, P. Nova genera parasitantia Copepodorum et Platyhelminthum. *Lunds Univ. Årsskrift*, **6**, 6 pp.

(1870) Metschnikoff, E. Embryologisches über *Gyrodactylus*. *Bull. Acad. Sci. St-Pétersb.* **14**, 61–5.

Beneden, E. van. Recherches sur la composition et la signification de l'œuf. *Mém. Acad. R. Belg. Cl. Sci.* **34**, 283 pp.

Stieda, L. Ueber den Bau des *Polystoma integerrimum*. *Arch. Anat. Physiol.* pp. 660–78.

Beneden, P. J. van. Les poissons des côtes de Belgique, leurs parasites et leurs commensaux. *Mém. Acad. R. Belg. Cl. Sci.* **38**, 100 pp.

Schmarda, L. K. *Zoologie*. Wien. 372 pp.

(1871) Stieda, L. Ueber den angeblichen inneren Zusammenhang der männlichen und weiblichen Organe bei Trematoden. *Arch. Anat. Physiol.* pp. 31–40.

Willemoes-Suhm, R. v. Vorläufiges über die Entwicklung des *Polystoma integerrimum* Rud. *Nachr. Ges. Wiss. Göttingen*, pp. 181–5.

Willemoes-Suhm, R. v. Ueber einige Trematoden und Nemathelminthen. *Z. wiss. Zool.* **21**, 175–203.

(1872) (*a*) Zeller, E. Untersuchungen über die Entwicklung und den Bau des *Polystoma integerrimum*. *Z. wiss. Zool.* **22**, 1–28.

(*b*) Zeller, E. Untersuchungen über die Entwicklung des *Diplozoon paradoxum*. *Z. wiss. Zool.* **22**, 168–80.

Willemoes-Suhm, R. v. Zur Naturgeschichte des *Polystoma integerrimum* und des *P. ocellatum* Rud. *Z. wiss. Zool.* **22**, 29–39.

(1873) Linstow, O. v. Ueber die Entwicklungsgeschichte des *Distomum nodulosum* Zeder. *Arch. Naturgesch.* 39th Year, **1**, 1–7.

Linstow, O. v. Einige neue Distomen und Bemerkungen über die weiblichen Sexualorgane der Trematoden. *Arch. Naturgesch.* 39th Year, **1**, 95–108.

Beneden, P. J. van. Les parasites des chauves-souris de Belgique. *Mém. Acad. R. Belg. Cl. Sci.* **40**, 42 pp.

Willemoes-Suhm, R. v. Helminthologische Notizen. III. *Z. wiss. Zool.* **23**, 331–45.

(1874) Zeller, E. Ueber *Leucochloridium paradoxum* Car. und die weitere Entwicklung seiner Distomenbrut. *Z. wiss. Zool.* **24**, 564–78.

Cobbold, T. S. *The internal parasites of our domesticated animals*. London, 160 pp.

Chatin, J. Études sur les helminthes nouveaux ou peu connus. *Ann. Sci. nat.* **65**, 1, 18 pp.

(1874–7) Moore, D. On *Bucephalus haimeanus* and other allied organisms. *J. Quekett Micr. Club*, **4**, 50–7.

Woods, W. F. On the relation of Bucephalus to the cockle. *J. Quekett Micr. Club*, **4**, 58–66.

(1875) McConnell, J. F. P. Remarks on the anatomy and pathological relations of a new species of liver fluke. *Lancet*, **2**, 271–4; also *Veterinarian*, **48**, 772–80.

(**1876**) LEWIS, T. R. & MCCONNELL, J. F. P. A new parasite affecting man. *Proc. Asiatic Soc. Bengal*, pp. 182–6.

SONSINO, P. Intorno ad un nuovo parassito de lue. *R.C. Accad. Napoli*, 15th Year, pp. 84–7.

OLSSON, P. Bidrag till skandinaviens helminthfauna. *Kgl. svenska vetensk. Acad. Handl. Stockholm*, **14**, 35 pp.

(**1877**) COBBOLD, T. S. Note on *Gastrodiscus*. *Veterinarian*, London, p. 326.

LINSTOW, O. v. Helminthologica. *Arch. Naturgesch.* **43**, **1**, 1–18, 173–98; also *Zool. Jb., Syst.* **3**, 104–5.

WIERZEJSKI, A. Zur Kenntnis des Baues von *Calicotyle Kröyeri* Dies. *Z. wiss. Zool.* **29**, 550–82.

(**1878**) LEUCKART, R. Bericht über die wissenschaftlichen Leistungen in der Naturgeschichte der niederen Thiere während der Jahre 1876–1879. II. Platodes. *Arch. Naturgesch.* **44**, 563–714.

LINSTOW, O. v. *Compendium der Helminthologie...etc.* (Hannover, 382 pp.).

LORENZ, S. Ueber die Organisation der Gattungen *Axine* und *Microcotyle*. *Arb. Zool. Inst. Univ. Wien*, **1**, 405–36.

VILLOT, A. Organisation et développement de quelques espèces des trématodes endoparasites marins. *Ann. Sci. nat.* (*Zool.*), 6th Ser. **8**, 40 pp.

KERBERT, C. Zur Trematoden-Kenntnis. *Zool. Anz.* 1st Year, pp. 271–3.

TASCHENBERG, E. O. Helminthologisches. *Z. ges. Naturwiss.* **51**, 562–77.

(**1879**) TASCHENBERG, E. O. Zur Systematik der monogenetischen Trematoden. *Z. ges. Naturwiss.* **52**, 232–65.

TASCHENBERG, E. O. *Didymozoon*, eine neue Gattung in Cysten lebender Trematoden. *Z. Naturwiss.* **52**, 606–17.

TASCHENBERG, E. O. Weitere Beiträge zur Kenntnis ectoparasitischer mariner Trematoden. *Festschr. naturf. Ges. zu Halle zur Feier ihrer* 100 *jähr. Bestehens.* Halle, pp. 25–76.

VILLOT, A. Sur une nouvelle forme de ver vésiculaire, à bourgeonnement exogène. *C.R. Acad. Sci. Paris*, **91**, 938–40.

WRIGHT, R. R. Contributions to American helminthology. *J. Proc. Canad. Inst.* N.S. **1**, 54–75.

COBBOLD, T. S. *Parasites; a Treatise on the Entozoa of Man and Animals, etc.* (London, 510 pp.).

(**1879–93**) BRAUN, M. Platyhelminthes. I. Trematodes. Bronn's *Klassen und Ordnungen des Thier-Reichs* 4, *Vermes*, Abt. 1*a*, pp. 567–925.

(**1880**) TASCHENBERG, E. O. Ueber *Tristomum molae* Blanch. *Zool. Anz.* 3rd Year, pp. 17–18.

FRAYPONT, J. Recherches sur l'appareil excréteur des trématodes et des cestodes. *Arch. Biol.* **1**, 415–56. (Also *Arch. Biol.* **2**, 1881, 1–40.)

BALFOUR, F. M. *A Treatise on Comparative Embryology* (London).

LANG, A. Untersuchungen zur vergleichenden Anatomie und Histologie der Nervensystems der Platyhelminthen. II. Ueber das Nervensystem der Trematoden. *Mitt. Zool. Stat. Napoli*, **2**, 28–52.

SOMMER, F. Zur Anatomie des Leberegels, *Distomum hepaticum* L. *Z. wiss. Zool.* **34**, 539–640.

(**1881**) LEVINSEN, G. M. R. Bidrag til Kundskab om Gröndlands Trematodfauna. *K. Danske Vidensk. Selsk. Forhdl.* no. 1, 49–84.

PAGENSTECHER, H. A. *Allgemeine Zoologie oder Grundsätze des thierschen Baues und Lebens*, Part IV (959 pp.).

GROBBEN, C. *Doliolum* und sein Generationswechsel, etc. *Arb. zool. Inst. Univ. Wien*, **4**, 201–98.

ERCOLANI, G. Dell' adattamento delle specie all' ambiente, nuove ricerche sulla storia genetica dei Trematodi. I. *Mem. R. Accad. Bologna*, 4th Ser. **2**, 237–334.

LEUCKART, R. Zur Entwicklungsgeschichte des Leberegels. *Zool. Anz.* 4th Year, pp. 641–6.

THOMAS, A. P. Report of experiments on the development of the liver fluke. *J. Roy. Agric. Soc.* **17**, 1–30; also *J. Roy. Micr. Soc.* **1**, 740–1.

(**1882**) ERCOLANI, G. Memoria. II. *Mem. R. Accad. Bologna*, 4th Ser. **3**, 43–111.

LEUCKART, R. *Zool. Anz.* 5th Year, pp. 524–8; also *Arch. Naturgesch.* 48th Year, **1**, 80–119.

THOMAS, A. P. Second report. *J. Roy. Agric. Soc.* **18**, 439–54; also *Nature, Lond.* **26**, 606–8.

(1883) THOMAS, A. P. The life history of the liver fluke. *Quart. J. Micr. Soc.* **23**, 90–133.

JACKSON, W. H. Note on the life history of *Fasciola hepatica*. *Zool. Anz.* 6th Year, pp. 248–50.

ZIEGLER, H. E. *Bucephalus* and *Gasterostomum*. *Z. wiss. Zool.* **39**, 537–71; also *Zool. Anz.* 6th Year, pp. 476–92.

POIRIER, J. Descriptions d'helminthes nouveaux du *Palonia frontalis*. *Bull. Soc. philom. Paris*, 7th Ser. pp. 73–80.

LOOSS, A. Zur Frage nach der Natur des Körperparenchyms bei den Trematoden. *Sächs. Ges. Wiss. math.-phys. Kl. Sitz.* 9. Jan.

LOOSS, A. Ist der Laur'sche Kanal der Trematoden eine Vagina? *Zbl. Bakt.* I. Orig. **13**, 808–19.

STOSSICH, M. Brani di elmintologia tergestina, Ser. 1. *Bol. Soc. Adr. Sci. Nat. Trieste*, **8**, 111–21.

(1884) CARUS, J. V. *Prodromus faunae mediterraneae, etc.* **1** (Stuttgart, xi+524 pp.).

BIEHRINGER, J. Beiträge zur Anatomie und Entwicklung der Trematoden. *Arb. Zool.-zootom. Inst. Würzburg*, **7**, 1–28.

(1885) BERGH, R. S. Die Excretionsorgane der Würmer. *Kosmos*, hrsg. v. Vetter, **17**, 97–122.

LOOSS, A. Beiträge zur Kenntnis der Trematoden. *Z. wiss. Zool.* **41**, 390–446.

POIRIER, J. Contribution à l'histoire des Trématodes. *Arch. Zool. exp. gén.* B, **3**, pp. 464–624.

(1886) LEUCKART, R. *Die Parasiten des Menschen und die von ihnen herrührenden Krankheiten*, 2. Aufl. **1**, 3. Lief., 2. Abteil., 1–96.

POIRIER, J. Trématodes nouveaux ou peu connus. *Bull. soc. philom. Paris*, 7th Ser. **10**, 20–41.

POIRIER, J. Sur les Diplostomidae. *Arch. Zool. exp. gén.* 2nd Ser. **4**, 327–46.

STOSSICH, M. *I distomi dei pesci marini e d' acqua dolce.* Lavoro Monografico (Trieste, pp. 1–66).

BELL, F. J. A new species of Distomum. *Ann. Nat. Hist.* (5), **19**, 116.

(1887) CUNNINGHAM, J. T. On *Stichocotyle nephropsis*, a new trematode. *Trans. R. Soc. Edin.* **33**, 273–86.

LINSTOW, O. v. Helminthologische Untersuchungen. *Zool. Jb., Syst.* **3**, 97–114.

LINTON, E. Notes on a trematode from the white of a newly laid hen's egg. *Proc. U.S. Nat. Mus.* **10**, 367–9.

PARONA, C. Elmintologia sarda, contributione allo studio dei vermi parassiti in animali di Sardegna. *Ann. Mus. Stor. nat. Genova*, Ser. 2, **4**, 275–384.

PARONA, C. Res. ligusticae. II. *Ann. Mus. Stor. nat. Genova*, Ser. 2, **4**, 483–501.

ZSCHOKKE, F. Helminthologische Bemerkungen. *Mitt. zool. Sta. Neapel*, **7**, 264–71.

(1888) LINSTOW, O. v. Helminthologisches. *Arch. Naturgesch.* 54th Year, pp. 235–46.

MONTICELLI, F. S. *Saggio di una morfologia dei Trematodi* (Napoli, 130 pp.).

MONTICELLI, F. S. *Cercaria setifera* Müller. In *Bol. Soc. Nat. Napoli*, **2**, 193–9.

KÜCHENMEISTER, F. & ZÜRN, F. A. *Die Parasiten des Menschen* (Leipzig).

ZELLER, E. Ueber den Geschlechtapparat von *Diplozoon paradoxum*. *Z. wiss. Zool.* **46**, 233–9.

BRANDES, G. Ueber das Genus *Holostomum*. *Zool. Anz.* 11th Year, pp. 424–6.

BRANDES, G. *Die Familie der Holostomeae, ein Prodromus zu einer Monographie derselben.* In Diss. Leipzig, 72 pp.

BRANDES, G. Helminthologisches. *Arch. Naturgesch.* 54th Year, pp. 247–51.

STOSSICH, M. *Appendice al mio lavoro i Distomi dei pesci marini e d' acqua dolce* (Trieste, 14 pp.).

VOELTZKOW, A. *Aspidogaster conchicola*. *Arch. Zool.-zootom. Inst. Würzburg*, **8**, 249–92.

(1889) JUËL, H. O. Beiträge zur Anatomie der Trematodengattung *Apoblema* Duj. *Vet.-Ak. Handl. Stockholm*, **15**, 46 pp.

LEUCKART, R. *Die Parasiten des Menschen..., etc.* **1**, 4. Lief. 2. Aufl., 97–440; with 131 woodcuts.

ZSCHOKKE, F. Erster Beitrag zur Parasitenfauna von *Trutta salar*. *Verh. Naturf. Ges. Basel*, **8**, III, 761–95.

BLANCHARD, R. *Traité de Zoologie médicale*, **1** (Paris, 808 pp.).

LINSTOW, O. v. *Compendium der Helminthologie, Nachtrag. Die Litteratur der Jahre* 1878–1889 (Hannover, 151 pp.).

MONTICELLI, F. S. Notes on some entozoa in the collection of the British Museum. *Proc. Zool. Soc. Lond.* **3**, 321–5.

PARONA, C. & PERUGIA, A. Di alcuni trematodi ectoparassiti di pesci marini nota preventiva (Res. ligusticae 8). *Ann. Mus. Stor. nat. Genova*, Ser. 2, **7**, 740–7.

HECKERT, A. *Untersuchungen über die Entwicklungs- und Lebensgeschichte des* Distomum macrostomum (Cassel, 66 pp., Bibliotheca zool. Leuckart und Chun, Heft 4).

LOOSS, A. *Ueber Degenerations-Erscheinungen im Thierreich, besonders über die Reduktion des Froschlarvalschwanzes und die im Verlauf derselben auftretenden histolytischen Processe* (Leipzig).

STOSSICH, M. *Distomi degli Anfibi*. Lavoro monografico (Trieste, 14 pp.).

(1890) HUBER, J. C. Zur Litteraturgeschichte der Leberegelkrankheit. *Dtsch. Z. Thiermed. vergl. Pathol.* **17**, 77–9.

SONSINO, P. Studie notizie elmintologiche. *Mem. Soc. tosc. Sci. nat.* **6**, 273–85; also *Mem. Soc. tosc. Sci. nat.* **7**, 99–114.

GOTO, S. On *Diplozoon nipponicum* n.sp. *J. Coll. Sci. Tokyo*, **4**, 151–92.

LINSTOW, O. v. Ueber den Bau und die Entwicklung des *Distomum cylindraceum* Zed. *Arch. mikr. Anat.* **36**, 173–91.

BRANDES, C. Die Familie der Holostomiden. *Zool. Jb., Syst.* **5**, 549–604.

CREUTZBURG, N. *Untersuchungen über den Bau und die Entwicklung von* Distomum ovocaudatum *Vulp.* (Diss. Leipzig, 33 pp.).

CUNNINGHAM, J. T. *A treatise on the Common Sole*. Plymouth, 148 pp.

MONTICELLI, F. S. Note elmintologiche. *Boll. Soc. Nat. Napoli*, Ann. 4, fasc. 2, 189–208.

(*a*) PARONA, C. & PERUGIA, A. On *Gastrocotyle*. *Atti Soc. ligust. Sci. nat. geogr.* **1**, no. 3, 225–42.

(*b*) PARONA, C. & PERUGIA, A. Nuove osservazioni sull' *Amphibdella torpedinis* Chatin. *Ann. Mus. Stor. nat. Genova*, Ser. 2, **9**, 363–7.

PERUGIA, A. & PARONA, C. Di alcuni trematodi ectoparassiti di pesci adriatici. *Ann. Mus. Stor. nat. Genova*, Ser. 2*a*, **9** (19), 16–32.

ST REMY, G. Synopsis des Trématodes monogénèses. *Rev. Biol. Nord. France*, **3**, 405–16, 449–57.

ZSCHOKKE, F. Erster Beitrag zur Parasitenfauna von *Trutta salar*. *Verh. naturf. Ges. Basel*, **8**, 761–95.

(1891) BRAUN, M. Verzeichniss von Eingeweidewürmern aus Mecklenburg. *Arch. Ver. Frde. Naturg. Meckl.* pp. 97–117.

BELL, F. J. Description of a new species of *Tristomum* from *Histiophorus brevirostris*. *Ann. Mag. Nat. hist.* (7), **7**, 534–5.

FRÖLICH, J. A. Beyträge zur Naturgeschichte der Eingeweidewürmer. *Abh. naturf. Ges. Halle*, **25**, 52–113.

(1891–2) ST REMY, G. Synopsis des Trématodes monogénèses. *Rev. Biol. Nord. France*, **4**, 1–21, 90–107.

(1892) BRANDES, G. Zum feineren Bau der Trematoden. *Z. wiss. Zool.* **53**, 558–77.

BRANDES, G. Revision der Monostomiden. *Zbl. Bakt.* I. Orig. **12**, 504–11.

LOOSS, A. *Schmarotzertum in der Tierwelt* (Leipzig: R. Freese).

LOOSS, A. Ueber *Amphistomum subclavatum* Rud. und seine Entwicklung. *Festschr. z. 70. Geburtstage R. Leuckarts, Leipzig*, pp. 147–67.

CUÉNOT, L. Commensaux et parasites des Echinodermes. *Rev. Biol. Nord. France*, **5**, pp. 1–23.

SONSINO, P. Studi sui parassiti molluschi di aqua dolce dintorni di Cairo in Egitto. *Festschr. z. 70. Geburtstage R. Leuckarts, Leipzig*, pp. 134–47.

Lutz, A. Zur Lebensgeschichte des *Distoma hepaticum*. *Zbl. Bakt.* I. Orig. **11**, 783–96.

Monticelli, F. S. *Cotylogaster Michaelis* n.g., n.sp. e revisione degli Aspidobothridae. *Festschr. z.* 70. *Geburtstage R. Leuckarts, Leipzig*, pp. 168–214.

Monticelli, F. S. Dei Monostomum del *Box salpa*. *Atti Accad. Sci. Torino*, p. 271.

Lang, A. Ueber die Cercariae von *Amphistomum subclavatum*. *Ber. naturf. Ges. Freiburg i. B.* **6**, p. 81.

Parona, C. & Perugia, A. Note sopra trematodi ectoparassiti (Res. ligusticae 17). *Ann. Mus. Stor. nat. Genova*, Ser. 2, **12**, 86–102.

(1893) Kowalewski, M. Studya helminthologiczne. III. *Bilharzia polonica* n.sp. *Rozpr. Wydz. mat.-przy. Ak. Um.* (1), **31**, 146.

Monticelli, F. S. Studii sui trematodi endoparassiti, etc. *Zool. Jb.* Suppl. **3**, Fischer, Jena, 229 pp.

Walter, E. Untersuchungen über den Bau der Trematoden. *Z. wiss. Zool.* **56**, 51 pp.

Linstow, O. v. Helminthologische Studien. *Jena. Z. Naturw.* **28**, 328–52.

Braun, M. Platyhelminthes. II. Bericht über thierische Parasiten. Trematodes. *Zbl. Bakt.* (1) **13**, 176–90.

(1894) Goto, S. Studies on the ectoparasitic trematodes of Japan. *J. Coll. Sci. Tokyo*, **8**, 273 pp.

Braun, M. Zur Entwicklungsgeschichte der Holostomiden. *Zool. Anz.* **17**, 165; *Zbl. Bakt.* **15**, 680.

Cerfontaine, P. Sur un nouveau tristomien, *Merizocotyle diaphanum* (n.g., n.sp.). *Bull. Acad. Belg. Cl. Sci.* **64**, 936–48.

Looss, A. Die Distomen unserer Fische und Frösche. Neue Untersuchungen über Bau und Entwicklung des Distomenkörpers. *Biblioth. Zool.* Part 16, 293 pp.

Looss, A. *Uber den Bau der* Distomum heterophyes *v. Sieb. und* Distomum fraternum *n.sp.* (Kassel, Fischer, 59 pp.).

Stiles, C. W. The anatomy of the large American fluke *Fasciola magna*, and a comparison with other species of the genus. *J. Comp. Med.* **15**, 161, 225, 299.

(1895) Kathariner, L. Die Gattung *Gyrodactylus* v. Nordm. *Arb. Zool.-zootom. Inst. Würzburg*, **10**, 125–64.

Coe, W. R. Notizen über den Bau des Embryos von *Distomum hepaticum*. *Zool. Jb., Anat.* **9**, 561–70.

Stafford, J. Anatomical structure of *Aspidogaster conchicola*. *Zool. Jb., Anat.* **9**, 477–542.

Parona, C. & Perugia, A. Sopra due nuove specie di trematodi ectoparassiti di pesci marini (*Phylline monticellii* e *Placunella vallei*). *Atti Soc. ligust. Sci. nat. geogr.* **6**, 84–7.

Pelseneer, P. Un trématode produisant la castration parasitaire chez *Donax truncatulus*. *Bull. sci. Fr. Belg.* **27**, 357–63.

(1896) Looss, A. Recherches sur la faune parasitaire de l'Égypt. I. *Mém. Inst. Égypt*, **3**, 296 pp.

Kowalewski, M. Études helminthologiques. III. *Bull. Acad. Sci. Cracovie*, pp. 63–72, 146.

Kowalewski, M. Studya helmintologiczne. IV. *Bilharzia polonica* sp.nov. *Rozpr. Akad. Kraków*, **30**, 145–8, 265–6, 345–56.

Cerfontaine, P. Contribution à l'étude des Octocotylides. I–III. *Arch. Biol.* **14**, 1895–6, 497–560.

Jägerskiöld, L. A. Ueber *Monostomum lacteum* n.sp. *Festskr. Lilljeborg*, pp. 167–77.

Mühling, P. Beiträge zur Kenntnis der Trematoden. *Arch. Naturgesch.* **62**, 243–79.

Olsson, P. J. Sur *Chimaera monstrosa* et ses parasites. *Mém. Soc. Zool., Paris*, **9**, 499–512.

Gamble, F. W. Platyhelminthes and Mesozoa. In *Cambridge Natural History*, **2**, 1–96.

(1897) Bettendorf, H. Über Muskulatur und Sinneszellen der Trematoden. *Zool. Jb., Anat.* **10**, 307–58.

Hausman, L. Ueber Trematoden der Süsswasserfishe. *Rev. suisse Zool.* **5**, 1–42.

(1898) St Remy, G. Complement du Synopsis des Trématodes monogénèses. *Arch. Parasitol.* **1**, 521–71.

Cerfontaine, P. Contribution à l'étude des Octocotylides. IV. *Arch. Biol.* **15**, 301–28.

Cerfontaine, P. Le genre *Merizocotyle* (Cerf.). *Arch. Biol.* **15**, 329–66.

Stossich, M. *Saggio di una fauna elmintologica di Trieste e provincie contermini* (Trieste, pp. 63–5) (*Program. civ. scuola r. sup.* 162 pp.).

Linton, E. Notes on trematode parasites of fishes. *Proc. U.S. Nat. Mus.* **20**, 507–48.

Jägerskiöld, L. A. *Distoma lingua* Creplin, ein genitalnapftragendes Distomida. *Bergens Mus. Aarborg*, No. 11, 7 pp.

Pratt, H. S. A contribution to the life-history and anatomy of the Appendiculate Distomes. *Zool. Jb., Anat.* **11**, Heft 3.

Mühling, P. Die Helminthenfauna der Wirbeltiere Ostpreussens. *Arch. Naturgesch.* **64**, 1–118.

Brandes, G. Die Gattung *Gastrothylax*. *Abh. naturf. Ges. Halle*, **26**, 33 pp.

(1899) Cerfontaine, P. Les Onchocotylinae. *Arch. Biol.* **16**, 345–478.

Brown, A. W. On *Tetracotyle metromyzontis*, a parasite of the brain of *Ammocoetes*. *Quart. J. Micr. Soc.* **41**, 489.

Goto, S. Notes on some exotic species of ectoparasitic trematodes. *J. Coll. Sci. Tokyo*, **12**, 263–95.

Hofmann, K. Beiträge zur Kenntniss der Entwicklung von *Distomum leptosomum* Olsson. *Zool. Jb., Syst.* **12**, 174–204.

Looss, A. Weitere Beiträge zur Kenntniss der Trematoden-Fauna Aegyptens, zugleich Versuch einer natürlichen Gliederung des Genus *Distomum* Retzius. *Zool. Jb., Syst.* **12**, 521–784.

Monticelli, F. S. Di una nuova specie del genere *Plectanocotyle*. *Atti Accad. Sci. Torino*, **34**, 1045–53.

Jacoby, S. Beiträge zur Kenntnis einiger Distomen. *Arch. Naturg.* Year 66 (1900), **1**, 1–30.

D. 1900 *onwards* (*alphabetical arrangement*)

Abernathy, C. (1937). Notes on *Crepidostomum cornutum* (Osborn). *Trans. Amer. Micr. Soc.* **56**, 206–7.

Ackert, J. E. (1942). Natural resistance to helminthic infections. *J. Parasitol.* **28**, 1–24.

Africa, C. M. (1930). *Pleurogenes loossi* sp.nov. from the small intestine of water frogs (*Rana esculenta*). *Zbl. Bakt.* (i), **115**, 448–51.

Agersborg, H. P. K. (1924). Studies on the effects of parasitism upon the tissues. I. With special reference to certain gastropod molluscs. *Quart. J. Micr. Soc.* **68**, 361–401.

Allison, L. N. (1943). *Leucochloridiomorpha constantiae* (Mueller) (Brachylaemidae), its life cycle and taxonomic relationships among digenetic trematodes. *Trans. Amer. Micr. Soc.* **62**, 127–68.

Almarza, N. (1935). Die Leberegel des Schafes. Beschreibung neurer Arten. *Z. InfektKr. Haustiere*, **47**, 195–202.

Alvey, C. H. (1936). The morphology and development of the monogenetic trematode *Sphyranura oligorchis* (Alvey, 1933) and the description of *Sphyranura polyorchis* n.sp. *Parasitology*, **28**, 229–53.

Ameel, D. J. (1934). *Paragonimus*, its life history and distribution in North America and its taxonomy (Trematoda: Troglotrematidae). *Amer. J. Hyg.* **19**, 279–317.

Ameel, D. J. (1937). The life history of *Crepidostomum cornutum* (Osborn). *J. Parasit.* **23**, 218–20.

Ameel, D. J. (1938). The morphology and life cycle of *Euryhelmis monorchis* n.sp. (Trematoda) from the mink. *J. Parasit.* **24**, 219–24.

ANDERSON, M. G. (1935). Gametogenesis in the primary generation of a digenetic trematode, *Proterometra macrostoma* Horsfall, 1933. *Trans. Amer. Micr. Soc.* **54**, 271–9.

ANDRÉ, J. (1910). Zur Morphologie des Nervensystems von *Polystomum integerrimum* Froel. *Z. wiss. Zool.* **95**, 191–220.

ARIOLA, V. (1902). Contributo per una monografia dei *Didymozoon*. I. *Didymozoon* parassite del Tonno. *Arch. Parasit., Paris*, **6**, 99–108.

ASHWORTH, A. W. & BANNERMANN, I. C. W. (1927). On a *Tetracotyle* (*T. phoxini*) in the brain of the minnow. *Trans. Roy. Soc. Edinb.* **55**, 159.

AZIM, M. A. (1933). On *Prohemistomum vivax* (Sonsino, 1892) and its development from *Cercaria vivax* Sonsino, 1892. *Z. Parasitenk.* **5**, 432–6.

AZIM, M. A. (1936). On the life history of *Lecithodendrium pyramidum* Looss, 1896, and its development from a xiphidiocercaria, *C. pyramidum* sp.nov., from *Melania tuberculata*. *Ann. Trop. Med. Parasit.* **30**, 351–4.

BABASKIN, A. (1928). Die Trematoden des Amurlachses (Keta), *Brachyphallus amuriensis* n.sp. *Zbl. Bakt.* (ii), **75**, 213–18.

BAER, J. G. (1924). Description of a new genus of Lepodermatidae (Trematoda) with a systematic essay on the family. *Parasitology*, **16**, 22–31.

BAER, J. G. (1928). Contribution à la faune helminthologique de Suisse. *Rev. Suisse Zool.* **35**, 27–41.

BAER, J. G. (1931*a*). Un nouveau genre de trématode provoquant des lésions dans le rein de la Taupe (Note préliminaire). *Verh. schweiz. naturf. Ges.* **112**, 337–8.

BAER, J. G. (1931*b*). Helminthes nouveaux parasites de la musaraigne d'eau, *Neomys fodiens* Pall (Note préliminaire). *Verh. schweiz. naturf. Ges.* **112**, 338–40.

BAER, J. G. (1932). Contribution à la faune helminthologique de Suisse (Deuxième partie). *Rev. suisse Zool.* **39**, 1–58.

BAER, J. G. (1933). Note sur un nouveau trématode, *Clinostomum lophophallum* sp.nov., avec quelques considérations générales sur la famille des Clinostomidae. *Rev. suisse Zool.* **40**, 317–42.

BAKER, F. C. (1922). Fluke infections and the destruction of the intermediate host. *J. Parasit.* **8**, 145.

BALOZET, L. & CALLOT, J. (1938). Trématodes de Tunisia. *Arch. Inst. Pasteur*, **27**, 18–29.

BARKER, F. D. (1911). The trematode genus *Opisthorchis*. *Arch. Parasit., Paris*, **14**, 513–61.

BARLOW, C. H. (1925). Life cycle of *Fasciolopsis buski* (human) in China. *Chin. Med. J.* **37**, 453–72. See also *Amer. J. Hyg.* Monogr. Ser. **4**, 98 pp.

BAUDET, E. A. R. F. (1929). *Tracheophilus sisowi* Skrjabin. *Tijdschr. Diergeneesk.* Jahrg. 10; quoted by Szidat, *Zool. Anz.* (1932).

BAYLIS, H. A. (1922). Notes on the collection and preservation of parasitic worms. *Parasitology*, **14**, 402–8.

BAYLIS, H. A. (1926). On the correct use of host-names. An appeal to parasitologists. *Parasitology*, **18**, 203–5.

BAYLIS, H. A. (1927). Notes on three little-known trematodes. *Ann. Mag. Nat. Hist.* (9), **19**, 426–33.

BAYLIS, H. A. (1928*a*). A new species of *Notocotylus* (Trematoda) with some remarks on the genus. *Ann. Mag. Nat. Hist.* (10), **2**, 582–5.

BAYLIS, H. A. (1928*b*). Records of some parasitic worms from British vertebrates. *Ann. Mag. Nat. Hist.* (10), **1**, 329–43.

BAYLIS, H. A. (1929). *A Manual of Helminthology, Medical and Veterinary.* London.

BAYLIS, H. A. (1931). The names of some molluscan hosts of the Schistosomes parasitic in man. *Ann. Trop. Med. Parasit.* **25**, 369–72.

BAYLIS, H. A. (1932*a*). What is *Psilochasmus lecithosus* Otte? *Ann. Mag. Nat. Hist.* (10), **9**, 124–5.

BAYLIS, H. A. (1932*b*). A list of worms parasitic in Cetacea. *Discovery Rep.* **6**, 393–418.

BAYLIS, H. A. (1938*a*). On two species of the trematode genus *Didymozoon* from the mackerel. *J. Mar. Biol. Ass. U.K.* N.S. **22**, 485–92.

BAYLIS, H. A. (1938*b*). Helminths and evolution. In *Evolution*, ed. de Beer, Oxford, 249–70.

BAYLIS, H. A. (1939*a*). Further records of parasitic worms from British vertebrates. *Ann. Mag. Nat. Hist.* (11), **4**, 473–98.

BAYLIS, H. A. (1939*b*). A larval trematode (*Diplostomum volvens*) in the lens of the eye of a rainbow trout. *Proc. Linn. Soc. Lond.* **151**, 130.

BAYLIS, H. A. & CANNON, H. G. (1924*a*). A new trematode from the grass-snake. *Ann. Mag. Nat. Hist.* (9), **13**, 194–9.

BAYLIS, H. A. & CANNON, H. G. (1924*b*). Further note on a new trematode from the grass-snake. *Ann. Mag. Nat. Hist.* (9), **13**, 558–9.

BAYLIS, H. A. & IDRIS JONES, E. (1932–3). Records of parasitic worms from marine fishes at Plymouth. *J. Mar. Biol. Ass. U.K.* N.S. **18**, 627–34.

BAYLIS, H. A. & MONRO, C. C. B. (1941). *Instructions for Collectors. No.* 9A. *Invertebrate Animals other than Insects.* (73 pp. + 12 plates.) London: Brit. Mus. Nat. Hist.

BEAUCHAMP, P. DE (1912). *Isancistrum loliginis* n.g., n.sp. trématode parasite du calamar. *Bull. Zool. Soc. Fr.* **37**, 96–9.

BEAVER, P. C. (1929). Studies on the development of *Allassostoma parvum* Stunkard. *J. Parasit.* **16**, 13–23.

BEAVER, P. C. (1937*a*). Experiments on regeneration in the trematode *Echinostomum revolutum*. *J. Parasit.* **23**, 423–4.

BEAVER, P. C. (1937*b*). Experimental studies on *E. revolutum* (Froel.), a fluke from birds and mammals. *Illinois Biol. Monogr.* **15**, 1–96.

BEAVER, P. C. (1939). The morphology and life history of *Psilostomum ondatrae* Price, 1932 (Trematoda: Psilostomidae). *J. Parasit.* **25**, 383–93.

BEAVER, P. C. (1941). Studies on the life history of *Euparyphium melis* (Trematoda: Echinostomatidae). *J. Parasit.* **27**, 35–43.

BEAVER, P. C. (1942). *Aequistoma* nom.nov. for *Pseudechinostomum* Shchupakov, 1936, preoccupied by *Pseudechinostomum* Odhner, 1911. *Proc. Helminth. Soc. Wash.* **9**, no. 1, p. 31.

BENHAM, W. B. (1901). *The Platyhelmia, Mesozoa and Nemertini.* In Oxford Treatise, Part, 4, 1–195.

BENNETT, H. J. (1936). The life history of *Cotylophoron cotylophorum*, a trematode from ruminants. *Ill. Biol. Monogr.* **14**, no. 4, 119 pp.

BENNETT, H. J. & HUMES, A. G. (1939). Studies on the pre-cercarial development of *Stichorchis subtriquetrus* (Trematoda: Paramphistomidae). *J. Parasit.* **25**, 223–31.

BENNETT, H. J. & TOBIE, J. E. (1936). New records of the prevalence and distribution of some Telorchiinae from *Pseudemys elegans* Wied. *Proc. Helminth. Soc. Wash.* **3**, 62–3.

BERGHE, L. VAN DEN (1937). A morphological study of bovine Schistosomes. *J. Helminth.* **15**, 125–32.

BEVELANDER, G. (1933*a*). Response to light in the cercariae of *Bucephalus elegans*. *Physiol. Zoöl.* **6**, 289–305.

BEVELANDER, G. (1933*b*). The relation between temperature and frequency of contraction in the tail-furcae of *Bucephalus elegans*. *Physiol. Zoöl.* **6**, 509–20.

BHALERAO, G. D. (1936*a*). Studies on the helminths of India. Trematoda. I. *J. Helminth.* **14**, 163–80.

BHALERAO, G. D. (1936*b*). Studies on the helminths of India. Trematoda. II. *J. Helminth.* **14**, 181–206.

BHALERAO, G. D. (1936*c*). Studies on the helminths of India. Trematoda. III. *J. Helminth.* **14**, 207–28.

BHALERAO, G. D. (1937*a*). Studies on the helminths of India. Trematoda. IV. *J. Helminth.* **15**, 97–124.

BHALERAO, G. D. (1937*b*). On *Pneumotrema travassosi* gen. et sp.n., and two other trematode parasites from the animals dying in the Zoological Society's Garden during 1936–7. *Proc. Zool. Soc. Lond.* B, **107**, 365–9.

BITTNER, H. (1923). *Schistogonimus rarus* (Braun), ein seltener Trematode in der Bursa Fabricii einer an *Tetrameres*-Invasion gestorben Hausente. *Arch. wiss. prakt. Tierheilk.* **50**, 253–61.

BITTNER, H. (1927). Parasitologische Beobachtungen. *Z. InfektKr. Haustiere*, **30**, 213–27.

BITTNER, H. & SPREHN, C. (1928). Trematodes. In P. Schulze, *Biologie der Tiere Deutschlands*, Lief. 27, Teil 5, 1–133. Berlin.

BONHAM, K. & GUBERLET, J. E. (1938). Ectoparasitic trematodes of Puget Sound Fishes. *Acanthocotyle*. *Amer. Midl. Nat.* **20**, 590–602.

BONNE, C. (1941*a*). Zoetwatermosselen en echinostomiasis. *Natuurk. Tijdschr. Ned.-Ind.* **101**, 176–9.

BONNE, C. (1941*b*). Vier echinostomen van den mensch in Nederlandsch-Indie. *Geneesk. Tijdschr. Ned.-Ind.* **81** (25), 1343–57.

BOSMA, N. J. (1934). The life history of the trematode *Alaria mustelae* Bosma, 1931. *Trans. Amer. Micr. Soc.* **53**, 116–53.

BRACKETT, S. (1940). Pathology of schistosome dermatitis. *Arch. Derm. Syph., Chicago*, **42** (3), 410–18.

BRACKETT, S. (1942). Five new species of avian Schistosomes from Wisconsin and Michigan, with life cycle of *Gigantobilharzia gyrauli* (Brackett, 1940). *J. Parasit.* **28**, 25–42.

BRAND, T. v. & WEISE, W. (1932). Beobachtungen über den Sauerstoffgehalt der Umwelt einiger Entoparasiten. *Z. vergl. Physiol.* **18**, 339–46.

BRAUN, M. (1900). Trematoden der Chiropteren. *Ann. naturh. (Mus.) Hofmus. Wien*, **15**, 217–36.

BRAUN, M. (1901*a*). Die Arten der Gattung *Clinostomum* Leidy. *Zool. Jb., Syst.* **14**, 1–48.

BRAUN, M. (1901*b*). Zur Revision der Trematoden der Vögel. I. *Zbl. Bakt.* **29**, 560–8.

BRAUN, M. (1901*c*). Zur Revision der Trematoden der Vögel. II. *Zbl. Bakt.* **29**, 896–7, 941–8.

BRAUN, M. (1901*d*). Zur Kenntnis der Trematoden der Säugethiere. *Zool. Jb., Syst.* **14**, 311–48.

BRAUN, M. (1901*e*). Trematoden der Chelonier. *Mitt. zool. Mus. Berl.* **2**, 7–58.

BRAUN, M. (1902*a*). Fascioliden der Vögel. *Zool. Jb., Syst.* **16**, 1–162, 411–894.

BRAUN, M. (1902*b*). Ueber *Distomum goliath* P. J. v. Ben. 1858. *Zbl. Bakt.* **32**, 800–3.

BRINKMANN, A. JR. (1940). Contribution to our knowledge of the Monogenetic Trematodes. *Bergens Mus. Årbok*, 1939–40, *Naturvitensk*, no. 1, 117 pp., 58 figs.

BRINKMANN, A. JR. (1942*a*). On some new and little-known *Dactylocotyle* species, with a discussion on the relations between the genus *Dactylocotyle* and the 'family' Diclidophoridae. *Medd. Göteborgs Musei Zool. Andeln. Göteborgs Kungl. Vetensk.- och Vitterhets-Samhälles Handlingar Sjätte Följden*, Ser. B, **1**, no. 13, 1–32.

BRINKMANN, A. JR. (1942*b*). On '*Octobothrium*' *leptogaster* F. S. Leuckart. *Medd. Göteborgs Musei Zool. Andeln. Göteborgs Kungl. Vetensk.- och Vitterhets-Samhälles Handlingar Sjätte Följden*, Ser. B, **2**, no. 3, 1–29.

BRINKMANN, A. JR. (1942*c*). A new trematode, *Pachytrema paniceum* n.sp., from the gall-bladder of the Lesser Black-backed Gull (*Larus fuscus* L.). *Medd. Göteborgs Musei Zool. Andeln. Göteborgs Kungl. Vetensk.- och Vitterhets-Samhälles Handlingar Sjätte Följden*, Ser. B, **2**, no. 2, 1–19.

BROOKS, F. G. (1928). The germ-cell cycle of trematodes. *Science*, **68**.

BROOKS, F. G. (1930). Studies on the germ-cell cycles of trematodes. *Amer. J. Hyg.* **12**, 299–340.

BROOKS, G. L. (1934). Some new ectoparasitic trematodes (Onchocotylinae) from the gills of American sharks. *Parasitology*, **26**, 259–67.

BROWN, E. M. (1929). On a new species of monogenetic trematode from the gills of *Pagellus centrodontus*. *Proc. Zool. Soc. Lond.* pp. 67–83.

BROWN, F. J. (1926). Some fresh-water larval trematodes, with contributions to their life histories. *Parasitology*, **18**, 21–34.

BROWN, F. J. (1927). On *Crepidostomum farionis* O. F. Müll. (= *Stephanophiala laureata* Zeder), a distome parasite of the trout and grayling. I. The life history. *Parasitology*, **19**, 86–99.

BROWN, F. J. (1931). Some fresh-water larval trematodes from Cheshire. *Parasitology*, **23**, 88–98.

BROWN, F. J. (1933). On the excretory system and life history of *Lecithodendrium* (Mehl.) and other bat trematodes, with a note on the life history of *Dicrocoelium dendriticum* (Rud.). *Parasitology*, **25**, 317–28.

BRUMPT, E. (1931). *Cerc. ocellata* déterminant la dermatite des nageurs, provient d'une bilharzie des canards. *C.R. Acad. Sci., Paris*, **193**, 612.

BRUMPT, E. (1936). *Précis de Parasitologie*. 2139 pp. Paris.

Brumpt, E. (1941). Observations biòlogiques diverses concernant *Planorbis* (*Australorbis*) *glabratus*, hôte intermédiare de *Schistosoma mansoni*. *Ann. Parasit. hum. comp.* **18**, 9–45.

Butler, J. B. & Humphries, A. (1932). On the cultivation in artificial media of *Catenaria anguillulae*, a Chytridiacean parasite of the ova of the liver fluke, *Fasciola hepatica*. *Sci. Proc. R. Dublin Soc.* N.S. **20**, 301–24.

Bychowsky, B. J. (1929). *Distomum kessleri* Grebnizsky, 1872, seine systematische Stellung und Synonymik. (In Russian). *Russ. hydrobiol. Z. Saratow*, **8**, 321–4.

Bychowski, B. J. (1931). Neue *Dactylogyrus*-Arten aus dem Aral See. *Zool. Anz.* **95**, 233–40.

Bychowsky, B. J. (1932*a*). Die russischen *Pneumonoeces*-Arten und ihre geographische Verbreitung. *Z. Parasitenk.* **5**, 51–68.

Bychowsky, B. J. (1932*b*). Die Amphibientrematoden aus der Umgegend von Kiew. (In Russian with German summary.) *J. Cycle Bio-Zool. Kiew*, **3**, 23–35, 35–8.

Bychowsky, B. J. (1933*a*). Die Amphibientrematoden aus der Umgegend von Kiew. *Zool. Anz.* **102**, 44–58.

Bychowsky, B. J. (1933*b*). Beitrag zur Kenntnis neuer monogenetischer Fischtrematoden aus dem Kaspisee nebst einigen Bemerkungen über die systematik der Monopisthodiscinea Fuhrmann. *Zool. Anz.* **105**, 17–38.

Bychowsky, I. & Bychowsky, B. (1934). Über die Morphologie und die Systematik des *Aspidogaster limacoides* Diesing. *Z. Parasitenk.* **7**, 125–37.

Byrd, E. E. (1941). The excretory system of *Paragonimus*. *J. Parasit.* Suppl. Abstr. **17**, p. 101.

Cable, R. M. (1931). Studies on the germ-cell cycle of *Cryptocotyle lingua*. I. Gametogenesis in the adult. *Quart. J. Micr. Sci.* **74**, 563–89.

Cable, R. M. (1934). Studies on the germ-cell cycle of *Cryptocotyle lingua*. II. Germinal development in the larval stages. *Quart. J. Micr. Sci.* **76**, 573–614.

Cable, R. M. (1937). The resistance of the herring gull, *Larus argentatus*, to experimental infections of the trematode *Parorchis acanthus*. *J. Parasit.* **23**, 559.

Cable, R. M. (1938). Studies on larval trematodes from Kentucky, with a summary of known related species. *Amer. Midl. Nat.* **19**, 440–64.

Cable, R. M. & Hunninen, A. V. (1938). Observations on the life history of *Spelotrema nicolli* n.sp. (Trematoda: Microphallidae), with the description of a new microphallid cercaria. *J. Parasit.* **24**, Suppl. Abstr. pp. 29–30.

Cable, R. M. & Hunninen, A. V. (1940). Studies on the life history of *Spelotrema nicolli* (Trematoda: Microphallidae) with the description of a new microphallid cercaria. *Biol. Bull. Woods Hole*, **78**, 136–57.

Cable, R. M. & Hunninen, A. V. (1941*a*). Studies on the life history of *Siphodera vinaledwardsii* (Linton) (Trematoda: Cryptogonimidae). *J. Parasit.* **27**, Suppl. Abstr. p. 13.

Cable, R. M. & Hunninen, A. V. (1941*b*). The systematic position of the genus *Deropristis* Odhner, with respect to a proposed revision of the Trematode families Acanthocolpidae and Allocreadiidae. *J. Parasit.* **27**, Suppl. Abstr. p. 14.

Cable, R. M. & Hunninen, A. V. (1942*a*). Studies on *Deropristis inflata* (Molin); its life history and affinities to trematodes of the family Acanthocolpidae. *Biol. Bull. Woods Hole*, **82**, 292–312.

Cable, R. M. & Hunninen, A. V. (1942*b*). Studies on the life history of *Siphodera vinaledwardsii* (Linton) (Trematoda: Cryptogonimidae). *J. Parasit.* **28**, 407–22.

Cameron, T. W. M. (1931). Experimental infection of sheep with *Dicrocoelium dendriticum*. *J. Parasit.* **9**, 41–4.

Cameron, T. W. M. (1935). Immunity against animal parasites. *12th Internat. Vet. Congr.* **3**, 44–65.

Cameron, T. W. M. (1939). Parasitism and its importance. *Canad. J. Comp. Med.* **3**, 175–81.

Carrère, P. (1936). Sur le cycle évolutif d'un *Maritrema* (Trématodes). *C.R. Acad. Sci., Paris*, **202**, 244–6.

Carrère, P. (1937). Sur quelques Trématodes des poissons de la Camargue. *C.R. Acad. Sci., Paris*, **125**, 158–60.

CARY, L. R. (1909). The life history of *Diplodiscus temperatus* Stafford with especial reference to the development of the parthenogenetic eggs. *Zool. Jb., Anat.* **28**, 595–659.

CAWSTON, F. G. (1921*a*). Some South African cercariae. *S. Afr. J. Nat. Hist.* **3**, 199–204.

CAWSTON, F. G. (1921*b*). South African larval trematodes and their intermediate hosts. *Trans. Amer. Micr. Soc.* **11**, 119–30.

CAWSTON, F. G. (1922). Some notes on the differentiation of closely allied Schistosomes. *Parasitology*, **14**, 245–7.

CAWSTON, F. G. (1941). A consideration of the resistance of molluscs and mammals to the attacks of larval trematodes. *J. Trop. Med.* (*Hyg.*), **44**, 8.

CHANDLER, A. (1936). *Introduction to Human Parasitology*, 5th ed. 661 pp. New York and London.

CHEN, H. T. (1936). A study of the Haplorchinae (Looss, 1899) Poche, 1926 (Trematoda: Heterophyidae). *Parasitology*, **28**, 40–55.

CHEN, PIN-DJI (1937). The germ-cell cycle in the trematode *Paragonimus kellicotti* Ward. *Trans. Amer. Micr. Soc.* **56**, 208–36.

CHU, H. J. (1940*a*). Studies on *Clonorchis sinensis* in vitro. Part III. Survival period in relation to certain dyes. *Chin. Med. J.* Suppl. **3**, 255–9.

CHU, H. J. (1940*b*). Studies on *Clonorchis sinensis* in vitro. Part IV. Combined effect of gentian violet and X-rays. *Chin. Med. J.* Suppl. **3**, 260–6.

CIUREA, J. (1914). Recherches sur la source d'infection de l'homme et des animaux par les distomes de la famille des opisthorchidées. *Bull. Sect. sci. Acad. roum.* **2**, 201–5.

CIUREA, J. (1917). Die Auffindung der Larven von *Opisthorchis felineus*, *Pseudamphistomum danubiense* und *Metorchis albidus*, etc. *Z. InfektKr. Haustiere*, **18**, 301–33, 345–57.

CIUREA, J. (1924). Hétérophyides de la faune parasitaire de Roumanie. *Parasitology*, **16**, 1–21.

CIUREA, J. (1933*a*). Les Vers parasites de l'homme, des mammifères et des oiseaux provenant des poissons du Danube et de la mer noire. I[e] Mém. *Arch. Roum. Path. exp. Microbiol.* **6**, no. 2, 5–134.

CIUREA, J. (1933*b*). Sur quelques larves des Vers parasites de l'homme, des mammifères et des oiseaux ichtyophages, trouvés chez les poissons des grands lacs de la Bessarabie, du Dniester et de son liman. *Arch. Roum. Path. exp. Microbiol.* **6**, no. 2, 161–70.

CLAPHAM, P. A. (1935). Some helminth parasites from partridges and other English birds. *J. Helminth.* **13**, 139–48.

CLAPHAM, P. A. (1936). Further observations on the occurrence and incidence of helminths in British partridges. *J. Helminth.* **14**, 61–8.

CLAPHAM, P. A. (1938). New records of helminths in British birds. *J. Helminth.* **16**, 47–8.

COHN, L. (1903). Zur Kenntnis einiger Trematoden. *Zbl. Bakt.* **34**, 35–42.

COLE, H. A. (1935). On some larval trematode parasites of the mussel (*Mytilus edulis*) and the cockle (*Cardium edule*). *Parasitology*, **27**, 276–80.

COLE, H. A. (1938). On some larval trematode parasites of the mussel (*Mytilus edulis*) and the cockle (*Cardium edule*). Part II. A new larval *Gymnophallus* (*Cercaria cambrensis*) sp.nov. from the cockle. *Parasitology*, **30**, 40–3.

CONYNGHAM, H. (1904). A new trematode of man (*Amphistoma watsoni*). *Brit. Med. J.* **2**, 663.

COOPER, A. R. (1915). Trematodes from marine and fresh-water fishes, including one species of ectoparasitic turbellarian. *Trans. Roy. Soc. Can.* Sect. 4, Ser. 3, **9**, 181–205.

CORRINGTON, J. D. (1940). Frog flukes. *Nature Mag.* **33**, 310–11.

CORT, W. W. (1914). Larval trematodes from North American fresh-water snails. *J. Parasit.* **1**, 65–84.

CORT, W. W. (1915*a*). Some North American larval trematodes. *Ill. Biol. Monogr.* **1**, no. 4, 87 pp.

CORT, W. W. (1915*b*). Egg variation in a trematode species. *J. Parasit.* **2**, 25–6.

CORT, W. W. (1915*c*). *Gordius* larvae parasitic in a trematode. *J. Parasit.* **1**, 198–9.

CORT, W. W. (1917). Homologies of the excretory system of the forked-tailed cercariae. *J. Parasit.* **4**, 48–57.

CORT, W. W. (1921). Sex in the trematode family Schistosomatidae. *Science*, N.S. **53**, 226.

CORT, W. W. (1922). A study ot the escape of cercariae from their snail hosts. *J. Parasit.* **8**, 177–84.

CORT, W. W. (1928). Schistosome dermatitis in the U.S. (Michigan). *J. Amer. Med. Ass.* **90**, 1027. Also, Further observations, etc. *Science*, **68**, 388.

CORT, W. W. (1942). Human factors in parasitic ecology. *Amer. Nat.* **76** (763), 113–28.

CORT, W. W. & BRACKETT, S. (1938). A new strigeid cercaria which produces a bloat disease of tadpoles. *J. Parasit.* **24**, 263–71.

CORT, W. W. & OLIVIER, L. (1941). Early developmental stages of strigeid trematodes in the first intermediate host. *J. Parasit.* **27**, 493–504.

CORT, W. W. & OLIVIER, L. (1943). The development of the sporocysts of a schistosome, *Cercaria stagnicolae* Talbot, 1936. *J. Parasit.* **29**, 164–76.

CORT, W. W., OLIVIER, L. & BRACKETT, S. (1941). The relation of physid and planorbid snails to the life cycle of the strigeid trematode, *Cotylurus flabelliformis* (Faust, 1917). *J. Parasit.* **27**, 437–48.

CRAIG, J. F. & DAVIES, G. O. (1937). *Paramphistomum cervi* in sheep. *Vet. Rec.* **49**, 1116–17. Also *Helm. Abstr.* **6**, 3, p. 109.

CRAIG, C. F. & FAUST, E. C. (1940). *Clinical Parasitology*, 772 pp. London.

CRAM, E. B. (1925). The egg-producing capacity of *Ascaris lumbricoides*. *J. Agric. Res.* **30**, 977–83.

CRAWFORD, W. W. (1937). A further contribution to the life history of *Alloglossidium corti* (Lamont) with especial reference to Dragonfly naiads as second intermediate hosts. *J. Parasit.* **23**, 389–99.

CROFTON, H. D. (1940). A note on *Microcotyle fusiformis* Goto, a fish trematode new to Britain. *Parasitology*, **32**, 318–19.

CROFTON, H. D. (1941). A record of Trematode parasites from *Mola* and *Raniceps raninus* (Linn.). *Parasitology*, **33**, 209–10.

CULBERTSON, J. T. (1941). *Immunity against Animal Parasites*, 274 pp. New York: Columbia Univ. Press.

DADAY, E. v. (1906). In südamerikanischen Fischen lebende Trematodenarten. *Zool. Jb., Syst.* **24**, 185–228.

DARR, A. (1902). Über zwei Fasciolidengattungen. *Z. wiss. Zool.* **71**, 644–701.

DAVIES, E. (1932). On a trematode, *Itygonimus lorum* (Duj., 1845), with notes on the occurrence of other trematode parasites of *Talpa europaea* in the Aberystwyth area. *Parasitology*, **24**, 253–9.

DAVIES, E. (1934). On the anatomy of the trematode *Petasiger exaeretus* Dietz, 1909, from the intestine of *Phalacrocorax carbo*. *Parasitology*, **26**, 133–7.

DAWES, BEN (1936*a*). Sur une tendance probable dans l'évolution des trématodes digénétiques. *Ann. Parasit. hum. comp.* **14**, 177–82.

DAWES, BEN (1936*b*). On a collection of Paramphistomidae from Malaya, with revision of the genera *Paramphistomum* Fischoeder, 1901, and *Gastrothylax* Poirier, 1883. *Parasitology*, **28**, 330–54.

DAWES, BEN (1940*a*). *Hexacotyle extensicauda* n.sp., a monogenetic trematode from the gills of the tunny (*Thunnus thynnus* L.). *Parasitology*, **32**, 271–86.

DAWES, BEN (1940*b*). Notes on the formation of the egg capsules in the monogenetic trematode *Hexacotyle extensicauda* Dawes, 1940. *Parasitology*, **32**, 287–95.

DAWES, BEN (1941*a*). On *Styphlodora elegans* n.sp. and *Styphlodora compactum* n.sp., trematode parasites of *Python reticulatus* in Malaya, with a key to the species of the genus *Styphlodora* Looss, 1899. *Parasitology*, **33**, 445–58.

DAWES, BEN (1941*b*). On *Multicotyle purvisi*, n.g., n.sp., an Aspidogastrid trematode from the river turtle, *Siebenrockiella crassicollis*, in Malaya. *Parasitology*, **33**, 300–5.

DAWES, BEN (1941*c*). Pigmentary changes and the background response in Amphibia. *Nature, Lond.* **147**, 806.

DAWES, BEN (1942*a*). Use of the camera lucida for transcribing diagrams. *Parasitology*, **149**, 140.

DAWES, BEN (1942*b*). A new name for *Styphlodora compactum* Dawes, 1941, a criticism of the proposed new genus *Paurophyllum* Byrd, Parker & Reiber, 1940, and a revised key for the separation of species of the genus *Styphlodora* Looss, 1899. *Parasitology*, **34**, 266–77.

DAWES, BEN (1947). *The Trematoda of British Fishes.* Ray Society, London.

De Beer (1930). *Embryology and Evolution*. Oxford.

De Beer (1940). *Embryos and Ancestors*, 108 pp. Oxford.

Dietz, E. (1910). Die Echinostomiden der Vögel. *Zool. Jb.* Suppl. **12**, 256–512.

Dingler, M. (1910). Über die Spermatogenese des *Dicrocoelium lanceolatum* Stil. et Hass. *Arch. Zellforschg*, **4**, 672–712.

Dinulesco, G. (1939). *Echinoparyphium recurvatum* Linstow. Conditions de son développement larvaire chez *Paludina vivipara* L. *Trav. Stat. zool. Wimereux*, **13**, 215–24.

Dobrovolny, C. G. (1939*a*). Life history of *Plagioporus sinitsîni* Mueller and embryology of new cotylocercous cercariae (Trematoda). *Trans. Amer. Micr. Soc.* **58**, 121–55.

Dobrovolny, C. G. (1939*b*). The life history of *Plagioporus lepomis*, a new trematode from fishes. *J. Parasit.* **25**, 461–70.

Dogiel, V. & Bychowsky, B. (1934). Die Fischparasiten des Aral-Sees. (In Russian.) *Mag. Parasit., Leningr.*, **4**, 241–344. (German summary, pp. 345–6.)

Dollfus, R. P. (1912). Une métacercaire margaritigène parasite de *Donax vittatus* Da Costa. *Mém. Soc. zool. Fr.* **25**, 85–114.

Dollfus, R. P. (1914). *Cercaria pachycerca* Diesing et les cercaires à queue dite 'en moignon'. *IX Congr. internat. Zool. Monaco*, pp. 683–5.

Dollfus, R. P. (1919). Continuité de la lignée des cellules germinales chez les trématodes Digenea. *C.R. Acad. Sci., Paris*, **168**, 124–7.

Dollfus, R. P. (1923). Remarques sur le cycle évolutif des Hémiurides. *Ann. Parasit. hum. comp.* **1**, 345–51.

Dollfus, R. P. (1924). Polyxénie et progénèse de la larve métacercaire de *Pleurogenes medians* Olss. *C.R. Acad. Sci., Paris*, July, p. 28.

Dollfus, R. P. (1925*a*). Liste critique des cercaires marines à queue setigère signalées jusqu'à présent. *Trav. Stat. zool. Wimereux*, **9**, 43–65.

Dollfus, R. P. (1925*b*). Distomiens parasites de Muridae du genre *Mus*. *Ann. Parasit. hum. comp.* **3**, 85–102.

Dollfus, R. P. (1926). Sur l'état actuel de la classification des Didymozoonidae Monticelli, 1888 (= Didymozoidae Franz Poche, 1907). *Ann. Parasit. hum. comp.* **4**, 148–61.

Dollfus, R. P. (1927*a*). Monorchisme accompagné ou non d'anomalies multiples chez les distomes normalement diorchides. *C.R. Soc. Biol., Paris*, **91**, no. 17, pp. 1349–52.

Dollfus, R. P. (1927*b*). Parasitisme chez un pagure d'une larve de distome de tortue. *C.R. Soc. Biol., Paris*, **91**, no. 17, 1352–5.

Dollfus, R. P. (1929*a*). Sur le genre *Telorchis*. *Ann. Parasit. hum. comp.* **7**, 116–32.

Dollfus, R. P. (1929*b*). Existe-t-il des cycles évolutifs abrégés chez les trematodes digénétiques? Le cas de *Ratzia parva* Stossich, 1904. *Ann. Parasit. hum. comp.* **7**, 196–203.

Dollfus, R. P. (1931). Amoenitates helminthologicae. I. A propos de la création de *Lecithodendrium laguncula*, Ch. W. Stiles et M. O. Nolan, 1931. *Ann. Parasit. hum. comp.* **9**, 483–4.

Dollfus, R. P. (1932*a*). Métacercaire progénétique chez un planorbe. *Ann. Parasit. hum. comp.* **10**, 407–13.

Dollfus, R. P. (1932*b*). Trématodes. Résultats scientifiques du voyage aux Indes Orientales Néerlandaises de LL.AA.RR. le Prince et la Princesse Léopold de Belgique. *Mém. Mus. R. Hist. Nat. Belg.* Brussels Hors. Sér. 2 (**10**), 18 pp.

Dollfus, R. P. (1934). Sur quelques *Brachylaemus* de la faune française récoltés principalement à Richelieu (Indre-et-Loire). *Ann. Parasit. hum. comp.* **12**, 6, 551–75.

Dollfus, R. P. (1935*a*). Sur quelques *Brachylaemus* de la faune française récoltés principalement à Richelieu (Indre-et-Loire) (continued). *Ann. Parasit. hum. comp.* **13**, 1, 52–79.

Dollfus, R. P. (1935*b*). Les distomes des Stylommatophores terrestres (excl. Succineidae). Catalogue par hôtes et résumé des descriptions. *Ann. Parasit. hum. comp.* **13**, 5, 176–7, 259–78, 369–85, 445–95.

Dollfus, R. P. (1935*c*). Sur quelques parasites de poissons récoltés à Castiglione (Algérie). *Bull. Sta. Aquic. Pêche Castiglione*, Year 1933, 2nd Fasc. 199–279.

DOLLFUS, R. P. (1935*d*). Sur *Crocodicola* et autres Hémistomes de Crocodiliens. *Arch. Mus. Hist. nat., Paris*, 6th Ser. **12**, 637–46.

DOLLFUS, R. P. (1936*a*). Amoenitates helminthologicae. III. Le rejet du genre *Ostiolum* H. S. Pratt, 1903. *Ann. Parasit. hum. comp.* **14**, 3, 302.

DOLLFUS, R. P. (1936*b*). Présence d'un *Pseudamphistoma* chez la loutre à Richelieu (Indre-et-Loire). *Ann. Parasit. hum. comp.* **14**, 5, 520–2.

DOLLFUS, R. P. (1936*c*). Trématodes de sélaciens et de cheloniens (Parasitologia Mauritanica Helmintha. III). *Bull. Com. A. O. F.* **19**, 397–519.

DOLLFUS, R. P. (1937). Les trématodes Digenea des selaciens (Plagiostomes). Catalogue par hôtes. Distribution géographique. *Ann. Parasit. hum. comp.* **15**, 55–73, 164–76, 259–81.

DUBOIS, G. (1928*a*). Études des cercaires de la région de Neuchâtel. *Bull. Soc. neuchâtel. Sci. nat.* N.S. **1**, Year 1927, **22** of collection, pp. 14–32.

DUBOIS, G. (1928*b*). Descriptions de nouveaux trématodes d'oiseaux du genre '*Hemistomum*'. *Bull. Soc. neuchâtel. Sci. nat.* N.S. Year **1**, 1927, **22** of collection, pp. 33–44.

DUBOIS, G. (1929). Les cercaires de la région de Neuchâtel. *Bull. Soc. neuchâtel. Sci. nat.* **53**, N.S. **2**, Year 1928, pp. 3–177.

DUBOIS, G. (1930). Deux nouvelles espèces de Clinostomidae. *Bull. Soc. neuchâtel. Sci. nat.* **54**, N.S. **3**, Year 1929, pp. 61–72.

DUBOIS, G. (1933). Notes sur deux espèces de Strigeidae et sur une espèce d'Hemiuridae. *Rev. Suisse Zool.* **40**, 1–10.

DUBOIS, G. (1938). Monographie des Strigeida (Trematoda). *Mém. Soc. neuchâtel. Sci. nat.* **6**, 1–535.

ECKMANN, F. (1932*a*). Über zwei neue Trematoden der Gattung *Aspidogaster*. *Z. Parasitenk.* **4**, 395–9.

ECKMANN, F. (1932*b*). Beiträge zur Kenntnis der Trematoden-familie Bucephalidae. *Z. Parasitenk.* **5**, 94–111.

ECKMANN, F. (1934). Rectification de Nomenclature. *Ann. Parasit. hum. comp.* **12**, 256.

EDWARDS, E. E. (1927). On the anatomy of the trematode *Paryphostomum radiatum* Dietz, 1909. *Parasitology*, **19**, 245–59.

EICHLER, W. (1940). Korrelationen in der Stammesentwicklung von Wirten und Parasiten. *Z. Parasitenk.* **12**, 94.

EJSMONT, L. (1926). Morphologische, systematische und entwicklungsgeschichtliche Untersuchungen an Arten des Genus *Sanguinicola* Plehn. *Bull. int. Acad. Cracovie*, 877–964.

EJSMONT, L. (1927). Nouvelles recherches sur les trématodes des hématophages (*Spirhapalum polesianum* n.g., n.sp., le trématode du sang d'*Emys orbicularis* L.). *C.R. I. Cong. Polon. Acad. Zool. Varsovie*, Year 1926, 1 p.

EJSMONT, L. (1928). Sur les deux genres de schistosomatides des oiseaux. *C.R. Acad. Cracovie*, **8**, 6–7.

EJSMONT, L. (1929). Über zwei Schistosomatidengattungen der Vögel. *Bull. int. Acad. Cracovie*, pp. 389–403.

EJSMONT, L. (1930*a*). *Astiotrema emydis* n.sp., trématode d'*Emys orbicularis*. *C.R. Acad. Cracovie*, **7**, 4–5.

EJSMONT, L. (1930*b*). *Astiotrema emydis* n.sp., ein Trématode aus *Emys orbicularis* L. *Bull. int. Acad. Cracovie*, B, **2**, 405–17.

EJSMONT, L. (1931). Über die Identität von *Prohystera rossittensis* Korkhaus und *Tanaisia fedtschenkoi* Skrjabin, nebst einigen Bemerkungen über Trematoden mit verbundenen Darmschenkeln. *Bull. int. Acad. Cracovie*, B, **2**, 531–47.

EJSMONT, L. (1937). *Opisthorchis tenuicollis* (= *O. felineus*) en Pologne. Cas observés chez l'homme. *Ann. Parasit. hum. comp.* **15**, 507–17.

ENIGK, K. (1932). *Leucochloridium paradoxum* in *Succinea oblonga*. *S.B. Ges. naturf. Fr., Berl.*, pp. 444.

ERHARDT, A. (1935). Systematik und geographische Verbreitung der Gattung *Opisthorchis* R. Blanchard, 1895, sowie Beiträge zur Chemotherapie und Pathologie der Opisthorchiasis. *Z. Parasitenk.* **8**, 188–225.

ERICKSON, A. B. (1940). *Euparyphium melis* (Trematoda: Echinostomidae) from the snowshoe hare. *J. Parasit.* **26**, 334.

FANTHAM, H. B. (1938). *Lecithostaphylus spondyliosomae* n.sp., a trematode parasite of the hottentot fish, *Spondyliosoma blochii*, found in South African waters. *Trans. Roy. Soc. S. Afr.* **26**, 387–93.

FAUST, E. C. (1917). Life-history studies on Montana trematodes. *Illinois Biol. Monogr.* **4**, no. 1, pp. 1–121.

FAUST, E. C. (1918*a*). Studies on Illinois cercariae. *J. Parasit.* **4**, 93–110.

FAUST, E. C. (1918*b*). Two new cystocercous cercariae from North America. *J. Parasit.* **4**, 148–53.

FAUST, E. C. (1919*a*). The excretory system in Digenea. I–III. *Biol. Bull. Woods Hole*, **36**, 315–44.

FAUST, E. C. (1919*b*). Notes on South African cercariae. *J. Parasit.* **5**, 164–75.

FAUST, E. C. (1921). Notes on South African larva; trematodes. *J. Parasit.* **8**, 11–21.

FAUST, E. C. (1922). Notes on larval flukes from China. *Parasitology*, **14**, 248–67.

FAUST, E. C. (1924*a*). Notes on larval flukes from China. II. Studies on some larval flukes from the central and south coast provinces of China. *Amer. J. Hyg.* **4**, 241–300.

FAUST, E. C. (1924*b*). The reactions of the miracidium of *Schistosoma japonicum* and *S. haematobium* in the presence of their intermediate hosts. *J. Parasit.* **10**, 199–204.

FAUST, E. C. (1926). Further observations of South African larval trematodes. *Parasitology*, **18**, 101–26.

FAUST, E. C. (1930*a*). Larval flukes associated with the cercariae of *Clonorchis sinensis* in Bithynoid snails in China and adjacent territory. *Parasitology*, **22**, 145–55.

FAUST, E. C. (1930*b*). *Human Helminthology*, 616 pp. Philadelphia.

FAUST, E. C. (1932*a*). The role of aquatic molluscs in the spread of human trematode infections. *China J.* **16**, 350–3.

FAUST, E. C. (1932*b*). The excretory system as a method of classification of digenetic trematodes. *Quart. Rev. Biol.* **7**, 458–68.

FAUST, E. C. & HOFFMAN, W. A. (1934). Studies on schistosomiasis mansoni in Puerto Rico. *Puerto Rico J. Publ. Hlth*, **10**, 1–47.

FAUST, E. C. & KHAW, O.-K. (1927). Studies on *Clonorchis sinensis* (Cobbold). *Amer. J. Hyg.* Monogr. Ser. no. 8, 284 pp.

FAUST, E. C. & MELENEY, H. E. (1924). Studies on schistosomiasis japonica. *Amer. J. Hyg.* Monogr. Ser. no. 3, 399 pp.

FAUST, E. C. & TANG, C.-C. (1935). New Aspidogastrid species, with a consideration of the systematic position of the group. *J. Parasit.* **21**, 435.

FAUST, E. C. & TANG, C.-C. (1936). Notes on new Aspidogastrid species, with a consideration of the phylogeny of the group. *Parasitology*, **28**, 487–501.

FERGUSON, M. S. (1940*a*). Excystment and sterilization of metacercariae of the avian Strigeid trematode, *Posthodiplostomum minimum*, and their development into adult worms in sterile cultures. *J. Parasit.* **26**, 359–72.

FERGUSON, M. S. (1940*b*). Further studies on the sterile culture of stages in a trematode life cycle. *J. Parasit.* **26**, Suppl. Abstr. **6**, 38–9.

FERGUSON, M. S. (1943). Development of eye flukes of fishes in the lenses of frogs, turtles, birds and mammals. *J. Parasit.* **29**, 136–42.

FISCHOEDER, F. (1901). Die Paramphistomiden der Säugetiere. *Zool. Anz.* **24**, 367–75.

FISCHOEDER, F. (1902). Die Paramphistomiden der Säugetiere. Inaug. Diss. Königsberg, pp. 59.

FISCHOEDER, F. (1903). Die Paramphistomiden der Säugetiere. *Zool. Jb., Syst.* **17**, 485–660.

FISCHOEDER, F. (1904). Beschreibung dreier Paramphistomiden-Arten aus Säugethieren. *Zool. Jb., Syst.* **20**, 453.

FISCHTHAL, J. H. (1942*a*). A *Paragordius* larva (Gordiacea) in a trematode. *J. Parasit.* **28**, 167.

FISCHTHAL, J. H. (1942*b*). Three new species of *Phyllodistomum* (Trematoda: Gorgoderidae) from Michigan fishes. *J. Parasit.* **28**, 269–75.

FISCHTHAL, J. H. (1942*c*). *Phyllodistomum etheostomae* n.sp. (Trematoda: Gorgoderidae) from percid fishes. *J. Parasit.* **28**, Suppl. Abstr. p. 18.

FISCHTHAL, J. H. (1943). A description of *Phyllodistomum etheostomae* Fischthal, 1942 (Trematoda: Gorgoderidae) from percid fishes. *J. Parasit.* **29**, 7–9.

FISCHTHAL, J. H. & ALLISON, L. N. (1941). *Acolpenteron ureteroecetes* Fischthal & Allison, 1940, a monogenetic trematode from the ureters of the black basses, with a revision of the family Calceostomatidae (Gyrodactyloidea). *J. Parasi.* **27**, 517–24.

FOGGIE, A. (1937). An outbreak of parasitic necrosis in turkeys caused by *Plagiorchis laricola* (Skrjabin). *J. Helminth.* **15**, 35–6.

FOLDA, F. (1928). *Megalocotyle marginata*, a new genus of ectoparasitic trematodes from the rock fish. *Publ. Puget Sound Biol. Sta. Seattle*, **6**, 196–206.

FRANKENBERG, G. v. (1935). Trematodencysten in Turbellarien. *Zool. Anz.* **112**, 237–42.

FRANKENBERG, G. v. (1937). *Phyllodistomum folium* als Parasit des Stichlings. *Mikrokosmos*, **30**, 66–7.

FREEMAN, A. E. & ACKERT, J. E. (1937). *Metorchis albidus*, a dog fluke new to North America. *Trans. Amer. Micr. Soc.* **56**, 113–15.

FREITAS, J. F. T. DE (1940). '*Cathaemasoides callis*' n.g., n.sp., trematodeo parasito de '*Euxenura galeata*' (Molina). *Mem. Inst. Osw. Cruz*, **35**, 589–92.

FRIEDHEIM, E. A. H. & BAER, J. G. (1933). Untersuchungen über die Atmung von *Diphyllobothrium latum...*, etc. *Biochem. Z.* **265**, 329–37.

FRIEND, G. F. (1939). Gill-parasites of brown trout in Scotland. *Scot. Nat.* pp. 123–6.

FUHRMANN, O. (1925). Deux nouvelles espèces de *Gorgodera*. *Bull. Soc. neuchâtel. Sci. nat.* **49**, 431–7.

FUHRMANN, O. (1928). Trematoda. In Kükenthal und Krumbach, *Handbuch d. Zool.* **2**, Part 2, pp. (2) 1–140.

GALLIEN, L. (1932*a*). Sur la reproduction néoténique chez *Polystomum integerrimum* Froelich. *C.R. Acad. Sci., Paris*, **194**, 1852–4.

GALLIEN, L. (1932*b*). Sur l'évolution de la génération issue des formes néoténiques de *Polystomum integerrimum* Froelich. *C.R. Acad. Sci., Paris*, **195**, 77–9.

GALLIEN, L. (1933). Transformations histologiques correlative du cycle sexual chez *Polystomum integerrimum* Froelich. *C.R. Acad. Sci., Paris*, **196**, 426–8.

GALLIEN, L. (1935). Recherches expérimentales sur le dimorphisme évolutif et la biologie de *Polystomum integerrimum* Froelich. *Trav. Sta. zool. Wimereux*, **12**, 1, 1–183.

GALLIEN, L. (1937). Recherches sur quelques trématodes monogénèses nouveaux ou peu connus. *Ann. Parasit. hum. comp.* **15**, 9–28, 146–64. With rectification of nomenclature. *Ann. Parasit. hum. comp.* **15**, Part 4, p. 383.

GILLE, K. (1914). Untersuchungen über die Eireifung, Befruchtung und Zellteilung von Gyrodactylus. *Arch. Zellforsch.* **12**, 415–56.

GOHAR, N. (1934*a*). Les trématodes parasites du milan égyptien *Milvus migrans* avec description d'une nouvelle espèce et remarques sur les genres *Haplorchis* Looss, 1899 et *Monorchotrema* Nishigori, 1924. *Ann. Parasit. hum. comp.* **12**, 218–27.

GOHAR, N. (1934*b*). Liste des trématodes parasites et de leurs hôtes vertébrés signalés dans la vallée du Nil. *Ann. Parasit. hum. comp.* **12**, 322–31.

GOHAR, N. (1935). Liste des trématodes parasites et de leurs hôtes vertébrés signalés dans la vallée du Nil (continued). *Ann. Parasit. hum. comp.* **13**, 80–90.

GOLDBERGER, J. (1911*a*). On some new parasitic worms of the genus *Telorchis*. *Hyg. Lab. Bull.* no. 71, pp. 36–47.

GOLDBERGER, J. (1911*b*). A new trematode (*Styphlodora bascaniensis*) with a blind Laurer's canal. *Proc. U.S. Nat. Mus.* **40**, 233–9.

GOLDSCHMIDT, R. (1902). Untersuchungen über die Eireifung, Befruchtung und Zellteilung bei *Polystomum integerrimum* Rud. *Z. wiss. Zool.* **71**, 397–444.

GOLDSCHMIDT, R. (1905). Eireifung, Befruchtung und Embryonalentwicklung des *Zoogonus mirus* Looss. *Zool. Jb., Anat.* **21**, 604–54.

GONDER, F. (1910). *Itygonimus lorum* (Dujardin). *Zbl. Bakt.* (1) **53**, 160–74.

GOODCHILD, C. G. (1939). *Cercaria conica* n.sp. from the clam *Pisidium abditum* Haldeman. *Trans. Amer. Micr. Soc.* **58**, 179–84.

GOTO, S. & KIKUCHI, H. (1917). Two new trematodes of the family Gyrodactylidae. *J. Coll. Sci. Tokyo*, **34**, Art. 4.

GOWER, W. C. (1936). New sporocyst of *Leucochloridium* from Louisiana. *J. Parasit.* **22**, 375–8.

GOWER, W. C. (1939). Modified stain and procedure for trematodes. *Stain Tech.* **14** (1), 31–2.

GRAHAM, R., TORREY, J. P., MIZELLE, J. D. & MICHAEL, V. M. (1937). Internal parasites of poultry. *Circ. Ill. Agric. Exp. Sta.* no. 469, 50 pp.

GRAY, J. (1931). *A Text-book of Experimental Cytology*, 516 pp. Cambridge.

GRÉGOIRE, V. (1909). La réduction dans le *Zoogonus mirus* et la Primärtypus. *Cellule*, **25**.

GRIEDER, H. (1937). In der Schweiz selten vorkommende Helminthen als pathogene Wirbeltierparasiten. *Z. Parasitenk.* **9**, 145–50.

GUBERLET, J. E. (1933). Notes on some Onchocotylinae from Naples with a description of a new species. *Publ. Sta. zool. Napoli*, **12** (3), 323–36.

GUBERLET, J. E. (1936). Two new ectoparasitic trematodes from the sting ray, *Myliobatus californicus*. *Amer. Midl. Nat.* **17**, 954–64.

GUBERLET, J. E. (1937). Trematodos ectoparasitos de los peces de los costas del Pacifico. *An. Inst. Biol. Univ. Méx.* **7**, 457–567.

GUBERLET, J. E. & BONHAM, K. (1938). Ectoparasitic trematodes of Puget Sound fishes—*Acanthocotyle*. *Amer. Midl. Nat.* **20**, 590–602.

GUBERLET, J. E., HANSEN, H. A. & KAVANAGH, J. A. (1927). Studies on the control of *Gyrodactylus*. *Univ. Wash. Publ.* **2** (2), 1–13.

HADLEY, C. E. & CASTLE, R. M. (1937). A trematode of the genus *Maritrema* (Nicoll) parasitic in the barnacle and the ruddy turnstone. *Anat. Rec.* Suppl. p. 139.

HADLEY, C. E. & CASTLE, R. M. (1940). Description of a new species of *Maritrema* Nicoll, 1907, *Maritrema arenaria*, with studies of the life history. *Biol. Bull. Woods Hole*, **78**, 338–48.

HALL, M. C. (1929). Arthropods as intermediate hosts. *Smithson. Misc. Coll.* **81**, no. 15, pp. 1–77.

HALL, M. C. (1936). *Control of Animal Parasites. General Principles and their Application*, 162 pp. Evanston, Ill.

HARDING, J. P. (1941). Simple modifications of the camera lucida for making larger drawings. *Nature, Lond.*, **148**, 754–5.

HARKEMA, R. (1942). *Pharyngostomoides procyonis* n.g., n.sp. (Strigeidae), a trematode from the raccoon in North Carolina and Texas. *J. Parasit.* **28**, 117–22.

HARNISH, O. (1932). Untersuchungen über den Gaswechsel von *Fasciola hepatica*. *Z. vergl. Physiol.* **17**, 365–86.

HARPER, W. F. (1929). On the structure and life histories of British fresh-water larval trematodes. *Parasitology*, **21**, 189–219.

HARPER, W. F. (1931). On the structure and life history of British fresh-water Furcocercariae. *Parasitology*, **23**, 310–24.

HARPER, W. F. (1932). On some British larval trematodes from terrestrial hosts. *Parasitology*, **24**, 307–17.

HARRAH, E. C. (1922). North American monostomes primarily from fresh-water hosts. *Ill. Biol. Monogr.* **7**, 1–106.

HARRISON, L. (1914). The Mallophaga as a possible clue to bird phylogeny. *Aust. Zool.* **1**, 7–11.

HARRISON, L. (1915). Mallophaga from *Apteryx*, and their significance. *Parasitology*, **8**, 88–100.

HARWOOD, P. D. (1932). The helminths parasitic in the Amphibia and Reptilia of Houston, Texas, and vicinity. *Proc. U.S. Nat. Mus.* **81**, 1–71.

HARWOOD, P. D. (1936). Notes on Tennessee helminths. III. Two trematodes from a kingfisher. *J. Tenn. Acad. Sci.* **11**, 251–6.

HARWOOD, P. D (1939). Notes on Tennessee helminths. IV. North American trematodes of the sub-family Notocotylinae. *J. Tenn. Acad. Sci.* **14**, 332–40.

HAZARD, F. O. (1941). The absence of Opalinids from the adult green frog, *Rana clamitans*. *J. Parasit.* **27**, 513–16.

HEDRICK, L. R. (1943). Two new large-tailed cercariae (Psilostomidae) from Northern Michigan. *J. Parasit.* **29**, 182–6.

HEGNER, R., ROOT, F. M., AUGUSTINE, D. L. & HUFF, C. G. (1938). *Parasitology, with special reference to man and domesticated animals*, 812 pp. New York and London.

HEIDEGGER, E. & MENDHEIM, H. (1939). Beiträge zur Kenntnis der Gattung *Platynosomum*. I. *Platynosomum fallax* n.sp., ein neuer Dicrocoeliine aus den Gelbwangenkakadu (*Cacatua sulfurea*). *Z. Parasitenk.* **10**, 94–107.

HEIDEGGER, E. & MENDHEIM, H. (1940). Beiträge zur Kenntnis der Gattung *Platynosomum*. II. Missbildungen bei *Platynosomum ventroplicatum* (Heidegger & Mendheim, 1937). *Z. Parasitenk.* **11**, 435–56.

HEINEMANN, E. (1936). Über ein Massensterben von Seeschwalben (*Sterna hirundo* und *S. paradisea*) und sein Ursache. *Dtsch. Jagd. Neudamm*, **4**, 2 pp.

HEINEMANN, E. (1937). Über den Entwicklungskreislauf der Trematodengattung *Metorchis* sowie Bemerkungen zur Systematik dieser Gattung. *Z. Parasitenk.* **9**, 237–60.

HEITZ, A. (1918). *Salmo salar* Lin., seine Parasitenfauna und seine Ernährung im Meer und im Süsswasser. *Arch. Hydrobiol. Plankt.* **12**, 311–72, 485–561.

HEITZ-BOYER (1935). Un cas de bilharziose vésicale d'origine européenne. *Bull. Soc. franç. Urol.* pp. 65–7.

HENKEL, H. (1931). Untersuchungen zur Ermittlung des Zwischenwirts von *Dicrocoelium lanceatum*. *Z. Parasitenk.* **3**, 664–712.

HENNEGUY, L. F. (1906). Recherches sur le mode de formation de l'œuf ectolecithe du *Distomum hepaticum*. *Arch. Anat. Micr.* **9**, 47.

HERBER, E. C. (1939). Studies on the biology of the frog amphistome, *Diplodiscus temperatus* Stafford. *J. Parasit.* **25**, 189–95.

HERBER, E. C. (1942). Life history studies on two trematodes of the subfamily Notocotylinae. *J. Parasit.* **28**, 179–96.

HESS, W. N. (1928). The life history and control of *Dactylogyrus* sp. *J. Parasit.* **15**, 138–9.

HESS, W. N. (1930). Control of external fluke parasites on fish. *J. Parasit.* **16**, 131–6.

HESSE, A. J. (1923). A description of two cercariae found in *Limnaea peregra* in Scotland. *J. Helminth.* **1**, 227–36.

HESSE, R., ALLEE, W. C. & SCHMIDT, K. P. (1937). *Ecological Animal Geography*, 597 pp. New York.

HICKMANN, V. V. (1934). On *Coitocaecum anaspidis* sp. nov., a trematode exhibiting progenesis in the fresh-water crustacean *Anaspides tasmaniae* Thomson. *Parasitology*, **26**, 121–8.

HIERONYMI, E. & SZIDAT, L. (1921). Ueber eine neue Hühnerenzootie, bedingt durch *Prosthogonimus intercalandus* n.sp. *Zbl. Bakt.* (i), **86**, 236–41.

HILARIO, J. S. & WHARTON, L. D. (1917). *Echinostoma ilocanum* Garrison, a Report of Five Cases and a contribution to the anatomy of the fluke. *Philipp. J. Sci.* B, **12**, 203–11.

HINTON, M. A. C., TUCKER, B. W., PARKER, H. W. & NORMAN, J. R. (1935). *List of British Vertebrates*. Brit. Mus. Nat. Hist.

HOCKLEY, A. R. (1937). An investigation of holostomiasis in Avon coarse fish. *Rep. Avon Biol. Res., Southampton*, **4**, 103–6.

HOEPPLI, R. (1927). Über Beziehungen zwischen dem biologischen Verhalten parasitischer Nematoden und histologischen Reaktionen des Wirbeltierkörpers. *Arch. Schiffs- u. Tropenhyg.* **31** (Beiheft 3), 88 pp.

HOEPPLI, R. (1929). Histologische Beiträge zur Biologie der Helminthen. *Virchow's Arch.* **271**, 356–65.

HOEPPLI, R. (1933). On histolytic changes and extracellular digestion in parasitic infections. *Lingnan Sci. J.* **12**, Suppl. May, 1–11.

HOEPPLI, R. & CHU, H. J. (1937). Studies on *Clonorchis sinensis* in vitro. *Festschr. B. Nocht 80. Geburtstag, Hamburg*, pp. 199–203.

HOEPPLI, R., FENG, J. C. & CHU, H. J. (1938). Attempts to culture helminths of vertebrates in artificial media. *Chin. Med. J.* Suppl. **2**, 343–74.

HOFFMAN, W. A. & JANER, J. L. (1936). Miracidial twinning in *Schistosoma mansoni*. *Proc. Helminth. Soc. Wash.* **3**, 62.

HOHORST, W. (1937). Die 'Fühler-Made' (*Leucochloridium* sp.) der Bernstein-Schnecke, ein für die Umgebung von Frankfurt-a.-M. neuer Saugwurm. *Natur. u. Volk*, **67**, 123–32.

HOLL, F. J. (1929). The phyllodistomes of North America. *Trans. Amer. Micr. Soc.* **48**, 48–53.

HOLLANDE, A. C. (1920). Réactions des tissues du *Dytiscus marginalis* L. au contact des larves de Distome encystées aux parois du tube digestif de l'insecte. *Arch. Zool. exp. gén.* **19**, 543.

HOLLIS, M. B. (1941). Revision de los generos *Diplodiscus* Diesing, 1836, y *Megalodiscus* Chandler, 1923 (Trematoda: Paramphistomoidea). *An. Inst. Biol. Univ. Méx.* **12**, 127–46.

HOPKINS, S. H. (1931). Studies on *Crepidostomum*. II. The '*Crepidostomum laureatum*' of A. R. Cooper. *J. Parasit.* **18**, 79–91.

HOPKINS, S. H. (1933*a*). Note on the life history of *Clinostomum marginatum* (Trematoda). *Trans. Amer. Micr. Soc.* **52**, 147–9.

HOPKINS, S. H. (1933*b*). The morphology, life histories and relationships of the papillose Allocreadiidae (Trematodes). *Zool. Anz.* **103**, 65–74.

HOPKINS, S. H. (1934). The papillose Allocreadiidae. *Ill. Biol. Monogr.* **13** (2), 1–80.

HOPKINS, S. H. (1937). A new type of Allocreadiid cercaria. The cercaria of *Anallocreadium* and *Microcreadium*. *J. Parasitol.* **23**, 94–6.

HOPKINS, S. H. (1940). The excretory systems of *Tergestia* Stossich, 1899 and *Cornucopula adunca* (Linton, 1905) (Trematoda). *Trans. Amer. Micr. Soc.* **59**, 281–4.

HOPKINS, S. H. (1941*a*). The excretory systems of *Helicometra* and *Cymbephallus* (Trematoda), with remarks on their relationships. *Trans. Amer. Micr. Soc.* **60**, 41–4.

HOPKINS, S. H. (1941*b*). New genera and species of the family Monorchiidae (Trematoda), with a discussion on the excretory system. *J. Parasit.* **27**, 395–407.

HORSFALL, M. W. (1934). Studies on the life history and morphology of the cystocercous cercariae. *Trans. Amer. Micr. Soc.* **53**, 311–47.

HSIAO, S. C. T. (1941). Melanosis in the common cod, *Gadus callarias* L., associated with trematode infection. *Biol. Bull. Woods Hole*, **80**, 37–44.

HSÜ, H. F. (1940). Studies on the food and the digestive system of certain parasites. V. On the food of liver flukes. *Chin. Med. J.* **56**, 122–30.

HSÜ, H. F. & CHOW, C. Y. (1938). Studies on helminths of fowls. II. Some trematodes of fowls in Tsingkiangpu, Kiangsu, China. *Chin. Med. J.* Suppl. II, p. 449.

HSÜ, H. F. & LI, S. V. (1940). Studies on the food and the digestive system of certain parasites. VI. On the food of certain helminths living in the digestive tract of Vertebrates. *Chin. Med. J.* **57**, 559–67.

HUBBS, C. L. (1928). The related effects of a parasite on a fish. *J. Parasit.* **14**, 75.

HÜBNER, F. (1939). Über *Echinostomum anceps* (Molin, 1859) Dietz, 1909. *Zool. Anz.* **128**, 176–87.

HUGHES, W. K. (1928). Some trematode parasites on the gills of Victorian fishes. *Proc. Roy. Soc. Vict.* **41**, N.S. 1, 45–54.

HUGHES, R. C., HIGGINBOTHAM, J. W. & CLARY, J. W. (1942). The trematodes of reptiles. Part I. Systematic Section. *Amer. Midl. Nat.* **27**, 109–34. Part II. Host Catalogue. *Proc. Okla. Acad. Sci.* for 1941. Part III. Conclusion (Index of specific names). *Proc. Okla. Acad. Sci.* **21**, 90–114.

HUNNINEN, A. V. & CABLE, R. M. (1940). Studies on the life history of a new species of *Anisoporus* (Trematoda: Allocreadiidae). *J. Parasit.* **26**, Suppl. Abstr. **6**, p. 33.

HUNNINEN, A. V. & CABLE, R. M. (1941*a*). The life history of *Podocotyle atomon* (Rud.) (Trematoda: Fam. Opecoelidae). *J. Parasit.* **27**, Suppl. Abstr. **8**, pp. 12–13.

HUNNINEN, A. V. & CABLE, R. M. (1941*b*). Studies on the life history of *Lecithaster confusus* Odhner (Trematoda: Hemiuridae). *J. Parasit.* **27**, Suppl. Abstr. **9**, p. 13.

HUNNINEN, A. V. & CABLE, R. M. (1941*c*). Studies on the life history of *Anisoporus manteri* Hunninen & Cable, 1940 (Trematoda: Allocreadiidae). *Biol. Bull. Woods Hole*, **80**, 415–28.

HUNNINEN, A. V. & CABLE, R. M. (1943*a*). The life history of *Podocotyle atomon* (Rudolphi) (Trematoda: Opecoelidae). *Trans. Amer. Micr. Soc.* **62**, 57–68.

HUNNINEN, A. V. & CABLE, R. M. (1943*b*). The life history of *Lecithaster confusus* Odhner (Trematoda: Hemiuridae). *J. Parasit.* **29**, 71–9.

HUNNINEN, A. V. & WICHTERMAN, R. (1938). Hyperparasitism; a species of *Hexamita* (Protozoa: Mastigophora) found in the reproductive systems of *Deropristis inflata* (Trematoda) from marine eels. *J. Parasit.* **24**, 95–101.

HUNTER, G. W. iii (1941). Studies on host parasite reactions. VI. An hypothesis to account for pigmented metacercarial cysts in fish. *J. Parasit.* **27**, Suppl. Abstr. **71**, p. 33.

HUNTER, G. W. iii & BANGHAM, R. V. (1932). Studies on fish parasites of Lake Erie I. New trematodes (Allocreadiidae). *Trans. Amer. Micr. Soc.* **51**, 137–52.

HUNTER, G. W. iii & HAMILTON, J. M. (1941). Studies on host parasite reactions to larval parasites. IV. The cyst of *Uvulifer ambloplitis* (Hughes). *Trans. Amer. Micr. Soc.* **60**, 498–507.

HUNTER, G. W. iii & HUNTER, W. S. (1934). The life cycle of the yellow grub of fish, *Clinostomum marginatum* (Rud). *J. Parasit.* **20**, 325.

HUNTER, W. S. (1928). A new strigeid larva, *Neascus wardi*. *J. Parasit.* **15**, 104.

HUNTER, W. S. & HUNTER, G. W. iii (1935). Studies on *Clinostomum*. II. The miracidium of *C. marginatum* (Rud.). *J. Parasit.* **21**, 186–9.

HUSSEY, K. L. (1941*a*). Partial twinning in a stylet cercaria. *J. Parasit.* **27**, 92–3.

HUSSEY, K. L. (1941*b*). Comparative embryological development of the excretory system in digenetic trematodes. *Trans. Amer. Micr. Soc.* **60**, 171–210.

HUXLEY, J. S. (1932). *Problems of Relative Growth*, 276 pp. Methuen.

IDRIS JONES, E. (1933*a*). Fertilisation and egg formation in a digenetic trematode *Podocotyle atomon*. *Parasitology*, **24**, 545–7.

IDRIS JONES, E. (1933*b*). On *Ceratotrema furcolabiata* n.g. et n.sp. and *Hemipera sharpei* n.sp., two new digenetic trematodes of British fishes. *Parasitology*, **25**, 248–54.

IDRIS JONES, E. (1933*c*). Studies on the Monogenea (Trematoda) of Plymouth. I. *Microbothrium caniculae* (Johnstone, 1911). *Parasitology*, **25**, 329–32.

IDRIS JONES, E. (1933*d*). Studies on the Monogenea of Plymouth. *Gastrocotyle trachuri* v. Ben. & Hesse, 1863. *J. Mar. Biol. Ass. U.K.* **19**, 227–32.

IDRIS JONES, E. & ROTHSCHILD, M. (1932). On the sporocyst and cercaria of a marine distomid trematode from *Nucula*. *Parasitology*, **24**, 260–4.

INGLES, L. G. (1932). Four new species of *Haematoloechus* (Trematoda) from *Rana aurora draytoni* from California. *Univ. Calif. Publ. Zool.* **37**, 189–201.

ISSAITSCHIKOW, I. M. (1919). Trématodes nouveaux du genre *Lyperosomum* Looss. *Izvest. Donsk. Vet. Inst. Novotscherkassk*, **1**, 1, 16 pp.

ISSAITSCHIKOW, I. M. (1920). Über einen neuen Vertreter der Gattung *Eurytrema* Looss. *Izvest. Donsk. Vet. Inst. Novotscherkassk*, **1**, II, 1–11.

ISSAITSCHIKOW, I. M. (1923*a*). On the parasitic worms of Cyprinoid fishes of the rivers of the Kuban. (Russian.) *Trud. Inst. eksp. Vet.* **2**, 12 pp.

ISSAITSCHIKOW, I. M. (1923*b*). Zur Kenntnis der Helminthenfauna der Amphibien Russlands. II–IV. *Zbl. Bakt.* (ii), **59**, 19–26.

ISSAITSCHIKOW, I. M. (1924). Parasitic worms from the Sea of Azov. (Russian.) *Trudy Sibir. Vet. Inst. Omsk*, **6**, 1–28.

ISSAITSCHIKOW, I. M. (1926*a*). Zur Kenntnis der Helminthenfauna der Amphibien Russlands. I. Parasitische Würmer aus *Bufo viridis* Laur. in der Krim. (Russian.) *Trudy Sibir. Vet. Inst. Omsk*, **7**, 1–99.

ISSAITSCHIKOW, I. M. (1926*b*). Zur Fauna der parasitischen Würmer des *Erinaceus europaeus* L. des Artjemowschen Bezirks (Don). *Results 25th Helminth. Exp. to Artemow Distr., Moscow*, pp. 82–91.

ISSAITSCHIKOW, I. M. (1926*c*). Contributions to the fauna of parasitic worms of *Rana arvalis altaica* Kascht. in the Omsk province. *Izvest. W. Siberian Sect. Russ. Geogr. Soc. Omsk*, **5**, 219–24.

ISSAITSCHIKOW, I. M. (1927). A new trematode of the family Heterophyidae. (Russian.) *Festschr. Knipovitch, Moscow*, pp. 260–9.

ISSAITSCHIKOW, I. M. (1928). Zur Kenntnis der parasitischen Würmer einiger Gruppen von Wirbeltieren der russischen Arktis. (Russian.) *Ber. wiss. Meeresinst. Moscow*, **3**, 2, 1–79.

ISSAITSCHIKOW, I. M. (1933). Contributions to parasitic worms of some groups of vertebrates from the Russian Arctic. (Russian.) *Trans. Oceanogr. Inst. Moscow*, **3**, 1–36; German summary, pp. 37–44.

ISSAITSCHIKOW, I. M. & ZAKHAROW, N. P. (1929). Ueber die parasitäre Helminthenfauna von *Rana esculenta* im Dongebiet. (Russian.) *Russ. hydrobiol. Z.* **8**, 49–53.

ISHII, N. (1935*a*). Studies on avian trematodes. I. *J. Exp. Med.* **19**, 5, 475–6.

ISHII, N. (1935*b*). Notes on the family Didymozooidae (Monticelli, 1888). *Jap. J. Zool.* **9**, 279–335.

JACOB, E. (1940). Zur Behandlung einiger parasitärer Fischkrankheiten. *Berl. tierärztl. Wschr.* 161–2.

JÄGERSKIÖLD, L. A. (1900*a*). *Diplostomum macrostomum* n.sp. *Zbl. Bakt.* I. Orig. **27**, 33–7.

JÄGERSKIÖLD, L. A. (1900*b*). *Levinsenia* (*Distomum*) *pygmaea* Levinsen, ein genitalnapftragendes Distomum. *Zbl. Bakt.* I. Orig. **27**, 732.

JÄGERSKIÖLD, L. A. (1901). *Tocotrema expansum* (Crepl.) (= *Monostomum expansum* Crepl.), ein genitalnapftragende Distomide. *Zbl. Bakt.* I. Orig. **30**, 979–83.

JÄGERSKIÖLD, L. A. (1907). Zur Kenntnis der Trematodengattung *Levinseniella*. *Zool. Stud. tillägnade Prof. T. Tullberg, Uppsala*, pp. 135–54.

JÄGERSKIÖLD, L. A. (1908). Kleine Beiträge zur Kenntnis der Vogeltrematoden. *Zbl. Bakt.* (ii), **48**, 302–17.

JAHN, T. L. & KUHN, L. R. (1932). The life history of *Epibdella melleni* MacCallum, 1927, a monogenetic trematode parasitic on marine fishes. *Biol. Bull. Woods Hole*, **62**, 89–111.

JAMESON, H. L. & NICOLL, W. (1913). On some parasites of the Scoter duck (*Oidemia nigra*) and their relation to the pearl inducing trematode in the edible mussel (*Mytilus edulis*). *Proc. Zool. Soc. Lond.* **12**, 53–63.

JANER, J. L. (1941). Miracidial twinning in *Schistosoma mansoni*. *J. Parasit.* **27**, 93.

JANICKI, C. v. (1903). Beziehungen zwischen Chromatin und Nucleolen während der Furchung des Eies von *Gyrodactylus elegans* von Nordm. *Zool. Anz.* **26**, 241–5.

JEGEN, G. (1916). *Collyriclum faba* (Bremser) Kossack. Ein Parasit der Singvögel, sein Bau und seine Lebensgeschichte. *Z. wiss. Zool.* **117**, 460–553.

JEPPS, M. W. (1933). Miracidia of the liver fluke for laboratory work. *Nature, Lond.*, **132**, 171.

JOHNSON, J. C. (1920). The life cycle of *Echinostoma revolutum* (Froelich). *Univ. Calif. Publ. Zool.* **19**, 335–88.

JOHNSTON, S. J. (1912). On some trematode parasites of Australian frogs. *Proc. Linn. Soc. N.S.W.* **37**, 2, 285–362.

JOHNSTON, S. J. (1913*a*). On some trematode parasites of Marsupials and of a Monotreme. *Proc. Linn. Soc. N.S.W.* **37**, 4, 727–40.

JOHNSTON, S. J. (1913*b*). On some Queensland trematodes, with anatomical observations and descriptions of new species and genera. *Quart. J. Micr. Sci.* **59**, 361–400.

JOHNSTON, S. J. (1916). On the trematodes of Australian birds. *J. Proc. Roy. Soc. N.S.W.* **50**, 204–6 and **1**, 1916 (1917), 187–261.

JOHNSTON, T. H. (1929). Remarks on the synonymy of certain Tristomatid genera. *Trans. Proc. Roy. Soc. S. Aust.* **53**, 71–8.

JOHNSTON, T. H. (1930). A new species of trematode, of the genus *Anoplodiscus*. *Aust. J. Exp. Biol. Med. Sci.* **7**, 108–12.

JOHNSTON, T. H. (1931). New trematodes from the Subantarctic and Antarctic. *Aust. J. Exp. Biol. Med. Sci.* **8**, 91–8.

JOHNSTON, T. H. (1934*a*). New trematodes from South Australian Elasmobranchs. *Aust. J. Exp. Biol. Med. Sci.* **12**, 25–32.

JOHNSTON, T. H. (1934*b*). Notes on some monocotylid trematodes. *Proc. Linn. Soc. N.S.W.* (251–2), **59**, 62–5.

JOHNSTON, T. H. (1938). Larval trematodes from Australian terrestrial and fresh-water molluscs. Part III. *Leucochloridium australiense*, n.sp. *Trans. Roy. Soc. S. Aust.* **62**, 25–33.

JOHNSTON, T. H. & CLELAND, E. R. (1937). Larval trematodes from Australian terrestrial and fresh-water molluscs. Part I. A Survey of literature. *Trans. Roy. Soc. S. Aust.* **61**, 191–201.

JOHNSTON, T. H. & TIEGS, O. W. (1922). New Gyrodactyloid trematodes from Australian fishes, together with a reclassification of the superfamily Gyrodactyloidea. *Proc. Linn. Soc. N.S.W.* **37**, 2, 83–131.

JOHNSTONE, J. (1905–14). Internal parasites and diseased conditions of fishes. *Rep. Lpool Sea Fish. Investig.* Years 1904–13.

JONES, D. O. (1943). The anatomy of three digenetic trematodes, *Skrjabiniella aculeatus* (Odhner), *Lecithochirium rufoviride* (Rud.) and *Sterrhurus fusiformis* (Lühe) from *Conger conger* (Linn.). *Parasitology*, **35**, 40–57.

JOYEUX, C. (1922). Recherches sur les Notocotyles. *Bull. Soc. Path. exot.* **15**, 331–43.

JOYEUX, C. & BAER, J. G. (1927*a*). Note sur les Cyclocoelidae (Trematodes). *Bull. Soc. zool. Fr.* **52**, 416–34.

JOYEUX, C. & BAER, J. G. (1927*b*). Recherches sur le cycle évolutif du trématode *Opisthioglyphe rastellus* Olss. *Bull. Biol.* **61**, 4, 359–73.

JOYEUX, C. & BAER, J. G. (1934). Note sur une nouvelle espèce de Trématode, *Gorgoderina capsensis* n.sp. *Rev. suisse Zool.* **41**, 197–201.

JOYEUX, C. & BAER, J. G. (1936). Quelques helminthes nouveaux et peu connus de la Musaraigne, *Crocidura russula* Herm. (I[e] Part). *Rev. suisse Zool.* **43**, 25–50.

JOYEUX, C., BAER, J. G. & CARRÈRE, P. (1934). Recherches sur le cycle évolutif d'*Euryhelmis squamula*. *C.R. Acad. Sci., Paris*, **199**, 1067–8.

JOYEUX, C., BAER, J. G. & TIMON-DAVID, J. (1932*a*). Le développement du trématode *Brachylaemus* (*Brachylaemus*) *nicolli* (Witenberg). *C.R. Soc. Biol., Paris*, **109**, 464–6.

JOYEUX, C., BAER, J. G. & TIMON-DAVID, J. (1932*b*). Recherches sur le cycle évolutif des trématodes appartenant au genre *Brachylaemus* Dujardin (Syn. *Harmostomum* Braun). *C.R. Acad. Sci., Paris*, **195**, 972–3.

JOYEUX, C., BAER, J. G. & TIMON-DAVID, J. (1934). Recherches sur les trématodes du genre *Brachylaemus* Dujardin (Syn. *Harmostomum* Braun). *Bull. Biol.* **68**, 385–418.

JOYEUX, C. & FOLEY, H. (1930). Les helminthes de *Meriones shawi* Rozet dans le nord de l'Algérie. *Bull. Soc. Zool. Fr.* **55**, 353–74.

JOYEUX, C. & HOUDEMER, E. (1928). Recherches sur la fauna helminthologique de l'Indochine (Cestodes et Trématodes) (suite et fin). *Ann. Parasit. hum. comp.* **6**, 27–58.

KATHARINER, L. (1904). Über die Entwicklung von *Gyrodactylus elegans* v. Nordmann. *Zool. Jb.* Suppl. **7**, 519–51.

KAY, M. W. (1942). Notes on the genus *Merizocotyle* Cerfontaine, with a description of a new species. *Trans. Amer. Micr. Soc.* **6**, 3, 254–60.

KEILIN, D. (1925). On cytochrome, a respiratory pigment common to animals, yeast and higher plants. *Proc. Roy. Soc.* B, **98**, 312–39.

KEMNITZ, G. A. v. (1913). Eibildung, Eireifung, Samenreifung und Befruchtung von *Brachycoelium salamandrae* (*B. crassicolle* (Rud.)). *Arch. Zellforsch.* **10**, 470–506.

KHALIL, M. (1937). The life history of the human trematode parasite, *Heterophyes heterophyes*. *C.R.* 12 *Congr. internat. Zool., Lisbon*, **3**, 1889–99.

KHAN, M. H. (1935). On eight new species of the genus *Cyclocoelum* Brandes from North Indian snipes. *Proc. Acad. Sci. U.P. India*, **4**, 342–70.

KOBAYASHI, H. (1918). Studies on cercariae from Korea. *Korean Med. J.* **21**, 62 pp.

KOBAYASHI, H. (1922). A review of Japanese cercariae. *Mitt. Med. Akad. Keijo*, 27 pp.

KOLLATH, W. & ERHARDT, A. (1936). Lebensdauer, Redoxlage und Fuadinwirkung bei *Opisthorchis in vitro*. *Biochem. Z.* **286**, 287–8.

KOMIYA, Y. (1939). Die Entwicklung des Exkretionssystems einiger Trematodenlarven aus Alster und Elbe, nebst Bemerkungen über ihren Entwicklungszyklus. *Z. Parasitenk.* **10**, 340–85.

KOSSACK, W. (1910). Neue Distomen. *Zbl. Bakt.* (i), **56**, 114–20.

KOURÍ, P. & NAUSS, R. W. (1938). Formation of the egg-shell in *Fasciola hepatica* as demonstrated by histological methods. *J. Parasit.* **24**, 291–310.

KOURÍ, P., BASNUEVO, J., ALVARÉ, L., LESCANO, O. & SIMON, R. (1936). Nota previa sobre la genesis del huevo de *Fasciola hepatica*. *Rev. Parasit. Clin. Lab.* **2**, 173–4.

KRAUSE, R. (1914). Beitrag zur Kenntnis der Hemistominen. *Z. wiss. Zool.* **112**, 93–238.

KRULL, H. (1935). Anatomische Untersuchungen an einheimischen Prosobranchieren und Beiträge zur Phylogenie der Gastropoden. *Zool. Jb., Anat.* **60**, 399–464.

KRULL, W. H. (1930). The life history of two North American lung flukes. *J. Parasit.* **16**, 207–12.

KRULL, W. H. (1931). Life-history studies on two frog lung flukes, *Pneumonoeces medioplexus* and *Pneumobites parviplexus*. *Trans. Amer. Micr. Soc.* **50**, 215–77.

KRULL, W. H. (1934). Some observations on the cercaria and redia of a species of *Clinostomum*, etc. *Proc. Helminth Soc. Wash.* **1**, 34–5.

KRULL, W. H. (1935). Studies on the life history of *Halipegus occidualis* Stafford, 1905. *Amer. Midl. Nat.* **16**, 129–43.

KRULL, W. H. (1940). Notes on *Typhlocoelum cymbium* (Diesing, 1850) Cyclocoelidae. *Trans. Amer. Micr. Soc.* **59**, 290–3.

KRULL, W. H. (1941). The number of cercariae of *Fasciola hepatica* developing in snails infected with a single miracidium. *Proc. Helminth. Soc. Wash.* **8**, 2, 55–8.

KRULL, W. H. & PRICE, E. W. (1932). Studies on the life history of *Diplodiscus temperatus* Stafford from the frog. *Occ. Pap. Mus. Zool. Univ. Mich.* no. 237, p. 1.

KU, C. T. (1937*a*). On a new trematode parasite from the Peking duck. *Peking Nat. Hist. Bull.* **12**, 39–41.

KU, C. T. (1937*b*). Two new trematodes of the genus *Notocotylus* with a key to the species of the genus. *Peking Nat. Hist. Bull.* **12**, 113–22.

KU, C. T. (1940). Studies on the genus *Prosthogonimus* of the domestic duck in Kunming. *Peking Nat. Hist. Bull.* **15**, 119–31.

KULWIÉC, Z. (1927). Untersuchungen an Arten der Gattung *Dactylogyrus* Diesing. *Bull. Internat. Pol. Sci. et Lett. Cl. Sci. Math. et Nat.*, B, Sci. Nat. Cracovie, 113–44.

LAL, M. B. (1936). A new species of the genus *Parorchis* from *Totanus hypoleucos* with certain remarks on the family Echinostomatidae. *Proc. Indian Acad. Sci.* **4**, 27–35.

LAL, M. B. (1939). Studies in helminthology. Trematode parasites of birds. *Proc. Indian Acad. Sci.* **10** B, 111–200.

LANDER, C. H. (1904). The anatomy of *Hemiurus crenatus* (Rud.) Lühe, an appendiculate trematode. *Bul. Mus. Comp. Zool.* **45**, 1–28.

LAPAGE, G. (1929). *Parasites*. Benn's Sixpenny Library, no. 76.

LA RUE (1926*a*). Studies on the trematode family Strigeidae (Holostomidae). No. I. *Pharyngostomum cordatum* (Diesing) Ciurea. *Trans. Amer. Micr. Soc.* **45**, 1–8.

LA RUE (1926*b*). Studies on the trematode family Strigeidae (Holostomidae). No. II. Taxonomy. *Trans. Amer. Micr. Soc.* **45**, 11–19.

LA RUE (1926*c*). Studies on the trematode family Strigeidae (Holostomidae). No. III. Relationships. *Trans. Amer. Micr. Soc.* **45**, 265–81.

LA RUE (1938). Life history studies and their relation to problems in taxonomy of digenetic tremates. *J. Parasit.* **24**, 1–11.

LA RUE, G. R. & AMEEL, D. J. (1937). The distribution of *Paragonimus*. *J. Parasit.* **23**, 382–8.

LAYMAN, E. M. (1922). Zur Charakteristik neuer *Lyperosomum*-Arten. *Zbl. Bakt.* (ii), **56**, 568–72.

LAYMAN, E. M. (1926*a*). Parasitic worms from fishes of the Murmansk coast. *Rab. parazit. Lab. moskovsk. Univ.* pp. 27–37.

LAYMAN, E. M. (1926*b*). Parasitic worms from the skate *Raja radiata*. *Rab. parazit. Lab. moskovsk. Univ.* pp. 9–26.

LAYMAN, E. M. (1926*c*). Parasitic worms from Murman birds. *Rab. parazit. Lab. moskovsk. Univ.* pp. 38–46.

LAYMAN, E. M. (1930). Parasitic worms from the fishes of Peter the Great Bay. *Bull. Pacif. Fish. Res. Sta.* **3**, 1–120.

LAYMAN, E. M. (1933*a*). Les vers parasitaires des poissons du lac Bajkal. *Trav. Sta. limnol. Lac. Bajkal*, **4**, 5–93.

LAYMAN, E. M. (1933*b*). Einige neue Tatsachen über die Ökologie der Froschtrematoden. *Zool. Anz.* **101**, 199–201.

LEBOUR, M. V. (1905). Notes on Northumbrian trematodes. *Rep. Northumb. Sea Fish. Comm.* 1905, p. 6.

LEBOUR, M. V. (1907). On three mollusk-infesting trematodes. *Ann. Mag. Nat. Hist.* (7), **19**, 102–6.

LEBOUR, M. V. (1908*a*). Fish trematodes of the Northumberland coast. *Rep. Northumb. Sea Fish. Comm.* 1907, 23–67.

LEBOUR, M. V. (1908*b*). Trematodes of the Northumberland coast. No. II. *Trans. Nat. Hist. Soc. Northumb.* **3**, N.S. 28–45.

LEBOUR, M. V. (1909). A preliminary note on *Echinostephilla virgula*, a new trematode in the turnstone. *Trans. Nat. Hist. Soc. Northumb.* **3**, N.S. 440–5.

LEBOUR, M. V. (1910). *Acanthopsolus lageniformis* n.sp., a trematode in the catfish. *Rep. Northumb. Sea Fish. Comm.* 1909 *to* 15 *June* 1910, pp. 29–35.

LEBOUR, M. V. (1911). A review of the British marine cercariae. *Parasitology*, **4**, 416–56.

LEBOUR, M. V. (1913). A new trematode of the genus *Lechriorchis* from the dark green snake (*Zamenis gemonensis*). *Proc. Zool. Soc. Lond.* pp. 833–6.

LEBOUR, M. V. (1914). Some larval trematodes from Millport. *Parasitology*, **7**, 1–11.

LEBOUR, M. V. (1916–18*a*). Medusae as hosts for larval trematodes. *J. Mar. Biol. Ass. U.K.* N.S. **11**, 57–9.

LEBOUR, M. V. (1916–18*b*). Some parasites of *Sagitta bipunctata*. *J. Mar. Biol. Ass. U.K.* N.S. **11**, 201–6.

LEBOUR, M. V. (1916–18*c*). A trematode larva from *Buccinum undatum* and notes on trematodes from post-larval fishes. *J. Mar. Biol. Ass. U.K.* N.S. **11**, 514–17.

LEBOUR, M. V. (1923). Note on the life history of *Hemiurus communis* Odhner. *Parasitology*, **15**, 233–5.

LEBOUR, M. V. (1935). *Hemiurus communis* in *Acartia*. *J. Mar. Biol. Ass. U.K.* **20**, 371–2.

LEBOUR, M. V. & ELMHIRST, R. (1922). A contribution towards the life history of *Parorchis acanthus* Nicoll, a trematode in the herring gull. *J. Mar. Biol. Ass. U.K.* **12**, 829–32.

LEE, C. U. & CHU, H. J. (1935). Simple technique for studying schistosome worms *in vitro*. *Proc. Soc. Exp. Biol., N.Y.*, **32**, 1397–1400.

LEIPER, J. W. G. (1938). The longevity of *Fasciola hepatica*. *J. Helminth.* **16**, 173–6.

LEIPER, R. T. (1913). Observations on certain helminths of man. *Trans. R. Soc. Trop. Med. Hyg.* **6**, 265–97.

LEIPER, R. T. (1915–18). Researches on Egyptian bilharziosis. *J. R. Asiatic Soc.* **25**, 1915, 1–85; **26**, 147–92, 253–67; **27**, 1916, 190–210; **30**, 1918, 235–60.

LEIPER, R. T. & ATKINSON, E. L. (1915). Observations on the spread of Asiatic Schistosomiasis. *Brit. Med. J.* 30 Jan., 16 pp.

LENT, H. & FREITAS, J. F. T. DE (1940*a*). Estado atual de trés espécies do gênero '*Cephalogonimus*' Poirier, 1886 (Trematoda). *Mem. Inst. Osw. Cruz*, **35**, 515–24.

LENT, H. & FREITAS, J. F. T. DE (1940*b*). Sur la position systématique de *Distoma arrectum* Molin, 1859. *Ann. Acad. bras. Sci.* **12**, 319–23.

LEPESKIN, V. D. (1914). Zur Ovogenese des *Zoogonus mirus*. *Moskva J. Sect. Zool. Soc. Nat.* **2**, 1–85.

LEWIS, E. A. (1926). Helminths of wild birds found in the Aberystwyth area. *J. Helminth.* **4**, 7–12.

LEWIS, F. J. (1935). The trematode genus *Phyllodistomum* Braun. *Trans. Amer. Micr. Soc.* **54**, 103–17.

LINDNER, E. (1914). Über die Spermatogenese von *Schistosomum haematobium* Bilh. mit besonderer Berücksichtigung der Geschlechtchromosomen. *Arch. Zellforsch.* **12**, 516–38.

LINSTOW, O. VON (1907). Zwei neue *Distomum* aus *Lucioperca sandra* der Wolga. *Ann. Mus. Zool. St Petersburg*, **12**, 201–2.

LINTON, E. (1901). Parasites of fishes in the Woods Hole region. *Bull. U.S. Fish. Comm.* 1900, pp. 405–92.

LINTON, E. (1905). Parasites of fishes of Beaufort, North Carolina. *Bull. U.S. Bur. Fish.* for 1904, **24**, 409.

LINTON, E. (1910). Helminth fauna of the Dry Tortugas. II. Trematodes. *Pap. Tortugas Lab.* **4** (*Publ. Carneg. Instn*, no. 113, pp. 11–98).

LINTON, E. (1914). Notes on a viviparous Distome. *Proc. U.S. Nat. Mus.* **46**, 551–5.

LINTON, E. (1915). *Tocotrema lingua* (Creplin), the adult stage of a skin-parasite of the cunner and other fishes of the Woods Hole region. *J. Parasit.* **1**, 128–34.

LINTON, E. (1928). Notes on trematode parasites of birds. *Proc. U.S. Nat. Mus.* **73**, Art. 1, p. 21.

LINTON, E. (1940). Trematodes from fishes mainly from the Woods Hole region, Massachusetts. *Proc. U.S. Nat. Mus.* (3078), **88**, 1–172.

LITTLE, P. A. (1929*a*). The trematode parasites of Irish marine fishes. *Parasitology*, **21**, 22–30.

LITTLE, P. A. (1929*b*). *Trochopus gaillimhe* n.sp., an ectoparasitic trematode of *Trigla hirundo* or *Trigla lucerna*. *Parasitology*, **21**, 107–19.

LITTLE, P. A. (1929*c*). The anatomy and histology of *Phyllonella solaea* Ben. & Hesse, an ectoparasitic trematode of the sole, *Solea vulgaris* Quensel. *Parasitology*, **21**, 324–37.

LITTLE, P. A. (1930). A new trematode parasite of *Acipenser sturio* L. (Royal sturgeon), with a description of the genus *Dihemistephanus* Lss. *Parasitology*, **22**, 399–413.

LLEWELLYN, J. (1941*a*). A revision of the monogenean family Diclidophoridae Fuhrmann, 1928. *Parasitology*, **33**, 416–30.

LLEWELLYN, J. (1941*b*). The taxonomy of the monogenetic trematode *Plectanocotyle gurnardi* (v. Ben. & Hesse). *Parasitology*, **33**, 431–2.

LLEWELLYN, J. (1941*c*). A description of the anatomy of the monogenetic trematode *Choricotyle chrysophryi*. *Parasitology*, **33**, 397–405.

LLOYD, L. C. (1938). Some digenetic trematodes from Puget Sound fish. *J. Parasit.* **24**, 103–33.

LLOYD, L. C. & GUBERLET, J. E. (1932). A new genus and species of Monorchidae. *J. Parasit.* **18**, 232–9.

LOEWEN, S. L. (1929). A description of the trematode *Catoptroides lacustri* n.sp., with a review of the known species of the genus. *Parasitology*, **21**, 55–62.

LOOSS, A. (1900*a*). Recherches sur la faune parasitaire de l'Égypte. *Mém. Inst. Égypt*, **3**, 152.

LOOSS, A. (1900*b*). Nachträgliche Bemerkungen zu den Namen der von mir vorgeschlagenen Distomengattungen. *Zool. Anz.* **23**, no. 630, pp. 601–8.

LOOSS, A. (1901*a*). Ueber einige Distomen der Labriden des Triester Hafens. *Zbl. Bakt.* **29**, 398–405, 437–42.

LOOSS, A. (1901*b*). Ueber die Fasciolidengenera *Stephanochasmus, Acanthochasmus* und einige andere. *Zbl. Bakt.* **29**, 595–606, 628–34, 654–61.

LOOSS, A. (1902*a*). Zur Kenntnis der Trematodenfauna des Triester Hafens. I. Ueber die Gattung *Orophocotyle* n.g. *Zbl. Bakt.* **31**, 637–44.

LOOSS, A. (1902*b*). Zur Kenntnis der Trematodenfauna des Triester Hafens. II. Ueber *Monorchis* Montic. und *Haplosplanchnus* n.g. *Zbl. Bakt.* **31**, 115–22.

LOOSS, A. (1902*c*). Ueber neue und bekannte Trematoden aus Seeschildkröten. *Zool. Jb., Syst.* **16** (3–6), 411–894.

LOOSS, A. (1902*d*). Notizen zur Helminthologie Ägyptens. V. Eine Revision der Fascioliden Gattung *Heterophyes* Cobb. *Zbl. Bakt.* I. Orig. pp. 886–91.

LOOSS, A. (1902*e*). Die Distomenunterfamilie der Haploporinae. *Arch. Parasit.* **6**, 129–43.

LOOSS, A. (1905). Von Würmern und Arthropoden hervorgerufene Erkrankungen. *Mense's Handb. d. Tropenkr.*

LOOSS, A. (1907*a*). Beiträge zur Systematik der Distomen. *Zool. Jb., Syst.* **26**, 63–180.

LOOSS, A. (1907*b*). Zur Kenntniss der Distomenfamilie Hemiuridae. *Zool. Anz.* **19/20**.

LOOSS, A. (1907*c*). On some parasites in the museum of the School of Tropical Medicine, Liverpool. *Ann. Trop. Med. Parasit.* **1**, 123–53.

LOOSS, A. (1908). Beiträge zur Systematik der Distomen. Zur Kenntnis der Familie Hemiuridae. *Zool. J., Syst.* **26**, 64–180.

LOOSS, A. (1912). Über den Bau einiger auscheinend seltner Trematoden-Arten. *Zool. Jb.* Suppl. 15, **1**, 323–66.

LÓPEZ-NEYRA, C. R. (1941*a*). Compendio di Helmintologia Iberica. 1*a*. Introduccion historica. 2*a*. Sistematica de los helmintos. *Rev. Iberica Parasit.* **1**, 7–34.

LÓPEZ-NEYRA, C. R. (1941*b*). Compendio di Helmintologia Iberica. (Continued.) *Rev. Iberica Parasit.* **1**, 171–83.

LÓPEZ-NEYRA, C. R. (1941*c*). Sobre dos *Lyperosomum* nuevos para la fauna helmintologica iberica. *Rev. Iberica Parasit.* **1**, 35–43.

LOWE, P. R. & BAYLIS, H. A. (1934). On a flock of razorbills in Middlesex found to be infected with intestinal flukes. With a parasitological report. *Brit. Birds*, **28**, 188–90.

LÜHE, M. (1900). Über die Gattung *Podocotyle* (Duj.) Stoss. *Zool. Anz.* **23**, 487–92.

LÜHE, M. (1901*a*). Über Hemiuriden. *Zool. Anz.* **24**, 394–403, 473–88.

LÜHE, M. (1901*b*). Über *Monostomum orbiculare*. *Zbl. Bakt.* I. Orig. **29**, 49–60.

LÜHE, M. (1906). Report on the trematode parasites from the marine fishes of Ceylon. *Lond. R. Soc. Rep. Pearl Oyster*, **5**, 97–108.

LÜHE, M. (1908). Zur Systematik und Faunistik der Distomen. I. Die Gattung *Metorchis* Looss, nebst Bemerkungen über die Familie Opisthorchiidae. *Zbl. Bakt.* I. Orig. **48**, 428–36.

LÜHE, M. (1909). Trematodes. In *Die Süsswasserfauna Deutschlands*, Heft 17, 217 pp.

LUNDAHL, W. S. (1941). Life history of *Caecincola parvulus* Marshall & Gilbert (Cryptogonimidae, Trematoda) and the development of its excretory system. *Trans. Amer. Micr. Soc.* **60**, 461–84.

LUTZ, A. (1921). Zur Kenntnis des Entwicklungszyklus der Holostomiden. *Zbl. Bakt.* I. Orig. **86**, 124–9.

LYSAGHT, A. M. (1941). The biology and trematode parasites of the gastropod *Littorina neritoides* (L.) on the Plymouth Breakwater. *J. Mar. Biol. Ass. U.K.* **25**, 41–67.

LYSAGHT, A. M. (1943). The incidence of larval trematodes in males and females of the gastropod *Littorina neritoides* (L.) on the Plymouth Breakwater. *Parasitology*, **35**, 17–22.

LYUBINSKI, G. A. & KULAKOVSKAYA, O. P. (1940). L'intensité de l'invasion et variabilité de l'*Opisthorchis felineus* (Riv. 1884). *Med. Parasitol., Moscow*, **9**, 434–8.

MACCAGNO, T. (1934). Il *Clinostomum marginatum* Rud. in Europa. *Boll. Mus. Zool. Anat. comp. Torino*, **44**, 5–12.

MACCALLUM, G. A. (1913). Further notes on the genus *Microcotyle*. *Zool. Jb., Syst.* **35**, 389–402.

MACCALLUM, G. A. (1915). Some new species of ectoparasitic trematodes. *Zoopathologica*, pp. 393–410.

MACCALLUM, G. A. (1917). Some new forms of parasitic worms. *Zoopathologica*, **1**, 2, 45–75.

MACCALLUM, G. A. (1918). Notes on the genus *Telorchis* and other trematodes. *Zoopathologica*, **1**, 3, 81–98.

MACCALLUM, G. A. (1921). Studies in helminthology. *Zoopathologica*, **1**, 3, 137–284.

MACCALLUM, G. A. (1926*a*) Deux nouveaux trématodes parasites de *Carcharinus commersonii; Philura ovata* et *Dermophthirius carcharini*. *Ann. Parasit. hum. comp.* **4**, 162–7.

MACCALLUM, G. A. (1926*b*). *Dermophagus squali* n.g., n.sp. *Ann. Parasit. hum. comp.* **4**, 330–2.

MACCALLUM, G. A. (1927). A new ectoparasitic trematode *Epibdella melleni* sp. nov. *Zoopathologica*, **1**, 3, 291–300.

MACCALLUM, G. A. & MACCALLUM, W. G. (1913). Four species of *Microcotyle, M. pyragraphorus, mucroura, eueides* and *acanthophallus*. *Zool. Jb., Syst.* **34**, 223–44.

MACFARLANE, W. V. (1939). Life cycle of *Coitocaecum anaspidis* Hickman, a New Zealand digenetic trematode. *Parasitology*, **31**, 172–83.

MACHATTIE, C. & CHADWICK, C. R. (1932). *Schistosoma bovis* and *S. mattheei* in Irak, with notes on the development of the eggs of the *S. haematobium* pattern. *Trans. R. Soc. Trop. Med. Hyg.* **26**, 147–56.

MACLAREN, N. (1903). On trematodes and cestodes parasitic in fishes. *Rep. Brit. Ass. Adv. Sci.* **72**, 260–2.

MACLAREN, N. (1904). Beiträge zur Kenntnis einiger Trematoden (*Diplectanum aequans* Wagener und *Nematobothrium molae* n.sp.). *Jena. Z. Naturwiss.* **38**, 573–618.

MACY, R. W. (1936*a*). A new genus and species of trematode from the little brown rat and a key to the genera of Pleurogenetinae. *Proc. U.S. Nat. Mus.* **83**, no. 2986, pp. 321–4.

MACY, R. W. (1936*b*). Three new trematodes of Minnesota bats with a key to the genus *Prosthodendrium*. *Trans. Amer. Micr. Soc.* **55**, 352–9.

MACY, R. W. (1942). The life cycle of the trematode *Echinostomum callawayensis* Barker. *J. Parasit.* **28**, 431–2.

MADSEN, H. (1941). The occurrence of helminths and Coccidia in partridges and pheasants in Denmark. *J. Parasit.* **27**, 29–34.

MAGATH, T. B. (1917). The morphology and life history of a new trematode parasite, *Lissorchis fairporti* nov.gen. et nov.spec., from the buffalo fish, *Ictiobus*. *J. Parasit.* **4**, 58–69.

MAGATH, T. B. (1920). *Leucochloridium problematicum* n.sp. *J. Parasit.* **6**, 105–14.

MALEVITSKIJA, M. (1941). Neue Dactylogyrus-Arten (Trematoda) aus dem Dnjepr. *C.R. Acad. Sci. Moscow*, N.S. **30**, 269–71.

MANTER, H. W. (1926). Some North American fish trematodes. *Ill. Biol. Monogr.* **10**, 1–138.

MANTER, H. W. (1931). Some digenetic trematodes of marine fishes of Beaufort, North Carolina. *Parasitology*, **23**, 396–411.

MANTER, H. W. (1932). Continued studies on trematodes of Tortugas. *Yearb. Carneg. Instn*, **31**, 287–8.

MANTER, H. W. (1933). The genus *Helicometra* and related trematodes from Tortugas, Florida. *Pap. Tortugas Lab.* **28** (*Publ. Carneg. Instn*, no. 435, pp. 167–82).

MANTER, H. W. (1934). Some digenetic trematodes from deep-water fishes of Tortugas. *Pap. Tortugas Lab.* **28** (*Publ. Carneg. Instn*, no. 435, pp. 257–345).

MANTER, H. W. (1937*a*). The status of the trematode genus *Deradena* Linton, with a description of six species of *Haplosplanchnus* Looss (Trematoda). *Skrjabin Jub. Vol.* pp. 381–7.

MANTER, H. W. (1937*b*). Modifications of the acetabulum in trematodes. *J. Parasit.* **23**, 566.

MANTER, H. W. (1939). Digenetic trematodes of fishes of the Galapagos Islands and their relationships. *J. Parasit.* **35**, Suppl. Abstr., p. 25.

MANTER, H. W. (1940*a*). Digenetic trematodes of fishes from the Galapagos Islands and the neighbouring Pacific. *Rep. Coll. A. Hancock Pacific Exp.* 1932–8, **2** (14), 329–497.

MANTER, H. W. (1940*b*). The geographical distribution of digenetic trematodes of marine fishes of the tropical American Pacific. *Rep. Coll. A. Hancock Pacific Exp.* 1932–8, **2** (16), 531–47.

MANTER, H. W. (1940*c*). Gasterostomes (Trematoda) of Tortugas, Florida. *Pap. Tortugas Lab.* **33** (*Publ. Carneg. Instn*, no. 524, pp. 1–19).

MANTER, H. W. (1941). Observations on the geographical distribution of digenetic trematodes of marine fishes. *J. Parasit.* **27**, Suppl. Abstr., **72**, 33–4.

MANTER, H. W. (1942). Monorchidae (Trematoda) from fishes of Tortugas, Florida. *Trans. Amer. Micr. Soc.* **61**, 349–60.

MAPLESTONE, P. A. (1923). A revision of the Amphistomata of mammals. *Ann. Trop. Med. Parasit.* **14**, 113–212.

MARKELL, E. K. (1943). Gametogenesis and egg-shell formation in *Probilotrema californiense* Stunkard, 1935 (Trematoda: Gorgoderidae). *Trans. Amer. Micr. Soc.* **62**, 27–56.

MARKOWSKI, S. (1933*a*). Die Eingeweidewürmer der Fische des Polaischen Balticums (Trematoda, Cestoda, Nematoda, Acanthocephala). *Arch. Hydrobiol. Ichthyol. Suwalki*, **7**, 1–58.

MARKOWSKI, S. (1933*b*). Untersuchungen über die Helminthenfauna der Raben (Corvidae) von Polen. *Mém. Cl. Sci. Acad. polon.* no. 5, pp. 1–65.

MARKOWSKI, S. (1935). Die parasitischen Würmer von *Gobius minutus* Pall. des polnischen Balticums. *Bull. int. Acad. Cracovie*, B, **2**, 251–60.

MARTIN, W. E. (1936). A new second intermediate host of the trematode, *Gorgodera amplicava*. *Proc. Indiana Acad. Sci.* **46**, p. 253.

MARTIN, W. E. (1938*a*). Studies on the trematodes of Woods Hole. The life cycle of *Lepocreadium setiferoides* (Miller & Northup), Allocreadiidae, and the description of *Cercaria cumingiae* n.sp. *Biol. Bull. Woods Hole*, **75**, 463–74.

MARTIN, W. E. (1938*b*). The life cycle of *Stephanostomum tenue* (Linton), family Acanthocolpidae. *J. Parasit.* **24**, Suppl. Abstr. p. 27.

MARTIN, W. E. (1939*a*). The life cycle of *Monorcheides cumingiae* (Martin) (Trematoda: Monorchiidae). *J. Parasit.* **25**, Suppl. Abstr. p. 18.

MARTIN, W. E. (1939*b*). Studies on the trematodes of Woods Hole. II. The life cycle of *Stephanostomum tenue* (Linton). *Biol. Bull. Woods Hole*, **77**, 65–73.

MARTIN, W. E. (1940). Studies on the trematodes of Woods Hole. III. The life cycle of *Monorcheides cumingiae* (Martin) with special reference to its effect on the invertebrate host. *Biol. Bull. Woods Hole*, **79**, 131–44.

MASSA, D. (1906). Materiali per una revisione del genere *Trochopus*. *Arch. zool.* (*ital.*) *Napoli*, **3**, 43–71.

MASSINO, B. G. (1927). Bestimmung der Arten der Gattung *Plagiorchis* Lühe. *Samml. Helm. Arb. K. I. Skrjabin gewidmet, Moskau*, pp. 108–13.

MATHESON, C. (1930). Notes on *Cercaria elvae* Miller as the probable cause of an outbreak of dermatitis at Cardiff. *Trans. R. Soc. Trop. Med. Hyg.* **23**, 421.

MATHIAS, P. (1922). Cycle évolutif d'un trématode Holostomide (*Strigea tarda* Steenst.). *C.R. Acad. Sci., Paris*, **175**, 599–602.

MATHIAS, P. (1924*a*). Cycle évolutif d'un trématode échinostome (*Hypoderaeum conoideum* Bloch). *C.R. Soc. Biol., Paris*, **90**, 13–15.

MATHIAS, P. (1924*b*). Sur le cycle évolutif d'un trématode de la famille des Psilostomidae (*Psilotrema spiculigerum* Mühling). *C.R. Acad. Sci., Paris*, **178**, 1217–19.

MATHIAS, P. (1924*c*). Cycle évolutif d'un trématode Holostomide (*Strigea tarda* Steenst.). *C.R. Acad. Sci., Paris*, **175**, 599–602.

MATHIAS, P. (1924*d*). Contribution à l'étude du cycle évolutif d'un trématode de la famille des Pleurogenetinae Lss. (*Pleurogenes medians* Olss.). *Bull. Soc. zool. Fr.* **49**, 375–7.

MATHIAS, P. (1925). Recherches expérimentales sur le cycle évolutif de quelques trématodes. *Bull. Biol.* **49**, 1–123.

MATHIAS, P. (1926*a*). Sur le cycle évolutif d'un trématode de la famille Échinostomidae Dietz (*Echinoparyphium recurvatum* Linstow). *C.R. Acad. Sci., Paris*, **183**, 90–2.

MATHIAS, P. (1926*b*). Sur une nouvelle espèce de trématode *Peracreadium perezi* nov.sp. *Bull. Soc. Zool. Fr.* **51**, 353–6.

MATHIAS, P. (1930). Sur *Cercaria ocellata* La Valette. *Ann. Parasit. hum. comp.* **8**, 151.

MATHIAS, P. (1934). Sur quelques trématodes de poissons marins de la région de Banyuls. *Arch. Zool. exp. gén.* **75**, 567–81.

MATHIAS, P. (1937). Cycle évolutif d'un trématode de la famille Allocreadiidae Stossich (*Allocreadium angusticolle* (Hausmann)). *C.R. Acad. Sci., Paris*, **205**, 626–8.

MATHIAS, P. & VIGNAUD, R. (1935). Sur le cycle évolutif d'un trématode de la sous-famille des Pleurogenetinae Looss (*Pleurogenes claviger*). *C.R. Soc. Biol., Paris*, **120**, 397–8.

MATTES, O. (1933). Neues über den Entwicklungsgang des Lanzettegels (*Dicrocoelium lanceatum*). *Naturwissenschaften*, **21**, 237.

MATTES, O. (1934). Der Entwicklungsgang des Lanzettegels (*Dicrocoelium lanceatum*) vollständig aufgeklärt. *Naturwissenschaften*, **22**, 777–8.

MATTES, O. (1936). Der Entwicklungsgang des Lanzettegels *Dicrocoelium lanceatum*. *Z. Parasitenk.* **8**, 371–430.

MATTES, O. (1937). Abschliessender Bericht über die in den letzten Jahren im Marburger Zoologischen Institut durchgeführten Untersuchungen zur Aufdeckung des Entwicklungsganges des Lanzettegels. *S.B. Ges. ges. Naturw. Marburg*, **72**, 69–100.

MCCOY, O. R. (1928). Life-history studies on trematodes from Missouri. *J. Parasit.* **14**, 207–28.

MCCOY, O. R. (1929*a*). Notes on cercariae from Missouri. *J. Parasit.* **15**, 199–208.

MCCOY, O. R. (1929*b*). Observations on the life history of a marine lophocercous cercaria. *J. Parasit.* **16**, 29–34.

MCCOY, O. R. (1929*c*). The life history of a marine trematode *Hamacreadium mutabile* Linton, 1910. *Parasitology*, **21**, 220–5.

MCCOY, O. R. (1930). Experimental studies on two fish trematodes of the genus *Hamacreadium* (family Allocreadiidae). *J. Parasit.* **17**, 1–13.

MCCOY, O. R. (1935). Physiology of animal parasites. *Physiol. Rev.* **15**, 221–40.

MCFARLANE, S. H. (1936). A study of the endoparasitic trematodes from marine fishes of Departure Bay, B.C. *J. Biol. Bd Canada*, **2**, 335–47.

MCINTOSH, A. (1927). Notes on the genus *Leucochloridium* Carus (Trematoda). *Parasitology*, **19**, 353–64.

MCINTOSH, A. (1932). Some new species on trematode worms of the genus *Leucochloridium* Carus, parasitic in birds from Northern Michigan, with a key and notes on other species of the genus. *J. Parasit.* **19**, 32–53.

MCINTOSH, A. (1934*a*). A new blood trematode, *Paradeontacyclix sanguinicoloides* n.g., n.sp., from *Seriola lalandi*, with a key to species of the family Aporocotylidae. *Parasitology*, **26**, 463–7.

McIntosh, A. (1934*b*). Two new species of trematodes, *Scaphiostomum pancreaticum* n.sp. and *Postharmostomum laruei* n.sp. from the chipmunk. *Proc. Helminth. Soc. Wash.* **1**, 2–3.

McIntosh, A. (1937). Two new avian liver flukes, with a key to the species of *Athesmia* Looss, 1899 (Dicrocoeliidae). *Proc. Helminth. Soc. Wash.* **4**, 21–3.

McMullen, D. B. (1936). A note on the staining of the excretory system of trematodes. *Trans. Amer. Micr. Soc.* **55**, 513–15.

McMullen, D. B. (1937*a*). The life histories of three trematodes, parasitic in birds and mammals, belonging to the genus *Plagiorchis*. *J. Parasit.* **23**, 235–43.

McMullen, D. B. (1937*b*). A discussion of the taxonomy of the family Plagiorchiidae Lühe, 1901, and related trematodes. *J. Parasit.* **23**, 244–58.

Mehl, S. (1932). Die Lebensbedingungen der Leberegelschnecke. *Arb. bayer. Landesanst. PflBau.* **10**, 177 pp.

Mehra, H. R. (1931). A new genus (*Spinometra*) of the family Lepodermatidae Odhner (Trematoda) from a tortoise, with a systematic discussion and classification of the family. *Parasitology*, **23**, 157–78.

Mehra, H. R. (1935). New trematodes of the family Lecithodendriidae Odhner, 1911, with a discussion of the classification of the family. *Proc. Acad. Sci. U.P. India*, **5**, 99–121.

Mehra, H. R. (1936). A new species of the genus *Harmotrema* Nicoll, 1914, with a discussion on the systematic position of the genus and the classification of the family Harmostomidae Odhner, 1912. *Proc. Nat. Acad. Sci. India, Allahabad*, **6**, 217–40.

Meier, N. T. (1913). Wissenschaftlichen Mitteilungen. I. Einige Versuche über die Regeneration parasitierender Platodes und deren Züchtung in künstlichem Medium. *Zool. Anz.* **42**, 481–7.

Mellen, L. (1928). The treatment of fish diseases. *Zoopathologica*, **2**, 1–31.

Mendheim, H. (1939). Über eine zweckmässige Abänderung der Looss'schen Schüttelmethode nebst Bemerkungen zur helminthologischen Technik. *Z. Parasitenk.* **10**, 436.

Mendheim, H. (1940). Beiträge zur Systematik und Biologie der Familie Echinostomidae (Trematoda). *Nova Acta Leop. Carol.* **8**, 489–588; Abstr. in *Biol. Zbl.* **61**, 1941, p. 97.

Meserve, F. G. (1938). Some monogenetic trematodes from the Galapagos Islands and the neighbouring Pacific. *Rep. Allen Hancock Pacific Exp.* **2** (5), 31–88.

Meserve, F. G. (1943). *Phyllodistomum coatneyi* n.sp., a trematode from the urinary bladder of *Ambystoma* (sic) *maculatum* (Shaw). *J. Parasit.* **29**, 226–8.

Metcalf, M. M. (1929). Parasites and the aid they give in problems of taxonomy, geographical distribution, and paleogeography. *Smithson. Misc. Coll.* **81**, no. 8, pp. 1–36.

Meyerhof, E. & Rothschild, M. (1940). A prolific trematode. *Nature, Lond.*, **146**, 367.

Miller, E. L. (1936). Studies on North American cercariae. *Ill. Biol. Monogr.* **14**, no. 2, 125 pp.

Miller, H. M. (1923). Notes on some furcocercous larval trematodes. *J. Parasit.* **10**, 35–46.

Miller, H. M. (1925). Larval trematodes of certain marine gastropods from Puget Sound. *Publ. Puget Sd Mar. (Biol.) Sta.* **5**, 75–89.

Miller, H. M. (1926*a*). Behaviour studies on Tortugas larval trematodes, with notes on the morphology of two additional species. *Yearb. Carneg. Instn*, no. 25, pp. 243–7.

Miller, H. M. (1926*b*). Comparative studies on furcocercous cercariae. *Ill. Biol. Monogr.* **10**, no. 3, 112 pp.

Miller, H. M. (1927). Furcocercous larval trematodes from San Juan Island, Washington. *Parasitology*, **19**, 61–82.

Miller, H. M. (1928). Variety of behaviour of larval trematodes. *Science*, **68**, 117.

Miller, H. M. (1929*a*). A large-tailed Echinostome cercaria from North America. *Trans. Amer. Micr. Soc.* **48**, 310–13.

Miller, H. M. (1929*b*). Behaviour and reactions of a larval trematode especially to changes of light-intensity. *Anat. Rec.* **41**, 36.

Miller, J. N. (1939). Observations on the rate of growth of the trematode *Postharmostomum laruei* McIntosh, 1934. *J. Parasit.* **25**, 509–10.

MILLER, M. J. (1940). Blackspot in fishes. *Canad. J. Comp. Med.* **4** (11), 303–5.

MILLER, M. J. (1941*a*). A critical study of Stafford's report on 'Trematodes of Canadian Fishes' based on his trematode collection. *Canad. J. Res.* D, **19**, 28–52.

MILLER, M. J. (1941*b*). The life history of *Apophallus brevis* Ransom, 1920. *J. Parasit.* **27**, Suppl. Abstr. **2**, 10.

MILLER, H. M. & MAHAFFEY, E. E. (1930). Reaction of *Cercaria hamata* to light and to mechanical stimuli. *Biol. Bull. Woods Hole*, **59**, 95–103.

MILLER, H. M. & NORTHUP, F. (1926). The seasonal infestation of *Nassa obsoleta* (Say) with larval trematodes. *Biol. Bull. Woods Hole*, **5**, 490.

MIZELLE, J. D. (1936). New species of trematodes from the gills of Illinois fishes. *Amer. Midl. Nat.* **17** (5), 785–806.

MIZELLE, J. D. (1938). Comparative studies on trematodes (Gyrodactyloidea) from the gills of North American fresh-water fishes. *Ill. Biol. Monogr.* **7**, no. 1, pp. 81.

MIZELLE, J. D. (1940*a*). Studies on monogenetic trematodes. II. New species from Tennessee. *Trans. Amer. Micr. Soc.* **59**, 285–9.

MIZELLE, J. D. (1940*b*). Studies on monogenetic trematodes. III. Redescriptions and variations in known species. *J. Parasit.* **26**, 165–78.

MIZELLE, J. D. (1941). Studies on monogenetic trematodes. IV. *Anchoradiscus*, a new Dactylogyrid genus from the bluegill and the stump-knocker sunfish. *J. Parasit.* **27**, 159–63.

MÖDLINGER, G. (1930). Trematoden ungarischer Chiropteren. *Studia zool.* **1**, 191–203.

MOLA, P. (1912). Die Parasiten des *Cottus gobio*. *Zbl. Bakt.* **65**, 491–504.

MORGAN, D. C. (1924). A survey of helminthic parasites of domestic animals in the Aberystwyth area of Wales. *J. Helminth.* **2**, 89–94.

MORGAN, D. C. (1927). Studies on the family Opisthorchiidae Braun, 1901, with a description of a new species of *Opisthorchis* from a Sarus crane (*Antigone antigone*). *J. Helminth.* **5**, 89–104.

MORISHITA, K. (1924). On the trematodes of the genus *Cyclocoelum* obtained in Japan, with notes on the phylogeny of the monostomatous trematodes. *Zool. Mag., Tokyo*, **36**, 89–104.

MÖNNIG, H. O. (1922). *Über* Leucochloridium macrostomum (Leucochloridium paradoxum *Carus*), ein Beitrag zur Histologie der Trematoden, 61 pp. Jena.

MÖNNIG, H. O. (1934). *Veterinary Helminthology and Entomology*, 402 pp. London.

MONTGOMERIE, R. F. (1928). Observations on artificial infestation of sheep with *Fasciola hepatica*, etc. *J. Helminth.* pp. 167–74.

MONTGOMERIE, R. F. (1929). Recent research on the dosage of sheep infested with liver flukes. *Scot. J. Agric.* **12**, 6 pp.

MONTICELLI, F. S. (1902). A proposito di una nuova specie del genere *Epibdella*. *Boll. Soc. Nat. Napoli*, **15**, 137–45.

MONTICELLI, F. S. (1903). Per una nuova classificazione degli Heterocotylea. *Monit. zool. ital.* **14**, 334–6.

MONTICELLI, F. S. (1904). Il genere *Lintonia* Monticelli. *Arch. Zool.* **2**, 117–24.

MONTICELLI, F. S. (1905). Osservazioni intorno ad alcune specie di Heterocotylea. *Boll. Soc. Nat. Napoli*, **18**, 65–80.

MONTICELLI, F. S. (1907). Il genere *Encotyllabe* Diesing. *Atti Ist. Sci. nat. Napoli*, **59**, 23–35.

MONTICELLI, F. S. (1908). Il genere *Nitzschia* von Baer. *Annuario Museo zool. Napoli*, N.S. **2**, 27, 19 pp.

MONTICELLI, F. S. (1909). Identificazione di una n.sp. del genere *Encotyllabe* (*lintonii* Monticelli). *Boll. Soc. Nat. Napoli*, **22**, Ser. 2, 2, 86–88.

MONTICELLI, F. S. (1910). *Calinella craneola* n.g., n.sp. Trématode nouveau de la famille des Udonellidae provenant des campagnes de S.A.S. le Prince de Monaco. *Ann. Inst. océan., Monaco*, **1**, 1–9.

MONTICELLI, F. S. (1912). Ricerche sulla *Cercaria setifera* Müller. *R.C. Accad. Napoli*, fasc. 7–9.

MONTICELLI, F. S. (1914). Ricerche sulla *Cercaria setifera* di Joh. Müller. *Atti Accad. Sci. fis. mat. Napoli*, **15**, 2, no. 11, 48 pp.

MUELLER, J. F. (1930). The trematode genus *Plagiorchis* in fishes. *Trans. Amer. Micr. Soc.* **40**, 174–7.

MUELLER, J. F. (1936*a*). Studies on North American Gyrodactyloidea. *Trans. Amer. Micr. Soc.* **55** (1), 55–72.

MUELLER, J. F. (1936*b*). New gyrodactyloid trematodes from North American fishes. *Trans. Amer. Micr. Soc.* **55** (4), 457–64.

MUELLER, J. F. (1937). The Gyrodactylidae of North American fresh-water fishes. *Fish Culture*, **3** (1), 1–14.

MÜHLSCHLAG, G. (1914). Beitrag zur Kenntnis der Anatomie von *Otodistomum veliporum* (Creplin), *Distomum fuscum* Poirier und *Distomum ingens* Moniez. *Zool. Jb., Syst.* **37**, 199–252.

MÜLLER, W. (1923). Die Nahrung von *Fasciola hepatica* und ihre Verdauung. *Zool. Anz.* **57**, 273–81.

MURIKAMA, S. (1937). Über die Eischalenbildung bei den Trematoden. (Japanese with German summary.) *J. Okayama Med. Soc.* **49**, 706–68.

MURRAY, F. V. (1931). Gill trematodes from some Australian fishes. *Parasitology*, **23**, 492–506.

NAGANO, K. (1930). On the intermediate host of *Asymphlodora tincae* in Japan. *Trans. II. Ann. Meeting Parasit. Soc. Japan*, p. 24.

NAGATY, H. F. (1937). Trematodes of fishes from the Red Sea. Part I. Studies on the family Bucephalidae Poche, 1907. *Egypt. Univ. Faculty Med. Publ.* **12**, 1–172.

NAGATY, H. F. (1941). Trematodes of fishes from the Red Sea; the genus *Hamacreadium* Linton, 1910, family Allocreadiidae, with a description of two new species. *J. Egypt. Med. Ass.* **24**, 300–10.

NAKAGAWA, K. (1921). On the life cycle of *Fasciolopsis buski*. *Kitasato Arch.* **4**, 159–67.

NAKAGAWA, K. (1922). The development of *Fasciolopsis buski* Lankester. *J. Parasit.* **8**, 161–6.

NARABAYASHI, H. (1914). On prenatal infection with *Schistosoma japonicum*. *Verh. nap. path. Ges.* no. 4. (Japanese.)

NÄSMARK, K. E. (1937). A revision of the trematode family Paramphistomidae. *Zool. Bidr. Uppsala*, **16**, 301–566.

NEUHAUS, W. (1936). Untersuchungen über Bau und Entwicklung der Lanzettegelcercariae (*Cercaria vitrina*) und Klarstellung des Infektionsvorganges beim Endwirt. *Z. Parasitenk.* **8**, 431–73.

NEUHAUS, W. (1939). Der Invasionsweg der Lanzettegelcercariae bei der Infektion des Endwirtes und ihre Entwicklung zum *Dicrocoelium lanceatum*. *Z. Parasitenk.* **10**, 476–512.

NEVEU-LEMAIRE, M. (1936). *Traité d'helminthologie médicale et vétérinaire*, 1514 pp. Paris.

NICKERSON, W. S. (1902). *Cotylogaster occidentalis* n.sp. and a revision of the family Aspidobothridae. *Zool. Jb. Syst.* **15**, 597–624.

NICOLL, W. (1906). Some new and little known trematodes. *Ann. Mag. Nat. Hist.* (7), **17**, 513–26.

NICOLL, W. (1907*a*). A contribution towards a knowledge of the Entozoa of British marine fishes. Part I. *Ann. Mag. Nat. Hist.* (7), **19**, 66–94.

NICOLL, W. (1907*b*). A contribution towards a knowledge of the Entozoa of British marine fishes. Part II. *Ann. Mag. Nat. Hist.* (8), **4**, 1–25.

NICOLL, W. (1907*c*). *Parorchis acanthus*, the type of a new genus of trematodes. *Quart. J. Micr. Sci.* **51**, 345–55.

NICOLL, W. (1909). Studies on the structure and classification of the digenetic trematodes. *Quart. J. Micr. Sci.* **53**, 391–487.

NICOLL, W. (1910*a*). The bionomics of helminths. *Trans. Soc. Trop. Med.* **3**, 353–78.

NICOLL, W. (1910*b*). On the entozoa of fishes from the Firth of Clyde. *Parasitology*, **3**, 322–59.

NICOLL, W. (1912). On the new trematode parasites from the Indian cobra. *Proc. Zool. Soc. Lond.* pp. 851–6.

NICOLL, W. (1912). On two new trematode parasites from British food-fishes. *Parasitology*, **5**, 197–202.

NICOLL, W. (1913*a*). New trematode parasites from fishes of the English Channel. *Parasitology*, **5**, 238–46.

NICOLL, W. (1913*b*). Trematode parasites from food-fishes of the North Sea. *Parasitology*, **6**, 188–94.

NICOLL, W. (1914*a*). Trematode parasites of fishes in the English Channel. *J. Mar. Biol. Ass. U.K.* N.S. **10**, 466–505.

NICOLL, W. (1914*b*). Trematode parasites from animals dying in the Zoological Society's Gardens during 1911–1912. *Proc. Zool. Soc. Lond.* pp. 139–54.

NICOLL, W. (1914*c*). The trematode parasites of North Queensland. I. *Parasitology*, **6**, 333–50.

NICOLL, W. (1914*d*). The trematode parasites of North Queensland. II. Parasites of birds. *Parasitology*, **7**, 105–26.

NICOLL, W. (1915*a*). A list of the trematode parasites of British marine fishes. *Parasitology*, **7**, 339–78.

NICOLL, W. (1915*b*). A new liver fluke (*Platynosomum acuminatum*) from the kestrel. *Proc. Zool. Soc. Lond.* pp. 87–9.

NICOLL, W. (1923*a*). A reference list of the trematode parasites of British birds. *Parasitology*, **15**, 151–202.

NICOLL, W. (1923*b*). A reference list of the trematode parasites of British mammals. *Parasitology*, **15**, 236–52.

NICOLL, W. (1924*a*). A reference list of the trematode parasites of British fresh-water fishes. *Parasitology*, **16**, 127–44.

NICOLL, W. (1924*b*). A reference list of the trematode parasites of British reptiles. *Parasitology*, **16**, 329–31.

NICOLL, W. (1926). A reference list of the trematode parasites of British Amphibia. *Parasitology*, **18**, 14–20.

NICOLL, W. (1927). A reference list of the trematode parasites of man and the primates. *Parasitology*, **19**, 338–51.

NICOLL, W. (1934, 1935). Vermes Section in *Zool. Rec.*

NICOLL, W. & SMALL, W. (1909). Notes on larval trematodes. *Ann. Mag. Nat. Hist. Lond.* (**8**), 3, 237–46.

NIGRELLI, R. F. (1935*a*). On the effect of fish mucus on *Epibdella melleni*, a monogenetic trematode of marine fishes. *J. Parasit.* **21**, Suppl. Abstr. p. 438.

NIGRELLI, R. F. (1935*b*). Experiments on the control of *Epibdella melleni* MacCallum, a monogenetic trematode of marine fishes. *J. Parasit.* **21**, Suppl. Abstr. p. 438.

NIGRELLI, R. F. (1935*c*). Studies on the acquired immunity of the Pompano, *Trachinotus carolinus*, to *Epibdella melleni*. *J. Parasit.* **21**, Suppl. Abstr. pp. 438–9.

NIGRELLI, R. F. (1937). Further studies on the susceptibility and acquired immunity of marine fishes to *Epibdella melleni*, a monogenetic trematode. *Zoologica, N.Y.*, **22**, 185–92.

NIGRELLI, R. F. (1940). Mortality statistics for specimens in the New York aquarium, 1939. *Zoologica*, **25** (4), 525–52.

NIGRELLI, R. F. & BREDER, C. M. jun. (1934). The susceptibility of certain marine fishes to *Epibdella melleni*, a monogenetic trematode. *J. Parasit.* **20**, 259–69.

NIGRELLI, R. F. & STUNKARD, H. W. (1937). Giant trematodes from the Wahoo, *Acanthocymbium solandra*. *J. Parasit.* **23**, Suppl. Abstr., p. 567.

NOBLE, A. E. & NOBLE, G. A. (1937). *Accacladium nematulum* n.sp., a trematode from the Sunfish, *Mola mola*. *Trans. Amer. Micr. Soc.* **56**, 55–60.

NOBLE, A. E. & PARK, J. T. (1937). *Helicometrina elongata* n.sp. from the Gobiescoid fish *Caularchus meandricus*, with an emended diagnosis of the trematode genus *Helicometrina*. *Trans. Amer. Micr. Soc.* **56**, 344–7.

NÖLLER, W. (1929). Befunde bei Schnecken von Thüringer Schafweiden in einem Lanzettegelgebiete. *Tierärztl. Rdsch.* **35** (26), 485–9.

NÖLLER, W. & ENIGK, K. (1933*a*). Ein *Platynosomum* beim Steinhuhn. *S.B. Ges. naturf. Fr. Berl.* Year 1932, pp. 419–23.

NÖLLER, W. & ENIGK, K. (1933*b*). Weitere Cercarienbefunde bei Landschnecken. *S.B. Ges. naturf. Fr. Berl.* Year 1932, pp. 424–37.

NÖLLER, W. & WAGNER, O. (1923). Der Wasserfrosch als zweiter Zwischenwirt einer Trematoden von Ente und Huhn. *Berl. tierärztl. Wschr.* **39**, 463.

NUTTALL, G. H. F. (1940). *Notes on the Preparation of Papers for Publication, etc.*, 62 pp. Cambridge.

NYBELIN, O. (1924). *Dactylogyrus vastator* n.sp. *Ark. Zool. Stockh.* **16**, no. 28, 1–2.

NYBELIN, O. (1926). Zur Helminthenfauna der Süsswasserfische Schwedens. I. Phyllodistomen. *Göteborg. VetenskSamh. Handl.* (4), **31**, iii, 1–29.

NYBELIN, O. (1933). *Crepidostomum suecicum* n.sp., ein Trematode mit ungewöhnlich weiter morphologischer Variationsbreite. *Ark. Zool., Stockh.* **25**, B, 1, 1–6.

NYBELIN, O. (1936). *Bunocotyle cingulata* Odhner, ein halophiler Trematode des Flussbarsches und Kaulbarsches der Ostsee. *Ark. Zool., Stockh.* **28**, B, 10, 1–6.

NYBELIN, O. (1937). Kleine Beiträge zur Kenntnis der Dactylogyren. *Ark. Zool., Stockh.* **29**, A, 1–29.

NYBELIN, O. (1941). *Dictyocotyle coeliaca* n.g., n.sp., ein leibeshöhlebewohnender monogenetischer Trematode. *Medd. Göteborgs Mus. Zool. Andeln. Göteborgs Kungl. Vetensk. och Vitterhets-Samhälles Handlingar Sjätte Följden*, Ser. B, **1**, no. 3, 21 pp.

ODHNER, T. (1900*a*). *Aporocotyle simplex* n.g., n.sp., ein neuer Typus von Ectoparasitischer Trematode. *Zbl. Bakt.* i, **27**, 62–6.

ODHNER, T. (1900*b*). *Gymnophallus*, eine neue Gattung von Vogeldistomen. *Zbl. Bakt.* I. Orig. **28**, 12–23.

ODHNER, T. (1901). Revision einiger Arten der Distomengattung *Allocreadium* Looss. *Zool. Jb., Syst.* **14**, 483–520.

ODHNER, T. (1902*a*). Mitteilungen zur Kenntnis der Distomen. I. *Zbl. Bakt.* I. Orig. **31**, 56–69.

ODHNER, T. (1902*b*). Mitteilungen zur Kenntnis der Distomen. II. *Zbl. Bakt.* I. Orig. **31**, 152–62.

ODHNER, T. (1905). Die Trematoden des arktischen Gebietes. *Fauna arct., Jena*, **4**, 291–372.

ODHNER, T. (1907). Zur Anatomie der *Didymozoen*: ein getrenntgeschlechtlicher Trematode mit rudimentärem Hermaphroditismus. *Zool. Stud. tillagnade, Prof. T. Tullberg, Uppsala*, pp. 309–42.

ODHNER, T. (1910*a*). Nordafricanische Trematoden grösstenteils vom weissen Nil. *Res. Swed. Zool. Exp. to Egypt and White Nile*, 1901, no. 23 A, pp. 1–170.

ODHNER, T. (1910*b*). *Gigantobilharzia acotylea* n.g., n.sp., ein mit dem Bilharzien verwandter Blutparasit von enormer Länge. *Zool. Anz.* **35**, 380–5.

ODHNER, T. (1911*a*). *Echinostomum ilocanum* (Garrison), ein neuer Menschenparasit aus Ostasien. *Zool. Anz.* **38**, 65–8.

ODHNER, T. (1911*b*). *Sanguinicola* M. Plehn—ein digenetischer Trematode. *Zool. Jb., Syst.* **31**, 33–45.

ODHNER, T. (1911–13). Zum natürlichen System der digenen Trematoden. I. Angiodictyidae. *Zool. Anz.* **37**, 181–91. II. Zoogonidae. *Zool. Anz.* **37**, 237–53. III. Steringophoridae. *Zool. Anz.* **38**, 97–117. IV. Azygiidae. *Zool. Anz.* **38**, 523–31. V. Bilharzia-Typus. *Zool. Anz.* **41**, 54–71. VI. Die Ableitung der Holostomiden und die Homologien ihrer Haftorgane. *Zool. Anz.* **42**, 289–318.

ODHNER, T. (1912). Die Homologien der weiblichen Genitalwege bei den Trematoden und Cestoden. *Zool. Anz.* **39**, 327–51.

ODHNER, T. (1913). Noch einmal die Homologien der weiblichen Genitalwege der monogenen Trematoden. *Zool. Anz.* **41**, 558–9.

ODHNER, T. (1914*a*). *Cercaria setifera* Monticelli. Eine Larvenform von *Lepocreadium album* Stoss. *Zool. Bidr. Uppsala*, **3**, 247–55.

ODHNER, T. (1914*b*). Die Verwandtschaftsbeziehungen der Trematodengattung *Paragonimus* Brn. *Zool. Bidr. Uppsala*, **3**, 231–46.

ODHNER, T. (1924). Remarks on *Sanguinicola*. *Quart. J. Micr. Sci.* **68**, 403–11.

ODHNER, T. (1926*a*). *Protofasciola* n.g. Ein Prototypus des grossen Leberegels. *Ark. Zool., Stockh.*, **18**, 7.

Odhner, T. (1926*b*). Zwei neue Arten der Trematodengattung *Cathaemasia* Looss. *Ark. Zool., Stockh.* **18** B, no. 10, pp. 1–4.

Odhner, T. (1927). Über Trematoden aus der Schwimmblase. *Ark. Zool., Stockh.* **19** A, no. 15, pp. 1–9.

Odhner, T. (1928). Weitere Trematoden mit Anus. *Ark. Zool., Stockh.* **20** (B2), pp. 1–6.

Odlaug, T. O. (1939). Abnormal conditions in the reproductive system of the trematode *Gorgodera amplicava*. *Trans. Amer. Micr. Soc.* **58**, 67–72.

Odlaug, T. O. (1940). Morphology and life history of the trematode *Alaria intermedia*. *Trans. Amer. Micr. Soc.* **5** (4), pp. 490–510.

Olivier, L. (1940). Life-history studies on two strigeid trematodes of the Douglas Lake region, Michigan. *J. Parasit.* **26**, 447–77.

Olsen, O. W. (1937*a*). A systematic study of the trematode subfamily Plagiorchiinae Pratt, 1902. *Trans. Amer. Micr. Soc.* **56**, 311–39.

Olsen, O. W. (1937*b*). Description and life history of the trematode *Haplometrana utahensis* sp.nov. (Plagorchiidae) from *Rana pretiosa*. *J. Parasit.* **23**, 13–28.

Olsen, O. W. (1940). Two new species of trematodes (*Apharyngostrigea bilobata*: Strigeidae, and *Cathaemasia nycticorax*: Echinostomidae) from herons, etc. *Zoologica*, **25**, 323–8.

O'Roke, E. C. (1917). Larval trematodes from Kansas fresh-water snails. *Kansas Univ. Sci. Bull.* **10**, 161–80.

Ortner-Schönbach, P. (1913). Zur Morphologie des Glykogens bei Trematoden und Cestoden. *Arch. Zellforsch.* **11**, 413–49.

Osborne, H. L. (1903). On *Phyllodistomum americanum* (n.sp.), a new bladder distome from *Amblystoma punctatum*. *Biol. Bull. Woods Hole*, **4**, 252–8.

Ozaki, Y. (1928). Some gasterostomatous trematodes of Japan. *Jap. J. Zool.* **2**, 35–60.

Ozaki, Y. & Asada, J. (1928). On some trematodes with anus. *Jap. J. Zool.* **2**, 5–33.

Ozerska, W. N. (1926). Parasitic helminth fauna of Don region. (Russian.) *Trudy Gos. Inst. Eksper. Vet. Moskva*, **2**, 103.

Palombi, A. (1923). Sulle cercarie del genere *Gymnophallus* Odhner dei mitili. *Monit. Zool. ital. Firenze*, **35**, 21–3.

Palombi, A. (1924). Le cercarie del genere *Gymnophallus* Odhner dei mitili. *Pubbl. Staz. zool. Napoli*, **5**, 137–52.

Palombi, A. (1929*a*). Richerche sul ciclo evolutivo di *Helicometra fasciata* (Rud.). Revisione delle specie del genere *Helicometra* Odhner. *Pubbl. Staz. zool. Napoli*, **9**, 237–92.

Palombi, A. (1929*b*). Il ciclo biologico di *Diphterostomum brusinae* Stossich. (Trematode digenetico: Fam. Zoogonidae Odhner). *R.C. Accad. Lincei* (6), **10**, 274–7.

Palombi, A. (1930*a*). Il ciclo biologico di *Diphterostomum brusinae* Stossich. (Trematode digenetico: Fam. Zoogonidae Odhner). Considerazioni sui cicli evolutivi delle specie affini e dei trematodi in generale. *Pubbl. Staz. zool. Napoli*, **10**, 111–49.

Palombi, A. (1930*b*). Il ciclo evolutivo di *Diphterostomum brusinae* Stoss. *Riv. Fis. mat.-sci. nat. Napoli*, Ser. ii, **4**, 8, 1–3.

Palombi, A. (1931*a*). Il polimorfismo dei Trematodi. Richerche sperimentali sui *Helicometra fasciata* (Rud.). *Annu. Mus. zool. Univ. Napoli*, **6**, no. 5, 1–8.

Palombi, A. (1931*b*). Per una migliore conoscenza del Trematodi endoparassiti dei pesci del golfo di Napoli. I. *Steringotrema divergens* (Rud.) e *Haploporus benedeni* (Stoss.). *Annu. Mus. zool. Univ. Napoli*, **6**, 15 pp.

Palombi, A. (1932*a*). *Bacciger bacciger* (Rud.) Nicoll, 1914, forma adulta di *Cercaria pectinata* Huet,. 1891. *Boll. Soc. Nat. Napoli*, **44**, 217–20.

Palombi, A. (1932*b*). Rapporti genetici tra *Cercaria setifera* Monticelli (non Joh. Müller) ed alcune specie della famiglia Allocreadiidae. Brevi notizie sul ciclo biologico di *Podocotyle atomon* (Rud.). *Boll. Soc. Nat. Napoli*, **44**, 213–16.

Palombi, A. (1933). *Cercaria pectinata* Huet e *Bacciger bacciger* (Rud.). Rapporti genetica e biologia. *Boll. Zool. Torino*, **4**, 1–11.

Palombi, A. (1934*a*). Gli stadi larvali dei Trematodi del Golfo di Napoli. I. Contributo allo studio della morfologia, biologia e sistematica delle cercarie marine. *Pubbl. Staz. zool. Napoli*, **14**, 51–94.

PALOMBI, A. (1934*b*). *Bacciger bacciger* (Rud.). Trematode digenetico: Fam. Steringophoridae Odhner. Anatomia, sistematica e biologia. *Pubbl. Staz. zool. Napoli*, **13**, 438–78.

PALOMBI, A. (1937*a*). La cercariae di *Mesometra orbicularis* (Rud.) e la sua transformazione in Metacercaria. (Appunti sul ciclo evolutivo.) *Riv. Parasit., Roma*, **1**, 13–17.

PALOMBI, A. (1937*b*). Il ciclo biologico di *Lepocreadium album* Stossich sperimentalmente realizzato. *Riv. Parasit., Roma*, **1**, 1–12.

PALOMBI, A. (1938*a*). Gli stadi larvali dei trematodi del Golfo di Napoli. Secondo contributo allo studio della morfologia, biologia e sistematica delle cercarie marine. Il gruppo delle cercariae cotilocerche. *Riv. Parasit., Roma*, **2**, 189–206.

PALOMBI, A. (1938*b*). Metodi impiegati per lo studio dei cicli evolutivi dei trematodi digenetici. Materiale per la conoscenza della biologia di *Podocotyle atomon* (Rud.). *Livro Jub. L. Travassos, Rio de Janeiro*, pp. 371–9.

PALOMBI, A. (1940). Gli stadi larvali dei trematodi del Golfo di Napoli. 3° contributo allo studio della morfologia, biologia e sistematica delle cercarie marine. *Riv. Parassit., Roma*, **4**, 7–30.

PANDE, B. P. (1937). On the morphology and systematic position of a new bladder fluke from an Indian frog. *Ann. Mag. Nat. Hist.* **20**, 250–56.

PARK, J. T. (1936). Two new trematodes, *Sterrhurus magnatestis* and *Tubilovesicula californica* (Hemiuridae) from littoral fishes of Dillon's Beach, California. *Trans. Amer. Micr. Soc.* **55**, 477–82.

PARK, J. T. (1937*a*). A new trematode, *Genitocotyle acirrus* gen.nov., sp.nov. (Allocreadiinae) from *Holconotus rhodoterus*. *Trans. Amer. Micr. Soc.* **56**, 67–71.

PARK, J. T. (1937*b*). A revision of the genus *Podocotyle* (Allocreadiidae) with a description of eight new species from the pool fishes from Dillon's Beach, California. *J. Parasit.* **23**, 405–22.

PARONA, C. (1911, 1912). *L'Elmintologia italiana da' suoi primi tempi all' anno* 1910. *Bibliografia. Sistematica. Cronologia. Storia.* I (502 pp.). II (1912) (540 pp.).

PARONA, C. (1919). Il tonno e la sua pesca. *R. Comm. Talassograf. Ital. Mem. Venezia*, **68**, 265.

PARONA, C. & MONTICELLI, F. S. (1903). Sul genere *Ancyrocotyle* (n.g.). *Arch. parasit., Paris*, **7**, 117–21.

PATWARDHAN, S. S. (1935). Three new species of trematodes from birds. *Proc. Ind. Acad. Sci.* **2**, 21–3.

PATZELT, H. (1930). Dos trematodos parasitos de *Sardinella aurita* Cuv. y Val. *Aphanurus stossichi* Mont. y *Hemiurus appendiculatus* Rud. *Notas Inst. esp. oceanogr.* (2), **45**, 1–23.

PAUL, A. A. (1938). Life-history studies of North American fresh-water Polystomes. *J. Parasit.* **24**, 489–507.

PELSENEER, P. (1906). Trématodes parasites de mollusques marins. *Bull. sci. Fr. Belg.* **40**, 161–86.

PELSENEER, P. (1916). La proportion relative des sexes chez les animaux et particulièrement chez les mollusques. *Mém. Acad. Roy. Belg.* **8**, no. 11, pp. 1–258.

PELSENEER, P. (1928). Les parasites des mollusques et les mollusques parasites. *Bull. Soc. Zool. Fr.* **53**, 158–89.

PENNER, L. R. (1941). The possibilities of systematic infection with dermatitis-producing Schistosomes. *Science*, **93** (2414), 327–8.

PENNYPACKER, M. I. (1936). The chromosomes in the maturation of the germ cells of the frog lung fluke, *Pneumonoeces medioplexus*. *Arch. Biol.* **47**, 309–17.

PENNYPACKER, M. I. (1940). The chromosomes and extra nuclear material in the maturing germ cells of a frog lung fluke, *Pneumonoeces similiplexus*. *J. Morphol.* **66**, 481–95.

PEREIRA, C. & CUOCOLO, R. (1940). Trematoides vesicais de Anfíbios do Nordeste Brasileiro. *Arch. Inst. biol. Def. agric. anim., São Paulo*, **11**, 413–20.

PERKINS, M. (1928). A review of the Telorchiinae, a group of distomid trematodes. *Parasitology*, **20**, 336–56.

PIGULEWSKY, S. W. (1927). Un cas de *Fasciola gigantica* Cob. chez un enfant usbek en Vieux Taschkent. (Russian.) *Pensée Méd. Usbekistane Taschkent*, pp. 59–61 (Fr. résumé, p. 131).

PIGULEWSKY, S. W. (1928). Zwei Fälle von *Fasciola gigantica* Cob. beim Menschen in Russland. *Arch. Schiffs- u. Tropenhyg.* **32**, 511–12.

PIGULEWSKY, S. W. (1931). Neue Arten von Trematoden aus Fischen des Dnjeprbassins. *Zool. Anz.* **96**, 8–18.

PIGULEWSKY, S. W. (1932). Fischparasiten des Dnjeprbassins. (Russian.) *Ann. Mus. Zool. Acad. Leningrad*, **32**, 425–50 (German summary, pp. 451–2).

PIGULEWSKY, S. W. (1938). Zur Revision der Parasitengattung *Lecithaster* Lühe, 1901. *Livro Jub. Prof. L. Travassos, Rio de Janeiro*, pp. 391–7.

PLEHN, M. (1905). *Sanguinicola armata* und *inermis* (n.g., n.sp.), n.f. Rhynchostomatidae. *Zool. Anz.* **39**, 244.

PLEHN, M. (1908). Ein monozoischer Cestode als Blutparasit (*Sanguinicola armata* und *inermis* Pl.). *Zool. Anz.* **33**, 427–40.

POCHE, F. (1907). Über die Kennzeichnung in ihrem Verhältniss zur Gültigkeit eines Namens. *Zool. Anz.* **32**, 99–106.

POCHE, F. (1911). Über die wahre Natur der von Busch in Siphonophoren beobachteten 'Eingeweidewürmer'. *Zool. Anz.* **38**, 369–73.

POCHE, F. (1926). Das System der Platodaria. *Arch. Naturgesch.* Abt. A, Heft 2, pp. 1–240; Heft 3, pp. 241–458.

POPOFF, P. (1926). Zur Fauna der parasitisch lebenden Würmer des Don-Flusssystems. Die parasitischen Würmer des Brachsens (*Abramis brama*). *Russ. Hydrobiol. Z., Saratov*, **5**, 64–71 (Fr. résumé, p. 79).

PORTER, A. (1902). The life history of the African sheep and cattle fluke, *Fasciola gigantica*. *S. Afr. J. Sci.* **17**, 126–30.

PORTER, A. (1938). The larval trematoda found in certain South African Mollusca with special reference to schistosomiasis. *Publ. S. Afr. Inst. Med. Res.* **8**, 492 pp.

PRATT, H. S. (1900). Synopsis of Monogenea. *Amer. Nat.* **34**, 645–62.

PRATT, H. S. (1902). The trematodes. Part II. Aspidocotylea and Malacocotylea. *Amer. Nat.* **36**, 887, 910, 953, 979.

PRATT, H. S. (1910). *Monocotyle floridana*, a new monogenetic trematode. *Publ. Carneg. Instn*, no. 133, pp. 1–9.

PRATT, H. S. (1916). The trematode genus *Stephanochasmus* Looss in the Gulf of Mexico. *Parasitology*, **8**, 229–38.

PRATT, H. S. (1919). A new cystocercous cercaria. *J. Parasit.* **5**, 128–31.

PRELL, H. (1928). *Cryptocotyle lingua* Creplin als Darmparasit des Silberfuchses. *Pelztierzucht*, **4**, 117–21.

PRICE, E. W. (1929*a*). A synopsis of the trematode family Schistosomatidae with descriptions of new genera and species. *Proc. U.S. Nat. Mus.* **75**, Part 18, p. 1.

PRICE, E. W. (1929*b*). Two new species of trematodes of the genus *Parametorchis* from fur-bearing animals. *Proc. U.S. Nat. Mus.* **76**, art. 12, 1–5.

PRICE, E. W. (1930). Two new species of trematode worms of the genus *Eucotyle* from North American birds. *Proc. U.S. Nat. Mus.* **77**, art. 1, 1–4.

PRICE, E. W. (1931*a*). *Metagonimoides oregonensis*, a new trematode from a raccoon. *J. Wash. Acad. Sci.* **21**, no. 16, 405–7.

PRICE, E. W. (1931*b*). A new species of trematode of the family Heterophyidae, with a note on the genus *Apophallus* and related genera. *Proc. U.S. Nat. Mus.* **79**, art. 17, 1–6.

PRICE, E. W. (1932). The trematode parasites of marine mammals. *Proc. U.S. Nat. Mus.* **81**, 1–68.

PRICE, E. W. (1934*a*). A new term for the adhesive organs of trematodes. *Proc. Helminth. Soc. Wash.* **1**, 34.

PRICE, E. W. (1934*b*). Losses among wild ducks due to infestation with *Sphaeridiotrema globulus* (Rud.) (Trematoda; Psilostomidae). *Proc. Helminth. Soc. Wash.* **1**, 631–4.

PRICE, E. W. (1934*c*). New trematode parasites of birds. *Smithson. Misc. Coll.* **91**, no. 6, 1–6.

PRICE, E. W. (1934*d*). New monogenetic trematodes from marine fishes. *Smithson. Misc. Coll.* (3286), **91**, no. 18, 1–3.

PRICE, E. W. (1934*e*). New digenetic trematodes from marine fishes. *Smithson. Misc. Coll.* (3234), **91**, no. 7, 1–8.

PRICE, E. W. (1935). Descriptions of some heterophyid trematodes of the sub-fam. Centrocestinae. *Proc. Helminth. Soc. Wash.* **2**, 70–3.

PRICE, E. W. (1936*a*). A new heterophyid trematode of the genus *Ascocotyle* (Centrocestinae). *Proc. Helminth. Soc. Wash.* **3**, 31–2.

PRICE, E. W. (1936*b*). North American monogenetic trematodes. *Bull. Geo. Washington Univ. Wash.* (Summary of Doctoral theses), 10–13.

PRICE, E. W. (1937*a*). North American monogenetic trematodes. I. The superfamily Gyrodactyloidea. *J. Wash. Acad. Sci.* **27**, 114–30, 146–64.

PRICE, E. W. (1937*b*). Redescriptions of two exotic species of monogenetic trematodes of the family Capsalidae Caird from the MacCallum Collection. *Proc. Helminth. Soc. Wash.* **4**, 25–7.

PRICE, E. W. (1937*c*). A new monogenetic trematode from Alaskan salmonoid fishes. *Proc. Helminth. Soc. Wash.* **4**, 27–9.

PRICE, E. W. (1937*d*). Three new genera and species of trematodes from cold-blooded vertebrates. *Skrjabin Jubilee Vol.* pp. 483–90.

PRICE, E. W. (1938*a*). A new species of *Dactylogyrus* (Monogenea; Dactylogyridae) with the proposal of a new genus. *Proc. Helminth. Soc. Wash.* **5**, 48–9.

PRICE, E. W. (1938*b*). A redescription of *Clinostomum intermedialis* Lamont (Trematoda: Clinostomidae), with a key to the species of the genus. *Proc. Helminth. Soc. Wash.* **5**, 11–13.

PRICE, E. W. (1938*c*). North American trematodes. II. The families Monocotylidae, Microbothriidae, Acanthocotylidae and Udonellidae (Capsaloidea). *J. Wash. Acad. Sci.* **28**, 109–26, 183–98.

PRICE, E. W. (1938*d*). The monogenetic trematodes of Latin America. *Livro Jub. Prof. L. Travassos, Rio de Janeiro*, pp. 407–13.

PRICE, E. W. (1939*a*). North American monogenetic trematodes. III. The family Capsalidae (Capsaloidea). *J. Wash. Acad. Sci.* **29**, 63–92.

PRICE, E. W. (1939*b*). North American monogenetic trematodes. IV. Polystomatidae (Polystomatoidea). *Proc. Helminth. Soc. Wash.* **6**, 80–92.

PRICE, E. W. (1939*c*). A review of the trematode superfamily Opisthorchioidea. *J. Parasitol.* **26**, 6, Suppl. Abstr. pp. 9–10.

PRICE, E. W. (1940*a*). A review of the trematode superfamily Opisthorchioidea. *Proc. Helminth. Soc. Wash.* **7**, 1–13.

PRICE, E. W. (1940*b*). A redescription of *Onchocotyle emarginata* Olsson, 1876 (Trematoda: Monogenea). *Proc. Helminth. Soc. Wash.* **7**, 76–8.

PRICE, E. W. (1942*a*). A new trematode of the family Psilostomidae from the lesser scaup duck, *Marila affinis*. *Proc. Helminth. Soc. Wash.* **9**, 30–1.

PRICE, E. W. (1942*b*). North American monogenetic Trematodes. V. The family Hexabothriidae (Polystomatoidea). *Proc. Helminth. Soc. Wash.* **9**, 39–56.

PRICE, E. W. (1943*a*). North American monogenetic trematodes. VI. The family Diclidophoridae (Diclidophoroidea). *J. Wash. Acad. Sci.* **33**, 44–54.

PRICE, E. W. (1943*b*). VII. The family Discocotylidae (Diclidophoroidea). *Proc. Helminth. Soc. Wash.* **10**, 10–15.

PRICE, E. W. (1943*c*). The presence of the lancet fluke, *Dicrocoelium dendriticum* (Rudolphi, 1819), in cattle in the United States. *Vet. Med.* **38**, 3 pp.

PRICE, E. W. & MCINTOSH, A. (1935). A new trematode, *Lyperosomum monenteron*, n.sp. (Dicrocoeliidae), from a robin. *Proc. Helminth. Soc. Wash.* **2**, no. 1, p. 1.

PRICE, H. F. (1931). Life history of *Schistosomatium douthitti* (Cort.). *Amer. J. Hyg.* **13**, 685–72.

PURVIS, G. B. (1931). The species of *Platynosomum* in Felines. *Vet. Rec.* **11**, 9, 228–9.

PURVIS, G. B. (1933). The excretory system of *Platynosomum concinnum* (Braun, 1901): Syn. *P. fastosum* (Kossack, 1910) and *P. planicipitis* (Cameron, 1928). *Vet. Rec.* **13**, 24, 565.

PURVIS, G. B. (1937) The synonyms of the trematode genus *Pachytrema* Looss, 1907. *Ann. Trop. Med. Parasit.* **31**, 457–60.

QUINTARET, G. (1905). Note sur une cercaire parasite du *Barleeia rubra* (Adams). *C.R. Soc. Biol., Paris*, **58**, 724–5.

RANKIN, J. S. jr. (1937*a*). An ecological study of parasites of some North Carolina salamanders. *Ecol. Monogr.* **7**, 171–269.

RANKIN, J. S., jr. (1937*b*). New helminths from North Carolina salamanders. *J. Parasit.* **23**, 29–42.

RANKIN, J. S., jr. (1938). Studies on the trematode genus *Brachycoelium* Duj. I. Variation in specific characters with reference to the validity of the described species. *Trans. Amer. Micr. Soc.* **57**, 358–75.

RANKIN, J. S., jr. (1939*a*). Ecological studies on larval trematodes from Western Massachusetts. *J. Parasit.* **25**, 309–28.

RANKIN, J. S., jr. (1939*b*). Studies on the trematode family Microphallidae Travassos, 1921. I. The genus *Levinseniella* Stiles & Hassall, 1901, and description of a new genus, *Cornucopula*. *Trans. Amer. Micr. Soc.* **58**, 431–46.

RANKIN, J. S., jr. (1939*c*). Studies on the trematode family Microphallidae Travassos, 1921. III. The genus *Maritrema* Nicoll, 1907, with a description of a new species and a new genus, *Maritreminoides*. *Amer. Midl. Nat.* **22**, 438–51.

RANKIN, J. S., jr. (1939*d*). The life cycle of *Cornucopula nassicola* (Cable & Hunninen, 1938) Rankin, 1939 (Trematoda: Microphallidae). *J. Parasit.* **25**, Suppl. Abstr. p. 12.

RANKIN, J. S., jr. (1940*a*). Studies on the trematode family Microphallidae Travassos, 1921. II. The genus *Spelotrema* Jägerskiold, 1901, and description of a new species, *Spelotrema papillorobusta*. *Trans. Amer. Micr. Soc.* **59**, 38–47.

RANKIN, J. S., jr. (1940*b*). Studies on the trematode family Microphallidae Travassos, 1921. IV. The life cycle and ecology of *Gymnaecotyla nassicola* (Cable & Hunninen, 1938) Yamaguti, 1939. *Biol. Bull. Woods Hole*, **79**, 439–51.

RANKIN, J. S., jr. & HUGHES, R. C. (1937). Notes on *Diplostomulum ambystomae* n.sp. *Trans. Amer. Micr. Soc.* **56**, 61–6.

RANSOM, B. H. (1920). Synopsis of the trematode family Heterophyidae, with descriptions of a new genus and five new species. *Proc. U.S. Nat. Mus.* **57**, 527–73.

RAO, M. A. N. (1939). A brief review of the species of schistosomes of the domesticated animals in India and their molluscan hosts. *Ind. Vet. J.* **15** (4), 349–58.

REES, F. G. (1931). Some observations and experiments on the biology of larval trematodes. *Parasitology*, **23**, 428–40.

REES, F. G. (1932). An investigation into the occurrence, structure and life histories of the trematode parasites of four species of *Limnaea*. *Proc. Zool. Soc. Lond.* pp. 1–32.

REES, F. G. (1933). On the anatomy of the trematode *Hypoderaeum conoideum* Bloch, 1782, together with attempts at elucidating the life cycles of two other digenetic trematodes. *Proc. Zool. Soc. Lond.* pp. 819–26.

REES, F. G. (1934). *Cercaria patellae* Lebour, 1911, and its effects on the digestive gland and gonads of *Patella vulgata*. *Proc. Zool. Soc. Lond.* pp. 45–53.

REES, F. G. (1937). The anatomy and encystment of *Cercaria purpurae* Lebour, 1911. *Proc. Zool. Soc. Lond.* pp. 65–73.

REES, F. G. (1939*a*). *Cercaria strigata* Lebour from *Cardium edule* and *Tellina tenuis*. *Parasitology*, **31**, 458–63.

REES, F. G. (1939*b*). Studies on the germ-cell cycle of the digenetic trematode *Parorchis acanthus* Nicoll. Part. I. Anatomy of the genitalia and gametogenesis in the adult. *Parasitology*, **31**, 417–33.

REES, F. G. (1940). Studies on the germ-cell cycle of the digenetic trematode *Parorchis acanthus* Nicoll. Part II. Structure of the miracidium and germinal development in the larval stages. *Parasitology*, **32**, 372–91.

REES, F. G. & LLEWELLYN, J. (1941). A record of the trematode and cestode parasites of fishes from the Porcupine Bank, Irish Atlantic Slope and Irish Sea. *Parasitology*, **33**, 390–6.

REES, W. J. (1935). The anatomy of *Cercaria buccini*. *Proc. Zool. Soc. Lond.* pp. 309–12.

REES, W. J. (1936*a*). Note on the ubiquitous cercaria from *Littorina rudis*, *L. obtusata* and *L. littorea*. *J. Mar. Biol. Ass. U.K.* **20**, 621–4.

REES, W. J. (1936*b*). The effect of parasitism by larval trematodes on the tissues of *Littorina littorea* (Linne). *Proc. Zool. Soc. Lond.* pp. 357–68.

REMLEY, L. W. (1942). Morphology and life-history studies of *Microcotyle spinicirrus* MacCallum, 1918, a monogenetic trematode parasitic on the gills of *Aplodinotus grunniens*. *Trans. Amer. Micr. Soc.* **61**, 141–55.

RILEY, W. A. (1921). An annotated list of the animal parasites of foxes. *Parasitology*, **13**, 86–96.

ROBINSON, V. S. (1934). A new species of Accacoeliid trematode (*Accacladocoelium alveolatum* n.sp.) from the intestine of a sun-fish (*Orthagoriscus mola* Bloch). *Parasitology*, **26**, 346–51.

ROEWER, C. F. (1906). Beiträge zur Histogenese von *Cercariaeum helicis*. *Jena. Z. Naturw.* **41**, 185–228.

ROSS, I. C. (1928). Liver fluke disease in Australia: its treatment and prevention. *Pamphl. Coun. Sci. Industr. Res., Aust.*, no. 5, 23 pp.

ROSS, I.C. & GORDON, H. M. (1936). *The Internal Parasites and Parasitic Diseases of Sheep, their Treatment and Control*, 238 pp. Sydney.

ROSS, I. C. & MCKAY, A. C. (1929). The bionomics of *Fasciola hepatica* in New South Wales, etc. *Bull. Coun. Sci. Industr. Res., Aust.*, **43**, 62 pp.

ROTHSCHILD, M. (1935*a*). Note on the excretory system of *Cercaria ephemera* Lebour, 1907 (nec Nitzsch). *Parasitology*, **27**, 171–4.

ROTHSCHILD, M. (1935*b*). The trematode parasites of *Turritella communis* Lmk. from Plymouth and Naples. *Parasitology*, **27**, 152–70.

ROTHSCHILD, M. (1936*a*). Gigantism and variation in *Peringia ulvae* Pennant, 1777, caused by infection with larval trematodes. *J. Mar. Biol. Ass. U.K.* **20**, 537–46.

ROTHSCHILD, M. (1936*b*). The process of encystment of a cercaria parasitic in *Lymnaea tenera euphratica*. *Parasitology*, **28**, 56–62.

ROTHSCHILD, M. (1936*c*). Rearing animals in captivity for the study of trematode life histories. I. *Larus ridibundus* L., the black-headed gull. *J. Mar. Biol. Ass. U.K.* **21**, 143–5.

ROTHSCHILD, M. (1936*d*). Preliminary note on the trematode parasites of *Peringia ulvae* (Pennant, 1777). *Novit. Zool.* **39**, 268–9.

ROTHSCHILD, M. (1937). Note on the excretory system of the trematode genus *Maritrema* Nicoll, 1907, and the systematic position of the Microphallinae Ward, 1901. *Ann. Mag. Nat. Hist.* (10), **19**, 355–65.

ROTHSCHILD, M. (1938*a*). Preliminary note on the life history of *Cryptocotyle jejuna* Nicoll, 1907 (Trematoda). *Ann. Mag. Nat. Hist.* (11), **1**, 238–9.

ROTHSCHILD, M. (1938*b*). A further note on the excretory system of *Maritrema* Nicoll, 1907 (Trematoda). *Ann. Mag. Nat. Hist.* (11), **1**, 157–8.

ROTHSCHILD, M. (1938*c*). *Cercaria sinitsîni* n.sp., a cystophorous cercaria from *Peringia ulvae* (Pennant, 1777). *Novit. Zool.* **41**, 42–57.

ROTHSCHILD, M. (1938*d*). Notes on the classification of cercariae of the superfamily Notocotyloidea (Trematoda), with special reference to the excretory system. *Novit. Zool.* **41**, 75–83.

ROTHSCHILD, M. (1938*e*). Further observations on the effect of trematode parasites on *Peringia ulvae* (Pennant, 1777). *Novit. Zool.* **41**, 84–102.

ROTHSCHILD, M. (1938*f*). The excretory system of *Cercaria coronanda* n.sp. together with notes on its life history and the classification of cercariae of the superfamily Opisthorchioides Vogel, 1934 (Trematoda). *Novit. Zool.* **41**, 148–63.

ROTHSCHILD, M. (1938*g*). A note on the fin-folds of cercariae of the superfamily Opisthorchioidea Vogel, 1934 (Trematoda). *Novit. Zool.* **41**, 170–3.

ROTHSCHILD, M. (1939). Large and small flame-cells in a cercaria (Trematoda). *Novit. Zool.* **41**, 376.

ROTHSCHILD, M. (1940*a*). A note on the systematic position of *Cercaria coronanda* Rothschild, 1938. *Proc. Helminth. Soc. Wash.* **7**, 13–14.

ROTHSCHILD, M. (1940*b*). *Cercaria imbricata* Looss, 1896, nec 1893...a note on nomenclature. *Novit. Zool.* **42**, 215–16.

ROTHSCHILD, M. (1940*c*). Rearing animals in captivity for the study of trematode life histories. II. *J. Mar. Biol. Ass. U.K.* **24**, 613–17.

ROTHSCHILD, M. (1941*a*). The effect of trematode parasites on the growth of *Littorina neritoides* (L.). *J. Mar. Biol. Ass. U.K.* **25**, 69–80.

ROTHSCHILD, M. (1941*b*). The metacercaria of a pleurolophocerca cercaria parasitizing *Peringia ulvae* (Pennant, 1777). *Parasitology*, **33**, 439–44.

ROTHSCHILD, M. (1941*c*). Observations on the growth and trematode infections of *Peringia ulvae* (Pennant, 1777) in a pool in the Tamar Saltings, Plymouth. *Parasitology*, **33**, 406–15.

ROTHSCHILD, A. & ROTHSCHILD, M. (1939). Some observations on the growth of *Peringia ulvae* (Pennant, 1777) in the laboratory. *Novit. Zool.* **41**, 240–7.

ROTHSCHILD, M. & SPROSTON, N. G. (1941). The metacercaria of *Cercaria doricha* Rothschild, 1934, or a closely related species. *Parasitology*, **33**, 359–62.

RUSHTON, W. (1937). Blindness in fresh-water fish. *Nature, Lond.*, **140**, 1014.

RUSHTON, W. (1938). Blindness in fresh-water fish. *Nature, Lond.*, **141**, 289.

RUSZKOWSKY, J. (1922). Die postembryonale Entwicklung von *Hemistomum alatum*. *Bull. int. Acad. Cracovie*, p. 237–50.

SALOMON, S. (1932*a*). *Fasciola magna* bei deutschem Rotwild. *Abh. naturf. Ges. Görlitz*, **31**, 139–42.

SALOMON, S. (1932*b*). *Fascioloides magna* bei deutschem Rotwild. *Tierärztl. Wschr.* **48**, 627–8.

SANDGROUND, J. H. (1929). A consideration of the relation of host-specificity of helminths and other metacoan parasites to the phenomena of age resistance and acquired immunity. *Parasitology*, **21**, 227–55.

SCHELLENBERG, A. (1911). Ovogenese, Eireifung und Befruchtung von *Fasciola hepatica* L. *Arch. Zellforsch.* **6**, 443–84.

SCHEURING, L. (1920). Die Lebensgeschichte eines Karpfenparasiten (*Sanguinicola inermis* Plehn). *Allg. Fisch. Ztg*, **45**, 225–30.

SCHEURING, L. (1923). Der Lebenszyklus von *Sanguinicola inermis* Plehn. *Zool. Jb., Anat.* **44**, 264–310.

SCHREINER, A. & SCHREINER, K. E. (1908). Neue Studien über die Chromatinreifung der Geschlechtzellen. V. Die Reifung der Geschlechtzellen von *Zoogonus mirus*. *Vidensk. Selsk. Skrifter, I. Math. nat. Kl.* **8**.

SCHUBMAN, W. (1905). Über die Eibildung und Embryonalentwicklung von *Fasciola hepatica* L. *Zool. Jb., Anat.* **21**, 571–606.

SCHULZ, R. E. (1932). Trematoden der Gattung *Plagiorchis* Lühe der Nagetiere. (Russian.) *Rev. Microbiol., Saratov*, **11**, 53–9 (German summary, p. 60).

SCHULZ, R. E. & DOBROVA, M. J. (1933–4). Parasitic worms of water-rats, Lower Volga. (Russian.) *Rev. Microbiol., Saratov*, **12**, 329–31.

SCHUMACHER, W. (1939). Untersuchungen über den Wanderungsweg und die Entwicklung von *Fasciola hepatica* L. im Endwirt. *Z. Parasitenk.* **10**, 608–43.

SCHUMAKOWITSCH, E. E. (1932). Eine neue Trematode *Maritrema sachalinicum* n.sp. aus einer Möwe (*Larus argentatus*). *Zool. Anz.* **98**, 154–8.

SCOTT, A. (1901). Some additions to the fauna of Liverpool Bay. *Trans. Lpool Biol. Soc.* **15**, 344.

SCOTT, A. (1904). Some parasites found on fishes in the Irish Sea. *Trans. Lpool Biol. Soc.* **18**, 113–23.

SCOTT, A. (1906). Faunistic notes. *Trans. Lpool. Biol. Soc.* **20**, 47–57.

SCOTT, J. A. (1937). The incidence and distribution of the human Schistosomes in Egypt. *Amer. J. Hyg.* **25**, 566–614.

SCOTT, T. (1901). Notes on some parasites of fishes. 19*th Ann. Rep. Fishery Board Scotland*, Part 3, pp. 120–53.

SCOTT, T. (1902). Notes on some parasites of fishes. 20*th Ann. Rep. Fishery Board Scotland*, Part 3, pp. 288–303.

SCOTT, T. (1904). On some parasites of fishes new to the Scottish marine fauna. 22*nd Ann. Rep. Fishery Board Scotland*, Part 3, pp. 275–80.

SCOTT, T. (1909). Some notes on fish parasites. 26*th Ann. Rep. Fishery Board Scotland*, Part 3, pp. 73–92.

SCOTT, T. (1911). Some trematodes parasitic on British fishes. *Trans. Edinb. Fld Nat. Micr. Soc.* **6**, 344–53.

SCOTT, T. (1912). Notes on some trematode parasites of fishes. 28th *Ann. Rep. Fishery Board Scotland*, Part 3, pp. 68–72.

SEIFERT, R. (1926). *Cercaria limbifera*, eine neue echinostome Cercarie. *Zool. Anz.* **67**, 112–19.

SEMENOV, W. D. (1927). Vogeltrematoden des westlichen Bereiches der Union S.S.R. (Russian.) *Samml. Helm. Arb. K. I. Skrjabin genr. Moscow*, pp. 221–71.

SEŘKOVA, O. P. & BYCHOWSKY, B. E. (1940). *Asymphylodora progenetica* n.sp. nebst einigen Angaben über ihre Morphologie und Entwicklungsgeschichte. *Mag. Parasitol. Inst. Zool. Acad. Sci. U.R.S.S.* **8**, 162–75.

SEVERINGHAUS, A. G. (1928). Sex studies on *Schistosoma japonicum*. *Quart. J. Micr. Sci.* **71**, 653–702.

SEWELL, R. B. S. (1922). Cercariae Indicae. *Ind. J. Med. Res.* **10**, Suppl. No., pp. 370.

SEWELL, R. B. S. (1931). Cercariae Nicobaricae. *Ind. J. Med. Res.* **18**, 785–806.

SHAFFER, E. (1916). *Discocotyle salmonis* nov.spec., ein neuer Trematode an den Kiemen der Regenbogenforelle (*Salmo irideus*). *Zool. Anz.* **46**, 257–71.

SHCHUPAKOV, I. (1936). Parasites of the Caspian Seal. (Russian.) *Ann. Boubnoff State Univ. Leningrad* **7**, 134–43.

SHULZ, R. E. & DOBROVA, M. J. (1933). Parasitic worms of water-rats, Lower Volga. (Russian; German summary.) *Vestnik Microbiol., Epidemol. i Parasitol.* **12**, 229–331.

SINITSÎN, D. F. (1905). *Data on the Natural History of Trematodes. Distomes of Fish and Frogs in the Vicinity of Warsaw*, 210 pp. English translation.

SINITSÎN, D. F. (1907). Observations sur les métamorphoses des trématodes. *Arch. Zool. exp. gén.* **7**, 21–37.

SINITSÎN, D. F. (1909). Studien über die Phylogenie der Trematoden. 2. *Bucephalus* v. Baer und *Cercaria ocellata* de la Val. *Z. wiss. Zool.* **94**, 299–325.

SINITSÎN, D. F. (1910). Studien über die Phylogenie der Trematoden. *Biol. Z.* **1**, 1–63. Also *Z. wiss. Zool.* **94**, 299–325.

SINITSÎN, D. F. (1911). La génération parthénogénétique des trématodes et sa descendance dans les mollusques de la Mer Noire. *Mém. Acad. Sci. St-Pétersb.* (8), **30**, 1–127.

SINITSÎN, D. F. (1931*a*). Studien über die Phylogenie der Trematoden. IV. (The life histories of *Plagioporus siliculus* and *P. virens*, with special reference to the origin of the Digenea.) *Z. wiss. Zool.* **138**, 409–56.

SINITSÎN, D. F. (1931*b*). Studien über die Phylogenie der Trematoden. V. (Revision of Harmostominae in the light of new facts from their morphology and life history.) *Z. Parasitenk.* **3**, 786–835.

SIWÀK, J. (1931). *Ancyrocephalus vistulensis* sp.n., un nouveau trématode parasite du Silure (*Silurus glanis* L.). *Bull. int. Acad. Cracovie*, B, **2**, 669–79.

SKRJABIN, K. I. (1913*a*). Vogeltrematoden aus Russisch-Turkestan. *Zool. Jb., Syst.* **35**, 351–88.

SKRJABIN, K. I. (1913*b*). *Tracheophilus sisowi* n.g., n.sp. Ein Beiträg zur Systematik der Gattung *Typhlocoelum* Stossich und der verwandten Formen. *Zbl. Bakt.* **69**, 90–5.

SKRJABIN, K. I. (1915). Trematoden der Vögel des Urals. *Annu. Mus. Zool. Acad. St-Pétersb.* **20**, 270.

SKRJABIN, K. I. (1923). Die Trematoden des Hausgeflügels. Versuch einer Monographie. *Arb. Inst. exp. Vet. Moskau*, **1**, 1–64.

SKRJABIN, K. I. (1924). Nierentrematoden der Vögel Russlands. *Zbl. Bakt.* (2), **62**, 80–90.

SKRJABIN, K. I. (1926). Infestation simultanée d'un oiseau par 17 espèces d'helminths. *C.R. Soc. Biol. Paris*, **94**, 307–8.

SKRJABIN, K. I. (1928). Sur la faune des trématodes des oiseaux de Transbaikalie. *Ann. Parasit. hum. comp.* **6**, 80–7.

SKRJABIN, K. I. *et al.* (1927). The results of research of twenty-eight helminthological expeditions in U.S.S.R. (1919–25). *Helminth. Dep. State Inst. Exp. Vet. Med. Moscow*, 296 pp.

SKRJABIN, K. I. & BASKAKOW, W. T. (1925). Über die Trematodengattung *Prosthogonimus*. Versuch einer Monographie. *Z. Infekt. Kr. Haustiere*, **28**, 195–212.

SKRJABIN, K. I. & ISSAITSCHIKOW, I. M. (1927). Four new species of the family Dicrocoeliidae from the livers of birds. *Ann. Trop. Med. Parasit.* **21**, 303–8.

SKRJABIN, K. I. & LINDTROP, G. T. (1919). Trématodes intestinales des chiens du Don. *Nach. Don. Vet. Inst.* **1**, 11 pp.

SKRJABIN, K. I. & MASSINO, B. G. (1925). Trematoden bei den Vögeln des Moskauer Gouvernements. *Zbl. Bakt.* (2), **64**, 453–62.

SKRJABIN, K. I. & POPOFF, P. (1924). Bericht über die Tätigkeit der helminthologischen Expedition in Armenien 1923. (Russian.) *Russ. H. Trop. Med. Moscow*, **1**, 58–63.

SKRJABIN, K. I. & SCHULZ, R. E. (1933). Ein neuer Trematode *Ogmocotyle pygargi* n.g., n.sp. aus einem Reh (*Capreolus pygargus bedfordi* Thomas). *Zool. Anz.* **102**, 267–70.

SKRJABIN, K. I. & UDINZEW, A. N. (1930). Two new trematodes from the biliary ducts of birds from Armenia. *J. Parasit.* **16**, 213–19.

SKRJABIN, K. I. & ZAKHAROFF, N. P. (1920). Zwei neue Trematodengattungen aus den Blutgefässen der Vögel. (Russian.) *Isv. Donsk. Vet. Inst. Novogerkassk.* **2**, 6 pp. (German summary).

SKWORTZOFF, A. A. (1927). Ueber den anatomischen Bau des Saugwurmes des Sterlets des Wolga-Flusssystems, *Acrolichanus auriculatus* (Wedl, 1856). (Russian.) *Samml. Helm. Arb. K. I. Skrjabin gewidmet, Moskau*, pp. 276–86 (German summary).

SKWORTZOFF, A. A. (1928). Ueber die Helmintenfauna des Wolgasterlets. *Zool. Jb., Syst.* **54** (5–6), 557–77.

SKWORTZOFF, A. A. (1934). Zur Kenntnis der Helminthenfauna der Wasserratten (*Arvicola terrestris* L.) (Russian.) *Rev. Microbiol., Saratov*, **13**, 317–26 (German summary).

SMITH, S. (1932). Two new cystocercous cercariae from Alabama. *J. Parasit.* **19**, 173–4.

SOLOWIOW, P. T. (1912). Vers parasitaires des oiseaux du Turkestan. *Annu. Mus. Zool. Acad. St-Pétersb.* **17**, 86–115.

SOPARKAR, M. B. (1921 *a*). Notes on some furcocercous cercariae from Bombay. *Indian J. Med. Res.* **9**, 23–32.

SOPARKAR, M. B. (1921 *b*). The cercaria of *Schistosoma spindalis* (Montgomery). *Indian J. Med. Res.* **9**, 1–22.

SOPARKAR, M. B. (1924). A new cercaria from northern India, *Cercaria patialensis* nov.sp. *Indian J. Med. Res.* **11**, 933–41.

SOUTHERN, R. (1912). Clare Island Survey, Part 56, Platyhelmia. *Proc. R. Irish Acad. Dublin*, **31**, 18.

SOUTHWELL, T. (1934). Observations on certain human parasites. *James Johnstone Mem. Vol. Liverpool*, pp. 132–53.

SOUTHWELL, T. & KIRSHNER, A. (1937 *a*). A description of a new species of Amphistome, *Chiorchis purvisi*, with notes on the classification of the genera within the group. *Ann. Trop. Med. Parasit.* **31**, 215–44.

SOUTHWELL, T. & KIRSHNER, A. (1937 *b*). On some parasitic worms found in *Xenopus laevis* the South African clawed toad. *Ann. Trop. Med. Parasit.* **31**, 245–65.

SOUTHWELL, T. & KIRSHNER, A. (1937 *c*). Parasitic infections in a swan and in a brown trout. *Ann. Trop. Med. Parasit.* **31**, 427–34.

SPREHN, C. E. W. (1932). *Lehrbuch der Helminthologie*, 998 pp. Berlin.

SPREHN, C. E. W. (1933). Trematoda. In Grimpe u. Wagler, *Die Tierwelt der Nord- u. Ostsee*, Lief. 24, Teil 4*c*, 1–60. Leipzig.

SRIVASTAVA, H. D. (1934). On new trematodes of frogs and fishes of the United Provinces, India. Part III. On a new genus *Mehraorchis* and two new species of *Pleurogenes* (Pleurogenetinae) with a systematic discussion and revision of the family Lecithodendriidae. *Bull. Acad. Sci. Allahabad*, **3**, 239–56.

SRIVASTAVA, H. D. (1939 *a*). The morphological and systematic relationships of two new distomes of the family Haplosplanchnidae Poche, 1926, from Indian marine food-fishes. *Indian J. Vet. Sci.* **9**, 67–71.

SRIVASTAVA, H. D. (1939 *b*). The morphology and systematic relationship of a new genus of digenetic trematode belonging to the family Monadhelminae (Dollfus, 1937). *Indian J. Vet. Sci.* **9**, 97–9.

SRIVASTAVA, H. D. (1939*c*). Three new parasites of the genus *Acanthocolpis* Lühe, 1906 (family Acanthocolpidae). *Indian. J. Vet. Sci.* **9**, 213–16.

SRIVASTAVA, H. D. (1939*d*). Two new trematodes of the family Monorchidae Odhner, 1911, from Indian marine food-fishes. *Indian J. Vet. Sci.* **9**, 233–6.

STAFFORD, J. (1904). Trematodes from Canadian fishes. *Zool. Anz.* **27**, 481–95.

STAFFORD, J. (1905). Trematodes from Canadian vertebrates. *Zool. Anz.* **28**, 681–94.

STAFFORD, E. W. (1931). Platyhelminths in aquatic insects and crustacea. *J. Parasit.* **18**, 131.

STEELMAN, G. M. (1938*a*). A description of *Gorgoderina schistorchis* n.sp. *Trans. Amer. Micr. Soc.* **57**, 383–86.

STEELMAN, G. M. (1938*b*). A description of *Phyllodistomum caudatum* n.sp. *Amer. Midl. Nat.* **19**, 613–16.

STEEN, E. B. (1938). Two new species of *Phyllodistomum* (Trematoda: Gorgoderidae) from Indiana fishes. *Amer. Midl. Nat.* **20**, 201–10.

STEINBERG, D. (1931). Die Geschlechtorgane von *Aspidogaster conchicola* Baer und ihr Jahreszyklus. *Zool. Anz.* **94**, 153–70.

STILES, C. W. (1905). The new Asiatic blood fluke (*Schistosoma japonicum* Katsurada, 1904) in the Philippines. *Amer. Med.* **10**, 21, 854.

STILES, C. W. & GOLDBERGER, J. (1910). A study of the anatomy of *Watsonius* (n.g.) *watsoni* of man., etc. *Bull. U.S.A. Treasury Dept. Hyg. Lab.* **60**, 259 pp.

STILES, C. W. & HASSALL, A. (1908). Index-catalogue of medical and veterinary zoology. Subjects: Trematoda and trematode diseases. *Hyg. Lab. Bull.* no. 37, 401 pp.

STILES, C. W. & NOLAN, M. O. (1931). Trematode parasites of bats. Key-catalogue. *Nat. Inst. Hlth Bull.* **155**, 609–789.

STOLL, N. R., CORT, W. W. & KWEI, W. S. (1927). Egg-worm correlations in cases of *Fasciolopsis buski*, with additional data on the distribution of this parasite in China. *J. Parasit.* **13**, 166–72.

STOSSICH, M. (1900). Osservazione elmintologiche. *Boll. Soc. Adr. Sci. Nat. Trieste*, **19**, 97.

STOSSICH, M. (1902). Sopra una nuova specie delle Allocreadiinae. *Arch. parasit., Paris*, **5**, 578.

STOSSICH, M. (1903). Una nuova specie di *Helicometra* Odhner. *Arch. parasit., Paris*, **7**, 373.

STOSSICH, M. (1904). Alcuni Distomi della collezione elmintologica del Museo di Napoli. *Ann. Mus. zool. Napoli*, no. 23, pp. 14.

STROM, J. K. (1927). *Parorchis asiaticus* n.sp., ein neuer Trematode der Gattung *Parorchis* Nicoll, 1907. *Zool. Anz.* **72**, 249–55.

STROM, J. K. (1928*a*). Eine neue Art der Vogeltrematode *Oswaldoia pawlowskyi* n.sp. *Zool. Anz.* **77**, 184–9.

STROM, J. K. (1928*b*). Beiträge zur Systematik der Gattung *Xenopharynx* Nicoll, 1912 im Zusammenhang mit der Beschreibung einer neuen Art, *X. amudariensis* n.sp. *Zool. Anz.* **79**, 167–72.

STROM, J. K. (1940*a*). Notes on the classification of the Dicrocoeliinae (Trematoda). *Mag. Parasit. Inst. Zool. Acad. Sci. U.R.S.S.* **8**, 176–88 (Russian, with summary in English).

STROM, J. K. (1940*b*). On the fauna of trematode worms from wild animals of Kirghisia. *Mag. Parasit. Inst. Zool. Acad. Sci. U.R.S.S.* **8**, 189–224 (Russian, with summary in English).

STROM, J. K. (1940*c*). New species of trematode worms of the genus *Plagiorchis*. *Mag. Parasit. Inst. Zool. Acad. Sci. U.R.S.S.* **8**, 225–31 (Russian, with summary in English).

STSCHUPAKOW, I. (1936). Die parasitäre Fauna des Kaspischen Seehundes (*Phoca caspica*). *Sci. Mem. Leningrad St Boubnoff Univ.* **7**, 134–43.

STUNKARD, H. W. (1915). Notes on the trematode genus *Telorchis* with descriptions of new species. *J. Parasit.* **2**, 57–66.

STUNKARD, H. W. (1917). Studies on North American Polystomidae, Aspidogastridae and Paramphistomidae. *Ill. Biol. Monogr.* **3**, no. 3, pp. 1–114.

STUNKARD, H. W. (1924). On some trematodes from Florida Turtles. *Trans. Amer. Micr. Soc.* **43**, 97–113.

STUNKARD, H. W. (1925). The present status of the Amphistome problem. *Parasitology*, **17**, 137–48.

STUNKARD, H. W. (1928). Observations nouvelles sur les trématodes sanguinicoles du genre *Vasotrema* (Spirorchidae) avec description des deux espèces nouvelles. *Ann. Parasit. hum. comp.* **6**, 303–20.

STUNKARD, H. W. (1929*a*). The excretory system of *Cryptocotyle* (Heterophyidae). *J. Parasit.* **15**, 259–66.

STUNKARD, H. W. (1929*b*). Parasitism as a biological phenomenon. *Sci. Mon., N.Y.*, **28**, 349–62.

STUNKARD, H. W. (1930*a*). The life cycle of *Cryptocotyle lingua* (Creplin), with notes on the physiology of the metacercaria. *J. Morph. Physiol.* **50**, 143–83.

STUNKARD, H. W. (1930*b*). Morphology and relationships of the trematode *Opisthoporus aspidonectes* (MacCallum, 1917) Fukin, 1929. *Trans. Amer. Micr. Soc.* **49**, 210–19.

STUNKARD, H. W. (1930*c*). An analysis of the methods used in the study of larval trematodes. *Parasitology*, **22**, 268–74.

STUNKARD, H. W. (1931). Further observations on the occurrence of anal openings in digenetic trematodes. *Z. Parasitenk.* **3**, 713–25.

STUNKARD, H. W. (1932). Some larval trematodes from the coast in the region of Roscoff, Finistère. *Parasitology*, **24**, 321–43.

STUNKARD, H. W. (1935). A new trematode, *Probilotrema californiense*, from the coelom of the sting ray, *Myliobatis californicus*. *J. Parasit.* **21**, 359–64.

STUNKARD, H. W. (1937). The physiology, life cycles and phylogeny of the parasitic flatworms. *Amer. Mus. Novit.* no. 908, pp. 1–27.

STUNKARD, H. W. (1938*a*). The morphology and life cycle of the trematode *Himasthla quissetensis* (Miller & Northup, 1926). *Biol. Bull. Woods Hole*, **75**, 145–64.

STUNKARD, H. W. (1938*b*). *Distomum lasium* Leidy, 1891 (Syn. *Cercariaeum lintoni* Miller & Northup, 1926), the larval stage of *Zoogonus rubellus* (Olsson, 1868) (Syn. *Z. mirus* Looss, 1901). *Biol. Bull. Woods Hole*, **75**, 308–34.

STUNKARD, H. W. (1939). Determination of species in the trematode genus *Himasthla*. *Z. Parasitenk.* **10**, 719–21.

STUNKARD, H. W. (1940*a*). Life-history studies and the development of parasitology. *J. Parasit.* **26**, 1–15.

STUNKARD, H. W. (1940*b*). Life-history studies and specific determination in the trematode genus *Zoogonus*. *J. Parasit.* **26**, Suppl. Abstr. pp. 33–4.

STUNKARD, H. W. (1941). Specificity and host-relations in the trematode genus *Zoogonus*. *Biol. Bull. Woods Hole*, **81**, 205–14.

STUNKARD, H. W. (1943*a*). A new trematode, *Dictyangium chelydrae* (Microscaphidiidae = Angiodictyidae) from the snapping turtle, *Chelydra serpentina*. *J. Parasit.* **29**, 143–50.

STUNKARD, H. W. (1943*b*). The morphology and life history of the digenetic trematode, *Zoögonoides laevis* Linton, 1940. *Biol. Bull. Woods Hole*, **85**, 227–37.

STUNKARD, H. W. & ALVEY, C. H. (1930). The morphology of *Zalophotrema hepaticum*, with a review of the trematode family Fasciolidae. *Parasitology*, **22**, 326–33.

STUNKARD, H. W. & CABLE, R. M. (1932). The life history of *Parorchis avitus* (Linton), a trematode from the cloaca of the gull. *Biol. Bull. Woods Hole*, **62**, 328–38.

STUNKARD, H. W. & DUNIHUE, F. W. (1933). *Gyrodactylus* as a parasite of the tadpoles of *Rana catesbiana*. *J. Parasit.* **20**, 137.

STUNKARD, H. W. & NIGRELLI, R. F. (1930). On *Distomum vibex* Linton, with special reference to its systematic position. *Biol. Bull. Woods Hole*, **58**, 336–43.

STUNKARD, H. W. & NIGRELLI, R. F. (1934). Observations on the genus *Sterrhurus* Looss, with a description of *Sterrhurus branchialis* sp.nov. (Trematoda: Hemiuridae). *Biol. Bull. Woods Hole*, **67**, 534–43.

STUNKARD, H. W. & WILLEY, C. H. (1929). The development of *Cryptocotyle* (Heterophyidae) in its final host. *Amer. J. Trop. Med.* **9**, no. 2, pp. 117–28.

SYÔGAKI, Y. (1936). *Aspidogaster conchicola* Baer. Structure and occurrence, Japan. *Zool. Mag., Tokyo*, **48**, 56–9. (In Japanese, with summary in English.)

SZIDAT, L. (1924). Beiträge zur Entwicklungsgeschichte der Holostomiden. I. *Zool. Anz.* **58**, 299–314. Note. II. *Zool. Anz.* **61**, 249–66. III. *Zool. Anz.* **86**, 133–49.

SZIDAT, L. (1928*a*). Studien an einiger seltener Parasiten der Kurischen Nehrung. *Z. Parasitenk.* **1**, 231–344.

SZIDAT, L. (1928*b*). Ueber ein Massenfischsterben im Kurischen Haff. *Schr. phys.-ökon. Ges. Königsb.* **65**, 245–7.

SZIDAT, L. (1929*a*). Beiträge zur Kenntnis der Gattung *Strigea* (Abildg.). I. Allgemeiner Teil. Untersuchungen über die Morphologie, Physiologie und Entwicklungsgeschichte der Holostomiden nebst Bemerkungen über die Metamorphose der Trematoden und die Phylogenie derselben. *Z. Parasitenk.* **1**, 612–87.

SZIDAT, L. (1929*b*). Beiträge zur Kenntnis der Gattung *Strigea* (Abildg.). II. Spezieller Teil. Revision der Gattung *Strigea* nebst Beschreibung einer Anzahl neuer Gattungen und Arten. *Z. Parasitenk.* **1**, 688–764.

SZIDAT, L. (1929*c*). Die Parasiten des Hausgeflügels. 3. *Bilharziella polonica* Kow. ein im Blut schmarotzender Trematoden unserer Enten, seine Entwicklung und Übertragung. *Arch. Geflügelk.* **3**, 78–87.

SZIDAT, L. (1929*d*). Zur Entwicklungsgeschichte der Bluttrematoden der Enten *Bilharziella polonica* Kow. 1. Morphologie und Biologie der Cercarie von *B. polonica* Kow. *Zbl. Bakt.* I. Orig. **3**, 461–70.

SZIDAT, L. (1930*a*). Die Parasiten des Hausgeflügels. 4. *Notocotylus* Diesing und *Catatropis* Odhner, zwei, die Blinddärme des Geflügels bewohnende Monostome Trematodengattung, ihre Entwicklung und Übertragung. *Arch. Geflügelk.* **4**, 105–11.

SZIDAT, L. (1930*b*). *Gigantobilharzia monocotylea* n.sp. ein neuer Blutparasit aus Ostpreussischen Wasservögeln. *Z. Parasitenk.* **2**, 583.

SZIDAT, L. (1931). *Cordulia aenea* S. ein neuer Hilfswirth für *Prosthogonimus pellucidus* v. Linst., den Erreger der Trematodenkrankheit der Legehüner. *Zbl. Bakt.* (i), **119**, 289.

SZIDAT, L. (1932*a*). Zur Entwicklungsgeschichte der Cyclocoeliden. Der Lebenszyklus von *Tracheophilus sisowi* Skrj. 1923. *Zool. Anz.* **100**, 205–13.

SZIDAT, L. (1932*b*). Über Cysticerke Riesencercarien, insbesondere *Cercaria mirabilis* M. Braun und *Cercaria splendens* n.sp. und ihre Entwicklung im Magen von Raubfischen zu Trematoden der Gattung *Azygia* Looss. *Z. Parasitenk.* **4**, 477–505.

SZIDAT, L. (1932*c*). Parasiten aus Liberia und Französisch-Guinea. II. Teil. Trematoda. *Z. Parasitenk.* **4**, 506–21.

SZIDAT, L. (1933). Über die Entwicklung und den Injektionsmodus von *Tracheophilus sisowi* Skrj. eines Luftröhrenschmarotzens der Enten aus der Trematodenfamilie der Zyklozöliden. *Tierärztl. Rsch.* **39**, 95–9.

SZIDAT, L. (1936*a*). Parasiten aus Seeschwalben. I. Über neue Cyathocotyliden aus dem Darm von *Sterna hirundo* L. und *Sterna paradisea*. *Z. Parasitenk.* **8**, 285–316.

SZIDAT, L. (1936*b*). Studien zur Systematik und Entwicklungsgeschichte der Gattung *Leucochloridium* Carus. I. Bemerkungen zur Arbeit von G. Witenberg (1925). Versuch einer Monographie der Trematodenunterfamilie Harmostominae Braun. *Z. Parasitenk.* **8**, 645–53.

SZIDAT, L. (1937*a*). Über die Entwicklungsgeschichte von *Sphaeridiotrema globulus* Rud. 1814 und die Stellung der Psilostomidae Odhner im natürlichen System. I. *Z. Parasitenk.* **9**, 529–42.

SZIDAT, L. (1937*b*). *Archigetes* R. Leuckart, 1878, die progenetische Larve einer für Europa neuen Caryophyllaeidengattung *Biacetabulum* Hunter, 1927. *Zool. Anz.* **119**, 166–72.

SZIDAT, L. (1939*a*). Beiträge zum Aufbau eines natürlichen Systems der Trematoden. I. Die Entwicklung von *Echinocercaria choanophila* U. Szidat zu *Cathaemasia hians* und die Ableitung der Fasciolidae von den Echinostomidae. *Z. Parasitenk.* **11**, 239–83.

SZIDAT, L. (1939*b*). Über *Allocreadium transversale* Rud., 1802 aus *Misgurnus fossilis* L. *Z. Parasitenk.* **10**, 468–75.

SZIDAT, L. (1939*c*). *Pseudobilharziella filiformis* n.sp., eine neue Vogelbilharzie aus dem Höckerschwan, *Cygnus olor* L. *Z. Parasitenk.* **10**, 535–44.

SZIDAT, L. (1940). Über Wirtspezifität bei Trematoden und ihre Beziehung zur Systematik und Phylogenie der zugehörigen Wirtstiere. *Int. Congr. (3rd) Microbiol., New York* (Abstr.), pp. 445–6.

SZIDAT, L. & SZIDAT, U. (1933). Beiträge zur Kenntnis der Trematoden der Monostomidengattung *Notocotylus* Diesing. *Zbl. Bakt.* (i), **129**, 411–22.

SZIDAT, U. (1935). Weitere Beiträge zur Kenntnis der Trematoden der Monostomidengattung *Notocotylus* Diesing. *Zbl. Bakt.* (i), **133**, 265–70.

SZIDAT, U. (1936*a*). Weitere Beiträge zur Kenntnis der Trematoden der Monostomidengattung *Notocotylus* Diesing. 3. *Notocotylus linearis* (Rud. 1819?) n.sp. aus den Blinddärmen des Kiebitz (*Vanellus vanellus* L.). *Zbl. Bakt.* **136**, 231–5.

SZIDAT, U. (1936*b*). Über eine neue Echinostomidencercarie, *Cercaria choanophila* n.sp. *Zool. Anz.* **116**, 304–10.

TABUNŠČIKOVA, A. W. (1932*a*). Einiges über *Fasciola gigantea* Cobbold. *Zool. Anz.* **100**, 185–91.

TABUNŠČIKOVA, A. W. (1932*b*). Quelques données sur *Fasciola gigantea* Cobbold. (Russian.) *C.R. Acad. Sci. Leningrad*, pp. 95–105.

TAGLIANI, G. (1912). *Enoplocotyle minima* nov. gen., nov. sp. Tremotode monogenetico parassiti sulla cute di *Muraena helana* L. *Ric. anat. e sistem. Arch. Zool. Napoli*, **5**, 281–318.

TALIAFERRO, W. H. (1940). The mechanism of acquired immunity in infections with parasitic worms. *Physiol. Rev.* **20**, 469–92.

TAYLOR, E. L. (1934). The production of malformed eggs by immature *Fasciola hepatica*. *Trans. R. Soc. Trop. Med. Hyg.* **27**, 499–504.

TAYLOR, E. L. & BAYLIS, H. A. (1930). Observations and experiments on a dermatitis-producing cercaria and on other cercariae from *Limnaea stagnalis* in Great Britain. *Trans. R. Soc. Trop. Med. Hyg.* **24**, 219–44.

TENNENT, D. H. (1906). A study of the life history of *Bucephalus haemeanus* a parasite of the oyster. *Quart. J. Micr. Sci.* **49**, 635–90.

THAPAR, G. S. (1938). Progress of helminthology in India. *Livro Jub. Prof. L. Travassos, Rio de Janeiro*, pp. 459–65.

THAPAR, G. S. & DAYAL, J. (1934*a*). A new species of the genus *Helicometra* from the intestine of *Trigla gurnardus*. *Proc. Ind. Sci. Congr. Calcutta*, **21**, 261.

THAPAR, G. S. & DAYAL, J. (1934*b*). The morphology and the systematic position of a new trematode from the intestine of the golden orfe, *Leuciscus idus*, with a note on the classification of the family Allocreadiidae. *J. Helminth.* **12**, 127–36.

THOMAS, L. J. (1939). Life cycle of a fluke, *Halipegus eccentricus* n.sp. found in the ears of frogs. *J. Parasit.* **25**, 207–21.

TIMON-DAVID, J. (1933*a*). Contribution à l'étude du cycle-évolutif des zoogonides (Trématodes). *C.R. Acad. Sci. Paris*, **196**, 1923–4.

TIMON-DAVID, J. (1933*b*). Sur une nouvelle espèce de *Renicola* trématode parasite du rein des Larides. *Bull. Inst. océanogr. Monaco*, no. 616, pp. 1–16.

TIMON-DAVID, J. (1934). Recherches sur les trématodes parasites des oursins en Mediterranée. *Bull. Inst. océanogr. Monaco*, no. 652, pp. 1–16.

TIMON-DAVID, J. (1935). Sur les *Wedlia* parasites de l'estomac du thon (Trematodes, Didymozoonidae). *Bull. Inst. océanogr. Monaco*, no. 670, pp. 1–11.

TIMON-DAVID, J. (1936). Sur l'évolution expérimentale des métacercaires de *Zoogonus mirus* Looss, 1901 (Trématodes, Famille des Zoogonidae). *C.R. Ass. franç. Av. Sci.* pp. 274–6.

TIMON-DAVID, J. (1937*a*). Étude sur les trématodes parasites des poissons du golfe de Marseille (Première liste). *Bull. Inst. océanogr. Monaco*, no. 717, pp. 1–24.

TIMON-DAVID, J. (1937*b*). Les kystes à *Didymocystis wedli* du thon, étude anatomopathologique. *Ann. Parasit. hum. comp.* **15**, 520–3.

TIMON-DAVID, J. (1938). On parasitic trematodes in Echinoderms. *Livro Jub. Prof. L. Travassos, Rio de Janeiro*, pp. 467–73.

TOSH, J. R. (1905). On the internal parasites of the Tweed salmon. *Ann. Mag. Nat. Hist.* (7), **16**, 115–19.

TRAVASSOS, L. (1928). Sur la systématique de la famille des Clinostomidae Lühe, 1901. *C.R. Soc. Biol., Paris*, **98**, 643–4.

TRAVASSOS, L. (1930*a*). Pesquizas helminthologicas realisadas em Hamburgo. I. Genero *Haplometra* Looss, 1899. *Mem. Inst. Osw. Cruz*, **23**, 163–8.

Travassos, L. (1930*b*). Pesquizas helminthologicas realisadas em Hamburgo. IV. Notas sobre o genero *Opisthioglyphe* Looss, 1899, e generos proximos. *Mem. Inst. Osw. Cruz*, **24**, 1–18.

Travassos, L. (1930*c*). Pesquizas helminthologicas realisadas em Hamburgo. V. Genero *Prosotocus* Looss, 1899 (Trematoda: Lecithodendriidae). *Mem. Inst. Osw. Cruz*. **24**, 57–61.

Travassos, L. (1930*d*). Genero *Pleurogenoides* Travassos, 1921 (Trematoda: Lecithodendriidae). *Mem. Inst. Osw. Cruz*, **24**, 63–70.

Travassos, L. (1930*e*). Informações sobre o genero *Pleurogenes* Looss, 1896 (Nematoda (*sic*): Lecithodendriidae). *Mem. Inst. Osw. Cruz*. **24**, 251–6.

Travassos, L. (1934). Synopse dos Paramphistomoidea. *Mem. Inst. Osw. Cruz*, **29**, 19–178.

Travassos, L. (1941). Sobre o '*Lyperosomum rudectum*' Braun, 1901, un equivoco na descrição deste parasito. *Rev. Brasil. Biol.* **1**, 83–5.

Travassos, L., Artigas, P. & Pereira, C. (1928). Fauna helminthologica dos peixes agua doce de Brasil. *Arch. Inst. Biol. São Paulo*, **1**, 5–68.

Travassos, L. & Darriba, A. R. (1930). Pesquizas helminthologicas realisadas em Hamburgo. III. Trematodeos dos generos *Pneumonoeces* e *Ostiolum*. *Mem. Inst. Osw. Cruz*, **23**, 237–53.

Travassos, L. & Vogelsang, E. (1930). Pesquizas helminthologicas realisadas em Hamburgo. II. Sobre dois trematodeos parasitos de mammiferos. *Mem. Inst. Osw. Cruz*, **23**, 169–71.

Tseng, S. (1931). Douve trouvée dans un œuf de poule à Nankin et considérations sur les espèces du genre *Prosthogonimus*. *Bull. Soc. zool. Fr.* **56**, 468–78.

Tubangui, M.A. (1928). Larval trematodes from Philippine snails. *Philipp. J. Sci.* **36**, 37–54.

Tubangui, M. A. (1931). Trematode parasites of Philippine vertebrates. IV. Ectoparasitic flukes from marine fishes. *Philipp. J. Sci.* **45**, 109–17.

Tubangui, M. A. (1932). Observations on the life history of *Euparyphium murinum* Tubangui, 1931 and *Echinostoma revolutum* (Froelich, 1802) (Trematoda). *Philipp. J. Sci.* **47**, 497–513.

Tubangui, M. A. & Pasco, A. M. (1933). The life history of the human intestinal fluke, *Euparyphium ilocanum* (Garrison, 1908). *Philipp. J. Sci.* **51**, 581–606.

Tyzzer, E. E. (1918). A Monostome of the genus *Collyriclum* occurring in the European sparrow, with observations on the development of the ovum. *J. Med. Res.* N.S. **33**, no. 2, pp. 267–92.

Ujiie, N. (1936*a*). On the process of egg-shell formation of *Clonorchis sinensis*, a liver fluke. *J. Med. Ass. Formosa*, **35**, 8, no. 377, 1894–6.

Ujiie, N. (1936*b*). On the structure and function of Mehlis gland on the formation of the egg-shell of *Echinochasmus japonicus*. *J. Med. Ass. Formosa*, **35**, 5, no. 374, 1010.

Van Cleave, H. J. (1921). Notes on two genera of ectoparasitic trematodes from fresh-water fishes. *J. Parasit.* **8**, 33–9.

Van Cleave, H. J. & Mueller, J. F. (1932). Parasites of Oneida Lake fishes. Part I. Description of new genera and new species. *Roosevelt Wild Life Ann.* **3**, 5–71.

Van Cleave, H. J. & Mueller, J. F. (1934). Parasites of Oneida Lake fishes. Part III. A biological and ecological survey of the worm parasites. *Roosevelt Wild Life Ann.* **3**, 161–334.

Van Cleave, H. J. & Vaughn, C. M. (1941). The trematode genus *Otodistomum* in North America. *J. Parasit.* **27**, 253–7.

Van Cleave, H. J. & Williams, C. O. (1943). Maintenance of a trematode, *Aspidogaster conchicola*, outside the body of its natural host. *J. Parasit.* **29**, 127–30.

Vanni, V. (1937). Grandi figure della parassitologia italiana e suo odierno sviluppo. *Minerva Med., Turin*, **28**, 159–62.

Vaz, Z. & Pereira, C. (1930). Nouvel hémiuride parasite de *Sardinella aurita* Cuv. et Val., *Parahemiurus* n.g. *C.R. Soc. biol. Paris*, **103**, 1315–17.

Verma, S. C. (1936). Studies on the family Bucephalidae (Gasterostomata). Part I. Descriptions of new forms from Indian fresh-water fishes. *Proc. Nat. Acad. Sci. India*, **6**, 66–89.

VEVERS, G. M. (1923). Observations on the life history of *Hypoderaeum conoideum* (Bloch) and *Echinostomum revolutum* (Froel.), trematode parasites of the domestic duck. *Ann. App. Biol.* **10**, 134–6.

VEVERS, G. M. (1942). The progress of helminthology in the U.S.S.R. *Post-Grad. Med. J.* **18** (194), 11.

VIANA, L. (1924). Tentative de catalogação des especies brazileiras de trematódeos. *Mem. Inst. Osw. Cruz*, **17**, 95–227.

VICKERS, G. G. (1940). On the anatomy of *Cercaria macrocerca* from *Sphaerium corneum*. *Quart. J. Micr. Sci.* **82**, 311–26.

VLASENKO, P. V. (1931). Zur Helminthfauna der Schwarzmeerfische. *Trav. Sta. biol. Karadagh*, **4**, 88–126 (Russian) (German résumé, pp. 127–34).

VOGEL, H. (1929). Beobachtungen über *Cercaria vitrina* und deren Beziehung zum Lanzettegelproblem. *Arch. Schiffs- u. Tropenhyg.* **33**, 474–89.

VOGEL, H. (1930*a*). Hautveränderungen durch *Cercaria ocellata*. *Derm. Wschr.* **90**, 577.

VOGEL, H. (1930*b*). Zerkarien-Dermatitis in Deutschland. *Klin. Wschr.* **9**, 883.

VOGEL, H. (1932). Über den ersten Zwischenwirt und die Zerkarie von *Opisthorchis felineus*. *Arch. Schiffs- u. Tropenhyg.* **36**, 558–60.

VOGEL, H. (1934). Der Entwicklungszyklus von *Opisthorchis felineus* (Riv.) nebst Bemerkungen über die Systematik und Epidemiologie. *Zoologica, Stuttgart*, **33**, 1–103.

WALL, L. D. (1941). *Spirorchis parvus* (Stunkard), its life history and the development of its excretory system. (Trematoda: Spirorchiidae). *Trans. Amer. Micr. Soc.* **60**, 221–60.

WALLACE, F. G. (1939). The life cycle of *Pharyngostomum cordatum* (Diesing) Ciurea (Trematoda: Alariidae). *Trans. Amer. Micr. Soc.* **58**, 49–61.

WALLACE, H. E. (1941). Life history and embryology of *Triganodistomum mutabile* (Cort) (Lissorchiidae, Trematoda). *Trans. Amer. Micr. Soc.* **60**, 309–26.

WALTON, A. C. (1938). The trematodes as parasites of Amphibia. List of parasites. *Contr. Biol. Lab. Knox Coll. Galesburg, Illinois*, no. 62, 64 pp.

WARD, H. B. (1917). On the structure and classification of North American parasitic worms. *J. Parasit.* **4**, 1–12.

WARD, H. B. (1918). Parasitic flatworms. Trematoda. In *Freshwater Biology* (Ward and Whipple).

WARD, H. B. (1921). A new blood-fluke from turtles. *J. Parasit.* **7**, 114–28.

WARD, H. B. (1938). On the genus *Deropristis* and the Acanthocolpidae (Trematoda). *Livro Jub. Prof. L. Travassos, Rio de Janeiro*, pp. 509–21.

WARD, H. B. & FILLINGHAM, J. (1934). A new trematode in a toadfish from South-eastern Alaska. *Proc. Helminth. Soc. Wash.* **1**, 25–31.

WARD, H. B. & HIRSCH, E. (1915). The species of *Paragonimus* and their differentiation. *Ann. Trop. Med. Parasit.* **9**, 109–62.

WARD, H. B. & HOPKINS, S. H. (1931). A new North American Aspidogastrid, *Lophotaspis interiora*. *J. Parasit.* **18**, 69–78.

WASSERMANN, F. (1912*a*). Zur Eireifung von *Zoogonus mirus*, ein Beitrag zur Synapsisfrage. *Anat. Anz.* **41**, 47–58.

WASSERMANN, F. (1912*b*). Ueber die Eireifung bei *Zoogonus mirus* Lss. (vorl. Mitt.). *S.B. Ges. Morph. Physiol. München*, **27** (1911), 123–31.

WASSERMANN, F. (1913). Die Oogenese des *Zoogonus mirus* Lss. *Arch. mikr. Anat.* **83**, 1–140.

WATSON, E. E. (1911). The genus *Gyrocotyle*, etc. *Univ. Calif. Publ. Zool.* **6**, 353–468.

WEIDMAN, F. D. (1918). A contribution to the anatomy and embryology of *Cladorchis* (*Stichorchis*) *subtriquetus* Rudolphi, 1814 (Fischoeder, 1901). *Parasitology*, **10**, 267–79.

WEINLAND, E. & BRAND, T. v. (1926). Beobachtungen an *Fasciola hepatica*. *Z. vergl. Physiol.* **4**, 212–85.

WENYON, C. M. (1926, 1928). *Protozoology*, 1563 pp. London.

WERBY, H. J. (1928). On the trematode genus *Harmostomum*, with the description of a new species. *Trans. Amer. Micr. Soc.* **47**, 68–81.

WESENBERG-LUND, C. (1931). Contributions to the development of the Trematoda Digenea. Part I. The biology of *Leucochloridium paradoxum*. *D. Kgl. Dansk. Vidensk. Selsk. Skrifter*, Naturw. Math. Afd., Raekke 9; **4** (3), 90–142.

WESENBERG-LUND, C. (1934). Contributions to the development of the Trematoda Digenea. Part II. The biology of the fresh-water cercariae in Danish fresh-waters. *D. Kgl. Dansk. Vidensk. Selsk. Skrifter*, Naturw. Math. Afd., Raekke 9; **5** (3), 223 pp. (Also in *Mem. Acad. Roy. Sci. Lett. Danemark*, Sect. Sci. **9**, Ser. 5, pp. 1–223.)

WHARTON, G. W. (1939). Studies on *Lophotaspis vallei* (Stossich, 1899) (Trematoda: Aspidogastridae). *J. Parasit.* **25**, 83–6.

WHARTON, G. W. (1940). The genera *Telorchis, Protenes* and *Auridistomum* (Trematoda: Reniferidae). *J. Parasit.* **26**, 497–518.

WHARTON, G. W. (1941). The function of respiratory pigments of turtle parasites. *J. Parasit.* **27**, 81–7.

WHEELER, N. C. (1939). A comparative study on the behaviour of four species of pleurolophocercous cercariae. *J. Parasit.* **25**, 343–53.

WILDE, J. (1937). *Dactylogyrus macrocanthus* Wegener also Krankheitserreger auf den Kiemen der Schleie (*Tinca tinca*), etc. *Z. Parasitenk.* **9**, 203–36.

WILHELMI, R. W. (1939). Serological reactions of some helminths. *J. Parasit.* **25**, Suppl. Abstr. p. 31.

WILHELMI, R. W. (1940). Serological reactions and species specificity of some helminths. *Biol. Bull. Woods Hole*, **79**, 64–90.

WILLEM, V. (1906). Deux trématodes nouveaux pour la faune belge, *Acanthocotyle branchialis* nov.sp. et *Distomum turgidum* Brandes. *Bull. acad. Cl. Sci. Belg.* **8**, 522–3, 599–612.

WILLEY, C. H. (1930). A cystophorous cercaria, *C. projecta*, n.sp., from the snail, *Helisoma antrosa* N. America. *Parasitology*, **22**, 481–9.

WILLEY, C. H. (1941). The life history and bionomics of the trematode, *Zygocotyle lunata* (Paramphistomidae). *Zoologica, N.Y.*, **26**, 65–88.

WILLEY, C. H. & GODMAN, G. C. (1941). Gametogenesis in the trematode *Zygocotyle lunata*. *Anat. Rec.* **81**, Suppl. 78, Abstract.

WILLIAMS, C. O. (1942). Observations on the life history and taxonomic relationships of the trematode *Aspidogaster conchicola*. *J. Parasit.* **28**, 467–75.

WIŚNIEWSKI, L. W. (1932). Sur deux nouveaux trématodes progénétiques des Gammarides balkaniques. *C.R. Acad. Cracovie*, **7**, 9.

WIŚNIEWSKI, L. W. (1933*a*). Über zwei neue progenetische Trematoden aus dem balkanischen Gammariden. *Bull. int. Acad. Cracovie*, B, **2**, 1932, 259–76.

WIŚNIEWSKI, L. W. (1933*b*). O rodzajach rodziny Coitocaecidae (Trematoda). *Arch. Tow. nauk. Lwów* (3), **6**, ii, 1–13.

WITENBERG, G. (1922). *Orchipedum centorchis* nov.sp. *Zbl. Bakt.* **56**, 572–5.

WITENBERG, G. (1925). Versuch einer Monographie der Trematodenunterfamilie Harmostominae Braun. *Zool. Jb., Syst.* **51**, 167–254.

WITENBERG, G. (1926). Die Trematoden der Familie Cyclocoelidae Kossack, 1911, etc. *Zool. Jb., Syst.* **52**, 103–86.

WITENBERG, G. (1928*a*). Notes on Cyclocoelidae. *Ann. Mag. Nat. Hist.* (10), **2**, 410–17.

WITENBERG, G. (1928*b*). Reptilien als Zwischenwirte parasitischer Würmer von Katze und Hund. *Tierärztl. Rdsch.* **34**, 32, 603.

WITENBERG, G. (1929). Studies on the trematode family Heterophyidae. *Ann. Trop. Med. Parasit.* **23**, 131–239.

WITENBERG, G. (1930). Corrections to my paper 'Studies on the trematode family Heterophyidae'. *Ann. Mag. Nat. Hist.* (10), **5**, 412–14.

WITENBERG, G. (1932*a*). On the anatomy and systematic position of the causative agent of so-called Salmon poisoning. *J. Parasit.* **18**, 258–63.

WITENBERG, G. (1932*b*). Über zwei in Palästina in Hunden und Katzen parasitierende Echinochasmus-Arten (Trematoda). *Z. Parasitenk.* **5**, 213–16.

WITENBERG, G. & ECKMANN, F. (1934). Notes on *Asymphylodora tincae*. *Ann. Mag. Nat. Hist.* (10), **14**, 366–71.

WITENBERG, G. & ECKMANN, F. (1939). On the classification of the trematode genus *Prosthogonimus*. *Vol. Jub. pro Prof. S. Yoshida, Osaka*, **2**, 129–44.

WITENBERG, G. & YOFE, J. (1938). Investigation on the purification of water with respect to Schistosome cercariae. *Trans. Roy. Soc. Trop. Med. Hyg.* **31**, 549–70.

WOODHEAD, A. E. (1931). The germ-cell cycle in the trematode family Bucephalidae. *Trans. Amer. Micr. Soc.* **50**, 169–87.

WOODHEAD, A. E. (1932). The germ-cell cycle in trematodes. *J. Parasit.* **19**, Suppl. Abstr. p. 164.

WOODHEAD, A. E. (1935). The mother sporocysts of *Leucochloridium*. *J. Parasit.* **21**, 337–46.

WOODHEAD, A. E. (1936). An extraordinary case of multiple infection with the sporocysts of *Leucochloridium*. *J. Parasit.* **22**, 227–8.

WOOLCOCK, V. (1935). Digenetic trematodes from some Australian fishes. *Parasitology*, **27**, 309–31.

WOOLCOCK, V. (1936). Monogenetic trematodes from some Australian fishes. *Parasitology*, **28**, 79–91.

WRIGHT, W. R. (1927). Studies on larval trematodes from North Wales. Parts I and II. *Ann. Trop. Med. Parasit.* **21**, 41–56, 57–60.

WRIGHT, W. R. (1928). Note on the locomotion of the redia of *Fasciola hepatica*. *Parasitology*, **20**, 113–14.

WU, K. (1937). Susceptibility of various Mammals to experimental infection with *Fasciolopsis buski* (Trematoda: Fasciolidae). *Ann. Trop. Med. Parasit.* **31**, 361–72.

WU, K. (1938). Progenesis of *Phyllodistomum lesteri* sp.nov. (Trematoda: Gorgoderidae) in fresh-water shrimps. *Parasitology*, **30**, 4–19.

WUNDER, W. (1923). Bau, Entwicklung und Function des Cercarien-Schwanzes. *Zool. Jb., Anat.* **46**, 303–42.

WUNDER, W. (1924). Die Schwimmbewegung von *Bucephalus polymorphus* v. Baer. *Z. vergl. Physiol.* **1**, 289.

WUNDER, W. (1932). Untersuchungen über Pigmentierung und Encystierung von Cercarien. *Z. Morph. Ökol. Thiere*, **25**, 336.

YAMAGUTI, S. (1933). Studies on the helminth fauna of Japan. Part 1. Trematodes of Birds, Reptiles and Mammals. *Jap. J. Zool.* **5**, 1–134.

YAMAGUTI, S. (1934*a*). Studies on the helminth fauna of Japan. Part 2. Trematodes of Fishes. I. *Jap. J. Zool.* **5**, 249–541.

YAMAGUTI, S. (1934*b*). Studies on the helminth fauna of Japan. Part 3. Avian trematodes. II. *Jap. J. Zool.* **5**, 543–83.

YAMAGUTI, S. (1935*a*). Studies on the helminth fauna of Japan. Part 5. Trematodes of birds. III. *Jap. J. Zool.* **6**, 159–82.

YAMAGUTI, S. (1935*b*). Über die Cercarie von *Clonorchis sinensis* (Cobbold). *Z. Parasitenk.* **8**, 183–7.

YAMAGUTI, S. (1936). Studies on the helminth fauna of Japan. Part 14. Amphibian trematodes. *Jap. J. Zool.* **6**, 551–76.

YAMAGUTI, S. (1937). *Hexacotyle dissimilis* n.sp. on gills of *Thynnus thynnus*. (Studies on the helminth fauna of Japan. Part 19.) *Jap. J. Zool.* **7**, 20–2.

YAMAGUTI, S. (1938). Studies on the helminth fauna of Japan. Part 24. Trematodes of fishes. V. *Jap. J. Zool.* **8**, 15–74.

YAMAGUTI, S. (1939*a*). Studies on the helminth fauna of Japan. Part 26. Trematodes of fishes. VI. *Jap. J. Zool.* **8**, 211–30.

YAMAGUTI, S. (1939*b*). Studies on the helminth fauna of Japan. Part 25. Trematodes of birds. IV. *Jap. J. Zool.* **8**, 129–210.

YAMAGUTI, S. (1939*c*). Zur Entwicklungsgeschichte von *Notocotylus attenuatus* (Rud., 1809) und *N. magniovatus* Yamaguti, 1934. *Z. Parasitenk.* **10**, 288–92.

YAMAGUTI, S. (1939*d*). Zur Entwicklungsgeschichte von *Centrocestus armatus* (Tanabe) mit besonderer Berücksichtigung der Cercarie. *Z. Parasitenk.* **10**, 293–5.

YAMAGUTI, S. (1939*e*). Über die Ursache der sog. 'Schwarzen Winterflecke' der japanischen Süsswasserfische. *Z. Parasitenk.* **10**, 691–3.

YAMASHITI, J. (1937). Studies on the family Echinostomatidae. Part II. Trematode parasites of Reptiles, Birds and Mammals arranged systematically. *Trans. Sapporo Nat. Hist. Soc.* **15**, 82–95.

YOUNG, R. T. (1938). The life history of a trematode (*Levinseniella cruzi*?) from the shore birds (*Limosa fedoa* and *Catoptrophorus semipalmatus inornatus*). *Biol. Bull. Woods Hole*, **74**, 319–29.

YUMOTO, Y. (1936). On the minute structure of the egg-shells of *Clonorchis sinensis*, and on its abnormal eggs. *J. Med. Ass. Formosa*, **35** (no. 377), 1845–6.

ZAKHAROW, N. P. (1920). *Prosthogonimus skrjabini* nov.sp. *Izvestia Donsk. Vet. Inst.* **1**, ii.

ZANDT, F. (1924). Fischparasiten des Bodensees. *Zbl. Bakt.* I. Orig. **92**, 225–71.

ZSCHOKKE, F. (1904). Die Darmcestoden der amerikanischen Beuteltiere. *Zbl. Bakt.* I. Orig. **36**, 51–61.

ZSCHOKKE, F. (1933). Die Parasiten als Zeugen für die geologische Vergangenheit ihrer Träger. *Forsch. Fortschr. dtsch. Wiss.* **9**, 466–7.

ZSCHOKKE, F. & HEITZ, A. (1914). Entoparasiten aus Salmoniden von Kamtschatka. *Rev. Suisse Zool. Genève*, **22**, 195–256.

INDEX

An asterisk before a name denotes that a diagnosis is given in the text. Where two or more names are given in the same line, those on the left of the dash are synonyms of those on the right of it. Figures are indicated in bold-face type. n=footnote.